INTRODUCTION TO MANUFACTURING PROCESSES

McGraw-Hill Series in Mechanical Engineering and Materials Science

INTRODUCTION TO MANUFACTURING PROCESSES

Third Edition

John A. Schey
Department of Mechanical Engineering
University of Waterloo, Ontario

Boston   Burr Ridge, IL   Dubuque, IA   Madison, WI   New York   San Francisco   St. Louis
Bangkok   Bogotá   Caracas   Lisbon   London   Madrid
Mexico City   Milan   New Delhi   Seoul   Singapore   Sydney   Taipei   Toronto

McGraw-Hill Higher Education

A Division of The McGraw·Hill Companies

INTRODUCTION TO MANUFACTURING PROCESSES
International Editions 2000

10 9 8 7 6 5 4 3
20 9 8 7 6 5 4 3 2 1 0 9
UPE   ANL

Library of Congress Cataloging-in-Publication Data

Schey, John A.
  Introduction to manufacturing processes / John A. Schey – 3rd ed.
     p.   cm. – (McGraw-Hill series in mechanical engineering and materials science)
  Includes index.
  ISBN 0-07-031136-6
1. Manufacturing processes.   I. Title.   II. Series.
TS183.S33   2000
670–dc21   99-30381
                                                CIP

www.mhhe.com

When ordering this title, use ISBN 0-07-116911-3

Printed in Singapore

To Gitta, for years of help and support

ABOUT THE AUTHOR

Dr. John A. Schey was educated in his native Hungary. He received his Dipl. Met. Ing. degree from the Jozsef Nador Technical University, Sopron, in 1946, and was awarded the Cand. Tech. Sci. (Ph.D.) degree by the Academy of Sciences, Budapest, in 1953. He received the degree Dr.-Ing. honoris causa in 1987 from the University of Stuttgart, Germany, and in 1989 from his alma mater (now the University of Miskolc).

He held positions as chief technologist at the Metal Works in Csepel, Budapest (1947–1951), Reader at the Technical University of Miskolc, Hungary (to 1956), department head at the Research Laboratories of the British Aluminium Co. Ltd., England (to 1962), Senior Metallurgical Advisor at IIT Research Institute, Chicago (to 1968), Professor of Metallurgical Engineering at the University of Illinois at Chicago (to 1974), and Professor in the Department of Mechanical Engineering at the University of Waterloo, Ontario, where he now has the title Distinguished Professor Emeritus.

He is Member of the National Academy of Engineering of the USA, Foreign Member of the Hungarian Academy of Sciences, Honorary Member of the Hungarian Academy of Engineering, Fellow of ASM International and of the Society of Manufacturing Engineers, and a Certified Manufacturing Engineer.

In 1984 he was awarded the Dofasco Award by the Canadian Institute of Mining and Metallurgy, in 1974 the Gold Medal Award by the Society of Manufacturing Engineers, and in 1966 the W.H.A. Robertson Award and Medal by the Institute of Metals, London, England.

He is the author of numerous books and research publications on manufacturing processes, metalworking processes, and the tribology of metalworking, including the monograph *Tribology in Metalworking: Friction, Lubrication, and Wear* (American Society for Metals, 1983). He has acted as a consultant to more than sixty industries, and he has eight patents.

PREFACE

In the more than twenty years that have passed since the publication of the first edition of this text, the study of manufacturing has found its rightful place in the engineering curriculum. Courses in manufacturing are being offered at various levels, from review courses given to incoming freshmen to intensive courses designed for juniors and seniors. There are also a growing number of graduate programs aimed at educating professionals capable of managing enterprises on a sound technological basis. To serve these diverse demands, various institutions have adopted or are in the midst of developing a variety of new techniques. This time period has also seen a spectacular growth in information technology, and this has had a major effect not only on manufacturing technology but also on the way manufacturing education is delivered. Every year, dozens of innovative approaches, many of them using the Web for enrichment or distance education, are presented in conferences devoted to manufacturing education.

There has also been a marked shift in the attitudes of industry toward its clients: the loop between producer and consumer is indeed closing. Concurrent engineering, with input from the consumer, has become reality in many companies.

All of this has implications for the student and practicing engineer. A vast amount of information is now available, not just in the printed form but also on the Internet, and there is an increasing abundance of software that helps in designing products and processes. None of these can, however, be effectively utilized unless there is a sound understanding of the physical fundamentals of processes. This understanding must, therefore, be the aim of any course on manufacturing, irrespective of the level of its presentation.

The second edition of this text aimed at furthering this understanding by incorporating background material that some students may not have. The present edition continues to emphasize the physical basis and its relation to actual processes, but with some important changes. Mindful of the highly varied background of students and practicing engineers, the number of chapters has been increased to allow separation of the background material into chapters of its own. Those with only high school chemistry and physics can gain from these a sufficient foundation for the subsequent treatment of processes; those with a knowledge of materials and mechanics can read them as refreshers, with an eye on their implications for manufacturing processes and product quality and properties. These chapters on background material are followed by a treatment of processes. Even a brief description of individual processes could fill several volumes; therefore, the emphasis here is on physical principles that are often common to several seemingly unrelated processes and can be applied to make reasoned judgments on the feasibility of a proposed solution.

While common threads are woven through the text, it is recognized that—more often than not—the instructor will have to select a limited number of topics and present them in an order different from that in the text. To facilitate this, copious cross-references are made to essential elements preceding a given discussion. With these considerations in mind, the subject matter is divided into three broad groups.

Chapters 1–5 prepare the subject: Chap. 1 offers a general view of the importance of manufacturing to humankind; Chap. 2 outlines the interaction of design and manufacturing and introduces basic concepts of process control; Chap. 3 deals with geometry, dimensions, and surface quality; and Chap. 4 with the service properties expected of a manufactured product. The last two of these chapters also discuss the measurement techniques that will be used for manufacturing control. Chapter 5 is new, narrowing the focus to the interactions between product design, material selection, and process choice in concurrent engineering. Recycling and health and environmental concerns are addressed in this chapter and throughout the text.

Chapters 6–20 deal with processes in the sequence to which a material is usually subjected. Basic concepts and their applications are first demonstrated on metals: Chap. 6 discusses solidification in preparation for casting (Chap. 7) and welding (Chap. 18), and Chap. 8 serves as preparation for deformation processes such as bulk deformation (Chap. 9) and sheet metalworking (Chap. 10). Particulate technologies are presented first for metals (Chap. 11) then, after some introduction to the structure and properties of ceramics, to the processing of ceramics (Chap. 12). The background to plastics (Chap. 13) lays the groundwork for a much expanded treatment of plastics processing (Chap. 14), and all the preceding material is drawn upon in the new Chap. 15 on composites. Metal removal by chip forming (Chap. 16) and non-traditional techniques (Chap. 17) is followed by the additive processes of joining (Chap. 18) with their application to solid free-form fabrication. A new chapter on surface treatments (Chap. 19) brings these highly varied technologies together. A brief introduction to solid-state electronics is followed by a discussion of processes for the manufacture of solid-state devices, and this lays the foundations of microfabrication for the production of microelectromechanical systems, surely one of the growth areas in manufacturing.

The last two chapters are devoted to the organization and competitive aspects of manufacturing, including an exploration of competition between processes discussed in the main section.

New developments are highlighted in all process chapters. Thus, the student will get a feel for cutting-edge technologies and precision manufacturing. Examples drawn from recent applications demonstrate the relevance of principles and techniques. Wherever justified, quantitative treatments are included, often with the use of spreadsheets.

For the practice of concurrent engineering, the manufacturing engineer must be able to interact with product designers, and the product designer must have at least a basic feel for the process consequences of a design decision. For this reason, the design implications of processes are emphasized throughout. Thus, the design of a product can be developed and judged not on the basis of sterile rules but with a fuller understanding of the reasoning behind these rules. To aid a better comprehension of process relationships, a classification of processes is given in each process chapter, and summarized in tables. Together with chapter-end summaries, they can be used to give an appreciation of technologies that had to be skipped because of time constraints.

Problems have been expanded and are presented in three groups: simple review questions; problems calling for reasoned judgement; and problems requiring quantitative answers. Several of them are suitable for examinations.

A comprehensive Instructor's Manual is provided to adopters of the text. Since the material presented here cannot possibly be covered in a one-semester course, suggestions are given to help in selecting sections for courses serving different aims, with due regard for the level of preparation of students. There are full solutions to all problem statements, even to review questions. Since all too often more than one alternative solution is acceptable, answers are given with explanations and hints, and sufficient detail is provided to allow marking by teaching assistants. Derivations of text equations are shown in a form suitable for reproduction and distribution to students. Suggestions for teaching aids are included.

I am greatly indebted to many colleagues who critically reviewed specific sections, particularly H.W. Kerr, A. Plumtree, C. Tzoganakis, R. Varin, and M. Worswick of the University of Waterloo and D. Edelstein, of IBM T. J. Watson Research Center. In addition to the companies and individuals specifically acknowledged in the text, I received helpful information from P.H. Abramowitz and D.A. Yeager (Ford), T. Altan (Ohio State University), R.A. Crockett (Lockheed Martin), K.F. Hens (Thermat), T.E. Howson (Wyman-Gordon), M.L. Devenpeck and H.R. Zonker (Alcoa), S.R Larrabee and C.J. Rogers (Modine), F. Norris (Howmet), J.D. Schreiber (American Superconductor), J. Stump (GE Aircraft Engines), and A.J.K. Tubman (Tubman Marketing).

I benefited greatly from the helpful comments and criticisms of the reviewers of the manuscript, L.R. Cornwell (Texas A&M University), A.S. El-Gizawy (University of Missouri at Columbia), J.G. Lenard (University of Waterloo, Ontario), D.G. Tomer (Rochester Institute of Technology), and A.A Tseng (Arizona State University). Valuable suggestions were made by the reviewers of the revision plan, X.D. Fang (Iowa State University), J.K. Gershenon (University of Alabama), D. Hall (Louisiana Technological University), D.W. Radford (Colorado State University) and J. Warner (Milwaukee School of Engineering).

I was most fortunate indeed to have the support of McGraw-Hill personnel, in particular, Jonathan Plant, editor, and Kristen Druffner, editorial assistant; Kimberly Moranda, project manager, and Rose Range, supplements coordinator. I am indebted also to John Corrigan and Debra Riegert who initiated this revision. As with the previous editions, my wife Gitta shared the task and provided support through many long months.

Waterloo, Ontario, May 1999

John A. Schey

CONTENTS

Chapter 8

Plastic Deformation of Metals 250

Chapter 9

Bulk Deformation Processes 302

Chapter   10

Sheet-Metalworking Processes   382

Chapter   11

Powder Metallurgy   441

Chapter   12

Processing of Ceramics   474

Manufacturing creates the goods for our existence, from the enormous to the miniature. The C-5 cargo plane is close to 76 m long and 20 m high, has a wingspan of 68 m, and can take off at 380 Mg mass, including 120 Mg of military or disaster-relief supplies. (*Courtesy Lockheed Martin, Atlanta, Georgia.*). In contrast, the Pentium II chip has 7.5 million transistors on an area measuring only a few millimeters on its sides. (*Courtesy Intel, Santa Clara, California.*)

1

Introduction to Manufacturing

In this chapter we will examine some important questions:

What is manufacturing?

Does manufacturing benefit humankind?

Is manufacturing important for the economy?

Do we really need manufacturing at all in the information age?

Manufacturing is a human activity that pervades all phases of our life. The products of manufacturing are all around us: Everything we wear, we live in, we travel on, even most of what we eat, has gone through some manufacturing process. The word manufacturing is derived from the Latin (*manus* = hand, *factus* = made), and is defined by dictionaries as "the making of goods and articles by hand or, especially by machinery, often on a large scale and with division of labor." We shall see that this definition is not necessarily complete, but we can use it to gain an understanding of the role of manufacturing in human development.

1-1 HISTORICAL DEVELOPMENTS

The history of manufacturing is marked by gradual developments, but the cumulative effects have been of such substantial social consequences that they can rightly be regarded as revolutionary.

1-1-1 Early Developments

Manufacturing has been practiced for several thousand years, beginning with the production of stone, ceramic, and metallic articles. The Romans already had factories for the mass production of glassware, and many activities, including mining, metallurgy, and the textile industry have long employed the principle of division of labor. Nevertheless, much of manufacturing remained for centuries an essentially individual activity, practiced by artisans and their apprentices. The ingenuity of successive generations of artisans led to the development of many processes and to a great variety of products (Table 1–1), but the scale of production was necessarily limited by the available power. Water power supplemented muscle power only in the Middle Ages, and then only to the extent allowed by the availability of swift water; this limited the location of industries and the rate of growth of industrial production.

1-1-2   The First Industrial Revolution

At the end of the eighteenth century, the development of the steam engine made power available in large quantities and at many locations. This spurred advances in manufacturing processes (Table 1–1) and facilitated the growth of production, providing an abundance of goods and, with the mechanization of agriculture, of agricultural products. As a result, society was also transformed, and later these developments came to be recognized as the industrial revolution. It was characterized by mechanical power supplementing the physical power of the worker. Many machines were driven by belts from a common drive shaft, and the scope for mechanization was limited.

Toward the middle of the nineteenth century, some functions of the worker were taken over by machines in which mechanical components such as cams and levers were ingeniously arranged to perform relatively simple and repetitive tasks. Such mechanization, or "hard automation," eliminated some jobs, but the workers thus displaced—together with those made redundant in agriculture—usually found jobs in the expanding manufacturing and service sectors of the economy. Around the turn of the twentieth century, development was further aided by the introduction of electric power: Machines could now be individually driven, and controls based on electric circuits allowed a fair degree of sophistication.

1-1-3   The Second Industrial Revolution

Beginning with the second half of the twentieth century, further developments have taken place. Computers have begun to offer hitherto undreamed of computational power, and solid-state electronics—growing out of the transistor—permitted the fabrication of devices of great versatility at ever-decreasing cost. In the early 1970s the availability of the microchip, with thousands of electronic components crammed onto a tiny silicon wafer, made it possible to perform computational, control, planning, and management tasks at high speeds, very often in real time (i.e., while the process to be controlled takes place) and at low cost. The consequences have been far-reaching in every facet of our life, and there still seem to be no limits to development. It is evident, however, that the social

Table 1-1 Historical development of manufacturing unit processes

Year	Casting	Deformation	Joining	Machining	Ceramics	Plastics	Machine and Controls
4000 BC	Stone, clay molds	Bending, forging (Au, Ag, Cu)	Riveting	Abrasion: stone, emery, garnet, flint	Earthenware, hand building	Wood, natural fibers	Wedge, manual control
2500	Lost wax (bronze)	Shearing, sheet forming	Soldering, brazing	Drilling, sawing	Glass beads, potter's wheel		Wheel, cord drill
1000		Hot forging (iron), wire drawing (?)	Forge welding (iron), glueing	Iron saws, turning (wood)	Glass pressing, glazing		Lever, pulley, cord lathe
AD 0		Screw press, coining (brass), forging (steel)	Diffusion bonding (jewelry)	Filing	Glass blowing		Crank
1000		Wire drawing			Stoneware, porcelain (China)	Protein glues	Waterwheel
1400	Sand casting, cast iron	Water hammer		Sandpaper, clock making	Majolica, crystal glass		Connecting rod, flywheel
1600	Permanent mold	Tinplate can, rolling for coinage			Slip casting		Cams, wheel lathe
1800	Flasks	Deep drawing, rolling (steel), extrusion (Pb)		Boring, turning, screw cutting	Extrusion, plate glass, porcelain (Germany)		Steam engine, boring mill, drill press, punch cards
1850	Centrifugal casting, molding machine	Steam hammer, tinplate rolling Mg, Al, Ni		Shaping, milling, chemical milling	Window glass from slit cylinder	Vulcanization	Mechanization, copying lathe, milling machine
1875	Bessemer steel	Rail rolling, continuous rolling			SiC, vitrified grinding wheel	Rubber extrusion, molding, celluloid	Turret lathe, universal mill
1900	Aluminum	Tube piercing, extrusion (Cu)	Oxyacetylene, arc, electrical resistance welding	Hobbing, high-speed steel	Automatic bottle making, synthetic SiC, Al_2O_3		Electric motor, automatic screw machine, geared lathe, GO–NOT GO gage

Table 1-1 (Continued)

Year	Casting	Deformation	Joining	Machining	Ceramics	Plastics	Machine and Controls
1920	Die casting	W wire from powder, precipitation hardening of Al, Co superalloy, Be	Coated electrode, gas-W arc, automatic welding, thermal spraying		WC, rolled glass	Casting, cold molding, injection molding, Bakelite, vinyl acetate	Hard automation (electrical), production line
1940	Lost wax (engineering), resin-bonded sand	Extrusion (steel), powder metal parts, Ti; Ni superalloys, isostatic pressing	Roll bonding, submerged-arc, stud, and inert-gas welding, covered electrodes, structural adhesives	ECM		Transfer molding, foaming PVC, PE, acrylics, polystyrene, nylon, polyesters, synthetic rubber	Roughness measurement
1950	Ceramic mold, nodular iron	Cold extrusion (steel)	TIG, MIG, electroslag welding	EDM	Glass ceramics	ABS, silicones, fluorocarbons, polyurethane	Numerical control (NC)
1960	Rapid solidification	Hydrostatic extrusion, superplastic forming, HIP	Plasma-arc, electron-beam welding, wave soldering	Synthetic diamond	Float glass	Acetals, polycarbonate, polypropylene, cyanoacrylate	CNC, CAD, CMM, group technology robot, automatic tool changer
1970	Vacuum casting, directional solidification	Isothermal forging, HIP, powder forging	Laser	Coated tools	CBN, technical ceramics, optical fibers	Polyimide, aramids, polybutylene	CAD/CAM, adaptive control, programmable controller
1980	Single crystal, squeeze casting	Thixoforming, powder injection molding	Seam tracking, surface-mount soldering	High-speed machining	Nanoscale powders	Composites, reaction injection molding	CIM, flexible manufacturing, AGV, artificial intelligence
1990		Nanophase powders		Hard machining	High-temperature superconductor		Lean manufacturing, Agile manufacturing

consequences of these changes will be as fundamental as those wrought by the nineteenth-century industrial revolution, and most observers now agree that we are in the midst of the second industrial revolution.

A characteristic of this industrial revolution is that, in addition to the possibility of replacing most physical labor, it is now also feasible to enhance and sometimes even replace mental effort. Some consequences of these developments are already evident: Many dangerous, physically demanding, or boring jobs are performed by machines or robots controlled by computers; product variety is increasing; quality is improving; productivity—as expressed by output per unit labor—is rising; demand on natural resources is decreasing.

One of the most pervasive features of development is the dramatic increase in our capability for gathering and processing information, and it is commonly accepted that we have entered the information age. There are those who believe that we are in the process of developing into a "postindustrial" society in which manufacturing will wither and the service sector based on information processing will generate wealth. To explore the reality of such beliefs, it is necessary to consider the role of manufacturing in the economy.

1-2   THE ECONOMIC ROLE OF MANUFACTURING

Manufacturing has often been cast as the villain on the stage of human development. Indeed, the first industrial revolution began with little concern for the very people who made this revolution possible. Yet the factory was the alternative willingly chosen by the masses seeking to escape a rural existence burdened with famine and disease. Modern demographic studies show that the misery of rural life prompted people to crowd into cities even before the first industrial revolution. Since then, the excesses of the early industrial revolution have been moderated and the growth of manufacturing has led to undeniable advances, not only in providing an abundance of material possessions, but also in creating the economic basis for genuine improvements in the quality of life.

There are no universal measures to express quality of life but, in the absence of better measures, the *gross national product* (GNP; the sum of the value of all goods and services produced in a national economy) can be taken as a measure of material well-being. (Even for this, it is an imperfect measure because it excludes the value of work performed in the home, in voluntary organizations, and so forth. Thus, it presents a distorted picture in favor of industrially developed nations.)

If one analyzes the components of the GNP, it is evident that material wealth comes from only two substantial, basic sources: material resources and the knowledge and energy that people apply in utilizing these resources. Agriculture and mining are of prime importance, yet they represent only 3–8% of the GNP of industrially developed nations. Manufacturing claimed the largest single share until the 1950s. Since then, much of the growth has taken place in the service sector, and recent data (Fig. 1–1) would suggest that—at least in highly developed

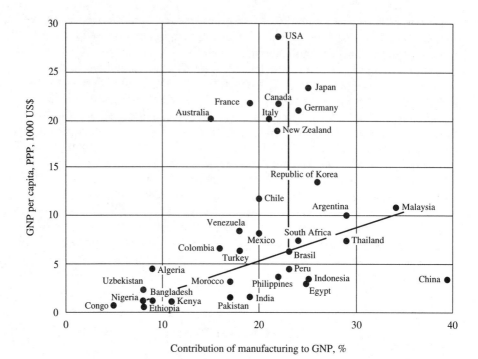

Figure 1-1   The standard of living, as expressed by the per-capita output of the economy, increases with the development of manufacturing. At a higher stage of industrial development, output reflects the increased sophistication of manufacturing. To account for differences in purchasing power, GNP for 1997 is presented on the basis of purchasing power parity (PPP). (*Data from World Development Report 1998/99, World Bank, 1998.*)

economies—material wealth is independent of the contribution of manufacturing to the GNP. This is, however, an illusion. What the numbers fail to show is that increasing wealth is based on an increasingly sophisticated manufacturing sector; this in turn creates the need for many similarly sophisticated supporting activities such as research, design, and financial services, distribution, maintenance, and field service of products, and even the hospitality and travel industry connected with manufacturing. For statistical purposes, all these supporting activities are classified as services. Yet, unless a nation is exceptionally well endowed with natural resources, a strong service sector can exist only if there is a similarly strong manufacturing sector. Only the interactions of the two can secure competitive advantages in a global economy where the simpler tasks migrate to low-wage environments. It is often said that, in the information age, knowledge is the most valuable commodity. This is quite true, but it is also true that knowledge itself can be bought relatively cheaply. Wealth is generated most abundantly by *producing tradable articles in which knowledge is embodied.*

No nation exists in isolation any more, and international trade has grown to the point where the economies of all nations are interconnected. The flow of goods and services is increasingly liberated from the many restrictions of earlier times. For the vital manufacturing industries to grow and survive, they must be competitive on the global scale. Manufacturing thus occupies a central position in the economy of nations and, indeed, in the economy of the world.

Many economic activities provide essential inputs to manufacturing; at the same time, manufacturing creates all the products that are needed for the conversion of energy and raw materials, construction, transportation, communication, health care, entertainment, and leisure. These industries and businesses, together with the individual consumer, dictate the range of products that manufacturing has to provide. Competitiveness implies a need not just for lowest cost, but also for developing and producing truly world-class products, and an essential step in reaching this goal is the education of engineers and technologists for the manufacturing industries.

Since the scope of manufacturing is so enormous, we will narrow it for our purpose to the manufacture of "hardware," durable articles for consumption and for machineries of production. A distinction is often made between high-technology (high-tech) and low-technology products; as we shall see, the distinction can be deceptive.

1-3   MANUFACTURING AS A TECHNICAL ACTIVITY

To recognize the challenges faced by manufacturing engineers and technologists, consider but a few examples taken from everyday experience.

Example 1-1

The jet engine is a machine designed with the most advanced knowledge of thermal and fluid engineering principles, using all available computer models for design and performance evaluation. Its manufacture too demands the most advanced techniques, not least because components are subjected to extreme conditions in operation. A turbofan engine (Fig. 1–2) consists of essentially three major sections. First, air is compressed in the low- and high-pressure compressor sections. Second, fuel is introduced into the compressed air in the combustor, and the fuel-air mixture is ignited. The resultant hot gases are discharged from the combustor through the high-pressure and low-pressure turbines, which extract energy from the gases to drive the high-pressure and low-pressure compressor rotors. Third, a large fan at the front of the engine increases thrust by increasing the mass of air displaced: in a high-bypass turbofan engine the bypass ratio is 6:1 or higher, which means that six times as much air passes around the engine as passes through it. Additional benefits of the fan include reduced fuel consumption and lower engine noise.

Thermodynamic efficiency increases with increasing turbine entry temperature, and fuel consumption drops with an increasing compression ratio (Fig. 1–3a). Compression raises the temperature of air and, in recent engines, the final-stage compressor blades are required to run red hot (Fig. 1–3b); we shall see throughout this text that new materials and technologies had to be developed for making compressor blades.

An even more demanding application is the turbine where entry temperatures are limited by the temperature capability of blade and vane materials (Fig. 1–4a). Developments in

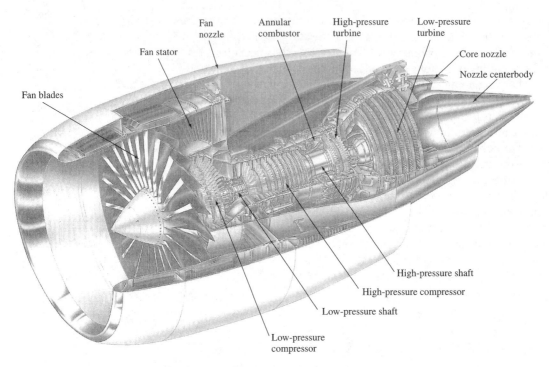

Figure 1–2   Today's jet engine is a highly sophisticated manufactured product; many of its parts must operate at high temperatures and stress levels, demanding advanced methods of manufacturing. The GE90 engine is one of the engines approved for the Boeing 777 aircraft. It is 5.3 m long, has a fan of 3.1-m diameter, and is rated at 410-kN thrust at a bypass ratio of 9:1. (*Courtesy General Electric Company, Cincinnati, Ohio.*)

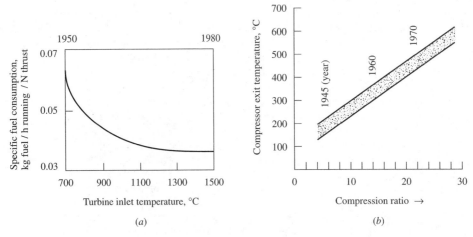

Figure 1–3   Higher operating temperatures result in higher efficiency and lower fuel consumption. Hence, (*a*) specific fuel consumption decreases with increasing turbine entry temperatures and (*b*) higher compression ratios lead to higher temperatures even in the compressor section. Both stages require new materials and manufacturing techniques. [(*a*) *Reprinted with permission from M. F. Ashby and D. R. H. Jones, Engineering Materials, Pergamon Press, 1980; (b) based on data by G. W. Meetham, The Metallurgist and Materials Technologist,* **8**(11):589–593, 1976.]

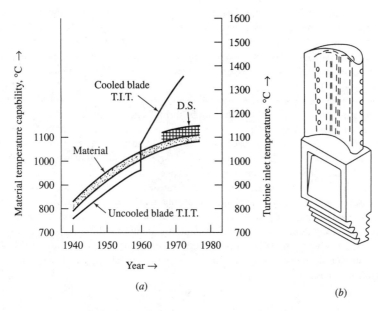

Figure 1–4   Turbine inlet temperatures (T.I.T.) can be increased by cooling the blades (a); for this, new techniques of making holes into hard materials had to be developed. D. S.: directional solidification. (*After M. F. Ashby and D. R. H. Jones, as in Fig. 1–3a.*). In the example shown (b) the holes are molded into a precision-cast turbine blade.

superalloys allowed a gradual increase in operating temperature but a jump in performance came from cooling the blades by passing uncombusted compressor air through small holes in the airfoils (Fig. 1–4b). Thus, in addition to developing new manufacturing processes for increasingly difficult-to-manufacture alloys, techniques had to be found for making very fine, deep holes in very hard materials. Even though few people know much about the jet engine, everybody recognizes it as a high-tech product.

Automobiles used to be regarded as typically middle- or even low-technology products, and there were those who suggested that their manufacture should be left to less-developed economies. No knowledgeable person would do so now. The design of automobiles involves the same advanced principles and techniques as that of aircraft, and many changes had to be made by the automobile industry to satisfy new demands regarding safety, pollution level, gasoline consumption, durability, and quality of the product. These changes have affected the choice of materials and manufacturing techniques. For example, early automobile bodies had a steel frame to which wooden panels were attached. Soon the wood was replaced with an all-steel body that was, however, still secured to a heavier frame (a chassis). The desire for weight reduction led to unibody construction; the all-welded, frameless steel bodies were made of low-carbon steel which has highly desirable forming properties. As we shall see, further efforts at weight reduction and increased corrosion resistance led to the introduction of galvanized steel, high-strength low-alloy steels, and aluminum alloys specially designed for automotive applications. Much had to be learnt about the forming behavior of these materials

Example 1-2

so that their ability to form complex shapes could be predicted. Polymers have also been used, first as fiberglass-reinforced epoxies and, more recently, as mass-produced body parts attached to a precision-machined drivable steel space frame. The car itself, in its entirety, has become a much more sophisticated product, with several on-board computers performing critical control functions. Auto production now has all the characteristics of a high-tech industry.

Example 1-3

Bioengineering offers examples where manufacturing benefits humanity directly. One of the most frequently performed orthopedic surgeries is the replacement of arthritic hip joints with surgical implants (Fig. 1–5). Materials had to be found that could be implanted into the body without adverse reactions and that could withstand the substantial dynamic loading (of millions of cycles per year) imposed by ever younger and more active patients. Materials for these applications show the benefit of technology transfer: Several alloys are direct descendants of alloys used in jet engines. The shape of the spherical head is critical, and techniques had to be developed for high-precision machining. Better ways of fastening the replacement in the bone also had to be found, since adhesive joints frequently fail after some years of service. A more recent approach uses the regenerative capacity of the bone to establish the bond, and manufacturing processes had to be developed to provide intricate channels in the surface of the part into which new tissue can grow.

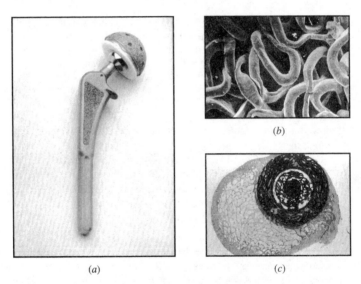

(a)   (b)   (c)

Figure 1–5   Hip joint replacements are advanced manufactured products.
(a) The titanium-alloy stem is implanted into the femur
(thighbone) and the cup into the acetabulum (socket of the
hipbone), with a wear-resistant polymer lining providing the
sliding surface. (*Courtesy Zimmer Inc., Warsaw, Indiana*)
(b) Pure titanium wire of 0.25-mm diameter is sintered into a
porous mass to (c) provide room for the ingrowth of fresh
bone tissue which fixes the implant in the bone of an
experimental animal. (*Courtesy Dr. W. Rostoker, University of
Illinois at Chicago.*)

Microelectronics, which is at the heart of the second industrial revolution, had its origins in physical phenomena that could be exploited only by adapting old and creating new manufacturing technologies for new materials. We shall become acquainted with advances in manufacturing processes that have made it possible to create millions of components on a single silicon chip. As a result of these advances, performance has increased dramatically while cost decreased substantially, a combination never paralleled.

Example 1-4

1-4   SCOPE AND PURPOSE OF BOOK

In this book we will explore manufacturing processes with an emphasis on physical principles. With the move into the information age, there has also been a proliferation of computer applications in manufacturing. It would thus be tempting to conclude that knowledge of physical principles is losing significance and that information processing—taken in the narrower sense of data processing—will become the central activity in manufacturing. Nothing could be further from the truth. As is often said, data do not equal information, and information is not knowledge. Information processing, while of great importance, is still only a tool, a tool that on its own cannot ensure competitiveness. We shall also see that the computer can be called upon for both developing and controlling processes. Yet, even the ultimate computerization of a process will be of no avail if a more original mind meanwhile develops a new process that wipes out the advantages of the old one. Existing processes can be effectively controlled and improved and new processes can be developed only if the role of process variables is understood and if, at least, an elementary yet physically sound model of the process can be formulated. For all these reasons, this book is devoted to developing an understanding of the physical background of unit processes used for the manufacture of parts. One must, of course, recognize that parts will have to be assembled into finished products, but these assembly operations will be touched upon only briefly.

General Approach   In choosing the particular approach adopted in this volume, the guiding principle was that fundamental, general principles are more important than details. The number of manufacturing processes in existence defies enumeration, let alone description, in a single volume. There are already a number of encyclopedic books available in which details on individual processes can be found. Thus, the purpose is not to give detailed information, but to *impart a knowledge of principles* which can then be used to consider process capabilities and their implications for product design, improve existing processes, create new ones, and interpret the information presented in books and, increasingly, computer databases. To quote Sherlock Holmes: "A man should keep his little brain attic stocked with all the furniture that he is likely to use, and the rest he can put away in the lumber room of his library, where he can get it if he wants it." (A. Conan Doyle, *Adventure of the Five Orange Pips*, Crown Publishers, New York, 1976.)

In dealing with principles, it is recognized that many engineers and technologists are needed to make up a manufacturing team. Some team members may be specialists in manufacturing, but others are experts in materials, mechanical, industrial, or systems engineering and technology. Some students may embark on their exploration of manufacturing after taking a course in the properties of materials and in the strength of materials; others may have no more preparation than high-school physics and chemistry. The book is constructed to cater to both groups. Those who have the appropriate preparation may simply skip the background material or read it as a refresher; others may use the background material given here as a jumping-off point for their studies. For transparency, background material is presented in separate chapters or sections. The aim is that, at the minimum, a student should acquire a knowledge of process principles to the level where useful interaction with specialists is possible. At a higher level, the foundations for specialization will be laid.

In general, an attempt is made to resolve the dilemma of depth versus breadth of treatment. A fully quantitative treatment would require excessive length and could easily obscure the larger issues; a fully qualitative treatment would give little guidance to intelligent product design or process selection. The compromise adopted here aims to retain the sound scientific principles that must be brought to bear on the subject; in particular, it emphasizes the mutual constraints exerted by materials and processes on each other and on design possibilities. Many processes can now be modeled in substantial detail with numerical methods; this is truly the domain of the specialist, and only the possibilities can be indicated here. Some processes are, however, amenable to simple quantitative treatments which also serve to clarify the physical foundations, and such treatments will be included, but without proofs or derivations that could—and do—fill up a library of specialized books. Derivations of the most important equations are given in the Instructor's Manual and can be obtained from the instructor.

The material is organized to make study at different levels easier. A general understanding of processes and underlying physical principles can be gained by reading the nonquantitative parts of the text. Basic process development becomes possible with the inclusion of quantitative aspects. In the spirit of viewing manufacturing as a total system in which design and analysis are integral parts, the design implications of process capabilities and restraints are emphasized whenever possible.

Examples and Problems   These are designed to further this understanding. In most chapters, problems are presented in three groups:

A: *Testing your knowledge*. This calls for a straightforward account of material presented.

B: *Applying your knowledge*. Problems of this kind could arise in the course of a discussion with a group of other people and require reasoning and logical presentation.

C: *Making quantitative judgments*. To solve these problems, data have to be looked up in this book (and, in limited instances, in reference books commonly available) and some calculations have to be performed.

Further Reading   The reader interested in further details or in the theories of processes will find ample material in the readings suggested at the end of each chapter. In addition to books, there is an ever-growing variety of computer programs and instructional videos available. The Web is a rich and up-to-date source of information on books and on publications by professional societies, trade associations, and manufacturers.

While not absolutely necessary, access to some basic reference volumes encourages the use of the literature. The most comprehensive single-volume sources of data are: *Metals Handbook Desk Edition*, 2d ed. (1998); *Engineered Materials Handbook Desk Edition* (1995); and *ASM Handbook*, vol. 20, *Materials Selection and Design* (1997), all published by ASM International, Materials Park, Ohio.

Standards   Standards govern many engineering activities. In the United States, standards are issued by the American National Standards Institute (ANSI), American Society for Testing and Materials (ASTM), American Society of Mechanical Engineers (ASME), and other professional and government bodies. There are corresponding institutions in other countries, and international standards are developed by the International Organization for Standardization (known as ISO). From time to time, standards are revised and the year of revision is then shown. ANSI maintains a database of worldwide standards (*A Global Standards Network*) on the Web (http://www.nssn.org).

Units of Measurements   In most engineering activities, measurements are now carried out in units of the Système International d'Unités (SI; see ASTM Standard E380, Standard Metric Practice and National Institute of Standards and Technology *NIST Special Publication 814: Interpretation of the SI for the United States and Metric Conversion Policy for Federal Agencies*). Technical publications now use SI units almost exclusively, and any engineer not conversant with the system will soon be unable to keep up with developments. Non-SI units are not even allowed on labels or in documentation in the European Union, including the United Kingdom. Nevertheless, the inch system is retreating only slowly in the United States, and a conversion table for the most frequently used units will be found in Appendix A and on the front inside cover. In everyday usage, approximate conversions are often made. Such approximations are discussed in Sec. 3-2-1 for dimensions and in Sec. 4-1-1 for stresses. For mass, we will use g, kg, or Mg (1 metric tonne = 1.102 short tons). A ton force (tonf) equals roughly 10 MN. Note that whenever the term *billion* is used, it refers to 10^9.

1-5   SUMMARY

Manufacturing is the making of goods and articles. For the last two centuries, it clearly has been the engine of economic development. Now that we have moved into the information age, it might appear that it is losing its importance. Nothing could be further from reality. True, manufacturing seems to be stuck at some 20% of the gross national product in developed nations, while the service sector is steadily growing. Much of the service sector exists, however, only because it supports an ever more sophisticated manufacturing sector. Manufacturing still provides the goods for our existence, for our housing, transportation, leisure, health needs, and even for information technology.

Globalization and fierce competition make the incentive for the study of manufacturing, if anything, stronger than ever and the present book aims to lay the physical foundations for such study. No single book can cover all topics in their entirety, and suggestions are given for further study.

FURTHER READING

History

McNeil, I. (ed.): *An Encyclopaedia of the History of Technology*, Routledge, London, 1990.
Singer, C., et al. (eds.): *A History of Technology*, vols. 1–8, Oxford University Press, 1963–1984.

General Manufacturing Textbooks

Alting, L.: *Manufacturing Engineering Processes*, 2d ed., Dekker, 1994.
Black, S.C., V. Chiles, A.J. Lissaman, and S.J. Martin: *Principles of Engineering Manufacture*, Arnold, 1996.
Brown, J.: *Modern Manufacturing Processes*, Industrial Press, 1991.
Bruce, R.G. (ed.): *Modern Materials and Manufacturing Processes*, Prentice Hall, 1997.
DeGarmo, E.P., J.T. Black, and R.A. Kohser: *Materials and Processes in Manufacturing*, 8th ed., Prentice-Hall, 1997.
Doyle, L.E., C.A. Keyser, J.L. Leach, G.F. Schrader, and M.S. Singer: *Manufacturing Processes and Materials for Engineers*, 3d ed., Prentice-Hall, 1985.
Edwards, L., and M. Endean (eds.): *Manufacturing with Materials*, Butterworths, London, 1990.
Fellers, W.O., and W.W. Hunt, *Manufacturing Processes for Technology*, Prentice Hall, 1995.
Ghosh, A., and A.K. Mallik: *Manufacturing Science*, Wiley, 1986.
Groover, M.P.: *Fundamentals of Modern Manufacturing*, Prentice Hall, 1996.
Hitomu, K.: *Manufacturing Systems Engineering*, 2d ed., Taylor and Francis, 1996.
Kalpakjian, S.: *Manufacturing Processes for Engineering Materials*, 3d ed. Addison-Wesley, 1997.
Kalpakjian, S.: *Manufacturing Engineering and Technology*, 3d ed., Addison-Wesley, 1995.
Lindberg, R.A.: *Processes and Materials of Manufacture*, 4th ed., Allyn and Bacon, 1990.
Ludema, K.C., R.M. Caddell, and A.G. Atkins: *Manufacturing Engineering: Economics and Processes*, Prentice Hall, 1987.
Ostwald, P.F., and J. Munoz: *Manufacturing Processes and Systems*, 9th ed., Wiley, 1997.
Tanner, J.P.: *Manufacturing Engineering*, 2d, Dekker, 1990.
Todd, R.H., D.K. Allen, and L. Alting: *Fundamental Principles of Manufacturing Processes*, Industrial Press, 1994.
Todd, R.H., D.K. Allen, and L. Alting: *Manufacturing Processes Reference Guide*, Industrial Press, 1994.
Waters, T.F.: *Fundamentals of Manufacturing for Engineers*, University College London, 1996.

Reference Books

Tool and Manufacturing Engineers Handbook, 4th ed., 9 volumes to 1998, Society of Manufacturing Engineers, 1983 onward.

Avallone, E., and T. Baumeister (eds.): *Marks' Standard Handbook for Mechanical Engineers*, 10th ed., McGraw-Hill, 1996.

Beitz, W., and K.H. Kuttner: *Handbook of Mechanical Engineering*, Springer, 1994.

Fuchs, J.H.: *The Prentice Hall Illustrated Handbook of Advanced Manufacturing Methods*, Prentice-Hall, 1988.

Koshal, D. (ed.): *Manufacturing Engineer's Reference Book*, Butterworth-Heinemann, 1993.

Krar, S.F.: *Illustrated Dictionary of Metalworking and Manufacturing*, McGraw-Hill, 1998.

Kreith, F. (ed.): *Handbook of Mechanical Engineering*, CRC Press, 1998.

Kutz, M. (ed.): *Mechanical Engineer's Handbook*, Wiley, 1998.

Kverneland, K.O.: *Metric Standards for Worldwide Manufacturing*, ASME, 1996.

The Making, Shaping and Treating of Steel, 11th ed., 5 vols., Association of Iron and Steel Engineers, 1999–2000.

Machinery's Handbook, 25th ed., Industrial Press, 1996.

Kirk-Othmer Encyclopedia of Chemical Technology, 4th ed., Wiley-Interscience, 1991–1998.

Walker, J.M. (ed.): *Handbook of Manufacturing Engineering*, Dekker, 1996.

Walsh, R.A.: *McGraw-Hill Machining and Metalworking Handbook*, McGraw-Hill, 1994.

Proceedings of Periodic Conferences (Up-to-Date Coverage)

Tobias, S.A., F. Koenigsberger, J.M. Alexander, B.J. Davies (eds.): *Advances in Machine Tool Design and Research* (proceedings of annual conferences; since 1960, Pergamon, Oxford; from 1972, Macmillan, London).

Proceedings of the North American Metalworking (since 1983: *Manufacturing*) *Research Conference* (proceedings of annual conferences: McMaster University, 1973; University of Wisconsin-Madison, 1974; subsequent volumes published by Society of Manufacturing Engineers).

Annual Review of Materials Science (includes up-to-date reviews of manufacturing technologies).

Selected Journals with General Coverage

Advanced Manufacturing Technology
Artificial Intelligence
Advanced Materials & Processes
American Machinist
Canadian Machinery and Metalworking
Computer and Industrial Engineering
Computer-Aided Engineering
Computers in Mechanical Engineering
International Journal of Advanced Manufacturing Technology
International Journal of the Japan Society for Precision Engineering
International Journal of Machine Tool Design and Research
International Journal of Machine Tools and Manufacturing
International Journal of Production Research
Journal of Engineering for Industry (Trans. ASME)

Journal of Engineering Materials and Technology (Trans. ASME)
Journal of Manufacturing Science and Engineering
Journal of Manufacturing Systems
Journal of Materials and Manufacturing
Machine and Tool Blue Book
Machinery
Machinery and Production Engineering
Manufacturing Engineering
Manufacturing Engineering Transactions
Manufacturing Review
Materials and Manufacturing Processes
Metal Progress
Metals Forum
Microtechnic
Precision Engineering
Production
Production Engineer (contains abstracts of papers in all fields)
Robotics and Computer-Integrated Manufacturing
SAMPE Quarterly
SME Transactions

Abstract Journals

Applied Mechanics Reviews
Applied Science and Technology Index
Engineering Index
Metals Abstracts

Trade, Business, and Commercial Organizations (USA)

Abrasive Engineering Society (AES), P.O. Box 3157, Butler, PA 16003.
The Aluminum Association (TAAI), 900 19th St. NW, Washington, DC 20006.
Aluminum Extruders Council (AEC), 1000 N. Rand Rd., Wauconda, IL 60084.
American Bureau of Metal Statistics (ABMS), 400 N. Main St., Manahawkin, NJ 08050.
American Foundrymen's Society, 505 Main St., Des Plaines, IL 60016-8399.
American Gear Manufacturers Association (AGMA), 1500 King St., Alexandria, VA 22314-2730.
American Iron and Steel Institute (AISI), 1101 17th Street NW, Washington, DC 20036.
American National Standards Institute (ANSI), 11 West 42d St., New York, NY 10036.
American Plastics Council, 1275 K Street NW, Washington, DC 20005.
Computer Aided Manufacturing International (CAM-I), 3301 Airport Freeway, Bedford, TX 76021.
Copper Development Association (CDA), 260 Madison Ave., New York, NY 10016.
Ductile Iron Society (DIS), 28938 Lorain Rd., North Olmstead, OH 44070.
Electronic Industries Association (EIA), 2500 Wilson Blvd., Arlington, VA 22201-3834.
Forging Industry Association (FIA), 25 Prospect Ave. West, Cleveland, OH 44115.
Grinding Wheel Institute (GWI), 30200 Detroit Rd., Cleveland, OH 44145-1967.
International Copper Research Association (INCRA), 708 Third Ave., New York, NY 10017.
International Lead Zinc Research Organization (ILZRO), Research Triangle Park, NC 27709-2036.

International Organization for Standardization (ISO), 1 Rue de Varembe, CH-1211 Geneva 20, Switzerland.

Investment Casting Institute (ICI), 8350 N. Central Expressway, Dallas, TX 75206-1602.

Metal Powder Industries Federation (MPIF), 105 College Rd. E., Princeton, NJ 08540-6692.

Steel Founders' Society of America (SFSA), 455 State St., Des Plaines, IL 60016.

Professional and Technical Societies (USA)

American Ceramic Society (ACerS), P.O. Box 6236, Westerville OH 43086-6136.

American Foundrymen's Society (AFS), 505 State Street, Des Plaines, IL 60016-8399.

American Institute of Mining, Metallurgical and Petroleum Engineers (AIME), 3 Park Ave., New York, NY 10016-5998.

American Society for Metals (ASM) (see ASM International).

ASM International, Materials Park, OH 44073-0002 (formerly American Society for Metals (ASM), Metals Park, OH 44073).

American Society for Nondestructive Testing (ASNT), 1711 Arlingate Lane, Columbus, OH 43228-0518.

American Society for Precision Engineering (ASPE), P.O. Box 10826, Raleigh, NC, 27605-0826.

American Society for Quality Control (ASQ), 611 E. Wisconsin Ave., Milwaukee, WI 53201-3005.

American Society for Testing and Materials (ASTM), 100 Barr Harbor Rd., West Conshohocken, PA 19428-2959.

American Society of Lubrication Engineers (ASLE), 840 Busse Highway, Park Ridge, IL 60068-2376.

American Society of Mechanical Engineers (ASME), 3 Park Ave., New York, NY 10016.

American Welding Society (AWS), 550 N.W. LeJeune Rd., Miami, FL 33126.

Association of Iron and Steel Engineers (AISE), 3 Gateway Center, Pittsburgh, PA 15222-1004.

The Clay Minerals Society (CMS), P.O. Box 4416, Boulder, CO 80306.

Federation of Materials Societies, 1899 L Street NW, Washington, DC 20036.

International Institution for Production Engineering Research (CIRP), 10, rue Mansart, F-75009 Paris, France.

International Society for Measurement and Control (ISA), 67 Alexandra Drive, Triangle Research Park, NC, 27709.

Institute of Electrical and Electronic Engineers (IEEE), 3 Park Ave., New York, NY 10016-5997.

Institute of Industrial Engineers (IIE), 25 Technology Park/Atlanta, Norcross, GA 30092.

Iron and Steel Society of AIME (ISS), 410 Commonwealth Dr., Warrendale, PA 15086-7512.

Materials Research Society, 506 Keystone Drive, Warrendale, PA, 15086-7573.

The Minerals, Metals and Materials Society (TMS), 420 Commonwealth Dr., Warrendale, PA 15086-7514.

Society for the Advancement of Material and Process Engineering (SAMPE), 1161 Parkview Drive, Covina, CA 91724-3748.

SAE International (formerly Society of Automotive Engineers, SAE), 400 Commonwealth Dr., Warrendale, PA 1509-0001.

Society of Manufacturing Engineers (SME), One SME Dr., Dearborn, MI 48121-0930.

Society of Plastics Engineers (SPE), 14 Fairfield Dr., Brookfield Center, CT 06804-0403.

The Welding Institute (WI), North American Office, P.O. Box 5268, Hilton Head Island, SC 29928.

The newest Boeing airplane, the twin-engine 777, was designed with the fullest use of concurrent engineering and is being built with international cooperation. (*Courtesy The Boeing Company, Seattle, Washington.*)

chapter

2

Manufacturing

In this chapter we will find a general introduction to manufacturing, including a quick review of:

Manufacturing as a system

The product realization enterprise

Steps in designing a product and process

Concurrent engineering for developing products faster and with higher quality

Integration of manufacturing with the aid of the computer

Control strategies and numerical control

The definition of manufacturing as the making of goods and articles reveals little about the complexity of the problem. A more specific definition is given by CAM-I (Computer Aided Manufacturing International, Arlington, Texas): "A series of interrelated activities and operations involving design, materials selection, planning, production, quality assurance, management, and marketing of discrete consumer and durable goods." This recognizes that, from the simple beginnings when an artisan provided all the necessary mental and physical input, manufacturing has grown to become a *system* with many components that interact in a dynamic manner.

2-1 THE MANUFACTURING ENTERPRISE

A manufacturing entity (a company or a branch of a larger corporation) usually possesses some special strengths, such as specific technology, knowledge, or equipment. On the basis of these strengths, new products must be developed to maintain or gain market share. For this, the appropriate markets are identified, their magnitudes estimated, and the existing and potentially emerging competition appraised. After the market and its future development are projected, new products are identified. The process of *product realization* involves a number of interconnected activities (Fig. 2–1).

2-1-1 Specification Development

In many ways, this is the most important phase. Needs are defined in terms of function, performance, time constraints, cost, and other criteria. Specifications that fail to meet the customer's needs lead to failure of the product in the market, but unnecessarily tight specifications lead to inflated cost and lack of competitiveness. Therefore, neither excess performance nor excess life is needed, and performance must be optimized. In general, it is found that a product satisfying minimum requirements can be produced at some minimum cost. Performance can often be increased—and thereby the selling price substantially raised—with relatively little increase in cost. Further improvements may lead to much higher manufacturing cost and only marginally increased customer appeal, thus the sell-

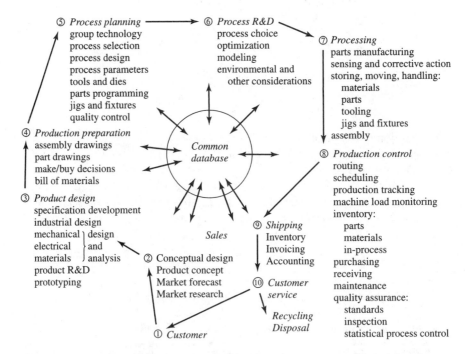

Figure 2–1   Product realization involves a large variety of activities, many of which have become specialties on their own. More properly, manufacturing is regarded as a system with interdependent activities. Interaction can be strengthened by maintaining a common, computerized database.

ing price cannot be raised proportionately. Consequently, there is always a point beyond which performance cannot be economically improved. To assure that the product will indeed succeed against the competition, it is examined against the best in the field (*benchmarking*). Sometimes an existing product, for which drawings and specifications are not available, will be reproduced by *reverse engineering*. The technique is used also for the evaluation of competitive products: the product is disassembled and its good features noted.

2-1-2   Conceptual Design

This is the most creative phase: The product is designed in broad outlines to fulfill its intended function, that is, to operate satisfactorily over its expected life and to fill the needs of the customer. A significant change from past practices is that the input of the customer is (or should be) eagerly sought. Decisions will be influenced by the anticipated size of the market. Developments in manufacturing technology have led to the situation where the consumer often expects personalized products at mass-production prices. When this is the case, design must make the product suitable for manufacture in flexible manufacturing systems. Detailed drawings are not needed at this stage; it is sufficient to make conceptual sketches that show the parts and their relation to each other. Preliminary material choices are made, and, since material always has an influence on process, production processes are tentatively identified.

2-1-3   Product Design

The product, whether it be a machine tool, household machine, building product, computer, automobile, aircraft, chemical processing plant, power station, oil drilling rig, cookware, or beverage container, is then designed to satisfy several criteria. This has led to *design for X* (DFX), where *X* stands for a growing list of criteria, a few of which are given below.

1. *Industrial designers* strive to create a visually appealing, functional product that the customer will be willing to buy.

2. Most products are assembled from a number of components; *design for assembly* (DFA) strives to make this as simple as possible by eliminating superfluous parts, combining functions of parts, and designing parts in a way that facilitates their assembly. Large assemblies may be broken down into *subassemblies* or *modules*.

3. *Mechanical and electrical designers and analysts* ensure that the product will properly function. This requires the choice of appropriate materials, often in cooperation with materials specialists. Most phases of product design can take place on a computer. With the aid of *geometric modeling*, the designer can explore a number of options which can then be analyzed with the aid of software packages (including those for *finite-element analysis*, FEM). Design can be optimized in a much shorter time, rapid design changes become possible, and changing consumer demands can be satisfied. Both assemblies and parts can be designed with assurance that they will properly fit together. Beyond this, *failure mode and effect analysis* (FMEA) can be carried out to examine possible failure modes and assess reliability of systems and components. A library of standard components can be built up. In some instances physical models will still have to be built, increasingly by *rapid prototyping* techniques, and a *prototype* product may also be made and tested. Computer-aided analysis often eliminates or minimizes the need for physical testing. For example, computer simulation of the behavior of an automobile body in a collision minimizes the need for actual crash tests on prototypes. Thus, activities indicated in blocks 3 and 4 of Fig. 2–1 are performed in what has become known as CAD (*computer-aided design*).

4. The product must serve the customer, with due regard to the physical capabilities and limitations of operators or consumers. These aspects are the subject of the science of *ergonomics* which takes a general, holistic approach to the relationship between people and machines.

5. The product must be easy to *maintain* over its intended life (*design for maintenance*); for this, parts needing service must be accessible and, if unavoidable, disassembly must be simple.

6. At the end of its life, *reuse*, *recycling* or, if unavoidable, *disposal* of the product must be possible in a safe and ecologically acceptable way. This too calls for ease of disassembly (*design for disassembly*) or separation by some mechanical or chemical means. Recycling (or, where possible, reuse) has assumed great significance in recent years for several reasons. It often requires less energy than production from raw materials; reduces waste volume and thus reduces pressure on landfill sites or incinerators; reduces pollution of the soil and atmosphere; and reduces the need for raw materials. With the rapidly increasing consumption of material goods, *life-cycle engineering* or *green engineering* is becoming more important.

7. Most important, all the above criteria must be satisfied while also ensuring ease of manufacture (DFM: *design for manufacture*). This requires not only close cooperation between industrial, mechanical, and electrical designers and manufacturing specialists, but also demands that designers should have some awareness of the manufacturing consequences of their decisions. Seemingly minor changes may often present (or remove) enormous manufacturing problems, thus affecting the cost, quality, and reliability of the product.

8. We shall see that all manufacturing processes are subject to variations. *Design for quality* (DFQ) leads to design and process choices that reduce the magnitude of these variations and minimize their impact on product performance.

9. The process of design often reveals areas where research and development are needed and, vice versa, research often leads to ideas for new products. Thus, *product research and development* (product R&D) is an integral part of manufacturing in an increasingly competitive world.

Product design is a subject of enormous importance and is treated in an ever-growing number of books and publications. Increasingly, software is available that embodies much practical experience, theory, and models for both design and manufacture.

2-1-4   Make or Buy

Once a product is designed, *production drawings* (or *computer databases*) are prepared of the assembly and of all parts other than standardized, mass-produced components such as screws, rivets, dowel pins, and bearings. Decisions can then be made on what parts should be bought from outside suppliers and what parts should be produced in house. As a general rule, it is almost always more economical to buy components and modules that are available as standard products (motors, clutches, valves, cylinders, etc.). A *bill of materials* is prepared which, in many ways, is central to the manufacturing process.

2-1-5   Process Design

For components produced in house, process design is carried out. As with product design, process design is not an isolated activity.

1. The best *process is selected*, and process parameters are chosen to optimize the quality and properties of the finished product and to facilitate inspection for quality control. Processes too are measured through benchmarking against the best in the field.

2. *Dies* are designed, tooling is chosen and, if the tool must follow a prescribed path, a *tool path* is selected and programmed. Beginning with the 1950s, computers have been increasingly used for this. Information contained in drawings is transformed into digital form for the *numerical control* (NC) or *computer numerical control* (CNC) of machines. When the part geometry is created by CAD, the database already exists and can be directly employed.

3. To facilitate processing and assembly, *fixtures* are designed to hold the workpiece in the correct position in relation to the machine tool or to hold several workpieces in the correct position relative to each other. *Jigs* perform a similar function but also incorporate guides for the tool.

4. Beginning with the 1970s, the computer has been used also for process optimization and control, materials management, material movement (including transfer lines, robots, etc.), scheduling, monitoring, etc. This entire field is now generally called *computer-aided manufacturing* (CAM); it encompasses blocks 5 to 8 in Fig. 2–1. (This group of activities is often described as manufacturing engineering in the narrower sense.)

5. A strong competitive position requires also that new processes be developed and existing ones improved through *process research and development*.

New processes often make it possible to develop new products, thus further increasing competitiveness. Process development on the production scale can be very expensive. Therefore, the fundamentals of processes are often explored in the laboratory or office. *Process models* can be used to explore the influence of process parameters. Two approaches are possible:

a. In *physical modeling* the process is conducted on a reduced scale or simulating materials are used that are easier and cheaper to work with than the real materials.

b. In *mathematical modeling* equations are set up that express the response of the process to changes in process parameters. Such models often require lengthy computations which are made *off line* (in the office), although with increasingly powerful computers and modeling techniques some *on-line* (real-time) modeling is becoming possible.

Whichever modeling approach is used, a sound understanding of physical realities is essential. Modeling is an enormous topic on its own; the present book aims to provide the physical background needed for even the simplest model.

6. In some instances the practical problem is so complex that only experts with long experience can solve it. The knowledge, logic, and judgment of the expert can be captured in *expert programs*, developed in cooperation between the expert and system specialists (also called knowledge engineers). The program contains facts generally available to experts in the particular field; rules of thumb (heuristics), which allow the expert to make educated guesses even when data are incomplete; and inferences, i.e., rules of good judgment. Special programs are available that reduce the programming effort needed for building expert systems; nevertheless, expert programs tend to be lengthy and are expensive to produce. Once completed, they allow a less-experienced person to find the solution to the problem by interacting with the program through if-then sequences.

7. In choosing and developing processes, their impact on the *environment* (air and water pollution, noise, vibration, etc.) and on the *safety and health* of operators and other people must be considered. Manufacturing often involves high temperatures, molten metal, highly stressed tooling, flammable or toxic liquids, and activities that generate noise, smoke, fumes, gases, or dust. It is imperative that appropriate precautions and remedial measures be taken. Beyond the social responsibilities of the engineer and technologist, there are also legal requirements, such as the regulations of the Environmental Protection Agency (EPA) and the Occupational Health and Safety Act (OSHA) in the United States and corresponding measures in other countries.

8. For components produced by a vendor, many of these functions are performed by the vendor, ideally in cooperation with the purchaser. In the past, the main producer (original equipment manufacturer, OEM), such as an automotive company, would design all details of the product and the outside vendor had to produce to these plans. Orders were often awarded purely on the basis of lowest cost. There is an increasing trend toward a cooperative scheme where only the functional specifications and spatial constraints are given and the vendor is responsible for design, manufacture, and quality. This approach demands close partnership between the vendor and final producer, from the early stages of design onward, since the expertise of the vendor may help in improving the basic design itself.

2-1-6   Production

The actual process of production takes place on the workshop floor. Once a product is established, customer orders are fed into the system at this point.

1. *Plant layout* is chosen to fit the characteristics of production.

2. *Process monitoring* is set up to observe critical characteristics of processes, check the dimensions, quality, etc. of parts and, when needed, activate process control to take corrective action.

3. *Material movement* is the most important auxiliary function. Raw materials, partly finished parts, tooling, and jigs and fixtures must be made available on a timely basis. Large stocks used to be maintained to assure continuity of production; this has now largely been abandoned in favor of *just-in-time* (JIT) delivery.

4. *Assembly* of manufactured and purchased parts is the final phase. After checking, the products are ready for packaging and shipping. Inventory control feeds information back to the production process on the basis of sales performance.

5. The complex sequences of production require a strong *manufacturing organization*. The status of production must be known. Formal methods of quality assurance must be established, together with a plan for preventive maintenance of equipment. An up-to-date inventory of parts in process, combined with inventories of purchased materials and parts, must be maintained to ensure that no shortages develop that could delay production and assembly. For a running analysis of performance, machine loading (utilization) and machine and labor performance are monitored. Many of these activities are in the domain of industrial engineering, while others are regarded primarily as management tasks. Superior organization, even of existing technology, can lead to substantial competitive advantage.

2-1-7   Customer Relations

In many ways, this is the central task of manufacturing.

1. Unless the customer provided input into the design process, contact begins with delivery of the product. An important part is provision of well-written user's guides and service manuals.

2. Field service ensures the continuing performance of products delivered to the customer. It is here that the information loop is finally closed: Feedback is valuable in sharpening up production practices and, if necessary, changing the design (this might be regarded as a failure of concurrent design but can be unavoidable in the most advanced products).

2-2 SEQUENTIAL MANUFACTURING

Even this very brief and in many ways incomplete discussion of the manufacturing system indicates the complexity of the problem. Efficient and competitive manufacturing requires close cooperation between the various activities, so that they become truly parts of an *interacting, dynamic system*. Unfortunately, for many years this ideal had seldom been reached.

The problem was—and to some extent still is—that the traditional approach views manufacturing essentially as a sequential process, following the direct arrows from activity to activity in Fig. 2–1. Companies are then organized into departments with clearly separated functions, in which highly specialized people make decisions with little regard for the consequences of their decisions. In effect, virtual walls are erected between departments. Nowhere has this been more harmful than in the wall between design and manufacturing. Design dictates, to a large extent, process possibilities and thus the cost of future production. As shown in Fig. 2–2, in *sequential manufacturing* the *actual* expenditure on design can be quite low, but the cost *committed* to production is high. Design changes

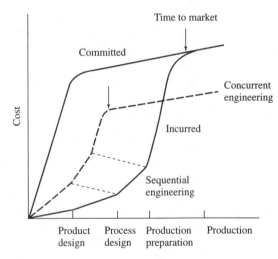

Figure 2–2   Sequential engineering leads to isolation of activities; problems arising in production force expensive design and process changes. Concurrent engineering considers all aspects from the beginning; the product is on the market sooner and its cost is lower.

are forced by problems that surface during production or, worse yet, in the service of the product. Since even a small change may affect many other components or functions, the consequences are steeply increasing costs and a long product maturation cycle. Therefore, companies practicing sequential engineering find themselves at a disadvantage against their more nimble competitors.

2-3   CONCURRENT OR SIMULTANEOUS ENGINEERING

Recognition of these problems has led to the introduction—or, rather, reintroduction—of *concurrent* or *simultaneous engineering*. As the name implies, activities are no longer isolated nor do they follow each other; instead, they overlap or take place simultaneously. The concept is not new. The artisan practiced simultaneous engineering, as did the Wright brothers. Even Henry Ford did it: His team was small, everyone was aware of the customer's needs, available materials, and the capabilities and limitations of processes, and design fitted the processes then available. The task became more difficult as products grew more complex, the variety of materials and processes proliferated, knowledge expanded, and individuals became more and more specialized. Concurrent engineering at this level demands a team of experts with maximum interaction. The team is often brought together in a common location, although computer-aided telecommunication can allow cooperation from several geographic locations.

Concurrent engineering starts with the recognition that design is always an iterative process and that product design has manufacturing consequences. At early stages of design, changes are made easily and at a low cost, and the product can be designed to assure manufacturing at low cost and highest quality. The actual cost of design will be higher, but the cost committed to manufacturing is lower, and the total time required to reach mature production is reduced (Fig. 2–2). When properly practiced, the results of concurrent engineering can be dramatic. For example, American automakers used to take 6 years to develop a new car model. With concurrent engineering, this has been halved and may even be further reduced. Shorter product development time is important not only because of reduced cost but also because the life of products is becoming shorter. In some industries, particularly electronics, development time is now longer than product lifetime. With the development of powerful three-dimensional (3-D) visualization software, digital prototypes of entire assemblies such as automobiles can be constructed. This allows exploration of more alternatives and, by reducing the number of physical prototypes required, greatly accelerates product development.

The newest member of the Boeing aircraft family, the midsize, twin-engine 777, offers an example of concurrent engineering in the fullest sense. Boeing held intensive discussions—including many group sessions—with a number of airlines, to define and develop the new airplane's configuration and specifications. For example, the design allows positioning of galleys and lavatories in 25.4-mm (1-in) increments within "flexibility zones" of the fuselage,   **Example 2-1** allowing reconfiguration of the plane in 3 days instead of the usual 2 to 3 weeks. With wings of 60.9-m span and advanced design, the plane can climb and fly faster and at higher altitudes. Airlines can specify General Electric, Pratt & Whitney, or Rolls Royce engines of 340-kN thrust, each offering low noise level and high fuel efficiency. Much of the structure is made of advanced versions of aluminum alloys of high strength and corrosion resistance. Vertical and horizontal tails and the floor beams of passenger cabins are made of carbon-fiber-reinforced plastics. Composites and adhesives account for 9% of the structural weight.

The engineers of some key customers worked side by side with Boeing designers to ensure that the plane will fill their needs. Computer simulation was used to design and electronically assemble the entire airplane, increasing accuracy and improving quality. Similar efforts were made by many suppliers of parts. Firms in the United States, Canada, Europe, and Asia-Pacific (particularly Japan) contribute to manufacturing. Simulated fly cycles supplemented actual flight testing, and the plane earned Federal Aviation Administration (FAA) approval to fly extended-range twin-engine operation at service entry. (Source: Boeing Commercial Airplane Group.)

Simultaneous engineering succeeds only when the team has a clear mandate and the full support of management. Within the team, each member has different expertise, but the team functions best if all members have at least a general appreciation of problems and solutions. Smaller companies may not even have a sufficient number of experts to create formalized teams, and then a small group or even an individual may have to provide all input. In view of the vast range of manufacturing processes available, this may seem an impossible task. Fortunately, there are physical principles that can be invoked across a wide range of processes, and one of the aims of the present volume is to address these principles.

2-4   COMPUTER-INTEGRATED MANUFACTURING

The benefits of CAD and CAM can be fully realized only if an effective interface is established between them, creating what is usually referred to as CAD/CAM. Information flow in both directions ensures that parts and assemblies will be designed with the capabilities and limitations of materials and manufacturing processes in mind. Superior products can be created and tremendous competitive advantages gained. Effort spent on design and process changes is reduced by ensuring that such changes are entered in the *common database* and are thus immediately recognized at all stages of design and production. An important benefit is that the introduction of CAD/CAM forces a review and improvement of existing design and manufacturing practices and production planning. CAD/CAM is an important tool also in concurrent engineering.

A logical extension is *computer-integrated manufacturing* (CIM), in which all actions take place with reference to a common database. *Database management* is a complex but not insurmountable task. Drawings and computer models serve only to help people visualize the geometry of parts; no changes are allowed on them. If changes in design, process, scheduling, bill of materials, quality standard, etc. are to be made, they are made in the database; thus, they reflect throughout the organization. The database is continuously updated by most recent information on production, sales, etc. For many industries, total CIM is still in the future, but beginnings have been made.

2-5 CONTROL OF MANUFACTURING PROCESSES

We shall discuss this topic again in Chap. 21, after gaining a closer acquaintance of processes. However, we have to clarify a few terms at this point, so that the potential of process control can be pointed out throughout the discussion of individual processes.

2-5-1 Control Strategies

The different approaches to control can be best explained by reference to a simple example, that of turning a cylindrical component on a lathe. The principles apply to any process.

Manual Control   To understand what task a control system is expected to perform, we must first examine some of the actions of a skilled lathe operator. Let us assume that the part to be machined has been mounted in the chuck and the cutting speed and the feed (the axial movement of the tool during each revolution of the workpiece) have been set. The task to be followed is that of maintaining the diameter of the finished part between specified minimum and maximum values and to ensure that the surface finish satisfies specifications.

A trained and experienced operator possesses knowledge that allows a decision on the depth of cut (the thickness of layer removed in one cut). Operator knowledge is often supplemented or even replaced by instructions provided on the basis of past experience or published data. The important point is that information is stored in some form.

Next, the operator checks the setting of the cross slide by reading the micrometer dial (Fig. 2–3a). In other words, the current status of the machine is sensed.

The operator then determines what changes are needed, makes logical decisions, and communicates these decisions to the system by actuating the screw to set the position of the cross slide. At this point, a correct part will be made, assuming that the dial is correctly calibrated and machine deflections are negligible.

A skilled operator will go further and stop the machine after the beginning of the cut, check the part diameter, and make necessary adjustments.

A highly skilled operator will observe the surface produced, listen to the sound of the machine, and, generally, sense changes that are often difficult to

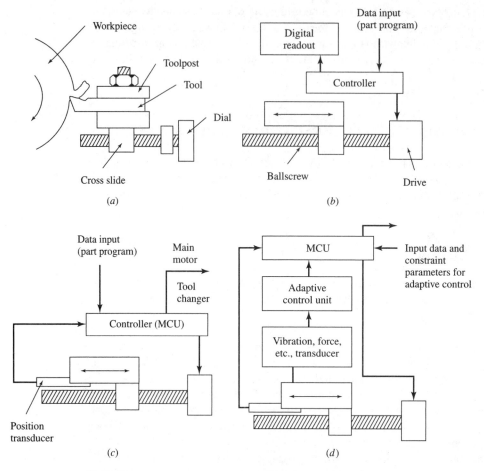

Figure 2–3   All manufacturing processes must be controlled. The example is for cutting on a lathe, under (*a*) manual, (*b*) open-loop, (*c*) closed-loop, and (*d*) adaptive control. Adaptive control takes actions in a manner that a highly skilled machinist would.

describe accurately. For example, under given conditions, vibration (chatter) may develop which causes the surface finish to vary in a systematic manner, resulting in an objectionable surface finish. The operator will then change cutting conditions (speed, feed, support of the part or tool) until the undesirable condition disappears. The operator will also compensate for tool wear, change the tool when needed, and will make sure that the machine tool is not overloaded.

A control system will take over several or all of the functions of the operator.

Open-Loop Control   In this, actions are taken without verifying the results of the action. Actuators may be mechanical (cam, lever, linkage), electromechanical (dc or ac motor, stepping motor), or hydraulic or pneumatic (motor or cylinder). For example, the cross slide of the lathe may be moved by a cam, stepping motor, or hydraulic cylinder to a predetermined position. The setting will be repeated for each part, but it still takes an operator or setup person to confirm that the part is within tolerance and, if not, to reset the cam, mechanical stop, microswitch, or change the program instruction (Fig. 2–3b).

Closed-Loop Control   The control loop is closed when sensors provide *feedback* to the system. In the simplest case, a high-resolution position transducer is added to confirm that the intended position of the cross slide has indeed been reached (Fig. 2–3c). The signal from the transducer is processed by a *comparator* that compares it with the control signal and then issues an error signal to correct the position. In other applications, the control maintains speed or other parameter at a set level. (The oldest example of closed-loop control is a purely mechanical device, the centrifugal governor invented by Watt in 1788 for maintaining a preset speed on a steam engine, irrespective of the load imposed on it. Its modern counterpart is the cruise control of automobiles.) A simple closed-loop control system is ignorant of possible secondary inputs to the system and will go on producing parts even with a worn or broken tool or under conditions of chatter.

Adaptive Control   This is a higher level of control which, in its fullest development, can replace the operator entirely. Sensors are used to provide *feedback of secondary inputs* (in the case of the lathe, in-process measuring devices check the diameter of the part, load cells measure forces, vibration transducers give signals characteristic of the existing cutting conditions, etc.). The feedback signal is then processed so that the control unit can take appropriate corrective action (Fig. 2–3d). Obviously, the corrective action will accomplish its intended purpose only if the effect of process variables on the finished part are known. Interrelations between process variables can be extremely complex, and full adaptive control can be successful only if a sufficiently quantitative model of the process can be formulated. Even if a simple model is used, the constraints of the process or system (maximum force, speed, etc.) must be obeyed (*adaptive control with constraints*, ACC). A more complex model allows optimization (*adaptive control with optimization*, ACO), for example, for maximum production rate.

Artificial Intelligence (AI)   In this, the power of the computer is used to endow the control with some measure of intelligence. As the name implies, the control program is designed to solve the problem the way humans solve it; it is capable of some reasoning, can learn from experience, and, ultimately, can do some self-programming. Alternatively or additionally, elements of expert programs may be incorporated in the control system.

2-5-2   Automation

The word *automatic* is derived from the Greek and means self-moving or self-thinking. The word *automation* was coined to indicate aspects of manufacturing in which production, movement, and inspection are performed or controlled by self-operating machines without human intervention. In general, one may distinguish between several levels of automation. Here we will make a distinction between:

Mechanization   This means that something is done or operated by machinery and not by hand. Feedback is not provided, thus one deals with open-loop control. An example would be the use of a cam to move the cross slide in Fig. 2–3a. A different cam would have to be used (or its position reset) if a cylinder of different diameter had to be machined.

Automation   This will imply closed-loop control and, in its advanced form, adaptive control. Automation utilizes programmable devices, the flexibility of which can be quite different:

1. *Hard automation* refers to methods of control that require considerable effort to reprogram for different parts or operations. Watt's centrifugal governor falls in this category.

2. *Soft or flexible automation* implies ease of reprogramming, usually simply by changing the software.

An important aspect of automation in manufacturing is automation of material movement. We shall come back to this topic in Sec. 21-1; however, we have to give here brief definitions: *Manipulators* are mechanical devices for the movement of materials, tools, and parts, and *robots* are programmable manipulators.

2-5-3   Numerical Control

Machine control has long been practiced with analog devices, for example, by comparing the voltage generated by a transducer to the control voltage. However, the greatest advances in manufacturing control have been made by the introduction of *numerical control* (NC). In the most general sense, NC is the use of symbolically coded instructions for the automatic control of a process or machinery. Various forms of NC have been developed:

Numerical Control   The hardware for basic NC includes the *machine control unit* (MCU, Fig. 2–3) which contains the logic required to translate information into appropriate action; *actuators*; and, if control is closed-loop, *feedback devices* and associated circuits. The plan of action is provided to the MCU in the form of a program on a punched tape or magnetic tape or disk. Programs are usually prepared by a programmer or the machine-tool operator, and are read into the MCU, incrementally, by a tape reader. The MCU is hard-wired to perform various functions. For example, the machine tool or other mechanical device may be expected to move from one point to another. This may be accomplished in several ways:

1. If the machine tool is equipped with two actuators arranged in x-y coordinates, the simplest MCU moves first the x then the y actuator by the prescribed distances, without controlling the motion itself (*point-to-point* or *positioning system*, Fig. 2–4a); when the programmed position is reached, the operation is performed (say, a hole is drilled). A slightly more complex system also moves first in one and then the other direction, but this time with full control of the rate of movement (*straight-cut system*, Fig. 2–4b) while an operation such as cutting, milling, or welding takes place.

2. NC is particularly valuable when a complex contour is to be followed (Fig. 2–4c). In *contouring systems* the MCU is programmed to break up the contour into shorter segments and to interpolate between the end points of segments. Linear interpolation approximates the curved profile in small straight lengths; better approximation is obtained with circular paths, splines, and, especially, nonuniform rational B-splines (NURBS).

Information is read in blocks, and a buffer memory (buffer register) prevents discontinuity of operation which, in the case of machining, welding, etc., would result in visible stop marks on the surface.

Computer Numerical Control (CNC)   In this, the functions of the MCU are partly or fully taken over by a dedicated computer (a mini- or microcomputer assigned to the machine tool, Fig. 2–5a). The entire program is read into memory. Since computers can be readily reprogrammed, much greater flexibility of operation is obtained. For example, it is possible to trace a complex curve without any breaks in continuity, and thus attain the closest approximation to the desired contour. Furthermore, programs can be added that provide technological functions, perform adaptive control, and incorporate some elements of a process model. The microprocessors used in place of the hard-wired NC circuits are more reliable and can have self-diagnostic features. In general, the part or process program is still received on tape or disk, although many CNC systems allow direct programming.

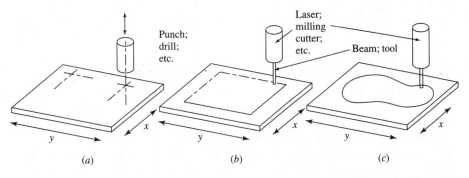

(a)	(b)	(c)

Figure 2–4   Control methods may provide (a) simple position or point-to-point control or (b) control over tool or workpiece movement in a straight cut or (c) along a contour.

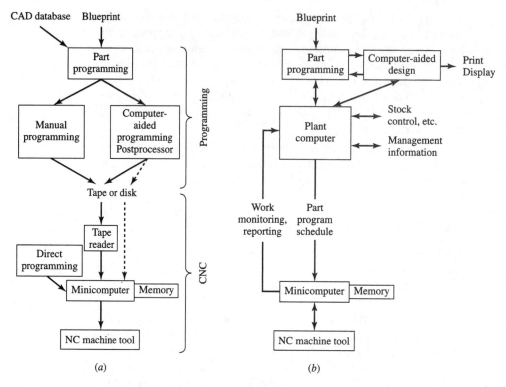

Figure 2–5   Many numerical control (NC) machines, with structures similar to those shown in Fig. 2–3, are now controlled by (a) a dedicated microcomputer (CNC) or (b) a hierarchy of computers (distributed NC, DNC).

The computer has sufficient memory to serve not only as a buffer but also to store the programs necessary for extended operation.

Both NC and CNC raise productivity and reproducibility, thus raising accuracy, quality, and reliability of the end product. CNC minimizes the errors introduced by the tape reader since the tape is read only once; it also reduces overhead relative to NC.

Direct Numerical Control (DNC)   In this, several machine tools are connected to one larger, central computer which stores all programs and issues the NC commands to all machines (Fig. 2–5b). This approach has been largely abandoned in favor of *distributed numerical control* in which each machine has its own computer and the central computer is used only to store, download, edit, and monitor programs, and to provide supervisory and management functions. With such hierarchical control, even very complex tasks can be broken down into manageable elements. The task of real-time computation and sensory processing is allocated to the first-level computers. The NC units may be of the conventional hard-wired type, with the tape reader replaced by a direct communication line to the central computer (*behind-the-tape-reader systems*), or specialized units which, like CNC units, use a minicomputer as the MCU. Obviously, the latter allows much greater flexibility.

Programmable Logic Controllers   The control of many processes requires sequencing, timing, counting, logic, and arithmetic functions which used to be satisfied by relay logic circuits. These had to be rewired if the logic was to be changed. Their place has now been taken by *programmable controllers* (PC); to avoid confusion with personal computers, they are now called *programmable logic controllers* (PLCs). Their great advantage is that the memory can be readily reprogrammed with a programming panel or a computer, in the "ladder logic" familiar to people versed in relay circuits. They are often used in combination with microcomputers to perform simple sequential tasks, very rapidly, in real time.

It should be noted that, to exploit all the benefits of computer control, it is usually necessary to improve the mechanical performance of the system. The integration of mechanical and electronic aspects is often termed *mechatronics*.

NC Programming   Programming of the machine tool has been greatly simplified over the years and has spread from machining to other processes. Programming starts by defining the optimum sequence of operations and the process conditions for each operation. The geometric features of the part are then used to calculate the tool path. The resulting program can be quite general and must be converted, with the aid of a program called the postprocessor, into a form acceptable for the particular machine tool control. The output is a punched tape or other storage medium. An important step is tape verification which reveals programming errors and ensures the production of correct parts. Basically there are four approaches:

1. *Manual programming*: All elements of the program are calculated by a skilled parts programmer who puts them into standardized statements. Programming is laborious and is now largely limited to simple point-to-point programs.

2. *Computer-assisted programming*: The programmer communicates with a software system in a special-purpose language that uses English-like words. The most comprehensive of these languages, APT (automatically programmed tools) was developed in the 1950s at the Massachusetts Institute of Technology, Cambridge, under U.S. Air Force sponsorship, and was expanded in the 1960s, under sponsorship of a consortium of users, at IIT Research Institute, Chicago, Illinois, and then at CAM-I. Many simplified languages and languages designed for specific processes have since been developed. Programming languages translate the input into a form understandable to the computer so that it can perform the necessary computations, including compensation for tool dimensions (cutter offset in machining). Tape verification must be done on the machine tool or a drafting machine.

3. *CAD/CAM*: When parts are designed by CAD, the numerical database can be used to generate the program on the graphics terminal, either by a programmer or by the designer of the part with the aid of the CAD/CAM software. The program can be immediately verified by viewing on a VDT (video display terminal) the path of the tool relative to the part. Programming is fast and relatively inexpensive, justifying its use even for single parts or, as it is often called, for *one-off production*.

4. *Manual data input*: Many CNC machine tools are equipped with a VDT display and powerful software that prepares the part program. In response to queries, the operator enters data to define part geometry, material, and tooling. Standard English words are used, and the software does the rest. The technique is highly economical when it allows programming while another job is running.

With the spread of CNC and manual data input, the trend is to entrust most programming to the machine-tool operator; however, conventional computer-assisted and CAD/CAM programming is still performed in programming departments.

The graphic database can be exchanged between different systems through standard formats such as IGES (Initial Graphics Exchange Specification, ANSI Y14.26M). Additional product information can be shared through ISO 10303 (Industrial Automation Systems and Integration—Product Data Representation) and ANSI/CAM-I 101-1995 (Dimensional Measurement Interface Standard).

2-6 SUMMARY

Manufacturing is an essential part of any industrialized economy. It is the mainspring of economic development and has been recognized as such by most nations, resulting in fierce international competition. Manufacturing is central to the activities of all engineers and technologists, because most research, development, design, and management activity finally results in some manufactured product. If an industrial unit (company) or nation is to be successful in the worldwide competition (and, indeed, if humankind is to be best served by plentiful, high-quality, high-value manufactured products), it is essential to recognize some very general features of manufacturing:

1. Manufacturing involves many steps from market research through the development, design, analysis, and control of products and processes to the delivery, service and, finally, recycling or disposal of the manufactured product. Gradually, the many activities associated with these stages have become specialized, compartmentalized, and disjointed. Isolated activities in sequential manufacturing resulted in long lead times, great inefficiencies, and high product cost.

2. Manufacturing must be viewed as a system, with all parts of the system interacting in an organic manner. This is reflected in the practice of concurrent engineering in which various activities overlap or proceed simultaneously; thus, lead times are reduced, quality improved, and expensive changes late in the product cycle are avoided.

3. Subsystems such as CAD and CAM have been based on the computer for some time now, with many benefits in improved productivity and quality.

The full benefits require integration of all manufacturing actions. This is facilitated by a common computer database, essential for the development of CIM.

4. Computers and other microelectronic devices such as programmable controllers have been used extensively for the control of production processes and machinery with the aid of NC, CNC, and DNC. A better understanding of processes and the development of appropriate transducers has allowed control in the adaptive mode, responding to changes in process conditions the same way or better than a highly skilled operator could.

5. The application of the computer to an outdated or basically defective process cannot solve the root problems. Therefore, if anything, it has become even more important to acquire a sound understanding of the physical principles upon which process control can be based. Knowledge of these principles is essential also if an interface is to be built between mechanical equipment and electronic devices.

FURTHER READING

Design

ASM Handbook, vol. 20, *Materials Selection and Design*, ASM International, 1997.

Bakerjian, R., and P. Mitchell (eds.): *Tool and Manufacturing Engineers Handbook*, 4th ed., vol. 6, *Design for Manufacturability*, Society of Manufacturing Engineers, 1992.

Andreasen, M.M., S. Kahler, and T. Lund: *Design for Assembly*, Springer, 1983.

Backhouse, C.J. and N.J. Brookes (eds.): *Concurrent Engineering*, The Design Council, 1996.

Badiru, A.B.: *Expert Systems Applications in Engineering and Manufacturing*, Prentice Hall, 1992.

Bedworth, D.D., M.R. Henderson, and P.M. Wolfe: *Computer Integrated Design and Manufacturing*, McGraw-Hill, 1991.

Boothroyd, G.: *Assembly Automation and Product Design*, Dekker, 1991.

Boothroyd, G., P. Dewhurst, and W. Knight, *Product Design for Manufacture and Assembly*, Dekker, 1994.

Bralla, J.G.: *Design for Excellence*, McGraw-Hill, 1996.

Bralla, J.G.: *Design for Manufacturability Handbook*, 2d ed., McGraw-Hill, 1998.

Chapman, W.L.: *Engineering Modeling and Design*, CRC Press, 1992.

Crabb, H.C.: *The Virtual Engineer*, Society of Manufacturing Engineers, 1998.

Cross, N.: *Engineering Design Methods*, 2d ed., Wiley, 1994.

Dieter, G.E., Jr.: *Engineering Design: A Materials and Processing Approach*, 3d ed., McGraw-Hill, 1999.

Dixon, J.R., and C. Poli: *Engineering Design and Design for Manufacturing*, Field Stone Publishers, 1995.

Dorrf, R.C., and A. Kusiak (eds.): *Handbook of Design, Manufacturing and Automation*, Wiley, 1995.

Helander, M., and M. Nagamachi: *Design for Manufacturability*, Taylor and Francis, 1992.

Hannam, R.: *Computer Integrated Manufacturing: From Concepts to Realization*, Longman, 1998.

Ingle, K.A.: *Reverse Engineering*, McGraw-Hill, 1994.

Kusiak, A. (ed.): *Concurrent Engineering: Automation, Tools, and Techniques*, Wiley, 1993.

Lindbeck, J.R.: *Product Design and Manufacturing*, Prentice Hall, 1995.

Maus, R., and J. Keys (eds.): *Handbook of Expert Systems in Manufacturing*, McGraw-Hill, 1991.

Mills, A.: *Collaborative Engineering*, Society of Manufacturing Engineers, 1998.

Mital, A., and S. Anand (eds.): *Handbook of Expert Systems in Manufacturing: Structures and Rules*, Chapman and Hall, 1994.

Prasad, B.: *Concurrent Engineering Fundamentals*, vol. 1, 1996; vol. 2, 1997, Prentice Hall.

Rhyder, R.F.: *Manufacturing Process Design and Optimization*, Dekker, 1997.

Roosenburg, N.F.M., and J. Eekels: *Product Design: Fundamentals and Methods*, Wiley, 1995.

Salomone, T.A.: *What Every Engineer Should Know about Concurrent Engineering*, Dekker, 1995.

Swift, K.G., and J.D. Booker: *Process Selection: From Design to Manufacture*, Wiley, 1997.

Ullman, D.G.: *The Mechanical Design Process*, 2d ed., McGraw-Hill, 1997.

Ulrich, K.: *Product Design and Development*, McGraw-Hill, 1995.

Woodson, W.E. (ed.): *Human Factors Design Handbook*, McGraw-Hill, 1991.

Products Liability

Enghagen, L.K.: *Fundamentals of Product Liability Law for Engineers*, Industrial Press, 1992.

Henley, E.J., and H. Kumamoto: *Designing for Reliability and Safety Control*, Prentice Hall, 1985.

Weinstein, A.S., A.D. Twerski, H.R. Piehler, and W.A. Donaher: *Products Liability and the Reasonably Safe Product*, Wiley, 1978.

CAD/CAM and Control

Amic, P.J.: *Computer Numerical Control Programming*, Prentice Hall, 1997.

Groover, M.P.: *Automation, Production Systems, and Computer-Integrated Manufacturing*, Prentice Hall, 1987.

Johnson, C.D.: *Process Control Instrumentation Technology*, 5th ed., Prentice Hall, 1997.

McMahon, C., and J. Browne: *Cadcam: Principles, Practice, and Manufacturing Management*, Addison-Wesley, 1998.

Nanfara, F., T. Uccello, and D. Murphy: *The CNC Workbook: An Introduction to Computer Numerical Control*, Addison-Wesley, 1995.

O'Sullivan, D.: *Manufacturing Systems Redesign: Creating the Integrated Manufacturing Environment*, Prentice Hall, 1995.

Rehg, J.A.: *Computer-Integrated Manufacturing*, Prentice Hall, 1994.

Soloman, S.: *Sensors and Control System in Manufacturing*, McGraw-Hill, 1994.

Soloman, S.: *Sensors Handbook*, McGraw-Hill, 1998.

Turbide, D.A.: *Computers In Manufacturing*, Industrial Press, 1991.

Vajpayee, S.K.: *Principles of Computer Integrated Manufacturing*, Prentice Hall, 1994.

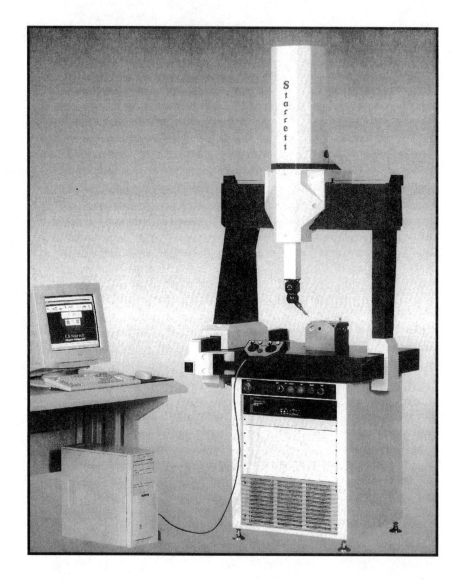

The touch probe of a coordinate measuring machine (CMM) sends its signal directly to a computer to build a file that fully describes the dimensions of the part. (*Courtesy The L. S. Starrett Co., Athol, Massachusetts.*)

3

Geometric Attributes of Manufactured Parts

Manufactured components come in all sizes and in a bewildering variety of shapes. In this chapter we shall see:

How to create order from chaos by group technology

What machine tools have to do to generate shapes

How to make sure parts will fit together

How to make a drawing that truly expresses design intent

Methods of measuring dimensions, statically and "on the fly"

What a good surface is and how it can be quantified.

The first impression we receive of a manufactured product is its shape and size. Both have esthetic connotations, and it is the industrial designer's task to create a pleasing product. Shape and dimensions are critical also to the function of the product. In an assembly, many parts need to fit together, and this requires that allowable deviations in dimensions (dimensional tolerances) be specified and not exceeded. This leads to the need for measuring techniques and procedures. Our impressions are greatly influenced also by the surface appearance of a product. Again, the industrial designer will specify a finish for visible parts but there are also strict technical requirements to be met if two mating parts are to function properly. Therefore, objective measures of surface quality must be found and appropriate measuring techniques employed.

3-1 SHAPE

The *shape* of a part is dictated, first of all, by its function. Complexity of this shape often determines what processes can be considered for making it, and, in the most general sense, increasing complexity narrows the range of applicable processes and increases the cost of design and manufacture. A cardinal rule of design is, therefore, to keep the shape as simple as possible. This rule may, however, be broken if a more complex shape allows consolidation of several parts and/or elimination of one or more manufacturing steps.

3-1-1   Shape Classification

There is no universally accepted shape classification system. The groupings in Fig. 3–1 are designed for identifying process capabilities and will be used in this book throughout. Products of uniform cross section (spatial complexity = 0) are two-dimensional, all others are three-dimensional. With increasing spatial complexity, definition of the shape requires additional geometric parameters; it can be said that the shape has greater *information content*. A small increment in information content may, however, have significant manufacturing consequences. For example, moving from the solid shape R1 to the hollow shape T1 adds only one dimension (the diameter of the hole), yet it immediately excludes some processes or necessitates extra operations in others. In contrast, adding a third outside diameter to the round product R1 would result in the same increase in information content without imposing the same limitations on process choice. We shall see that limitations on shape are tightened by properties of the material and by interactions with the tooling. It is, therefore, important not to finalize the part configuration too early in the design process, otherwise the most economical manufacturing process may be excluded. In each chapter dealing with processes, tables will be incorporated to show what shapes are most suitable for a given process. The aim is, generally, to produce a "net-shape" part ready for assembly; if this is not feasible, a "near net-shape" part will need only minor finishing, usually by machining.

3-1-2   Group Technology

Group technology (GT) is a powerful tool in designing for manufacturability. Its essence is the recognition that many problems have similar features and, if these problems are solved together, great efficiency and economy result. In applying the concept to manufacturing, individual parts are analyzed in terms of commonalities of design features as well as manufacturing processes and process sequences. This way, *families of parts* can be identified and economies are assured:

1. In design, the task of repetitive design is eliminated. It has been estimated that 40% of all design is simple duplication, 40% requires only some modification of existing design, and only 20% calls for original design. The designer who chooses a standard bolt or other component practices GT at the most elementary level.

2. In manufacturing, programs required for making families of parts can be optimized and retained for the future when the part is to be produced again.

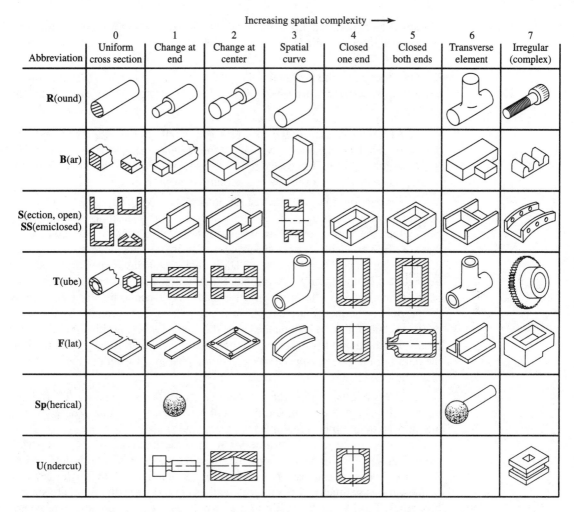

Increasing spatial complexity ⟶

Abbreviation	0 Uniform cross section	1 Change at end	2 Change at center	3 Spatial curve	4 Closed one end	5 Closed both ends	6 Transverse element	7 Irregular (complex)
R(ound)								
B(ar)								
S(ection, open) **SS**(emiclosed)								
T(ube)								
F(lat)								
Sp(herical)								
U(ndercut)								

Figure 3–1   The choice of possible manufacturing methods is aided by classifying shapes according to their geometric features.

Because parts that are geometrically similar often require the same production sequence, GT is the first step also in reorganizing a production facility (see Sec. 21-2-4).

3. In production planning, cycle time estimation is accelerated, workpiece movement is rationalized, process design is simplified. Cost estimation is facilitated too.

Introduction of the computer has made GT particularly attractive, because programs relating to the design of standard elements such as solid and hollow cylinders, rectangular blocks, and cones, can be retained in memory and easily combined and modified for a large variety of part configurations. Similarly, process details can be filed away for later use, with modifications, if necessary.

Part Classification   The first step in GT is the *classification of parts* into families. Several approaches can be taken on the basis of design and/or manufacturing attributes:

1. *Experience-based judgment.* This works only in the simplest cases. The part is classified into a family by visual judgment of its shape, and the classification is further refined from a knowledge of the usual production sequence. There is no assurance that such a sequence is actually the optimum one.

2. *Production-flow analysis* (PFA). Information relating to the sequence of operations in an existing plant is contained in routing sheets or routing cards from which the flow of parts through various operations can be extracted. Parts that are made by identical operations form a family. Good engineering judgment will tell whether parts on which some additional operations are performed should be included in the family. A critical examination may also reveal that some parts falling outside a family could be made more economically by adopting the production sequence typical of the family. Parts that are made by the same processes but in different sequences may still logically be classified into the same family, but the flexibility of the production system will have to be greater to allow the return of the part to a previous operational position. Computer-aided process planning (CAPP, Sec. 21-4-1) creates the basis for PFA.

3. *Classification and coding.* This is a more formal exercise. There is no universally accepted system, and there will perhaps never be one. Some systems are more suitable for design, others for parts made by specific processes (casting, forging, machining, etc.), yet others aim at some universality. They all start from a classification of basic workpiece shapes (something similar is done in Fig. 3–1). Part codes are usually made up of several (sometimes up to 30) digits which define various geometrical features as well as composition and requisite surface properties. Further digits may be added to define processes, process parameters, and processing sequences. Some computer-based systems facilitate coding by guiding the operator through the necessary steps in a conversational mode, others use the database generated by CAD to help in assigning code numbers.

4. *Engineering database.* The most versatile system is based on engineering databases that contain—in addition to all information found on an engineering drawing—information about the part, its use, and manufacture. When structured as relational databases, they can be searched according to different attributes and thus groups can be formed to satisfy specific criteria.

3-1-3   Machine Tool Movement and Control

Shape complexity has a profound influence on the process controls needed. This is most obvious in machining, where the cutting tool has to follow an exact path to generate the required shape.

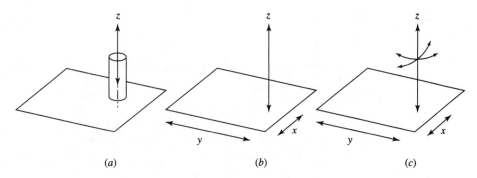

Figure 3–2 Tools and workpieces may be moved and control may be exercised along (a) one; two; (b) three; or (c) several axes.

When the movement of tool or workpiece is restricted to a single axis, one speaks of *one-axis* or *single-axis* (usually denoted as the *z* axis) *movement* or *control* (as, for example, when drilling a hole in a clamped workpiece, Fig. 3–2a). Table movement requires *two-axis* control (usually referred to as the *x* and *y* axes, Fig. 3–2b); programmed movement in the *z* axis makes it into a *three-axis* machine. (When movement in the *z* direction is simply on-off and proceeds at some preset rate, one usually speaks of two-and-a-half axis control.) Swiveling the tool (or table) would add the fourth and fifth axes (Fig. 3–2c). Every joint in the tool holder or table adds a further freedom (axis) of movement and permits more complex shapes to be made, but at the expense of more complex and expensive machinery and control. This again points to the need for a design that facilitates manufacture and assembly with minimum complexity, an issue of paramount interest for automation and robotics. The human body has dozens of freedoms of movement and, while it might be possible to build machines and robots of similar versatility, *it is easier and cheaper to accommodate the limitations of machinery by appropriate design of the parts and assemblies that the automated equipment will have to handle.* Humans can always assemble what robots can, but the converse is not true.

There are some shape features that immediately set certain limitations:

1. Axial symmetry is, in many ways, the simplest because a two-dimensional shape can be generated by rotating the part or tool (around the *z* axis) while moving the tool in a straight path. Two or more outer diameters (shapes R1 and R2) require a second axis of control.

2. Parts of nonrotational symmetry call for a minimum of two-axis control, and spatial curvature can be followed only with three- (or more) axis control.

3. A surface in line with the tool movement (Fig. 3–3a) can be made with one-axis control although, if the tool is difficult to withdraw, a *draft angle* (Fig. 3–3b) may be necessary. Undercut shapes (Fig. 3–3c and d) require control in more than one axis, and five-axis control is now frequently encountered.

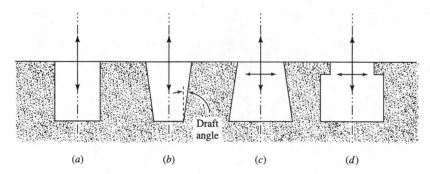

Figure 3–3   (a) A straight-walled pocket is often easy to produce but a (b) draft angle may be required to withdraw a tool. (c), (d) Undercut shapes require multiaxis control or complex tooling.

3-2   DIMENSIONS

We are accustomed to see manufactured products of enormous size range, from a pin to a jumbo airplane. Individual parts also have a wide size range and not all processes are suitable for making all of them. Minimum size is often limited by laws of nature, while maximum size may also be set by the availability of equipment. All these constraints must be taken into account in design, and tables of process capabilities are included in each process chapter to give guidance in this respect.

3-2-1   Dimensional Units

The SI unit of length is the meter (m); smaller dimensions are expressed in millimeters (mm) or micrometers (μm, 10^{-6} m, colloquially referred to as micron). Some products (products of *nanotechnology*) are so small that their dimensions are expressed in nanometers (nm, 10^{-9} m). For atomic dimensions, the non-SI unit of Angstrom (Å) has been widely used (10 Å= 1 nm). Unless otherwise specified, *dimensions in this text are given for 20°C.*

In the U.S. customary system (USCS) the unit of length is the inch (in). Smaller dimensions are referred to in units of 10^{-3} in (colloquially, mil or thou). Yet smaller dimensions are given in microinches (μin, 10^{-6} in). As shown in the conversion table of App. A and the inside covers, 1 in = 25.4 mm and 1 μin = 25.4 nm. Hence, 1 mm = 0.03937 in (for conversational purposes, 1 mm = 40 thou and 1 μm = 40 μin).

3-2-2   Dimensional Tolerances

The artisan made individual products in which each part was tailored to fit into the assembly. When repair or replacement of a part was necessary, it had to be custom made and fitted. Mass production requires interchangeability of parts and, for this, dimensions must be controlled. From beginnings in the nineteenth century, dimensional control has been progressively tightened with the help of rapidly developing measurement techniques (Table 1–1). Again, different processes have inherently different capacity for making parts to controlled dimensions, as will be seen in chapters on processes.

While dimensions must be controlled, it is neither possible nor necessary to make parts to exact dimensions. Therefore, maximum and minimum *limits of dimensions* (length or angle) are specified with two goals in mind:

1. The limits must be set close enough to *allow functioning of the assembled parts* (including interchangeable parts).

2. The *limits must be set as wide as functionally possible*, because tighter limits usually call for more expensive processes or process sequences. The single most important cause of excessive production costs is the specification of unnecessarily close dimensional limits. All too often tight tolerances are specified even when there is no mating part.

The designer specifies dimensions and the *allowance*, i.e., the difference in dimensions necessary to ensure proper functioning of mating parts (the allowance is called also the *functional dimension* or *sum dimension*). This is best illustrated on the example of a shaft fitting into a hole (Fig. 3–4). The *basic size* of one of the mating parts is first defined, from tables of preferred sizes if at all possible, so that standard shafts or tools can be used. In principle, the basic size could be assigned to either the hole or the shaft. In practice, holes are often manufactured with some special tool (drill, reamer, punch) and are, furthermore, difficult to measure while the hole is being made; therefore, the *hole-basis system* is generally used. The allowance (the minimum clearance or maximum interference) is then specified to satisfy functional requirements.

The position of the *tolerance zone* relative to the basic dimension defines the type of fit. *Clearance fits* allow sliding or rotation (Fig. 3–4a). *Transition fits* provide accurate location with slight clearance or interference (Fig. 3–4b). *Interference fits* ensure a negative clearance (interference) and are designed for rigidity and alignment, or even to develop a specified pressure (shrink pressure) on the shaft (Fig. 3–4c).

The next step is determination of the *tolerance*, that is, the permissible difference between maximum and minimum limits of size. Tolerance can be expressed with respect to the basic size as deviation in both upper and lower directions (*bilateral tolerancing*) or in only one direction, if the consequences of inaccuracy in that direction are less dangerous (*unilateral tolerancing*).

Experience has taught that, in most manufacturing processes, dimensional inaccuracies are proportional to the cube root of the absolute size (denoted D for diameter, in units of mm or in). The American National Standards Institute (ANSI) standard ANSI B4.1-1967, R1972, gives tables for 8 classes (comprising 34 subclasses) of fits, ranging from loose fit to force fit. The International Organization for Standardization (ISO) standard *System of Limits and Fits*, ISO

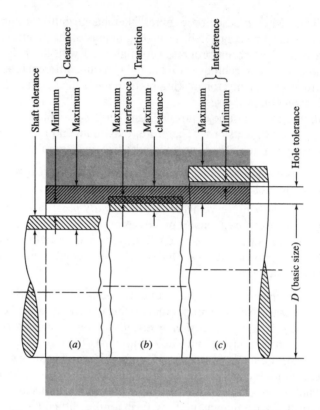

Figure 3–4   In the hole-basis system, the diameter of the hole is chosen from a table of preferred sizes and then the tolerances are applied to create (*a*) clearance, (*b*) transition, or (*c*) interference fits.

286-1:1988, is based on the tolerance unit *i*

$$i = 0.45D^{1/3} + 0.001D \tag{3-1}$$

and the grade of tolerance is expressed as the standardized multiple of *i* (within the grades 5–16, the tightest tolerance IT5 implies a standard tolerance of 7*i*, while the loosest, IT16, implies 1000*i*). The actual value of tolerance may be obtained from tables. There are also computer-aided techniques for assigning dimensions and tolerances.

The example given in Fig. 3–5 is for an assembly that is suitable for running with a viscous lubricant (it is a hydrodynamic bearing, such as that on an automotive crankshaft). Capital letters show the position of the tolerance zone relative to the basic dimension of the hole and lowercase letters show it for the shaft.

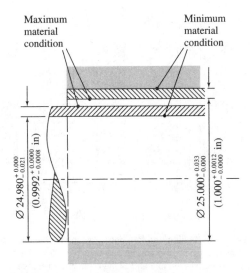

Figure 3–5   Example of a close-running fit, according to American National Standard preferred hole-basis metric clearance H8/f7.

3-3   SHAPE AND LOCATION DEVIATIONS

For a part to function properly with respect to other components, it is often necessary to place further restrictions on the location (position) of geometric features and on geometric properties such as concentricity, runout, flatness, parallelism, and perpendicularity. This information is conveyed on engineering drawings (or computer files). Two basic methods are employed:

Coordinate dimensioning is faster and would seem to be ideally suited for machining on NC machine tools, yet it can give rise to ambiguities. Geometric attributes must be spelled out and can easily be misinterpreted.

Geometric dimensioning and tolerancing (GD&T) takes more effort but gives a clearer expression of the intent of design, which in turn helps the choice of the most appropriate manufacturing method. It also indicates how the part should be inspected. It achieves all this with the use of symbols expressing geometric attributes, as embodied in ASME Y14.5 M-1994 and the corresponding ISO standards.

The subject is beyond the scope of our discussions, but a simple example of a shaft will illustrate some advantages. The shaft is of 15.00 ± 0.05-mm diameter and is 50 mm long. This tells us nothing about its shape; it could be severely bowed. We know that it will mate with another part and therefore specify that bow should not exceed 0.2 mm over 50 mm. This would be spelled out in

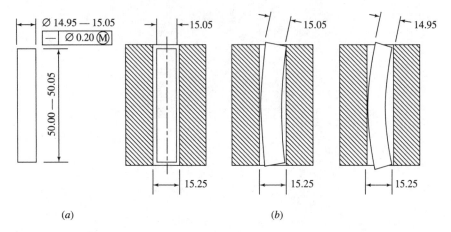

(a)   (b)

Figure 3–6   Geometric dimensioning and tolerancing expresses the intent of design and the mode of measurement and can lead to a gain in allowable tolerances. In this instance, the shaft at the low end of the diametral tolerance is allowed to bow an extra 0.10 mm without affecting the function. (*Adapted from J.D. Meadows, Geometric Dimensioning and Tolerancing, Dekker, 1995, p. 237, Fig. 11–9. By courtesy of Marcel Dekker, Inc*).

words on the traditional drawing. With GD&T the drawing would appear as in Fig. 3–6a. The symbol $\varnothing$ denotes diameter. Below the diameter dimensions, a *feature control frame* contains a straight line (the symbol for straightness) with $\varnothing$ indicating that the axis of the body (the derived median line) must be within 0.2 mm. At first glance, this is no different from the traditional drawing. Consider, however, the implications. The shaft must fit within a cylindrical envelope of $15.05 + 0.20 = 15.25$ mm. This can be checked by a gage containing the envelope (a *functional* or *receiver* gage, Fig. 3–6b). It is evident that, if the shaft is at minimum material condition (of 14.95-mm diameter), it will still fit if bow is $15.25 - 14.95 = 0.30$ mm. Thus, we gain 0.10 mm in extra (*bonus*) tolerance. This shows one of the advantages of GD&T: Parts are now saved that are perfectly functional but would have been rejected if conventional dimensioning and tolerancing had been used.

3-4   ENGINEERING METROLOGY

Metrology is the science of physical measurement, applied to variables such as dimension, surface finish, and mechanical and electrical properties. The narrower field of *engineering metrology* (or *industrial metrology*) concentrates on the measurement of dimensions, including those of length and angle. It is of prime importance for the control of quality by in-line (in-process and postprocess) and off-line inspection, and as such it will again be discussed in Sec. 21-2-4 as an element of manufacturing organization. Our concern here is with the techniques of measurement.

3-4-1   Principles of Measurement

Measurement must be performed with a device of sufficient accuracy and precision. No measurement can be repeated perfectly; readings (as well as the variables subject to measurement) will always show *dispersion* (Fig. 3–7).

Accuracy expresses the degree of agreement between measured dimension and true value. The difference between measured value and true value is the *error*; since the true value can never be known, the error can be established only

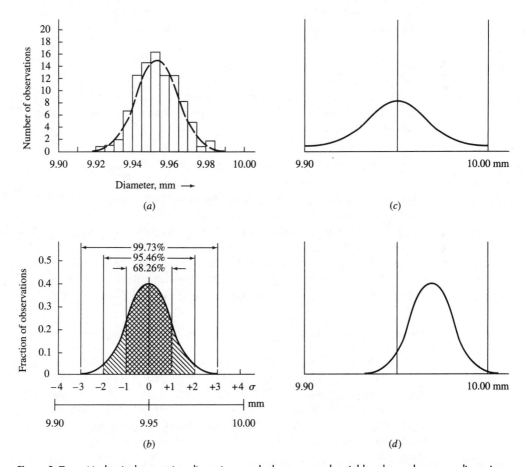

Figure 3–7   Mechanical properties, dimensions, and other measured variables always show some dispersion. In the example, 100 shafts were turned to 9.95 ± 0.05-mm diameter. On grouping the actual measurements into narrow groups (*a*), the distribution proves both accurate and precise (*b*); in another instance it may be accurate but imprecise (*c*) or precise but inaccurate (*d*).

by checking against a standard. For example, working standards (working gages) used for length measurement in the workplace are checked against reference standards which, ultimately, are checked against national standards. Other standards are based on physical phenomena; thus, the meter is now defined by the vacuum wavelength of the orange line of krypton 86, with a precision of 1 part in 10^9.

Precision is the degree of repeatability of measurement. Tight bunching of data indicates high precision (Fig. 3–7a and b) but not necessarily high accuracy (Fig. 3–7d); precision is poor when the data are widely dispersed (Fig. 3–7c).

In many instances, the distribution is bell-shaped and approximates the so-called *normal distribution* (the curve of which can be mathematically derived). Such a curve has two important characteristics: one is the mean, the other is the dispersion of data (Fig. 3–7b).

Accuracy can be judged from the *statistical average* or *mean* $\bar{x}$. This is the center line of the bell and is simply the sum of all measured values x divided by the number of measurements n (or, for smaller sample sizes, by $n - 1$):

$$\bar{x} = \frac{\sum x}{n - 1} \tag{3-2}$$

The *dispersion of data* can simply be characterized by the *range R* (the difference between maximum and minimum values). For statistical analysis the *standard deviation* σ is most useful:

$$\sigma = \left[\frac{\sum (x - \bar{x})^2}{n - 1} \right]^{1/2} \tag{3-3}$$

The value of σ is a measure of the width of the bell. The area under the normal curve is a measure of the number of values that fall within specified limits. Figure 3–7b shows that approximately 68% of the parts will be within $\pm\sigma$, 95% within $\pm 2\sigma$, and 99.73% within $\pm 3\sigma$ of the average. In general industrial practice it used to be acceptable if the $\pm 3\sigma$ limit fitted within the specified tolerance. This means, however, that 27 readings in 10 000 will still be outside the tolerance range, and recent trends aim at tighter control (Sec. 21-3-4).

Measuring Instruments   These have to possess several attributes:

1. *Sensitivity* is the smallest variation that the device can detect. It is called *resolution* when reading is digital or is made against a scale. A scale subdivided into increments smaller than the device can detect gives only spurious resolution; the accuracy of the device should be several times better than the smallest graduation of the readout. A rough rule of thumb is that precision of the measuring instrument should be 10 times better than the precision of the dimension to be measured.

2. *Linearity* affects readings over a specified measurement range. Even if an instrument is set (calibrated) against a standard at some point in the range, nonlinearity affects other points in the range.

3. *Repeatability* determines the possible highest precision that can be achieved under well-controlled conditions. The instrument must be capable of repeating readings to the same accuracy to which it can be read.

4. *Stability* expresses resistance to drift that would reduce both accuracy and precision and would necessitate frequent recalibration.

5. *Speed of response* is critical when a transient variable is to be measured, usually during production.

6. *Feasibility of automation* is important in many applications, especially now with the spread of 100% in-process inspection.

Variations of Measurements   Repeated readings may suffer from errors:

* *Assignable (systematic) errors* are measurable and often controllable. In addition to errors inherent in the device, *temperature variation* is the main source of systematic error. If tolerances are tight, the temperature of the part must be uniform and known so that an allowance can be made for thermal expansion. In postproduction measurements this is best ensured by taking the part to a climate-controlled room and allowing it to equalize with the temperature of the measuring device; this may take hours or days.

* *Random errors* stem from human error (inaccurate scale reading, excessive force applied to a contacting gage; improper setup, etc.) and from sources such as dust and rust. Again, a climate-controlled room helps with its filtered air and controlled humidity.

Measurements are often made with reference to a datum surface such as a flat, hole, or shaft. It must be chosen with due regard to the method of manufacture and inspection.

An AISI 1020 steel shaft of 100.00-mm diameter is made by turning on a lathe. The part heated up to 70°C during cutting. Can the dimension be measured to the nearest 0.01 mm without allowing for the temperature increase?

    The steel is a carbon steel of 0.2% C content. The coefficient of linear thermal expansion is 11.7 μm/m·°C. Hence the diameter will increase by $(70 - 20)(0.100)(11.7) = 58.5$ μm, which is 6 times the desired precision of measurement.

| **Example 3-1** |

A 100-mm dimension is checked on a Zn-alloy part (coefficient of linear thermal expansion = 27.4 μm/m·°C) with a gap gage used as a GO–NOT GO gage. Both are at 38°C. Is there a potential problem?

    Relative to 20°C, the part expands by $(0.1)(38 - 20)(27.4) = 49.32$ μm. The gage expands by only $(0.1)(18)(11.7) = 21.06$ μm; it will reject good parts.

| **Example 3-2** |

3-4-2   Gages

In the broader sense, a *gage* is an instrument that measures some variable. In the narrower sense used here, the term gage refers first of all to bodies of hardened steel, tungsten carbide, ceramic (glass), etc. that are manufactured to close tolerances. They can be fixed or adjustable. Once set, an adjustable gage can also be used as a fixed gage. There are several types:

1. *Gage blocks* still are the primary length gages in many applications. They are made in sets that allow building up any dimension by *wringing* (a sliding-twisting motion) of several blocks (Fig. 3–8*a*). Adsorbed moisture or oil films on the mating measuring surfaces have negligible thickness but provide sufficient adhesion to handle the built-up column as one unit; a sliding motion is again needed for separating the blocks. Gage blocks come in several grades. Tolerances (expressed in μm) are as follows: Grade 3 gages, used directly in production, +0.15, −0.05; Grade 2 sets, used as inspection and toolroom standards, +0.10, −0.05; Grade 1 laboratory gage blocks, for the calibration of other gages and indicating instruments, +0.05, −0.05; Grade 0.5 reference gages, used only in work of the highest precision, +0.03, −0.03.

2. *Angle blocks* (Fig. 3–8*b*) are constructed according to the same principles as gage blocks. *Sine bars* (Fig. 3–8*c*) are used in conjunction with gage blocks to create any angle.

3. Other length gages include *length bars* (measuring rods, Fig. 3–9*a*) and fixed (Fig. 3–9*b*) and adjustable (Fig. 3–9*c*) *gap gages*.

4. *Plug and ring gages* are used for the measurement of diameters (Fig. 3–10*a*). They are usually of the GO–NOT GO type. The GO limit gage is the negative (the reverse replica) of the dimension at the *maximum material condition* (see Fig. 3–5), indicating that the mating parts can be assembled. The NOT GO limit gage is made to the dimension of the *minimum material condition* and rejects parts that are outside the tolerance. There are three problems with these

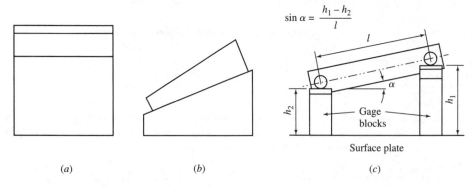

$$\sin \alpha = \frac{h_1 - h_2}{l}$$

(*a*)   (*b*)   (*c*)

Figure 3–8   Hardened steel (*a*) gage blocks, (*b*) angle blocks, and (*c*) sine bars are extensively used for comparative gaging purposes.

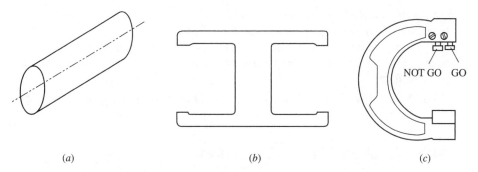

(a)   *(b)*   *(c)*

Figure 3–9   Comparative measurement of length dimensions is possible with (*a*) length bars, (*b*) fixed gap gages, or (*c*) adjustable gap gages.

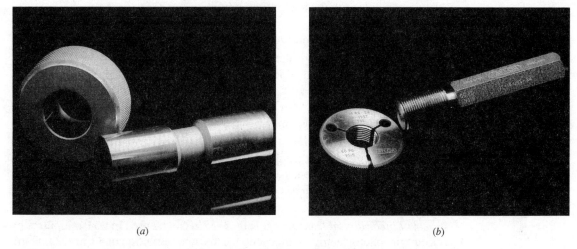

(a)   *(b)*

Figure 3–10   (*a*) Diameters of holes can be checked with plug gages and diameters of bars with ring gages. (*b*) More complex configurations are checked with special gages such as thread plugs.

gages: first, they themselves can be made only to certain tolerances, resulting in the rejection of good parts or passing of bad ones; second, they are subject to operator judgment; third, they give no information on the variations of part dimensions within the limits, and are thus of limited use for statistical production control. (Also, in the form shown, they violate Taylor's principle: Only the GO gage should be of full form to check *both* size and geometric features, whereas the NOT GO gage should check only one linear dimension. Thus, a GO ring gage should go fully over a shaft but the NOT GO gage should measure only the diameter and should not fit over the shaft anywhere.)

5. Multiple-diameter gages—*thread plugs* (Fig. 3–10*b*) and rings, spline gages, etc.—check the combined effect of several parameters. *Contour gages* or *templates* (including straight edges and radius gages) test the coincidence of shapes by visual observation or optical magnification. In the broader sense, *surface plates* also come in this category; they are often used for setting up other gaging elements and are made of some very stable material, such as granite, to specified flatness.

6. *Assembly gages* test not only dimensions but also alignment and coaxiality.

3-4-3   Graduated Measuring Devices

These devices allow the reading of a dimension against a scale. Some have a zero point, others read only relative displacement. Relative to fixed gages, their great advantage is that information on the *distribution of dimensions* within a batch is obtained. For best results, Abbe's principle should be observed: The line of scale should coincide with the line of measurement.

1. *Line-graduated rules and tapes* limit reading to the nearest division.

2. The use of a *vernier* increases the sensitivity of *caliper gages* (Fig. 3–11a) to 25 μm (0.001 in) and that of *micrometers* to 3 μm (0.0001 in). Verniers (Fig. 3–11b) are often replaced by digital readouts (Fig. 3–11c). Abbe's principle is satisfied in measuring with a micrometer, but with caliper gages the line of measurement (between the jaws) is separated from the scale.

3. When two *diffraction gratings* (closely spaced parallel lines on a glass surface) are superimposed at a slight inclination, they produce interference fringes, the location of which depends on the relative position of the gratings (Fig. 3–12a). The number of fringes can be counted electronically, to give a sensitivity of 5 μm (0.0002 in).

4. *Linear digital transducers* (Fig. 3–12b) can be used to transmit pulses by electronic, photoelectric, or magnetic means, to a resolution of 4 μm (0.0002 in). A rotary pulse-generation encoder can be used for angular measurements and, with a rack-and-pinion or slide-wire movement, also for linear measurements.

5. *Numerical encoding disks* (Fig. 3–13) provide, with the appropriate interface, direct readout or, if desired, input to NC controls. (The gray scale resolves 2^n different values, where n is the number of bits assigned to each pixel.)

6. Solid-state electronic devices that convert light to an electrical signal [*photodetector diodes* and *charge-coupled devices* (CCDs)] sense the presence or absence of light and, arranged in a linear array, offer resolutions of 3 μm (0.0001 in) or better when used alone or in TV cameras. All electronic gages can be used for on-line measurement.

7. *Toolmaker's microscopes* are optical microscopes equipped with cross-slide stages driven by micrometers, eyepieces with cross hairs for length measurement, and protractor eyepieces for angle measurement. They can also be used for checking and measuring the form (shape) of parts.

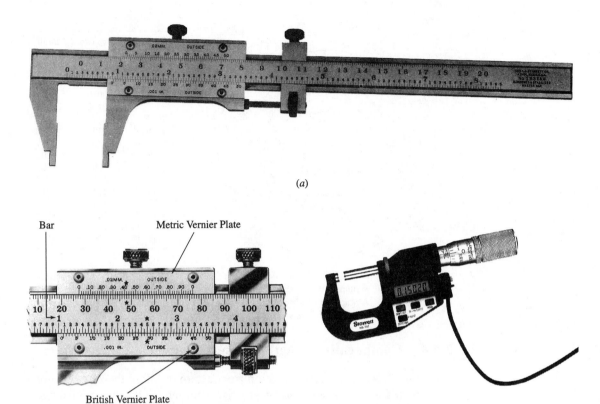

(a)

(b)　　　　　　　　　　　　(c)

Figure 3–11　A vernier allows reading to some fraction of the smallest division on the main scale (a). To read, the line of the vernier scale which coincides with a line on the major scale is read, and the reading is added to the basic reading on the major scale of the vernier caliper (b). There are no verniers on digital-readout instruments such as this micrometer (c). (*Courtesy The L. S. Starrett Co., Athol, Massachusetts.*)

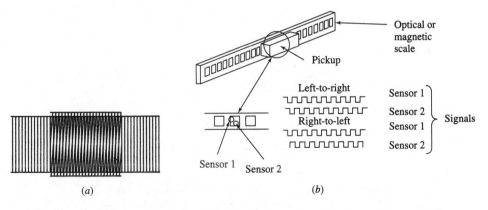

(a)　　　　　　　　　　　　(b)

Figure 3–12　(a) Length may be measured by counting the number of interference fringes. (b) The direction of displacement of an optical or magnetic scale is sensed by two transducers.

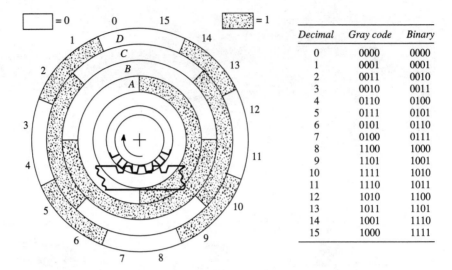

Decimal	Gray code	Binary
0	0000	0000
1	0001	0001
2	0011	0010
3	0010	0011
4	0110	0100
5	0111	0101
6	0101	0110
7	0100	0111
8	1100	1000
9	1101	1001
10	1111	1010
11	1110	1011
12	1010	1100
13	1011	1101
14	1001	1110
15	1000	1111

Figure 3–13   A numerical encoding disk provides a digital signal for control purposes. Ambiguous readings are avoided in this 4-bit encoder by the use of the gray code, which is then converted into binary code.

3-4-4   Comparative Length Measurement

Indicators measure only the deviation from a zero position; the zero position is set up with a setting gage (master gage) that is chosen to give the nominal size of the part (Fig. 3–14). If an indicator of sufficient sensitivity is used and a suitable

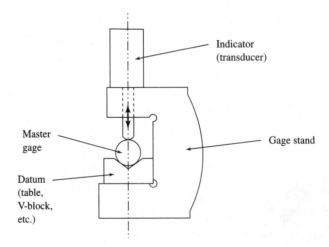

Figure 3–14   Dimensions may be read by gages which are equipped with an indicator or some form of position transducer.

datum (reference surface) is provided, much relevant information can be obtained not just on length and its variation from part to part but also on runout, alignment, etc. The indicator can be of several kinds:

1. *Dial indicators* are purely mechanical devices that convert linear displacement into rotation (e.g., with a rack-and-pinion movement) and amplify it with a gear train to increase sensitivity to 1 μm (50 μin). Some gages have built-in electrical contacts which activate signal lights, making them into GO–NOT GO gages.

2. *Electronic gages* (transducers) transform mechanical movement into an electrical signal according to various principles. Frequently used is the differential transformer (Fig. 3–15a), the output of which is zero when the movable core is exactly centered; the output is proportional to displacement elsewhere. Other transducers convert the deflection of a leaf spring into an electrical signal: Strain gages (resistance wire loops) are attached to the spring with an adhesive and interconnected to form a Wheatstone bridge circuit (Fig. 3–15b). Upon deformation, the resistance of the wire changes and the bridge is unbalanced to give an output proportional to deflection (this is also the principle upon which many load cells operate). With suitable contact components, they can be used on line.

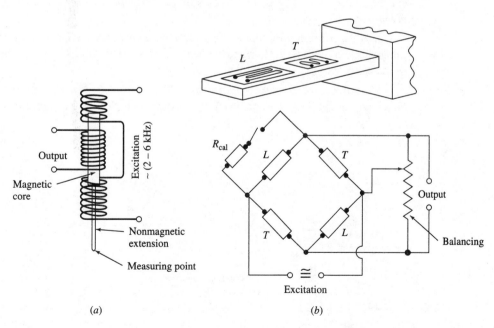

(a)   (b)

Figure 3–15   Displacements may be obtained from (a) the position of a differential transformer or (b) the deflection of a beam to which strain gages, interconnected to form a Wheatstone bridge, have been attached.

3. *Pneumatic gages* measure the back pressure generated when air emerging from the orifice of the gage head impinges upon the surface of the part (Fig. 3–16). Within a narrow dimensional range, pressure change is proportional to the size of the gap between gage head and workpiece surface.

4. *Capacitive gages* measure the distance between the part and a reference surface. Both these and pneumatic gages are suitable for on-line measurement.

5. *Ultrasonic gages* determine thickness from the time delay between first (top) and second (bottom) surface reflections. Because the velocity of sound propagation is material-dependent, calibration against another device is necessary. This gage too is suitable for process control or on-line measurements.

6. *Profile tracing* is based on the techniques of surface roughness measurement (Sec. 3-5-3) and reveals deviations from a reference surface.

3-4-5 Optical Devices

Light waves have a number of characteristics that can be exploited for engineering metrology.

1. Instead of being observed in a microscope, the magnified shape of a part can be projected onto a screen on which dimensions as well angles and shapes can be measured. These instruments are called *optical projectors* or *optical comparators*.

2. A *light-sectioning microscope* projects a narrow band of light obliquely (at an angle of 45°) to the surface of the part. The reflected light gives an outline of the cross section, suitably magnified.

3. Visible light has wavelengths from 400 nm (violet end) to 760 nm (red end). When an optical flat (glass or fused quartz disk with parallel flats, true to within

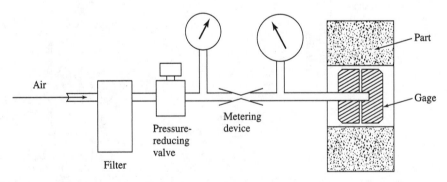

Figure 3–16   Air gages give a measure of the distance between gage head and workpiece surface.

50 nm) is placed at a slight angle to the workpiece surface and monochromatic light is beamed on it, light and dark bands (*interference fringes*) become visible to the eye (or a photodetector). The reason is that light rays from the monochromatic light source are reflected from both the bottom surface of the optical flat and the surface of the workpiece (Fig. 3–17a). The two reflected rays interact; the ray reflected from the workpiece surface travels a path that is longer by the distance *CDE*. If this distance equals a wavelength λ (or an integer multiple of it, $n\lambda$), the two rays reinforce each other (Fig. 3–17b) and the observer sees a light band. Conversely, if the distance is $\lambda/2$ (or $\lambda/2 + n\lambda$), the rays cancel and a dark band appears (Fig. 3–17c). Since *DE* is practically equal to *CD*, the fringes repeat every time the height between flat and workpiece surface changes by $\lambda/2$. By counting the number of fringes, the total distance (or the height of the workpiece from a reference plane) can be measured. (Note that refraction and phase changes are ignored here because they do not affect the argument.) Helium-neon lasers are increasingly used as the light source.

Interferometry is also useful for checking the flatness of surfaces: Fringes are straight, parallel, and evenly spaced when the surface is flat.

4. Highly collimated *laser beams* can be used for the noncontacting (and on-line) measurement of dimensions. In one approach, the workpiece is placed in the light path between source and photodetector. The beam sweeps at a preset rate, hence the length of time for which the light is cut off is a measure of the dimension. In another approach, the beam is split, and interference with the beam reflected from the surface of the part gives the distance on a digital display, to a resolution of 2.5 μm (0.0001 in) or better.

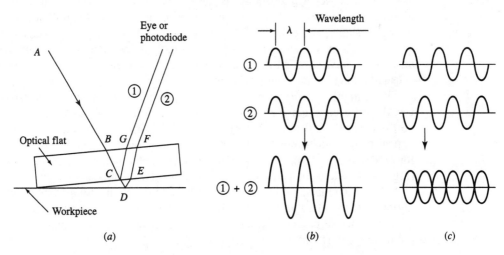

(a)   (b)   (c)

Figure 3–17   Flatness of a surface is obtained from (a) interference fringes produced with the aid of an optical flat. Light is split into two beams: (b) when in phase, they reinforce each other and a light band appears; (c) when out of phase, they cancel and a dark band appears.

Figure 3–18   Laser *x–y* coordinate measuring machine allows mapping of the dimensions of a part to submicron accuracy. (*Courtesy microVu, Windsor, California.*)

Many of the above techniques are used also for measuring flatness, straightness, circularity, and perpendicularity.

3-4-6   Measuring Machines

The term *measuring machine* is used to denote structures built with extreme care to provide supports for transducers relative to a reference surface or reference axes. In a sense, a micrometer is also a measuring machine, but of limited accuracy. In the more commonly applied sense, measuring machines are made to be highly stable, and contain high-precision movements to permit measurement along a single axis or along two or three mutually perpendicular axes (*coordinate measuring machines*, CMMs). Some machines also measure angles. Calibration is usually done by laser interferometry.

With highly sensitive touch probes driven by hand or computer program, CMMs are used extensively for the tracing of complex surfaces. Resolutions on the order of 250 nm (10 μin) are possible with mechanical, electronic, or optical readout instruments. The readout is usually supplied in a digital form for processing by computer. For noncontact measurement, laser-beam scanning (Fig. 3–18), video-image processing, and optical transducers are available.

Measuring machines can be used for layout prior to machining and for checking dimensions after machining. They can be linked to computers to perform automatic measurements, sometimes in conjunction with a computer-controlled production cell (flexible manufacturing cell, Sec. 21-2-4).

3-5 SURFACE TOPOGRAPHY

Few surfaces are smooth and flat (or of cylindrical or other pure geometrical shape). On the microscopic scale, surfaces exhibit *waviness* and *roughness*.

3-5-1   Roughness and Waviness

The surface profile can be measured and recorded. For easier visualization, recordings are usually made with a larger gain on the vertical axis (Fig. 3–19). This gives a distorted image with sharp peaks and steep slopes; in reality the peaks (*asperities*) have gentle slopes of typically 5–20° inclination (as in Fig. 4–21*b*). The traces or, more frequently, the signal obtained from the profilometer may be processed electronically or, after digitization, in a computer, to derive various values for a quantitative characterization of the surface profile. Of the various measures given in ANSI B46.1-1978, R1995, the following are most frequently used:

1. R_t is the *maximum roughness height* (the height from maximum peak to deepest trough). It is important when the roughness is to be removed, for example, by polishing. Often a more meaningful figure is obtained by taking the average height difference between the 5 highest peaks and 5 deepest valleys within the sampling length (10-point height, R_z).

2. A line, drawn in such a way that the area filled with material equals the area of unfilled portions, defines the centerline or mean surface. The average deviation from this mean surface is called the *centerline average* (CLA) or *arithmetical average* (AA), denoted also as R_a:

$$R_a = \frac{1}{l} \int_0^l |y| \, dl \quad \text{or} \quad R_a = \frac{y_1 + y_2 + y_3 + \cdots + y_n}{n} \qquad \textbf{(3-4)}$$

3. The *root mean square* (RMS) value R_q is frequently preferred in practice and also in the theory of contacting surfaces

$$R_q = \left[\frac{1}{l} \int_0^l y^2 \, dl \right]^{1/2} \quad \text{or} \quad R_a = \left(\frac{y_1^2 + y_2^2 + y_3^2 \cdots + y_n^2}{n} \right)^{1/2} \qquad \textbf{(3-5)}$$

R_q is closely related to $R_a (R_a = 1.11 \, R_q$ for a sine wave) and, for technical

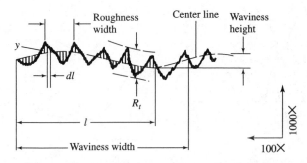

Figure 3–19   The roughness of technical surfaces can be revealed by various techniques; typical recordings are made with a larger magnification in the direction perpendicular to the surface.

surfaces, the relationship between various values is fairly well defined (Table 3–1).

Table 3–1   Approximate relationship of surface roughness values

Type of Surface	R_q / R_a	R_t / R_a
Turned	1.1	4–5
Ground	1.2	7–14
Lapped	1.4	7–14
Random	1.25	8

4. *Skewness* expresses the distribution of roughness heights and is a quantitative measure of the "fullness" of the surface (Fig. 3–20). The Abbot curve shows the load-bearing area available when cuts are taken at various levels from the top of the profile.

Convenient units of measurement are the micrometer (μm) or nanometer (nm) and the microinch (μin).

$$1 \; \mu in = 0.025 \; \mu m \; = 25 \; nm$$
$$1 \; \mu m = 40 \; \mu in$$

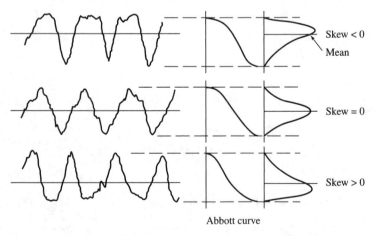

Abbott curve

Figure 3–20   For the same peak-to-valley roughness height, surfaces may have very different profiles, resulting in a skewing of the roughness-height distribution. (*After ANSI B46.1-1978, R1995.*)

The finer details of surface roughness are superimposed on larger-scale periodic or nonperiodic variations (waviness, Fig. 3–19). In measuring the surface roughness, the waviness is usually filtered out by electronic processing of the signal, although the allowable waviness is specified and measured (in units of millimeters or inches) when it is functionally important.

On drawings, *roughness limits* are given by a check mark written over the line to which the roughness designation applies (Fig. 3–21). A single roughness number indicates an upper limit, below which any roughness is acceptable; if a minimum roughness is required, two limits are shown. Waviness, when important, is limited by a number over the horizontal line of the check mark. Surfaces usually exhibit a topography characteristic of the finishing process. The characteristic *directionality (lay)* is indicated by a symbol placed under the check mark.

It must be recognized that the same numerical R_q or R_a values may be obtained on surfaces of greatly differing profiles, and that highly localized troughs add very little to the average values. Therefore, averages are inadequate to describe surfaces for specific applications. In general, surface characterization remains a challenge. Nevertheless, the manufacturing process must be capable of providing a surface suitable for the intended function of the part, and various quantitative or descriptive terms can be and often are used to elaborate on the required finish.

3-5-2　Surface Finish and Tolerances

We already stated that unnecessarily tight tolerances and surface finish specifications are a major course of excessive manufacturing costs. There is a close

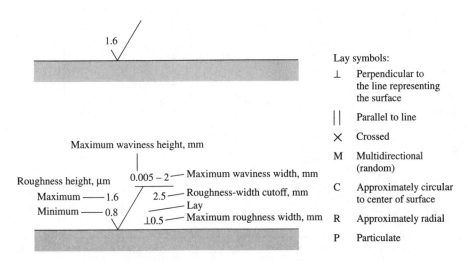

Lay symbols:

⊥　Perpendicular to the line representing the surface

||　Parallel to line

✕　Crossed

M　Multidirectional (random)

C　Approximately circular to center of surface

R　Approximately radial

P　Particulate

Figure 3–21　Characteristics of the surface finish are described by standard symbols (the example is given in SI units, with roughness in μm R_a).

relationship between roughness and tolerances. A good rule of thumb is that the maximum roughness height R_t (and waviness, if any) should be about one-third to one-half the tolerance, unless the fit is a forced fit and the surface roughness can be at least partially smoothed out in the fitting process. Since $R_t = 10R_a$, a roughness value of 3.2 μm (125 μin) R_a would be too coarse for a tolerance of 0.025 mm (0.001 in), and a roughness of maximum 0.8 μm (32 μin) R_a should be specified for such a tight tolerance.

Each manufacturing process is capable of producing a part to a certain surface finish and tolerance range without extra expenditure. Some general guidance is given in Fig. 3–22. The surface finish and tolerances *usually* attainable in the process are indicated by heavy lines adjacent to the name of the process. The capabilities of some processes overlap; for example, shell casting at its best can compete with plaster casting, but can never match the best plaster casting results. When ranges are common to several processes, the names of these processes are separated by commas; for example, the same tolerance and surface finish are obtained in drilling or punching a hole. Plastic parts usually bear the surface finish of the mold or die in which they were made, hence they can be produced to any finish (although fiber-reinforced parts may be quite rough).

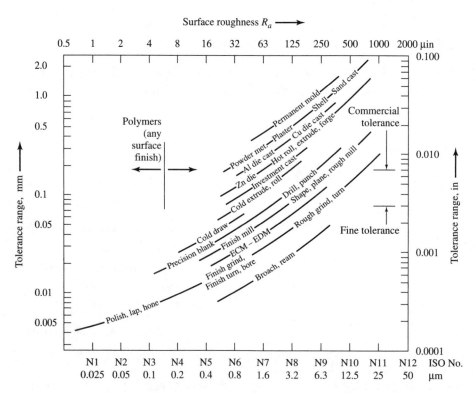

Figure 3–22   Under typical conditions, each manufacturing process is capable of producing parts to some characteristic tolerance and surface finish. See text for interpretation of graph.

The tolerances given apply to a 25-mm (1-in) dimension. For larger or smaller dimensions, they do not necessarily increase or decrease linearly. In a production situation it is best to take the recommendations published by various industry associations (see Chap. 1) or individual companies.

Surface roughness in Fig. 3–22 is given in terms of R_a [arithmetic average, Eq. (3-4)]. In many applications the texture (lay) of the surface is also important, and for a given R_a value, different processes may result in quite different finishes (Fig. 3–23).

It used to be believed that cost tends to rise exponentially with tighter tolerances and surface finish (Fig. 3–24). This is true only if these tolerances are achieved by the use of a process sequence involving processes and machine tools of limited capability. There are, however, processes and machine tools of inherently greater accuracy and better surface finish (Sec. 16-9-2), and then higher-quality products can be obtained with little extra cost and, if the application justifies it, certainly with greater competitiveness. Still, a cardinal rule of the cost-conscious designer is to specify the loosest possible tolerances and coarsest surfaces that still fulfill the intended function. The specified tolerances should, if possible, be within the range obtainable by the intended manufacturing process (Fig. 3–22) so as to avoid separate finishing operations.

From what was said above, it is obvious that the limits indicated in Fig. 3–22 are not inflexible. Indeed, various manufacturing industries have responded to competitive pressures by tightening tolerances, improving surface topography, and generally improving quality without necessarily raising the cost of their product.

An infant respirator relies on a small-volume, fast-cycle air pump. There is a PTFE seal on the piston, and it has been found that the walls of the pump cylinder must be finished to between 0.1 and 0.2 μm R_a. A rougher finish results in loss of compression because of rapid wear of the seal. A very smooth finish results in higher friction between seal and cylinder; heat builds up, the PTFE melts locally, and the respirator fails. (Source: M. F. DeVries, M. Field, and J. F. Kahles, *Annals of CIRP*, **25**: 569–573, 1976).

Example 3-3

3-5-3   Surface Roughness Measurement

The most common surface-roughness measuring instrument is based on the principle of the phonograph (Fig. 3–25). An arm with a reference rest is drawn across the surface, while a stylus follows the finer surface details. The surface profile can be recorded (as in Fig. 3–19) and various roughness characteristics computed. Portable instruments give a readout of a few parameters (typically, R_a or R_q) directly.

For plant use there are also collections of standard sample surfaces (*replica blocks*) available, with R_q marked for each sample. By drawing a fingernail

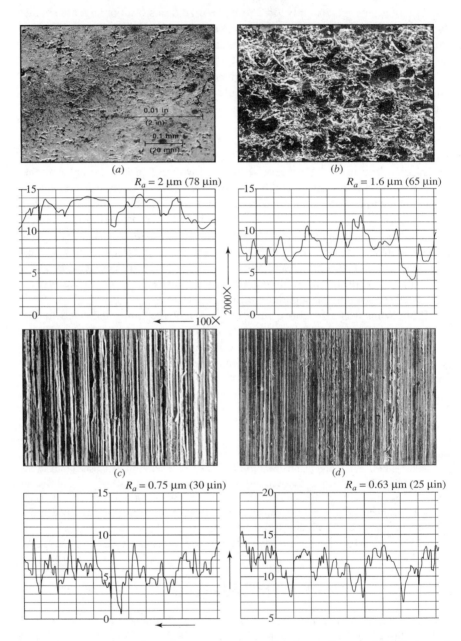

Figure 3–23   Scanning electron microscope images and surface profile traces (all 100×) reveal that quite different detail features may exist on surfaces of similar roughness averages. Random surfaces: (a) permanent-mold cast and (b) shotblasted. Directional surfaces: (c) cold-rolled and (d) ground.

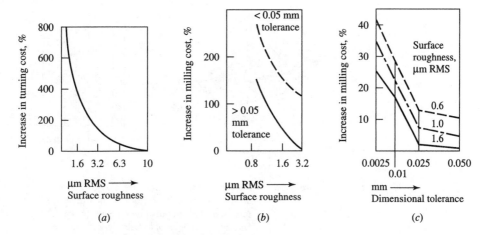

Figure 3–24   Smoother surfaces and tighter tolerances can be produced only at increased cost, whether in (a) turning, (b) milling, or (c) surface grinding. (L. J. Bayer, ASME Paper 56-SA.9, 1956. With permission of the American Society of Mechanical Engineers, New York.)

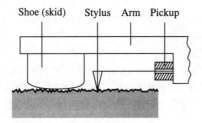

Figure 3–25   Surface features are revealed by drawing a stylus, attached to a pickup, across the surface.

across the sample and the production part, remarkably close estimates of R_q can be obtained.

Other inspection devices are based on the measurement of capacitance, optical interference (Fig. 3–26), diffraction, and air-pressure drop.

In some instances, surface attributes are difficult to quantify, and then comparison specimens, chosen to represent acceptable and reject qualities, are used.

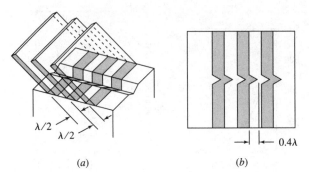

$\lambda/2$ $\lambda/2$

(a)

0.4λ

(b)

Figure 3–26   Light-interference microscopy is used to observe deviations from a flat surface; in the example a scratch of 0.4-μm depth is revealed. (*From F.T. Farago, Handbook of Industrial Measurement, 2d ed., Industrial Press, New York, 1982, p. 394, Fig. 15.21. With permission.*)

3-6   SUMMARY

Manufactured parts must have closely defined shapes, dimensions, and surface roughness.

1. The shape of parts affects function and manufacture; classification into groups according to shape and manufacturing sequence (group technology) offers many savings.

2. Dimensions and their tolerances are prime attributes and are expressed in engineering drawings. Geometric dimensioning and tolerancing helps to communicate the intent of the design and facilitates production and quality control.

3. Dimensions are checked by various gages based on hard bodies or on various electronic and optical techniques. The outputs of many of them can be digitized, thus made available for production and quality control. Because of the importance of making parts right the first time, engineering metrology and, especially, in-process measurement assume a central role in manufacturing.

4. Surface appearance is important for esthetic appeal. Surface topography, including roughness and lay, is critical for contacting surfaces. Numerous measures of surface topography can be obtained with various contacting and noncontacting instruments.

5. Maintenance of specified tolerances is vital to the function of assemblies and makes interchangeability possible. Excessively tight tolerance and surface finish specifications do lead, however, to excessive manufacturing costs.

PROBLEMS 3A

3A-1 Define allowance, tolerance, and fit.

3A-2 In measuring the diameter of a shaft, give two examples each for (*a*) assignable errors and (*b*) random errors.

3A-3 Define for a measuring instrument (*a*) sensitivity, (*b*) resolution, (*c*) repeatability, and (*d*) stability.

3A-4 Make a sketch to show a typical surface roughness recording one would obtain from a stylus instrument traversing perpendicular to the lay of a turned surface. (*a*) Indicate relative magnifications; (*b*) show the centerline; (*c*) show the roughness width; (*d*) show the waviness width and height; (*e*) define R_a, R_q, and R_t (in words or equations).

3A-5 Make sketches to show (*a*) two-axis, (*b*) three-axis, and (*c*) five-axis movement.

3A-6 In a drawing similar to Fig. 3–5, show a shaft with interference fit in a hole. Assuming a hole-basis assembly, show tolerance bands and maximum and minimum material condition.

3A-7 Draw the pattern of interference fringes when a flat part has a scratch in the surface.

PROBLEMS 3B

3B-1 Standards are available that list preferred sizes. Have they been developed for reasons of technology, economy, or both? Explain.

3B-2 The diametral tolerance of a shaft of 20.00-mm diameter and 200.0-mm length is ±0.10 mm. Coordinate dimensioning and tolerancing was used. (*a*) Make a dimensioned sketch of a geometrically perfect cylinder. (*b*) Make sketches of two other bodies that satisfy the specifications.

3B-3 What noncontacting technique would be suitable for automatic, 100% inspection of shafts produced to the tolerances shown in Fig. 3–5? List at least 2 options and indicate what factors will affect the choice.

3B-4 Is tolerancing in Fig. 3–5 unilateral or bilateral? Explain why this particular system of tolerancing was chosen.

3B-5 The diameters of turned cylindrical bodies are measured with a micrometer. (*a*) Draw a diagram with the bell curve within ±3σ of the tolerance zone. Superimpose a curve one would expect if (*b*) measurement was done by an unskilled operator or (*c*) by a skilled operator but if all cylinders were oversize. (*d*) In case (*a*), would all bodies fall within the tolerance zone?

3B-6 A batch of components is returned to the manufacturer with the claim that all parts are outside tolerance, even though their dimensions are closely clustered. In-process, 100% inspection had been made with a pneumatic gage. Can the complaint be correct? Justify.

3B-7 Suggest at least 2 in-line, in-process measuring techniques for the (*a*) shaft and (*b*) bearing in Fig. 3–5.

3B-8 The surface finish of a part is specified as 1.6 μm R_a maximum. A hand-held, stylus-type instrument gives readings in μm RMS. Should the maximum permissible limit be greater or smaller than 1.60 μm?

3B-9 A customer complains that a part is not usable because it was received with a scratch that could not be removed by the usual polishing. Inspection records show that the part left the plant with a surface finish that satisfied the specified R_a roughness. (*a*) Can the customer's claim be valid? Justify. (*b*) If the answer is yes, how would you change your in-house specification for surface topography?

3B-10 Make a sketch of the pattern of interference fringes produced when: (*a*) a ball-bearing ball and (*b*) a roller-bearing roller is placed on an optical flat.

3B-11 The roughness of surfaces similar to those shown in Fig. 3–23 is measured. State for each of the four surfaces whether the roughness value R_a will be greater in the horizontal or vertical direction of the picture.

PROBLEMS 3C

3C-1 The shaft and bearing shown in Fig. 3–5 are checked for dimensions. (*a*) What precision should the measuring device have and (*b*) what would be the smallest scale division on such an instrument?

3C-2 Make dimensioned sketches of GO–NOT GO gages used for checking the (*a*) shaft and (*b*) bearing of Fig. 3–5.

3C-3 Hydrodynamic bearings are typically made with a clearance of 0.1% of diameter. Check whether the bearing of Fig. 3–5 satisfies this criterion.

3C-4 The shaft of Fig. 3–5 is made of AISI 410 stainless steel. There is a change of 10°C in room temperature; what will be its effect on the diameter?

3C-5 A flywheel (with a hole diameter of 50.000–50.025 mm) is fitted on a shaft (of diameter 50.070–50.086 mm). The thermal expansion coefficient of the steel is 12.4 μm/m · K. (*a*) Is the assembly dimensioned in the hole-basis or shaft-basis system? Why? (*b*) Calculate the fit for maximum and minimum material condition. (*c*) The shaft and flywheel are assembled by heating the flywheel; calculate the necessary temperature. (*d*) Alternatively, the shaft could be cooled in liquid nitrogen. Determine whether this is feasible.

3C-6 After grinding the shaft of Problem 3C-5, the diameter is measured with a gap gage (C-gage). The shaft is at 80°C, the gage at room temperature (20°C). (*a*) Calculate the magnitude of measurement error. (*b*) Can the shaft diameter be measured with the required accuracy? If not, what should be done? (*c*) Indicate the maximum value of desirable R_t and (*d*) R_a.

3C-7 A 2024 aluminum block of 200 mm length was milled on all six sides and is now at 70°C. Vernier calipers, made of steel and calibrated at 20°C, are used to check the dimensions. They are now at 25°C. Calculate the error if the 200-mm dimension is specified at 20°C.

3C-8 In Prob. 3C-7 both the aluminum block and the vernier calipers now are at 30°C. What is the error?

3C-9 The aluminum block of Problem 3C-7 is still at 70°C but the the vernier caliper is at 16°C. What is the error?

3C-10 An optical flat is laid upon a polished but wavy surface. Make a sketch showing the interference fringes produced by a monochromatic light if the wavelength is one quarter of the depth of the wave.

Further Reading

Wick, C., and R. Veilleux (eds.): *Tool and Manufacturing Engineers Handbook*, vol. 4, *Quality Control and Assembly*, Society of Manufacturing Engineers, 1987.

Bosch, J.A.: *Coordinate Measuring Machines and Systems*, Dekker, 1995.

Farago, F.T., and M.A. Curtis: *Handbook of Dimensional Measurement*, 3d ed., Industrial Press, 1994.

Henzhold, G.: *Handbook of Geometrical Tolerancing*, Wiley, 1995.

Krulikowski, A.: *Fundamentals of Geometric Dimensioning and Tolerancing*, Delmar, 1997.

Medows, J.D.: *Geometric Dimensioning and Tolerancing*, Dekker, 1995.

Medows, J.D.: *Measurement of Geometric Tolerances in Manufacturing*, Dekker, 1998.

Murphy, S.D.: *In-Process Measurement and Control*, Dekker, 1990.

Puncochar, D.E.: *Interpretation of Geometric Dimensioning and Tolerancing*, 2d ed., Industrial Press, 1997.

Sydenham, P.H.: *Transducers in Measurement and Control*, 3d ed., Hilger, Bristol, 1985.

The automobile exemplifies the multitude of properties a manufactured product must satisfy: it carries static and dynamic loads, protects its occupants in a collision, resists corrosion, has pleasing appearance, and does all this at a highly competitive price. [Generic auto of UltraLight Steel Auto Body (ULSAB) project.] (*Courtesy American Iron and Steel Institute, Southfield, Michigan.*)

chapter
4

Service Attributes of Manufactured Products

This chapter reviews the most important properties of engineering materials. It serves not only as a refresher for those who have had a course in materials, but also highlights principles with manufacturing implications. You will learn about:

Mechanical properties and their determination

Ductile and brittle behavior of materials

The harmful effects of internal defects, inclusions, and minute surface defects

Methods of neutralizing these effects: the importance of cleanliness, hydrostatic pressure, and residual stresses

Elastic deformation of test and production equipment

Tribological, physical, and chemical properties.

The properties that make manufactured products valuable are called service attributes. They are important also for design, because service properties often dictate the choice of materials or, at least, narrow the choice of alternative materials that can be considered. This has immediate repercussions for manufacturing for several reasons:

1. *Optimum processes are different for different materials* and the choice of manufacturing process is affected.

2. Processes can be directed to change the properties of materials, often by influencing the structure or stress state in the material. The *sequence* of manufacturing processes must be chosen for any given material so that the *desired end properties will be reached at minimum cost.*

3. The acceptability of the finished product is judged on the basis of tests in which conformance to specifications is checked. *The same tests must be em-*

ployed during manufacturing to assure that the properties of the final product will meet specifications. They are often supplemented with technological tests, i.e., tests that simulate the conditions imposed on the material during manufacture. Technological tests will be described in the individual chapters dealing with processes.

4-1   MECHANICAL PROPERTIES IN TENSION

A most obvious property of manufactured products is that they are capable of supporting loads. *Loads (forces)* may be of many kinds; accordingly, there are many test methods designed with the specific aim of reproducing loading in service. In many applications the load is static, i.e., constant and stationary, and several tests are conducted at such low speeds that the application of force can be regarded as static. Yet other tests aim to establish behavior at elevated temperatures or at controlled speeds of load application. We shall see in Chap. 5 that materials of construction can be broadly classified as metals, ceramics, plastics, and composites, and different tests may be applicable to different materials. Test results are affected by the test method itself; therefore, tests must be conducted in conformance to standards.

4-1-1   The Tension Test

Most structures impose tensile stresses on components; therefore, properties are routinely checked in the *tension test* (also called *tensile test*), subject to ASTM Standard E8.

Test Setup   The test specimen is machined with larger heads at its ends to ensure secure gripping. There are standard specimen geometries for round and flat (sheet) specimens. On imposing a load, the weaker part of uniform cross section (the *gage length*) deforms (Fig. 4–1). The gage length is usually accurately marked on the surface.

The specimen is held in self-aligning heads to ensure that only pure tensile loads (and no bending) will be imposed. The *test machine* is essentially a press in which a moving crosshead is displaced in a controlled manner (such as a preset speed) by an actuator. In Fig. 4–1 the actuator is a hydraulic cylinder, but it could be a screw-and-nut or other mechanism too. The movement of the crosshead develops a force P which is balanced by the reaction force P. The magnitude of P is measured with an instrument called a *dynamometer*. Most machines are equipped with a load cell that gives an electrical signal proportional to the applied load. All load cells are calibrated against another load cell of known accuracy.

Extension of the specimen is measured by attaching an *extensometer* to the gage length. Transducers give an electrical output proportional to the elongation Δl. Noncontacting optical methods may also be used.

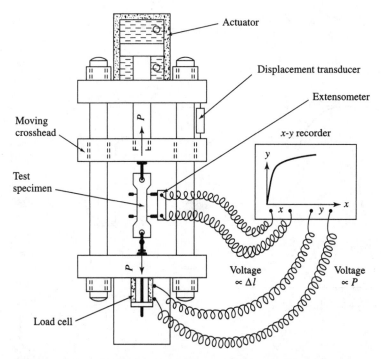

Figure 4–1   Universal testing machines can be used for tension, compression, and bending tests. A recorder or data-acquisition system is used to obtain values of force and displacement; the latter may be obtained from an extensometer attached to the specimen or from a displacement transducer attached to the moving crosshead.

In the course of testing, both load and extension change continuously. Most conveniently, transducer outputs are used to drive an x-y recorder so that a force (dependent variable) versus extension (independent variable) recording is obtained (Fig. 4–2). Outputs can be directly digitized with a data-acquisition system linked to a computer, thus the analysis of results is speeded up. Even so, there is merit in a visual recording that often reveals features potentially obscured by numerical processing.

Stress–Strain Curve   The force-displacement diagram shown in Fig. 4–2*a* is typical of ductile metals such as copper tested at room temperature. If specimens of a larger cross section were tested, different curves would be obtained, simply because it takes a greater force to deform a larger specimen. Therefore, results can be normalized by dividing the force P by the area A over which the force acts. In general, stress is defined as the internal force per unit area in an object subjected to external forces. A *normal stress* acts perpendicular to the cut surface

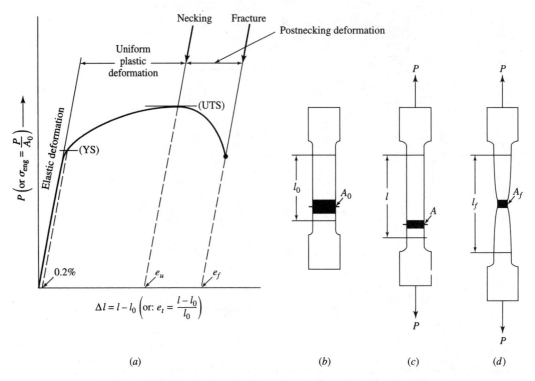

Figure 4-2 (a) The force–displacement (or engineering stress–strain) curve obtained on testing a ductile material reflects the sequence of events: (b) a specimen of A_0 initial cross section first suffers elastic deformation, then (c) deforms plastically—more or less uniformly within the gage length—and (d) subsequently necks and finally fractures.

and is denoted by σ. Its value is

$$\sigma = \frac{P}{A} \tag{4-1}$$

The SI unit of stress is N/m^2 (also called pascal, Pa); this represents a very small stress, therefore, MN/m^2 or MPa is often used. The unit MPa is numerically equal to N/mm^2, which is more convenient for many calculations. In the U.S. Conventional System (USCS) the unit is lbf/in^2, written as psi. This too is a small unit, and a thousandfold value (often denoted as ksi but, more logically, written as kpsi in this book) is more customary. (For a quick conversion, 1 kpsi = 7 MPa). In the old metric system, the unit was kg/mm^2, which roughly equals $10\ N/mm^2$ (the kg stands for kg force).

In the course of the tension test the specimen is forcefully elongated. To a first approximation, most engineering materials are incompressible. While there is a small change in volume during elastic deformation, their volume V remains virtually constant during plastic deformation. This is expressed as the *principle of constancy of volume*:

$$V = A_0 l_0 = A_1 l_1 = A l \qquad \textbf{(4-2)}$$

where A and l are instantaneous cross-sectional area and length, respectively. The subscript 0 refers to starting dimensions, the subscript 1 to final dimensions. We shall come back repeatedly to this principle.

Because of constancy of volume, there must be a reduction in cross-sectional area to compensate for the increase in length (compare Fig. 4–2c with Fig. 4–2b). However, the exact value of the cross-sectional area is not immediately known and, for convenience, the convention has been adopted to calculate force P divided by the original cross-sectional area A_0. By definition, stress is force acting on unit area; since here we divide force by an area that does not exist any more, the result is distinguished from a true stress by calling it *nominal, conventional,* or *engineering stress* (σ_{eng} or S):

$$\sigma_{\mathrm{eng}} = \frac{P}{A_0} \qquad \textbf{(4-3)}$$

Elongation too can be normalized by taking the change in length and dividing it by the original length; this is usually termed *engineering tensile strain* e_t,

$$e_t = \frac{l - l_0}{l_0} \qquad \textbf{(4-4a)}$$

where $l - l_0 = \Delta l$, the change in length. For conversational convenience, a percentage value is often quoted:

$$e_t(\%) = \frac{l - l_0}{l_0} 100 \qquad \textbf{(4-4b)}$$

It should be noted that data-acquisition systems can be added to all test equipment, and data can be processed by a computer. Combined with computer control of the test equipment itself (for example, by servohydraulic control), testing can be automated.

4-1-2   Process/Equipment Interactions

For quality-control purposes the test method can often be simplified: No extensometer is used, and the force-displacement curve is recorded simply from the moving crosshead. The recording thus obtained (Fig. 4–3, broken line), is similar to the one obtained with an extensometer (Fig. 4–3, full line) but with one significant difference: the initial slope of the curve is now much lower. Inspection of Fig. 4–1 will show that the force P deforms not only the gage length (and the rest of the specimen), but also the machine: the stationary crossheads are bent and the columns are compressed. Even though the machine is much sturdier than the specimen, the length over which deformation occurs (the *elastic loading path*) is much longer: the machine behaves as a very long spring, attached in series with

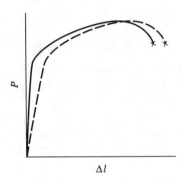

Figure 4–3   The initial slope of a force–displacement curve obtained from an extensometer (full line) gives the elastic modulus of the material; a curve obtained from crosshead displacement (broken line) includes the elastic deformation of the machine.

the short spring representing the specimen. In consequence, the deformation of the machine is added to the deformation of the specimen, and the initial slope of the force-displacement curve represents the sum of the two.

This is an important observation because, in most manufacturing processes, elastic deformation of the machine is large enough to affect dimensional control of the parts produced. We will have numerous occasions to refer to the principle of *elastic deformation of production equipment.*

4-1-3   Strength in Tension

Inspection of the engineering stress–strain curve shows a number of critical points that can be used to characterize a material.

Elastic Modulus   At the beginning of the test, the force increases rapidly and proportionately to strain: the stress–strain curve obeys *Hooke's law*

$$\sigma = E e_t \tag{4-5}$$

The proportionality constant (the slope of the curve) is called the *elastic modulus* or *Young's modulus E*

$$E = \frac{\sigma}{e_t} \quad \text{(MPa or psi)} \tag{4-6}$$

If the specimen is unloaded in this range, it will return to its original length, i.e., *all deformation is elastic.* Most structures are designed so that they should never suffer permanent deformation and E then determines the change in the length of a component for a given load. The elastic modulus reflects the basic structure and bond strength of materials. To gain a general feel for the order of magnitudes encountered, Chap. 5 has tables with typical values for some metals, ceramics, plastics, and composites. Note that, while the elastic modulus is little affected by processing, other properties often have a wide range for a given material, and one of the purposes of this book is to show how properties can be tailored by manufacturing control.

A completely *brittle material* deforms only elastically. At some critical stress, separation (fracture) occurs suddenly (Fig. 4–4a), usually in a plane perpendicular to the axis of load application (Fig. 4–4b). Fracture often originates from some minute crack that locally raises the stress (see Sec. 4-1-6). Brittle behavior, identifiable by zero elongation, is typical of a few metals, almost all ceramics, and thermosetting polymers, and plastic deformation is minimal with some alloys.

A tension-test specimen of 12.7-mm diameter and 50-mm gage length was cast of a Zn-12Al alloy. With an extensometer attached, the line shown in Fig. 4–4a (broken line) was recorded. Determine the elastic modulus. | **Example 4-1**

Draw a line parallel to elastic line at $\Delta l = 0.05$ mm.

From Eq. (4-3), $\sigma = P/A_0 = 1900 \text{ N}/(12.7^2 \pi/4) \text{ mm}^2 = 15 \text{ N/mm}^2$; from Eq. (4-4a), $e_t = 0.05/50 = 0.001$; from Eq. (4-6), $E = 15 \text{ MPa}/0.001 = 15 \text{ GPa}$.

In determining the elastic modulus, we must be careful to take the slope only from a tensile test curve produced with an extensometer. If displacement is taken from the crosshead, the initial slope represents the combined deformation of specimen and machine (Sec. 4-1-2).

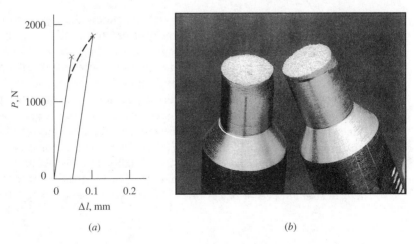

(a)   (b)

Figure 4–4   A brittle material shows (a) little or no evidence of plastic deformation in the tension test and (b) fracture often occurs along grain boundaries or other weakening features. (*The example shown is a Zn-12Al alloy. Courtesy Dr. P. Niessen, University of Waterloo.*)

Some materials (e.g., gray cast iron, some plastics) gradually yield from the beginning and Hooke's law does not hold. An arbitrary modulus (*secant modulus*) is determined by connecting the origin with a specified point (e.g., one-fourth the tensile strength, or an arbitrarily chosen strain).

Yield Strength   When ductile materials are tested, at some critical stress the slope of the curve changes, and this stress is termed the *proportional limit*. However, its determination is quite difficult; therefore, it is customary to choose a point at which the specimen deforms permanently. The corresponding engineering stress is called the *yield strength* YS or S_y. For most metallic materials, 0.2% permanent deformation is taken as the threshold because it is relatively easily measured, and then the yield strength is denoted as $\sigma_{0.2}$ (or $S_{0.2}$)

$$\sigma_{0.2} = \frac{P_{0.002}}{A_0} \tag{4-7}$$

Note that, by definition, the strain of 0.002 is all *plastic* (permanent) *strain*, therefore, the corresponding force $P_{0.2}$ is found by drawing a line from $e_t = 0.002$ *parallel to the elastic line*. If the specimen were unloaded at this point, all deformation would be recovered, at a slope equal to the initial slope of the force-displacement curve. By drawing the parallel line, the contribution of elastic deformation to total strain is eliminated (Fig. 4–2a).

Yield strength is an important design quantity. To prevent even the slightest plastic deformation of an engineering structure, the design stress is often kept to some fraction of $\sigma_{0.2}$ by the use of a safety factor, or design is made to some lower value such as $\sigma_{0.02}$.

Tensile Strength   On further loading and elongation, the gage section of the specimen elongates (and its cross section reduces) uniformly along its entire length (Fig. 4–2c) yet the force gradually increases. For reasons to be explained in Sec. 8-1-4, the material becomes stronger with deformation (it strain-hardens). At some critical deformation level typical of the material and its processing history, strain hardening cannot counterbalance the loss of strength resulting from the ever-decreasing cross-sectional area, and a neck forms at the weakest point. Since the cross section is now locally reduced, the force sustained by this weakened section is less, and the force P declines while deformation is concentrated in the already necked zone (Fig. 4–2d). Finally, fracture occurs.

The engineering or conventional stress at the maximum load is called the *tensile strength* (TS or S_u) or often also *ultimate tensile strength* (UTS),

$$\text{TS} = \frac{P_{\max}}{A_0} \tag{4-8}$$

The TS is not a true stress (because force is divided by an area that does not exist at this point), but it has great practical value for quality-control purposes. It is also a measure of the maximum force a component can sustain before catastrophic failure.

An aircraft component, made of 7075-T6 aluminum alloy, can be represented as a bar of diameter 20 mm and length 400 mm. It is loaded in pure tension. Calculate: (a) the extension of the bar under an imposed load of 80 kN, (b) the load at which the bar suffers permanent deformation, and (c) the maximum load the bar can take without fracture. | **Example 4-2**

From Table 5–2, $E = 70$ GPa; YS $= 500$ MPa; TS $= 570$ MPa.

(a) The cross-sectional area of the bar is $A_0 = 20^2 \pi / 4 = 314$ mm^2. The imposed tensile stress is $80\,000/314 = 255$ N/mm^2 ($= 255$ MPa) and is thus less than the YS: deformation will be purely elastic. From Eq. (4-5), $e_t = \sigma / E = 255/70\,000 = 0.0036$ or 0.36%.

(b) YS $= \sigma_{0.2} = 500$ N/mm^2. From Eq. (4-7), $P_{0.002} = (\sigma)(A_0) = (500)(314) = 157$ kN.

(c) From Eq. (4-8), $P_{\max} = (TS)(A_0) = (570)(314) = 179$ kN.

4-1-4   Ductility in Tension

The engineering stress–strain curve also provides information on the ductility of the material, i.e., its ability to deform without fracture.

1. **Uniform elongation.**   Prior to necking, the cross section reduces roughly uniformly along the gage length. Therefore, the engineering strain sustained at the point of maximum load is called *uniform elongation*, denoted by e_u

$$e_u = \frac{l_u - l_0}{l_0} \qquad \textbf{(4-9a)}$$

where l_u is the specimen length at the point of necking. Uniform elongation will be important for some deformation processes, but it has little application as a service property and is seldom quoted in databases.

2. **Elongation.**   More frequently, the total *elongation to fracture* (also called *total elongation* or simply, and somewhat misleadingly, *elongation*) is measured, most often by placing the broken parts of the specimen together and measuring the distance l_f between the gage marks

$$e_f = \frac{l_f - l_0}{l_0} \qquad \textbf{(4-9b)}$$

Alternatively, the length at fracture is taken from the extensometer output. Note that, if l_u or l_f (or any length during deformation) is measured from a recording, the elastic contribution to elongation must be taken out *by drawing a line parallel to the elastic loading line* (Fig. 4–2a).

As visible from Fig. 4–2a and d, e_f is the sum of uniform elongation and elongation in the neck. Thus, it is sensitive to gage length: a shorter gage length will make the same material appear to have a larger elongation. For this reason, the gage length must always be stated, otherwise total elongation—a readily measurable quality-control indicator—would lose its meaning. In this text, unless otherwise stated, elongation (abbr.: el.) is always measured on a 50-mm (or, what is practically the same, 2-in) gage length, where the effect of gage length becomes fairly minor for specimens of standard geometry.

Example 4-3	Scribe marks were made every 5-mm distance on a standard 50-mm gage-length test specimen of 9.5-mm diameter of hot-rolled 1018 steel. After testing to fracture, the change in length was measured over different gage lengths, always including the necked and fractured portion. Elongation was calculated from Eq. (4-9*b*):

l_0, mm	l_1, mm	e_f	el., %
15.0	22.8	0.52	52.0
25.0	36.7	0.468	46.8
50.0	63.5	0.27	27.0

Evidently, elongation must always be stated with reference to gage length.

Reduction in Area   The most sensitive measure of the ductility of materials is the *reduction in area* measured at fracture. To understand this, it is necessary to consider the stress state.

We saw that initial deformation is uniform along the gage length; in a cylindrical specimen, the diameter within the gage length becomes uniformly smaller (Fig. 4–5*a*). It may appear that there must be compressive stresses acting in the radial direction; this is, however, not the case. The material simply obeys the principle of constancy of volume [Eq. (4-2)]: to compensate for the increase in length, the cross section of the specimen must reduce. The only stress acting is in the axial (pulling) direction: the stress state is that of *uniaxial tension*.

All this changes at the onset of necking. The neck is the weakest part of the specimen, hence deformation is concentrated there. The material in the neck cannot, however, deform freely because the adjacent nondeforming material exerts a restraint. This sets up radial tensile stresses; thus, within the neck, the stress state changes to *triaxial tension* (Fig. 4–5*b*).

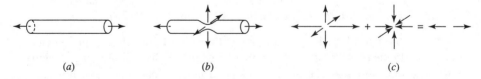

(a)   (b)   (c)

Figure 4–5   In the tension test, the stress state is (a) uniaxial during uniform extension but (b) becomes triaxial in the neck zone. Superposition of a hydrostatic pressure cancels the triaxial tension and (c) suppresses void formation.

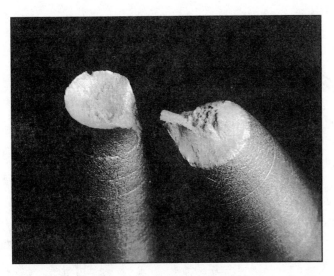

Figure 4–6   A ductile material undergoes plastic deformation before and beyond necking, and the fractured surface shows the formation and interlinking of voids and a characteristic cup-and-cone configuration.

Triaxial tensile stresses literally tear the material apart. First, cavities (voids) open in the center of the neck. On further straining, voids interlink, roughly on a plane perpendicular to the axis. Once the remaining annular cross section is insufficient to carry the load, the specimen fails (Fig. 4–6). The minimum cross-sectional area of the fractured test specimen, A_f, can be measured and reduction in area q (or RA) can be calculated as

$$q = \frac{A_0 - A_f}{A_0} \qquad \textbf{(4-10)}$$

Toughness   The area under the stress–strain curve has the dimension of force times distance, i.e., *work*. Thus it can be regarded as a measure of *toughness*, i.e., the *energy absorbed by the material prior to fracture*. Evidently, ductile materials, such as low-carbon steels and many aluminum and copper alloys have much greater toughness than brittle materials (Fig. 4–4).

We are now ready to evaluate the results of an actual tension test. A test specimen of 6.35-mm thickness and 6.38-mm width was machined from an annealed 80Cu-20Ni alloy plate. The gage length of $l_0 = 25.0$ mm was lightly marked with a scriber. The test was performed on a 10 000-kgf (98-kN) capacity testing machine, with an extensometer attached to the gage length. The curve shown as a full line was recorded. To obtain a better resolution at low strains, the

Example 4-4

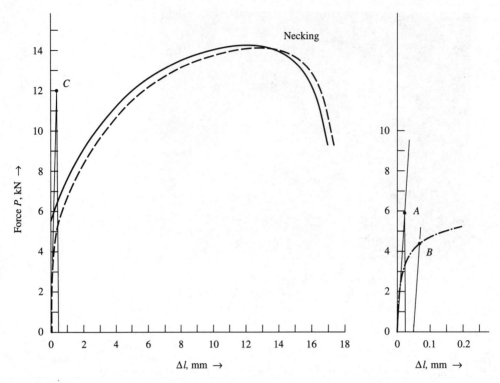

Figure Ex. 4-4

test was repeated with a 20 times higher gain on the extension axis (dash-dot line) (Fig. Ex. 4-4). The fractured specimen halves were placed together and the distance between the scribe marks was $l_f = 40.2$ mm. The fractured cross section was 2.85 mm × 3.50 mm. Calculate: (a) Young's modulus; (b) $\sigma_{0.2}$; (c) TS; (d) el.; and (e) RA.

(a) For the elastic modulus, draw a straight line through the elastic range; select a convenient point. In the example chosen (point A), $P = 5.7$ kN and $A_0 = (6.35)(6.38) = 40.5$ mm$^2 = 40.5(10^{-6})$ m^2. Extension is $\Delta l = 0.025$ mm, hence $e_t = 0.025/25.0 = 0.001$. From Eq. (4-6),

$$E = 5700/((40.5)(10^{-6})(0.001)) = 141 \text{ GPa}$$

For the rest, it is convenient to set up a spreadsheet for repetitive calculations. Such a spreadsheet can then serve as a template for all tension tests. Constant: $l_0 = 25.00$ mm. Since our test results are in the form of a recorded curve, we need to find:

(b) For the yield strength, the force where strain $e_t = 0.2\% = 0.002$. From Eq. (4-4a), $\Delta l = 0.002(25.0) = 0.05$ mm. For the dash-dot curve, draw a line from this point, parallel to the elastic line. It intersects the recording at point B, where $P = 4.4$ kN.

(c) For TS, we need the maximum load: $P_{\max} = 14.2$ kN.

(d) For elongation, we use $l_f = 42.2$ mm measured on the specimen. From the recording (full line), $(l_f - l_0) = 17.0$ mm and $l_f = 17.0 + 25.0 = 42$ mm, which is slightly less than the measured l_f because perfect fitting of the broken halves is difficult.

(e) Reduction of area is from measured values. $A_f = 2.85 \times 3.5 = 9.975 \text{ mm}^2$.

A	B	C	D	E	F	G	H	I	J
Spec.	A0	P0.002	YS	Pmax	TS	lf	el. (25 mm)	Af	q
	mm^2	N	N/mm^2	N	N/mm^2	mm	%	mm^2	%
			Eq. (4-7)		Eq. (4-8)		Eq. (4-9b)		Eq. (4-10)
1	40.51	4400	108.2	14200	350.5	42.2	68.8	9.975	75.4

It is always advisable to check results against published data. For this alloy, handbooks give, for a soft material, YS = 90 MPa, TS = 340 MPa, and el. = 40%. Thus, agreement is acceptable, except that the measured elongation is much higher. Note, however, that databases refer to a gage length of 50 mm, whereas the *gage length was only 25 mm in the present test* (see Example 4-3).

Example 4-5

To explore the effect of deformation of the test machine, the tension test of Example 4-4 was repeated, but this time the recording was made from the crosshead movement (broken line in recording). Calculate the spring constant of the system.

Taking a convenient point at, say, $\Delta l = 0.4$ mm (point C), the force is $P = 12.0$ kN. Thus, the overall spring constant $K = P/\Delta l = 30$ kN/mm. The total elastic deformation is the sum of deformations in the specimen and the machine; the contribution of each can be calculated if the spring constant of the machine is known or the spring constant of the specimen is calculated.

4-1-5   Assuring Increased Ductility

Reduction of area and thus also toughness are highly sensitive to material condition and are seldom given in material specifications. They are, however, of great importance for both service and processing, and substantial efforts are often made to increase ductility:

1. It can be expected that cavities form earlier if there are points of weakness in the material. Therefore, one of the principal aims of manufacturing control is to assure *freedom from internal defects*. We shall see in Sec. 6-3-4 that inclusions can create such defects, and *cleanliness* is a prime requisite for improved ductility. As seen in Fig. 4–7, strength is not necessarily affected, but postnecking deformation (which determines reduction in area) increases. Very clean ductile materials may reduce to a point before separation, which is then sometimes called *rupture*.

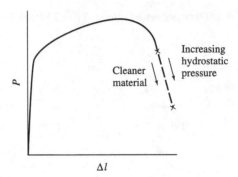

Figure 4–7   Postnecking deformation can be increased by (a) assuring greater cleanliness or by (b) imposing hydrostatic pressure.

2. If triaxial tensile stresses are responsible for opening up weak points, fracture can be delayed by superimposing compressive stresses. When the three stresses are equal, one speaks of *hydrostatic pressure*. This can be achieved, for example, by conducting the test in a pressurized fluid. Superposition of hydrostatic pressure neutralizes the tensile stresses (Fig. 4–5c) and deformation will continue to higher strains (Fig. 4–7). More broadly, this observation leads to a very important principle, applicable to both service and manufacturing: *fracture in a part can be delayed or prevented if a sufficiently high compressive stress state prevails.*

4-1-6   Notch Effects

We saw that internal defects or inclusions reduce the ductility of metals. Even more harmful can be surface defects, particularly notches. Notches cause *stress concentration*, i.e., a local increase in stress to $\sigma_{\max}$. The *stress concentration factor K* is the ratio of $\sigma_{\max}$ to the stress σ that would prevail in a smooth body (Fig. 4–8) and can reach very high values when the notch radius is small. When the maximum stress or strain reaches some critical value, a crack develops and propagates at high speed through the part. Thus, the presence of cracks on the surface or inside the body may severely reduce the tensile stress that a material can withstand without fracture. This *fracture stress* σ_{fr} can be shown to depend on the crack radius r_c and crack depth (crack length) a as

$$\sigma_{\mathrm{fr}} = \left(C\frac{r_c}{a} \right)^{1/2} \tag{4-11a}$$

where C is a material constant. For truly brittle materials, r_c is on the order of atomic radii and then Eq. (4-11a) reduces to the *Griffith criterion*

$$\sigma_{\mathrm{fr}} = \left(C\frac{1}{a} \right)^{1/2} = \left(\frac{2E\gamma_s}{\pi a} \right)^{1/2} \tag{4-11b}$$

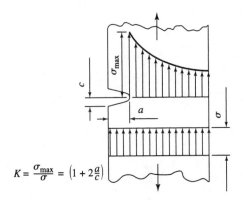

$$K = \frac{\sigma_{max}}{\sigma} = \left(1 + 2\frac{a}{c}\right)$$

Figure 4–8   A notch on the surface of a body results in a sharp increase of stresses: It causes stress concentration.

where γ_s is the surface energy of the crack surfaces. Cracks tend to have random distribution; therefore, all mechanical properties that are influenced by cracks or other stress raisers are also subject to scatter. It is necessary to conduct repeated tests and treat the results by statistical methods.

4-1-7   Bending Tests

Brittle materials are frequently used in situations where tensile stresses are imposed either in pure tension or in bending. Testing in tension is difficult because the slightest misalignment in the jaws imposes bending which increases stresses in an unknown manner. There are specially designed jaws available that facilitate testing of very carefully prepared specimens but much testing is still conducted in pure bending (Fig. 4–9).

The specimen is supported at two points (ASTM F417). In the *three-point test* (Fig. 4–9a) a force P is applied at the center. The specimen bends, and the outer (lower) half is put into tension, whereas the inner half is put into compression. Tensile stresses reach their maximum at the outer surface, midway between the supports. Failure (fracture) occurs when the maximum tensile stress reaches a critical value, often called the *rupture strength* (or *flexural strength* or *modulus of rupture*). For a rectangular beam

$$\sigma_B = \frac{3}{2}\frac{Pl}{bh^2} \qquad \textbf{(4-12a)}$$

For a round specimen

$$\sigma_B = \frac{8Pl}{\pi d^3} \qquad \textbf{(4-12b)}$$

Alternatively, the deflection for a given load (or load for a specified deflection) is given.

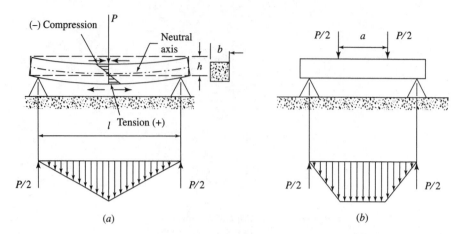

Figure 4–9   Less-ductile materials are often subjected to (*a*) three-point or (*b*) four-point bending tests. Tensile stresses peak at the center in the three-point test but are distributed uniformly between the two loading points in the four-point test.

The *four-point test* (Fig. 4–9*b*) generates uniform tensile stresses between the loading points. If $a = l/3$, the modulus of rupture for a rectangular specimen is

$$\sigma_B = \frac{Pl}{bh^2} \qquad \text{(4-12c)}$$

Less-ductile materials—such as some alloys, most ceramics, and many filled polymers—may have minute defects, cracks in the surface or body of the specimen (Sec. 4-1-6). Rupture strength is then a function of testing method and is highest— and shows the greatest scatter—in three-point bending because there is a low probability of a defect residing at the point of maximum stress. The uniform stress distribution in the four-point test makes it more likely that a defect will be found, hence rupture strength is lower but more consistent. Note that rupture strength is not equal to tensile strength.

Example 4-6   **A** high-technology ceramic (hot-pressed silicon nitride, Si_3N_4) was tested by bending 3.2-mm-thick, 6.4-mm-wide specimens loaded over a 38-mm span. Fracture occurred at a load of 1070 N in three-point bending and at 1250 N in four-point bending. Calculate the maximum stresses in each case.

From Eq. (4-12*a*), three-point test $\sigma_B = 3(1070)(38)/(2)(6.4)(3.2)^2 = 930$ MPa.

From Eq. (4-12*c*), four-point test $\sigma_B = (1250)(38)/(6.4)(3.2)^2 = 725$ MPa.

We shall see in Sec. 12-1-2 that almost all ceramics have minute imperfections, cracks in the surface. The probability of finding such a crack increases with an increasing length over which a high stress is developed, hence the measured strength is lower in four-point bending than in three-point bending. (Data for this example were taken from D.W. Richerson, *Modern Ceramic Engineering*, Dekker, 1982.)

4-2   IMPACT ENERGY AND FRACTURE TOUGHNESS

In Sec. 4-1-4 we mention that the energy per unit volume is sometimes used as a measure of toughness. It is found, however, that some normally ductile, tough materials suffer brittle fracture when they are in the form of a notched specimen or component and are exposed to sudden loading (impact force), especially below some critical *ductile-to-brittle transition temperature*. This can be a problem, for example, in arctic service of welded structures such as ships, drilling rigs, and pipelines that may contain planar welding defects and also residual stresses.

Several standard *impact tests* (ASTM E23) exist, each using different test geometry and loading method. In impact tests a load is suddenly applied, for example, by a swinging pendulum (Fig. 4–10a). The impact energy absorbed by the specimen (the energy lost from the pendulum) is reported (in units of joule). At the transition temperature the energy absorbed drops more or less suddenly. Also, the appearance of the fracture surface changes.

It will be noted from Fig. 4–10b that the impact specimen is notched, thus a stress concentration is set up (Sec. 4-1-6) to make the test more severe. Because of the great sensitivity of impact test results to specimen geometry and preparation, the impact energy quoted is only a comparative value between materials tested under identical conditions. It is a very useful quality control indicator but cannot be used for purposes of design calculations.

In a given material system, highest strength can usually be attained only at the expense of ductility and, thus, increased sensitivity to brittle fracture. This is true, for example, of high-strength aluminum alloys used in aircraft construction

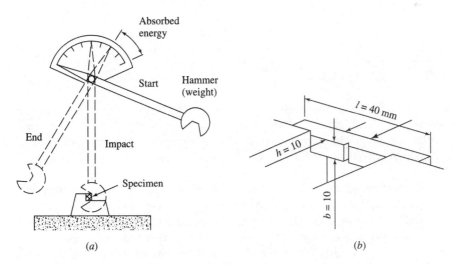

(a)   (b)

Figure 4–10   (a) The Charpy impact test is one of the tests for determining the fracture toughness of a material; (b) to induce a stress concentration, the specimen is notched.

and of the highest-strength steels. The need to design with this danger in mind has led to the development of a *linear elastic fracture mechanics* approach. Special tests (ASTM E813) are used to determine the *plane-strain fracture toughness* K_{Ic}

$$K_{Ic} = \alpha\sigma\sqrt{\pi a_c} \qquad (4\text{-}13)$$

where α is a factor depending on specimen and crack geometry, σ is stress or a function of the stress field, and a_c is the critical crack length below which fracture will not occur. Thus a structure can be designed to the allowable stress if the likely crack length is known, or the maximum allowable crack length may be specified for a given design stress.

In Sec. 4-1-5 we saw that ductility can be increased by removing internal defects or by imposing a compressive stress state. The same principles apply to improving fracture toughness. Cracks or notches can be a problem in all but the most ductile materials, and one of the aims of manufacturing processes is to *prevent the formation of cracks*. If this is not possible, *cracks must be kept in compression* during the service of the part. This can be achieved by design that allows only compressive loading, or by manufacturing processes that induce compressive residual stresses in the surface of the part (Sec. 4-7).

Example 4-7

During the Second World War, a large number of transport ships (the Liberty ships) were constructed. The traditional riveted structure was abandoned in favor of welding, thus greatly speeding up the rate of production. Of the over 2700 ships built, about 24 had serious cracking, and about a dozen broke into two in the cold waters of the Northern Atlantic. Fractures were of the brittle type, even though the steel was ductile in room-temperature tension and impact tests. Research into the causes of the problem did much to shed light on ductile-to-brittle transition and has led to the specification and manufacture of steels with guaranteed impact energies at low temperatures.

4-3   COMPRESSION

For reasons to be explained later, some materials, such as gray cast iron and concrete, are weak in tension but strong in compression (in contrast, some composites are weaker in compression). When design of the structure ensures that only compressive loads will be imposed, *compression testing* (ASTM E9) is most relevant.

The test equipment is again a press (or a universal testing machine), this time arranged so that the specimen is compressed between two well-lubricated, flat, parallel, hardened platens. Elastic deflections in the press and tooling would cause significant errors, therefore, deformation of the specimen is measured between the platens (Fig. 4–11a). The principle of constancy of volume [Eq. (4-2)] again applies: The cross-sectional area of the specimen (perpendicular to its axis) must increase to compensate for the decrease in height. Thus, the recorded force rises not only because of strain hardening (if present) but also because of the increasing cross-sectional area (Fig. 4–11b). The instantaneous cross-sectional area A can

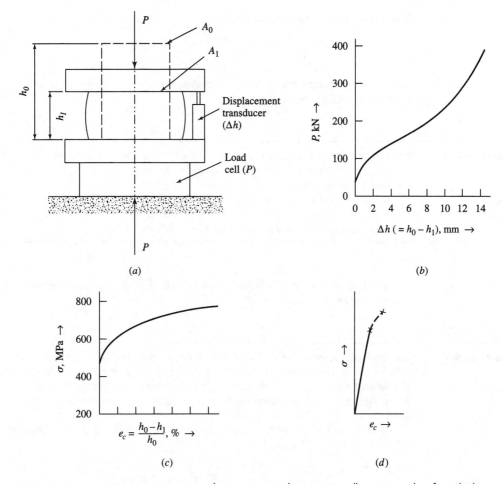

Figure 4–11   In (a) compression testing, the cross-sectional area continually increases, therefore, (b) the recorded force increases even if the material does not harden with deformation. (c) The derived stress–strain curve in this instance shows strain hardening. (d) Brittle materials fracture after initial elastic compression although some plastic deformation is sometimes observed (broken line).

be calculated from the instantaneous height h obtained from the output of the displacement transducer ($h = h_0 - \Delta h$)

$$A = \frac{A_0 h_0}{h} = \frac{V}{h} \qquad \textbf{(4-14)}$$

At any point of the press stroke, the *die pressure p* is force divided by area; we shall see in Sec. 9-2-1 that, *if friction effects are negligible*, the stress state is uniaxial compression, and the die pressure equals the (true) *compressive strength*

$$\sigma = \frac{P}{A} \qquad \textbf{(4-15)}$$

The engineering *compressive strain* is

$$e_c = \frac{h_0 - h}{h_0} = \frac{A - A_0}{A} \qquad \textbf{(4-16)}$$

Note that *the compressive strain thus defined is numerically different from the equivalent tensile strain.*

Example 4-8 | As far as the material is concerned, elongating a specimen from a length of $l_0 = 30$ mm to a length of $l_f = 60$ mm is the same as compressing from $h_0 = 60$ mm to $h_f = 30$ mm. Calculate the tensile and compressive strains.

From Eq. (4-4b), $e_t = [(60 - 30)/30]100 = 100\%$.
From Eq. (4-16), $e_c = [(60 - 30)/60)]100 = 50\%$.
The problem is resolved if true strains are used (Sec. 8-1-1).

From the measured force and displacement (Fig. 4–11b), a stress–strain curve can be plotted (Fig. 4–11c).

Example 4-9 | The recording shown in Fig. 4–11b was made by compressing, at room temperature, a steel cylinder of diameter 15.00 mm and height 22.5 mm, made of hot-rolled AISI 1020 steel. A graphited grease was used to reduce friction. Force P readings at six points are given below together with the instantaneous heights h. Calculate the true stress σ and compressive strain e_c.

It is again best to set up a spreadsheet. The volume of the specimen is $V = (15.0^2 \pi/4)(22.5) = 3976$ mm^3.

	A	B	C	D	E	F	G
	Point No.	h	P	A0	sigma	ec	epsilon
		mm	kN	mm^2	N/mm^2	%	
				Eq. (4-11)	Eq. (4-12)	Eq. (4-13)	Eq. (8-3)
	0	22.5		177			
	1	20.5	115	194	593	8.9	0.09
	2	17.5	158	227	695	22.2	0.25
	3	14.5	200	274	729	35.6	0.44
	4	12.5	235	318	739	44.4	0.59
	5	10.5	290	379	766	53.3	0.76
	6	8.5	370	468	791	62.2	0.97

The resulting plot is given in Fig. 4–11c. Note that there is no decline in stress, since there is no necking.

Engineering components or structures are seldom allowed to deform substantially, therefore, the compressive stress corresponding to some small strain (say, 0.2% or 0.5%) is usually taken as the basis of design. On further compression, the specimen may assume a barrel shape. The stress state changes and the stress calculated from Eq. (4-15) is no longer a true compressive strength.

Since compressive stresses keep the material together (Sec. 4-1-5), it might appear the deformation should continue indefinitely. Yet fracture is still possible:

1. Barreling in the presence of friction generates tensile stresses on the surface and cracks develop if the material has limited ductility (see Sec. 9-2-4).

2. Sooner or later, the ductility of many materials is exhausted and fracture sets in (Fig. 4–11d), often on a 45° diagonal. Again, fracture may be delayed by imposing hydrostatic pressure.

3. Brittle materials fail suddenly on reaching a critical stress.

The importance of hydrostatic pressure was recognized some 80 years ago, and tests in pressurized fluid have shown that ductility of most metals increases almost linearly. A more sudden transition from brittle to ductile behavior is typical of gray cast iron; it is, however, necessary to enclose the specimen in a rubber sleeve to prevent penetration of pressurized fluid into surface cracks. A brittle Zn-5Al alloy (similar to the one shown in Fig. 4–4b) deforms plastically with almost 100% reduction of area when the pressure is raised to 130 MPa.

Example 4-10

4-4 HARDNESS

The resistance of a material to deformation is most conveniently tested by indentation (Fig. 4–12). For reasons to be explained in Sec. 9-2-2, the specimen must be large enough to keep deformation highly localized, so that the indentor pushes the displaced material up around the indentation *but does not deform the entire thickness of the specimen.* A great advantage is that a relatively small local indentation may be permissible even on a full-size part; thus, there is no need to destruct the part to obtain a reading. Tests are standardized, including the geometry and dimensions of the indentor, the magnitude of the applied load, and the rate of load application.

1. In the *Brinell hardness test* (ASTM E10) the indentor is a steel (or, for harder materials, tungsten carbide) ball (Fig. 4–12a). After the load is applied, the mean diameter of the impression is measured. Force divided by the *surface area* of the indentation gives the *Brinell hardness number* (HB or BHN), which is still quoted in the old metric units of kg/mm^2. Since surface area is not a linear function of impression diameter, tables are available to simplify the calculation. Very deep indentations must be avoided; hence the load is reduced for softer materials to keep the indentation diameter between 2.50 and 4.75 mm.

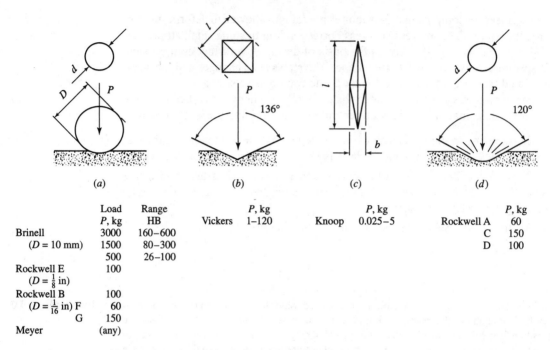

	Load P, kg	Range HB		P, kg		P, kg		P, kg
Brinell	3000	160–600	Vickers	1–120	Knoop	0.025–5	Rockwell A	60
(D = 10 mm)	1500	80–300					C	150
	500	26–100					D	100
Rockwell E	100							
(D = $\frac{1}{8}$ in)								
Rockwell B	100							
(D = $\frac{1}{16}$ in) F	60							
G	150							
Meyer	(any)							

Figure 4–12   Hardness tests have the advantage that information on the compressive strength can be obtained by localized deformation, without destroying the workpiece.

Fundamentally more correct is *Meyer's hardness test* (Fig. 4–12a) in which the load on the ball is divided by the *projected area* of the indentation; unfortunately, the test has not become popular.

2. Impressions made by a pyramid (Fig. 4–12b) remain geometrically similar independent of load, therefore they can be used for a wide range of hardnesses. In the *Vickers hardness test* (ASTM E92) the hardness number (HV or VHN, in kg/mm^2) is again obtained on dividing the force by the surface area, calculated from the diagonal of the impression.

3. *Microhardness tests* (ASTM E384) are used to explore localized variations in hardness within a body and close to the edges. Loads have to be very small, hence the surface must be prepared by polishing, taking care not to cause local deformation which would increase the hardness. In the *Knoop test* (Fig. 4–12c) hardness is calculated from the long diagonal of the indentation. Because of elastic recovery at low loads, the hardness (HK, in units of kg/mm^2) is not a linear function of the diagonal.

4. For quality-control purposes, the *Rockwell hardness test* (ASTM E18) is most widespread because of the convenience of using the test apparatus. The indentor is a ball or a diamond cone (Fig. 4–12a and d). After preloading to minimize surface roughness effects, the main load is applied. The apparatus automatically measures the depth of indentation and gives a readout on arbitrary scales of which the A, B, and C scales are used most frequently (reported as HRA, HRB, HRC, etc.). Conversion to other units is possible, particularly for materials of low strain hardening such as heat-treated steels. Conversion tables are found in ASTM E140 and in handbooks; a nomograph is given in App. B.

5. The hardness of large parts can be measured with a *scleroscope* (ASTM E448), a portable instrument which relates hardness to the rebound of a small weight (hammer) dropped from a standard height; a diamond indentor is attached to the hammer.

6. The hardness of brittle materials is measured in a comparative *scratch test* and is reported on the Mohs scale, which is based on the scratch resistance of selected minerals (see App. B.).

7. Because elastomers recover their shape upon unloading, their hardness is measured with a *durometer* by indentation under a specified load.

For reasons to be explained in Sec. 9-2-2, the hardness of materials is approximately 3 times their TS (*but only if both are expressed in consistent units*). The relationship works best for materials of low strain hardening (such as heat-treated steels) and for HV, less well for HB.

A cold-drawn steel bar has a Brinell hardness of HB = 190. What TS is to be expected? | **Example 4-11**

$$TS = 190/3 = 63.3 \text{ kg/mm}^2 = 620 \text{ N/mm}^2$$

Converting to conventional units, we get 620 N/mm^2 = 90 kpsi = 90 000 psi. Hence the convenient conversion: multiply HBN by 500 to obtain TS (psi).

A large rolling-mill roll was supposed to be heat-treated to a hardness of HRC 55. How could | **Example 4-12**
one check if this has indeed been done?

The roll is too large to be placed in a Rockwell hardness tester, and no specimen can be cut from it. Therefore, a Model C Shore scleroscope is used. It gives a reading of 78. From conversion tables available in many handbooks or from App. B, this corresponds to HRC 58, thus the heat treatment has indeed been carried out.

4-5 FATIGUE

In many instances, materials are subjected to repeated load applications. Even though each individual loading event is insufficient to cause permanent deformation and, even less, fracture, the repeated application of stress can lead to *fatigue failure*. Fatigue is the result of cumulative damage, caused by stresses much smaller than the tensile strength. Fatigue failure begins with the generation of small cracks, invisible to the naked eye, which then propagate on repeated loading until brittle fracture occurs or the remaining cross-sectional area is too small to carry the load. Fractured surfaces bear evidence of this sequence of events (Fig. 4–13a).

The suitability of a material can be judged from experiments (ASTM E206) in which a specimen is exposed to a preset level of stress S until fracture occurs after N cycles (Fig. 4–13b and c). The results are reported in fatigue diagrams or S–N diagrams (Fig. 4–14a). In some environments, materials such as steel may sustain some minimum stress level indefinitely; this is called the *fatigue limit* or *endurance limit*. It is better, however, to specify the stress which can be sustained for a given number (say, two million) of cycles.

Because fatigue involves the propagation of cracks under an imposed tensile stress, the number of cycles to failure (or the stress sustained for a given number of cycles) is greatly reduced if there are preexisting cracks (Fig. 4–14a), internal defects, or inclusions of a brittle nature. The surface roughness produced in some processes acts similarly, therefore, fatigue strength is reduced if the surface is rough, especially in high-strength materials that are less ductile (Fig. 4–14b).

Repeated loading at elevated temperatures—such as is caused by differential expansion and contraction of the surface of a part—may lead to *thermal fatigue*. It is particularly troublesome in forging tools, casting dies, and glass molds because cracking (crazing) of the surface is reproduced on the surface of the part.

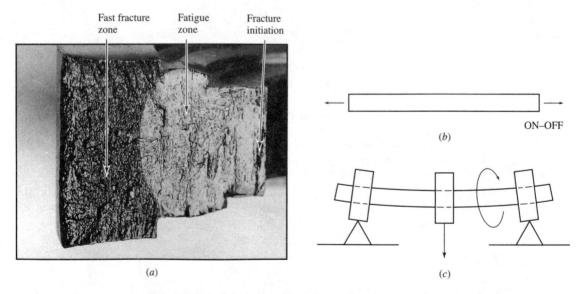

(a)   *(b)*   *(c)*

Figure 4–13   Repeated application of even relatively small stresses can result in fatigue; (a) the fractured surface shows evidence of crack initiation and propagation. Materials are tested by subjecting specimens to (b) cyclic tension, tension and compression, or (c) bending in rotation. The example shown is of a high-strength steel sealing ring subjected to fluctuating internal pressure. (*Courtesy of Dr. D. J. Burns, University of Waterloo.*)

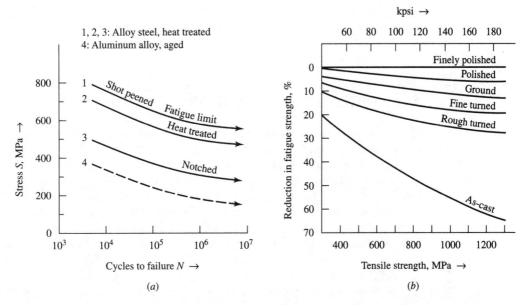

Figure 4–14   With an increasing number of loading cycles, (a) the stress at which fracture occurs drops, although some materials show an indefinite life at some stress level, the so-called fatigue limit or endurance limit. Fatigue strength is greatly impaired by the presence of surface cracks or notches and (b) even by a rough surface. [(a) From various sources; (b) from E. S. Burdon, SCRATA Proceedings 1968 Annual Conference, Steel Castings Research and Trade Association, London, 1968, Paper No. 3, with permission.]

A special form of failure occurs when certain materials are exposed to a chemically aggressive (corrosive) environment. Surface cracks form and, in combination with the applied stress, lead to *stress-corrosion cracking*. If there are residual tensile stresses on the surface of the part (as in Fig. 4–18), cracks develop in a part even in the absence of external loads.

Because of the scatter of data (Sec. 4-1-6), fatigue life and limiting stress are often expressed as probabilities.

The first commercial jet-powered plane was the British Comet. Two planes disintegrated in flight with the loss of all lives. Fatigue failure due to cyclic hoop stresses, generated by repeated pressurization of the cabin, was suspected. Therefore, a complete airframe, which had been through 1230 flights, was submerged in water in a test tank and was subjected to pressure cycles. After 1830 cycles the cabin failed by fatigue cracks that grew at the corners of cabin windows. The lesson was well learned; since then, great advances have been made in the science of designing for materials of limited fracture toughness as well as in manufacturing methods for improved fracture toughness.

Nevertheless, planes are still regularly inspected for evidence of cracks; they are designed so that cracks too small to be detected in one inspection will not grow so fast that they would

Example 4-13

result in catastrophic failure before the next inspection. (Data taken from J. K. Williams, in *Fatigue Design Procedures*, E. Gassner and W. Schutz (eds.), Pergamon, Oxford, 1969.)

4-6 HIGH-TEMPERATURE PROPERTIES

Many components are expected to operate at some elevated temperature. Our perception of temperature is conditioned by our own response to it; similarly, there is a temperature scale for each material that is much more relevant for it than any of our temperature scales. Using a Greek derivation (*homos* = the same, *legein* = to speak) this is termed the *homologous temperature* scale, signifying that it corresponds to specific points of relevance for each material. Not surprisingly, one of the end points is absolute zero, the other is the melting point T_m (expressed in kelvins). Below roughly $0.5T_m$, most metals (and many thermoplastic polymers) show a "cold" behavior (Fig. 4–15): strength is high, ductility is relatively low. Above $0.5T_m$, they exhibit typically "hot" properties: strength is lower, ductility is higher. Substantial deformation may occur after necking, with the neck spreading over the entire length of the specimen (Fig. 4–16). The structural reasons for this behavior will become evident in Sec. 8-1-5 for metals and in Sec. 13-2-4 for polymers. It should be noted here that $0.5T_m$ is a very rough dividing line; alloying of metals and changes in the structure of polymers can push the onset of hot behavior to much higher temperatures. Nevertheless, in some plastics the transition to hot behavior may occur below room temperature, whereas very high temperatures are needed to induce ductility in ceramics (Chap. 12).

At this point it is important to note that deformation in the hot temperature range involves substantial rearrangement of atoms (in metals) or molecules (in polymers). These processes take time. Therefore, properties are also a function of the rate of load application or, more correctly, of the imposed *strain rate* $\dot{\epsilon}$ which in the tension test is simply

$$\dot{\epsilon} = \frac{v}{l} \qquad \textbf{(4-17)}$$

where v is the crosshead velocity (see Fig. 4–1) and l is the *instantaneous deforming length* (the gage length prior to necking, but *the length of the necked portion after necking*). Low strain rates allow more time for atomic or molecular rearrangement, hence stresses are lower and ductility is higher.

This also means that, over a long period of time at elevated temperatures, some deformation may occur even if the applied load is quite small. We say that the material suffers *creep*. This is very important when deformation is unacceptable. The growth of a turbine blade would cause damage in a jet engine, and we all have experience with products that fail to function after some time because of distortion of one of the plastic parts.

In the typical *creep* test (ASTM E139), a tension-test specimen is exposed to a preset (constant) load at a constant temperature. There is rapid initial extension

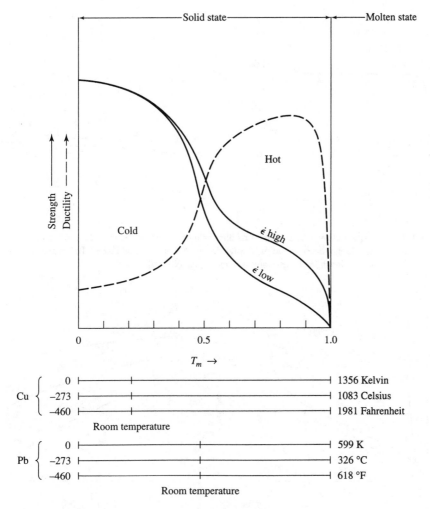

Figure 4–15   Materials such as metals have their own, built-in, homologous temperature scale. In the "cold" regime they are strong but less ductile, whereas in the "hot" regime they are less strong but more ductile. In the hot regime their strength is greater at higher rates of loading.

(*primary creep*), followed by slower deformation at a constant rate (*secondary creep*) and, finally, when structural damage occurs, creep accelerates and the part fails (*tertiary creep*, Fig. 4–17a). For parts expected to give long service, design is based on the stress that produces a linear creep rate of 1% per 10 000 h (e.g., for jet engine components) or per 100 000 h (e.g., for steam turbine components). Alternatively, the minimum creep rate is plotted against stress (or vice versa) on log-log paper.

To accelerate testing and to obtain design data for components that may be allowed to creep but must not fracture, tests are conducted at higher stresses to

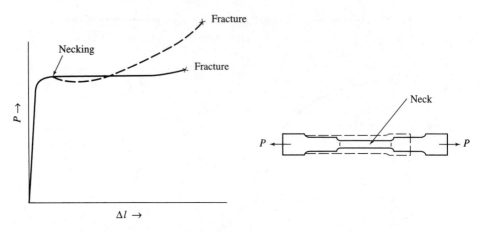

Figure 4–16   In the hot regime, a neck forms after little deformation, yet total deformation is substantial because the neck spreads over the entire gage length. The broken line is characteristic of tough polymeric materials (plastics).

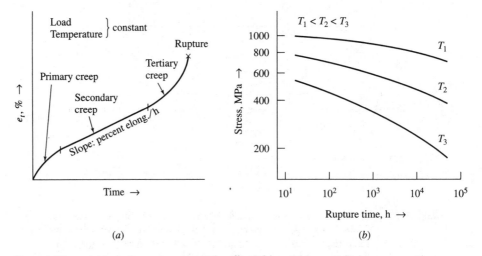

Figure 4–17   (a) In the hot regime, materials suffer deformation even under low stresses: They creep. Ultimately, fracture occurs. (b) Fracture (rupture) sets in faster at higher stress levels and temperatures.

total failure. In *stress-rupture tests* the time required for rupture is determined at various stress levels (Fig. 4–17b).

The onset of hot behavior in plastics is often determined in a bending test. The temperature at which a 125-mm bar deflects by 0.25 mm under a stress of 0.44 and/or 1.82 MPa is reported as the *heat-deflection temperature* (or *deflection temperature under load*, DTUL).

Temperatures in the turbine stage of jet engines are limited by creep deformation of the turbine blade. From 1940 to 1960, improvements in superalloys permitted gradually increasing temperatures; after 1960, a jump increase in temperature became possible with the introduction of internally cooled blades. First, cooler air was ducted from the compressor stage through holes provided in the length of the blade; later, cool air was passed over the surface of the blade to give a cooling boundary layer (Fig. 1–4b). The creep resistance of the blade material was improved too by novel manufacturing techniques such as directional solidification (Sec. 7-5-3). Yet higher temperatures can be obtained with ceramics (Chap. 12) and thermal barrier coatings (Secs. 19-4-4 and 19-6-1).

Example 4-14

4-7   RESIDUAL STRESSES

Structures and components are designed to sustain the externally imposed stresses. These are, however, not necessarily the only active stresses. As a result of manufacturing operations, there may also be stresses, called *internal stresses* or *residual stresses*, locked into the part or structure.

To understand how internal stresses arise, consider a cylindrical component. Assume that it has been made by joining a shorter tube and a longer, closely fitting core (Fig. 4–18a). Also assume that while joining was performed, the core was compressed to the length of the tube (Fig. 4–18b). Upon completion of the joint the core was released, whereupon the cylinder assumed a new length: The core wanted to expand to its original length, while the tube also wished to retain its original length. The mutually exerted forces must reach a balance. Since core and tube are of the same material and were chosen to have the same cross-sectional area, the cylinder will take up a length halfway between the original lengths of tube and core (Fig. 4–18c). The tube will be extended relative to its original length and will thus be subjected to (residual or internal) tensile stresses, while the core will be compressed and subjected to compressive stresses. Even though the cylinder is solid and sound, its surface is in tension. When the component is then put in tension, the applied stress is added to the surface residual stress. This would be dangerous for a material of limited ductility, because any surface defects that may be present would propagate much earlier, and the cylinder would fail in tension or succumb to fatigue at lower loads than a cylinder free of internal stresses. Stress corrosion may also occur.

Internal stresses can be reduced by heating to some higher temperature (*stress-relief anneal*). As visible in Fig. 4–15, the strength of materials drops at higher temperatures; therefore, internal stresses are reduced to the YS prevailing at the stress-relief anneal temperature. This may have undesirable consequences. Take, for example, a manufactured part of rectangular shape which has a high residual surface compressive stress on one surface, balanced by a much lower tensile stress in the bulk. The shape of the part will remain stable as long as the force balance is maintained (Fig. 4–19a). On stress-relief annealing, the surface compressive stress is reduced but the tensile stress, being lower than the YS at the annealing

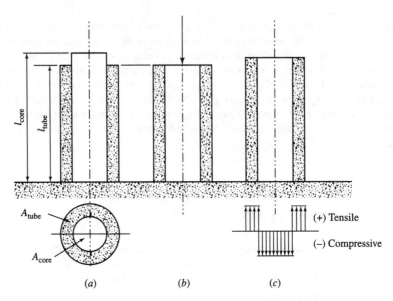

Figure 4-18   If a bar (a), longer than a tube into which it fits, is (b) joined to the tube while compressed to the same length as the tube, the assembly—released from compression—will (c) occupy an intermediate length, and surface tensile stresses will be generated.

temperature, remains unchanged. Thus, the force balance within the part is upset: the bulk, originally subjected to tension, now shrinks, and a new force balance must be established (Fig. 4–19b). Physically, this means that the part curves (warps).

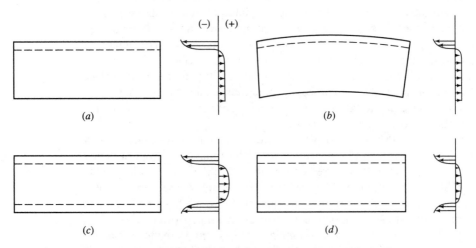

Figure 4-19   A rectangular part, (a) produced with a residual compressive stress on one of the surfaces, (b) will distort when subjected to stress-relief anneal; in contrast, (c) a workpiece with equal residual stresses on both surfaces (d) will retain its shape.

Residual stresses may be eliminated by mechanical means, i.e., by deforming the body to induce a stress exceeding the yield strength. Small deformations, including vibrations, may suffice. If the residual stresses are nonsymmetric, distortion may occur when mechanical loading or vibration changes the force balance (as in Fig. 4–19*b*).

Manufacturing processes or sequences of processes are often directed at either minimizing residual stresses or introducing a favorable stress distribution. Remembering that compressive stresses delay fracture, compressive residual stresses are induced in the surface of a part so that tensile and, particularly, fatigue strength increases (Fig. 4–14*a*). If compressive stresses are equal on both surfaces (Fig. 4–19*c*), the shape remains unchanged even if stresses are partially relieved (Fig. 4–19*d*).

Residual stresses can be determined by drilling out the center of or removing surface layers from the part and measuring the resulting dimensional changes. Nondestructive methods based on x-rays are also available.

4-8   NONDESTRUCTIVE TESTING (NDT)

In critical components, the presence of cracks and other defects is checked by various *nondestructive testing* (NDT) techniques. Many of them can be linked to computers for rapid data acquisition and processing and can be used for 100% inspection of parts during processing. Some techniques are suitable also for gaging.

1. *Liquid-penetrant inspection* reveals surface defects. Penetrants (dyes) are applied as sprays or by immersion to a thoroughly cleaned and dried surface. After wiping off the excess, the penetrant trapped in defects is drawn out and made visible by an absorbent developer. Some dyes are fluorescent and make the defect highly visible in ultraviolet light. A great advantage is that the process can be applied to all materials.

2. *Magnetic particle inspection* is limited to ferromagnetic workpieces. When the workpiece is magnetized, cracks lying more or less perpendicular to the field interrupt the magnetic field and become visible when fine ferromagnetic particles are dusted onto the surface.

3. *Eddy current inspection* can be carried out on any conductive material. A probe supplied with a high-frequency current induces an electric field in the part; the field changes in the presence of surface or near-surface defects. These changes show up on instruments. The technique is noncontacting and is suitable for on-line inspection, measurement of the thickness of surface coatings, and changes in metallurgical condition.

4. *Ultrasonic inspection* is based on the observation that a beam of ultrasonic (typically, 1–25 MHz) energy (high-frequency acoustic energy) passes through a solid structure with little loss but is partially reflected from internal surfaces;

therefore, cracks and cavities show up on a VDT. Moving the transducer or specimen in an x-y pattern allows mapping, and the time delay of the reflected wave gives the depth dimension. Good coupling between the transducer and workpiece is ensured by a coupling fluid (couplant). The technique is the most important method for polymer-matrix composites.

5. *Radiographic inspection* using x-rays, gamma rays, or neutrons is capable of revealing internal defects as well as surface cracks, which reduce the absorption of penetrating radiation and show up as darker areas. Also, absorption of x-rays and γ-rays increases with increasing atomic number and density, hence internal structures (as in semiconductor devices) are seen. Neutrons are absorbed by some light elements, including hydrogen, making the technique also suitable for plastics. In conventional radiography a two-dimensional image is created on a film and the location of defects or features within the depth of the body is not known. This disadvantage is eliminated by *computed tomography* (CT), originally developed for medical purposes. A highly collimated, fan-shaped x-ray beam is passed through the part and its absorption is measured by a linear array of photodetectors. Measurements are repeated while the part is rotated and moved (translated) (Fig. 4–20). Computer algorithms construct a three-dimensional (3-D) image of the part, with all details of the inside. Spatial resolution is better than 50 μm and dimensional accuracy better than 10 μm; thus the technique is suitable for reverse engineering of components having complex shape and internal cavities. The output can be directly converted into a solid model.

6. *Electromagnetic sorting* is used to separate ferromagnetic components according to their hardness, composition, or compositional change in surface layers (such as occur in case hardening). Sorting is based on the effects of these variables on magnetic properties.

7. *Holography* constructs a 3-D image of the part. Optical holography shows surface defects; acoustical holography, using ultrasonic waves, reveals internal flaws.

8. *Acoustic emission* is of great value for monitoring processes and machin-

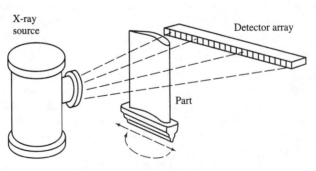

Figure 4–20   In computed tomography (CT) a 3-D image is constructed by passing an x-ray through the part while rotating and translating it.

ery. Internal processes such as fracture and bulk plastic deformation and surface processes such as shearing and sliding all result in the release of short bursts of elastic energy, which can be detected with transducers attached to the surface. Analysis of the emission spectrum gives valuable clues regarding the process and, in some instances, can be used for closed-loop control.

Interpretation of NDT readings requires considerable skill and judgment. Personal bias is minimized when computer graphics is drawn upon for displaying and interpreting signals. It is then possible to obtain a complete map of imperfections in a large workpiece.

4-9   PHYSICAL PROPERTIES

Physical properties other than strength are often of great importance and have to be satisfied by manufactured parts.

4-9-1   Density

We know from everyday experience that a component weighs much less when made of aluminum rather than steel: the density of aluminum is roughly one third that of steel. Density is mass per unit volume. The SI unit of mass is the gram or megagram (Mg; metric tonne); in the U.S. conventional system, the pound (lb). For a quick conversion, 1 kg = 2.2 lb. Thus density is in units of Mg/m^3 (= g/cm^3) or, in the U.S. conventional system, lb/in^3. If an aluminum component has the same strength as the steel, it will have 3 times the strength-to-weight (or strength-to-mass) ratio. This explains why subsonic airplanes are constructed mostly of aluminum alloys. There are, however, steels that are much stronger than any aluminum alloy, and the most highly stressed parts will be made of steel (or titanium alloy). In yet other instances, high density is of benefit. For example, balancing weights attached to the rim of automobile wheels are of lead, and lead walls or skirts are used by radiologists to protect against x-ray radiation. Thus, density will be one of the material selection factors in Chap. 5.

4-9-2   Tribological Properties

Tribology is the science, technology, and practice relating to interacting surfaces in relative motion. The term was coined in Britain in 1966 from the Greek (*tribein* = to rub), in recognition of the great importance of this interdisciplinary subject. It encompasses several fields:

Adhesion   When two bodies are brought together into such intimate contact that atoms come within interatomic distances, strong bonds may develop; in the language of tribology, *adhesion* occurs, and it takes a measurable force to separate the two bodies. Adhesion between two solids may result in the formation of a strong joint (a *pressure weld*), and manufacturing control aims at controlling the strength of joint. High strength is desirable when the purpose is to make a composite structure, such as the nickel-clad copper used in U.S. coinage. It is undesirable when low friction and wear is to be secured in either manufacturing or service, as in contact between die and workpiece or in bearings.

Adhesion can be reduced by an appropriate choice of the contacting materials. Generally, materials of greater hardness show less adhesion, and some materials show inherently low adhesion (e.g., lead in contact with other metals, or PTFE in contact with metals or plastics). Alternatively, a contaminant film may be interposed that prevents atomic bonding. Some contaminant films are provided by nature: materials processed in the normal terrestrial atmosphere have surface films formed in contact with air. At the least, there are adsorbed films of gases and water vapor. On many surfaces, chemical reactions also occur: most metals oxidize in air (Fig. 4–21a) and some polymers and ceramics undergo an irreversible change on contact with humid air. Thus, technical surfaces are never absolutely clean. Nevertheless, adhesion may still occur when relative sliding causes surface films to be broken through, and when temperatures are high enough to cause migration (diffusion) of atoms from one body into the other.

Friction   Mechanical components often slide against another body. The normal force P exerts a normal stress, which is usually called an *interface pressure* and is denoted p (instead of σ). The force required to move the body parallel to the surface is called a shear force F (Fig. 4–22); on dividing F by the surface area A, a *shear stress* τ_i is obtained (the subscript i signifies the interface). By definition, the *coefficient of friction* μ is

$$\mu = \frac{F}{P} = \frac{\tau_i}{p} \tag{4-18}$$

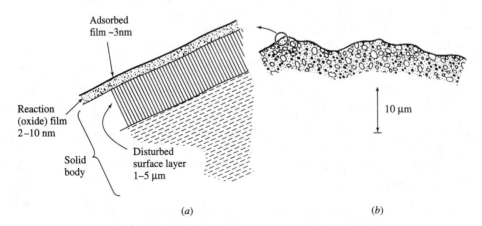

Adsorbed
film ~3nm

Reaction
(oxide) film
2–10 nm

Solid
body

Disturbed
surface layer
1–5 μm

10 μm

(a)   (b)

Figure 4–21   (a) The surface of materials differs from their bulk, showing evidence of prior processing and reactions with the atmosphere and other media. (b) Very few surfaces are truly smooth; most show peaks (asperities) and valleys.

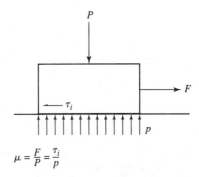

$$\mu = \frac{F}{P} = \frac{\tau_i}{p}$$

Figure 4–22   When two bodies are in contact, it takes a finite force to move them relative to each other. This allows us to walk, but it also accounts for much loss of energy.

On the micro scale, surfaces are not perfectly smooth but show hills (*asperities*) and valleys (Fig. 4–21b). Friction arises from the interaction of these asperities and from adhesion. In many applications it is necessary to minimize μ, either by the use of a lubricant, or by selecting materials that show inherently low friction, or both. Material pairs that show low adhesion usually—but not always—also give low friction. Manufacturing techniques can be directed to produce an internal structure in a component that is favorable for low friction (see Sec. 11-6). The surface texture (roughness and its orientation) of the part, which is controlled by the manufacturing process, also assumes prime importance.

Wear   Economic losses due to wear are enormous. *Wear* is the progressive loss of substance from the operating surface of components. It is usually a consequence of the simultaneous action of several mechanisms, with one mechanism dominating. The most important ones are the following:

1. *Adhesive wear* occurs when a pressure-welded joint is stronger than one of the contacting bodies, and rips out a particle from that body (Fig. 4–23a).

2. *Abrasive wear* is caused by hard particles, whether they are within one of the contacting bodies (*two-body wear*, Fig. 4–23b) or are interposed between the two components (*three-body wear*, Fig. 4–23c).

3. *Fatigue wear* occurs when the repeated passage of a component over the surface of the other component leads to the separation of small particles from the surface, as in ball bearings (Fig. 4–23d).

4. *Chemical wear* is caused by chemical attack accelerated by the pressure and rubbing prevailing in tribological contacts.

Numerous wear-evaluation techniques are available; they usually simulate, as closely as possible, the conditions encountered in service. Materials have been developed for high wear resistance. Alternatively, wear resistance can be increased by coating the surface or by transforming the surface into a form of

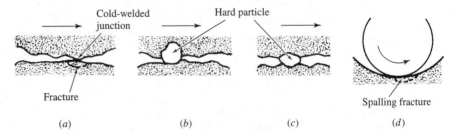

Figure 4–23   Wear is the progressive loss of material. It may be caused by: (a) the formation of adhesive junctions; (b) rubbing (abrasion) by a hard particle embedded in one of the mating surfaces; (c) abrasion by a hard particle trapped between surfaces; or (d) fatigue resulting from repeated loading.

greater wear resistance (Chap. 19). Controlled, accelerated wear is intentionally induced in some manufacturing processes (Sec. 16-8).

Lubrication   The purpose of lubrication is to reduce or, more accurately, to control both friction and wear. In addition to choosing material pairs that show low adhesion and friction, a separate substance (*lubricant*) is often interposed between the contacting surfaces. Lubricants may be grouped according to their mode of action:

1. Viscous fluids (such as mineral oils) introduced into a converging gap between moving surfaces (Fig. 4–24a) may build up a thick enough film to separate the two surfaces. Such *hydrodynamic lubrication* virtually eliminates wear, and friction is very low.

2. *Boundary lubricants* are organic substances (such as fatty acids) that adsorb on the surfaces of the contacting bodies and prevent adhesion, even

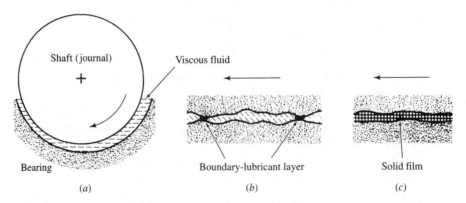

Figure 4–24   Friction and usually also wear may be reduced by (a) viscous fluids, (b) boundary lubricants attached to the surface by physical or chemical adsorption, or (c) solid films.

when the fluid film thins out to the point where asperity contact takes place (Fig. 4–24b). Natural oils, fats, soaps, and waxes also possess this property to some extent.

3. *EP* (extreme-pressure) *lubricants* are chemicals (usually organic materials with S, Cl, or P content) that react at elevated temperatures with metals to protect them from adhesion and rapid wear; often they also reduce friction. Environmental concerns have led to the development of so-called passive EP additives.

4. *Solid lubricants* (such as graphite and molybdenum disulfide, MoS_2) separate the two surfaces with a layer of low shear strength (Fig. 4–24c). They lubricate even when sliding speeds are low or temperatures are high.

Lubrication is of critical importance in many manufacturing operations and in the service of mechanical devices. The successful operation of such devices demands very close control of dimensions and surface finish. This does not necessarily mean a very smooth finish; for example, operation of an internal combustion engine hinges on the controlled, crosshatched roughness produced on the cylinder bore.

4-9-3   Electrical Properties

While there is some relation between mechanical and tribological properties, electrical properties can be quite independent of either.

Electric current is conducted in most solids by the movement of electrons. In order to move, the electron must be given extra energy by the imposition of an electric field.

Metals can be visualized as consisting of positively charged centers (ions) bonded by freely moving electrons. Thus, metals are *conductors*, with *resistivities* on the order of 20×10^{-9} $\Omega \cdot m$ (or 20 n$\Omega \cdot$m). *Conductivity* (the reciprocal of resistivity) is often reported as a percentage of IACS (International Annealed Copper Standard), which is obtained upon dividing 1724.1 by the electrical resistivity in n$\Omega \cdot$m. Any alloying elements and crystal imperfections make passage of electrons more difficult, and maximum conductivity can be attained only if the manufacturing process sequence and the final condition of the part are closely controlled.

Some materials become *superconductive* at some critical temperature: Their resistivity drops to zero. The phenomenon used to be little more than a scientific curiosity, since ductile materials had to be cooled close to absolute zero. Metallic NbTi and Nb_3Sn, cooled by liquid helium (boiling at 4 K), were the first superconductors in practical applications, to power highly stable magnets for magnetic resonance imaging and high-energy accelerators. In 1986, the discovery of "high-temperature" superconductors based on copper oxide ushered in a new era. Superconducting temperatures are well above the boiling point (77 K) of cheap and plentiful liquid nitrogen, and composite metal-ceramic wires (Sec.

15-5) will play an increasing role in electromagnets, motors, and electric power generation and distribution.

Insulators are materials in which all—or virtually all—electrons are tied down in covalent, ionic, or molecular bonds. A large energy is required to break loose an electron (there is a large energy gap). Therefore, their resistivities are greater than 10^8 $\Omega \cdot$m. They lose their insulating quality only at some critical field intensity, the *dielectric strength*.

Some materials (chiefly ceramics but also polyvinylidene fluoride plastic) exhibit *piezoelectricity*: subjected to mechanical loading, the material generates a potential difference. Thus, it can be used as a force transducer; in the reverse mode, an applied potential difference causes a dimensional change which can be exploited in ultrasonic transducers and force generators. The bulk resistivity of *piezoresistive* materials (chiefly semiconductors) changes significantly upon the imposition of a stress. Yet other materials (ceramic crystals) are *pyroelectric*: They develop a voltage in response to a temperature difference.

Of great technical significance are solids that are normally insulators but become conductors when an electric field is applied. They form the basis of the *semiconductor* industry and will be discussed in Chap. 20.

4-9-4   Magnetic Properties

Many materials are *ferromagnetic*: They contain *magnetic domains*. When these are readily reoriented under the influence of imposed magnetic fields, one speaks of *magnetically soft materials* (e.g., the core sheets in transformers or motors). In contrast, magnetically hard materials are difficult to remagnetize, and *permanent magnets* retain the magnetic orientation imposed during manufacture (e.g., magnets of loudspeakers). Some materials can be magnetized repeatedly, opening opportunities for magnetic recording and data storage. There is also an analogous effect to piezoelectricity: When an external magnetic field is imposed, *magnetostrictive* materials change their dimension.

In all instances, not only composition but also manufacturing technologies must be closely controlled to obtain the desired properties.

4-9-5   Thermal Properties

Thermal properties such as the coefficient of thermal expansion (CTE), specific heat, and latent heat of fusion and evaporation are important in many manufacturing processes and service situations, and their values may be found in handbooks. A selection is given in Table 4–1.

For our purpose, the most important is melting point. In service it signifies total loss of load bearing capacity; also, we saw in Sec. 4-6 that it affects the onset of hot behavior. It will also determine to a large extent the ease of producing a casting.

Thermal expansion assumes great significance in composite structures: A large difference in thermal expansion coefficient leads to high stresses and possibly failure. Thermal expansion decreases with increasing bond strength, hence

Table 4–1   Thermal properties of some engineering materials

Material	Thermal Conductivity, W/m·K	Coefficient of Thermal Expansion at 20°C, μm/m·K	Material	Thermal Conductivity, W/m·K	Coefficient of Thermal Expansion at 20°C, μm/m·K
Silver	428	19.7	ABS		60–130
Copper	390	16.5	Nylon 66	0.24	80
Gold	318	14.2	Glass-filled	0.2–0.5	15–20
Aluminum	240	23.6	Polycarbonate	0.2	70
Iron	74	11.7	LDPE	0.33	110–220
Invar (Fe-36Ni)	11	0.6–0.3	HDPE	0.48	60–110
Kovar (Fe-28Ni-18Co)	16.7	4.4	PMMA		50–90
304 stainless steel	15	16.5	Polyimide		20–50
410 stainless steel	24	10.0	Polypropylene	0.12	80–100
			Polystyrene	0.12	50–80
Alumina	17	6.6	PTFE, glass-reinforced	0.3–0.4	77–100
Beryllia	218	8.5	PVC, rigid		50–100
E-glass	1.7	6.0	Flexible		70–250
Fused silica	1.4	0.6–0.9	Epoxy	0.17	45–65
Silicon	1.5	2.6–3.6	Glass-filled		11–50
			Silver-filled	0.8–1.3	33–53
			Phenolic resin		30–45
			Polyimide		20–50
			Polyurethane	0.2	100–200

it is highest in polymers, lower in metals, and lowest in ceramics. Differential expansion or contraction in a body sets up stresses that lead to warping in a ductile material and fracture in a brittle one (thermal shock).

In common with other thermal properties, *thermal conductivity* is an intrinsic, structure-independent material property. However, *heat transfer* in a structure often depends, in addition to conduction through the structure, also on the movement of some heated material, such as gas or other fluid (convective heat transfer), and on radiation. The purpose of manufacturing is often the production of a composite structure in which heat transfer by these means is either promoted or hindered.

Example 4-15

Internal combustion engines, in common with all heat engines, become more efficient at higher operating temperatures. However, temperature limits are set by lubricants and the materials of construction. Therefore, most engines are cooled with a water-base (aqueous) circulating fluid, from which heat is extracted with the aid of a sophisticated manufactured product, the radiator. In this, the coolant is pumped through parallel tubes from which heat is extracted through fins. Fins are designed and manufactured into often complex shapes so that air flowing over them removes heat most efficiently. Heat exchangers are vital to the operation of refrigerators, air conditioners, industrial and domestic furnaces, solar collectors, and heat sinks for computers; all of these products represent different manufacturing challenges. At the other end of the spectrum, heat transfer is minimized by insulating structures such as fiberglass mats, foamed plastics, and furnace refractories.

4-9-6   Optical Properties

Manufacturing processes are controlled to endow manufactured parts with desirable optical attributes, for both esthetic reasons (appearance) and technical function.

The surface appearance of parts is controlled by manufacturing techniques to reflect light in a desirable manner. A very smooth finish reflects light at the same angle as the angle of incidence (*specular reflection*, as given by a mirror-finish surface), whereas a rough surface reflects light randomly (*diffused reflection*, as given by a matte finish).

Some materials absorb light and are *opaque* (not transparent). Others, such as amorphous polymers, glasses, and ceramics are *transparent*. If by appropriate manufacturing techniques internal reflecting surfaces are created, the same materials become *translucent* (partially transparent) or opaque (see Sec. 12-5-1).

4-10   CHEMICAL PROPERTIES

Many manufactured structures are expected to survive for prolonged periods of time while being exposed to the atmosphere or other gases or liquids. Their deterioration by chemical or electrochemical action (*corrosion*) is governed primarily by the choice of materials, but is affected also by the method of manufacture.

The aim is usually to avoiding harmful situations. For example, residual stresses can lead to accelerated corrosion and also to stress-corrosion cracking (Sec. 4-5); steel screws corrode when used for joining brass sheet; some stainless steels lose their corrosion resistance if slowly cooled from the welding temperature. On the positive side, steps can be taken to protect a structure from corrosion. For example, most auto bodies are now constructed from zinc-coated sheet.

Corrosion resistance may be undesirable in manufacturing when the function of a lubricant requires a chemical reaction to take place. For example, the corrosion resistance of stainless steel necessitates the use of special lubrication techniques in forming processes.

4-11   SUMMARY

The term *manufacturing*, as used for the present purpose, refers to the production of durable articles that must meet a number of service requirements. Processes and process sequences must be chosen and controlled to give an optimum combination of properties. For this reason, one of the important activities in manufacturing is the measurement of properties both during manufacture and in the finished product.

1. Mechanical properties are determined under conditions designed to simulate loading in service. Thus, tests conducted in tension, bending, or compression provide information on strength and ductility.

2. Fracture in ductile materials is initiated by triaxial tensile stresses at points of weakness. Therefore, many manufacturing processes aim at the generation of compressive stresses and the elimination of weak points by increasing the cleanliness of the material.

3. Notches and cracks in the surface of parts cause stress concentrations which can lead to early fracture and random variations of properties. Therefore, one of the aims of processing is to prevent the formation of cracks or neutralize their harmful effect. These measures also increase resistance to sudden loading, fast fracture, and fatigue in repeated loading.

4. Forces acting on the test specimen deform the machine also. Indeed, elastic deformation of equipment is of major consequence in many manufacturing operations.

5. Mechanical properties of all materials are a function of temperature. In metals and some thermoplastic polymers a transition to "hot" behavior occurs at $0.5T_m$ on the homologous temperature scale; strength then becomes a function of strain rate.

6. Prolonged service in the hot temperature regime calls for creep resistance and long stress-rupture life, properties that are again greatly improved by the absence of weakening internal features such as inclusions, cracks, and voids.

7. All the above properties are structure-sensitive, i.e., they change—for a given material—with the internal structure of the part. These features can be changed, in a controlled manner, by manufacturing techniques.

8. A powerful aid to quality improvement is nondestructive testing for the detection and quantification of surface and internal defects, residual stresses, and deviations from specified material conditions.

9. Among physical properties, density is structure-insensitive. All others (tribological, electrical, magnetic, and optical properties) are structure-sensitive and are controlled by manufacturing techniques.

10. Chemical properties such as corrosion resistance are vital in many applications and can be controlled by manufacturing techniques, including special treatments of the surface.

It should be noted that data-acquisition systems can be added to all test equipment, and data can be processed by a computer. Combined with computer control of the test equipment itself, testing can be fully automated. Indeed, there are installations in which masses of bar-coded specimens are loaded by special-purpose robots, moved through a succession of test units, and results are immediately processed, all without operator intervention.

PROBLEMS 4A

4A-1   Define (a) yield strength and (b) tensile strength.

4A-2   Name and define two measures of ductility obtained from the tension test.

4A-3   Name two methods for increasing reduction of area for a metal. Justify.

4A-4   Define toughness; draw a sketch to clarify your answer.

4A-5   Explain why hydrostatic pressure increases reduction of area in the tension test. To make your explanation clear, make a sketch of a tensile specimen with the stresses acting on it.

4A-6   Define (a) rupture strength, (b) flexural strength, and (c) modulus of rupture. (d) Make sketches to show how these properties are determined.

4A-7   Make three sketches, showing the engineering stress–engineering strain tensile test curves for (a) a ductile material, (b) for a material of very limited ductility, and (c) for a completely brittle material. (d) On the coordinates, give the definitions of stress and strain. On the diagrams identify (e) $\sigma_{0.2}$; (f) TS; (g) total elongation; and (h) toughness.

4A-8   Define strain rate in the tension test.

4A-9   Define "high temperature" for a metal. Draw a curve to show tensile deformation at high temperature versus time; identify the primary, secondary, and tertiary creep regimes.

4A-10   (a) In a single diagram, draw two curves that describe the dependence of strength and ductility of most pure metals as a function of homologous temperature (draw separate curves for low and high strain rate, where required). (b) Show the effect of alloying on the onset of the hot temperature regime.

4A-11   Define homologous temperature.

4A-12   A steel part is subjected to cyclic tensile loading. (a) Draw a diagram to show the stress at which failure occurs as a function of the number of cycles. Mark this line (a). Draw additional lines to indicate the failure stress if the part (b) has a rougher surface, (c) has compressive residual stresses or (d) tensile residual stresses on the surface.

4A-13   Make a sketch to define the coefficient of friction.

4A-14   Make sketches to show (a) adhesive wear and (b) abrasive wear.

4A-15   Define adhesion.

PROBLEMS 4B

4B-1   The specimen specified in Example 4-4 is tested on a machine of 20-kN capacity. Recording is made from the crosshead of the machine. Would you expect the initial slope of the recording to be steeper for the smaller machine? Justify.

4B-2   In the construction industry, either steel or reinforced concrete is used in many applications where tensile stresses are generated (e.g., beams in bridges and road overpasses). What method of testing would you recommend to establish the safe tensile stress for (a) steel and (b) reinforced concrete?

4B-3   Tensile tests are conducted on a ceramic. The results show extreme scatter. Suggest three possible causes.

4B-4   To characterize an alumina ceramic, one sample is tested in three-point bending ($\sigma_B = 420$ MPa) and one sample in four-point bending ($\sigma_B = 485$ MPa). Subject the results to a critique: (a) Can σ_B be larger in four-point bending? Justify. (Draw sketches to clarify your answer.) (b) Are the data suitable for truly characterizing the tensile properties of the material? Why? (c) What test or tests would be better?

4B-5   A steel is to be used in the construction of an offshore drilling platform for the Beaufort Sea. What test would you recommend for quick checks on its susceptibility to brittle fracture?

4B-6   Forgings made for aircraft applications must be free of laps, seams, and cracks. List all NDT techniques you would consider if the material is: (*a*) steel, (*b*) aluminum alloy, or (*c*) titanium alloy or if (*d*) the forging is replaced by a fiber-reinforced polymer-based composite. Indicate which technique is suitable for surface defects or internal defects.

4B-7   It is often found that a given alloy has higher fatigue strength in a fine-grained than in a coarse-grained form. Explain why this should be so.

4B-8   There are specially developed materials that are described as *creep-resistant*. Does this mean that they do not creep, or something else?

4B-9   Some books quote data for the compressive strength of metals. (*a*) Define what is meant by compressive strength. What is the meaning of such data for (*b*) brittle and (*c*) ductile materials (to clarify your answer, draw typical force–displacement curves).

4B-10   A critical aerospace component is made by casting. What technique is most suitable for determining the size and location of potential porosity? Why?

4B-11   A colleague claims that all effort should be made to eliminate residual stresses in a metal part subject to fatigue loading in service. Subject the claim to a critique. Justify your opinion.

4B-12   A steel part is subjected to cyclic compressive stresses during service. The stress is equal to one-third of the yield strength. Do you expect fatigue failure? Justify.

4B-13   In research projects the samples are often too small to cut standard test specimens of 50-mm gage length, and smaller specimens with 25- or even 10-mm gage length are pre-pared (the diameter may be proportionately reduced). Are the test results comparable to results on standard specimens as far as (*a*) YS, (*b*) TS, (*c*) el., and (*d*) RA are concerned? Justify.

4B-14   The elastic modulus of a large number of specimens is to be determined; to save time, it is proposed to record the force–extension curves directly from the crosshead of the machine. Do you agree with the proposition? Why?

4B-15   It is argued that low adhesion is always desirable in manufacturing. Do you agree? Justify.

PROBLEMS 4C

4C-1   Steel wire used in steel-belted radial tires has YS = 2100 MPa. How much elastic extension is possible before permanent deformation sets in?

4C-2   A company specification calls for a steel component to have a minimum TS of 180 kpsi (1240 MPa). Tension tests are conducted on selected samples, but all components are also subjected to Rockwell C hardness testing. What HRC is the minimum acceptable value?

4C-3   If the components of Prob. 4C-2 were too large to be tested in a Rockwell hardness tester, what method would one use to keep a running check on hardness?

4C-4   Calculate the $0.5T_m$ temperature for Zn, Cu, and Ni. Explain on this basis whether significant creep of these metals should be expected at 200°C.

4C-5   Hardness tests are routinely performed to check the mechanical properties of materials. To assess the validity of this procedure, (*a*) check whether hardness = $3\times$ TS for the brass sheets of Example 8-7; (*b*) explain any difference that may exist (*hint*: make sure to convert hardness values into MPa); (*c*) explain why the particular test methods are used and why the Brinell test should *not* be used.

4C-6   Draw a spring model of the test setup for Fig. 4–1.

4C-7   A metal bar is tested in tension; reduction in area at fracture is 45%. The same metal is tested in compression; surface cracks appear after 87% reduction in height. (*a*) Define reduction in area. (*b*) Using the principle of constancy of volume, convert into equivalent tensile strain. (*c*) Convert the tensile strain into equivalent compressive strain. (*d*) Compare the equivalent compressive strain with the strain to fracture in the compression test. (*e*) Explain why the material can take so much higher strain in compression than in tension.

4C-8   Steel cables used in mine-shaft elevators have YS = 2100 MPa. What is the length of a single, 1000-m-long strand if the load imposed develops a stress equal to one-third the YS?

4C-9   A part has two notches. One is 2 mm deep with a 0.1-mm radius, the other is 1 mm deep with 0.05-mm radius. Which is more dangerous?

4C-10   A gray cast-iron cylinder of 25-mm diameter and 25-mm height is tested in compression. Fracture occurs after very little deformation at a load of 354 kN. Determine the compressive strength of the material.

4C-11   The same cast iron is tested in tension on specimens of 22.4-mm diameter. Fracture occurs at a load of 81.2 kN. (*a*) Find the tensile strength. (*b*) Compare with the result from Prob. 4C-10; explain the difference, if any.

4C-12   Two laboratories test the same copper in tension at 600°C. One reports a much higher strength than the other. What is the first question you would ask to clarify the situation? Justify.

4C-13   Hot-rolled plates from a single batch of steel were tested in tension on specimens of 50-mm gage length. Calculate the YS, TS, and el. Determine the mean and standard deviation of these properties.

No.	w_0, mm	h_0, mm	$P_{0.002}$, kg	P_{max}, kg	el., mm
1	12.64	3.16	1031	1407	21.39
2	12.60	3.22	1022	1340	19.50
3	12.64	3.22	964	1349	22.09
4	12.59	3.25	959	1376	21.95
5	12.57	3.21	956	1351	21.54
6	12.51	3.10	964	1352	21.64

4C-14   An 1100 Al cylinder of 50-mm diameter and 75-mm height is compressed with a highly effective grease lubricant. Force readings are taken at four points in the stroke. Obtain the compressive stress at each point and plot against stroke.

Point	h, mm	P, kN
0	75	0
1	72	128.7
2	65	195.1
3	55	279.7
4	45	387.3
5	40	458.9

FURTHER READING

ASM Handbook, vol. 10, *Materials Characterization*, 1986; vol. 11, *Failure Analysis and Prevention*, 1986; vol. 12, *Fractography*, 1987; vol. 18, *Friction, Lubrication, and Wear Technology*, 1992, ASM International. (Also on CD-ROM.)

Bever, M.B. (ed.): *Encyclopedia of Materials Science and Technology* (8 vols.), Pergamon, 1986.

Davis, J.R. (ed.): *ASM Materials Engineering Dictionary*, ASM International, 1992.

Davis, J.R. (ed.): *Metals Handbook Desk Edition*, 2d ed, ASM International, 1998.
Engineered Materials Desk Edition, ASM International, 1995.

Introductory Texts on Materials

Anderson, J.C., K.D. Leaver, R.D. Rawlings, and J.M. Alexander: *Materials Science*, 4th ed., Chapman and Hall, 1990.
Askeland, D.R.: *The Science and Engineering of Materials*, 3d ed., PWS Engineering, 1994.
Ashby, M.F., and D.R.H. Jones: *Engineering Materials 1: An Introduction to Their Properties and Applications*, Pergamon, 1980; *2: An Introduction to Microstructures, Processing and Design*, Pergamon, 1986.
Budinski, K.: *Engineering Materials: Properties and Selection*, 5th ed., Prentice Hall, 1996.
Callister, W.D. Jr.: *Materials Science and Engineering*, 4th ed., Wiley, 1997.
Carter, G.F., and D.E. Paul: *Materials Science and Engineering*, ASM International, 1991.
Flinn, R.A., and P.K. Trojan: *Engineering Materials and Their Applications*, 3d ed., Houghton Miffin, 1990.
John, V.B.: *Introduction to Engineering Materials*, 3d ed., Industrial Press, 1992.
Schaffer, J., A. Saxena, S. Antalovich, T. Sanders, and S. Warner: *The Science & Design of Engineering Materials* (*with Materials in Focus CD-ROM*), 2d ed., McGraw-Hill, 1999.
Shackelford, J.F.: *Introduction to Materials Science for Engineers*, 4th ed., Macmillan, 1995.
Smith, W.F.: *Principles of Materials Science and Engineering*, 3d ed., 1996.
Van Vlack, L.H.: *Elements of Materials Science and Engineering*, 6th ed., Addison-Wesley, 1989.

Mechanical Behavior

Booser, E.R.: *Tribology Data Handbook*, CRC Press, 1998.
Boyer, H.E. (ed.): *Atlas of Creep and Stress-Rupture Curves*, ASM International, 1986.
Boyer, H.E. (ed.): *Atlas of Fatigue Curves*, ASM International, 1986.
Boyer, H.E. (ed.): *Atlas of Stress-Strain Curves*, ASM International, 1986.
Caddell, R.M.: *Deformation and Fracture of Solids*, Prentice Hall, 1980.
Courtney, T.H.: *Mechanical Behavior of Materials*, McGraw-Hill, 1990.
Dieter, G.E., Jr.: *Mechanical Metallurgy*, 3d ed., McGraw-Hill, 1986.
McClintock, F.A., and A.S. Argon: *Mechanical Behavior of Materials*, Addison-Wesley, 1966.
Meyers, M.A., and K.K. Chawla: *Mechanical Metallurgy: Principles and Applications*, Prentice Hall, 1984.

Testing

ASM Handbook, vol. 8, *Mechanical Testing*, 1985; vol. 17, *Nondestructive Evaluation and Quality Control*, 1989; ASM International.
Cartz, L.: *Nondestructive Testing*, ASM International, 1995.
Cheremisinoff, N.P., and P.N. Cheremisinoff: *Handbook of Advanced Materials Testing*, Dekker, 1994.
McMaster, R.C. (ed.): *Nondestructive Testing Handbook*, 2d ed., American Society for Nondestructive Testing, Columbus, Ohio, 1982.
Pohlandt, K.: *Material Testing for the Metal Forming Industry*, Springer, 1989.
Whitestone, D.J.: *Handbook of Surface Metrology*, Institute of Physics Publishing, Philadelphia, 1994.

Cooperation between steel companies, auto companies, and design teams created an UltraLight Steel Auto Body (ULSAB) which has superior performance yet 36% lower mass than bodies of present-day mid-sized cars. (*Courtesy American Iron and Steel Institute, Southfield, Michigan.*)

chapter

5

Materials in Design and Manufacturing

In this chapter we revisit concurrent engineering, this time narrowed to the interactions between product design, material selection, and process choice. We will look at:

The design of parts that will be united in an assembly

Material groups available to the designer

The sources of primary products from which manufactured goods will be made

The impact of design and process decision on the environment

Recycling as a means of minimizing harmful effects

We saw in Chapter 2 that manufacturing is an activity that requires the input of many specialists. Now we are narrowing our focus to the activities of the manufacturing engineer, never forgetting the broader picture. Our emphasis will be on manufacturing processes; these cannot, however, stand on their own. Even in the narrowly focused view, interactions between material selection, process choice, and part design cannot be separated. Indeed, the term *concurrent* or *simultaneous engineering* is often applied to this group of activities.

5-1   DESIGN

The narrower focus on concurrent engineering may be illustrated by the interactions shown in Fig. 5–1. The emphasis here is on *unit processes*, a term often applied to the production of parts which will then be assembled into a functional product. Even though design is acknowledged to be an iterative process, in concurrent engineering each step is mirrored in all three fields of Fig. 5–1. The following list of actions is only representative and not exhaustive.

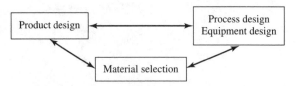

Figure 5–1   Concurrent engineering in the narrower sense concentrates on interactions between product and process design, as influenced by materials selection.

1. Determine the functions the part will have to satisfy, with due regard to operating conditions, safety aspects (including fail-safe characteristics), regulatory requirements, product liability implications, ease of maintenance, packaging requirements, service life, and environmental impact (storage and disposal).

2. Determine the configuration that will fulfill the requisite functions, and assign dimensions.

3. Analyze the design for loads and stresses, possible failure modes, and aspects of reliability. Consider the use of standard designs and components of known reliability.

4. Choose a material that satisfies all service criteria. Voluminous handbooks list properties of thousands of materials, usually classified according to composition. There are generally applicable guidelines that allow the designer to consider the broadest possible group or groups of materials without prematurely restricting the choice and thus limiting the possibilities of manufacture. Material choice is facilitated by increasingly sophisticated computerized databases.

5. Optimize the material choice by considering alternative materials. Depending on application, various factors become important: minimum cost for a given strength (load-bearing capacity), minimum weight for a given strength (specific strength = strength/density) or stiffness (specific modulus = elastic modulus/density). A formalized procedure is given by Ashby.[1] Computerized materials selectors incorporate much of the logic necessary for making a sound decision.

6. Assign the widest possible tolerances and roughest surface finish allowable for the given function.

7. Choose an appropriate process or process sequence, with due regard to the cost of processing and assembly and the number of parts to be produced. Savings can often be made by combining several parts (a subassembly) into a single part. Consider ease of production (producibility), inspection (inspectability), and testing (testability). The cost of these functions can exceed, by a wide margin, the cost of the starting material. Establish acceptance and rejection criteria. We shall see that some processes are not suitable for parts below or above certain sizes or cannot produce very thin or very thick walls, and that a process that may be economical for a few parts may be noncompetitive in mass production.

---

[1]M. F. Ashby: *Materials Selection in Mechanical Design*, Pergamon, 1992.

8. Optimize the design by an iterative refinement of steps 2 to 7. Consider the total cost implications; while a material may fulfill the required function, it may also present substantial manufacturing difficulties. The mutual constraints imposed by materials and processes are considered by Dieter.[2]

9. Industrial production has generated and continues to generate enormous amounts of waste materials. Disposal is increasingly difficult, and there is also a genuine desire of husbanding the limited resources of our Earth. Hence, serious efforts are being made to reduce, reuse, or recycle materials.

10. At different times and in different places, some other considerations assume overriding significance. For example:

a. The energy consumed in manufacturing varies greatly for various materials (Table 5–1). This is always an important consideration but, in times of energy shortages, it may become critical.

Table 5–1   Production data and energy consumption for selected materials of manufacturing*

Material	World Production, 10^6 Mg[†]		Energy Consumption, MJ/kg	
	1972	1994	From ore	From scrap
Iron (steel)	634	750	35	14
Aluminum	11	19.4	240	13
Copper	7	11.5	120	20
Zinc	5.2	7.1	70	20
Lead	3.6	5.4	30	10
Nickel	0.6	0.9	150	16
Magnesium	0.26	0.34	380	10
Titanium	0.06	0.1	550	
Plastics		130	170	
Plywood			10	

*From various sources.
[†]Mg = 1000 kg = metric tonne = 2200 lb.

b. Many raw materials are found only in some parts of the world. Their supply may become critical in periods of upheaval, and substitution may then require different approaches to design and manufacture.

The UltraLight Steel Auto Body (ULSAB) project is representative of concurrent engineering in the more restricted sense. Intense competitive pressures from aluminum, plastics, and space-frame construction prompted a consortium 35 steel companies from 18 nations to demonstrate

Example 5-1

---

[2]G. E. Dieter, Jr.: *Engineering Design: A Materials and Processing Approach*, 3rd ed., McGraw-Hill, 1999.

the feasibility of developing a unibody that saves weight, outperforms present structures, can be manufactured without the need to develop new technology, and does all this at a potentially lower cost. Porsche Engineering Services, Inc. was contracted to perform engineering and manufacturing management. The users (auto companies) were consulted from the beginning, and a baseline was established by benchmarking 32 midsize four-door sedans from around the world. Structural benchmarks were based on nine selected cars. The goal was set to outperform not only the current benchmark but also a projected improved structure. Steel producers and manufacturing engineers worked with designers to find the optimum solution. Computer simulation was extensively used for design optimization, crash simulation, and metal forming simulation. The final structure contains 94 major parts (in total, 158 parts versus the more than 200 in the existing structures). All parts were built and assembled by welding to prove feasibility of manufacture. We return to these aspects in later chapters. Physical prototypes were tested to validate the static and dynamic performance modeling. The performance criteria were fully met or exceeded (torsional and bending rigidity increased and vibrational response improved) with a significantly lower mass, as shown below:

	Benchmark Structure	Future Reference Structure	ULSAB Structure
Static torsional rigidity, N·m/degree	11,531	13,000	20,800
Static bending rigidity, N/mm	11,902	12,200	18,100
Vibrational response (first body structure mode), Hz	38	40	60
Mass, kg	271	250	203

(SOURCE: *UltraLight Steel Auto Body Final Report*, American Iron and Steel Institute, Washington, D.C., 1998.)

Example 5-2

Bumpers of automobiles give another illustration how changing tastes and functional requirements have affected design and manufacturing. Cars of the 1950s had bumpers of relatively heavy steel pressings, chrome-plated for corrosion protection and maximum reflectivity. They had a complex, wraparound shape which sometimes incorporated raised features for turn-signal lights, and manufacturing difficulties were substantial. Bumpers were attached to the frame of the automobile through rigid supports that transmitted the force of the smallest collision to the chassis. Bumpers of recent cars are entirely different, complex structures. With the shift in tastes, they often have a plastic surface layer for durability, color-coordinated with the body for visual appeal. For resistance to impact, the plastic is usually backed up by a metal (steel or aluminum alloy) pressing of relatively simple shape, which in turn is attached to the body through energy-absorbing devices designed to minimize damage in small collisions. Thus, design has changed to satisfy esthetic demands, offer superior protection and durability, and reduce manufacturing problems.

5-2   MAJOR CLASSES OF ENGINEERING MATERIALS

Manufacturing, in the sense used here, is concerned with parts and assemblies made of materials capable of carrying loads or fulfilling other technical functions (conduct electricity, insulate, etc.), as discussed in Chapter 4. Thus, our book focuses on the transformation of input materials into usable articles. Most of these materials are products of prior manufacturing operations (*primary processes*). The input material may often be obtained through a number of alternative routes, some of them much shorter than others. It would, however, be too hasty to conclude that the more complex processes are necessarily more expensive. Very often, economy is a matter of scale; thus, one can buy steel strip at a lower price than powder, partly because of the vast quantities produced in strip form.

5-2-1   Metals

Metals are still the most generally employed engineering materials, and the growth of their production (and especially that of steel) has often been taken as an indicator of industrial development. With the increasing sophistication of many products and with the growth of plastics and microelectronics, these relationships have ceased to be valid, particularly in industrialized nations. In the United States, steel consumption declined since 1950 per unit of GDP and since 1980 even per capita. Steel still represents an overwhelming portion of total metal production (Table 5–1) but other metals offer unique properties and remain indispensable. Thus, the low density of magnesium and the high strength-to-mass ratio of titanium have led to increasing use despite the high energy outlay.

Ores, usually oxides or sulfides, are the major source of metals. Various techniques are used to enrich them and make them more suitable for further processing. Metals are then extracted on a very large scale, in dedicated plants, by several methods (Fig. 5–2).

1. In *pyrometallurgy* the ores are reduced with carbon (coke, oil, or gas) in furnaces (*smelting*). For example, iron ores are charged into blast furnaces with coke and fluxes (principally limestone) to produce high-carbon *pig iron* and slag. Production is typically in excess of one million tons per year per furnace. Impurities may be removed by *fire refining*; in the case of iron, by blowing oxygen through molten pig iron in a basic oxygen furnace. Other metals, principally copper and zinc, are often refined by electrolysis (*electrorefining*) in which the impure metal forms the *anode* and a high-purity metal is deposited on the *cathode*.

2. *Direct reduction* (without melting) of some ores yields a high-purity powder.

3. *Hydrometallurgy* involves dissolution (*leaching*) of the ore in an acid. The metal may be *precipitated* or it may be deposited on an electrode (*electrowinning*). Low-grade ores and slag heaps may be leached in situ.

4. *Electrolysis* of a high-temperature melt also yields relatively pure metal but in a liquid form, as in the electrolysis of alumina (gained from bauxite) to produce aluminum.

An important attribute of metals is that they can be recycled without degradation of properties but the value of *scrap* depends greatly on quality. As shown in broken lines in Fig. 5–2, highly mixed or contaminated scrap may have to be put through the entire production sequence together with the ores; less contaminated

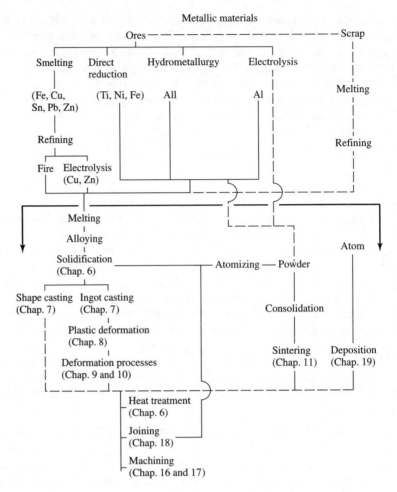

Figure 5–2   Metallic products are made through a sequence of preparatory steps, not discussed in this volume, and subsequent processes in which engineering parts are produced. These processes, shown below the heavy line, are the subject of our interest.

scrap may need only refining; and carefully separated scrap can substitute for new (virgin) metal.

Example 5-3 | Copper has been a key metal for millenia (Table 1–1) but competitive materials have made inroads in many traditional applications. Thus, aluminum has replaced copper in high-voltage power lines and many heat exchangers; fiber optics replaced it for signal transmission. Still, its use is growing, even in cars, because of the increasing number of servomotors installed. Some 60% of total consumption is from recycled scrap, saving substantial energy. The specific energy consumption in MJ/kg is: 110 from ore (mined, concentrated, smelted, refined); 50 from alloy scrap (fire- and electrorefined); 20 from Cu scrap (electrorefined and remelted).

Thus, energy consumption is considerably higher if scrap needs to be introduced at earlier stages of the full cycle. (Data from K. Gluckmann, *CIM Bulletin*, **85**(3), 1992: 150–156.)

We will be concerned only with the subsequent processing of metals, shown in Fig. 5–2 below the horizontal heavy line.

Pure metals have specific applications (e.g., Cu or Al for conductor wire) but alloys are much more often used. Most alloys are processed by the melt route: some will be cast into parts of complex shape (Chap. 7) but the large majority is cast into simple forms (Chap. 7) suitable for deformation processing (Chap. 9). The resulting forgings, construction sections, wire, tube, or sheet may be directly used but some will be further deformed (Chap. 10) into more complex shapes such as automotive body panels or beverage cans.

An entirely different route is followed when metal *powder* (either from primary processes or from "atomization" of a melt) is consolidated (Chap. 11). In yet another approach, atoms (or rather, ions) are deposited in a controlled manner to make *coatings* or *electroformed* parts (Chap. 19).

Products may be subjected to improvement of properties by heat treatment (Chap. 6). Machining (Chaps. 16 and 17) creates special shape features and improves dimensional tolerances and surface finish of cast or worked parts, or it may be used to produce parts directly from simple preforms.

All these processes not only impart the desired shape, dimensions, and surface finish, but also affect mechanical and other properties. Table 5–2 gives an idea of the enormous range of room-temperature mechanical properties that can be obtained. Changing the composition (alloying) is not the only way of changing properties; note that—for a given composition—properties can range widely depending on the manufacturing process. Thus, many alloys can be treated to secure, for example, a high specific strength, and this accounts for the extensive use of aluminum alloys in aircraft, heat-treated steel or titanium alloys for highly stressed components, and magnesium alloys where mass is of primary concern. Note also that some materials are much stronger in compression than in tension (see Sec. 4-3). An extension of data to higher temperatures would show an expanded range of possibilities. One of the aims of the present text is to show how these changes can be achieved in a controlled manner. Mechanical properties are not the only attributes of importance (Chap. 4), and other considerations—such as corrosion resistance—may be overriding in material choice and will influence also manufacturing.

Example 5-4

Aluminum–lithium alloys have long been investigated because Li reduces density and increases the elastic modulus. It was necessary to develop special processing techniques to allow the development of alloys of high strength, fatigue resistance, and toughness at cryogenic temperatures. Thus they are used in the aerospace industry. For example, in a 47-m-long, 8.4-m-diameter liquid hydrogen tank for the new Space Shuttle, they reduce weight by 3400 kg. [Source: P.S. Fielding and G.J. Wolf, *Adv. Mater. Proc.*, 1996(10): 21–23.]

Table 5-2   Room-temperature properties of metallic materials

Material		Density, g/cm³*	E, GPa	YS, MPa	TS, MPa	el., %
Designation	**Composition**					
AM60B	Mg-6Al-0.3Mn	1.8	45	130	220	6
	Be-38Al	2.1	200	190–310	260–380	7–2
A1100	99.5% Al	2.71	70	35–150	90–165	35–5
A7075-T6	Al-5.5Zn-1.6Cu-2.5Mg	2.8	70	105–500	230–570	17–11
	Ti-6Al-4V	4.43	120	920	1000	16
AC41A	Zn-4Al-1Cu-0.04 Mg	6.6			330	7
AISI 1008	Fe-0.08C	7.87	200	160–700	260–700	45–2
AISI 4140	Fe-0.4C-1Cr-0.2Ni	7.82	200	420–1700	650–1900	25–8
AISI 304	Fe-0.08C-19Cr-9Ni	7.9	193	205–2000	515–2200	
Gray cast iron	Fe-3C-2Si, tension	7.15	N.A.†	N.A.†	150–430	0.5
	Compression				570–1300	
C10100	99.99% Cu	8.9	130	70–365	220–255	55–4
C26000	70Cu-30Zn brass	8.53	110	75–450	300–900	66–3
	Ni	8.9	205	110–620	340–660	50–4
	W (at 500°C)	19.3	405	110	300	50

SOURCE: From various sources, chiefly *ASM Handbook*, vols. 1 and 2, *Properties and Selection*, ASM International, 1990 and 1991.
*Identical to Mg/m³ (multiply by 1000 to get kg/m³).
†Does not obey Hooke's law.

5-2-2   Ceramics

Ceramics are inorganic materials, often a metal oxide, boride, carbide, or nitride. They are characterized by low density and high temperature capability, and are always processed at high temperatures. Some are based on raw materials that occur in nature and are processed into mostly clay-based products such as bricks, tiles, tableware, etc. These *traditional ceramics* are produced in vast quantities. Our concern is primarily with *engineering ceramics*, or *advanced structural ceramics*, derived from specially processed or manufactured starting materials (Fig. 5–3). They often fulfill critical requirements such as stiffness, toughness, strength at low or high temperatures, abrasion resistance, and corrosion resistance. Thus, they are used as cutting tools, engine components, and products for the chemical industries. Because they are brittle, they are often characterized by fracture toughness (Table 5–3) and much design is based on statistical treatment. A prime purpose of manufacturing process development is narrowing the scatter in properties (Chap. 12).

Glasses form a special class of ceramics in that they are first melted and then shaped by various techniques (Chap. 12). While the greatest quantities

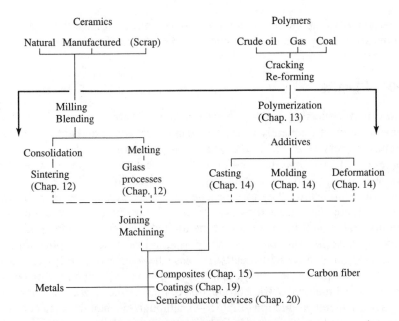

Figure 5-3 Processing of ceramics and polymers involves preparatory steps but the emphasis of our discussion is on subsequent processes, shown below the heavy line, aimed at producing engineering parts.

Table 5–3 Properties of selected engineering ceramics

Material	Density, g/cm³	E (tensile), GPa	Tensile Strength, MPa	Compressive Strength, MPa	Flexural Modulus, MPa	Fracture Toughness, MPa·m^{1/2}
SiO₂ fused		73	70	700–1400	100	
Al₂O₃	3.96	380	310	3800	300–1000	2.7–4.2
Si₃N₄	3.18	304	580	>1200	400–1000	5–7
Zirconia, PS	5.75	210	460	1760	630	9
Glass (E-glass)	2.5	75	500–5000	1200	55	0.9
Glass ceramic	2.7	60–140		120–560	60–100	1.6–2.4
Low-voltage porcelain	2.3	48	10–17	170–350	25–40	1
Zirconia porcelain	3.6	140–210	70–100	550–1050	140–240	

SOURCE: From various sources, chiefly *Engineered Materials Handbook Desk Edition*, ASM International, 1995. Properties vary widely with method of manufacture.

are found in traditional applications such as window glass, bottles, and lamp bulbs, smaller quantities of special glasses fill vital roles in corrosion-resistant containers, electrical insulators, etc. Glass is the only ceramic that can be recycled in large proportions.

Carbon and graphite fibers are not truly ceramics but have the high elastic modulus and temperature (but not oxidation) resistance of ceramics.

5-2-3   Plastics

Plastics are products in which *polymers* (Chap. 13) are the major ingredients. Polymers are most frequently based on a carbon backbone and are thus organic molecules of very large molecular weight, derived from oil, gas, or coal (Fig. 5–3). Some of them may be used on their own but most of them contain also fillers, stabilizers, plasticizers, colorants, and other additives. On the basis of the method of manufacture, they are classified as *thermoplastic* and *thermosetting* plastics. All are characterized by low density. Their relatively low temperature capability (Table 5–4) has for a long time limited their engineering application, but *advanced engineering plastics* can be used in load-bearing applications and have become competitive with metals. Because they can be relatively easily made into parts of complex shape, they are increasingly encroaching on the markets of more traditional materials. Total production of plastics has now outstripped, on a volume basis, metal production (Table 5–1), although much of this is in consumer goods that are still mostly discarded (and not often recycled) after use. A special class of nonmetallic materials is *elastomers* which have the capability of large deformations yet return to their original shape after the load is removed.

Table 5–4   Properties of selected engineering plastics

Material	Density, g/cm^3	E (tensile), GPa	TS, MPa	Flexural Strength, MPa	Impact Strength, J/m	el., %
Thermoplastics						
ABS	1.05	1.8–2.5	20–70	55–75	50–400	1–45
HDPE	0.96	1.1	20–35		20–200	10–1200
Nylon 6/6	1.14	1.6–3.8	55–95	110	30–60	10–130
30% glass	1.2–1.4	9	170	280	85–240	2–30
Polycarbonate	1.2	2.3	60–75	75–105	650–850	110–125
PET	1.56	9	60–160	240	100	
30% glass	1.68	8.9	150	235	95	2
PMMA	1.18	2.2–3.3	60–70	110	20	2–5
Thermosets						
Polyester (cast)	1.22	2.8–3.5	40–75	85–130		1.5–3.5
40% glass	1.60	5.5–11.5	125–195	160–240	570–640	3
Epoxy	1.22	2.7–3.4	40–80	100–130	70–210	1.2–5.7
Phenolics	1.36	0.8	30–60			
Filled	1.3–2.1	7–21	100–120	70–140	15–800	
Polyimide	1.32	3.9	40	175	53	

SOURCE: From various sources, chiefly *Engineered Materials Handbook Desk Edition*, vol. 2, *Engineering Plastics*, ASM International, 1988.

In many applications, plastics are competitive with metals. The reason for change is often the potential to replace complex assemblies with fewer parts of greater shape complexity. For example, concurrent engineering allowed Ford Motor Company to replace a 22-piece sheet-metal front-end module with two plastic parts made of a sheet molding compound, saving 22% in mass and 14% in assembly cost. [Source: *Manufacturing Engineering*, 1992(11):71]

Example 5-5

5-2-4   Composite Structures

Material choice is often the art of making compromises, since few materials can fulfill all requirements. Optimum properties may sometimes be more closely approximated by combining two or more chemically different materials in such a way that the benefits are retained but limitations are circumvented. By definition, there is a distinct interface between the components, which may be metallic, ceramic, or polymeric (Fig. 5–3).

Composites   By convention, the term is applied to structures in which one of the components (typically, fibers or particles) is surrounded by a continuous matrix of the other component (Chap. 15). The matrix may be a polymer, metal, ceramic, or carbon. Data in Table 5–5 indicate the scope of potential improvement.

Table 5–5   Properties of selected composites*

Material	Density, g/cm³	E (tensile), GPa	TS, MPa	Compressive Strength, MPa	Toughness, MPa·m$^{1/2}$	Flexural Strength, MPa
Epoxy	1.2	3	40			
With axial Kevlar fiber	1.4	76	1400	280		
Polyester (cast)	1.35	2	50	85		
With axial glass fiber	1.7	28	580	490		
With glass fabric	1.7	12	180	220		
6061 Al	2.7	69	300			
41% C fiber	2.44	320	620			
50% SiC	2.93	230	1480			
380 Al, 24% Al$_2$O$_3$		120	340			
Glass						
10% Si$_3$N$_4$ whiskers					6.5–9.5	400–500
C with 55% HTU fiber[†]		125	600	285	70	1250–1600
55% HMS fiber[‡]		220	575	380	20	850–1000

*Properties depend greatly on fiber orientation and method of manufacture.

[†]HTU = high-tensile strength, untreated-surface fiber.

[‡]HMS = high-modulus, surface-treated fiber.

Surface Treatments   Their aim is to impart special properties to the surface (or near-surface region) of a part. Some are applications of traditional techniques, for example, surface hardening by heat treatment (Chap. 6); deposition of wear-resistant overlays by welding (Chap. 18); glazing of a steel bath tub (Chap. 15); or galvanizing a steel sheet (Chap. 10). More specifically, the term surface treatment is applied to special processes (Chap. 19), frequently related to the manufacture of semiconductor devices.

Semiconductor devices   These are highly complex composite structures (Chap. 20) in which special semiconductors are combined with metals, ceramics, and plastics into devices capable of performing a great variety of functions, such as electrical conduction, isolation, storage, and light emission and reception.

5-2-5   Joining

Most parts made by any of the above techniques are subsequently assembled into larger, more complex structures. The assembly process itself is a specialized subject beyond the scope of our inquiry, but the joints that hold together the assembly are properly within the domain of unit processes (Chap. 18).

Nonpermanent joints make disassembly easy. We are all familiar with snap joints, especially for plastic parts, and with screw joints for parts made of all materials. The durability of such joints is very much a function of design and the manufacturing process.

Permanent joints are designed to enable the assembly to behave like a single part. They are made by techniques derived from other manufacturing processes, such as the solidification of alloys or polymerization of polymers.

5-3   ENVIRONMENTAL ASPECTS

Over many centuries, humans exploited the often limited resources of the Earth and in doing so have also placed an enormous burden on the environment in the form of air, water, and ground pollution. In the last few decades, there has been an increasing awareness of harmful consequences, and this has led to new approaches in both design and manufacturing, collectively denoted as *green engineering*.

5-3-1   Impact on Design

Attempts at environmentally responsible design have frequently foundered on a too narrow view, usually limited to a consideration of energy consumption. It is now recognized that the impact of a product must be evaluated from cradle to grave. In an automobile, for example, the cost and impact of operating the vehicle must be added to the cost and impact of raw material production, manufacturing, and disposal. This is formally done with *life cycle analysis* (LCA), a very demanding task that requires accurate accounting for all inputs and outputs in the system (Fig. 5–4). The outcome is highly sensitive to assumptions about manufacturing processes and methods and efficiencies of pollution control, and may even depend on geographic location. In a location with low gasoline cost, the initial (raw material and manufacturing costs) will weigh much more heavily than the cost of operation, although the cost of pollution remains the same.

The new approach to design is exemplified by the Partnership for a New Generation of Vehicles (PNGV), formed in 1993 by the U.S. government, Chrysler, Ford, General Motors, several universities, laboratories, and businesses. Its aim is to develop a midsize car that, compared to a 1993 vehicle, uses only one-third the fuel (3 L/100 km or 80 mpg), has the same performance, meets all safety and emission requirements, is 80% recyclable, and costs the same. The major means of achieving these goals is through reducing the mass of the vehicle by 40%. The ULSAB project (Example 5-1) is an offshoot of this, and throughout this text we shall see other examples of the potential contributions of manufacturing. Chapter 22 will tie together some of the solutions.

Example 5-6

5-3-2   Impact on Manufacturing

It is undeniable that manufacturing is a major user of resources and a significant contributor to pollution. Not long ago, industrial furnaces belched thick smoke, loaded with toxic compounds; liquid by-products were discharged into rivers; and solids were landfilled with abandon. All this has changed in the last decades, especially in industrially developed nations where a proliferation of local and national laws has forced or encouraged measures to curb pollution.

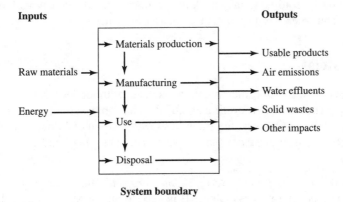

Figure 5–4   Life-cycle analysis evaluates what impact a manufactured product has on the environment. It encompasses the entire life of the product, from raw material and energy consumption through manufacturing to use and disposal. (*After J. L. Sullivan and S. B. Young, Adv. Mater. Proc., Feb. 1995:37–40. With permission of ASM International.*)

Limits are set on hazardous air pollutants (HAPs), volatile organic compounds (VOCs), and nitrogen–oxygen compounds (NO_x), and maximum achievable control technologies (MACT) are often mandated. The International Organization for Standardization has developed the ISO 14000 series of Standards for Environmental Management Systems. Like the ISO 9000 standards (Sec. 21-3-1), they aim at promoting continuous improvement. They too are voluntary, yet one can hope that competitive pressures will assure worldwide compliance. The culmination of all these efforts is what is often called *green manufacturing*.

Example 5-7

The State of Wisconsin banned landfilling or burning of containers as of January 1, 1995. In response, John Deere's Horicon Works switched from corrugated paper boxes to returnable containers for the shipment of its parts. As a result, parts are better protected, inventory tracking has improved, freight costs decreased, health and safety in the plant improved, and landfilling or incineration of boxes ceased. [Source: S. L. Buchholz, *Manufacturing Engineering*, 1993(10): 136.]

5-4   RECYCLING

Materials have always been recycled, if for nothing else, because it made economic sense. There are, however, additional pressures. Landfill sites are difficult to find, and the cost of disposal is increasing. Some products and many by-products of manufacturing are classified as hazardous materials which require special and expensive treatment or containment. Local and national legislation set limits to disposal methods. The most vivid example is Germany, where the manufacturer is required to take back the products once the consumer has no use for them. Thus there are pressures that increase the need for recycling.

5-4-1   Metals

As indicated in Sec. 5-2-1, metals can be recycled without degradation of properties, although the route of recycling greatly depends on the quality of scrap. Since recycling has a long history, the trade has developed an elaborate classification system for various grades. For our purpose, there are a few grades with distinct properties:

1. *New scrap* (*in-process scrap, home scrap, in-house scrap*) is generated in the manufacturing process itself. It is kept totally separated and substitutes for pure metal. Nevertheless, all effort should be made to minimize it, since its generation and recycling involves energy, labor, and other costs. Thus, our aim will be to maximize the yield of manufacturing processes.

2. *Segregated scrap* consists of returns from further processing. A typical example is trimmings from press shops and chips from machine shops. If carefully segregated, it is of equal value to new scrap. Beverage cans constitute a special class of segregated postconsumer scrap. In North America alone, some 110×10^9 aluminum cans [amounting to 1300×10^9 g or 1.3 million Mg (metric tonnes) of metal] are produced, of which over 50% is recycled. Since their composition is known, they can be treated as a special class of mixed scrap (see Sec. 7-4-1).

3. *Mixed scrap* consists of returns from further processing plants that fail to keep alloys fully separated. This includes press shops that mix bare and galvanized sheet and many machine shops that mix all steels or all aluminum alloys.

4. *Old scrap* (*postconsumer scrap*) is from discarded components of unknown composition. Mixed chips from machine shops also fall into this category.

The processes for dealing with various groups of scrap depend on metal and can be illustrated on the example of automobile recycling.

Vast numbers of automobiles end up at automotive dismantlers (not just "junkyards" anymore). Components in reasonable condition are removed for resale, a most important form of reuse. Batteries are drained and the lead sold: It can be readily refined. Tires are removed; unless a special use can be found for them (as in adding to asphalt for road building), they are of no value and are symptomatic of the problems that polymers in general present. Wiring harnesses are removed; the insulation is burnt off—a process in which pollution must be controlled—and the wire, which is always very pure, can be added to a melt or a refinery charge. Brass radiators have a Pb–Sn solder on them and will be smelted as ores would be. Catalytic converters with their valuable components are removed. Aluminum heat exchangers are removed, the fluorocarbons recovered from the air conditioning system, and the metal is sold. The engine block is removed, drained, and crushed in a hammer mill. Magnetic and hand sorting separates cast iron, sold to foundries. Aluminum is either melted into large blocks (sows) or sold to processors whose secondary smelters produce carefully controlled alloys for the casting industry, in the form of ingots or liquid metal transferred directly to the casting shop. The body is ripped apart by slowly rotating wheels and the ripped material is pounded in a hammer mill. Thus separated, ferrous materials are taken out by electromagnets. Large blowers remove lighter material such as glass, plastics, carpet etc., into *fluff* or ASR (*automotive shredder residue*); since it can be reprocessed only at great expense, it is often landfilled and represents a financial burden. The remaining nonferrous metals are sold for sorting and recycling. At present, the reclaimed metals pay for the entire operation and subsidize the disposal of ASR.

Example 5-8

5-4-2   Ceramics

Recycling of in-process materials will be discussed in Chap. 12. Fired ceramics can be crushed and used in limited quantity to supplement starting materials.

Glass is produced through the melt stage, hence it can be and is extensively recycled. Postconsumer scrap, chiefly nonreturnable containers, replaces raw materials and can save about 3% of energy for every 10% scrap in the charge. It does involve, however, substantial effort to sort by color, and contaminants (metal rings, paper labels, etc.) must be removed by techniques similar to those used for minerals.

5-4-3 Plastics

From modest beginnings, recycling of plastics has grown steadily but is still far behind that of metals. Segregation of various kinds is critical and determines the process:

Primary recycling   In-house thermoplastic scrap is ground up and added to new (virgin) materials for processing.

Secondary recycling   This involves mechanical separation by polymer type, followed by grinding, washing, and return to the original application. A problem is the multitude of plastics. For consumer goods, sorting is helped by molding into the product the code numbers (Table 5–6) of the Society of Plastics Industries (SPI), and automated sorting lines with x-ray or infrared sensors are increasingly used. More than half of PET bottles and a significant portion of HDPE are reprocessed. Much more extensive coding systems have been developed by various organizations (Society of Automotive Engineers J1344, ISO 1043, etc.). Mixed scrap is ground up and processed into lower-value "plastic lumber" products.

Tertiary recycling   Plastics produced by step-growth polymerization can be thermally or chemically decomposed and used as feedstock for polymerization or as a fuel.

Table 5–6   Recycling codes for commercial plastics

Code	Abbreviation	Polymer Name	Use of New Material	Recycled Products
1	PETE	Polyethylene terephthalate	Beverage containers, bottles	Fiberfill, soft-drink bottles, skis, boats
2	HDPE	High-density polyethylene	Beverage containers, toys	Drain pipes, recycling bins, toys
3	V	Polyvinyl chloride	Food packaging, shampoo bottles	Floor mats, hoses, mud flaps
4	LDPE	Low-density polyethylene	Grocery bags, bread bags	Grocery bags, garbage bags
5	PP	Polypropylene	Margarine tubs, bottles	Paint buckets, ice scrapers
6	PS	Polystyrene	Coffee cups, videocassette cases, fastfood containers	Drainage pipes, flower pots, food service trays
7	Others	Others		

SOURCE: American Plastics Council.

Quaternary recycling   This is not really recycling: Scrap is incinerated to recover its energy content. The amount of energy recovered depends on the plastic; it is around 40% for polyethylene but only half that for PVC. Pollution control is important.

5-4-4   Composites

Composites present some of the greatest challenges. Polymer-matrix composites can be recycled to some extent. Thus, parts made of sheet molding compound can be shredded and milled; the fine powder is added to virgin material. Tertiary recovery produces monomers that can be used as feedstock.

5-5   SUMMARY

Concurrent engineering in the narrower sense encompasses product design, material selection, and process choice and their interactions.

1. The designer must be cognizant of the wide range of materials available, including metals, ceramics, plastics, and composites.

2. Manufacturing processes, in the sense used here, transform primary products into parts suitable for assembly into functioning devices. A knowledge of the source of these primary products helps to understand the implications of many manufacturing decisions.

3. Life cycle analysis provides a cradle-to-grave evaluation of the impact of process and product on the environment, material and energy consumption, and cost to the user and society.

4. Recycling is one way of minimizing the harmful impact of industrial activity and will be one of the recurring themes in this book.

5. Many manufacturing operations use or create potentially hazardous products and by-products; in many cases, ways of dealing with them assume overriding importance. While the field is huge and cannot be treated here, indications will be given at appropriate points in our discussions.

Further Reading

Materials Data

ASM Handbook, Properties and Selection, vol. 1, *Irons, Steels, and High-Performance Alloys*, 1990; vol. 2, *Nonferrous Alloys and Special-Purpose Materials*, 1991, ASM International. (Also on CD-ROM.)
ASM Handbook, vol. 20, *Materials Selection and Design*, ASM International, 1997.

Alloy Finder (CD-ROM), 2d ed., ASM International, 1996.

Davies, J.R. (ed.): *Metals Handbook Desk Edition*, 2d ed., ASM International, 1998.

Engineered Materials Handbook Desk Edition, ASM International, 1995.

Wick, C., and R. Veilleux (eds.): *Tool and Manufacturing Engineers Handbook*, 4th ed., vol. 3, *Materials, Finishing and Coating*, Society of Manufacturing Engineers, 1985.

Bauccio, M.L.: *ASM Metals Reference Book*, 3d ed., ASM International, 1993.

Bauccio, M.L.: *ASM Engineered Materials Reference Book*, 2d ed., ASM International, 1994.

Brady, G.S., H.R. Clauser, and J.A. Vaccari: *Materials Handbook*, 14th ed., McGraw-Hill, 1996.

Brandes, E.A. and G.B. Brook (eds.): *Smithells Metals Reference Book*, 7th ed., Butterworth-Heinemann, 1997.

Frick, J. (ed.): *Woldman's Engineering Alloys*, 8th ed., ASM International, 1994.

Rahoi, D. (ed.): *Alloy Digest*, ASM International (continuing series; also on CD-ROM).

Wegst, C. G.: *Stahlschluessel* (*Key to Steel*), 18th ed., Verlag Stahlschluessel, Marbach, 1998. (Also on CD-ROM.)

Worldwide Guide to Equivalent Irons and Steels, 3d ed., ASM International, 1993.

Worldwide Guide to Equivalent Nonferrous Metals and Alloys, 3d ed., ASM International, 1996.

Materials Selection

Ashby, M.F.: *Materials Selection in Mechanical Design*, Pergamon, 1992.

Datsko, J.: *Materials Selection for Design and Manufacturing*, Dekker, 1997.

Dieter, G.E., Jr.: *Engineering Design: A Materials and Processing Approach*, 3d ed., McGraw-Hill, 1999.

Farag, M.M.: *Materials Selection for Engineering Design*, Prentice Hall, 1997.

MacDermott, C.P., and A.V. Shenoy: *Selecting Thermoplastics for Engineering Applications*, 2d ed., Dekker, 1997.

Mangonon, P.C.: *The Principles of Materials Selection for Engineering Design*, Prentice Hall, 1999.

Wroblewski, A. J., and S. Vanka: *MaterialTool: A Selection Guide of Materials and Processes for Designers*, Prentice Hall, 1997.

Environment and Recycling

Ayres, R.U., and L.W. Ayres: *Industrial Ecology*, Edward Elgar, 1996.

Hartinger, L.: *Handbook of Effluent Treatment and Recycling for the Metal Finishing Industry*, ASM International, 1994.

Kirk-Othmer Encyclopedia of Chemical Technology, 4th ed., Wiley-Interscience, vol. 20 (1996) 1075–1134; vol. 21 (1997) 1–46; Supplement (1998) 460–473.

Lewis, R.J.: *Hazardous Chemicals Desk Reference*, 4th ed., International Thomson Publication, 1997.

Misra, K.B. (ed.): *Clean Production*, Springer, 1996.

Mustafa, N.: *Plastics Waste Management: Disposal, Recycling, and Reuse*, Dekker, 1993.

Porter, R., and T. Roberts (eds.): *Energy Savings by Wastes Recycling*, Elsevier, 1985.

Tibor, T.: *ISO 14000: A Guide to the New Environmental Management Standards*, Irwin, 1995.

When viewed in a metallographic section, only the white areas are visible; the dendrites lurking below the surface are revealed when the matrix is etched away. (Primary cobalt dendrites are in a Co–Sa–Cu alloy matrix.) (*Courtesy Dr. W. Kurz, Swiss Federal Institute of Technology, Lausanne.*)

chapter

6

Solidification and Heat Treatment of Metals

Solidification and heat treatment are covered in introductory materials courses, but this chapter should serve as a useful review of topics such as:

Crystal structure of metals

Equilibrium diagrams as road maps to structures in alloys

Diffusion and its role in nonequilibrium solidification

Reactions in the iron–iron carbide system

The influence of structure on mechanical properties

The role of interface strength in two-phase structures

Methods of changing properties by heat treatment

Attaining a desirable combination of properties in steel parts by modifying a surface layer.

Solidification is a process of wide applicability for metals and alloys. The principles to be discussed here apply equally to casting, powder metallurgy, and welding processes, and they will also lay the foundations for an understanding of deformation behavior. An extension of these principles helps to rationalize changes that take place in heat treatments designed to facilitate processing and improve service properties.

6-1 SOLIDIFICATION

Solid metals are crystalline materials characterized by the metallic bond, reasonable strength and ductility, and good electrical conductivity. If their atoms, complete with electrons, are visualized as tiny spheres (of diameter around 0.2 nm) these spheres occupy strictly prescribed positions in space. The arrangement of points representing the center of atoms is called a *lattice*. Atoms vibrate about their lattice position; vibration is minimum at absolute zero. When the solid is heated, atoms vibrate at ever-increasing amplitudes; at a critical temperature—the melting point, T_m (Sec. 4-6)—the solid melts, turns into a liquid. The *long-range* crystalline order of the solid is largely lost, although some *short-range* order, extending to a few atoms, may exist. Thus, on melting, the *crystalline* solid changes into an *amorphous* liquid.

6-1-1 Pure Metals

We may follow the solidification of a *pure metal* by inserting a thermocouple into a melt contained in a small crucible and recording the change in temperature with time (Fig. 6–1a). If no heat is supplied, the melt gradually cools by releasing sensible heat or internal energy (A in Fig. 6–1b) until at T_m very small crystalline bodies, *nuclei*, form at several points in the melt. Temperature now

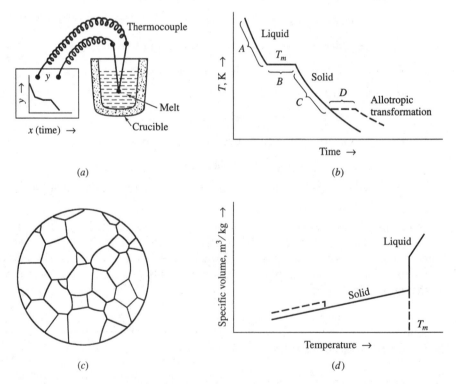

Figure 6–1   Solidification of a pure metal may be observed by (a) inserting a thermocouple and (b) recording the temperature as a function of time. (c) A micrograph of the resulting structure shows only grain boundaries. (d) The volume shrinks on solidification but may increase with an allotropic transformation.

remains constant while nuclei grow by the deposition of further atoms in the same crystallographic orientation, and the heat of fusion (B) is removed. When all melt is solidified, temperature drops again (Fig. 6–1b), and the solid releases its sensible heat energy (C).

The solidified body is *polycrystalline*, i.e., it consists of many randomly oriented crystals (usually called *grains*). Mechanical and other properties of a single crystal are *anisotropic*, i.e., a function of the direction of testing relative to lattice orientation. In contrast, a polycrystalline body consisting of a large number of randomly oriented grains is *isotropic* (has the same properties in all directions), with properties representing a mean of all crystallographic directions.

Since adjacent grains have different orientations, the *grain boundary* is a zone of disorder. To reveal grain boundaries, the solidified body may be cut and the surface ground, polished, and etched with a suitable reagent. Because of the higher chemical energy of atoms on the grain boundary, they are preferentially attacked and the groove appears as a dark line under the optical microscope (Fig. 6–1c).

In the liquid state, the randomly spaced, highly agitated atoms occupy much room, hence the *specific volume* (volume per unit mass) is large (Fig. 6–1d). During the cooling of the melt, thermal excitation becomes less violent, and specific volume drops gradually until the melting point is reached. Here the atoms occupy their lattice sites which are more closely spaced, and the specific volume drops substantially; the *solidification shrinkage* is typically 2.5–6.5%. This means that, if a casting free of cavities is to be produced, melt will have to be supplied to make up for solidification shrinkage, and this will be one of the major challenges in casting processes. Diminishing thermal excitation during cooling in the solid state results in further shrinkage, as given by the thermal expansion coefficient. Typically, metals shrink about 1% per 1000°C temperature drop. Even in a fully solidified metal there will be some unoccupied atomic sites, *point defects*, called *vacancies*.

The packing arrangement of atoms is characteristic of the metal and can be described by the *unit cell* (the smallest volume that fully defines the atomic arrangement). For practical engineering metals, three lattice types are of importance: *face-centered cubic* (fcc) with atoms at each corner and in the middle of the face of a cube (Fig. 6–2a); *body-centered cubic* (bcc) with atoms at each corner and in the middle of a cube (Fig. 6–2b); and *hexagonal close-packed* (hcp) with an atom at each corner and the center of the end face (*basal plane*) and at three sites in the middle of the body (Fig. 6–2c and d). Structure plays important roles in solidification and plastic deformation.

Some metals undergo, in the solid state, a change in crystal structure (*allotropic transformation*) at some critical temperature, releasing the latent heat of transformation (D in Fig. 6–1b). For convenience, different crystallographic forms of the same metal are denoted by Greek letters. Thus, on cooling, the bcc δ-iron changes to the fcc γ-iron at 1400°C, which again changes to the bcc α-iron at 906°C. The hcp β-titanium changes on cooling to the bcc α-titanium at 880°C. Allotropic transformations are often accompanied by a volume change (Fig. 6–1d) which may result in sufficient internal stresses to cause cracking.

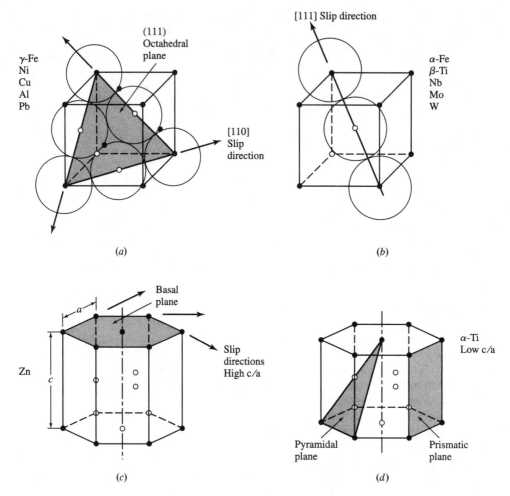

Figure 6–2   Lattice sites, slip planes, and slip directions in (*a*) face-centered cubic, (*b*) body-centered cubic, and (*c*) and (*d*) hexagonal close-packed structures.

6-1-2   Solid Solutions

Most technically important metals are not pure metals but contain a number of other metallic or nonmetallic elements which are either added intentionally (*alloying elements*) or are present because they could not be removed economically (*minor elements* or *contaminants*). Under favorable conditions, the alloying element may be uniformly distributed in the base metal, forming a *solid solution*.

Types of Solid Solutions   There are two possibilities:

1. The alloying element (*solute*) has a crystal structure similar to the base metal (*solvent*), has a similar (within 15%) atomic radius, and satisfies some criteria of compatibility in electronic structure. Then solute atoms can replace solvent atoms to give a *substitutional solid solution* (Fig. 6–3a). Some metals can form solid solutions over the entire composition range (e.g., copper and nickel, with atomic radii of 0.128 and 0.125 nm, respectively).

2. The solute atoms are much smaller (< 60%) than the solvent atoms and can fit into the spaces existing in the crystal lattice of the solvent metal, to form an *interstitial solid solution* (Fig. 6–3b, e.g., C and N in iron; also H and O).

Diffusion   It is important to realize that atoms are not immovably tied to their lattice position. If, for example, there is a vacancy, one of the adjacent atoms may move in, and the previously occupied site now becomes vacant (Fig. 6–3c). By a repetition of these events, atoms can move, diffuse within the lattice. The case quoted above is called *vacancy diffusion* (or *substitutional atom diffusion*). An interstitial solute atom can also move into an adjacent space between the solvent atoms by *interstitial diffusion* (Fig. 6–3d); since no vacancy is required, diffusion is fast.

If the solute atoms are not distributed evenly in a solid solution, they will diffuse until concentration gradients are eliminated. According to Fick's first law, the *flux of atoms J* (the number of atoms passing through a plane of unit area, in unit time), in units of atoms/m^2·s, is proportional to the concentration gradient ΔC (the change in concentration) over a Δx distance

$$J = -D\frac{\Delta C}{\Delta x} \tag{6-1}$$

where D is *diffusivity* or *diffusion coefficient*. The value of D is larger at higher temperatures

$$D = D_0 \exp\left(-\frac{Q}{RT}\right) \tag{6-2}$$

where D_0 is a constant for a given material pair, Q is the *activation energy* (the energy required to overcome the energy barrier involved in moving atoms through the lattice), and R is the gas constant (7.937 J/mol·K).

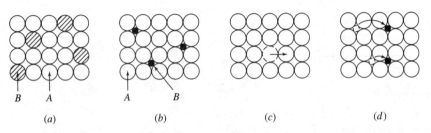

B A A B

(a)   (b)   (c)   (d)

Figure 6–3   Alloying elements may be accommodated in (a) substitutional or (b) interstitial solid solutions. Migration of atoms may take place by (c) vacancy (substitutional) or (d) interstitial diffusion.

Diffusion is a most important mechanism not only in solidification but also in many other phases of manufacture. There are two important points to remember:

1. Diffusion is greatly accelerated by high temperatures.

2. Because diffusion takes time, the distance over which diffusion can take place is much reduced if insufficient time is available.

Solidification of Solid Solutions   The events occurring during the solidification of solid solutions under equilibrium conditions may be followed by making up different melts of, say, copper and nickel, with a Ni content of 0, 50, and 100 weight percent (wt% or, in this text, simply %). The melts with 0% Ni (100% Cu) and 100% Ni are pure metals, and their cooling curves are the same as in Fig. 6–1b. The melt of $C_0 = 50\%$ Ni is different (Fig. 6–4). Solidification begins at 1315°C with the formation of nuclei with 68% Ni content. Temperature drops gradually; alloy less rich in Ni solidifies unto the nuclei until, at 1270°C, all melt disappears. If solidification was very slow and Cu atoms could diffuse into the already solidified crystals, the composition will be uniform everywhere at 50% Ni. At some intermediate temperature T_1, the alloy is in a *mushy state*: by drawing a horizontal *tie line*, we see that solid crystals (of composition C_S) coexist with a liquid (of composition C_L). Their relative quantities are given by the *inverse*

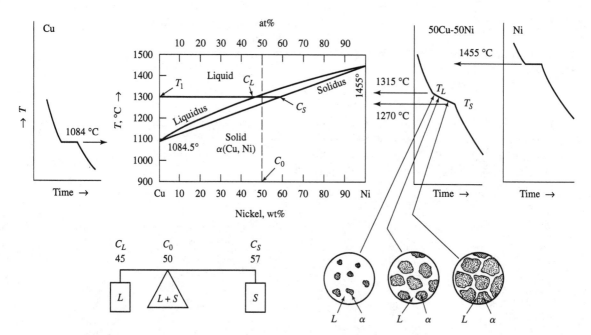

Figure 6–4   The copper–nickel equilibrium diagram shows complete solid solubility of the two elements in each other. Solidification of a solid solution takes place at gradually dropping temperatures and the proportion of solid, and liquid phases may be found from the inverse lever arm rule.

lever arm rule; the weight fraction of solid S is proportional to the horizontal distance (lever arm) between the nominal composition C_0 and the composition of the liquid phase C_L:

$$\frac{S}{S+L} = \frac{C_0 - C_L}{C_S - C_L}100 \quad (\%) \qquad \text{(6-3a)}$$

Similarly, the weight fraction of liquid L present is

$$\frac{L}{S+L} = \frac{C_S - C_0}{C_S - C_L}100 \quad (\%) \qquad \text{(6-3b)}$$

By repeating the experiment at other concentrations, lines defining complete melting (the *liquidus*, T_L) and solidification (the *solidus*, T_S) are defined and an *equilibrium diagram* of temperature versus composition is obtained. Obviously, the quantity of solid is vanishingly small at all points on the liquidus, and solid crystals gradually grow during cooling to the solidus.

Because the solvent atoms are uniformly distributed in the solute, each grain in a polycrystalline body will appear *homogeneous* and will look like the grains of a pure metal (Fig. 6–1c). It is usual to denote solid solutions by a Greek lowercase letter.

Example 6-1

For an alloy containing 50% Cu and 50% Ni, calculate the amounts of solid S and liquid L present at 1300°C. Show that the total nickel content in the solid and liquid phases adds up to 50%.

The equilibrium diagram (Fig. 6–4) shows that, for $C_0 = 50\%$ Ni, at 1300°C a liquid of composition $C_L = 45\%$ Ni is in equilibrium with a solid of composition $C_S = 57\%$ Ni. We know that mass must be conserved, hence for a batch of 100-g mass,

$$L + S = 100 \text{ g}$$

We also know that the masses must balance, i.e., the total amount of Ni (C_0) must reside in the liquid and solid phases

$$C_L L + C_S S = 100 C_0$$

substitution of $L = 100 - S$ results in Eq. (6-3a). Then

$$\frac{S}{S+L} = \frac{50 - 45}{57 - 45}100 = 42\%$$

and

$$L = 100 - 42 = 58\%$$

Amount of nickel in 100 g alloy:

$$
\begin{aligned}
\text{Solid} &= (57)(0.42) & = 23.9 \text{ g} \\
\text{Liquid} &= (45)(0.58) & = 26.1 \text{ g} \\
\text{Total} & & = 50.0 \text{ g or } 50.0\%
\end{aligned}
$$

6-1-3   Eutectic Systems

Generally, elements that exhibit a greater than 15% difference in atomic radii or have a different crystal structure are soluble in each other only up to a certain limit. When this limit of solid solubility is exceeded, the excess solute atoms are rejected into a *second phase* (phase means a chemically and stucturally homogeneous part of the system), which may again be a solid solution. The equilibrium diagram shows the temperatures and concentrations at which a given phase can exist. Thus, the equilibrium diagram is like a political map that reveals what phases to expect; therefore, it is also called a *phase diagram*.

An example is the silver-copper system (with atomic radii of 0.1444 and 0.1278 nm, respectively). The phase diagram (Fig. 6–5) shows that the maximum solubility of Ag in Cu is 7.9% and that of Cu in Ag 8.8%. A solid of overall compositions between these limits will consist of a two-phase mixture.

A unique point exists at 71.9% Ag. An alloy of this composition cools until it solidifies, like a pure metal, at a constant temperature T_E. However, T_E (779°C) is below T_m of both Cu (1083°C) and Ag (961°C); therefore, this low-melting composition is called, from the Greek, the *eutectic composition*. The temperature of its solidification (or melting) is termed the *eutectic temperature* T_E. An examination of the microstructure of the solidified eutectic shows that two phases can be distinguished: One is a solid solution of 7.9% Ag in Cu, and can be conveniently called α solid solution, while the other, β solid solution contains 91.2% Ag (i.e., it is a solid solution of 8.8% Cu in Ag). The two phases

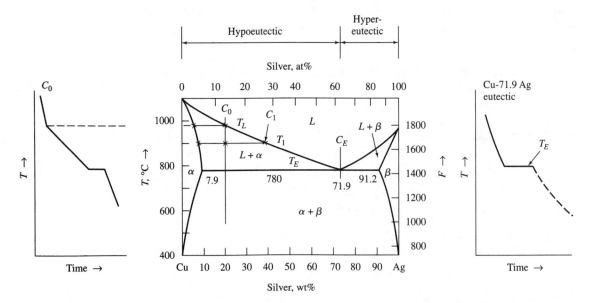

Figure 6–5   Limited solid solubility may result in eutectic solidification.

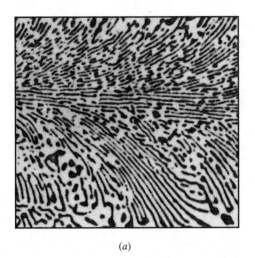
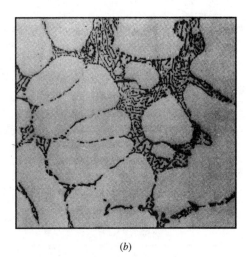

(a)   (b)

Figure 6–6   (a) The structure of a eutectic alloy is often lamellar. (b) A proeutectic alloy contains primary α grains surrounded by the eutectic. [(a) *Courtesy Dr. H. Kerr, University of Waterloo;* (b) *from Metals Handbook Desk Edition, ASM International, 1985, p. 6.50. With permission.*]

frequently appear as parallel plates; therefore, the eutectic shown in Fig. 6–6a is called *lamellar*. Because the *eutectic transformation* can occur only at a given composition and temperature, it is called an *invariant reaction*.

If the alloy contains 20% Ag, solidification begins at T_L with the formation of solid-solution nuclei, of approximately 6% Ag. At some intermediate temperature T_1, more α coexists with a liquid of C_1 composition. On reaching the eutectic temperature T_E, the remaining liquid is of the eutectic composition and solidifies as a eutectic. Thus, the microstructure consists of α solid-solution crystals embedded in the eutectic (Fig. 6–6b). Because the α crystals formed prior to eutectic solidification, we may also say that the structure consists of *proeutectic* α in a eutectic *matrix*. The eutectic itself is a two-phase structure, yet it is often regarded as a single constituent, especially from the point of view of its effects on mechanical properties.

Calculate the relative proportions of phases in a Cu–Ag alloy of eutectic composition, just below the eutectic temperature. | **Example 6-2**

The inverse lever arm rule, Eq. (6-3), can again be applied to find the proportion of α

$$\frac{\alpha}{\alpha + \beta} = \frac{C_\beta - C_E}{C_\beta - C_\alpha} = \frac{91.2 - 71.9}{91.2 - 7.9} 100 = 23.2\%$$

the proportion of $\beta = 100 - 23.2 = 76.8\%$.

Example 6-3

Calculate the relative proportions of phases in a solidified copper–silver alloy of 20% Ag content just below the eutectic temperature. First calculate the total weight percent α and β, and then the relative proportions of proeutectic α and eutectic E.

The total weight percent α is, by analogy to Example 6-2,

$$\frac{\alpha}{\alpha + \beta} = \frac{C_\beta - C_0}{C_\beta - C_\alpha} = \frac{91.2 - 20}{91.2 - 7.9} 100 = 85.5\%$$

and that of $\beta = 100 - 85.5 = 14.5\%$. This tells us little about the structure. To find the proportion of proeutectic α, the calculation is carried out for a temperature just above the eutectic.

$$\frac{\alpha}{\alpha + \beta} = \frac{C_\beta - C_0}{C_\beta - C_\alpha} = \frac{71.9 - 20}{71.9 - 7.9} 100 = 81.1\%$$

The α crystals will be surrounded by $100 - 81.1 = 18.9\%$ matrix of eutectic composition (in which 23.2% is α, Example 6-2).

There is a further piece of important information provided by the phase diagram: When the temperature drops below T_E, the mutual solubilities of Cu and Ag diminish. The compositions of the α and β phases are given by the *solvus* lines. Excess solute atoms will be rejected and this will provide a powerful strengthening mechanism in some alloy systems (Sec. 6-4-2).

6-1-4   Peritectic Systems

Phase diagrams of practical alloy systems may show further features. We may generalize the discussion by calling one of the metals A and the other B.

When the melting points of two metals are greatly different, the invariant reaction is often of the *peritectic* type, illustrated in Fig. 6–7 on the example of the Cu–Zn system. An alloy of peritectic composition C_P (36.9% Zn) begins to solidify with the formation of α solid-solution (32.5% Zn) crystals. At the invariant temperature T_P (903°C) all the remaining liquid must disappear and the entire solid must transform into a β solid solution (38.0% Zn). This can be achieved only by the circumferential diffusion of Zn atoms into the already solidified α crystals, hence the name *peritectic diffusion reaction*.

6-1-5   Intermetallic Phases

In many alloy systems, phases with distinct properties are formed. Their composition is characterized by a more or less fixed ratio of the two elements.

In intermetallic compounds the ratio is stoichometric (of fixed value) and can be denoted as $A_m B_n$. The two atomic species A and B occupy fixed sites in the often very large unit cell. Bonds may have predominantly covalent characteristics, with electrons shared between atoms. Even though some electrical conductivity is retained, these compounds are often brittle, hard, and of a high melting point

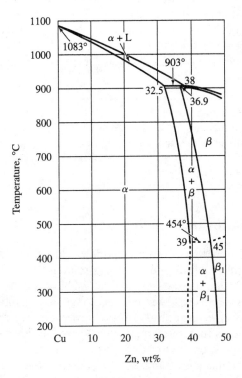

Figure 6–7   Large differences in melting points often lead to peritectic solidification, as in the Cu–Zn system.

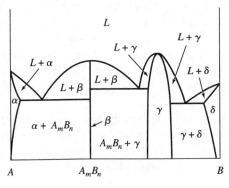

Figure 6–8   Many elements form intermetallic compounds ($A_m B_n$) or intermediate phases (γ).

(β in Fig. 6–8). Compounds may also form between metallic and nonmetallic elements. The most important example is Fe_3C in steels (Sec. 6-2-1).

In some cases the intermetallic can exist over an extended composition range and then one speaks of an *intermediate phase* (γ in Fig. 6–8) which often has an *ordered crystal structure* in which atoms are at specific locations in the lattice.

Example 6-4 | **S**ome intermetallics may be used as materials of construction. The titanium aluminides TiAl and Ti_3Al are ordered solid solutions (with solute atoms in regular sequence) in which dislocation propagation is difficult and requires a high energy, thus offering high-temperature strength with reasonable ductility, and they are used experimentally for turbine components. Other intermetallics are Ni and Fe aluminides.

6-1-6   Nonequilibrium Solidification

We assumed until now that cooling conditions during solidification allow the attainment of complete equilibrium. This is seldom the case, because cooling rates in most solidification processes are relatively fast (on the order of a fraction of a degree to a few degrees per second in casting and much faster in particulate processing and fusion welding) and diffusion processes are, in general, too slow. Therefore, solidified structures typically show nonequilibrium microstructural features, particularly when the freezing range is wide.

In a system like that shown in Fig. 6–9, solidification of a melt of composition C_0 begins with the rejection of α solid-solution crystals of composition C_1. On further cooling, the crystals not only grow but, according to the equilibrium phase diagram, their composition would also have to become enriched in the B element as dictated by the solidus. If time is insufficient to allow Mg atoms to diffuse into the already solidified core, the core remains leaner in the alloying element. The excess Mg atoms are retained in the melt, and solidification does not end when the T_S equilibrium solidus temperature is reached; instead it continues by the gradual deposition of richer and richer layers. The nonequilibrium solidus shown as a broken line in Fig. 6–9 represents the average composition of the solid. Because the centers (cores) of crystals grown during nonequilibrium solidification have a different composition (Fig. 6–9b), it is usual to refer to this phenomenon as *microsegregation* or *coring*. This has practical consequences for a finished part; it may become visible if, for example, an Al-5Mg alloy is anodized (Sec. 19-3).

An alloy of wider solidification range (*long freezing range*, $T_L - T_S$) is more prone to coring. The freezing range may be normalized in relation to T_L:

$$\text{Freezing range } = \frac{T_L - T_S}{T_L} \tag{6-4}$$

Microsegregation may be undesirable for a number of reasons and, if it cannot be prevented, may be partially or fully eliminated by subsequent homogenization heat treatment (see Sec. 6-4-1).

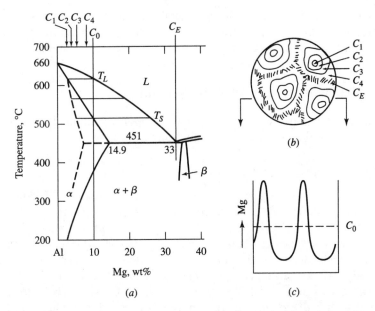

Figure 6–9 At usual rates of cooling, a solid solution of C_0 composition (a) will show coring (b), with a higher concentration of the alloying element on the grain boundaries (c), as in the Al–Mg system.

In the case shown in Fig. 6–9 a liquid phase remains until the eutectic temperature is reached, when the remaining liquid finally solidifies along the grain boundaries as a eutectic. This can have highly undesirable consequences:

Hot shortness According to the equilibrium phase diagram, the alloy is a solid solution and should thus be readily deformable. However, when heated above the T_E eutectic temperature, it will suffer fracture by separation at the grain boundaries where the nonequilibrium, low-melting eutectic is present. Hot-short fracture is readily identified by its ragged appearance as it follows the grain boundaries. Sometimes the presence of an unsuspected contaminant which forms a low-melting eutectic may totally destroy ductility. A striking example is sulfur in excess of 0.004% (i.e., 40 parts per million) in nickel or high-nickel superalloys.

Example 6-5

A 5056 Al alloy was cast into a 0.5-m-thick, 1.2-m-wide, 2.4-m-long ingot. Since it contains 5% Mg, it was known to have microsegregation (Fig. 6–9) and was placed into a bottom-blown air-circulating furnace for homogenization at 440°C. After several hours it was found that the bottom part of the ingot melted. Investigation showed that the bottom of the furnace was 15°C hotter than the temperature measured at midheight, thus the eutectic temperature was exceeded and the nonequilibrium eutectic melted.

Brittleness If the B element (which could also be an $A_m B_n$ intermetallic compound) is brittle, the solidified structure will also be brittle, even though from the equilibrium diagram we would judge it to be a ductile solid solution.

6-1-7 Nucleation and Growth of Grains

The account of solidification given above is still highly simplified. In reality, the processes goes through a sequence of nucleation and growth.

Nucleation There are two ways in which nuclei can form:

1. *Homogeneous nucleation* occurs only in very clean melts. The nucleus is formed by the ordering of atoms into positions corresponding to the crystal lattice. Such ordering exists also in the melt but only over short distances. Below the melting point, longer-range ordering is possible but much of it is only temporary. Atoms are in a highly agitated condition at this temperature and embryonic nuclei continually form and disappear. Only nuclei that have reached a critical size are stable and able to grow, and then only at temperatures considerably below T_m (the degree of such *undercooling* can be expressed as a fraction of the melting point and is around $0.2T_m$ in pure metals); since few nuclei form, grain size is coarse.

2. *Heterogeneous nucleation* is typical of most practical metals. The number of nuclei is greatly increased, grain size reduced, and the need for undercooling is reduced or eliminated by nucleation on the solid surface of *nucleating agents*. These may be residual impurities or finely divided substances (often, intermetallic compounds), intentionally added to the melt just before pouring. If they have a compatible crystal structure with little difference in lattice spacing and if they are wetted by the melt, atoms can easily deposit on them to form crystals at less than $5°C$ undercooling.

Growth of Crystals Once nucleated, crystals grow in essentially two ways:

1. *Planar growth* occurs when heat extraction is through the solid phase and a smooth solid/liquid interface moves into the liquid (Fig. 6–10a).

2. *Dendritic growth* is typical of solid-solution alloys. Crystals again grow in the direction of heat extraction but, as shown in Sec. 6-1-6, solidification begins with a leaner solid solution whereas the remaining solid is enriched. Coupled with local undercooling in the liquid, this leads to the formation of a branched crystal skeleton which resembles a tree (Fig. 6–10b), and is, therefore, called *dendrite* (from the Greek *dendron*, tree). At higher cooling rates or in the presence of nucleating agents, grains are refined and, more importantly for mechanical properties, the *secondary dendrite arm spacing* is also reduced.

Grain Size Nucleation and growth of grains occur simultaneously, but at different rates. Nucleation rate is maximum at substantial undercooling, whereas growth rate peaks close to the solidus temperature. Therefore, grain size depends on

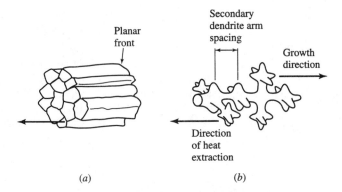

Figure 6–10   Solidification proceeds with (a) planar front in pure metals but (b) with the growth of dendrites in solid solutions.

residence time at a given temperature, which in turn depends on *cooling rate.* At low cooling rates, there is time for the few nuclei formed to grow and the structure will be coarse-grained. At high rates of cooling, a high nucleation rate gives many sites on which growth can take place and grain size will be small. At extremely high rates of cooling crystallization may be suppressed and a noncrystalline (*amorphous*) body obtained (Sec. 11-2-1).

Nuclei that preexist in the melt (either as homogeneous or heterogeneous nuclei) are dissolved by overheating the melt. Therefore, with increasing superheat, grain size increases too. The magnitude of *superheat* is usually expressed as the difference between melt temperature and liquidus temperature

$$\text{Superheat} = T_{\text{melt}} - T_L \qquad \textbf{(6-5a)}$$

There is not much information, but it is reasonable to assume that, for materials of different melting points, the effects of superheat may be rationalized by relating it to the melting point

$$\text{Superheat} = \frac{T_{\text{melt}} - T_L}{T_L} \qquad \textbf{(6-5b)}$$

Freshly formed crystals are extremely weak and easily broken to provide more nuclei. Thus grain size is refined by thermal currents or mechanical agitation of the solidifying melt, provided that the superheat is low, and crystal fragments are not remelted but survive as nuclei (*grain multiplication*).

6-2   SOLID-STATE REACTIONS

The solid alloy may undergo further changes as the temperature drops. We saw one such event in Fig. 6–5 where the phase boundary (*solvus*) of the *terminal solid solution* (i.e., solid solution at the end of the phase diagram) indicated that less of the solute species is kept in solution with decreasing temperature. The excess solute must then separate (*precipitate*) into a second phase, and this can be a powerful mechanism for controlling the properties of solid alloys, to be discussed in Sec. 6-4-2.

There are also other possibilities: As the temperature drops, the stability of various phases may change too and, at some critical temperature, transformations similar to those occurring in the liquid-to-solid transition may occur. To distinguish them from their counterparts, their names are formed by the ending *-oid*. Thus, when a homogeneous solid solution decomposes into two phases, one speaks of a *eutectoid transformation*.

6-2-1 The Iron–Iron Carbide System

Some transformations are best illustrated on the example of the Fe–Fe_3C system (solid lines, Fig. 6–11).

Above 2% C, the alloy is called a *cast iron*. In the absence of other alloying elements and at fast cooling rates, solidification between 2% and 4.3% C begins with the rejection of γ solid-solution crystals and ends with the formation of a γ–Fe_3C eutectic matrix. At 4.3% C, solidification is eutectic; at higher carbon contents, Fe_3C is embedded in the eutectic matrix (Fig. 6–12a). Because of the industrial importance of different phases in the system, each phase is also given a name. Thus, the eutectic is called *ledeburite*, and is composed of *austenite* (the γ phase) and *cementite* (Fe_3C). Since the Fe_3C is formed during solidification, it is called *primary cementite*.

Steels are alloys of less than 2% carbon in iron. There is a peritectic reaction at 1495°C which we may ignore for our purposes. At temperatures of practical importance, the C occupies interstitial sites, and at higher temperatures forms the fcc γ solid solution (austenite). With dropping temperatures, the solubility of C in austenite decreases and, in steels with over 0.8% C, the excess is rejected in the form of Fe_3C (*secondary cementite*).

Interstices in bcc iron are much smaller than in fcc iron, and the sudden drop in solubility leads to the eutectoid decomposition of austenite, at 727°C (also called the A_1 transformation temperature), into cementite and a solid solution of C in α-iron (*ferrite*). A eutectoid of lamellar structure (Fig. 6–12b) nucleates at austenite grain boundaries. Because it is formed in the solid state, the short diffusion paths make the lamellae much finer than in eutectics. It is called *pearlite* because the Fe_3C lamellae cause a fresh fracture surface to shimmer in a pearl-like manner.

Not shown in Fig. 6–11 is that the solubility of C in ferrite decreases (from 0.0218%) with dropping temperatures and a small amount of *tertiary cementite* is rejected.

Example 6-6 | Various countries adopt their own standards for the designation of steels. In the U.S. and Canada, the four-digit system of the American Iron and Steel Institute (AISI) and Society of Automotive Engineers (SAE) is widely accepted. The first two digits show what kind of alloying

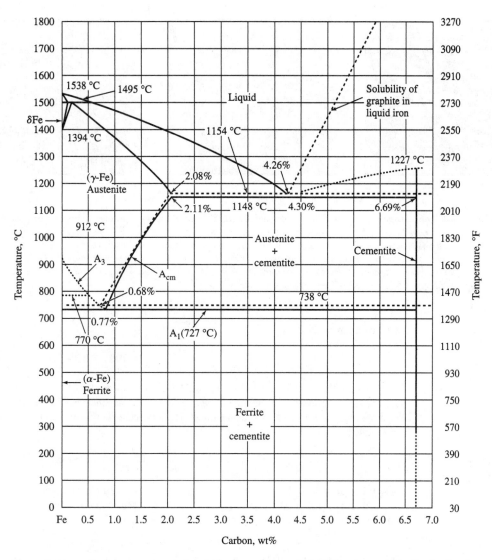

Figure 6–11   Carbon may occur in iron in the form of cementite (solid lines) or graphite (broken lines). (*From Metals Handbook Desk Edition, ASM International, 1985, p. 28.2. With permission.*)

elements are added to iron, and the last two digits give the carbon content in hundredths of percent. Thus, AISI 1040 is a plain carbon steel of 0.4% C content. (A complete description of the classification and designation of steels is given in MHDE *Metals Handbook Desk Edition*.) Calculate the metastable equilibrium phases present in this steel at (*a*) 1000°C and (*b*) room temperature.

   (*a*) At 1000°C, the structure is 100% austenite, with 0.4% C in interstitial solid solution in fcc iron.

   (*b*) If the proportion of the eutectoid is denoted as ED, the proportion of proeutectoid α at room temperature can be obtained, as in Example 6-3, by taking a temperature just above

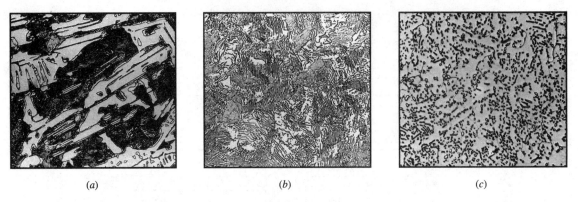

(a)   (b)   (c)

Figure 6–12   In the absence of graphite-stabilizing elements such as Si, carbon usually separates in the form of cementite. (a) Proeutectic cementite is found in a matrix of eutectic in a white cast iron. (b) Eutectoid decomposition of austenite results in pearlite (0.8% C) in which cementite and ferrite platelets alternate. (c) By spheroidizing heat treatment, the cementite can be brought into a spherical form in a 0.4% C steel. (*Courtesy Dr. G. F. VanderVoort, Carpenter Technology Corporation. Also in Metals Handbook Desk Edition, ASM International, 1985, pp. 35.37, 35.42, 27.28. With permission.*)

the transformation temperature:

$$\frac{\alpha}{\alpha + ED} = \frac{C_{ED} - C_0}{C_{ED} - C_\alpha} = \frac{0.77 - 0.40}{0.77 - 0.02}100 = 49.3\%$$

Eutectoid pearlite will constitute $100 - 49.3 = 50.7\%$ of the volume. The two phases will show up in the same proportions in the microstructure (a two-dimensional slice through the body). The proportions of ferrite (F) and cementite (CM) in the eutectic pearlite can be obtained, as in Example 6-2:

$$\frac{F}{F + CM} = \frac{C_{CM} - C_{ED}}{C_{CM} - C_F} = \frac{6.67 - 0.77}{6.67 - 0.02}100 = 88.8\%$$

Thus, $100 - 88.8 = 11.2\%$ cementite platelets (by weight) alternate with platelets of ferrite. We could also calculate the total weight fraction of α present in both the pearlite and ferrite from the composition just below the eutectoid temperature:

$$\frac{F}{F + CM} = \frac{C_{CM} - C_0}{C_{CM} - C_F} = \frac{6.67 - 0.4}{6.67 - 0.02}100 = 94.3\%$$

This will, however, tell us little about the properties of the steel.

6-3   STRUCTURE–PROPERTY RELATIONSHIPS

The aim of most manufacturing processes is to produce parts that not only have the correct size and shape, but also possess the best possible properties. For load-bearing components, this calls for high strength coupled with acceptable ductility, properties that are greatly influenced by structure.

6-3-1   Metals and Single-Phase Alloys

We saw in Sec. 4-1-1 that metals subjected to loading deform at some critical stress. When the deformed specimen is observed under an optical microscope, deformation appears to have taken place by the *slip* of adjacent zones (Fig. 6–13a). At high magnifications, each *slip zone* appears composed of many small steps, indicating that displacement must have taken place along *preferred slip planes* in each crystal (Fig. 6–13b).

Calculations can be made to show that slip by the massive movement of entire adjacent crystal zones would take much higher stresses than actually observed. Indeed, it is found that slip takes place by the movement of *line defects* (*dislocations*) along preferred slip planes in the crystal lattice: Shear stresses on these planes must reach a critical value before deformation can commence. This *critical shear stress* depends on the metal, crystal structure, and shear direction. In the simplest view, a dislocation could be regarded as an extra line or plane of atoms inserted into the structure (*edge dislocation*, Fig. 6–13c); thus, it is only necessary to dislodge this extra line of atoms along the slip plane instead of moving hundreds of thousands of atoms of a slip plane at the same time. Many of the deformation characteristics of metals can be interpreted by contemplating the ease with which these dislocations can move and by considering obstacles that may impede or arrest their movement.

Pure Metals   One might expect that dislocation movement (slip) should be easier on planes that give the smoothest movement, the least bumpy ride: If atoms are

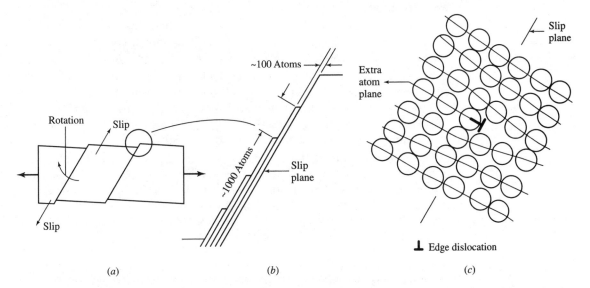

(a)   (b)   (c)

Figure 6–13   A single crystal subjected to tension deforms by (a) slip of adjacent zones. (b) At higher magnification, closely spaced slip planes are discovered. (c) Within each slip plane, there are numerous line defects, dislocations.

visualized as touching spheres (Fig. 6–2), one finds that slip takes place most readily in the most *closely packed planes* in the *closest-packed crystallographic directions*.

1. In the fcc structure (Fig. 6–2*a*) there are four equivalent closely packed planes (the {111} octahedral planes) with three equivalent slip directions ⟨110⟩, giving a total of 12 independent *slip systems* (i.e., combinations of slip planes and directions). If slip is limited on one plane because dislocations are arrested, there is always a likelihood that some other slip system will be oriented in the direction of the maximum deforming shear stress. Because of the close atomic packing, the critical shear stress is relatively low. In a polycrystalline body adjacent grains prevent free deformation and five independent slip systems must be available if grain-boundary fracture is to be avoided. This is amply satisfied in fcc structures, and fcc metals are readily deformable, essentially at all temperatures. Indeed, this is characteristic of Pb, Al, Cu, Ni, and γ-iron.

2. In the bcc structure, one cannot readily identify obviously close-packed planes, but a clearly closest-packed direction is found in the body diagonal (Fig. 6–2*b*). Therefore, these crystals slip in systems containing various planes that have the body diagonal ⟨111⟩ as a common slip direction, rather like a bunch of pencils. This co-called *pencil slip* allows extensive deformation, for example, in α-iron and β-titanium. However, because packing is not as close as in fcc metals, the critical shear stress may be higher.

3. The deformation of hcp structures is governed by the ratio of height-to-side dimensions, the *c/a ratio* (Fig. 6–2*c*). In an ideal hcp structure this ratio would be 1.633.

a. Some metals show a larger *c/a* ratio; i.e., the basal planes are more widely separated. Slip then occurs only in the basal planes along the three equivalent closest-packed directions (*basal slip*, Fig. 6–2*c*, as in zinc, $c/a = 1.856$).

b. When the *c/a* ratio is less than the ideal, the atoms of the basal planes are effectively squashed into each other and slip is now prevented here; the material will choose slip planes either along the side of the prism or on a pyramidal surface (*prismatic* or *pyramidal slip*, Fig. 6–2*d*). The prime example of this behavior is α-titanium ($c/a = 1.587$, at temperatures below 880°C).

c. A metal of close to the theoretical *c/a* ratio cannot slide readily along any of these planes and it is usually necessary to raise the temperature somewhat so that the increased freedom of atomic movement brings a number of slip systems into play. This is most clearly evidenced by magnesium ($c/a = 1.624$) which can take very little deformation at room temperature but deforms readily when heated to 220°C.

Frequently, deformation in hcp materials is aided by *twinning* (which occurs when a part of the crystal flips over into a mirror-image position), bringing more slip planes into a favorable direction relative to the maximum shear stress.

Solid Solutions   Solid-solution alloys have the structure of the solvent metal. Substitution of solute atoms of slightly different size distorts the lattice and makes dislocation propagation on the slip planes more difficult; thus strength increases without necessarily reducing ductility. The higher concentration of alloying elements associated with microsegregation can result in further strengthening; smaller dendrites (and, even more so, smaller secondary dendrite arm spacing) contribute to higher strength. Interstitial elements play a similar role in impeding dislocation mobility although they can also have an embrittling effect if they entirely block the movement of dislocations.

The benefits of solid-solution alloying are clearly visible in the Cu–Zn system. In the annealed condition, there is a useful increase in strength with increasing zinc content and even ductility increases over that of pure copper.

Example 6-7

Composition, %	YS, MPa	TS, MPa	el., %
Cu	70	220	55
Cu-5Zn	70	235	45
Cu-15Zn	70	270	55
Cu-30Zn	75	300	65
Cu-35Zn	97	320	65
Cu-40Zn	145	370	52

Thus, Cu-30Zn is suitable for heavy deformation, such as is needed for making cartridge cases—hence the traditional name *cartridge brass*. The solubility limit is exceeded over 35% Zn (Fig. 6–7) and the presence of a second phase results in a sudden increase in strength and concomitant drop in ductility.

6-3-2   Two-Phase Materials

In considering the properties of two-phase structures, it is necessary to recognize that the presence of two phases immediately implies the existence of an *interface* between them. The strength of bond in this interface is critical not just in two-phase alloys but also in all products composed of different entities (Chap. 15).

Interface Strength   Any interface, even a grain boundary in a pure metal, is a site of many unsatisfied, broken interatomic bonds that add up to an excess energy, the *interfacial energy* γ. The magnitude of interfacial energy is larger when the mismatch between adjacent atomic groupings is greater. Thus, the interfacial energy between the vapor and solid γ_{SV} of a substance, say metal, is much larger than the interfacial energy between its liquid and solid phases γ_{SL}. The relative magnitudes of interfacial energies between two dissimilar materials are readily judged by placing a liquid drop of one on top of a flat, solid surface of the other. The liquid drop sits in place (hence the name *sessile drop*) but is free to change its shape until *surface tensions* establish a force equilibrium (Fig. 6–14):

$$\gamma_{SV} = \gamma_{SL} + \gamma_{LV} \cos \theta \qquad \text{(6-6)}$$

When $\theta < 90°$, the surface is *wetted* and the drop spreads out; when $\theta > 90°$, the surface is *not wetted* and the liquid forms, in the limit, a spherical droplet.

Wetting, then, is an indication of relative surface energies and, through these, a measure of the strength of the interface. Wetting is a sign of reasonable match between the atomic lattices of the contacting phases, and one can expect a wetted interface to resist stresses that might pull the phases apart. A nonwetted interface, on the other hand, can behave as a preexisting crack and thus impair mechanical properties (Sec. 4-1-6). It should be remembered that surface tension is indeed a surface effect; therefore, even the minutest amounts of a contaminant, segregated at the interface, can substantially reduce wetting.

Mechanical Properties   With these preliminaries in mind, one would intuitively expect that the properties of a two-phase structure must depend on a number of factors, such as the properties, quantity, distribution, shape, and size of individual phases and the nature of interfaces between them. Some basically different situations can be envisaged (Fig. 6–15):

1. *Both phases are ductile and wet each other.* The material behaves like a homogeneous body: Dislocations pass freely through both phases and properties can be estimated from the relative volumes of the two phases (this is the case for Cu–Ag alloys).

2. *One of the phases is ductile, the other is brittle and wetted.* In this case, the properties, relative quantity, shape (*morphology*), and location of the brittle phase become dominant, because dislocations will pass freely through the ductile phase but will be blocked by the brittle phase.

a. If the brittle phase is the matrix (i.e., the other phase is embedded in it), the resulting structure will be brittle; it may have a high compressive strength, but tensile strength and ductility are low because cracks initiated in the brittle phase will easily propagate through the material (an example of this is white cast iron).

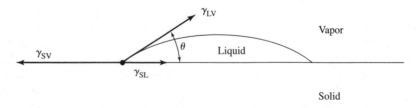

Figure 6–14   A liquid wets the solid surface when the angle θ is small.

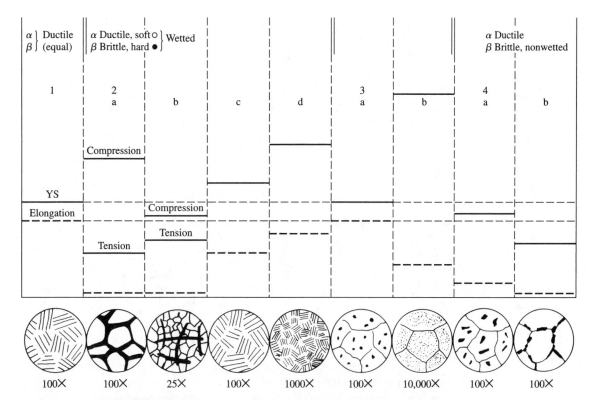

Figure 6–15   Properties of two-phase structures depend on the properties, wetting, shape, size, and distribution of the two phases. Note the different magnifications for the schematic microstructures.

b. If the matrix is ductile, but coarse plates (*lamellae*) or needles (*aciculae*) of the brittle phase weave through it, the brittle phase causes stress concentrations on loading, cracks propagate through the brittle phase; the structure will be both weak and of low ductility, and it will have low fracture toughness (e.g., the effect of sulfides in Fig. 6–16).

c. If finer plates of a strong brittle phase are confined within a grain and are systematically aligned into a lamellar structure, they will substantially strengthen the matrix but usually at the expense of ductility, since the brittle plates fracture during deformation and the stress raisers terminate plastic deformation in the ductile matrix. An example of this is shown by pearlitic carbides in Fig. 6–16.

d. If the platelike structure is extremely fine and closely spaced so that the hard phase acts as a barrier to dislocation mobility, the composite structure may show high strength coupled with reasonable ductility, because rapid strain hardening delays necking (see Sec. 8-1-1).

3. *The hard constituent is wetted by the soft phase and is in a roughly spheroidal form.*

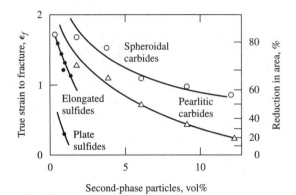

Figure 6–16   Ductility is highly sensitive to second-phase particles, especially if they are of unfavorable shape and of low ductility. (After T. Gladman, B. Holmes, and L. D. McIvor, in Effect of Second-Phase Particles on the Mechanical Properties of Steel, Iron and Steel Institute, London, 1971. With permission of The Metals Society.)

a. If the hard particles are relatively coarse and widely spaced, they have an only minor effect on the strength of the material. Ductility is impaired, but not excessively (spheroidal carbides in Fig. 6–16) because dislocations circumnavigate such large blocks of harder material, and the notch effect is minimal because of the large notch radius.

b. When the particle size becomes small enough to arrest or slow down dislocations, the structure will be strengthened. Particularly effective are very thin particles that still retain atomic registry with the matrix (*coherent precipitates* in Al–Cu alloys, Fig. 6–19). If the particles form a clearly defined second phase, the result depends on the total quantity and spacing of hard particles. Great strengthening can be obtained by loading the ductile matrix with masses of particles (until the matrix becomes little more than a ductile cement), but then ductility will greatly suffer. This is exploited for tool materials (Sec. 16-3-1).

c. Small particles located on grain boundaries and inside the grains are effective in blocking dislocation movement and grain boundary sliding, hence *dispersion-hardened* alloys show rapid strain hardening and good creep resistance.

4. *The hard phase is not wetted by the matrix.* The interface between phases acts as a premade crack.

a. When the nonwetted particles are inside ductile grains, they can be relatively harmless, although any cracks that form can coalesce readily, greatly impairing ductility. If the base material is relatively brittle, strength is reduced too because the interface acts as a notch.

b. Nonwetted particles are most harmful when located on the grain boundary, which tends to be less ductile in any event.

A feel for the effects of structure may be gained from typical mechanical property data of annealed C steels:

Example 6-8

AISI No.	YS, MPa	TS, MPa	el., %	RA, %	HB, kg/mm^2
1008	210	350	38	70	95
1015	285	385	37	70	110
1020	295	395	36	65	110
1040	350	520	30	57	150
1080	375	615	25	45	175
1090	380	655	13	20	192

Note the gradual increase in strength and drop in ductility with increasing pearlite content up to the eutectoid composition and the steep drop in ductility with the appearance of grain-boundary secondary cementite in 1090.

6-3-3   Ternary and Polycomponent Alloys

In practice, very few truly *pure metals* or *binary alloys* (i.e., alloys formed by two atomic species) are used. Even if only on the order of a few parts per million, contaminants are always present, while intentional additions may reach such high proportions that it becomes difficult to classify a material according to its base metal. Such *ternary*, *quaternary*, etc., alloys (*polycomponent systems*) still exhibit distinct phases found in binary alloys. Thus one can find solid solutions, eutectics, peritectics, intermetallics, and their various combinations. The behavior and properties of polycomponent systems can be derived by analogy to two-phase materials, especially if the phase diagram is known or at least a section of the phase diagram (a *pseudobinary diagram*, for constant percentages of the other alloying elements) has been established. Great strides have been made in using the computer to predict what phases to expect in polycomponent alloys, at any temperature and composition, on the basis of binary phase diagrams and a data bank of thermodynamic data.

The whole battery of strengthening mechanisms is drawn upon in Ni- and Co-based superalloys for high-temperature applications such as turbine blades. Cr, Mo, and W provide solid-solution strengthening, effective even at high temperatures; Al and Ti give precipitation hardening; and

Example 6-9

several elements (Ti, B, Zr, Ta, Cr, Mo, W) form very fine carbides, stable at high temperature, which impede dislocation movement as well as grain-boundary sliding. The Ni-based superalloy MAR-M 200 (Ni-9Cr-10Co-1Fe-5Al-2Ti-1Nb-12.5W-0.15C-0.015B-0.05Zr) has a melting range of 1315–1370°C and retains useful strength beyond $0.5T_m$:

Temperature, °C	YS, MPa	TS, MPa	el., %
21	840	930	7
540	850	945	5
650	855	950	4
760	840	930	3.5
870	760	840	4
980	470	550	4.5
1090		325	

6-3-4   Inclusions

The term *inclusion* is used to describe foreign particles in a metallic structure. They find their way into the alloy usually from the ore, during melting (for example, from the furnace lining, contamination of the charge, or even as a result of reaction—usually oxidation—with the surrounding atmosphere), or during pouring. As with all second-phase particles, their effect depends greatly on whether they are wetted by the matrix or not (Sec. 6-3-2).

If inclusions are wetted, strong, and perhaps even ductile, and are dispersed inside the grains in an approximately globular or fibrous form, they are harmless and sometimes even useful. Arranged along grain boundaries they are likely to be harmful, unless they are extremely small and well distributed. Brittle plates and, particularly, films (such as those formed by aluminum oxide) are detrimental, as are low-strength inclusions in elongated or platelike forms. Strength may not be greatly affected, but ductility (Fig. 4–7; also sulfides in Fig. 6–16), fatigue strength, and fracture toughness suffer. Hence our aim is generally that of producing clean metals, free from inclusions, except when inclusions of controlled size and shape are intentionally introduced to improve machinability (Sec. 16-2).

Nonwetted inclusions are almost always harmful, reducing the strength, ductility, and fatigue and impact properties of the material. If gases are present, they tend to congregate on the interface between inclusion and matrix and can build up such high pressures that a bubble (blister) is formed on inclusions close to the surface, particularly if the part is in high-temperature service or is heated during manufacture. Even inside the body, gases segregating on nonwetted interfaces aggravate the crack effect and lead to embrittlement (e.g., hydrogen embrittlement in steel, Sec. 18-4-2).

It is evident from the above discussion that even minor changes in structure may have an influence on mechanical properties. This also means that any material produced to a given specification will show a range of properties; it is extremely important to realize that clear distinction must be made between *typical properties* and the actual *range in properties*, as shown by the example of annealed, drawing-quality special-killed (DQSK) low-carbon steel sheet:

	YS, MPa	TS, MPa	el., %	HRB
Typical	170	290	42	40
Range	140–210	260–340	31–47	34–51

In this text, unless otherwise specified, data refer to typical properties.

6-3-5   Gases

Gases normally exist in a molecular form but, at higher temperatures and in contact with metal, a significant portion may dissociate into the atomic form and enter into the metal. They can be accommodated interstitially in the relatively loose, nonordered structure of melts. Thus, solubility of gases may be high above the melting point (Fig. 6–17) but it drops steeply as the melt solidifies. Some gas may be trapped in the solid in the atomic form, but much is rejected at the solid/liquid interface to combine into molecules. These molecules coalesce into gas bubbles

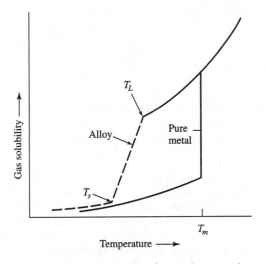

Figure 6–17   The solubility of gases drops greatly on solidification.

which rise in the melt or, if trapped during solidification, cause *gas porosity* (*pin-holes* or larger *blowholes*) in the structure. In contrast to interdendritic porosity, gas pores are generally round and, if they contain a neutral or reducing gas, they have a clean, bright surface. They too can be regarded as inclusions of zero strength, but their larger radius makes them less damaging to mechanical proper-ties. They can also cause blistering, as discussed in conjunction with nonwetted inclusions.

Not all gases are equally soluble in all metals. Hydrogen is soluble in practi-cally all metals because of the small size of its atoms. It may be introduced into the melt by dissociation of water from the air, the charge, or combustion products. It is particularly troublesome for aluminum and magnesium alloys. In contrast to hydrogen, nitrogen is soluble in iron but not in nonferrous metals. Oxygen is soluble in steel. Noble gases (of which argon is technically most significant) are completely insoluble.

The solubility S of any one gas in the melt increases (or decreases) with the square root of the *partial vapor pressure* p_g of that gas over the melt (Sievert's law)

$$S = k\sqrt{p_g} \tag{6-7}$$

where k, the equilibrium constant, drops sharply upon solidification (Fig. 6–17). It follows that the concentration of any gas in the melt can be reduced by either reducing the overall gas pressure (*vacuum degassing*) or by bubbling a nonsoluble *scavenging* gas through the melt just before pouring. Because the partial pressure of the offending gas is zero in the scavenging gas bubbles, the offending gas is drawn out of solution into the rising scavenging gas and is removed.

Some gases can be rendered harmless by combining them with other elements. A prime example is *deoxidation*, although care must be taken that the reaction product itself be harmless (Secs. 7-4, 8-3, and 10-1).

Example 6-11

The maximum equilibrium solubility of hydrogen in liquid magnesium is 26 cm^3 H/100 g; this drops to 18 cm^3 H/100 g upon solidification. What would be the porosity if liquid Mg saturated with H were allowed to solidify?

The volume of H rejected upon solidification is $26 - 16 = 8$ cm^3.

The volume of Mg is 100 g/(1.74 g/cm^3) = 57.5 cm^3.

The solid will look like Swiss cheese, and will have a total volume of $8 + 57.5 = 65.5$ cm^3, of which $8/65.5 = 0.122$ or 12.2 vol.% will be pores.

Example 6-12

When the partial pressure of hydrogen is 1 atm above a melt of aluminum, the equilibrium solubility of the gas is 0.7 cm^3/100 g Al. Maximum solubility in solid Al is 0.04 cm^3/100 g. What partial vapor pressure p_{H_2} should be maintained over the melt if a pore-free casting is to be obtained?

Substituting into Eq. (6-7) for the liquid, the equilibrium constant is obtained

$$0.7 \text{ cm}^3/100g = k\sqrt{1 \text{ atm}}$$

$$k = 0.7 \left(\text{cm}^3/100 \text{ g } \sqrt{\text{atm}}\right)$$

For the solid

$$0.04 = 0.7\sqrt{p_{H_2}}$$
$$p_{H_2} = (0.04/0.7)^2 = 0.0033 \text{ atm}$$

Note that a very low partial vapor pressure must be maintained.

6-3-6   Effects of Grain Size

In a polycrystalline material, at relatively low temperatures, individual grains can deform only by the propagation of dislocations. Grain boundaries represent defects in the structure and are thus sources of dislocations. At the same time, however, grain boundaries also present obstacles to dislocation propagation. Therefore, it is generally found that *the yield strength of a material increases with decreasing grain size* (Fig. 6–18) according to the Hall-Petch relationship[1]

$$\sigma = \sigma_0 + k_y d^{-1/2} \tag{6-8}$$

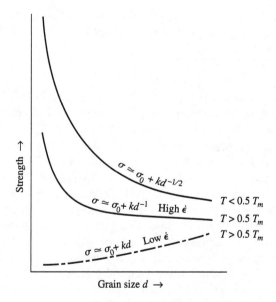

Figure 6–18   The strength of metals increases with diminishing grain size, except when deformation takes place at high homologous temperatures and low strain rates, with a massive diffusion of atoms.

[1] E.O. Hall, *Proc. Phys. Soc. London,* **64B**:747 (1951); N.J. Petch, *J. Iron Steel Inst. London,* **173**:25 (1953).

where d is the average grain size and σ_0 and k_y are material constants. In some metals (such as Al alloys) an even better agreement is found between strength and the *size of subgrains* (relatively strain-free subunits within the larger grain, arranged with a slight crystallographic misorientation).

| **Example 6-13** | As indicated in Example 6-7, cartridge brass is used for heavy deformations, most often in sheet form. In the annealed condition it is available with controlled grain size, which determines mechanical properties. The designation O5 indicates that annealing produced a prescribed average grain size, expressed in units of μm (thus O5025 indicates an average grain size of 25 μm). |

Designation	YS, MPa	TS, MPa	el., %	HRF
O5100	75	300	68	54
O5050	105	325	62	64
O5025	130	350	55	72
O5015	150	365	54	78

When the data are plotted on a log–log scale, the results conform reasonably well to a power law, although the exponent is slightly lower than called for by Eq. (6-8). Similar results are found for steels.

It should be noted that Eq. (6-8) holds only when dislocation propagation is the primary mechanism of deformation. This is true of deformation in the *cold temperature range* (Fig. 4–15), i.e., typically below $0.5T_m$ on the homologous temperature scale. At higher temperatures, in the *hot temperature range*, other deformation modes are possible at certain strain rates $\dot{\epsilon}$ [Eq. (4-17)]. At high strain rates, dislocation movement still dominates (although other mechanisms come into play too, Sec. 8-1-6) and strength decreases with increasing grain size (Fig. 6–18). However, at the very low strain rates typical of creep (Sec. 4-6) and temperatures close to T_m, there is time for substantial diffusion to take place. The part may deform by the sliding of grains as complete blocks relative to each other, or by the reshaping of individual grains in the loading direction. Both processes are easier if grain size is small, hence creep strength drops with decreasing grain size, and a *large grain size (and even a single grain) is preferable for materials destined for high-temperature service.*

6-4   HEAT TREATMENT

In later chapters we will see how conditions can be varied during solidification for controlling the properties of the finished part. Even with the best control, it may still not be possible to attain the desired properties, and then further treatment may be imposed on the already solidified part.

6-4-1   Annealing

Annealing is the process of heating a material to some elevated temperature, holding at that temperature, and cooling back to room temperature. The rate of heating and cooling may have to be controlled. Undesirable reactions—in particular, oxidation—may be significant at high temperatures and, if this is objectionable, annealing is carried out in vacuum or in an inert or reducing gas atmosphere. Annealing may serve several purposes:

1. We already discussed *stress-relief annealing* in Sec. 4-7; it is usually performed below the temperature at which recrystallization or phase transformation would occur.

2. In some applications, the compositional variations due to nonequilibrium solidification (Sec. 6-1-6) are objectionable. The part may then be subjected to *homogenization anneal*. The part is heated just below the solidus temperature (T_S) and is held, usually for several hours, until diffusion equalizes the concentration of the alloying element throughout the part. If the distances over which diffusion must take place (*diffusion paths*) are large, holding times may have to be increased to several days. Even so, some elements may diffuse too slowly to achieve complete homogenization. It must be kept in mind that, if solidification was nonequilibrium, a low-melting eutectic may be present and the part may distort or, in extreme cases, it may distort and even fall apart (Example 6-5).

3. The shape of second-phase particles may be changed by holding below the transformation temperature. The most important application is *spheroidizing* treatment of steels (Sec. 6-4-3).

4. The purpose of annealing may be to remove the effects of cold working; such *process annealing* will be discussed in Sec. 8-1-5.

6-4-2   Precipitation Hardening

In alloy systems where the solubility of an alloying element changes with temperature, there are opportunities for influencing mechanical properties by various heat treatments, especially if the excess solute is rejected in the form of an intermetallic compound. The example shown in Fig. 6–19 is for an Al-4Cu alloy.

Annealing   When the alloy cools slowly through the solvus, the rejected solute atoms combine with solvent atoms to form large, stable second-phase $A_m B_n$ particles (Fig. 6–19, Annealed). The relatively few and large particles (in this case, hard and brittle intermetallic $CuAl_2$) have little effect on strength or ductility.

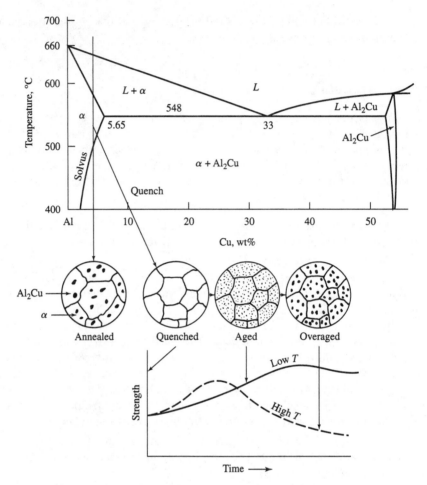

Figure 6–19   After solidification, some alloys may be substantially strengthened by precipitation-hardening heat treatment.

Precipitation Hardening   Substantial improvement may be obtained through a sequence of steps:

1. *Solution treatment.* The second-phase particles are completely dissolved by heating the alloy into the homogeneous α solid-solution temperature range. This is completed faster at temperatures closer to the solidus but the possible presence of nonequilibrium phases must again be considered. Also, high solution-treatment temperatures may allow some grains to grow at the expense of others, and the coarse-grained structure will have lower strength.

2. *Quenching.* The solid solution is preserved and diffusion of atoms and precipitation of second-phase particles is suppressed by rapid *quenching*, usually in water. The resulting solid solution (Fig. 6–19, Quenched) is *metastable*; i.e., it reverts to the stable two-phase structure if conditions are favorable for diffusion.

3. *Precipitation.* In many alloys precipitation proceeds at room temperature, and then one speaks of *natural aging*. Diffusion is limited to short distances, precipitation begins at many sites, and the precipitated particles are thin and coherent (Fig. 6–19, Aged). In other alloys, diffusion is too slow at room temperature and is accelerated by holding at some elevated temperature; this is called *artificial aging*. As shown in Fig. 6–15, strength increases greatly without undue loss of ductility, thus, the entire heat treatment sequence is also termed *precipitation hardening*.

4. *Overaging.* When a solution-treated part is subjected to higher temperatures, further diffusion results in fewer, coarser, noncoherent, stable precipitates and the strength of the structure (Fig. 6–19, Overaged) declines. Such overaging may also occur in service if the material is exposed to excessive temperatures.

No benefit is derived if the precipitates are not coherent. For this reason, Al–Mg alloys (Fig. 6–9) are not hardenable.

The aluminum alloy 2024 is extensively used in aircraft construction and for other high strength-to-weight ratio applications. It contains 3.8–4.9% Cu and is used most of the time in the heat-treated condition.

Example 6-14

Condition	Code	YS, MPa	TS, MPa	el., %	HB, kg/mm^2
Annealed	-0	75	185	20	47
Solution-treated and naturally aged	-T4	325	470	20	130
Solution-treated and artificially aged	-T6	415	480	13	

Note the tremendous increase in YS and TS in natural aging, with no loss in ductility. Precipitation is relatively rapid, with a noticeable increase in YS after only 1 h and almost full hardening after 10 h. Artificial aging confers much higher YS with some loss in ductility.

6-4-3   Heat Treatment of Steel

Alloy systems with solid-state transformations offer a variety of heat treatment possibilities that are best explored on the example of the Fe–Fe$_3$C system (Fig. 6–11).

Carbon Steels   In practice, carbon steels contain up to 1.7% C; the eutectoid composition lies at approximately 0.8% C. *Hypereutectoid steels* (i.e., those of

0.8–1.7% C) are hard but brittle because of the presence of secondary cementite, and thus have limited application. Most steels are *hypoeutectoid* (i.e., contain less than 0.8% C) and their structure consists of α solid solution and a pearlite eutectoid. The distribution of phases and the morphology of the eutectoid depend on cooling history: The pearlite is coarser in thicker sections where slower cooling allows diffusion over longer distances. Reheating the solid part into the homogeneous γ range (*austenitizing*) takes all C into solution; subsequently, the formation of Fe_3C can be controlled by choosing an appropriate cooling rate through the transformation temperature. The events are graphically summarized in *time-temperature-transformation* (TTT) diagrams, an example of which is given in Fig. 6–20 for steel of eutectoid composition.

Above 723°C, stable austenite exists. On cooling below 723°C, the austenite decomposes into the low-carbon ferrite (α) and cementite (Fe_3C). Since diffusion of C takes time, metastable austenite exists for some time. Transformation begins and is completed only after a certain time has elapsed, and this time is a function of temperature, giving the characteristic C-shaped or sigmoidal curves in the TTT diagram.

1. On slow cooling (Fig. 6–20, line 1) the transformation curve is crossed at a high temperature. Pearlite is nucleated at austenite grain boundaries; diffusion is fast, and the structure will consist of thick cementite platelets in a ferrite matrix. Such coarse *lamellar pearlite* is relatively soft but not very ductile. The heat treatment consisting of austenitization followed by slow cooling on air is called a *normalizing anneal*. The cementite may be changed to a spheroidal form by holding the steel just below the eutectoid temperature or by repeatedly heating and cooling just above and below this temperature. *Spheroidized* steels (Fig. 6–12c) have lower strength but significantly higher ductility.

2. On faster cooling (Fig. 6–20, line 2), the curve is crossed at lower temperatures. Diffusion is slower and the structure will consist of much finer but still lamellar pearlite, of higher strength (Fig. 6–15).

3. If the steel is cooled very rapidly and then held at an intermediate temperature, say around 300°C, the nose of the transformation curve is missed and transformation occurs *isothermally*, along line 3 in Fig. 6–20, with the formation of *bainite* (*austempering*). In this, the lack of diffusion time makes the carbide particles appear as extremely fine spheroids in a matrix of α solid solution. As expected from Fig. 6–15, such a structure possesses a desirable combination of strength and ductility. Isothermal transformation at around 400°C results in the formation of very fine pearlite (exploited in *patenting heat treatment*).

4. When cooling is again fast enough to miss the nose of the curve entirely (Fig. 6–20, line 4), but is now taken to room temperature (*quenching*), separation of the carbide phase is suppressed but the diffusionless transformation of iron from the fcc to the bcc form cannot be prevented. Transformation starts at the temperature marked M_s in Fig. 6–20 and is completed before room temperature is reached. Carbon atoms are retained in a supersaturated solid solution, distorting the bcc structure into a body-centered tetragonal lattice. This highly stressed structure (*martensite*) is very hard and brittle. Martensite formation results in

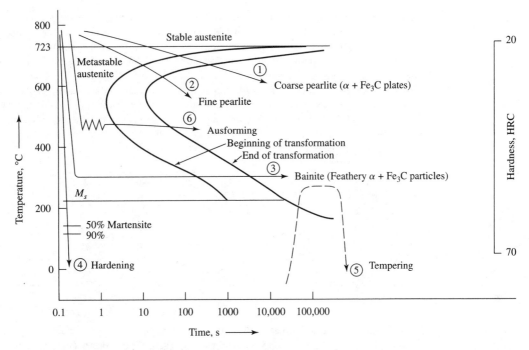

Figure 6–20   Depending on cooling rates, the eutectoid decomposition of austenite may result in a variety of structures. The example shown is for steel of 0.8% C content; the hardness of the resultant structures is given, approximately, on the right-hand ordinate.

volume increase and this can lead to *quench cracks* on the surface of a part: Martensite first forms on the surface; when the center transforms, its expansion puts the surface in tension and the brittle martensite cracks.

5. Ductility can be restored by reheating the martensite (Fig. 6–20, line 5) so that carbide can precipitate in a very fine form. The strength and hardness of such *tempered martensite* is somewhat lower but the sacrifice is well justified by the increased ductility and toughness. Heating at yet higher temperatures causes *overtempering*: The Fe_3C particles coarsen and hardness drops. (It should be noted that there are also martensites in other alloy systems, but not all martensites are necessarily hard.)

Carbon steels must have minimum 0.2% C to benefit from martensitic hardening. A disadvantage is that rapid quenching sets up steep temperature gradients; differential contraction results in the generation of high internal stresses which can lead to cracking and even disintegration of the part.

A part made of AISI 1040 steel is to be subjected to heat treatment for higher strength. We saw in Example 6-6 that there is sufficient carbon to form 50.7% pearlite, hence martensite will form on quenching. First, the casting must be heated into the austenitic temperature | **Example 6-15** range. From Fig. 6–11, all traces of ferrite and pearlite disappear at 780°C (actually, the transition temperature is 793°C on heating up). In practice, one heats to a somewhat higher temperature, say, by 50°C. Excessive temperatures cause coarsening of the austenite with some deterioration in final properties. Hence, we austenitize at 830–855°C and quench. Because the nose of the transformation diagram for this steel is similar to that in Fig. 6–20, water quench will be necessary. This could cause distortion and set up internal stresses leading to cracking. Furthermore, the depth to which martensite forms is also limited because, inside the part, cooling rates will not be high enough to avoid pearlitic transformation. (If through-hardening is required, an alloy steel must be chosen). The as-quenched hardness will be high, about HV 700. Tempering reduces hardness, but imparts ductility (toughness).

Tempering Temperature, °C	TS, MPa	YS, MPa	el., %	RA, %	HB, kg/mm^2
205	780	590	19	48	260
425	760	550	21	54	240
540	720	490	26	57	210
650	635	435	29	65	190

Alloy Steels   *Critical cooling rates* are reduced (the nose of the curve in Fig. 6–20 is shifted to the right) by the addition of alloying elements (usually several elements) to a carbon steel. Such steels still form a characteristic martensitic structure upon quenching but allow heat treatment of thicker sections; the required quenching rates are lower and quenching may be carried out in oil or even in air. Some elements serve multiple functions. For example, manganese is a solid-solution hardening element which is also effective in this respect, thus increasing the *hardenability* of the steel (i.e., the depth to which hardening is obtained upon quenching). Furthermore, it combines with sulfur in the form of inclusions, thus preventing the formation of iron sulfide which would cause hot shortness.

Phases that appear upon solidification can be modified too. Some alloying elements increase the stability of austenite (e.g., the solid-solution alloying element nickel), others that of ferrite (e.g., chromium). Alloying-element concentrations can be raised to the level where austenite is retained at room temperature in the stable form (*austenitic stainless steels*). In yet other steels alloying elements such as chromium, vanadium, and molybdenum are introduced that combine with carbon to form very stable carbides, so that the steels retain their hardness at temperatures where the martensite would be overtempered. Apart from their use in high-temperature service, such steels find application in manufacturing processes as molds, dies, and cutting tools.

6-4-4   Surface Treatment of Steel

In many applications—such as gears, shafts, rolling-mill rolls, components subjected to wear—it is desirable to have a high hardness on the surface combined with great toughness throughout the body of the part. One option is to apply a surface coating (Chap. 19); the other is to change properties in a surface layer of the steel part itself. Three fundamentally different approaches are available.

1. The steel has sufficient carbon and alloying-element concentration to form martensite upon quenching. *Surface hardening* is then possible by a two-step treatment: First, the part is heat-treated to obtain the toughness required in the core. Second, a surface layer of the part is rapidly heated (e.g., by induction, flame, or laser beam) and immediately quenched. Alternatively, the same result is obtained in a single step: A composition is chosen that will, upon controlled quenching from the austenitic temperature, transform into martensite on the surface but into pearlite in the core (*shell hardening*).

2. The steel has a low (typically 0.2%) carbon content and the surface is made hardenable by diffusing carbon into the surface—from a gas atmosphere, a liquid, or a solid pack—in the austenitic temperature range. Upon quenching, the carbon-enriched surface layer or *case* transforms into martensite, while the core remains tough. Hence the term *case hardening* or *carburizing*.

3. The steel is hardened by diffusing nitrogen into the surface (*nitriding*). Treatment is carried out below the A1 transformation temperature; hence, no quenching is required, distortion is less, but treatment times are longer. The danger of distortion is fully avoided by *ion nitriding*, i.e., the injection of N atoms into the surface (see Sec. 19-6-4).

6-5 SUMMARY

For the vast majority of metallic materials, solidification is the first step in manufacturing, and solidification is also a basic process in many joining techniques. Basic principles apply to all of these processes, and some of these principles have applications to other thermal treatments.

1. Pure metals undergo a phase change at the melting point T_m. Most practical metals solidify into fcc, bcc, or hcp crystals. Many alloys are solid solutions in which atoms of the solute element are accommodated by substitution or by fitting into interstices of the crystals.

2. Once limits of solubility are exceeded, solidification may occur by forming eutectics, peritectics, or intermetallics. Phase diagrams are roadmaps to equilibrium solidification for which perfect diffusion is necessary. Diffusion is fast at high temperatures but still requires time, therefore, lack of diffusion leads to nonequilibrium solidification at practical cooling rates.

3. Crystals may form by homogenous nucleation but in most instances existing solid particles initiate heterogeneous nucleation. While grain growth is faster at high temperatures, nucleation is faster at greater undercooling, hence grain size can be controlled by controlling cooling rates.

4. The effects of grain size depend on the deformation mechanism. In the cold temperature range smaller grain size gives greater strength; in the hot temperature range coarse grain is desirable.

5. Polymorphic metals undergo solid-state reactions, the primary example of which is the eutectoid decomposition in the Fe–Fe_3C system.

6. The mechanical properties of alloys are greatly affected by metallographic structure. Properties depend on the properties, size, shape, and distribution of individual phases and are vitally influenced by the strength of the interface. A week interface acts as a premade crack in the body. Interfaces between the matrix metal and inclusions often fall into this category, and this explains the need for inclusion control.

7. Properties of an already solidified part may be changed by thermal treatments such as annealing, precipitation hardening, and the special treatments made possible by the eutectoid transformation in steel.

PROBLEMS 6A

6A-1 Draw sketches of (*a*) fcc and (*b*) bcc cells; give an example of a metal of (*c*) fcc and (*d*) bcc structure.

6A-2 (*a*) Define allotropic transformation. (*b*) Give a practically important example.

6A-3 (*a*) Draw an equilibrium diagram showing complete solubility of metals *A* and *B*. Name the lines defining the temperature of (*b*) beginning and (*c*) end of solidification.

6A-4 (*a*) Draw a typical eutectic phase diagram with the eutectic composition at 60% *B* and 10% limited solubility at both ends. (*b*) Identify the phase fields. (*c*) Indicate the composition ranges for hypoeutectic and hypereutectic alloys.

6A-5 Draw a hypothetical phase diagram incorporating both (*a*) intermetallic compound and (*b*) intermediate phase.

6A-6 Draw a diagram showing strength as a function of grain size. Draw lines indicating the expected trends for (a) cold deformation and (b) fast and (c) slow hot deformation.

6A-7 (a) Draw the *A*-rich end of a eutectic phase diagram with limited solid solubility of *B*.

(b) Choose a composition resulting in a solid solution. (c) Draw a line showing the change in solidus due to nonequilibrium solidification. (d) Make a sketch of the expected microstructure.

6A-8 (*a*) Draw a part of a phase diagram suitable for explaining precipitation hardening. State (*b*) the processing steps needed for the development of maximum strength, (*c*) the structure in each step, and (*d*) what type of precipitate will give maximum strength.

6A-9 Define isotropy.

6A-10 (*a*) Draw the portion of the Fe–Fe_3C phase diagram that includes all steels (ignore the high-temperature peritectic transformation). (*b*) Give the approximate composition and temperature for the eutectoid transformation. (*c*) Identify phase fields and the associated crystal structure, giving also the customary name of each phase.

PROBLEMS 6B

6B-1 The alloy of Example 6-9 is to be used as the material of a tray for a high-temperature furnace. (*a*) Would you recommend large or small grain size? (*b*) Why?

6B-2   Anodized Al–Mg alloy sheet is often used as decorative trim. Concentration differences due to microsegregation impair appearance. (*a*) What remedy would you suggest? (*b*) What precautions should be observed in (*a*)?

6B-3   Under otherwise identical conditions, would a melt heated to $1.2T_L$ give a finer-grained casting than one heated to $1.1T_L$? Give two reasons for your answer.

6B-4   An Al-4Cu alloy part fails by tensile fracture in service. Fracture is ductile and investigation shows a structure with coarse second-phase particles. (*a*) State the likely cause of failure. (*b*) Suggest to the manufacturer a remedy to prevent future failures. (*c*) State the processing steps to be taken.

6B-5   An AISI 1045 steel is to be cold-formed. (*a*) What is the C content? (*b*) What structure is to be expected at room temperature? (*c*) What form of the structure would give greatest ductility?

6B-6   A steel component is examined and has fine lamellar colonies in a featureless matrix. (*a*) Suggest what the colonies and the matrix may be. (*b*) What is the likely production sequence that led to this structure?

6B-7   A steel billet, heated to 900°C, is rolled and the hardened steel rolls immediately develop severe surface cracking where contact with the billet was made. (*a*) What is the likely cause and (*b*) what remedy would you suggest?

6B-8   Steel shafting used with linear roller bearings must have a tough core and very hard surface. Name three methods by which such bars can be produced.

6B-9   A cast Al alloy has insufficient strength. Suggest two ways of increasing its strength without changing its composition.

PROBLEMS 6C

6C-1   Apply a material balance to Example 6-2 to show that 28.1% Cu is in the alloy.

6C-2   Would you expect significant coring in al-loys of (*a*) Cu-10Ni; (*b*) Cu-30Zn; (*c*) Al-5Mg? Support your judgment with quantitative statements.

6C-3   State the temperatures and sequence of precipitation hardening treatment in an Al-4Cu alloy.

6C-4   (*a*) Plot the property data given in Example 6-15 as a function of tempering temperature. Connect the points with a continuous curve; note the more rapid decline of strength properties at the highest temperature. (*b*) Check whether hardness is indeed three times TS.

6C-5   By taking σ_0 and k_y in Eq. (6-8) arbitrarily equal to unity, calculate σ for d ranging from 10^{-2} to 10^2.

6C-6   Al-4Cu alloy castings are solution-treated in a continuous, belt-type furnace at 545°C for 15 min. The furnace has three heating zones, each capable of holding temperature within ±10°C. Lately many castings have suffered severe distortion, sagging in the furnace. Suggest a reason or reasons for the problem.

6C-7   (*a*) AISI 1010 and 1045 steel bars are heated to 900°C; what phase of what crystal structure is present? (*b*) The bars are now quenched in water. One of the steels becomes hard, the other remains soft. Which one and why? (In your answer refer to the metallographic structure of each steel.)

6C-8   On the basis of the diagram drawn in Prob. 6A-6, state what grain size is desirable for (*a*) an aluminum alloy operating near room temperature and (*b*) a superalloy turbine blade operating at temperatures near the melting point.

6C-9   Handbooks state that 5056 Al alloy (5% Mg) is not heat-treatable. Inspect the equilibrium diagram. On this basis, does the statement mean that (*a*) no heat treatment can be applied at all or (*b*) no precipitation hardening is possible. (*c*) Explain your answers with reference to events in (*a*) and (*b*).

6C-10 Plot the YS and TS from Example 6-13 as a function of grain size and determine whether they follow the Hall-Petch relation [Eq. (6-7)]. (*Hint:* If plotted against $1/\sqrt{d}$, the slope equals k_y.)

6C-11 Continuing Problem 6A-4, choose a composition with 20% B. Calculate the proportion of α solid solution in the alloy.

FURTHER READING

ASM Handbook, vol. 3: *Alloy Phase Diagrams*, 1992; vol. 4, *Heat Treating*, 1991; vol. 9, *Metallography and Microstructures*, 1985; ASM International.

Alexiades, V.: *Mathematical Modeling of Melting and Freezing Processes*, Hemisphere, Washington, 1993.

Chalmers, B.: *Principles of Solidification*, Wiley, 1964.

Flemings, M. C.: *Solidification Processing*, McGraw-Hill, 1974.

Karlsson, L. (ed.): *Modeling in Welding, Hot Powder Forming and Casting*, ASM International, 1997.

Kurz, W., and D.J. Fisher: *Fundamentals of Solidification*, Trans Tech Publications, Switzerland, 1994.

Poirier, D.R.: *Heat Transfer Fundamentals for Metal Casting*, The Minerals, Metals, and Materials Society, 1992.

Poirier, D. R., and G.H. Geiger: *Transport Phenomena in Materials Processing*, The Minerals, Metals, and Materials Society, 1994.

Szekely, J.: *Fluid Flow Phenomena in Metals Processing*, Academic Press, 1979.

Totten, G.E., and M.A.H. Howes: *Steel Heat Treatment Handbook*, Dekker, 1997.

Precision integral turbine components are made by vacuum casting of superalloys. (*Courtesy Howmet Corporation, Greenwich, Connecticut.*)

chapter

7

Metal Casting

We now look at one application of solidification processes, namely, the casting of metals. We will explore:

The solidification of alloys in molds

Factors affecting the fluidity and quality of melts

Major classes of casting alloys

Casting of simple shapes for further processing by plastic deformation

Processes for casting shapes into expendable and reusable molds

Methods of improving the properties of castings

Choosing the casting process and designing parts for ease of casting

The principles of solidification are put to widest use in casting processes. To create a casting, the molten metal is poured into colder molds which extract heat. Pouring involves fluid flow; its interaction with solidification determines the suitability of an alloy for casting. Therefore, this interaction will be discussed before reviewing casting alloys and exploring casting processes.

7-1 STRUCTURE AND PROPERTIES OF CASTINGS

It is clear from the discussion in Chap. 6 that the properties of a solidified alloy are dependent not just on composition but also on grain size and the shape and distribution of phases. These factors may be controlled and modified in the course of solidification.

7-1-1   Solidification of Melts

When a melt is poured into a colder mold, metal in contact with the mold solidifies in the form of roughly equiaxed (of roughly equal dimensions in all directions) fine grains, because cooling rates are high (*chill zone*) and the mold wall induces heterogeneous nucleation. The latent heat of fusion released during solidification slows down the rate of solidification and the course of further solidification depends on the type of alloy being cast.

Pure Metals   Solidification proceeds by the growth of a few favorably oriented nuclei, in the direction of heat extraction. This leads to the often observed *columnar structure* (Fig. 7–1a) throughout the bulk of the casting. Because of the preferred growth of these large grains, the casting will have very anisotropic properties. (In a larger casting the central zone cools very slowly and heat extraction is almost omnidirectional; if solidification proceeds by heterogeneous nucleation, this results in an equiaxed structure, of grains much coarser than on the surface.)

Since most metals shrink on solidification (Fig. 6–1c), the liquid meniscus gradually drops and, if there is no supply of fresh liquid, a *shrinkage cavity* remains. A cavity of the geometry shown in Fig. 7–1a is called a *pipe* and always forms when a pure metal solidifies.

Eutectics   Eutectics, like pure metals, solidify at a constant (invariant) temperature (Fig. 6–5), and the solidification front is more or less planar (Fig. 7–1a).

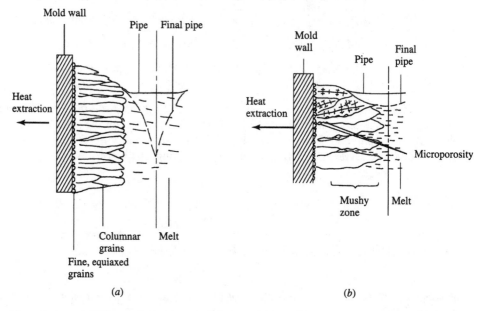

(a)   *(b)*

Figure 7–1   Solidification proceeds (*a*) with the growth of columnar grains in pure metals but (*b*) with the growth of dendrites in solid solutions.

Within each grain, there are several groups, *eutectic cells* or *colonies*. Properties of the casting can be influenced by various means:

1. Rapid cooling reduces cell size and, in lamellar eutectics, also interlamellar spacing and thus increases the strength of the casting.

2. Nucleating agents promote the formation of fine equiaxed eutectic grains of superior mechanical properties.

3. The lamellar structure is only one of the possible forms of eutectics. In certain instances the "natural" morphology of the eutectic may be changed by *modification*, with marked changes in properties. For example, the platelets of a lamellar eutectic may be changed into spheres (spheroidal structure) or rod-like particles. Such structures have distinctly different properties; typically, a spheroidal eutectic has higher ductility than a lamellar one (see Fig. 6–15).

Solid Solutions   Solid solutions solidify over the freezing range $T_L - T_S$ (Fig. 6–4), and this has significant effects on the structure. Crystals again grow in the direction of heat extraction but in the form of dendrites (Fig. 7–1b). When the melt finally solidifies, each grain contains one or more complete dendrite (*cellular dendritic structure*). Dendrite arms are initially very weak and can be easily broken by thermal and/or mechanical agitation to give, at low superheat, many nuclei and thus a fine grain size.

The intricate network of dendrite arms makes free movement of the remaining liquid difficult, and spaces formed between arms may be starved of the fluid necessary to make up for solidification shrinkage. Consequently, *microporosity*—characterized by the presence of holes with ragged edges (Fig. 7–2)—is typical of solid solutions. Such holes represent inclusions of zero strength and, because of the notch effect, are harmful for strength and ductility. In alloys (Fig. 7–3) the total shrinkage is similar to that of the constituent metals, but the pipe is much smaller and a large proportion of total shrinkage is in a *distributed* form (Fig. 7–3b).

Other Systems   Properties and porosity in a binary alloy system may be predicted with fair accuracy from the phase diagram. For example, in the eutectic system of Fig. 7–3, microporosity increases from A to B until the solubility limit of the α solid solution is reached, declines toward the eutectic composition, to rise again to the β solid solution. Strength (characterized here by the yield strength) rises by solid-solution alloying and changes little with the appearance of the eutectic. Ductility may rise or fall with solid-solution alloying; the effect of the eutectic depends greatly on its morphology (Fig. 7–3c).

Occasionally an alloying element cannot be dissolved even in the liquid metal, instead, it exists as a separate liquid phase (*limited liquid solubility or total immiscibility*). A prime example is lead, which is practically insoluble in many metals. Its effect on properties depends on its distribution. Since it is soft, it can act as a useful lubricant if trapped in interdendritic spaces within grains or in a globular

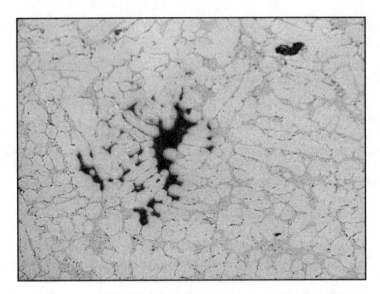

Figure 7–2   Microporosity in a 356.0 aluminum alloy due to a combination of interdendritic shrinkage and hydrogen gas evolution. (*Courtesy of Dr. J.E. Gruzleski, McGill University, Montreal, Quebec.*)

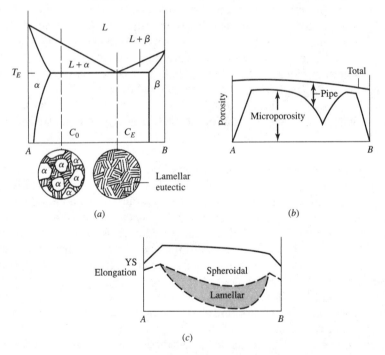

Figure 7–3   In a eutectic system (*a*), microporosity dominates when the structure consists primarily of a solid solution (*b*), and the mechanical properties reflect the effects of structure (*c*).

form on grain boundaries; it also improves machinability (Sec. 16-2). However, because of its low melting point, it causes hot shortness when it segregates on grain boundaries.

7-1-2   Macrosegregation

We already observed (Sec. 6-1-6) that lack of complete diffusion leads to microsegregation, i.e., compositional variations within a grain. The process of solidification can also result in compositional differences extending over long distances within a casting. There are basically three kinds of *macrosegregation*:

1. So-called *normal segregation* occurs when a more or less plane solidification front (as in Fig. 7–1a) drives the lower-melting constituent toward the center. A section taken through the solidified cross section will show a lower alloying-element concentration on the surface than in the center (Fig. 7–4a). If gases are liberated during solidification, they drive the richer fluid out from the solidifying zone and contribute to the segregation of alloying elements to the center.

2. *Inverse segregation* is typical of solid-solution alloys with a dendritic solidification pattern (Fig. 7–1b). Since dendrite arms form first and have a lower alloying-element concentration, the interdendritic spaces formed by solidification shrinkage must be filled in by a liquid of higher solute concentration. This liquid flows back in a direction opposite to the growth direction of dendrites; hence, the surface has a higher than average alloying element concentration (Fig. 7–4b).

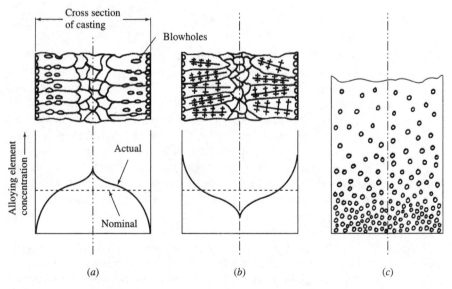

Figure 7–4   Solidification of almost pure metals leads to (a) "normal" macrosegregation, especially in the presence of gas evolution. (b) Dendritic solidification leads to "inverse" segregation. (c) High-density constituents that do not dissolve in the melt separate by gravity segregation.

3. *Gravity segregation* occurs when insoluble compounds, inclusions, or metals immiscible in the liquid have a density that is greatly different from the melt; therefore, they rise or sink (Fig. 7–4c). One of the attractions of manufacturing in space is that, in the absence of gravity, unusual alloys—consisting of metals of greatly different densities—may be solidified without segregation.

These forms of macrosegregation, if undesirable, must be prevented during solidification because diffusion distances are too large to equalize the composition by homogenization heat treatment.

7-2   CASTING PROPERTIES

Solidification characteristics combine with fluid properties to determine the suitability of various alloys for casting.

7-2-1   Viscosity

The pouring of the melt into a mold is essentially a problem in fluid flow, and as such, it is greatly affected by the resistance exerted by the fluid against flow. This resistance can be measured as a shear stress τ. If a fluid film of h thickness is sheared between two flat parallel plates, one of which moves at v velocity, the shear stress τ is the force per unit area acting on these plates (Fig. 7–5a)

$$\tau = \eta \frac{dv}{dh} = \eta \dot{\gamma} \tag{7-1}$$

where $\dot{\gamma}$ is the shear strain rate and η is the dynamic viscosity (in units of $N \cdot s/m^2$).

The laws governing the flow of substances are the subject of *rheology* (from the Greek *rheos* = current, flow). Many fluids (e.g., mineral oils used in machines) exhibit *Newtonian viscosity*, independent of $\dot{\gamma}$ (Fig. 7–5b, line A). Fluids in which solids are suspended shear readily at low strain rates, but the solid particles obstruct flow at high strain rates (*dilatant fluids*, line B). Substances in which the particles or molecules can orient themselves in the direction of flow shear readily at high shear strain rates (*pseudoplastic flow*, line C). When, at a given strain rate, viscosity declines with the time of exposure, the fluid is said to be *thixotropic*. An important group of materials begins to deform only after some minimum initial shear stress is applied, and then continues to shear in a viscous manner (*Bingham solids*, line D).

Above T_m (or T_L), most metals behave as Newtonian fluids whose viscosity is a function of free volume and, therefore, drops with superheat. While information is sketchy, one might generalize by saying that viscosity is a function of composition and of superheat as expressed on the homologous temperature scale [Eq. (6-5b)]. However, the nature of phases that are present is also important. For example, in eutectic systems one may find (Fig. 7–5c) that viscosity changes

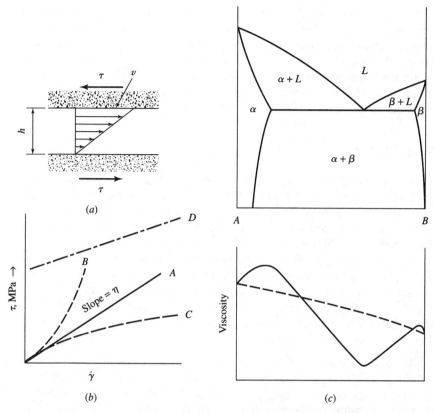

Figure 7–5   Shearing of a fluid requires (*a*) a shear stress which increases with increasing fluid viscosity. (*b*) Fluids may exhibit Newtonian (*A*), dilatant (*B*), pseudoplastic (*C*), or Bingham behavior (*D*). (*c*) In a eutectic system, viscosity may vary greatly.

linearly with alloy composition (broken line), but it could also show marked variations with phase boundaries (solid line: maximum viscosity at the limit of solid solubility, minimum at the eutectic composition).

Between T_L and T_S, the presence of the solid phase induces non-Newtonian effects. An *apparent viscosity*, which is a function of the quantity and structure of the solid phase, can be defined for a constant $\dot{\gamma}$. Equiaxed crystals hardly affect viscosity up to about 60% concentration by volume. Dendrites increase the apparent viscosity greatly, except when shear rates are large enough to break up the dendrites; then viscosity is low, similar to that found with equiaxed crystals.

7-2-2   Surface Effects

When the melt has to flow through small (typically, below 5 mm) channels, surface tension [Eq. (6-6)] becomes significant. A high surface tension makes it impossible to fill sharp corners.

On exposure to the atmosphere, the surface of many melts becomes rapidly coated with an oxide film, and the nature of this film greatly influences casting behavior. Thus, the extremely dense and tenacious oxide of aluminum (Al_2O_3) makes it flow as if it were inside a rather tough bag, and alloying elements that modify the oxide greatly affect the casting behavior of aluminum alloys. Aluminum as an alloying element in other metals usually oxidizes preferentially; an aluminum oxide skin forms which has the effect of increasing surface tension.

7-2-3 Fluidity

When a mold is filled, heat is extracted and solidification begins while flow is taking place. Therefore, *mold filling* depends on many factors, the exact effects of which may not be known. To characterize materials under complex conditions, it is customary to develop *technological tests* that allow a quantitative comparison of materials, *but only if the test conditions are carefully specified.*

The mold-filling ability of a metal is described as *fluidity*. It is a system property that is a function not only of the metal but also of the mold. Typically, a long spiral-shaped or thin plate-like cavity is made in the mold material of interest (Fig. 7–6) and fluidity is quoted as a *fluidity index* (length of the spiral or plate). Alternatively, the length of fill under vacuum (Fig. 7–6c) is quoted. Fluidity is affected by a number of factors:

1. Fluidity increases with increasing superheat because this lowers viscosity and delays solidification. However, excessive superheat may lead to undesirably large grain size, and may also be impractical because the melting furnace may not be able to withstand such high temperatures.

2. Fluidity increases with increasing mold temperature, because solidifi-

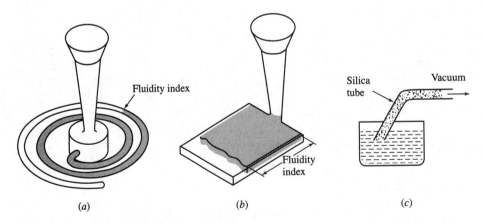

Figure 7–6   Fluidity is a technological property which is determined by pouring into a (a) spiral or (b) plate mold or (c) by pulling vacuum.

cation is slowed down. This benefit is, however, gained at the expense of a lower cooling rate which leads to coarser grain and may limit productivity (Fig. 7–7).

3. The type of solidification has a great effect. A solidification mechanism that allows orderly freezing, such as is found in pure metals and eutectics, is helpful (Fig. 7–8*a*). However, pure metals with their higher melting points tend to have lower fluidity than eutectics. Dendrite arms growing into the path of liquid supply slow down the flow and can cut off the supply of liquid entirely (Fig. 7–8*b*); therefore, the fluidity of alloys with a long freezing range is generally low. However, if the fluid is forced to flow (by a large gravity head or by externally applied pressure), dendrites are ripped off and broken up, and fluidity increases greatly. Flow stops when broken crystals freeze to form a "plug" at the meniscus (Fig. 7–8*c*).

4. Surface tension and the presence of oxide films have an effect.

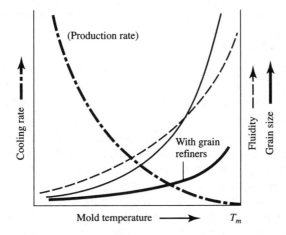

Figure 7–7   Mold temperature is a powerful factor in determining production rates, attainable shape complexity, and mechanical properties.

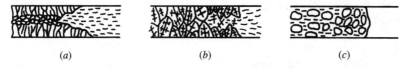

Figure 7–8   When a melt solidifies in a channel, communication with hot metal is (*a*) kept open longer with the frontal solidification of pure metals and eutectics than with (*b*) the dendritic solidification of solid solutions. (*c*) Debris of crystals chokes off the flow.

5. Mold material and mold dressing affect fluidity by influencing heat extraction and wetting of the mold surface.

In a rather loose sense, one also speaks of the *castability* of a metal. This term incorporates, in addition to the technological concept of fluidity, aspects that define the ease of producing a casting under average foundry conditions. Thus, an alloy is regarded as highly castable when it not only has high fluidity but is also relatively insensitive to accidental changes in process conditions, is more tolerant to the design of the fluid supply system, is less sensitive to wall thickness variations, and, in general, will produce castings of acceptable quality with less skill.

7-3   CASTING ALLOYS

With the exception of metals and alloys that are produced directly by powder metallurgy or electrolytic techniques, all metals and alloys must first go through the melting and casting stage (Fig. 5–2). It is, however, usual to distinguish between two broad classes:

1. *Wrought alloys* possess sufficient ductility to permit hot and/or cold plastic deformation (these will be discussed in Chaps. 9 and 10). They represent some 85% of all alloys produced, and are cast into simple shapes suitable for further working (ingot casting).

2. *Casting alloys* are selected for their good castability (such as the eutectics) or are materials of a structure that cannot tolerate any deformation (for example, alloys with unfavorably distributed hard or nonductile phases or with high proportions of intermetallic compounds and other hard constituents). These are cast directly into the final shape (shape casting). There are, of course, overlaps between the two groups, and the same material—because of its attractive service properties—may be produced in both wrought and cast forms.

Total quantities cast into shape have gradually declined in the industrially developed nations, but their value has gone up considerably because of the increased complexity and highly improved quality of products. Ferrous castings still represent the largest tonnages (Table 7–1) but nonferrous castings contribute much of the value. It should be noted that primary and secondary scrap constitutes a large part of the total quantity. Some properties of the most popular alloys are given in Table 7–2; general properties are discussed in the following.

7-3-1   Ferrous Materials

In its most familiar form, the iron–carbon diagram actually shows the equilibrium phases in the iron–carbide (Fe–Fe_3C) system (solid lines in Fig. 6–11). The compound Fe_3C is, however, metastable, and under certain conditions can revert

Table 7–1   Shipments of Castings
(United States)*

	Thousand Mg[†]	
Type	**1972**	**1997**
Gray iron	14 000	6 150
Malleable iron	860	200
Ductile iron	1 830	4 070
Steel	1 450	1 350
Aluminum alloys	850	1 700
Zinc alloys	460	360
Copper and brass	345	313
Magnesium alloys	21	40

*1972 data compiled from *Metal Statistics
1974*, American Metal Market, Fairchild Publications Inc., New York, 1974. 1997 data from
American Foundrymen's Society, Des Plaines,
Illinois.
[†]1Mg = 1000 kg = metric tonne = 2200 lb.

to the more stable carbon (*graphite*) form. Alternatively, melt composition and solidification conditions may be controlled to allow the carbon to separate in the form of graphite during solidification, at a somewhat higher eutectic temperature (Fig. 6–11, broken lines). Thus, several families of materials can be derived from the iron–carbon system.

Cast Steels   In carbon steels (up to 1.7% C) the carbon is always in the form of Fe_3C. Their high melting point and, above 0.15% C, the long freezing range make steels less suitable for casting purposes. However, they are ductile and have a high strength and fatigue resistance which, as discussed in Sec. 6-4-3, can be further increased by heat treatment and alloying. Since sulfide inclusions impair properties (Fig. 6–16), S content is reduced to 0.006–0.010% by adding elements such as Ca. Steel is deoxidized with Al, which forms Al_2O_3 inclusions. Most steels can be readily welded to build up components of unusually large size or complexity. Hence they have important applications, primarily for railroad equipment (wheels, truck frames, couplers); construction and mining equipment (track shoes, axle housings, hoist drums, buckets and bucket teeth, grinding balls); metalworking machinery (rolling mill, press, and hammer housings); and oil-field and chemical plant components (valve bodies, impellers, drill-rig parts).

Because of poor fluidity, wall thickness must be fairly large and castings tend to be of larger size (almost half of all castings are in the range of 200–500 kg mass). Stainless steels are indispensable in the food and chemical industries, but their high melting point and long freezing range present substantial technological challenges.

Table 7-2   Properties of Selected Casting Alloys*'

Name	ASTM No.	Typical Composition, wt%	Preferred Casting Method	Liquidus (Solidus), °C	Shrinkage Allowance,‡ %	TS MPa	YS MPa	el., 50 mm %	Hardness,§ HB
Ferrous:									
Cast steel	60-30	≤ 0.25C	Expendable mold		1.5–2	420	210	24	180
	175-145		Expendable mold		1.5–2	1200	1000	6	360
Gray iron	20	3.5C-2.4Si-0.4P-0.1S	Expendable mold	1180	1	140	(570)	<1	160
	60	2.7C-2.0Si-0.1P-0.1S-0.8Mn	Expendable mold	1290	1	420	(1300)	<1	300
Malleable iron	A47	2.5C-1.4Si-0.05P-0.1S-0.4Mn	Expendable mold	1140	1	350	220	10	150
Ductile iron	60-40-18	3.5C-2.4Si-0.1P-0.03S-0.8Mn	Expendable mold		0.8–1	420	280	18	160
Stainless steel	CF8	0.08C-19Cr-9Ni	Expendable mold		2.5	500	240	45	
Cu-based:									
Tin bronze	C90500	10Sn-2Zn	All	999 (854)	1.6	320	150	30	80
Leaded red	C83600	5Sn-5Pb-5Zn	All	1010 (854)	0.8–1.8	240	110	32	62
Bearing	C93700	10Sn-10Pb	All	926 (760)	1–2	220	110	20	60
Leaded yellow	C85400	1Sn-3Pb-29Zn	All	940 (925)	0.8–1.5	230	80	37	55
Nonferrous:									
Al-based	208.0	3Si-4Cu	Sand	627 (521)	1.5	150	100	2.5	55
	332.0	9.5Si-3Cu-1Mg	Permanent mold	582 (520)	1	250	195	1	105
	380.0	8Si-3.5Cu	Die	590 (520)	0.6	330	170	3	
	413.0	12Si	Die	577		295	145	2.5	
Mg-based	AZ91D	9Al-0.7Zn-0.2Mn	All	596 (468)	1.5 (die 0.6)	200	135	3	65
	EZ33A	2.7Zn-0.5Zr-3 rare earths	Sand and permanent mold	643 (543)	1.2	160	110	3	50
Ti-based	Ti-6Al-4V	6Al-4V	Investment	1330 (1157)		1000	900	8	
Ni-based	Inconel 718	19Cr-3Mo-5Nb-1Ti-0.5Al-18Fe	Investment			1200	1150	16	
Zn-based	AC41A	4Al-1Cu-0.04Mg	Die	386 (381)	0.3–0.6	330		10	82
	ZA12	11Al-1Cu-0.025Mg	Die	432 (377)	1.3	430	320	2	100
Pb-based	Babbitt	16Sb-1Sn-1As	Bearings	272 (248)	2.6				20
Sn-based	B 560	7.5Sb-1Cu	Permanent mold	295 (244)	2				24

*Data compiled from *ASM Handbook*, 10th ed., vols. 1 and 2, 1990. ASM International, Materials Park, Ohio.

†Minimum properties in the as-cast condition, except malleable and nodular cast iron (annealed) and 332.0 and EZ33A (precipitation-hardened).

‡Patternmakers' allowance.

§Load: 3000 kg for ferrous, 500 kg for nonferrous materials.

¶Compressive strength.

NOTE: To convert MPa into 1000 psi, divide by 7.

White Cast Irons   As discussed in Sec. 6-2-1, cast irons contain in excess of 2% C. The form in which the carbon solidifies depends on cooling rates as well as composition. Control is exerted primarily by the total C and Si (and also P) content, and their combined effect can be expressed by the *carbon equivalent* (C.E.):

$$\text{C.E. (\%)} = \text{C\%} + \frac{\text{Si\%} + \text{P\%}}{3} \qquad \text{(7-2)}$$

At C.E. < 3 and fast cooling (small section thickness; say, below 6 mm in sand casting) the entire cross section will solidify with a white microstructure, i.e., with all carbon in the form of Fe_3C; even lower C.E. must be maintained for thicker walls. Primary cementite in the eutectic makes these *white iron* castings hard and brittle, hence their use is limited to wear-resistant parts such as grinding balls, liners for ore-crushing mills, and some agricultural machinery parts. They are virtually unmachinable except by grinding.

Malleable Iron   The Fe_3C of white iron may be converted to the stable graphite by an annealing treatment in which cementite decomposes into graphite (*temper carbon*) and ferrite. The carbon appears in the form of irregular aggregates (Fig. 7–9*a*) embedded in a ferrite matrix, thus strength and ductility are similar to steel but, with a lower melting point and higher fluidity, castability is better. Since the casting must be white to begin with, only thin-walled products (max. 40 mm) can be cast. Replacement by nodular iron has shrunk the market to small electrical components and general fittings. The presence of graphite imparts good machinability.

Gray Iron   At relatively high C.E. and slower cooling rates, there is time for the iron to solidify in the stable form, with the carbon separating in the form of *graphite flakes* (Fig. 7–9*b*), making the fracture surface dull gray, hence the name *gray iron*. The formation of graphite counterbalances much of the solidification shrinkage, thus ensuring soundness and relative freedom from solidification porosity.

The graphite flakes reduce ductility to practically nil, and Young's modulus is also lower than that of pure iron (it ranges from 70–150 GPa). The size, shape, and distribution of flakes can be controlled to give low to medium strength (Table 7–2); properties are always better in compression because the graphite flakes act as incipient crack sites in tension. Too-fast cooling results in a *mottled structure* in which primary carbide is also present; rapidly cooled surface layers may be entirely white, creating problems in machining a material of otherwise excellent machinability.

Its low cost makes gray iron the preferred choice in all fields where ductility and high strength are not needed (weights; frames; motor, gear, and pump housings; pipe fittings). Its high damping capacity is an advantage for machine tool bases. High fluidity and good tribological properties have made it the traditional engine block material. Some markets have been lost to aluminum and even plastics.

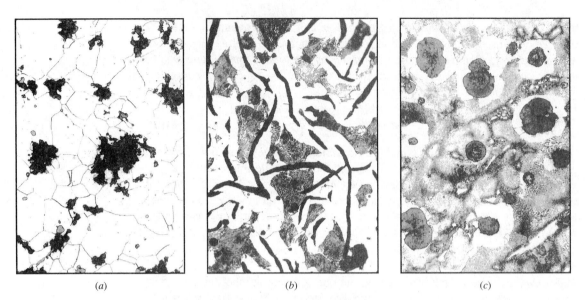

(a)   *(b)*   *(c)*

Figure 7–9   Carbon is present in different forms of graphite in (*a*) malleable iron (ferritic), (*b*) gray iron (50% ferrite, 50% pearlite), and (*c*) nodular iron (50% ferrite, 50% pearlite). (*From Metals Handbook Desk Edition, ASM International, 1985, p. 27.26. With permission.*)

Nodular Iron   A ductile cast iron is obtained when the graphite is brought into a less detrimental, globular form upon solidification. This is achieved by adding to the melt, just prior to casting or during pouring, a small amount of magnesium or cerium (introduced in the form of a ferroalloy) which, through a mechanism that is only partially understood, causes the graphite to separate in well-defined, roughly spherical (or elongated, *nodular*) particles, distributed in the α-iron or pearlite matrix (Fig. 7–9*c*). Since Mg promotes the formation of cementite, the iron is then inoculated with silicon. Sulfur interferes with the development of spheroids and is removed by desulfurization, and, like steels, the melt is deoxidized. *Nodular* (or *ductile* or *spheroidal*) *cast iron* combines the good castability and machinability of gray iron with some of the ductility of steel. Castings are usually heat-treated, and austempered iron can sometimes replace steel. *Compacted graphite* (*vermicular graphite*) *cast iron* is produced by inoculation with less Mg and has a structure and properties intermediate between nodular and gray cast irons.

Nodular iron has an extremely wide range of applicability, from automotive crankshafts and hypoid gears to pump housings, rolling mill rolls, and, in general, for parts subjected to impact loading or requiring a high elastic modulus ($E = 150$–175 GPa). Its use is still growing.

The matrix of cast irons may be produced with varying levels of carbon content; therefore, cast irons may be heat treated just as steels are. Also, cast irons can be alloyed for enhanced mechanical or chemical (corrosion-resistant) properties. Grain refinement is possible with calcium silicide and other nucleating agents.

7-3-2   Nonferrous Materials

The most important alloy groups are discussed here in order of increasing melting point, which also indicates increasing cost and difficulty of melting and superheating them to the appropriate temperature. Properties of selected alloys are given in Table 7–2.

Tin-Based Alloys   Of the widely used metals, tin has the lowest melting point (232°C). It is highly corrosion resistant and nontoxic, but its low strength precludes its use as a construction material.

An important application is for bearings where its low shear strength and low adhesion to other metals assures low friction even when lubrication fails. It must be backed by a stronger material or, if the bearing layer is to be thicker, it must be strengthened by creating a duplex structure in which a hard compound is dispersed in the soft tin matrix. This is achieved by adding Sb to form the intermetallic compound SbSn, in the shape of small, hard, cubic crystals (*cuboids*). These tend to rise to the surface of the melt; some copper is added to form a copper–tin intermetallic which solidifies as a spatial network of needles and traps the cuboids.

Old pewter contained lead, but modern pewter is free of lead (Table 7–2), and is suitable for decorative items and tankards.

Lead-Based Alloys   Lead too has a low melting point (327°C) and good corrosion resistance, but it is toxic and its use is limited to applications where human contact is avoided. Large sand or permanent-mold castings are used as x-ray and γ-ray shields.

The low strength of lead and its low solubility in other metals qualifies it as a bearing material, although of somewhat lower quality than tin. Strengthening is again obtained by alloying, usually with tin and antimony, so that the SbSn cuboids are dispersed in a matrix of ternary Sn–Pb–Sb eutectic. These ternary alloys are not only hard but also possess a high fluidity imparted by the presence of tin; therefore, they were also used as type metal that gave a clear, clean typeface.

Antimonial or calcium lead is extensively used for cast lead–acid battery grids. Great economy is ensured by recycling most of the used batteries (the majority of all lead used in the metallic form is recycled).

Zinc-Based Alloys   Zinc is the only low-melting (419°C) metal widely used as a structural casting material. Its major weakness is low creep strength. Also, its corrosion resistance is low in the presence of contaminants such as Cd, Sn, and Pb, which lead to intergranular corrosion. However, its high fluidity and low melting point make it eminently suitable for casting into steel dies. Strengthening is done by solid-solution alloying with approximately 4% Al and 1–2% Cu (the eutectic is at 5% Al in the Zn–Al system). By the use of 99.99% pure zinc and careful control of contaminants, good corrosion resistance is secured, making these alloys highly competitive (even with plastics) for thin-walled parts of intricate shape, such as instrument housings and automotive components and trim; for the last application, the excellent response to chromium plating is an advantage. Alloys with 11% Al offer high strength combined with good fluidity.

Magnesium-Based Alloys   The melting point of magnesium is substantially higher (649°C) but still low enough to allow casting by all techniques. Its low density (1.74 g/cm^3) and reasonable strength, coupled with corrosion resistance (except in marine environments) make it very attractive for structural applications including air-cooled automotive engine blocks, transmission housings, and wheels. The major barrier is cost. Casting alloys are solid-solution strengthened with up to 10% Al (the eutectic composition with 32% Al is too brittle to be practical), and some precipitation hardening may be obtained by adding Mn, Zr, or Zn. Fluidity is quite adequate because the oxide is not dense and does not hinder flow. The Mg–Zn–Zr–rare earth alloys are suitable for service up to 260°C. Grains are refined by adding Zr.

Aluminum-Based Alloys   Melting at an only slightly higher temperature (660°C), almost as light and considerably cheaper than magnesium, aluminum and its alloys represent (beside nodular iron) the fastest-growing segment of the casting industry. The corrosion resistance of aluminum is excellent (except to alkali) and its strength is readily improved through solid-solution and precipitation-hardening mechanisms. Increased recycling of secondary scrap (Sec. 5-4-1) has reduced the total energy consumed in making aluminum parts.

Pure aluminum is used for domestic utensils. High-conductivity, 99.6% pure aluminum is pressure die cast into squirrel-cage rotors for fractional horsepower motors and is used, as a permanent mold casting, for larger motors too.

The oxide film on the melt is dense and tough and, as already mentioned, reduces fluidity. The ease of casting is greatly affected by the influence of alloying elements on this oxide film. Silicon is most beneficial, making silicon alloys the most castable aluminum alloys. The eutectic composition (around 12% Si, with a melting point of 577°C) is, of course, the most favorable. Its properties are greatly improved by refining (*modifying*) the eutectic structure through rapid cooling or, more frequently, by the addition of a small quantity of sodium (or, more recently, strontium) to the melt just prior to pouring, whereupon the eutectic silicon is modified and separates in the form of fine rods or fibers instead of coarse flakes. Hypereutectic alloys contain the very hard and brittle silicon in a proeutectic form and are thus extremely wear-resistant. The structure is refined by the addition of 0.01% P to form aluminum-phosphide nuclei. The hard and brittle silicon of the eutectic limits the ductility of the alloy. Therefore, in the most popular casting alloys some Si is replaced with Cu which increases strength by solid-solution strengthening and also opens the door to precipitation hardening (Sec. 6-4-2). Castability is still high, especially in die casting where the dendrites are broken by the applied pressure. Magnesium is a useful solid-solution strengthening element but creates problems typical of a long freezing range.

Hydrogen would lead to porosity (Examples 6-11 and 6-12) and scavenging with argon or chlorine just prior to pouring is practiced. Grain refinement is achieved by adding Ti–B alloy nucleating agents.

The use of aluminum alloy castings is very wide ranging and is always worth considering, especially when high strength-to-weight ratio and corrosion resistance are desired. Typical applications include automotive transmission cases, pistons, engine blocks, and some aircraft components.

Copper-Based Alloys   The melting point of copper ($1083°C$) is too high for steel dies (unless protected by heavy coatings), but other casting methods are practiced because the metal has attractive color, good corrosion resistance, and high electrical conductivity.

A majority of castings are made of alloys that combine good fluidity with reasonably high strength. Because copper alloys have been around for such a long time, many of them were given proprietary names and sometimes misleading designations (some brasses are commonly called bronzes). Extensive past use of copper alloys now allows a substantial part (up to 35%) of the total consumption to be covered from secondary scrap. The technologically important features of alloys can be deduced from their composition and phase diagrams.

There are few copper alloy systems with useful eutectics; therefore, most casting alloys are hardened by adding solid-solution elements, up to the limit of solubility for maximum strength without undue embrittlement by excessive intermetallic particle content. Whenever the solidification range is wide, fluidity suffers.

Tin bronzes (Cu–Sn alloys) have a very long freezing range, and fluidity is increased by added phosphorous which forms a low-melting ternary eutectic (*phosphor bronzes*) and also deoxidizes the melt. Hydrogen porosity is prevented by purging with N_2. The addition of zinc with its low vapor pressure also increases fluidity. An 88Cu-10Sn-2Zn alloy still has high strength, making it suitable for gears, bearings, and pump parts. Lead is often added, primarily to improve machinability, but it also benefits fluidity. The 85Cu-5Sn-5Pb-5Zn alloy is the most castable of all and is extensively used for fixtures, pump bodies, and general castings. The high lead content in the 80Cu-10Sn-10Pb alloy reduces strength but makes the alloy suitable for bearing applications.

Aluminum bronzes have a short freezing range but the oxide reduces fluidity. They do give, however, high strength, especially in the heat-treated condition (with iron and other precipitation-hardening additions). Their excellent corrosion resistance makes them favorites for marine applications, worm gears, valves, and nonsparking tools.

Brasses are Cu–Zn alloys. They have a short freezing range (Fig. 6–7) and are especially suitable for fittings, plumbing fixtures, and other smaller parts. Many casting alloys also contain lead, which improves fluidity.

Beryllium-Based Alloys   Only slightly more dense (1.85 g/mm^3) than Mg but of a much higher melting point ($1277°C$), Be offers very high stiffness ($E = 280\,\text{GPa}$). Newly developed alloys (30Al-3Ag with Co, Ge, Si additions) allow investment casting of thin-walled parts for aerospace applications.

Nickel- and Cobalt-Based Alloys   The high melting points of nickel (1435°C) and cobalt (1495°C) and their corrosion resistance make them eminently suitable for many critical applications. Their strength and, particularly, hot strength can be greatly increased with solid-solution and precipitation-hardening alloying elements (Example 6-9). Some of these superalloys have such a high second-phase content that they are not deformable, and these cast superalloys can outperform other materials in high-temperature applications, particularly as gas turbine (jet engine) parts.

High-Temperature Materials   Some of the higher-melting alloys are used in only very specific cases to produce castings.

Titanium (melting point 1670°C) can be alloyed to give high elevated-temperature strength combined with low weight (high strength-to-weight ratio) and corrosion resistance. These alloys are thus used in chemical plants and in subsonic and, especially, in supersonic aircraft construction. The great affinity of Ti to oxygen, the high melting point, and the low fluidity place great demands on the skills of the foundry technologist. Properties of castings can be greatly improved by HIPing (Sec. 7-6-2).

Refractory metals are, by definition, resistant to heat and are difficult to melt. The most important ones are: molybdenum, Mo (melting point: 2610°C); niobium, Nb (also called columbium, Cb, 2470°C); and tungsten, W (3410°C). They oxidize extremely rapidly; therefore, special (vacuum arc or electron beam) melting and casting techniques are needed. They are indispensable in some applications, such as rocket motor nozzles.

7-4   MELTING AND POURING

Thus far we established the basic principles bearing on casting; we may now proceed to discuss actual techniques.

7-4-1   Melting

The first step is to prepare a melt of the correct composition. Figure 7–10a shows, schematically, the major elements of the system:

1. A *charge* is made up to yield, upon melting, the alloy of specified composition. It is seldom necessary, practicable, or even desirable to make up a charge entirely of metals obtained from ores (primary or virgin metals). Alloying elements of much higher melting point than the base metal would be slow in dissolving and would require excessive overheating, therefore, they are added in the form of a *master metal* (*temper alloy, hardener*), which contains a higher concentration of the alloying element in the base metal. For economy of operation, it is most important that as much *scrap* as possible should be added. The

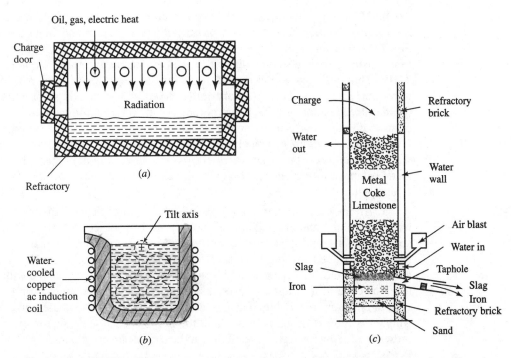

Figure 7–10 Some examples of melting systems: (*a*) Heat is supplied externally, through radiation from the furnace walls, in a reverbatory furnace. (*b*) Currents induced in the charge of an induction furnace assure thorough mixing. (*c*) Cast irons are usually melted in cupolas which often have water-cooled walls.

aim is to produce a melt of the composition specified by relevant standards while holding contaminants below the allowed maximum levels, and accomplish all this at the lowest possible cost. Computer programs are available to facilitate this task. Smaller plants often find it more economical to use prealloyed material cast into ingots purchased from specialized companies (*ingoted melting practice*).

2. The charge is loaded into a *furnace* which contains the melt and provides a source of heat. While the physical arrangement may vary substantially, there are some common elements:

a. The melt is contained with a material of substantially higher melting point than the metal while also minimizing contamination of the melt by inclusions or dissolved elements. The material is chosen so that its oxide is not reduced by the metal; otherwise it would be quickly attacked (e.g., Al is never melted in steel). The material may range from iron (for lead) to refractory-lined furnace structures (a *refractory* is a ceramic of high temperature resistance). The charge may also be contained in a graphite or refractory *crucible* (pot) placed into the furnace. Alternatively, the melt may be contained by maintaining a chilled outer zone that forms its own container.

b. *Heat* is provided externally (for example, radiation from the walls of a *reverbatory furnace* heated by gas or oil burners or electric heating elements, Fig. 7–10*a*) or internally (as in an electric *induction furnace*, Fig. 7–10*b*). Cast iron is usually melted semicontinuously in a vertical shaft furnace (*cupola*, Fig. 7–10*c*); lining of the cupola with a refractory is being abandoned in favor of water-cooled steel jackets. The charge is mixed with coke and some minerals (primarily limestone, $CaCO_3$), and hot air is blown through the column. Coke burns to give heat and is also a source of carbon for the cast iron. The liquid metal is tapped at the bottom, separately from the slag which is formed by the limestone with nonmetallic contaminants and metal oxides. In the *duplex process*, the liquid metal is tapped into an electric holding furnace where alloying and superheating is also practiced.

3. An inevitable factor is the presence of an atmosphere. This may be air which, with its humidity and various pollutants, is a source of N, H, and O gas absorption; it could be a *protective atmosphere* (such as argon gas); or even *vacuum*, produced at some expense. Combustion products including H_2O and H are also present in oil- and gas-fired furnaces. When the charge is mixed with fuel (such as coke in the cupola), reactions of the fuel and its combustion products with the melt are inevitable. Thus, interactions with the atmosphere range from simple dissolution of gases in the melt to reactions such as oxidation or, in the presence of reducing agents, reduction and even carbon enrichment.

4. The charge is covered or mixed with *fluxes*, various (usually inorganic) compounds that can be spread on the surface or mixed into the metal to react with the melt. They often have sophisticated formulations to perform specific functions: react with contaminants and nonmetallic elements; gather up inclusions; isolate the melt from the atmosphere; and reduce vapor losses of metals of low vapor pressure. The resulting *slag* floats to the surface of the melt. Metal lost in the slag and losses due to oxidation or evaporation represent a financial loss and the aim is their minimization, except when selective loss of contaminants is desired.

Many foundry operations generate fumes, gases, and dust. Techniques are available to minimize and even eliminate *environmental pollution*, and some of the most difficult and unpleasant jobs are now performed by robots and other mechanical devices.

Example 7-1	Aluminum beverage cans are highly sophisticated products. The can body is heavily deformed and is made of the highly formable 3004 alloy (nominally 1.2Mn-1Mg), with $T_L = 654°C$ and $T_S = 629°C$. The lid must be harder to stand internal pressure and facilitate tearing the pull tab, and is made of 5182 alloy (nominally 4.5Mg), with $T_L = 638°C$ and $T_S = 577°C$. For recycling, the cans are shredded, heated to burn off the lacquer, then heated to 600°C and shredded again. Alloy 5182 begins to melt at the grain boundaries and breaks into small chips which can be screened out from the large 3004 shreds. Alternatively, the cans are melted in their entirety and chlorine is used to remove Mg; the melt is then alloyed to 3004 specification.

7-4-2 Pouring

When the melt reaches the desired temperature and composition, it is *tapped*. A stationary furnace is tapped by breaking through a refractory plug placed in a hole close to the bottom of the furnace. As their name implies, tilting furnaces are tapped by tilting. Lower melting-point metals can be pumped or siphoned out of the furnace.

The melt may be transferred directly to the mold or tapped into a *ladle* (a refractory-lined vessel) which is then taken to the mold; metal is dispensed through a bottom orifice or by tilting the ladle. In some instances, melt is distributed from a central melting facility to several plants located at some distance.

There may be mismatch between the rates of melting and using material, and then *holding furnaces* are employed in which some treatment or alloying of the melt may also take place.

It is at the *pouring* stage where temperatures are finally adjusted. Highly volatile alloying elements that—because of their high vapor pressure—would be lost during melting, may be introduced (e.g., Mg into Al melts). Elements for deoxidation may also be added. In general, the aim is to keep the flowing metal free from turbulence that would cause entrapment of oxides and slag. Pouring rates and the quantity poured must be controlled too. Automated pouring, where economically possible, gives the most reproducible results.

7-4-3 Quality Assurance

Great strides have been made in improving the quality of castings, and many of these improvements are related to the melting and pouring stage.

Composition Alloy composition used to be controlled entirely by careful charge makeup. Rapid analytical methods, including high-speed spectrography, now provide analyses for the important elements within minutes so that adjustments can be made, to each charge, before pouring.

Inclusions We saw in Sec. 6-3-4 that inclusions can greatly impair mechanical properties, particularly impact properties (Sec. 4-2) and fatigue resistance (Sec. 4-5). Thus, the reliability of castings can be increased and the range of applications broadened by the introduction of techniques aimed at reducing the number of inclusions or, if this is not practicable, changing inclusion morphology and distribution to minimize harmful effects (these techniques are, of course, used also for wrought alloys). Many possibilities exist:

1. *Holding* the metal at a constant temperature allows lighter inclusions to separate. An active flux helps to gather up inclusions. Alternatively, the melt is passed through *filters* (especially for aluminum alloys). *Electroslag refining* (related to electroslag welding, Sec. 18-6-2) is applicable to steels and superalloys, and involves remelting of an electrode, previously cast to the specified composition, by drawing an arc submerged in the slag.

2. *Purging* (scavenging) of the melt with a gas (Sec. 6-3-5) reduces gas content. Reactions with the melt may also occur, as when using chlorine gas to remove hydrogen from magnesium or aluminum melts. Sometimes these effects are a by-product of another operation; e.g., in the oxygen blowing of steel the aim is to remove carbon, but nitrogen and hydrogen concentrations are also reduced. This is true also of the argon–oxygen blowing of stainless steel. Oxygen is later reduced by deoxidation.

3. Many inclusions are the products of *unwanted reactions* (usually oxidation) with a gas. Cleanliness is improved by reducing the partial vapor pressure (Eq. 6-7) by the application of vacuum during melting (vacuum induction melting, consumable-electrode remelting), after melting is complete (vacuum degassing in the furnace or ladle), or during pouring.

Gases   Gas bubbles form inclusions of zero strength (Sec. 6-3-5) and are, in most instances, unwanted. In addition to purging and the application of vacuum, there is the option of tying up gases in a solid reaction product. Thus, copper alloys are *deoxidized* with phosphorus; steels are deoxidized with aluminum, silicon, manganese, or calcium. The reaction products remain in the casting, and process controls aim to distribute them in the least harmful form. Because of its importance to wrought steels, we will return to deoxidation in Sec. 8-3-1.

Alloys (especially steels) of a given composition are commercially produced with various levels of impurities and inclusions, with the cleaner—and more expensive—versions used in more critical applications.

Example 7-2 | The benefit of the above measures can be gaged from the developments in casting the superalloys used in jet engines. The rupture strength (expressed as the temperature for a life of 100 h at 140 MPa) has increased at an average rate of 10°C/year from about 1940 to 1980, and some alloys now operate only 50°C below their solidus. Much of the improvement has come from reducing low-melting impurities (5 ppm Pb can reduce ductility by 75%) and eliminating inclusions by special melting techniques (vacuum casting, careful design of gating and risering, and the use of ceramic filters). As a result, fuel consumption has been halved and thrust-to-weight ratio increased tenfold. [Source: W.J. Molloy, *Adv. Mater. Proc.*, 1990(10):23–30.]

7-5   CASTING PROCESSES

The number of casting processes is very large, and only the major ones can be covered here.

7-5-1   Classification

From the technological point of view, casting processes can be grouped into two broad categories: casting into expendable and into permanent molds. The *expendable mold* is used only once and must be broken up to free the solidified casting, whereas a *permanent* mold is expected to last up to several hundreds or thousands of castings and must be of such a construction as to release the solidified casting.

An alternative classification is based on the purpose of casting:

1. *Ingot, slab, and billet casting.* The metal is a wrought alloy; in preparation for rolling, extrusion, or forging, it is cast into a simple shape suitable for further working. As mentioned, some 85% of all metals is processed this way, in specialized plants. It is essentially a primary manufacturing activity and will be discussed here only to the extent necessary in preparation for the discussion of Chaps. 9 and 10.

2. *Remelt ingots.* These are simple shapes, cast from melts of closely controlled and analyzed composition, for easy transport and loading into the furnaces of secondary manufacturers.

3. *Shape casting.* The melt is cast into the final shape which needs only cleaning and/or machining to produce a finished part. This is a typical secondary manufacturing process and will be the focus of our discussion.

7-5-2   Ingot Casting

Cast bodies of circular, octagonal, or round-cornered square cross sections are called *ingots* when their diameter or side dimension is about 200 mm or greater, and are called *billets* when smaller. Bodies of rectangular cross section are generally called *slabs*. Almost universally, they are poured into permanent molds by a variety of techniques:

1. *Ingot molds*, usually of iron or steel, may be used for static casting of all alloys (Fig. 7–11a). Solidification begins from the mold walls and proceeds toward the center, giving the solidification patterns shown in Fig. 7–1. Better quality (better surface, less slag and gas entrapment) is often obtained by bottom pouring (broken lines in Fig. 7–11a). Piping is avoided by feeding molten metal, either from the ladle or from a hot metal reservoir contained within a refractory-lined extension of the mold. The metal may be kept hot longer and the depth of pipe reduced by placing an insulating collar or an exothermic compound (a compound that ignites on contact with the hot metal and produces heat) on top of the melt (*hot top*). This gives a higher yield of sound metal.

2. *Water-cooled molds* of double-walled construction are employed mostly in the casting of copper-based alloys.

The smallest ingots cast by the first two techniques may be 25–50-mm thick in nonferrous metals and 150–200-mm thick in steels, and their mass ranges up to 20 Mg in nonferrous metals and 300 Mg in steels.

3. *Continuous casting* processes are used for casting the vast majority of slabs and billets.

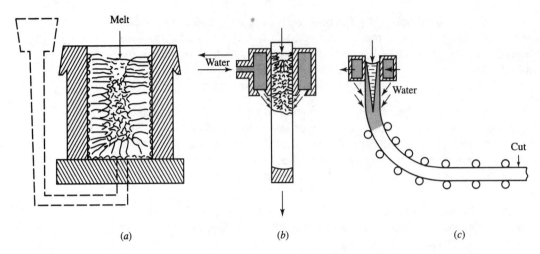

Figure 7–11   Metals destined for further working by plastic deformation are cast into simple shapes in permanent molds (a) or are subjected to semicontinuous casting in short lengths (b) or continuous casting in long lengths (c).

        a. *Semicontinuous* (or *direct chill*) *casting* (Fig. 7–11*b*) is used for nonferrous metals. Solidification is almost complete in the water-cooled copper mold. The casting is withdrawn gradually as solidification progresses and is further cooled with water sprays. The process is interrupted periodically to allow removal of an ingot. To prevent adhesion to the mold, a lubricant is applied. Contact with the mold is eliminated and surface quality greatly improved in casting aluminum by containing the molten zone with air pressure (gas-pressurized hot top) or with an electromagnetic field. An electromagnetic field is used also in the upward *levitation casting* of copper and its alloys.

        b. *Continuous casting*, also called *strand casting* (Fig. 7–11*c*), is used mostly for steels. Casting may go on for hundreds of heats and the slab or bar must be cut up during its movement with a flying saw or torch. A lubricant is again used, and further protection is obtained by oscillating the mold. Solidification extends to great distances beyond the mold, and the hot steel can be bent and unbent to reduce the height of the building. Simple shapes, such as preforms for the rolling of wide-flange beams, are also cast.

        c. *Strip casting* employs twin rolls or flexible bands to produce thin slabs or strips, thus bypassing the early stages of hot rolling.

        d. *Properzi* and *Southwire* processes solidify the metal in the space defined by a grooved roll and a flexible band; the billet is fed directly into a rolling mill, creating a completely continuous process.

7-5-3   Casting of Shapes

When the casting process aims at producing a component of complex shape, a mold is prepared with a cavity that defines the shape of the component, with due allowance for shrinkage after solidification. For all casting processes used (Fig. 7–12), some basic principles apply. First of these is that a sound casting is obtained only if the melt is brought to the cavity in an orderly fashion and solidifies in a planned manner.

Fluid Flow The fluid supply system of a mold is designed in accordance with principles of fluid flow. Ideally, flow should be laminar (smooth) but, in practice, turbulence cannot be entirely avoided. However, turbulence must be kept at minimum to avoid erosion of the mold and entrapment of slag, mold material, and gases. The mold system must be filled with metal under positive pressure, so that no gas is aspirated (sucked in) anywhere.

The fluid supply systems have some common features for all shape casting processes (Fig. 7–13):

1. The *pouring basin* is a receptacle large enough to accommodate the stream of metal, and is often shaped to ensure smooth flow of the melt. At the surface (level l, Fig. 7–13), hydrostatic pressure is nil and potential energy is maximum. *Dross* (oxides and other inclusions rising to the surface) may be held back with a *skimmer*, and heavier inclusions with a *weir*.

2. The fluid is transported down into the mold by the *sprue*. In doing so, potential energy is converted into kinetic energy and velocity increases. Because

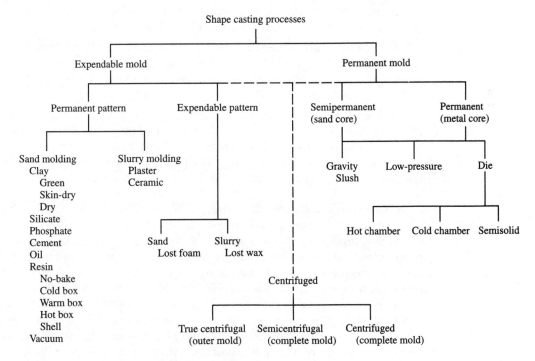

Figure 7–12 Classification of shape-casting processes.

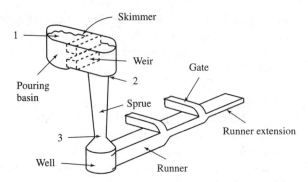

Figure 7–13   Orderly distribution of the melt to a mold cavity requires a well-designed rigging (running and gating) system.

the mass flow is constant, the stream will pull away from the sprue walls as the velocity increases and this will draw unwanted air into the mold. To avoid aspiration of air, a positive pressure differential must be maintained throughout, and for this the sprue is tapered downward (from level 2 to level 3, Fig. 7–13). A *well* of large cross section is at the base of the sprue; the sudden slowing of flow dissipates kinetic energy and helps to drop out inclusions that may have been washed in with the fluid stream. *Ceramic* or *wire mesh filters* may be placed at the sprue base to filter out dross and other large inclusions.

3. The melt is distributed through *runners* which are of larger cross section and often streamlined (Fig. 7–13) to slow down and smooth out the flow, and are designed to provide approximately uniform flow rates to various parts of the cavity.

4. The runners are connected by *gates* to the mold cavity. At the junction to the cavity, these gates are much reduced in thickness (*in-gates*) not only to allow easy separation from the solidified casting but also to choke the flow of metal and ensure quiet entrance into the cavity. The runner is often extended beyond the last gate to serve as a trap for inclusions carried into the runner by the first metal.

Castings are usually gated at the side, as shown in Figs. 7–13 and 7–15. Some parts are suitable for top gating, with the sprue serving also as a riser. Feeding from the bottom gives the quietest flow, although the top of the casting is then filled by the coldest metal. A particularly favorable situation exists when the metal is drawn up into a heated mold by vacuum (*suction casting*) because the mold fills without splashing and there are few, if any, mold gases to dissipate.

The dimensions of the various parts of the fluid distribution system (sometimes called *rigging*) may be calculated in an approximate manner by considering that melts are incompressible. Therefore, the *flow rate* (the volume passing through any given cross section in unit time) in any part of the system obeys the *equation of continuity*:

$$A_0 v_0 = A_1 v_1 \quad \text{etc.} \qquad \textbf{(7-3)}$$

where A is the cross-sectional area of a section and v is the velocity at that point. If $A_1 > A_0$, flow slows down and vice versa.

The velocity may be approximated from *Bernoulli's theorem* which states that, under steady, well-developed flow conditions, the total energy of a unit volume of material must be a constant at every part of the system. There are four components of energy: pressure energy due to the pressure p, which is the sum of external and hydrostatic pressure (pressure due to the weight of the melt); kinetic energy due to the velocity v; potential energy due to the height h above a reference plane; and energy losses f due to friction in the melt (this term may be taken to include energy lost in turbulence, directional change, and friction against the mold walls). Thus, total energy per unit volume is

$$p_0 + \frac{\rho v_0^2}{2} + \rho g h_0 = p_1 + \frac{\rho v_1^2}{2} + \rho g h_1 + f \qquad \textbf{(7-4)}$$

where ρ is density.

If velocities are too low, freezing occurs before the mold is completely filled; if they are too high, the mold will be washed out and inclusions will be swept into the mold cavity.

Positioning and dimensioning of the runner system vitally influence the soundness of the casting, because they determine the supply rates of the melt to various parts of the cavity, and thus also influence the solidification pattern. The subject has been greatly developed in recent years, both experimentally and theoretically. Commercial computer programs are usually based on a blend of basic theory and experience, and aid in the gating of castings in an interactive mode.

Example 7-3

Assuming negligible frictional losses, use Eqs. (7-3) and (7-4) to show that the areas of the top and bottom of the sprue must obey the following relation to avoid aspiration

$$\frac{A_2}{A_3} = \sqrt{\frac{h_3}{h_2}}$$

Let the pressure at the top of the sprue (Fig. 7–13, level 2) be p_2; at the bottom (level 3) p_3. To avoid aspiration, $p_3 > p_2$; for purposes of this example, take $p_3 = p_2$. From Eq. (7-4)

$$\frac{\rho v_1^2}{2} + \rho g h_1 = \frac{\rho v_2^2}{2} + \rho g h_2 \qquad \textbf{(7-4')}$$

The cross-sectional area of the pouring basin is very large, hence

$$v_1 = 0 \quad \text{at} \quad h_1 = 0 \quad \text{and} \quad v_2 = \sqrt{2 g h_2} \qquad \textbf{(a)}$$

similarly,

$$v_3 = \sqrt{2gh_3}$$ **(b)**

From Eq. (7-3),

$$v_2 = \frac{A_3/A_2}{v_3}$$ **(c)**

Substituting (c) into (a) and equating to (b) gives the relationship

$$\frac{A_2}{A_3} = \frac{v_3}{v_2} = \sqrt{\frac{h_3}{h_2}}$$

Heat Extraction and Solidification   Once the melt enters the mold, heat is extracted through mold walls and solidification begins. If no special measures are taken, heat is extracted all around, so that solidification occurs *progressively* from all surfaces inward.

Solidification time t_s is, as might be expected, directly proportional to volume V (which governs heat content) and inversely proportional to surface area A (over which heat extraction occurs). It can be shown that, for a large variety of shapes and sizes, the relationship is quadratic (*Chvorinov's rule*)[1]

$$t_s \propto (V/A)^2$$ **(7-5)**

where V/A is referred to as the modulus of the casting. Chunky portions of the casting freeze last; therefore, progressive solidification can lead to early freezing of thinner sections, denying access of liquid to thicker parts and leading to porosity and the formation of shrinkage cavities (Fig. 7–14a). Remedies may take different forms:

1. *Risers* (*feeder heads*) provide a reservoir of molten metal. Made with a high V/A ratio (Fig. 7–14b), they solidify last and feed enough liquid to heavy sections of the casting to make up for shrinkage before and during solidification. An actual example is shown in Fig. 7–15; it will be noted that the risers are placed between runners and casting, so that they are filled last and contain the hottest metal (*live* or *hot risers*). At other times, as in Fig. 7–14, a riser must be placed at the end of the casting (*dead* or *cold riser*). To ensure uninterrupted feeding, the junction between riser and casting can reach, in the casting of steel, 70–90% of the cross section to be fed.

Risers may be *open* to the atmosphere and then exothermic compounds may be placed on them; the *blind riser* shown in Fig. 7–14 loses less heat, but a porous ceramic *pencil core* must be inserted to equalize pressure in the shrinkage cavity. While often indispensable, risers reduce the yield and increase the amount of scrap to be recycled.

---

[1] N. Chvorinov, *Proc. Inst. Br. Foundryman,* **32**:229 (1938–39).

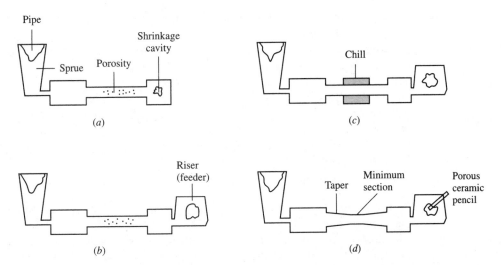

Figure 7-14 A casting may show shrinkage cavities and microporosity (*a*). Feeder heads or risers, removed after solidification, provide hot metal (*b*). Microporosity may be eliminated with directional solidification by incorporating a metal chill into the mold (*c*) or tapering the thinnest section (*d*).

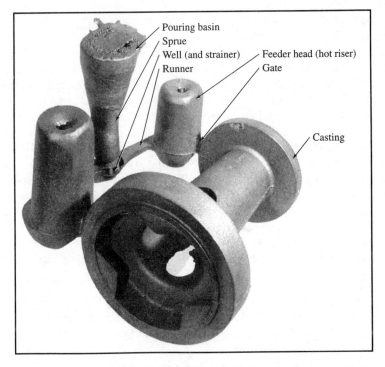

Figure 7-15 An example of a cored gray-iron casting showing sprue, runners, gates, and risers. Note the strainer configuration at the base of the sprue. (*Courtesy Massey-Ferguson Brantford Foundry, Brantford, Ontario.*)

2. Porosity in a thin section can be avoided by initiating freezing in that section and moving the solidification front toward thicker sections, i.e., by changing progressive solidification to *directional solidification*. In expendable molds, this can be aided by placing metal inserts (*chills*) into the refractory mold at points where maximum cooling is desired (Fig. 7–14c). In permanent molds localized cooling is achieved by placement of cooling fins or pins on external surfaces, or even by air or water cooling passages in the mold. In addition, it is necessary that liquid should be supplied to compensate for solidification shrinkage. A temperature gradient of minimum 1.5°C/cm is required to ensure good feeding and avoid microporosity.[2] Porosity may be eliminated also by tapering (Fig. 7–14d) although this requires extra material (*padding*).

Controlling Structure   Heat extraction and the mode of solidification affect not only the soundness of the casting but also the structure and grain size of the solidified metal. Indeed, one of the powerful ways of improving the properties of a casting is by controlling the grain size. We saw in Sec. 6-3-5 that this usually entails a small grain size, and nucleating agents together with controlled cooling rates are employed for the purpose (Sec. 7-1-1). There are, however, exceptions: In high-temperature, creep-resistant applications a coarse grain is preferable if the structure is equiaxed (Fig. 6–18). Better properties are obtained if grain boundaries transverse to the loading direction are eliminated. Thus, the properties of turbine blades were improved when the polycrystalline structure (Fig. 7–16a) was replaced by axially oriented grains (Fig. 7–16b); for this, the mold is placed on a water-cooled base and is slowly withdrawn from a heated enclosure. Even stronger—and, in the absence of grain boundaries, more corrosion-resistant—is a single grain (Fig. 7–16c). If a helical channel ("pigtail") is placed between chill and mold (Fig. 7–16e), only the most favorably oriented grain (in Ni-based superalloys, the ⟨001⟩ grain) can grow through the helix and into the mold. If a different orientation is needed for higher strength, a single-crystal seed is placed at the bottom of the mold (Fig. 7–16f). High hot strength can be developed in single-crystal blades because the absence of low-melting grain-boundary phases permits solutionizing at higher-temperatures. Yet further improvements are possible by the directional solidification of eutectic structures, so that fibers of a hard, strong phase (such as an intermetallic compound) become oriented in the direction of loading and provide integral reinforcement.

Mold Design   With the above principles in mind, mold design proceeds by the following steps:

1. The volume and weight of the casting are determined.

2. Based on volume and geometric configuration (long and narrow, or blocky, or of nonuniform cross sections, etc.), the size and number of risers are determined.

3. On the basis of theory and empirical relationships, the optimum pouring time is found.

---

[2]M.C. Flemings: *Solidification Processing*, McGraw-Hill, 1974.

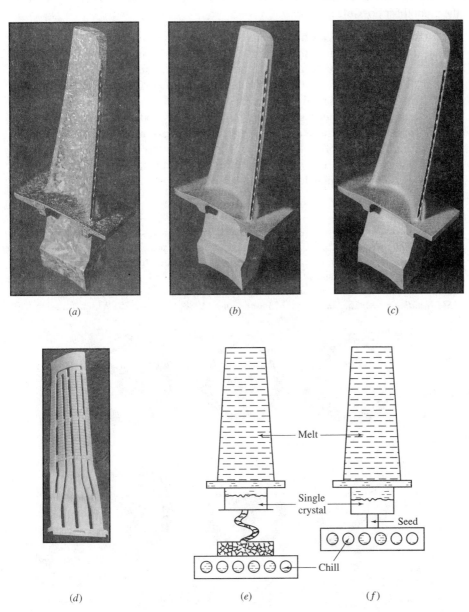

(a)   (b)   (c)

(d)   (e)   (f)

Figure 7–16   Jet engines may be operated at higher temperatures by changing the structure of
Ni-based superalloy turbine blades from (a) the equiaxed, polycrystalline structure
produced by conventional casting to (b) columnar-grain produced by directional
solidification and (c) to single crystals. Complex internal passages are produced with
(d) ceramic cores in these lost-wax castings. (*Courtesy Howmet Corporation,
Greenwich, Connecticut.*) The single-crystal blades are produced (e) by allowing only
the most favorable orientation to grow or (f) by placing a seed crystal in the mold.

4. The feeder system is designed to feed the mold, in the allowed time, in the smoothest possible manner.

Computer programs are helpful in many ways. The simpler programs take the mold designer through the steps outlined above, using simple theory and a great deal of empirical data. Mathematical models, based on analytical or numerical methods, have advanced to the point where mold filling can be observed on a screen, the gating and risering system can be designed, freezing times, and even microstructure and properties can be predicted to some extent. The influence of changes in casting conditions can be evaluated without extensive experimentation.

The solidification of melts can also be studied in the laboratory by the use of simulating materials (e.g., organic solutions whose crystallization may be observed in a transparent plastic mold).

Example 7-4 | Computer modeling of fluid flow reveals problems in casting a three-spoke wheel in the vertical position. With top gating, the metal cascades down freely (*a*), possibly eroding the sand, and cold shuts can develop at the hub (*b*). Bottom gating gives quieter fill (*d*) although cold shuts can develop in the rim (*e*) and (*f*). Modeling was confirmed by high-speed photography of filling in a sand mold, with one wall replaced by glass. (After W-S. Hwang and R. A. Stoehr, *ASM Handbook*, vol. 15, *Casting*, ASM International, 1988, pp. 874–875. With permission.)

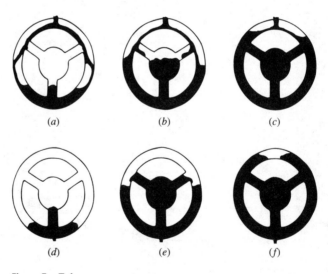

(*a*)   (*b*)   (*c*)

(*d*)   (*e*)   (*f*)

Figure Ex. 7-4

7-5-4   Expendable-Mold, Permanent-Pattern Casting

Expendable molds are prepared by consolidating a refractory material (*sand*) around a *pattern* that defines the shape of the cavity and also incorporates, in most instances, the gates, runners, sprue, and risers required to fill the mold (Fig. 7–17).

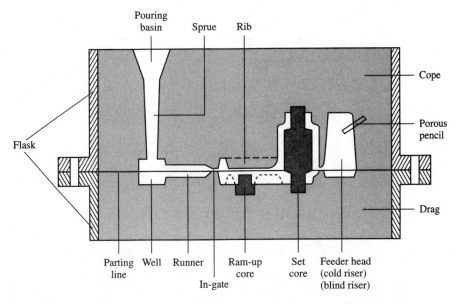

Figure 7–17 Some characteristic elements of a cope-and-drag sand mold. To avoid the "dead" feeder head (riser), the mold could be fed from the right, making material in the "live" riser the hottest.

Patterns Patterns differ from the finished part in some important respects. All dimensions are increased to account for the contraction (solid shrinkage) of the casting from the solidus to room temperature (not to be confused with solidification shrinkage). There are patternmaker's rules that are longer by the *shrinkage allowance* (Table 7–2); in CAD/CAM the allowance is preprogrammed. If the casting is to be machined, an appropriate thickness (*machining allowance*) is added.

Because permanent patterns are to be used repeatedly, they are made of wood or, for greater durability and dimensional stability, of a metal or strong plastic. The pattern must be easily removed from the consolidated mold; for this, molds will have to be made in two halves. Accordingly, a *parting plane* is selected that conveniently divides the shape into two parts (Fig. 7–18a). Surfaces parallel to the direction of withdrawal are given a draft (Fig. 7–18a) to allow removal of the pattern without damaging the mold.

Cavities, undercuts, and recesses in the cast shape must be formed by the insertion of *cores* (Figs. 7–16d and 7–17). Thus, greater complexity of shape is attainable, but at a higher cost. For the accurate location of cores, the pattern provides nesting holes (*core prints*).

The simplest pattern for producing the shape shown in Fig. 7–18 would be in *one piece* (but usually split along the parting plane), and gates, runners, and risers would be added during molding (*loose pattern molding*). This makes molding slow and labor-intensive. For higher productivity, elements of the feeding system

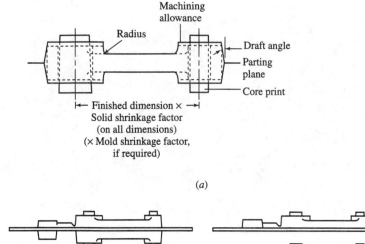

Machining
allowance

Radius

Draft angle
Parting
plane
Core print

|← Finished dimension × →|
Solid shrinkage factor
(on all dimensions)
(× Mold shrinkage factor,
if required)

(a)

(b)   *(c)*

Figure 7-18   A pattern must allow for (*a*) solid shrinkage and easy removal from the
mold. For faster production, it is fastened onto (*b*) a match plate or
(*c*) cope-and-drag plates.

are incorporated into the pattern, split along the parting line, preferably with the
runner in the drag and the in-gates in the cope; this way, the runner fills before
admitting metal to the cavity. The two halves are mounted either on the two
surfaces of the same plate (*match plate* Fig. 7–18*b*) or on separate upper and
lower plates (*cope half* and *drag half,* Fig. 7–18*c*). Production rates are further
increased if several pieces are molded and cast simultaneously in the same mold;
for this, *multipiece pattern plates* are prepared.

Large parts of fairly simple configuration are often molded by hand, using
skeleton patterns or, if the part is of rotational symmetry, by rotating a cross-
section board (a *sweep pattern*) in the sand.

The refractory will have to be contained around the pattern and the container is
traditionally called a *flask.* When split into two pieces to accommodate the upper
and lower halves of the pattern, one speaks of *cope* and *drag halves,* respectively.
Very large molds may be formed into a pit in the ground.

Cores, like the mold itself, are made of refractory materials. However, their
bond strength must be greater to allow handling but they must still be removable
after solidification. They are molded into *core boxes* made of wood or metal.
Cores may be made in halves (or several parts) and pasted together. Since cores
are often almost fully surrounded by the melt, they must be vented to the outside.

Sands   Of all refractory materials, silica sand (SiO_2) is of the lowest cost and, if
quality (composition and contaminants) is carefully controlled, it is satisfactory
for quite high casting temperatures, including that of steel. Some other refractories such as zircon ($ZrSiO_4$), aluminum silicate (Al_2SiO_5), chromite ($FeCr_2O_4$),

or olivine ($(MgFe)_2SiO_4$) are used for special purposes. Sand in itself flows freely and must be bonded temporarily. The *bond* must be strong enough to withstand the pressure of and erosion by the melt, yet it must be sufficiently weakened by the heat of the metal to allow shrinkage of the casting and, finally, removal of the sand without damage to the solid casting. However, the bond must not destroy the permeability of the sand so that gases—present in the melt or produced by the heat of the melt in the binder itself—can escape. The quality of the sand is routinely tested in the sand laboratory for properties such as grain size; compressive, shear, and tensile strength; hardness; permeability; and compactability (decrease in height of a column under a specified load). Cost is reduced and landfills minimized by reclaiming the used sand through washing, dry scrubbing, or, for organic binders, heating. A special case is *graphite*, used for metals (Ti, Zr, refractory metals) that react with silica.

Binders   Processes are often described according to the bonding agent (*binder*):

1. *Green sand molds* are the cheapest because they are bonded with clay. Clay is a hydrated aluminosilicate with a layered structure (Sec. 12-3-1). It is fairly strong, but brittle in the dry state. It becomes readily deformable when water is added: Water adsorbs on the platelets and allows their movement relative to each other. Some sands already contain the required few percent clay, but superior qualities are usually obtained when a quality clay (e.g., 6–8% bentonite) is added to pure quartz sand. With 2–3% water and thorough mixing (*mulling*), a readily transportable and moldable sand mix is obtained. When left in the damp condition, one speaks of a green sand mold. A great advantage is that used sand is readily reclaimed.

2. The clay-bonded sand may be partially dried around the cavity to improve the surface quality of the casting and reduce pinhole defects that may develop as a result of steam generation (*skin-dried sand mold*). Alternatively, the entire mold may be dried out, but such *dry sand molds* have largely been replaced by no-bake molds (see below).

3. *Silicate bonds* are produced by several techniques. In the *CO_2 process* sand is mixed with 3–5% water glass ($Na_2O{\cdot}xSiO_2 + nH_2O$), a liquid. On completion of molding, CO_2 is bubbled through to form the reaction products Na_2CO_3 and a gel of $xSiO_2 \cdot nH_2O$ composition. This gives a firmer sand mold with less wall movement and, therefore, larger, more accurate castings can be made. Another process uses hydrolyzed ethyl silicates.

4. *Phosphate bonding* uses a water-soluble phosphate with a powdered metal oxide hardener to produce an environmentally friendly bond.

5. *Cement molding* relies on the hydration of cement to form a gel of great strength. Some 10–15% cement is used as a binder, mostly for large steel castings molded in a pit. The sand is hard to break away from the finished casting.

6. *Oil sands* consist of sand mixed with a drying-type vegetable oil (such as linseed oil) and some cereal flour. These oils are unsaturated hydrocarbons (with double and/or triple bonds in the carbon chain) and form a polymer on heating to temperatures around 230°C. The process is fast and gives high strength, and is thus suitable also for cores.

7. *Resin-bonded sands* are bonded with thermosetting resins (polymers, Sec. 13-3). *No-bake binders* are thermosetting resins which cure upon combining two or more components and a catalyst. *Cold-box processes* use airborne catalysts. Liquid resins with catalyst are used in *warm-box* (over 150°C) and *hot-box* (over 230°C) processes. Originally developed for single-piece, strong cores, resins are increasingly also used for molding.

8. *Vacuum-molding* relies on the observation that sand is firmly lodged in place if air is removed. Patterns have small holes in them so that a thin, heated thermoplastic polymer sheet can be tightly drawn over their surface by vacuum. Clean, unbonded sand is then applied in a flask, the surface of the flask is sealed, and vacuum is drawn on the sand. The vacuum is now released on the pattern, the pattern is removed, the mold is assembled, and the metal is poured. The polymer sheet burns up and, once the casting is solidified, the vacuum is released and the loose sand falls off.

Compaction   Bonded sand is compacted by various techniques, chosen according to production rates and the total number of parts produced:

1. For only a few parts, the sand may be shoveled into the flask around a one-piece pattern and rammed by hand. It requires high skill to produce a mold of uniform packing.

2. For mass production, the sand is conveyed to the molding station and dropped or blown unto patterns surrounded by flasks. The sand is compacted in the flask by mechanical means, such as jolting or squeezing. A typical sequence of molding operations is given in Fig. 7–19.

3. High densities can be attained by dynamic compaction. The sand is slung by a fast-rotating impeller or the flask is evacuated and then a large valve opened through which sand is drawn in. Good results are obtained also using the pressure wave from the detonation of natural gas above the flask.

4. When pressures are high enough (around 7 MPa), a properly bonded sand acquires sufficient strength to maintain the integrity of the mold without a supporting flask (*flaskless molding*). Molds with vertically oriented parting lines are stacked end to end; production can reach 250–750 molds/h in a single production line. The high strength of molds also minimizes wall movement during casting and solidification and gives castings of higher accuracy.

In all but flaskless molding, weights must be placed on the cope; otherwise, the metallostatic pressure exerted by the melt would lift up the cope and a breakout would occur.

It is customary to use finer sand in the vicinity of the mold surface. Facing materials (coal or graphite) can be added to generate gases on contact with the hot metal, reducing metal penetration and adhesion to the sand (burning of the

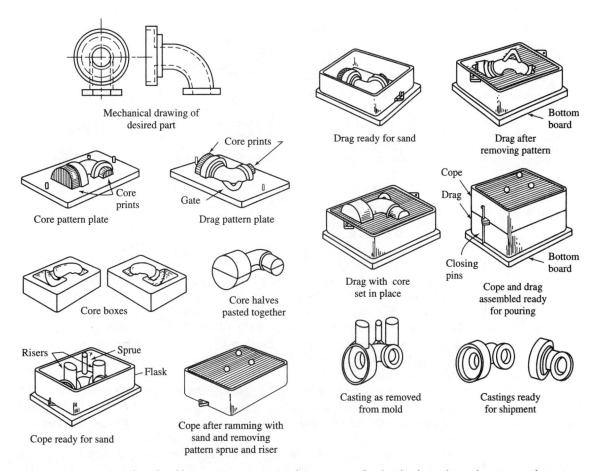

Mechanical drawing of
desired part

Drag ready for sand

Drag after
removing pattern

Bottom
board

Core prints

Core
prints

Gate

Core pattern plate

Drag pattern plate

Core boxes

Core halves
pasted together

Drag with core
set in place

Cope

Drag

Closing
pins

Bottom
board

Cope and drag
assembled ready
for pouring

Risers

Sprue

Flask

Cope ready for sand

Cope after ramming with
sand and removing
pattern sprue and riser

Casting as removed
from mold

Castings ready
for shipment

Figure 7–19 A typical sand molding sequence. (*From Steel Castings Handbook, 5th ed., Steel Founder's Society of America, Des Plaines, Illinois, 1980. With permission.*)

sand), thus giving a better surface finish, free of defects. Alternatively, various refractory materials can be suspended in a liquid and applied as a coating to the mold and core surfaces.

Cores greatly increase the variety of shapes that can be cast. If their weight cannot be supported, cores are placed on *chaplets* (small, often perforated metal supports that will melt into the casting alloy).

Shell Molding This is a variant of the resin-bonded sand technique. The sand is coated with a solid (B stage, Sec. 13-3) thermosetting resin. After coating with a parting agent, the pattern, heated to 200–260°C, is placed on top of a box that contains the sand (Fig. 7–20). On inverting the box, the sand settles on the pattern and a thin shell cures in situ, faithfully reproducing details of the pattern. Once the shell is thick enough, the box is turned back, whereupon excess, unbonded

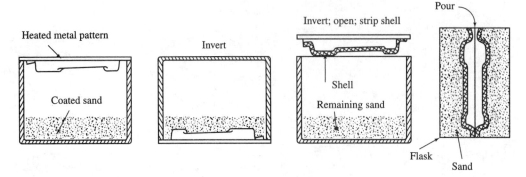

Figure 7–20   Typical sequence of shell molding.

sand drops back into the box. The shell is stripped, combined with the other half, placed in a flask, and backed with some inert material such as steel shot to provide support. The greater strength of the mold often allows forming of cores integrally with the mold. Separate cores are made by blowing sand into heated molds; because the cores are hollow, they give good venting. Compared to green sand casting, the surface finish and tolerances of parts are better, floor space and sand quantity in circulation are reduced, but recycling of the sand is more expensive.

Thinner walls can be achieved by partially immersing the resin-bonded mold into molten alloy held in a separate furnace and filling the cavity by drawing vacuum (*vacuum casting*, Fig. 7–21*a*). Yield is almost 100% because filling is through a "pin gate" which often feeds a riser. High-integrity structural parts can be made in all metals for the aerospace, automotive, and chemical industries. Alternatively, the mold is bottom fed by a pressurized melt (Fig. 7–21*b*).

Slurry Molding   Instead of compacting the sand by force, a finer-grained refractory may be made into a *slurry* with water and poured around the pattern. A smoother surface finish is obtained and, if so chosen, the refractory may be more heat-resistant than the bonded-sand variety. Since the shrinkage of mold and casting can be closely controlled, one often speaks of precision casting.

Plaster molding relies on the well-known ability of a plaster of paris slurry to flow around all details of a pattern. Various inorganic fillers are added to improve strength and permeability. After a rather complex baking sequence, the mold is assembled and the metal poured. In patented versions, steam pressure treatment or foaming agents are used to create controlled subsurface porosity and thus increase permeability. Since gypsum is destroyed at 1200°C, plaster molding is not suitable for ferrous castings. For other metals, plaster molding gives near–net shape parts of very good surface finish and tight tolerances.

Ceramic-mold casting is suitable for all materials, because the slurry is now made up of selected refractory powders, such as zircon ($ZrSiO_4$), alumina

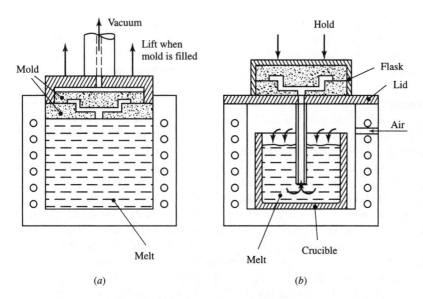

Figure 7-21   Excellent mold filling is secured by (a) vacuum casting and (b) pressure casting.

(Al_2O_3), or fused silica (SiO_2), with various patented bonding agents. The fine-grained ceramic slurry is applied as a thin facing to the pattern and is backed up with lower-cost fire clay (*Shaw process*) or coarse sand aggregate (*Unicast* process). The mold is fired at 1000°C and the melt is poured while the mold is still hot. Precision is assured by taking account of dimensional changes at each stage of mold making and casting, often by CAD/CAM. The higher cost is well justified by the quality of castings in highly alloyed, high-melting-point metals. Large constructional parts as well as forging and casting dies are cast to final shape and dimensions, often without the need for subsequent surface finishing.

7-5-5   Expendable-Mold, Expendable-Pattern Casting

An *expendable pattern* must be made of a material that can be either melted out before pouring or burnt up during casting. This way, the pattern may be left in the mold, and there is no need for parting planes, draft angles, or even cores. Shape limitations are few, and the only criterion is that the refractory can be shaken out or otherwise removed from all cavities and intricate details of the finished casting (the refractory is sometimes left in cavities of sculptures).

Expendable patterns are made by injecting the pattern material into the cavity of a pattern mold. Thus, there is the requirement that the pattern must be extracted from the mold; however, more complex shapes are easily produced by assembling the pattern from several simpler shapes. Small, complex internal passages may be formed by placing ceramic cores (usually of fused silica) into the pattern. In some instances, a rubber (usually silicone rubber) mold allows stripping from reentrant shapes. Shrinkage of the pattern material must be taken into account, and a draft must be provided if the cavity in the pattern mold is deep.

Investment Casting   *Investment casting*, also called the *lost-wax process*, was already used in ancient Egypt and China (Table 1–1), but it has found widespread industrial application only since the Second World War, with the need to produce precision parts in high-temperature materials for jet engines. It is capable of producing the most complex shapes, because the pattern is made of a carefully blended wax mixture (or, for thin sections, a plastic such as polystyrene) complete with the feeding system, and the refractory slurry is poured around this.

Wax patterns are readily produced in large quantities by injection molding into metal dies, usually made from highly machinable aluminum alloys. Ceramic cores may be incorporated, as for making complex internal passages in turbine blades (Fig. 7–16d). Individual patterns are assembled with wax sprues, runners, and gates into a so-called tree, simply by local melting of the wax, using a hot knife or blade between the two mating surfaces. Two approaches are then practical:

1. In *solid investment*, the tree is precoated by dipping in a refractory slurry, dusted with refractory sand, and placed in a flask where a thick, coarser refractory slurry is poured around it. When the slurry has gelled by drawing off excess water, the mold is dried in an oven in an upside-down position to allow the wax to run out. Before casting, the mold is fired at 700–1000°C; this imparts strength to the mold, eliminates the danger of gas formation from water during casting, increases the fluidity of the melt that will be poured, and ensures a good surface finish as well as close dimensional tolerances.

2. The cost of the mold may be reduced and the rate of production increased by dispensing with the solid mold. In the *ceramic shell-molding process* (Fig. 7–22) the tree is prepared as before, but is then covered with refractory in a *fluidized bed*. (When air is blown from the bottom of a container partially filled with powder, the powder is suspended in the air and flows like a fluid.) Several layers of gradually coarsening coats are applied to reach sufficient thickness. The repetitive operation can be entrusted to a robot. The shell mold is dried and fired, if necessary supported by a granular material, and the metal is cast.

Example 7-5	Investment casting offers exceptional freedom in shape complexity, even for high-temperature materials in fairly large, critical components. A transmission adapter for the Bell-Boeing V-11 Osprey vertical and short take-off and landing (VSTOL) aircraft consisted of 43 forged and sheet metal parts and 536 fasteners. It was replaced by three investment-cast Ti-6Al-4V alloy parts, HIPed (Sec. 7-6-2) and heat-treated, and 32 fasteners. The castings were developed with the aid of computer modeling, concurrent engineering, and rapid prototyping. Machining costs were reduced by 60%, and total part cost was slashed. [Data from B. C. Goodley et al., *Adv. Mater. Proc.*, 1997(6):43–46.]

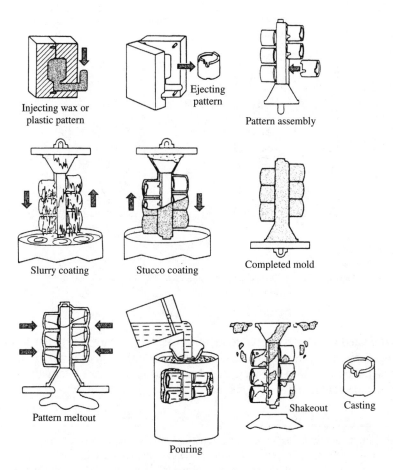

Injecting wax or
plastic pattern

Ejecting
pattern

Pattern assembly

Slurry coating

Stucco coating

Completed mold

Pattern meltout

Pouring

Shakeout   Casting

Figure 7–22   Typical sequence of ceramic shell molding. (*Courtesy Investment Shell Casting Institute, Dallas, Texas.*)

Significant improvements are obtained with vacuum pouring. The mold is placed over the melt (with a nozzle reaching into the melt); vacuum is then drawn on the mold so that the melt rises smoothly to fill the cavity (similar to Fig. 7–21*a*). The sprue and runner are made large enough to prevent solidification in them, and the liquid is allowed to flow back into the melt. Yields rise to 85–95%.

Lost Foam Casting   An interesting variant of sand casting utilizes an expendable mold with an expendable pattern made of expanded polystyrene foam, similar to that used in cups for hot beverages, but of carefully controlled density and surface quality. Very complex shapes can be built up and runners etc. can be attached with hot-melt resins or rubber cement. The process, also called *evaporative casting*, again allows great freedom in shape, without draft, because the pattern is left in the mold to evaporate and burn up during casting (Fig. 7–23). The plastic foam

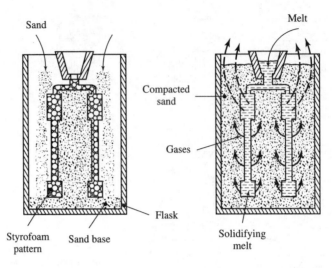

Figure 7–23   Parts of complex shape can be made by the lost-foam process.

is first coated with a permeable refractory wash. The foam is firm but would be damaged by high compaction pressures; therefore, it is backed up by dry sand compacted by vibration and weighted down during pouring. Alternatively, steel shot may be kept in place by a magnetic field (*magnetic molding*).

Example 7-6 | Saturn Corporation casts its aluminum engine blocks in house. Lost-foam molding allows more complex shapes to be cast, thus features such as alternator and oil-filter mounts can be integrated with the engine block, saving material and machining time. [Source: *Manufacturing Engineering*, 1993(5):53.]

7-5-6   Permanent-Mold Casting

While in the processes described above the mold was destroyed after the solidification of the casting, the mold is reused repeatedly in the permanent-mold casting processes.

Mold Materials   The mold material must have a sufficiently high melting point to withstand erosion by the liquid metal at pouring temperatures, a high enough strength not to deform in repeated use, and a high thermal-fatigue resistance to resist premature crazing (the formation of thermal fatigue cracks) that would leave objectionable marks on the finished casting. Finally, and ideally, it would also have low adhesion (Sec. 4-9-2) to the melt to prevent welding of the part to the mold.

The mold material may be cast iron, although alloy steels are the most widely used. For casting higher-melting alloys (brasses and ferrous materials), the mold steel must contain large proportions of stable carbides so that strength is retained at higher temperatures. Refractory metal alloys, particularly the precipitation-hardenable molybdenum alloy TZM (0.015C-0.5Ti-0.08Zr), find increasing application. Graphite molds can be used for steel although only for relatively simple shapes.

Coatings or dressings are chosen according to casting temperature. Graphite, MoS_2, silicone, and other films (*parting compounds*) reduce adhesion and facilitate ejection. Refractory coatings are sometimes built up to thicker layers to reduce temperature fluctuations on the mold surface. Uniform application is most important, and programmable robots are well suited for the repetitive task.

Die Design   All permanent molds (dies) have some common features:

1. A prime requirement is that the solidified casting be readily removed from the die cavity. Therefore, shapes that can be produced are more limited than in casting in an expendable mold, although great complexity is allowable when the mold is made in several parts (for example, housings for office machines, sewing machines, and chain saws, etc.).

2. Internal cavities are formed with the aid of fixed or movable *metal cores*. Undercut shapes require that cores be made in several interlocking parts which are withdrawn in a fixed sequence. In casting with gravity or low-pressure feed, sand, plaster, or graphite cores may be inserted into the permanent mold (*semipermanent molds*).

Semipermanent mold casting has a wide range of applications. A single casting of aluminum alloy 357.0 (7 Si-0.5Mg-0.2Ti) replaced a 17-part polymer/sheet metal assembly for a stowage-bin support frame in the Boeing 777 aircraft. Wall thickness is only 2.5 mm. [Source: *Adv. Mater. Proc.*, 1994(3):8.]

| **Example 7-7**

---

3. Nests in the die allow accurate location of *inserts* (threaded inserts, heating elements, etc.).

4. *Ejector pins* are necessary to remove the solidified casting, particularly if the process is mechanized. The number and location of ejector pins must be chosen to prevent objectionable surface marks or distortion of the casting. Early ejection is important when the casting would shrink onto bosses, but enough time must be allowed for solidification.

5. The mold material is not permeable; therefore *vents* must be provided to avoid trapping gases. Clearances along parting planes and around ejector pins may also serve as vents.

Solidification   The permanent mold works as a heat exchanger. At the start of a production run, the mold must be preheated to the desired temperature (typically 150–200°C for Zn, 250–275°C for Mg, 225–300°C for Al, and 300–700°C for Cu alloys). During steady-state production, heat given off by solidification is removed by means of radiating pins or fins, or internal water cooling channels. Evaporation of water from mold dressings, prior to closing the mold, also cools the mold face.

Close control of die temperatures and mold dressing composition and application allows casting with thinner walls, even though at the expense of slower solidification and a prolonged casting cycle (Fig. 7–7). Nevertheless, solidification rates are much higher than in refractory molds; therefore, output rates are high and grain size is small. However, the supply of melt to thicker sections of the casting may be cut off prematurely. It is thus imperative that proper feeding be provided; even so, porosity can be and usually is higher than in similar castings made in expendable molds. The permanent mold is always stronger than the solidifying casting, therefore, casting alloys prone to hot shortness (those of long freezing range, or containing a low-melting matrix) are avoided. There are several variants of permanent-mold casting, distinguished by the method and pressure of feeding.

Gravity-Feed Permanent-Mold Casting   The process is usually referred to simply as *permanent-mold casting* or *gravity die casting*. It builds on the same principles as expendable-mold casting, except that the mold is made of an appropriate permanent material. The casting machine is basically a bed that supports the stationary and movable mold halves. The halves may be hinged as pages in a book (*book mold*). Manually operated machines are equipped with a long handle and clamps; mechanized machines have hydraulic actuators (Fig. 7–24). In conjunction with split metal cores or collapsible sand cores, the process is very versatile.

The process is widely used for aluminum alloys (e.g., tens of millions of internal combustion engine pistons are cast in molds with five-piece movable cores) as well as magnesium and copper alloys. Smaller cast iron and steel castings can also be made. The mold is protected by a lubricant-type coating for aluminum and magnesium alloys, and with ceramic coatings of up to 1-mm thickness for copper-base alloys and gray iron.

Slush casting is a variant of permanent-mold casting, used mostly for non-structural, decorative products such as hollow lamp bases, candlesticks, and statuettes. The mold is filled with the melt and, after a short time, inverted to drain off most of the melt, leaving behind a hollow casting with a good outer surface but very rough inner surface.

Low-Pressure Permanent-Mold Casting   The mold is seated right above the melting or holding furnace, and metal is fed by air pressure through the bottom gate into the mold cavity (as in Fig. 7–21*b*), ensuring smooth filling. Solidification is directed from the top downward. Air pressure is released as soon as the cavity is filled with solid metal; thus, material losses are minimized. Thin coatings ensure an acceptable surface quality. The die halves must be held together under

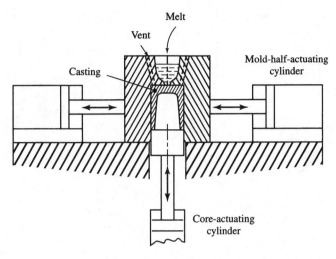

Figure 7–24   Schematic of permanent mold filled under gravity (press frame not shown).

sufficient force to resist the force generated by fluid pressure in the cavity. In a variant of the process, yet smoother filling is secured by drawing vacuum on the mold. Widest application is to aluminum alloys, but a variant with graphite molds is suitable for casting larger steel parts, such as railroad car wheels.

Die Casting   The term refers to processes in which the mold cavity is filled under moderate to high pressures, forcing the metal into intricate details of the cavity. Die halves are held together by a correspondingly high force; therefore, die-casting machines resemble hydraulic presses of two-, three-, or four-column construction, rated by the die-holding force (of up to 40 MN or 4000 tonf capacity). For greater rigidity, the dies are usually locked with toggle clamps while the melt is forced into the die—slowly at first and then at increasing speeds—by a separate plunger. Shot sizes range from a few grams to tens of kilograms, and production rates of up to 1000 shots/h can be achieved in the smaller sizes. There are two basic process variants:

1. In the *hot-chamber process* the liquid metal is transferred to the mold directly from the holding pot by a submerged pump (cylinder and plunger, Fig. 7–25a) at pressures up to 40 MPa. Turbulence is minimal, oxidation is prevented, and no heat is lost. Thus, castings of considerable complexity, such as instrument cases and automotive components can be made. The pump is, however, exposed to severe conditions. For zinc and magnesium alloys, pumps are constructed of steel, but there are still problems with the durability of ceramic pumps needed for aluminum alloys.

In a variant of the process, *direct injection* of Zn alloys takes place through a heated manifold and mininozzles (as in the injection molding of plastics, Sec. 14-3-3), so that gates and runner are eliminated and the yield is raised.

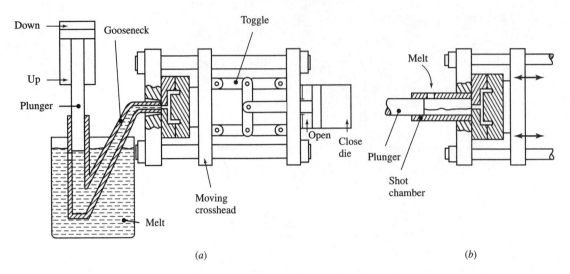

Figure 7–25   Higher pressures are applied in (*a*) hot-chamber and (*b*) cold-chamber die casting. The die is opened and closed by a toggle mechanism. Vents and ejectors are not shown.

2. The *cold-chamber process* uses separate melting facilities. Melt of a quantity sufficient for one shot is individually transferred, often automatically, to the cylinder (*shot chamber*) from which the plunger squirts it into the die cavity (Fig. 7–25*b*) at pressures up to 150 MPa. The plunger is exposed only briefly to high temperatures; therefore, higher-melting metals can be treated. The process has long been established for zinc, magnesium, and aluminum alloys (for example, for transmission and pump housings and rotors of squirrel-cage motors), and is finding increasing use for brass. Occasionally, steel is cast in TZM dies.

It is essential to apply a die lubricant to the mold surfaces between each shot. The lubricant is usually graphite or MoS_2 in an oily carrier, which is then dispersed in water. Evaporation of the water aids cooling.

The repetitive tasks of opening and closing the die, removing and quenching the casting, and insertion into a trim press can be performed by robots. Some die-casting machines use an indexing turret to move the part through flash removal and minor machining steps.

Process control is vital for success. In the past, this was the task of the die-casting machine operator. More recently, a better understanding of the role of process variables has led to extensive instrumentation so that measured variables can be fed into a microprocessor which then exercises control. A number of variables are of importance: melt temperature, dissolved gas content, die temperature and its distribution, plunger velocity and its variation during the stroke, chamber pressure, cavity pressure, and gas composition. In a machine of given capacity, the pressure exerted drops as the rate of metal delivery (pumping rate) increases. Much is now known about optimum pumping rates and pressures needed to make good, thin-walled castings. Computer programs are available to design gating for optimum velocity: Excessive velocity would result in mold erosion and gas entrapment due to turbulence, too slow filling in incomplete parts (*misruns*) and *cold shuts*. (Gate velocities can reach 40 m/s in casting zinc alloys.)

Most alloys are solid solutions and, even though the solidification pattern (Fig. 7–8b) would choke off the flow, pressures are high enough to rip off dendrites, and intricate cavities can be filled. With proper control, wall thickness can be reduced to the point where the entire cross section exhibits the fine structure typical of cast surfaces, thus increasing the strength of the product. With decreasing wall thickness, elastic deflections of die and press become significant, and forces on each press column are measured and equalized.

Porosity is a major problem in solid-solution alloys because solidification is too rapid for the proper feeding of heavier sections, and control generally aims at the generation of distributed microporosity. The other source of porosity is gas entrapped during casting. The difficulty lies in the need for the rapid expulsion of air from the die cavity and in the presence of air itself. One solution is based on the evacuation of the cavity, the other on the observation that in some alloys gas pores contain only nitrogen. If the cavity is purged prior to casting with pure oxygen, the gas reacts with the melt to form finely dispersed Al_2O_3 particles. This process, called *pore-free die casting*, produces better castings, although porosity from solidification or hydrogen entrapment is still possible. In *high-pressure casting* injection pressures are kept at a medium level until the casting is almost solidified; at this point pressures are raised to 150 MPa to consolidate the still mushy material.

Semisolid Processing   Microporosity due to dendritic solidification can be minimized or eliminated and grain size refined by several approaches:

1. *Squeeze-casting or melt forging.* A premeasured amount of melt is loaded into a die, allowed to cool below the liquidus temperature, and then the die is closed while solidification is completed. Using typical forging dies (as in Fig. 9–19b), the process represents a transition between die casting and hot forging, and gives highly refined grain structures and net or near–net shape parts.

Squeeze casting is highly competitive with other casting and forging processes. A squeeze-cast aluminum alloy steering knuckle replaced a ductile iron casting in the Ford Taurus and Mercury Sable automobiles, reducing weight by 50%, and also reducing machining costs. [Source: *Adv. Mater. Proc.*, 1997(6):14.] | **Example 7-8**

---

2. *Rheocasting.* The melt is subjected to shear (e.g., by a rotor submerged in the melt or by an electromagnetic field) during solidification. In consequence, the structure is greatly modified by the breakdown of dendrites and the semisolid acquires thixotropic properties. The partially solidified but still highly fluid mass can be injected into a die.

3. *Thixoforming.*   Billets produced by rheocasting are reheated into the mushy state and forged (also called *semisolid forging*). Wall thickness can be a fraction that of die castings, hence the process is used for aluminum automotive wheels and brake components.

4. *Thixomolding.* Pellets or chips of the alloy are loaded into a screw extruder (as in Fig. 14–3, but without screen pack and breaker plate), heated into the mushy range, sheared to break down the dendrites, and injection molded (chiefly for magnesium alloys).

7-5-7   Centrifugal Casting

When a mold is set in rotation during pouring, the melt is thrown out by the centrifugal force under sufficient pressure to assure better die filling. Solidification progresses from the outer surface inward; thus, porosity is greatly reduced and, since inclusions tend to have a lower density, they segregate toward the center (which in axially symmetric parts is often machined out anyway). Forced movement by shearing the melt results in grain refinement. Centrifuging can be applied to all casting processes if the mold is strong enough to withstand rotation. It is customary to distinguish between various processes according to the shape of the mold.

1. True *centrifugal casting* employs molds of rotational symmetry, essentially tubes, made of steel (protected with a refractory mold wash or even with a green- or dry-sand lining) or of graphite. The melt is poured while the mold rotates, resulting in a hollow product such as a tube or ring (Fig. 7–26a). By controlling flow rates and moving the pouring orifice along the axis, long and large tubes of very uniform quality and wall thickness can be cast. If desired, the outer contour of the casting can be varied, while the inside remains cylindrical. Surface quality is good outside but can be poor inside.

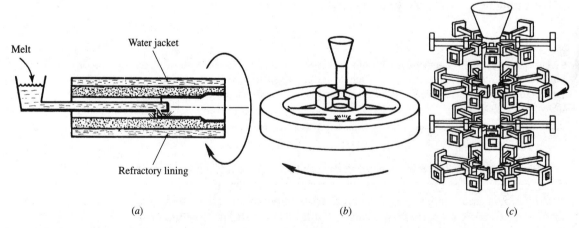

(a)   (b)   (c)

Figure 7–26   Good filling and fine grain can be obtained by rotating the mold in (a) centrifugal, (b) semicentrifugal, or (c) centrifuged casting. [(b) and (c) G. E. Schmidt, Jr., Massachusetts Institute of Technology.]

2. Centrifuging can be applied to all expendable and permanent molds. When only one piece of approximately rotational symmetry (e.g., a wheel with spokes and central hub) is cast, the term *semicentrifugal casting* is applied (Fig. 7–26*b*).

3. When odd-shaped parts are placed around a central sprue in a balanced manner (e.g., by investment molding), the term *centrifuged casting* is usual (Fig. 7–26*c*). Jewelry is *spin-cast* by placing the investment mold at the end of a rotating arm and placing the crucible, containing the molten metal, next to it. Centrifugal force transfers the melt and ensures good mold filling.

7-6   FINISHING PROCESSES

The solidified casting must be subjected to a number of auxiliary operations before it can be used.

7-6-1   Cleaning and Finishing

1. When casting is performed in expendable molds, the first step is to free the casting from the mold. For green and dry sand, *shaking* is a most effective procedure; the clay-bonded sand is then recycled and, with suitable additions, reused. This is one of the reasons for the survival of sand casting as the dominant process for making larger parts even in mass production. With other molding materials reclamation is a matter of economy because it often requires special equipment or processes.

2. Resin-bonded cores are removed (knocked out) by mechanical means, such as pulsating load, high-frequency vibration, or high-pressure water.

3. Residual sand is removed by *shot blasting* (*airless blasting*). Round particles (*shot*) are hurled on the surface of the casting by a fast-rotating paddle wheel. The shot is made of steel, malleable iron, or white iron for harder castings, and of mild steel, bronze, copper, or glass for the softer nonferrous materials.

4. Gates, runners, risers, and sprue are removed (before or after shotblasting), together with any fin (*flash*) that forms when melt flows out into gaps between two mold halves or at cores. In brittle materials, the excess material is simply broken off; in more ductile materials, sawing or grinding becomes necessary. Robots are often employed in these operations, especially in die casting.

5. The entire surface is cleaned by various processes, including shot blasting, tumbling in a dry or wet medium of some refractory material or steel shot, or chemical pickling.

6. If each part has sufficient value, any defects detected may sometimes be repaired by welding without jeopardizing the function of the finished part. Otherwise, defective parts are rejected and remelted.

7-6-2   Changing Properties after Casting

Many castings are subjected to heat treatment (Sec. 6-4), which reduces residual stresses or changes mechanical properties.

To heal existing defects, we have to reach back to the observation (Sec. 4-1-5) that hydrostatic pressure increases ductility while it is applied. If we combine it with high temperature, properties can be permanently improved. For *high-temperature isostatic pressing* (HIP), the casting is placed into a well-insulated furnace which is then loaded into a specially constructed pressure vessel (Fig. 7–27). The system is pressurized with a neutral gas such as argon. Pressures up to 200 MPa and temperatures up to 2000°C are possible, but 100 MPa and 1250°C are more typical of HIPing the superalloy and titanium alloy castings used in jet engines. The metal becomes soft enough for internal porosity to be closed under the imposed pressure and, if the surface of pores is clean, adhesion and solid-state welding take place. Since pressure is applied isostatically (from all directions), shape change is negligible but the properties of the casting—especially ductility, stress rupture, and fatigue properties—greatly improve.

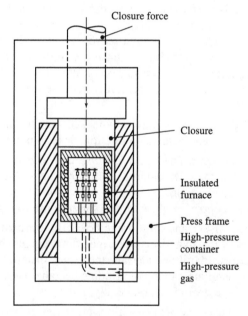

Figure 7–27   Hot isostatic pressing (HIP) imposes omnidirectional pressure and is a powerful method of improving the properties of castings.

7-7 QUALITY ASSURANCE

Quality control and inspection at all stages of production are vital to the success of the modern foundry.

7-7-1 Inspection

Mechanical properties are usually measured on separately cast test bars. It is essential that cooling rates are the same as in the casting, otherwise the results are irrelevant. Cooling rates can be obtained by direct measurements on the casting or from computer simulation.

Inspection for quality employs all techniques. Visual inspection has always been practiced, and is now aided by the use of penetrant dye and magnetic particle techniques. The greatest advances in the quality of castings have been made by the extensive use of nondestructive testing techniques, including ultrasonic, eddy-current, x-ray, and CT inspection for internal soundness (Sec. 4-8). While these measures can be costly and can account for half the cost of a casting, they have made castings competitive in applications hitherto reserved for forgings. Some of the prime examples are aircraft-quality parts and the crankshaft of the internal combustion engine. Nondestructive techniques are particularly important in detecting internal defects, whether they be due to solidification shrinkage, internal hot tearing, or gas porosity.

7-7-2 Casting Defects

The *International Atlas of Casting Defects* helps in classifying and identifying the causes of casting defects, and the American Foundrymen's Society has an ongoing program to develop an expert system. Here we describe only some of the most frequently encountered defects with reference to our earlier discussion.

1. Improper gating and risering or poor molding practices may cause sand erosion and embedment (*sand skin, scab*) in expendable-mold casting. In the extreme case, especially with steels, the melt penetrates into the sand and causes the metal to fuse into a mass. When sand drops off the surface, an *expansion scab* is formed.

2. Poorly controlled sand compaction may cause the dimensional tolerance limits to be exceeded by allowing too much movement of the mold walls (*swell*) in expendable-mold casting.

3. *Shifting* of mold halves and particularly of cores is a common cause of exceeding tolerance specifications.

4. The melt may break out from the flask if the mold is inadequately weighted or the mold wall is too thin. An incipient *breakout* shows up as a thin fin at the parting plane.

5. Insufficient cavity filling (*short run, misrun*) may occur with any process and can be caused by inadequate metal supply, improperly designed gating, or too low mold or melt temperatures.

6. Too low melt temperature can also lead to uneven die filling and visible folds (*cold laps*). *Cold shuts* form when two metal streams meet without complete fusion; this can be particularly troublesome when the metal flows inside an oxide cover, as do alloys that contain aluminum.

7. Surface or subsurface *pinholes* and *blowholes* are caused by gas liberated from the melt or are formed as a result of metal–mold reactions.

8. Shrinkage cavities and porosity may become exposed when gates and risers are removed.

9. Shrinkage on cooling from the solidus temperature requires that the mold should give. If this does not happen, the casting will be larger than expected, will have residual stresses, or will become distorted (thinner sections remaining longer). In the extreme case, when the material is hot-short, fracture occurs in the section that solidifies last.

7-8   PROCESS CAPABILITIES AND DESIGN ASPECTS

Drawing on the preceding discussion, it is now possible to concentrate on process choice and part design. We assume that a concurrent engineering approach has identified material and part geometry and that the decision has been reached to produce by casting.

7-8-1   Process Capabilities

Factors decisive in process choice are summarized in Table 7–3.

1. Some processes are immediately excluded if the melting point of the alloy is too high.

2. Among all metal processing methods, casting permits the greatest shape complexity, except when a permanent die is used: referring back to the classification in Fig. 3–1, some shapes cannot be made at all, others need an expendable core or a multipiece, collapsible permanent core.

3. Surface detail depends on process and, within that, alloy melting point, although some processes (e.g., investment casting) are less influenced by it.

4. Surface finish and tolerances are governed by process and alloy (Fig. 3–22). Indicated tolerances are across parting lines.

5. Size (mass) limits are fairly flexible, and are often defined by practical considerations. For example, the maximum weight of die castings decreases with

Table 7–3   General Characteristics of Shape Casting Processes

Characteristics	Sand		Lost Foam	Plaster	Investment	Permanent Mold	Die
	Green	Resin-Bonded					
Part							
Casting alloy	All	All	Al to cast iron	Zn to Cu	All	Zn to cast iron	Zn to Cu
Shape*	All	All	All	All	All	Not T3, 5, F5, U1, 5, 7 with solid core	Not T3, 5, F5, U1, 5, 7
Surface detail[†]	C	B	C	A	A	B–C	A–B
Mass, kg	0.01–300 000	0.01–100	0.01–100	0.01–100	0.001–100	0.1–100	<0.01–50
Min. section, mm	3–6	2–4	2–4	1	0.75	2–4	0.5–1.5
Min. core diam., mm	4–6	3–6	4–6	10	0.5–1	4–6	3 (Zn: 0.8)
Porosity[†]	C–E	D–E	C–E	D–E	E	B–C	A–C
Cost							
Equipment[†]	C–E	C	B–C	C–E	C–E	B	A
Die (or pattern)[†]	C–E	B–C	B–C	C–E	B–C	B	A
Labor[†]	A–C	C	C	A–B	A–B	C	E
Finishing[†]	A–C	B–D	C–D	C–D	C–D	B–D	C–E
Production							
Operator skill[†]	A–C	C	C	A–B	A–B	C	C–D
Lead time[†]	Days	Weeks	Weeks–months	Days	Hours–weeks	Weeks	Weeks–months
Rates (piece/h·mold)	1–20	5–50	1–20	1–10	1–1000	5–50	20–200
Min. quantity	~1–100	~100	~500	~10	~10–1000	~1000	~10 000

*From Fig. 3–1.
[†]Comparative ratings, with A indicating the highest value of the variable, E the lowest. For example, investment casting produces excellent surface detail, gives very low porosity, involves moderate to low equipment cost, medium to high pattern cost, high labor cost, medium to low finishing cost, and high operator skill. It can be used for low or high production rates and requires a minimum quantity of 10 to 1000 to justify the cost of the pattern mold.

increasing pouring temperature (from zinc to brass) because of thermal loading of the die. At the low end, limits may be set by physical factors, such as fluidity.

6. Minimum section thickness is dictated by fluidity and solidification and is, thus, a function of the distance over which the metal must flow (Fig. 7–28).

7. Porosity is closely related to solidification. It must be remembered, however, that porosity (and even larger cavities) can be tolerated in low-stress locations of the part, and distributed porosity may be allowable in many cases.

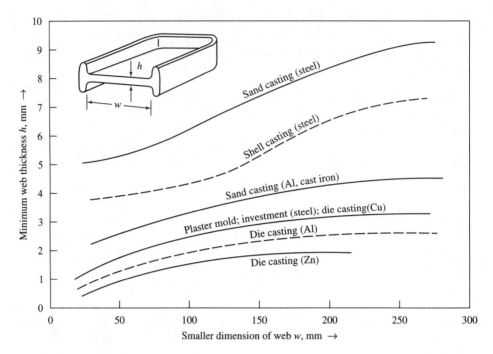

8. Minimum core diameter is limited by the strength of expendable cores and by the strength and heating of permanent cores.

9. Cost and production characteristics are important factors. Thus, in making a few prototype parts, low die and equipment cost will outweigh the high labor cost and skill requirement, and the converse will hold for mass production.

7-8-2   Part Design

Once a decision is reached on the process, the part configuration has to be optimized.

Dimensions   Unless the process yields a net-shape part, machining allowances must be applied to all close-tolerance surfaces. Allowances are typically 1.6 mm for a maximum dimension of 200 mm and increase by 0.8 mm for each additional 200 mm. Some processes, particularly ceramic mold, lost-wax, and die casting are suitable for near-net-shape production and tolerances can be reduced below those shown on Fig. 3–22. The terminology is diffuse but such products are often called precision castings.

Shape   The shape must allow release of solid bodies, such as the core from the core box and pattern from the sand or ceramic mold in expendable-mold casting; the pattern from the mold in expendable-pattern casting; and the casting from the mold in permanent-mold casting.

1. The parting plane, when needed, should be straight, if at all possible. Its location is critical because this determines the need for and location of cores and draft angles. Cores and the extra material required by the draft add to production cost. Therefore, draft should be chosen to contribute to load bearing, and it should not interfere with positive clamping and location in subsequent machining.

2. Undercuts, requiring extra cores in sand casting and movable cores in permanent mold casting, should be avoided if at all possible (Fig. 7–29).

3. Draft angles are larger on inside than on outside surfaces (Fig. 7–29*b*) and, to secure release of the part from the mold or of the core from the core mold, are larger for low height than for greater height. In sand casting, they range from 6° for low walls to 2° for high (over 50-mm) walls. In die casting, angles are lower for low-melting zinc alloys. Angles depend also on production method and must be agreed upon with the supplier. Very small angles are possible with lost-pattern methods.

4. Locations of drilled holes should be strengthened by bosses, shaped preferably so as to make the drill enter perpendicular to the cast surface. Where justified, inserts should be considered.

5. In permanent-mold casting, ejectors must be located where they cause no distortion and the print is not objectionable.

Directional Solidification   The shape of the casting should promote solidification from the remotest parts toward the feeding end, and should not close off access of melt to thicker sections. Otherwise, headers must be added to avoid shrinkage cavities and microporosity.

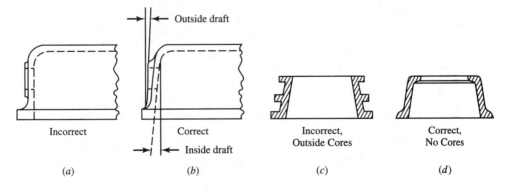

Incorrect	Correct	Incorrect, Outside Cores	Correct, No Cores
(*a*)	(*b*)	(*c*)	(*d*)

Figure 7–29   Design should aim at simplifying casting: Undercuts require (*a*) extra cores and (*b*) can be eliminated. Ribs on this part require (*c*) outside cores which (*d*) become superfluous with redesign. (*Steel Castings Handbook, 6th ed., ASM International, 1995, p. 7–11, Figs. 7–27 and 7–28, with permission.*]

1. Directional solidification is assured if the wall is tapered away from the point of feeding (Fig. 7–30a).

2. When wall-thickness variations are unavoidable, transition must be made by generous radii (Fig. 7–30a). Small radii or sharp corners act as stress raisers in the finished casting, create turbulence during pouring, and prevent proper feeding during solidification.

3. Localized heavy cross sections—such as result when a radius is applied only to the inside surface of a corner or when two ribs cross each other—create hot spots where shrinkage cavities form. As might be expected from Chvorinov's rule, the problem increases with increasing difference in inscribed circle diameter. Applying a radius to the outer surface (Fig. 7–30b) or offsetting the ribs (Fig. 7–30c) helps. Otherwise, it would be necessary to reduce the cross section by placing a core into the thickest section or accelerating cooling with external chills (Fig. 7–30c).

4. Thin mold parts break off or overheat, creating hot spots. Generous radii alleviate the problem (Fig. 7–30d).

Warping   The casting will warp if free contraction is restrained by the mold or if two sections of different thickness cool at different rates. Delayed shrinkage of the thicker section induces stresses in the already solidified thinner section (Fig. 7–31a).

Hot Tears   Materials susceptible to hot shortness (Sec. 6-1-6) can tear. They must be cast into simple shapes that do not develop tensile stresses during solidification,

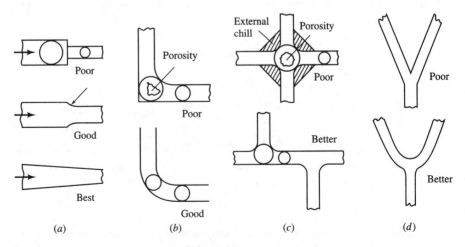

Figure 7–30   Design features help to alleviate casting problems due to (a) wall thickness variation, (b) hot spot in corner, (c) hot spot at cross ribs, and (d) hot spot at Y junction.

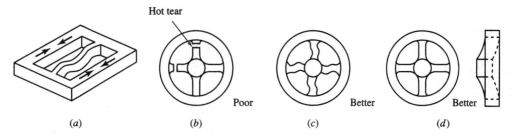

Figure 7–31   Differential shrinkage of unequal sections leads to warping (*a*), and shrinkage against an unyielding mold results in hot tearing (*b*), which can be prevented by redesign (*c*) and (*d*).

or in mold materials that collapse or give freely and allow shrinkage. The shape should permit deformation without moving large mold masses. Straight spokes of a wheel would tear even in a sand mold (Fig. 7–31*b*); S-shaped spokes can straighten by displacing relatively little sand and allow shortening during and after solidification (Fig. 7–31*c*), and a similar result is obtained when the hub is placed at a different level from the rim (Fig. 7–31*d*).

For many parts, there is a wide choice of manufacturing processes. In the concurrent engineering approach, a quick look is taken early on (even during conceptual design, Sec. 2-1-2), to see what processes can conceivably be used and what their implications are for design. Materials, part size, ratios of dimensions, and tolerances can have significant effects, and they must be considered in the next stage of deliberations. To illustrate some of these points, we shall consider two generic groups of parts throughout the text.

Example 7-9

One group of parts has rotational symmetry (Fig. 3–1, Group S3). To make changes in problem statements easier, geometry is shown in capital letters (*a*). Thus, for example, ID = inner diameter, mm; TID = tolerance for inner diameter, mm; SID = surface finish of internal diameter, μm R_a.

1. In the first stage of evaluation, general feasibility is considered. Assume that the shape is to be cast; from Fig. 3–1 and Table 7–3, all processes are suitable. If, however, the material is spheroidal cast iron, only various forms of sand casting and investment casting remain as feasible options (Table 7–3).

2. In the second stage of evaluation, dimensions and tolerances are needed. Assume OD = 100 ± 1 mm; ID = 60 ± 0.05 mm; FH = 20 ± 1 mm; FT = 5 ± 1 mm; CD = 70 ± 1 mm. SFT = SID = 0.8 μm R_a; other surfaces undefined.

    From Fig. 3–22, the ID tolerance can be barely met by investment casting. SFT and SID are achievable only with investment casting. Since machining is likely to be needed on the ID anyway, the cost of investment casting cannot be justified for such a simple part. For sand casting, apply the machining allowance to ID and FT. Most likely, four flanges would be cast at a time, arranged symmetrically around a central sprue, with gates on the side or flange surface.

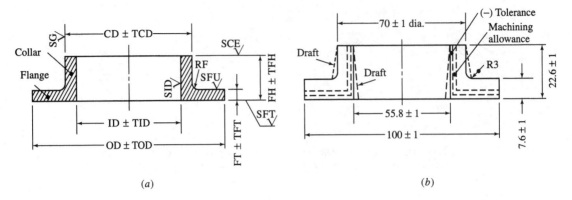

(a)   (b)

Figure Ex. 7-9

Thus, metal will flow in the circumference of the ring and fillet radius (RF) is not needed to aid flow. However, the sharp edge would be difficult to maintain without sand erosion; choose RF = 3 mm or more. For lowest mold cost, part at base of flange; this necessitates draft on the collar surfaces. A first approximation to the cast shape is shown in (b). More detailed evaluation will take into account production quantities, cost, etc.

Example 7-10

Another generic group of parts can be described as a section (an H beam, Fig. 3–1, Group S0.) A straight section (a) is suitable for loading in either compression or bending. For torsional stiffness, a cross rib is added (b). This configuration (Fig. 3–1, Group S6), repeated in the SL direction, is the basis of many structural parts subjected to combined stresses and is representative of many manufacturing challenges.

In the first evaluation, assume sand casting. Even without assigning dimensions, we can address some issues for the straight section (a). The most obvious choice is to part through the center of the web, thus no cores are needed. Draft must, however, be applied to the flanges (c), and this represents extra material that contributes little to strength in bending. The cavity will most likely be fed from the end; if fed from the bottom of the flange, rising melt streams will meet in the middle of the web; if fed from the middle of the web, melt will cascade down into the bottom flanges. Not immediately obvious is the possibility of casting in the vertical position. If external cores are used, internal draft can be close to zero (d). Bottom feeding gives smooth melt flow, and feeder heads can be placed at the junctions of web to flange. Further consideration must include material of casting, machining allowances, size and tolerances, production quantities, and relative cost.

With a cross rib (b), a horizontal parting line allows feeder heads to be placed at the intersections of flange and cross rib. In the vertical position, feeding the intersection would be difficult except that RR can be quite small and the hot spot can be minimized. It should be noted that in many applications web thickness WT can be reduced with little loss in load-bearing capacity. This can certainly be done in casting, provided that thickness is not reduced below the capabilities of the process (Fig. 7–28).

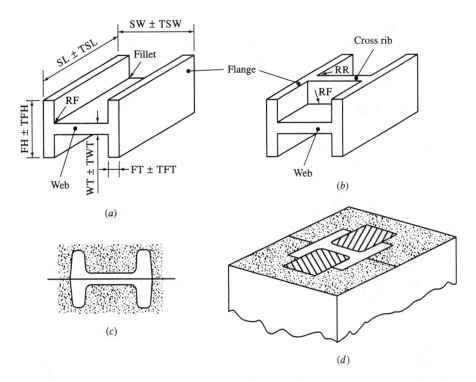

Figure Ex. 7-10

7-9   SUMMARY

Solidification processes are involved in producing the vast majority of metallic materials. Almost all wrought materials are cast before deformation, and a significant proportion of metals is cast into a shape directly. Cast components are used for applications as diverse as machine-tool bases, automotive engine blocks and crankshafts, turbine blades, plumbing fittings, and decorative hardware. With the appropriate molding process, parts of a complexity unmatched by any other process may be cast.

A sound casting results only if the limitations imposed by the solidification process are recognized in the design of the casting and of the process:

1. For solidification to occur, the mold must be colder than T_m of the metal. Problems of fluid flow and of heat transfer limit the minimum attainable wall thickness, especially if the alloy solidifies with the growth of dendrites which choke off fluid flow.

2. Gating and risering techniques must ensure smooth, complete filling of the die cavity followed by orderly solidification, with a liquid metal supply sufficient to feed the pipes that would otherwise form.

3. Heat transfer must be locally controlled to prevent starvation of late-solidifying portions of the casting and minimize porosity.

4. Grain-size control is one of the most powerful means of improving mechanical properties. One usually aims at a fine grain, but exceptionally (for high-temperature creep resistance) at coarse grain, in the limit, by directional solidification of a single dendrite.

5. Mechanical properties, particularly ductility and fatigue and impact strength, may be greatly improved by melting, pouring, and casting techniques that reduce the number and size of inclusions and bring inclusions into the least harmful shape and location.

6. Properties can be improved by the application of pressure in the mushy temperature range or by hot isostatic pressing of the solidified casting.

7. Desirable properties may be imparted also by heat treating the solidified casting to remove residual stresses (stress-relief anneal), homogenize the structure (homogenization), increase strength (solution treatment and aging of precipitation-hardenable alloys and quenching and tempering of steel), or develop the optimum structure (annealing of white iron to convert it into malleable iron, and annealing of ductile cast iron).

8. The dangers of handling molten metal, light and thermal radiation of high-temperature melts, and operation of foundry machinery require special training, personal protective outfits, and stringent safety measures. Noxious fumes and gases must be contained or treated for the protection of workers and the environment.

PROBLEMS 7A

7A-1   (a) Draw a sketch showing planar solidification in a cylindrical mold. (b) State what alloys solidify in this pattern. (c) Draw a sketch for dendritic solidification. (d) State what kind of alloys solidify in this pattern.

7A-2   (a) Define superheat. (b) Normalize superheat for alloys of different melting points.

7A-3   (a) Draw a diagram, showing the solubility of gases in pure metals and solid solutions as a function of temperature. (b) Give two methods suitable for assuring a low gas content in the solidified casting.

7A-4   Draw a diagram showing shear stress versus strain rate for (a) dilatant, (b) Newtonian, (c) pseudoplastic fluids and (d) Bingham substance.

7A-5   (a) Define thixotropy. (b) Give an example of a material that shows this behavior.

7A-6   Draw a diagram showing the effect of mold temperature on (a) cooling rate, (b) grain size, (c) productivity.

7A-7   State three possible functions of fluxes in melting.

7A-8   Define for sand casting: (a) pattern; (b) core; (c) core box; (d) core print.

7A-9   (a) State the components of green sand. (b) Define green-sand and dry-sand mold. (c) Which is stronger? (d) Which is more suitable for casting hot-short materials? Why?

7A-10   Make a sketch of the cross section of a sand mold, showing a part similar to a connecting rod (with larger sections at each end), and the required: (*a*) pouring basin, (*b*) sprue, (*c*) well, (*d*) runner, (*e*) gate, (*f*) in-gate, (*g*) hot riser, (*h*) cold riser, (*i*) feeding head, (*j*) dead riser, (*k*) live riser. Identify by letter each element (one element may have several names).

7A-11 Make a sketch showing the principal elements of a cold-chamber die-casting machine. (*a*) Label the major components of the machine and the critical features of he mold. (*b*) State what the major limitation is on the materials that can be cast by the technique.

7A-12 Repeat Prob. 7A-11 for hot-chamber die casting.

7A-13 Make simple sketches showing the principal distinguishing features of the four major types of permanent-mold casting processes.

7A-14 Make simple sketches showing the principal distinguishing features of (*a*) centrifugal casting, (*b*) semicentrifugal casting, and (*c*) centrifuging.

PROBLEMS 7B

7B-1 Sometimes it is preferred not to disturb the structure of a casting by machining its surface. Suggest a possible reason for this preference.

7B-2 An AISI 1020 steel is continuously cast into 200 mm × 200 mm billets. Make a sketch of the casting process, clearly showing the die and the shape of the solidification zone. Indicate in what part of the billet one might expect microporosity due to solidification shrinkage.

7B-3 A 1000-mm-wide, 400-mm-thick, and 2000-mm-long ingot is cast into a permanent mold. (*a*) Make a sketch of the 400 × 2000 mm cross section showing the grain size and grain distribution expected in casting a pure metal, free of gases. Show the shape and location of piping and microporosity, if any. (*b*) Draw another sketch as for (*a*), but this time for a solid-solution

alloy. Underneath draw a diagram showing macrosegregation; explain with the aid of the relevant section of an equilibrium diagram.

7B-4 A ring of 120 mm OD, 80 mm ID, and 20 mm height is to be cast in a sand mold. As a part of your preliminary design, make sketches to show at least four options for positioning the parting line. Indicate where draft and/or core may be needed.

7B-5 A part is claimed to have been made by thixoforming. (*a*) Is the material likely to be a pure metal, a solid solution, or a eutectic? (*b*) What is the likely microstructure? (*c*) What was the likely production sequence?

7B-6 Despite best efforts, a superalloy casting shows some porosity. You suggest that it be subjected to HIP. For the benefit of others: (*a*) Define HIP. (*b*) Explain what it does. (*c*) State what properties are improved most: strength, elongation, reduction in area, fatigue strength, impact strength, or hardness. (*d*) Explain whether it will always work.

7B-7 (*a*) Solid-solution alloys of wide freezing range are generally regarded as poor casting alloys; why? (*b*) They are extensively used in die casting. Provide an explanation.

7B-8 (*a*) Draw a sketch of a part with hot tearing. (*b*) Explain, with the aid of appropriate sketches, the events leading to the problem. (*c*) State what type of alloy is most prone to it.

7B-9 Continuing Prob. 7B-8, indicate whether hot tearing is a greater danger for a material of high thermal expansion or low thermal expansion.

7B-10 The flanged elbow fitting shown is to be cast of stainless steel. In a design sketch, (*a*) chose the parting line; (*b*) indicate the direction of feeding, and (*c*) suggest a design modification to assure directional solidification. (*d*) Consider all casting processes (Fig. 7–12 and Table 7–3 will help); eliminate all technically unsuitable processes (justify each decision with a brief sentence). (*e*) Suggest a casting orientation that gives the best fill.

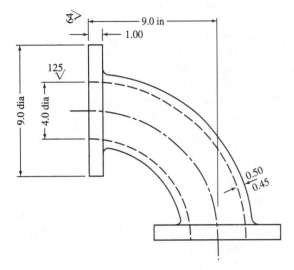

7B-11   A large, flat machine-tool base is to be cast with intersecting stiffening ribs on both sides of the flat. Show in a sketch the problem that may arise and suggest (in a sketch) at least one design modification to minimize the problem.

7B-12   A part resembling a connecting rod is cast in sand. Porosity is found in the shank and cavities in the heads. Show, with appropriate design sketches, (*a*) a remedy for the heads and (*b*) two possible remedies for the shank. Explain in one sentence each.

7B-13   A part of undercut shape (Fig. 3–1, shape U4) is to be cast. As part of your preliminary design, determine if it can be cast into (*a*) a sand mold and (*b*) a permanent mold. To justify your answers, make sketches to show the cores and the limitations on shape.

7B-14   In a meeting you are asked to describe, very briefly, the essential features of the four principal die-casting processes. There are no facilities to make sketches. (Make sure to clarify the method of filling the cavity.)

7B-15   In a meeting you are asked to describe, very briefly, the essential features of the three principal centrifugal casting processes. There are no facilities to make sketches.

7B-16   Automotive grills are sometimes made of Al–Si–Cu die-casting alloys. To explain the reasons, (*a*) draw one end of a typical binary eutectic equilibrium diagram with a terminal solid solution. (*b*) Choose a composition strengthened purely by a solid-solution mechanism; mark it C_s. (*c*) Choose a composition strengthened by both solid-solution and precipitation-hardening mechanisms, mark it C_p. (*d*) Show in a sketch the expected crystallization mode of C_p as it solidified in a thin-wall section. (*e*) Will the metal allow thin-wall pressure die casting? Why? (*f*) Do you expect to see microporosity in this casting? Why?

7B-17   A casting of very intricate shape and good surface finish is to be cast. Suggest some suitable methods if the casting is (*a*) solid and if it is (*b*) hollow, with a cavity of complex shape.

7B-18   It is argued that gas evolution during solidification causes inverse segregation. To settle the point, (*a*) define macrosegregation; (*b*) state the expected variation for inverse segregation, (*c*) state the most frequent cause of such segregation. (*d*) On this basis, what variation should one expect for gas-induced segregation?

7B-19   The part shown in Fig. 7–29c is sand cast. Assuming rotational symmetry, make design sketches to show (*a*) the cope and drag halves of the mold, (*b*) the cores required, if any. (*c*) Repeat for Fig. 7–29d.

7B-20   Repeat Prob. 7B-19 for Fig. 7–29a and b (assume a rectangular geometry with only two holes).

7B-21   Assume that the part of Fig. 7–29a is made in a permanent mold. Make a design sketch to show the mold halves and movable cores, if any.

7B-22 Draw a typical eutectic equilibrium diagram with limited solid solubility at each end. (a) Identify all phase fields. (b) Draw underneath a diagram to show variations of microporosity and pipe formation. (c) Mark in the equilibrium diagram the composition corresponding to pure metal A as C_A; a composition within the range of limited solid solubility as C_R; at the limit of solid solubility as C_L; the eutectic composition as C_E. (d) Define fluidity. (e) Set up a table as shown here and rank, marking the best 1 and the worst 4. Assume gravity-fed sand casting and, for fluidity, equal superheat.

	C_A	C_R	C_L	C_E
1. Fluidity				
2. Danger of microporosity				
3. Ease of superheating				
4. Castability				

7B-23 Suggest three methods of refining the grain size in a cast component.

7B-24 Explain the difference between vacuum molding and vacuum casting.

PROBLEMS 7C

7C-1 A part cast of aluminum alloy 413.0 breaks in service. Analysis confirms that the composition is correct. Metallographic examination shows large, angular interdendritic particles. What is the likely cause of the problem?

7C-2 Draw a diagram showing the changes in fluidity expected on moving in the Cu–Ag system from the Cu to the Ag end.

7C-3 Show that Bernoulli's equation [Eq. (7-4)] is dimensionally homogeneous and indeed represents energy per unit volume.

7C-4 Look up the Al–Si equilibrium diagram and select compositions with 2% Si and 12% Si. (a) What are the critical temperatures of the 2% Si alloy and what phases would you expect in what location in the structure at room temperature? (b) Repeat this for the 12% Si alloy. On this basis, which alloy is (c) more ductile in the equilibrium condition, (d) more prone to coring, (e) more susceptible to hot shortness, (f) of higher fluidity, (g) easier to feed, and (h) more prone to microporosity? (i) State which alloy is more favorable for casting.

7C-5 Liquid steel of 0.2% C with 0.01 wt% oxygen is to be strand cast. Assume no solubility of O in the solid and that the O will be rejected in the form of CO. (a) Calculate the amount of gas released in units of $cm^3/100$ g metal. (b) Recalculate as a percentage of the total (gas + metal) volume. (c) If the steel is continuously cast into a 100 mm × 100 mm square strand at 1.5 m/min, calculate the feed rate of 10-mm-diameter aluminum wire used to deoxidize the steel.

7C-6 A Cu-40Zn brass melt is poured into a sand mold. The metal level in the pouring basin is 200 mm above the centerline of the runner, which is taken as the zero level. The cross section of the runner is 10mm × 10mm. (a) Make a sketch. (b) Calculate, from Bernoulli's theorem [Eq. (7-4)], the velocity and rate of flow at the entry to the mold, ignoring friction losses (the pouring basin is so large that the velocity in it can be taken as zero).

7C-7 A wheel is cast of low-carbon steel with straight spokes (as in Fig. 7–31b). The length of each spoke is 100 mm. The mold is of a refractory material, unyielding, and changes its dimensions insignificantly during heating or cooling. The spoke cools from 1100 to 900°C in 10 min. To assess the feasibility of design, calculate (a) the strain, if the thermal expansion coefficient is 23 × 10^{-6} per °C; (b) the average tensile strain rate; (c) the flow stress at 1000°C (from Table 8–2). (d) Assuming that the material at this temperature behaves like an ideal elastic-plastic body, and Young's modulus is 60 percent of the room-temperature value, determine whether shrinkage will be accommodated by the development of elastic (residual) stresses or by plastic deformation. (*e*) Subject the above problem statement to a detailed critique regarding the validity of the simplifying assumptions.

7C-8   With the aid of Chvorinov's rule [Eq. (7-5)], calculate the relative solidification times for castings of identical volumes and of the following shapes: (*a*) sphere of diameter d; (*b*) cylinder with $h/d = 1$; (*c*) cylinder with $h/d = 10$; (*d*) cube; (*e*) right rectangular prism with $h/a = 10$; (*f*) flat plate of the same length as (*e*) but of one-third the thickness. (*g*) Plot the results to illustrate the effect of shape changes.

7C-9   The ring of Prob. 7B-4 is made of a Zn alloy by cold-chamber die casting. For your design, (*a*) choose the parting line halfway through the thickness of the ring. (*b*) Determine the size of the die-casting machine needed.

7C-10   From Chvorinov's rule, calculate the relative solidification times for risers of unit volume, of shapes defined in Prob. 7C-8.

7C-11   A charge of 2000 kg "jewelry bronze" (alloy C22600, 87.5Cu-12.5Zn) is to be made up for semicontinuous casting. Pure copper and zinc and 70/30 brass scrap are available. Design the charge makeup, allowing 0.5% Zn for loss by oxidation.

7C-12   A charge of 1000 kg leaded yellow brass (C85400 in Table 7-2) is to be made up for melting in an induction furnace. Materials available: 60/40 brass (sufficient to provide up to 60% of the charge); 70/30 brass; pure copper; and zinc. For economy, Sn is added in the form of bearing bronze (C93700 in Table 7-2). Design the charge makeup utilizing as much as possible of the 60/40 brass; the 70/30 brass scrap is more expensive but still cheaper than the pure metals. Allow 1% Zn loss.

7C-13   The connecting rod shown in Fig. 7–18*a* is cast with a solid end of diameter 80 mm and height 50 mm. The part is fed from the big end, and a cold riser is placed at the small end. To assure soundness of your design, compute the dimensions of a riser of $h = 1.2d$ ratio that takes 50% longer time to solidify than the small end.

FURTHER READING (see also Chap. 6)

ASM Handbook, vol. 15, *Casting*, ASM International, 1988.

Analysis of Casting Defects, American Foundrymen's Society, 1974.

Beeley, P.R.: *Foundry Technology*, Butterworths, 1972.

Campbell, J.: *Castings*, Butterworth-Heinemann, 1991.

Clegg, A.J.: *Precision Casting Processes*, Pergamon, 1991.

Davis, J.R. (ed.): *Cast Irons*, ASM International, 1996.

Davis, J.R. (ed.): *Tool Materials*, ASM International, 1995.

Elliott, R.: *Cast Iron Technology*, Butterworths, 1988.

Johns, R.: *Casting Design*, American Foundrymen's Society, 1987.

Miller, R.K.: *Robots in Industry: Applications for Foundries*, SEAI Institute, Madison, GA, 1982.

Rowley, M.T. (ed.): *International Atlas of Casting Defects*, American Foundrymen's Society, 1974.

Wieser, P.F. (ed.): *Steel Castings Handbook*, 6th ed., Steel Founders' Society of America and ASM International, 1995.

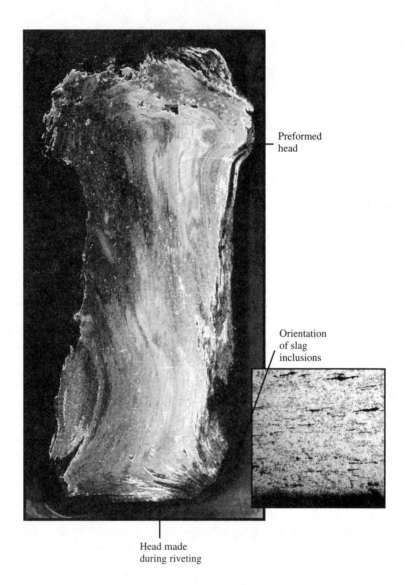

Preformed
head

Orientation
of slag
inclusions

Head made
during riveting

When the Titanic struck an iceberg, the steel plates of the hull and bulkheads—held
together by rivets—were ripped apart. A marine forensic panel concluded that slag
stringers in the rivets, oriented perpendicular to the direction of loading, were the
likely cause of the catastrophic damage. (*Courtesy Dr. Tim Foecke, National Institute
of Standards and Technology, Gaithersburg, Maryland.*)

8

Plastic Deformation of Metals

In preparation for our discussion of bulk deformation and sheet metal-working processes, we will critically review fundamentals, including:

Why YS is not flow stress

Anomalies in plastic flow: yield point, serrated yielding, and textures

The consequences of cold working and their removal by recovery and recrystallization

Exploiting strain-hardening and restoration mechanisms for the control of structure and mechanical properties

Elements of the mechanics of plastic deformation: effects of stress state, friction, and inhomogeneous deformation

Concepts of bulk workability and sheet formability

In Sec. 7-5 we mentioned that some 85% of all metals is cast into ingots, slabs, or billets for further working by plastic deformation (Table 8–1). Plastic deformation implies that the shape of the workpiece is changed without a change in volume or melting of the material. It is, obviously, essential that the material should be able to undergo plastic deformation without fracture but, because all deformation occurs in the solid state, die filling will not be as easy as it was in casting. Therefore, in the design of parts and of the metalworking processes themselves it will be necessary to consider not only the laws governing material flow (because they define whether the desired configuration can be obtained) but also the ductility of materials (because it sets a limit to the attainable deformation) and pressures, forces, and power requirements (because they determine the loading of tools and equipment).

The success of processes depends on interactions between material properties and process conditions, and the principles to be discussed here have universal applicability. For practical reasons it is, however, usual to divide metalworking processes into two groups. In *bulk deformation processes*

Table 8-1   Shipments of Wrought Products (U.S.A.)

Alloy group	Thousand Mg*	
	1972	1995
Steel		
Sections, rails	5 900	8 700
Plate	7 300	9 000
Hot-rolled sheet and strip	14 200	17 800
Cold-rolled sheet and strip	17 800	14 200
Galvanized sheet	4 900	14 600
Tinplate	5 000	2 600
Hot-rolled bar	11 800	11 900
Cold-finished bar	1 600	1 800
Wire	2 300	650
Tube, pipe	6 900	5 400
Forgings	1 200	
Copper and brass	2 600	3 300
Aluminum	4 100	6 500
Lead (incl. battery)	480	760
Magnesium	16	24

SOURCE: Compiled from *Metal Statistics* 1974 and *Metal Statistics* 1997, American Metals Market, Fairchild Publications Inc., New York, 1974 and 1997.
*Mg (= 1000 kg = metric tonne = 2200 lb).

(Chap. 9) the thickness, diameter, or other major dimension of the workpiece is substantially changed. In *sheet-metalworking processes* (Chap. 10) thickness change is incidental; furthermore, the sheet—which is the starting material—is the product of a bulk deformation process, namely rolling. Here our concern is with fundamentals that are equally applicable to bulk deformation and sheet metalworking processes.

8-1   MATERIAL PROPERTIES

In Secs. 4-1 and 4-3 we already discussed many properties of solid materials, but we did so with reference to the properties required in service. Now we need to reexamine these properties with emphasis on their relevance to deformation processing.

8-1-1   Flow Stress in Cold Working

For metalworking calculations, yield strength and tensile strength—the properties of primary interest for the design of products—are of secondary importance. Our first concern is the stress required to deform the workpiece material.

Flow Stress in Tension   The engineering stress (Eq. 4-3) conventionally calculated from the tension test is of little value for computations, although it is widely used in communication. We need a true stress which, by definition, is force P divided by instantaneous area A. We could measure the instantaneous cross-sectional area, but most of the time we compute it using the principle of constancy of volume. As long as elongation is uniform over the gage length (Fig. 4–2c),

$$A = A_0 \frac{l_0}{l} = \frac{V}{l} \tag{8-1}$$

where l is instantaneous length. Once necking begins, the minimum diameter—which is the only diameter of relevance—is unknown and no further points can be calculated.

Since our interest is in permanent deformation which begins at the point of yielding, the true stress is usually calculated from initial yielding to necking. Each calculated point defines the stress that must be applied to keep the material deforming, flowing; hence, we call it the *flow stress* σ_f

$$\sigma_f = \frac{P}{A} \tag{8-2}$$

where P is the instantaneous force. We could—and sometimes do—plot flow stress as a function of engineering tensile strain e_t [Eq. (4-4)], but for computational purposes the *true strain* ε (also called *natural* or *logarithmic strain*) is needed. By definition, it is obtained as the natural logarithm of the ratio of instantaneous length l to original length l_0

$$\varepsilon = \ln \frac{l}{l_0} = \ln \frac{A_0}{A} \tag{8-3}$$

The data derived from the tensile force–displacement curve may now be plotted to define the true stress–true strain curve (Fig. 8–1a). For comparison, the engineering stress–true strain curve is also shown in Fig. 8–1a in broken lines. [There is one point that can be calculated, even if only approximately, beyond necking: The fracture force P_f is available, and the corresponding minimum cross-sectional area A_f (Fig. 4–2d) can be measured on the broken specimen. For reasons to be explained later, the true stress thus calculated is somewhat high].

Flow stress curves of many materials have been determined, and an atlas of such curves can be—and has been—built up. However, a more condensed and, for computations, more convenient record can be kept. When σ_f is replotted against ε on log–log paper, a straight line frequently results (Fig. 8–1b), indicating that σ_f must be a power function of ε

$$\sigma_f = K\varepsilon^n \tag{8-4}$$

where K is the *strength coefficient* and n is the *strain-hardening exponent*. From the log–log plot, K is the stress at a strain of unity, and n is the slope of the line, *measured on a linear scale*. Alternatively, K and n are found by fitting a power-law curve to the data points (note that n is *not* the slope of the true stress–true strain curve!).

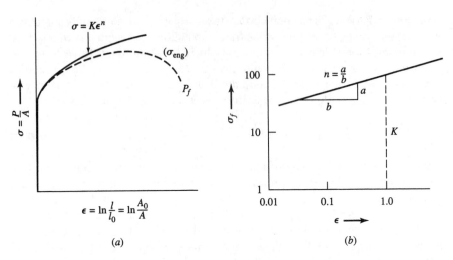

Figure 8–1 In the cold-working temperature range (a) many materials obey the power law of strain hardening as shown by (b) the linear log–log plot of flow stress versus true strain.

Example 8-1 From the force–displacement curve given in Example 4-4, calculate the flow stress of the material at several points. Plot to obtain the K and n values.

The task lends itself to a spreadsheet solution. Constants are: gage length 25.0 mm; thickness 6.35 mm; width 6.38 mm; and volume of the specimen $V = (6.35)(6.38)(25.0) = 1013$ mm^3. Note that Δl is always obtained by drawing a line, from the point of interest, *parallel to the elastic line*. The spreadsheet gives the following results:

A	B	C	D	E	F	G
Delta l	P	l	et	epsilon	A	sigma
mm	N	mm				N/mm^2
			Eq. (4-4)	Eq. (8-3)	Eq. (8-1)	Eq. (8-2)
2.0	9100	27.0	0.080	0.077	37.51	243
4.0	11200	29.0	0.160	0.148	34.93	321
6.0	12800	31.0	0.240	0.215	32.87	386
8.0	13500	33.0	0.320	0.278	30.69	440
10.0	14000	35.0	0.400	0.336	28.94	484
12.5	14200	37.5	0.500	0.405	27.01	519

(Note that true stress is always higher than engineering stress.)

A last point may be calculated from the fracture area: $A = 2.85 \times 3.5 = 9.98$ mm^2 and strain is based on the fracture strain $\varepsilon = \ln(A_0/A_f) = \ln(40.5/9.98) = 1.4$. The log–log plot of points defines a straight line, thus the material obeys Eq. (8-4). $K = 760$ MPa, and $n = 0.45$ (quite high but not unreasonable, since the material is a ductile solid solution).

Effect of Strain Hardening on Necking   An important observation is that strain hardening delays the onset of necking. This may be understood by considering the events involved in the formation of a neck. In the course of extension, an incipient neck may form anywhere along the gage length, generally at a point of inhomogeneity, i.e., where the material is, for any reason, weaker (because of a surface irregularity, an inclusion, or a large grain of weak orientation). If the n value is high, localized deformation in the incipient neck raises σ_f at this point [Eq. (8-4)]. Deformation will now continue in other, less strain-hardened parts of the specimen, until hardening can no longer keep up with the loss of load-bearing capacity due to reduced cross section; at this time, one of the necks stabilizes and continues to neck (Fig. 8–2a) while the applied force drops. It can be shown that, for a material that obeys the power law of strain hardening [Eq. (8-4)], the n value is numerically identical to the uniform (prenecking) strain expressed as true strain ε_u; therefore, a material of low n necks soon after initial yielding (Figs. 8–2a and b).

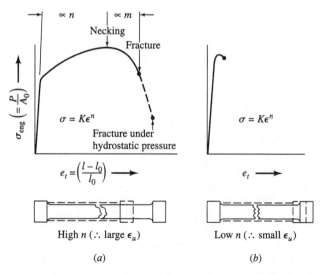

Figure 8–2   (a) A high strain-hardening rate, as expressed by a high n value, results in large uniform (prenecking) elongation; postnecking deformation increases with increasing strain-rate sensitivity or high m value, and fracture is delayed under hydrostatic pressure. (b) A material of low n necks early and, if m is low, fractures soon.

Example 8-2

Check whether $n = \varepsilon_u$ for the material of Example 4-4.

By the definition of Eq. (4-9a), uniform strain can be expressed as natural strain $\varepsilon_u = \ln(l_u/l_0)$. From the recording in Example 4-4, $l_u = 25 + 12.5 = 37.5$ mm. Thus, $\varepsilon_u = \ln(37.5/25.0) = 0.405$. This is less than the $n = 0.45$ found in Example 8-1. Good agreement between the two values can be expected for steels; agreement is often less good for nonferrous materials.

Flow Stress in Compression   A problem with the tension test is that necking limits the uniform strain that can be obtained, and the development of flow stress at higher strains is uncertain. Yet, many metalworking processes involve heavy deformation, and the compression test (Sec. 4-3) is then more useful. The instantaneous cross-sectional area is again calculated from constancy of volume [Eq. (4-2)], but now the length is more descriptively called the height h [Eq. (4-14)]. The true strain ε is, by definition,

$$\varepsilon = \ln\frac{h}{h_0} = \ln\frac{A_0}{A} \qquad \textbf{(8-5a)}$$

The calculation yields a negative number. As far as the material is concerned, compressive and tensile deformations cause the same metallurgical changes. Hence the convention is usually ignored and, to obtain a positive value, true strain is taken as the natural logarithm of the *ratio of the larger value to the smaller value*

$$\varepsilon = \ln\frac{h_0}{h} = \ln\frac{A}{A_0} \qquad \textbf{(8-5b)}$$

The true stress is again from Eq. (8-2). In reality, this is an interface pressure and may be regarded as the flow stress only if friction effects can be neglected (Sec. 9-2-1). From a log–log plot of σ_f versus ε, the K and n values can be extracted. Indeed, most published data (including those that will be given in Tables 8–2 and 8–3) have been determined in compression tests.

Example 8-3

Find the K and n values for the steel of Example 4-9.

When we plotted the true-stress/compressive-strain curve in Example 4-9, we already had all the relevant data. Only the true strain has to be calculated from Eq. (8-5b) (results are entered in the table of Example 4-9). From the log–log plot, $K = 800$ MPa and $n = 0.13$. Note that the strain-hardening capacity, while not as high as for the Cu–Ni alloy of Example 8-1, is still quite substantial for this interstitial solid solution of C in Fe. The test material was in the slightly cold-drawn condition, hence n is less than it would be for the same steel in the annealed condition (Sec. 8-1-4). (Note also that the engineering and natural strains are very similar at low reductions, but the numerical value of natural strain becomes progressively larger with increasing reductions.)

Example 8-4

As far as their effect on strain hardening is concerned, extension of a bar from $l_0 = 1$ unit length to $l = 2$ units should be the same as compressing a bar of the same material from $h_0 = 2$

units to $h = 1$ unit height. Calculate the corresponding engineering and true strains.

Engineering tensile strain [Eq. (4-4b)] $e_t = 100(2 - 1)/1 = 100\%$

compressive strain [Eq. (4-16)] $e_c = 100(2 - 1)/2 = 50\%$

Natural strain, tension [Eq. (8-3)] $\varepsilon = \ln(2/1) = 0.69$

compression [Eq. (8-5a)] $\varepsilon = \ln(1/2) = -0.69$

Note that it is very misleading to quote engineering strain without specifying whether it is tensile or compressive because, to calculate tensile strain, the *change in dimensions* is divided by the *smaller* dimension, whereas for compressive strain it is divided by the *larger* dimension (see also Example 4-8). The absolute value of natural strain is the same for tension and compression, correctly indicating that the two deformations are equivalent in their effects on the material.

8-1-2   Discontinuous Yielding

Not all metals and alloys show the smooth transition from elastic to plastic deformation discussed hitherto, and not all of them strain-harden in a continuous fashion. Such anomalies in plastic flow behavior have structural reasons.

Yield-Point Elongation   We mentioned in Sec. 6-1-2 the possibility of forming interstitial solid solutions in which solute atoms, much smaller than the solvent atoms, fit into the spaces existing between atoms in the basic lattice. These solute atoms often seek more comfortable sites where lattice defects have created voids in the structure. Most markedly, this is found with carbon and nitrogen in iron. Their atoms are small enough to fit into the lattice; nevertheless, they tend to migrate to dislocations where distortion of the lattice provides more room (just below the extra row of atoms in Fig. 6–13c). In a sense, the solute atoms form a *condensed atmosphere* that completes the lattice and immobilizes, *pins the dislocations*.

In the course of deformation, a larger stress must be applied before dislocations can break away from the condensed atmosphere of carbon or nitrogen atoms. This leads to the appearance of a *yield point* on the stress–strain curve of low-carbon steels (Fig. 8–3a; note, however, that the initial force peak is due to interactions with the elastic properties of the test equipment, therefore, the so-called upper yield point is not a true material parameter). After the dislocations have broken away from the pinning atoms, they multiply and move in large groups in the direction of maximum shear stress (very approximately, at 45° to the applied force). Under favorable conditions, such localized yielding becomes visible with the appearance of *Lüders lines* or *strain bands* on the surface of the specimen (Fig. 8–4) but results in objectionable *stretcher strain* marks on the surface of stretch-formed sheet-metal parts (Sec. 10-1-1). Successive generation of strain bands continues over the whole length of the specimen at a relatively

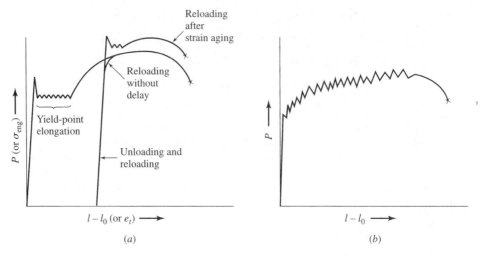

Figure 8–3   (*a*) The yield-point elongation typical of mild steel returns if a steel subject to strain aging is stored after initial deformation; (*b*) serrated yielding is typical of solid-solution alloys.

Figure 8–4   The yield point phenomenon results in the development of visible shear bands, Lüders' lines, on a polished mild-steel strip subjected to tension. (*Courtesy S. Kadela, University of Waterloo.*)

low stress, giving the familiar *yield-point elongation* (Fig. 8–3*a*). Once the strain bands cover the entire surface, normal strain-hardening behavior takes over.

If straining is interrupted and then immediately resumed, the original strain-hardening curve is rejoined. However, if sufficient time is allowed for the interstitial atoms to seek out new dislocation sites (so that the carbon and nitrogen atmospheres condense again), the steel is strengthened and the yield-point phenomenon returns (broken line in Fig. 8–3a). This behavior is described as *strain aging*.

Serrated Yielding   Discontinuous yielding during strain hardening is observed in some materials for causes related to negative strain-rate sensitivity rather than to dislocation pinning. When tested on a "soft" machine (of low spring constant), yielding is *stepwise*; on a hard machine, the force drops erratically and rapidly (*serrated yielding*) (Fig. 8–3b). Such behavior can be particularly troublesome with some substitutional aluminum alloys, because it again leads to the development of visible and objectionable marks on the surface (Sec. 10-1-2).

8-1-3   Textures (Anisotropy)

We saw in Sec. 6-3-1 (Fig. 6–13) that crystals deform by slip on preferred planes. If the crystal shown in Fig. 6–13a is to become longer, the slip planes must rotate into the direction of straining; in compressive deformation the slip planes rotate across the direction of straining. This has important consequences in polycrystalline materials, particularly when only a limited number of slip systems are available. Before deformation, properties will be *isotropic* (the same in all directions), representing the average properties of randomly oriented crystals. Deformation results in an elongation of grains and, within the grains, rotation of slip planes. This leads to a noticeable alignment (*preferred orientation* or *texture*) of crystallographic orientations (Fig. 8–5a). A polycrystalline material possessing a texture will show some of the directional properties typical of single crystals. This *directionality* or *anisotropy* of properties (dependence on the direction of testing) is evident in variations of the elastic modulus, YS, TS, elongation (Fig. 8–5b), and other properties. It can be exploited for special purposes. Thus, silicon steel (3Si-0.003C) sheets are processed to align the cube edge along the rolling direction to optimize magnetic properties for transformer cores. Most importantly for metalworking, the relative magnitudes of strains also change during deformation.

Anisotropy of Deformation   We can express the principle of constancy of volume [Eq. (4-2)] in terms of natural strain: the sum of the three principal *true* strains is equal to zero

$$\varepsilon_1 + \varepsilon_2 + \varepsilon_3 = 0 \tag{8-6}$$

Remember that true strain is the natural logarithm of new dimension divided by old dimension [Eqs. (8-3) and (8-5a)]. In a tension test the major strain ε_1 is positive (tensile) whereas the transverse strains ε_2 and ε_3 are negative (compressive). For convenience, it is usual to speak of length strain ε_l, width strain ε_w, and thickness strain ε_t (Fig. 8–5c). Then

$$\varepsilon_l + \varepsilon_w + \varepsilon_t = 0 \tag{8-7}$$

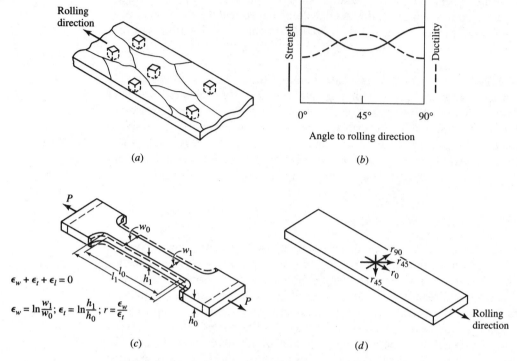

$$\epsilon_w + \epsilon_t + \epsilon_l = 0$$

$$\epsilon_w = \ln\frac{w_1}{w_0};\ \epsilon_t = \ln\frac{h_1}{h_0};\ r = \frac{\epsilon_w}{\epsilon_t}$$

(c)

(d)

Figure 8–5   The development of a texture (a) reflects in variations of mechanical properties (b). The effects of anisotropy on the deformation of a material are determined in tension tests (c), which are repeated in different directions relative to the rolling direction (d).

This relationship always holds, but ε_w and ε_t need not be equal in magnitude. By convention, the relative magnitudes of transverse strains are expressed by the *r value*, which is the ratio of width strain to thickness strain

$$r = \frac{\varepsilon_w}{\varepsilon_t} \tag{8-8}$$

Types of Anisotropy   Several possibilities exist:

1. When the material is *isotropic*, $\varepsilon_w = \varepsilon_t$, and $r = 1$. It does not matter whether the specimen is cut in the rolling direction, across it, or at an intermediate angle (Fig. 8–5d); in an isotropic material

$$r_0 = r_{90} = r_{45} = 1 \tag{8-9a}$$

2. It is conceivable that the r values vary in relation to the rolling direction.

$$r_0 \neq r_{90} \neq r_{45} \tag{8-9b}$$

This is denoted as *planar anisotropy* and leads to such problems as earing in deep drawing (Sec. 10-6-2).

3. If the r values measured in the plane of the sheet are identical in all directions but deviate from unity

$$r_0 = r_{90} = r_{45} \lessgtr 1 \qquad \textbf{(8-9c)}$$

we speak of *normal anisotropy*, because deformation of the test specimen in the thickness direction (normal to the sheet surface) is greater or smaller than in the width direction.

4. It is possible and indeed usual that normal and planar anisotropy occur simultaneously

$$r_0 \neq r_{90} \neq r_{45} \neq 1 \qquad \textbf{(8-9d)}$$

To separate the two kinds of anisotropy, we can define a mean r, denoted $\bar{r}$ or r_m

$$r_m = \frac{r_0 + r_{90} + 2r_{45}}{4} \qquad \textbf{(8-9e)}$$

as a measure of normal anisotropy (frequently, the symbol r is used loosely to denote $\bar{r}$ or r_m). A measure of planar anisotropy is Δr

$$\Delta r = \frac{r_0 + r_{90} - 2r_{45}}{2} \qquad \textbf{(8-9f)}$$

The supplier of a steel sheet provided the following values: $r = 1.70$ and $\Delta r = 0.64$. It is known that r is the same in the rolling and transverse directions and is the high value. Find r_0, r_{90}, and r_{45}. | **Example 8-5**

$$\Delta r = 0.64 = (2r_0 - 2r_{45})/2; \quad 0.64 = r_0 - r_{45}$$
$$r_0 = r_{90} = 1.700 + 0.32 = 2.02; \quad r_{45} = 1.70 - 0.32 = 1.38$$

Anisotropy of Sheet Materials   Development of texture greatly depends on crystal structure.

1. In hexagonal materials the limited number of slip systems leads to the development of a texture after relatively small (20–30%) deformations, with most of the basal planes aligned perpendicular to the application of the rolling force, that is, with basal planes almost parallel to the sheet surface. When a tension test specimen cut from such a sheet is elongated, deformation is highly anisotropic.

a. In hcp materials with a *high c/a ratio*, sliding is limited to the basal planes (Fig. 6–2c), thus the thickness of the sheet is reduced while its width is hardly affected, just as a card pack can be elongated by sliding the cards over each other (Fig. 8–6a). The r value becomes very small, typically 0.2 for zinc.

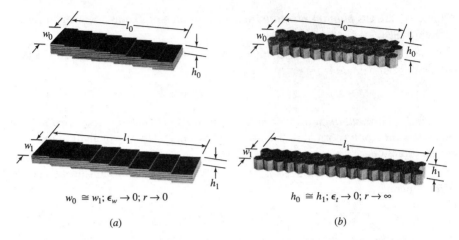

$$w_0 \cong w_1; \epsilon_w \to 0; r \to 0 \qquad\qquad h_0 \cong h_1; \epsilon_t \to 0; r \to \infty$$

(a) (b)

Figure 8–6   Deformation of hexagonal metals: (a) A high c/a ratio leads to basal slip and low r value; (b) a low c/a ratio results in prismatic slip and a high r value.

b. Deformation of a tensile specimen cut from an hcp material with a *low c/a ratio* shows a dramatically different behavior. Since slip now takes place on prismatic and/or pyramidal planes (Fig. 6–2d), sheet thickness is hardly reduced; instead, most of the deformation takes place by rearrangement of the hexagonal prisms, leading to a marked reduction in the width of the specimen (Fig. 8–6b). The *r* value could, theoretically, reach infinity, but in practice seldom exceeds 6, the value for titanium.

2. Metals of fcc structure possess many equivalent slip systems (Fig. 6–2a); therefore, only much later—typically after more than 50% reduction—do they develop a texture. A completely randomly oriented polycrystalline fcc material is nearly isotropic ($r = 1$). However, after deformation the *r* value may drop, and many aluminum alloys tend to have $0.4 < r < 0.8$.

3. The common slip direction in bcc materials (Fig. 6–2b) can be exploited by appropriate processing to give *r* values ranging from 0.8 to over 2.

Example 8-6 | **A** tension test is conducted on a sheet specimen (as in Fig. 8–5c) of $l_0 = 50.0$ mm, $w_0 = 6.0$ mm, and $h_0 = 1.00$ mm. The test is interrupted before the onset of necking; at this time, $l_1 = 60.0$ mm and $w_1 = 5.42$ mm (the thickness h_1 is difficult to measure with sufficient accuracy). Calculate the r value.

We may calculate the average thickness from constancy of volume, or obtain ε_t from Eq. (8-7):

$$\varepsilon_t = -\varepsilon_l - \varepsilon_w = -[\ln (60/50)] - [\ln (5.42/6.00] = -0.1823 + 0.1017 = -0.0806$$
$$r = (-0.1017)/(-0.0806) = 1.26$$

8-1-4 Effects of Cold Working

It is obvious from Fig. 8–1 that an ever-increasing true stress is needed for the continuing deformation of a metal. Because this is a direct consequence of working or straining, one speaks of *work hardening* or *strain hardening*. The reason for it is to be found in the mechanism of plastic deformation.

We saw in Sec. 6-3-1 that crystalline metals deform by slip and, on the atomic scale, by the propagation and multiplication of dislocations (Fig. 6–13). Slip occurs on close-packed planes in close-packed directions (Fig. 6–2); inspection of Fig. 6–2 will show that there are a number of equivalent *slip systems* in each crystal structure. As deformation proceeds, dislocations may begin to move on several systems. It takes a higher stress to move a succession of dislocations on the same plane, and a yet higher stress is needed to move them once dislocations propagating on different planes become entangled. This higher stress is the cause of the increase in flow stress. Distortion of the crystal lattice by foreign atoms further inhibits the free movement of dislocations and increases strain hardening; therefore, solid solutions have a higher n value. Since this gives a larger prenecking strain, *solid solutions have high ductility.*

A material subjected to cold working, for example, by rolling or drawing, strain-hardens too. Dislocation density increases and, when a tension test is performed on this strain-hardened material, a higher stress will be needed to initiate and maintain plastic deformation; thus, the YS rises. The TS rises too, although not as rapidly as the YS, and the TS/YS ratio approaches unity (Fig. 8–7). How-

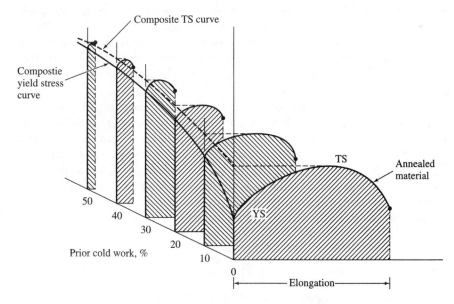

Figure 8–7   Tension tests conducted on previously worked material show that cold working increases strength and reduces ductility.

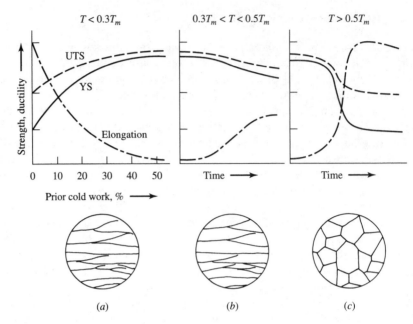

Figure 8–8   The effects of prior cold work (*a*) are partially removed by recovery (*b*) and the original, soft condition is fully reestablished by recrystallization (*c*).

ever, the ductility of the material—as expressed by total elongation and reduction of area—drops because of the higher initial dislocation density. Similarly, K rises and n drops. The microstructure changes too: Crystals (grains) become elongated in the direction of major deformation. These changes are summarized in Fig. 8–8*a*. The material may also develop directional properties, as discussed in Sec. 8-1-3.

Strain hardening is important for several reasons. Since many cold-worked materials retain a reasonable level of ductility, cold working offers the designer a low-cost method of obtaining higher-strength materials. There is, however, a price to pay: the increased flow stress can generate excessive tool pressures and the reduced ductility may lead to fracture of the workpiece. This can become a major problem when heavy reductions are to be taken or when the manufacture of products involves a succession of cold-working steps. It is then necessary to remove the effects of cold working by annealing.

Example 8-7 | The effects of cold working are well demonstrated in the properties of Cu-30Zn brass. As indicated in Example 6-7, this was the traditional material for cartridge cases, but it is used for many other purposes, mostly in sheet form. It is supplied in various rolled "tempers."

From *Metals Handbook*, 9th ed., vol. 2, p. 324:

Temper	Rolling Reduction, %	YS, MPa	TS, MPa	el., %	Hardness, HRB
H01 ($\frac{1}{4}$ hard)	10.9	275	370	43	55
H02 ($\frac{1}{2}$ hard)	20.7	360	425	23	70
H04 (hard)	37.1	435	525	8	82
H06 (extra hard)	50.1	450	595	5	83
H08 (spring)	60.5		650	3	91
H10 (extra spring)	68.6		680	3	93

8-1-5 Annealing

We defined annealing as heat treatment that involves heating to (and holding) at some elevated temperature (Sec. 6-4-1). When its purpose is the removal of the effects of cold working in the finished product, one speaks simply of *annealing*. When the purpose is softening of a workpiece for further cold working, one speaks of *process annealing*. Fundamentally, both involve the same metallurgical processes.

Recovery In Sec. 4-6 we introduced the concept of homologous temperature and indicated that above $0.5T_m$ the strength of many materials drops. One of the reasons is that larger thermal excursions allow atoms to move to vacant sites (Sec. 6-1-2) and thus to change places with relative ease. Even before this temperature is reached, increased atomic mobility allows the rearrangement of dislocations into regular arrays (typically, at temperatures of $0.3–0.5T_m$). Given enough time, such *recovery* restores some of the original softness without changing the visible grain structure (Fig. 8–8*b*). This offers some special benefits: In most metals, ductility drops rapidly with even a small degree of cold work (Fig. 8–8*a* and solid line in Fig. 8–9) while recovery increases ductility without greatly affecting strength (broken line in Fig. 8–9). Therefore, *recovery anneal* is a useful method for producing a material of higher strength yet reasonable ductility. It also restores electrical conductivity, important for electric wires.

Some aluminum alloys would gradually soften at room temperature and are stabilized by a low-temperature heat treatment (H3 temper). From *Metals Handbook*, 9th ed., vol. 2, p. 102:

Example 8-8

	YS, MPa	TS, MPa	el., %
5056-O (annealed)	152	290	35
5056-H18 (full hard)	407	434	10
5056-H38 (recovery annealed)	345	414	15

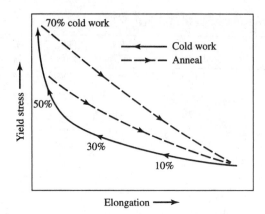

Figure 8–9 Cold working followed by partial annealing can give relatively high ductility combined with good strength. *(From J.A. Schey, in Techniques of Metals Research, R. F. Bunshah (ed.), vol. 1, pt. 3, Interscience, 1968, p. 1415. With permission.)*

The increase in ductility may appear very minor, yet it is often a recovery annealed material that allows a sheet metalworking process of critical difficulty to proceed.

Recrystallization Above $0.5T_m$ atoms can move, diffuse to form new, relatively dislocation-free nuclei which grow until all the cold-worked structure is recrystallized. Diffusion is greatly time- and temperature-dependent (Fig. 8–8c). An equiaxed structure normally results, with a grain size that is a function of prior cold work, annealing temperature, and time.

The driving force for *recrystallization* is provided by the increased energy content (*stored energy*) resulting from the higher dislocation density induced by cold working. Therefore, recrystallization begins at a lower temperature with increasing *prior cold work* (Fig. 8–10). We know that coarse-grained material has low strength (Sec. 6-3-6), hence the aim is generally that of producing finer grain. This can be achieved with increasing cold work because, for any given temperature, more nuclei form and grain size diminishes. Strength increases with little loss in ductility (see Example 6-13). There is, of course, no recrystallization possible if cold work is zero, and the original grain size is retained. The low dislocation densities induced by very light (say 2–4%) cold work results in the formation of only a few nuclei which can then grow to a large size. Such *critical cold work* is usually undesirable because of the poor mechanical properties of coarse-grained structures. At the opposite end of the spectrum, very fine grain—obtained by annealing a heavily cold-worked metal—gives high strength yet reasonable ductility (see Example 6-13). It should be noted that annealing does not necessarily restore isotropy; the deformation texture may simply be replaced with an annealing texture.

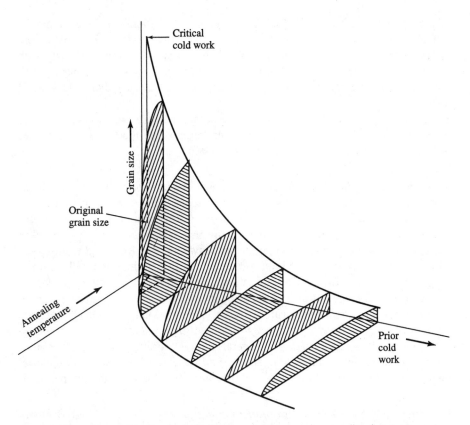

Figure 8–10   Recrystallization begins at lower temperatures and recrystallized grain size decreases with increasing prior cold work.

The temperature of $0.5T_m$ should be taken only as a very rough guide, since even minor amounts of alloying elements can substantially delay the formation of new grains and thus raise the recrystallization temperature. In alloys specifically designed for high-temperature service, such as the superalloys, heavy alloying pushes the onset of recrystallization to around $0.8T_m$ (see Example 6-9).

When a metal is held at temperature for a prolonged time, larger grains—which have a smaller surface area per unit volume and hence a lower surface energy—grow at the expense of smaller grains. Such *grain growth* is, in general, undesirable because strength and, if excessive grain growth occurs, even ductility suffers.

Recovery and recrystallization are collectively termed *softening processes* or *restoration processes*.

8-1-6   Hot Working

We have noted that temperatures above $0.5T_m$ greatly facilitate the diffusion of atoms. This means that an arrested dislocation has the option of climbing, and can thus move into another, unobstructed, atomic plane. If, therefore, deformation it-self takes place at such elevated temperatures, many dislocations can immediately disappear; in fact, one finds that softening processes work simultaneously with dislocation propagation. Material resulting from such *hot working* has a much lower dislocation density and, therefore, is less strain-hardened than cold-worked material.

In practice, hot working is conducted at higher temperatures, where softening processes are fast, but not at such high temperatures that there would be danger of incipient melting (typically between $0.7T_m$ and $0.9T_m$).

Mechanisms of Hot Working   Since $0.5T_m$ is also the temperature of recrystalliza-tion, it is often said that hot working is conducted above the recrystallization temperature. However, recrystallization *during* hot working (*dynamic recrystal-lization*) is by no means universal; in many materials *dynamic recovery* takes place during working, resulting in quite low flow stresses. Recrystallization may still occur on holding at or cooling from the hot-working temperature. Therefore, the distinctive mark of hot working is not a recrystallized structure, but the si-multaneous occurrence of dislocation propagation and softening processes, *with or without recrystallization during working*. The dominant mechanism depends on temperature, strain rate, and grain size, and may be conveniently shown on deformation mechanism maps. In general, the recrystallized structure becomes finer with lower deformation temperature and faster cooling rates, and material of superior properties is often obtained by controlling the finishing temperature.

Flow Stress in Hot Working   Since all softening processes require the movement of atoms, the time available for these processes is critical. This means that in hot working there is substantial *strain-rate sensitivity*. We already observed that strain rate should not be confused with deformation velocity (Sec. 4-6). In its simplest definition, strain rate is the instantaneous deformation velocity divided by the instantaneous length or height of the workpiece [Eq. (4-17)]. For compressive deformation (Fig. 4–11)

$$\dot{\varepsilon} = \frac{v}{h} \tag{8-10}$$

Again, $\dot{\varepsilon}$ is expressed in units of s^{-1}.

To find the flow stress of a metal, specimens are heated to a constant tem-perature and then compressed (or tested in tension) at a constant strain rate, on machines in which the crosshead velocity changes in a programmed manner so as to keep $\dot{\varepsilon}$ [Eq. (8-10)] constant.

Example 8-9 | The flow stress of metals is to be determined by compressing 20-mm-high cylinders at constant strain rates. Calculate the press speed needed for compression to 60% reduction in height at $\dot{\varepsilon} = 5 \, s^{-1}$.

From Eq. (8-10), the press must slow down as the height diminishes to the 8-mm final height.

Height, mm	Press Speed, mm/s
20	100
16	80
12	60
8	40

From recordings of force versus displacement, stress–strain curves are plotted which may show a number of trends (Fig. 8–11a):

1. After an initial peak, flow stress drops with increasing strain. Such *strain softening* is usually a sign of dynamic recrystallization.

2. The stress–strain curve may be fairly flat after initial yielding, indicating that strain hardening and softening processes roughly balance each other.

3. At yet higher strain rates, stresses increase with increasing strain, indicating that softening processes could not keep pace with strain hardening.

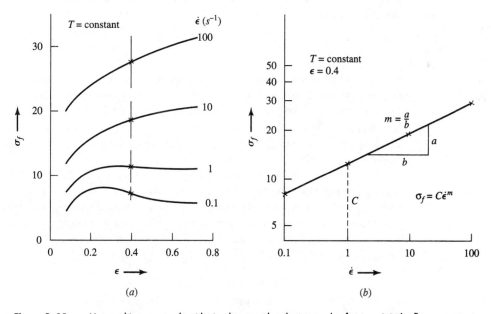

Figure 8–11   Hot working proceeds with simultaneous hardening and softening. (*a*) The flow stress is sensitive to strain rate and (*b*) for a given temperature and strain, it is often a power function of strain rate.

To a first approximation, hot working can be regarded as though it were governed purely by strain rate. Then flow stress values for a given strain may be extracted from the true stress–true strain curves (Fig. 8–11a) and replotted as a function of strain rate on a log–log scale (Fig. 8–11b). In the majority of instances, the line thus defined will be straight, indicating that hot-working flow stress is a power function of strain rate $\dot{\varepsilon}$:

$$\sigma_f = C\dot{\varepsilon}^m \tag{8-11}$$

where C is a *strength coefficient*, and m is the *strain-rate-sensitivity exponent*. The value of C is found at a strain rate of unity, and m is the slope of the line, again measured on a linear scale (Fig. 8–11b). Alternatively, a power-law function is fitted to the data points. It is evident from Fig. 8–11a that different C and m values will be found for different strains. Both C and m also change with temperature. Increasing temperature usually increases strain-rate sensitivity and thus m, but always decreases the flow stress and thus C. [*Note*: The full form of Eq. (8-11) would have $\dot{\varepsilon}/\dot{\varepsilon}_0$ in it; with $\dot{\varepsilon}_0 = 1$, the universally used form of Eq. (8-11) is obtained.]

Example 8-10 | Calculate C and m for the material shown in Fig. 8–11, assuming that σ_f is given in MPa.

From the graph in Fig. 8–11b, $C = 11.8$ MPa (remember to read on the log scale) and $m = 7.5/17 = 0.44$ (remember to read on the linear scale). This high m value indicates a superplastic material.

For computational purposes, experimentally determined C and m values (for example, from Tables 8–2 and 8–3) or flow stress curves must be used. It is worth noting, however, that time and temperature are equivalent in their effects on softening. Therefore, it is sometimes possible to express all hot-working flow stress values with a single curve that is a function of a *velocity*- (or *strain-rate-*) *modified temperature*.

In discussing cold-working flow stress [Eq. (8-4)] we made the tacit assumption that strain-rate effects could be ignored (i.e., $m = 0$). This is not entirely true; a fuller description of the response of metals would include both strain and strain rate. Strain-rate sensitivity increases with increasing homologous temperatures, and increases rather suddenly when the hot-working temperature is reached. Typical values of the strain-rate sensitivity exponent are

Cold working	$-0.05 < m < 0.05$
Hot working	$0.05 < m < 0.3$
Superplasticity	$0.3 < m < 0.7$
Newtonian fluid	$m = 1$

Warm Working   Deformation at $0.3T_m$ to $0.5T_m$ is often denoted as *warm work-ing*, and is characterized by reduced strain hardening, increased strain-rate sensitivity, and a somewhat lower flow stress relative to cold working.

Ductility   A high m value means that markedly higher forces are needed to deform the material at higher strain rates. This translates into greater total elongation for the following reason:

When, in the course of testing in tension, a neck begins to form, this incipient neck is the smallest cross section of the specimen. In a non-strain-rate-sensitive material, it would also be the weakest part and it would thin out and fracture. Events take a different turn in a material with positive m value. Since deformation is concentrated in the neck, the instantaneous deforming length in Eq. (4-17) suddenly drops (see Figs. 4–2 and 4–5). Strain rate in the neck becomes much higher than it was before necking, whereas it drops to zero outside the necked zone. Consequently [Eq. (8-11)], the flow stress of the material in the neck increases, and the neck resists further deformation. Instead, adjacent material deforms and further locations neck until the entire gage length is deformed (Fig. 4–16). Thus, we find that total elongation increases with higher n (strain-hardening exponent) which governs *prenecking strain* and a higher m (strain-rate sensitivity exponent) which governs *postnecking strain* (Fig. 8–2a). This will be particularly important in stretching-type operations (Secs. 10-5 and 14-4).

Superplasticity   In some extremely fine-grained materials (often alloys with a two-phase *microduplex structure*) high-temperature deformation takes place by extensive grain-boundary sliding and accompanying diffusion (essentially, by entire grains sliding past each other) or by mass diffusion which reshapes entire grains. Deforming forces can be very low and, as long as strain rates are kept within the limits that allow these deformation mechanisms to prevail (Fig. 8–12), the superplastic behavior is maintained and very large elongation values (up to several hundred and even thousand percent) are readily obtained. Thus, techniques developed for the forming of polymers (Sec. 14-4) can be applied to these metals.

After cooling from the superplastic temperature, many alloys develop substantial strength. However, the same mechanisms that allow superplastic deformation also account for the poor creep resistance of fine-grained materials (Sec. 6-3-6). Therefore, superplastically deformed parts may be made suitable for high-temperature service by a high-temperature anneal. The coarse grains thus formed have relatively little grain-boundary area and offer greater resistance to creep at low strain rates (Fig. 6–18). This process sequence is the basis of Gatorizing®[1] a patented process for making superalloy turbine disks.

---

[1] Registered trademark of Pratt & Whitney, United Technologies.

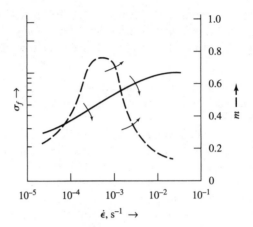

Figure 8–12   Some very fine-grained materials exhibit superplasticity, with very high *m* values within a limited strain-rate range. Arrows indicate effects of decreasing grain size or increasing temperature.

8-1-7   Interactions between Deformation and Structure

Up to now we tacitly assumed that the workpiece was homogeneous. This can be far from reality, and the interactions of deformation processes with structural features can be exploited to control the service properties of materials.

Destruction of Cast Structure   The structure of cast ingots or billets shows a number of undesirable features. Grains and dendrite arm spacing within grains tend to be large, thus strength is low; columnar grains (Fig. 7–1*a*) may be oriented in unfavorable directions, further reducing strength and ductility in some directions. Concentration gradients usually exist, as evidenced by microsegregation (coring, Fig. 6–9), and macrosegregation (Fig. 7–4). Microporosity, typical of dendritic solidification (Fig. 7–1*b*) is often present and there may even be gross piping (Fig. 7–1*a*). Pinholes and blowholes may remain as a result of gas evolution during solidification (Fig. 7–4*a* and Sec. 6-3-5).

Hot working is the most powerful method for eliminating harmful features because:

1. The forced movement of atoms favors recrystallization and the equalization of composition. Thus, grains are refined and homogenization is accelerated. It is generally found that minimum 75% reduction ($\varepsilon > 1.4$) is needed for destroying the cast structure. If necessary, the direction of deformation can be reversed to accumulate the necessary strain without overall shape change (*redundant work processing*).

Example 8-11

An electric generator rotor is forged from a cast ingot of 1500-mm diameter and 3000-mm length. To assure the integrity of the rotor, 75% hot work must be imparted, yet the shape of the rotor requires that the original dimensions be maintained. Find a method for accumulating the necessary work.

In Example 8–4 we already found that axially compressing a cylinder of 2 units height and 1 unit diameter to half its height imparts $\varepsilon = -0.69$ strain. Thus. we can compress to 1500 mm and then draw out (see Sec. 9-2-3) the workpiece to its original dimensions, again imposing $\varepsilon = 0.69$ strain. In terms of their effect on structure, compressive and tensile strains are equivalent, hence the total strain is $\varepsilon = 0.69 + 0.69 = 1.38$. (The example is highly simplified and ignores the complications introduced by inhomogeneous deformation.)

2. Pores are compressed until their walls touch; if pressures and temperatures are high enough, adhesion and solid-state welding effectively eliminate the pore as a defect (at least if its walls were originally free of contaminants). However, cracks oriented in the direction of force application are likely to open up rather than heal.

3. Deformation greatly extends oxides and other internal contaminant films. Consequences depend on the nature of the inclusions.

a. *Brittle* inclusions are broken up into small particles around which pressure welding can take place. Thus, even though the inclusions remain in the material, they may be rendered harmless from the point of view of mechanical properties. This is also true of oxide films that may be present on the internal surfaces of pores and pipes. Because intermetallic compounds are generally brittle, they may also be broken up.

b. *Ductile* inclusions will be stretched out and could considerably impair properties.

c. *Heavy oxides* and slag inclusions found in pipes prevent welding and cause laminations in the hot-worked product.

4. Cast ingots are usually subjected to a sequence of hot-working steps (*passes*), and recrystallization during or in between passes replaces the coarse cast grain with a fine, equiaxed structure of much better mechanical properties.

5. The more or less randomly distributed inclusions and second-phase particles become aligned and, to some extent, oriented in the direction of major deformation. This *mechanical fibering* gives rise to *anisotropy*, i.e., a variation of properties with testing direction, quite independently of any directionality which may be due to crystal structure (Sec. 8-1-3). Typically, in the direction of fiber orientation the properties of the matrix dominate, and strength and ductility are high (Fig. 8–13a). When the material is loaded (during manufacture or in service) in the transverse direction, inclusions serve as effective stress raisers (Fig. 8–13b). Therefore, the so-called *short-transverse properties* (such as strength and, even more so, impact strength, fatigue strength, and ductility) suffer (Fig. 8–13c).

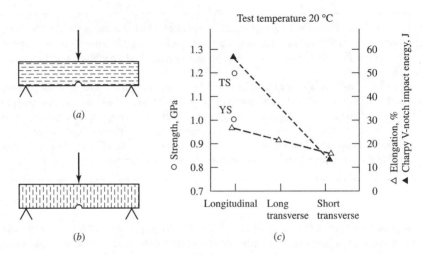

Figure 8–13   Fibering due to alignment of second-phase particles, inclusions, and segregation leads to directional properties, which (*a* and *b*) are revealed in impact testing and (*c*) are most clearly evident in ductility and impact energy. (*Data for Inconel 718 taken from Forging Design Handbook, American Society for Metals, 1972, p. 14.*)

6. Fibering can be revealed by deep *macroetching* (as opposed to the lighter microetching used to reveal the grain structure). The fibered structure developed in earlier passes is distorted on subsequent working, and etching a cross section to reveal *flow lines* is a most useful tool in studying material flow. Even in the absence of inclusions or second-phase particles, flow lines show up when homogenization is not perfect and traces of microsegregation remain. This is true of steels in which the large phosphorus atoms remain segregated even after heavy hot working, thus outlining the flow lines upon macroetching (as in Figs. 9–4*d* and 9–18).

Example 8-12 | A brittle inclusion embedded in a bar has a width of 5 mm, length of 12 mm, and thickness of 1 mm. It is oriented in the direction of elongation (elongation may be the result of any metalworking process that causes elongation by reducing the height of the workpiece). If the bar is reduced by 90% in thickness without a change in width, and if the inclusion breaks up without any change in thickness, over what length will the fragments be distributed?

A 90% reduction in thickness results in a strip with a height equal to $(100 - 90) = 10\%$ of the original thickness. From constancy of volume, $l_0 w_0 h_0 = l_1 w_1 h_1$. If $w_0 = w_1$, the extension is $l_1/l_0 = h_0/h_1 = 100/10$ or tenfold. Thus fragments of the inclusion of 12 mm original length will be scattered over a length of 120 mm. In between, over a distance of $120 - 12 = 108$ mm (or 90%), the bar material will be in full contact and re-weld.

Example 8-13 | When the Titanic was launched in 1912, it embodied the latest technology of the day and was dubbed "unsinkable" by the popular press. Yet, when it hit an iceberg, it suffered severe damage and sank in three hours. In 1996 a section of the hull plating and several rivets were recovered and turned over to the Marine Forensic Panel of the Society of Naval Architects and

Marine Engineers. The plate was found to have a ductile-to-brittle transition temperature much above the water temperature (recall the problem with Liberty ships, Example 4-7). The major damage was, however, caused by the fracture of wrought iron rivets, so that the hull plates were ripped apart. Wrought iron of 1912 contained about 2–3 vol.% iron-silicate slag, which became elongated during hot rolling. Slag stringers act as inclusions of zero strength, thus the transverse strength and ductility of the iron was only one-tenth the longitudinal strength and ductility. The recovered rivets had even more (over 9%) slag, rolled out into coarse stringers. This created no problem in the preformed head in which the stringers followed the shape of the head, but proved catastrophic in the head formed during riveting: The stringers were oriented transverse to the loading direction and the rivets popped off when the ship hit the iceberg (Source: T. Foecke: *Metallurgy of the RMS Titanic*, National Institute of Standards and Technology-IR 6118, Feb. 4, 1998.)

Thermomechanical Processing   Because plastic deformation involves the movement of atoms, it *accelerates* all processes that rely on diffusion or transformation. We saw that dislocations, multiplied and entangled during deformation, provide sites for recrystallization. They also provide sites for the nucleation of precipitate particles, thus increasing the number and decreasing the size of precipitates. Many possibilities exist of which only the major ones will be discussed here.

1. When a steel is alloyed so that the metastable austenite can exist for some reasonable time (the nose of the curve is pushed to the right in Fig. 6–20), there is time for working the metastable austenite. For this, the steel is austenitized and then rapidly cooled some 100 to 200°C below the transformation temperature where it is worked. The high dislocation density induced in the austenite results in a substantial refinement of transformation products, and such *ausformed steels* have high strength. If the austenite is worked at lower temperatures (*low-temperature thermomechanical* working, Fig. 6–20, line 6), strength increases further but at the expense of ductility.

2. By alloying, the starting temperature of martensitic transformation (M_s) can be depressed and a metastable austenitic structure retained at room temperature. When such a material is subjected to deformation, the greater mobility of atoms during deformation initiates the transformation to martensite. Therefore, in the course of tension testing, an incipient neck is stabilized by the transformation of austenite into the much stronger martensite, and the onset of localized necking is delayed until the entire volume of the specimen is transformed. Thus, *transformation-induced plasticity* (TRIP) provides a third means (besides increasing n and m values) of increasing the ductility of a metal while offering great strength. TRIP steels further benefit from a high dislocation density induced in the austenite by warm working.

3. We mentioned that not all martensites are hard. If the carbon content is very low, as in *maraging steels*, the martensite will be soft and readily worked, but subsequently can be greatly strengthened by the precipitation of intermetallic compounds (such as Ni_3Ti or Ni_3Mo) at the numerous sites of high dislocation density induced by cold working.

4. Precipitation-hardening materials (Sec. 6-4-2), such as aluminum and nickel alloys, may be worked while heated into the homogeneous solid-solution temperature range. Upon cooling, precipitates are refined because they begin to form at sites of dislocation concentrations. Alternatively, the material may be solution-treated and the supersaturated solid solution (which is quite ductile) cold-worked to introduce a high dislocation density which refines the precipitates on subsequent aging. Great increase in strength may result without loss of ductility.

5. An already aged material—and even a tempered martensite—may be cold-worked to take advantage of the great strengthening resulting from dislocation pile-up against finely distributed obstacles. Ductility usually suffers.

6. Further possibilities exist if the material undergoes an allotropic transformation. A material heated to the vicinity of the transformation temperature often shows low strength and high ductility, although not to the same degree as a superplastic material. (This is exploited in the hot working of titanium and its alloys around the α-to-β transformation temperature.) Typically, grains are also refined, and the morphology of transformation products may change too. For example, pearlite formed while working steel at the transformation temperature is much refined and can become globular (spheroidal). Steel that is worked right through the transformation temperature may show unusually high strength and reasonable ductility (*controlled hot working*).

Example 8-14   Some stainless steels are alloyed with small amounts of Cu, Al, P, Nb, or Ti. After cold forming, an age-hardening treatment causes precipitation of hard intermetallic compounds which increase strength. Steel 17-4 PH [17Cr-4Ni-4Cu-0.3(Nb+Ta)] is typical of these precipitation-hardening stainless steels:

Condition	YS, MPa	TS, MPa	el., %
Solution-annealed	275	900	35
Aged	1310	1415	9
Cold-worked and aged	1790	1825	2

(Data from R. Brucker, *Adv. Mater. Proc.*, 1995(12):25–27.)

Example 8-15   The aluminum alloy 6063 is provided in several forms:

Temper	Designation	YS, MPa	TS, MPa	el, %
Annealed	-0	48	90	
Solution-treated and naturally aged	-T4	90	172	22
Solution-treated and artificially aged	-T6	214	240	12
Solution-treated, cold-worked, and artificially aged	-T8	270	290	12

Note the marked increase in strength when the material is cold worked after solution treatment but prior to precipitation hardening.

Example 8-16

High-strength low-alloy (HSLA) steels contain minor amounts of carbide- and nitride-forming elements (V, Nb, Ti, Mo) which allow control of austenite grain size, recrystallization temperature, and ferrite grain size. Thus, high strength and toughness can be developed with an appropriate thermomechanical processing sequence. For example, after controlled cooling, hot-forged parts can be used in the as-forged condition. Elimination of conventional quench and temper heat treatment, straightening, and stress relieving saves cost and accelerates delivery schedules.

8-2   MECHANICS OF DEFORMATION PROCESSING

There is a great variety of plastic deformation processes but some principles can be applied to all of them. Without an understanding of these principles, no process can be intelligently designed or controlled.

Plastic deformation is conducted with the aid of tools (dies). All die materials have limited strength; therefore, a primary concern is the magnitude of pressure developed in the course of deformation. If pressure is too large, the process is not feasible; even if pressures are reasonable, the total deformation force may be too high for available equipment. Consequently, computation of pressures and forces is the primary preoccupation of books on deformation processes. The emphasis is often on the relative accuracy of various theories; for our purpose it is more important that any estimate of pressures and forces should be truly relevant. The simple approach presented here will be accurate to within ±20% and allows for an appropriate safety factor. In order to get a meaningful estimate, four points must be observed: (1) the stress state must be analyzed; (2) a relevant flow stress must be found; (3) the effects of friction must be judged; (4) inhomogeneous deformation must be taken into account.

8-2-1   Yield Criteria

The stress state is, in the general case, triaxial—that is, stresses act in all directions. Analysis is simplified if the coordinate system is oriented in such a way that shear stresses disappear and only three normal stresses act. These are then called principal stresses and are denoted σ_1, σ_2, and σ_3 (Fig. 8–14a).

For plastic flow to occur, the combination of stresses must satisfy the *yield criterion*. Yield criteria have been formulated to describe the beginning of plastic deformation by relating the principal stresses to the tensile or compressive yield strength of the material (Secs. 4-1-3 and 4-3). Our concern is with large plastic deformations; therefore, we will use the flow stress σ_f. (Thus we should really speak of flow criteria; however, the term *yield criterion* is so widely entrenched that we will retain it for our purpose.) For metals, two criteria are frequently used. The yield criterion due to Tresca can be written as

$$\frac{\sigma_{\max} - \sigma_{\min}}{2} = \frac{\sigma_f}{2} \tag{8-12}$$

where $\sigma_{\max}$ is the most positive and $\sigma_{\min}$ the most negative stress.

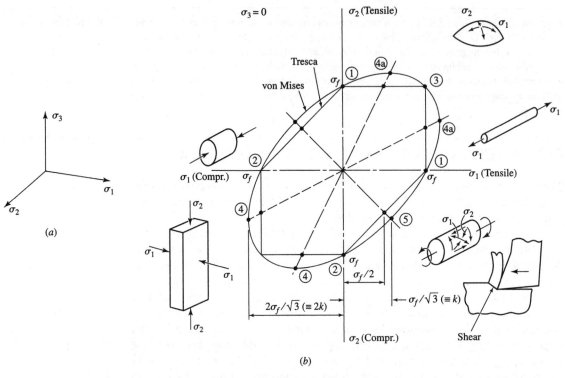

Figure 8–14 (a) The coordinate system may be rotated to obtain only principal stresses. (b) Under plane-stress conditions, some of the important stress states may be shown on the Tresca yield hexagon and von Mises yield ellipse. (From J.A. Schey, Tribology in Metalworking: Friction, Lubrication and Wear, ASM International, 1983, p. 12. With permission.)

The yield criterion according to von Mises is

$$(\sigma_1 - \sigma_2)^2 + (\sigma_2 - \sigma_3)^2 + (\sigma_3 - \sigma_1)^2 = 2\sigma_f^2 \qquad \textbf{(8-13)}$$

The significance of yield criteria is best illustrated by examining a simplified stress state in which $\sigma_3 = 0$ (*plane stress*). For ease of visualization, one may think of a plate in which the rolling direction is arbitrarily taken as the σ_1 direction and the width direction the σ_2 direction (Fig. 8–14b). Plastic flow can be initiated in many ways:

1. If a tensile specimen is cut in the rolling direction, flow occurs—according to both Tresca and von Mises—at the flow stress σ_f (points 1, corresponding to the two directions in the plane of the plate).

2. Shorter cylinders cut in the same directions can be tested in compression and will usually be found to flow at the same stress σ_f (points 2).

3. When the plate is bulged by a punch or a pressurized medium (as a balloon is blown up by air), the two principal stresses acting in the plane of the plate are equal (*balanced biaxial tension*) and must reach σ_f (point 3).

4. A technically very important condition is reached when deformation of the workpiece is prevented in one of the principal directions (*plane strain*) for one of two reasons:

  a. A die element keeps one dimension constant (Fig. 8–15a).

  b. Only one part of the workpiece is deformed, and adjacent nondeforming portions exert a restraining influence (Fig. 8–15b).

  In either case, the restraint imposes a stress on the material in that principal direction; the stress is the arithmetic average of the other two principal stresses (corresponding to points 4 in Fig. 8–14b). The stress required for deformation is still σ_f according to Tresca, who ignores the intermediate principal stress. However, according to von Mises, the stress required is higher, $1.15\sigma_f$, which value is often denoted as $2k$. It is also called the *plane-strain flow stress* or *constrained flow stress* of the material. Plane strain may be imposed also in tension, points 4a.

  5. If a cylinder is cut out and twisted (*torsion*), the two principal stresses on the surface of the cylinder are of equal magnitude but of opposing sign (points 5 in Fig. 8–14b). This is a condition of *pure shear*, and flow occurs at the shear flow stress τ_f, which is equal to $0.5\sigma_f$ according to Tresca and $0.577\sigma_f$ according to von Mises. The shear flow stress according to von Mises is often denoted as k. The important point is that when, in the course of deformation by compression, a transverse stress of opposing sign (a tensile stress) is imposed, *the stress required for compression will decrease*. This offers a powerful mechanism for reducing die pressures.

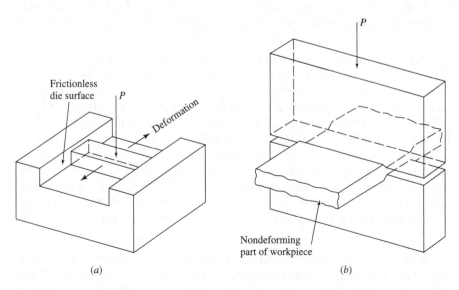

(a)   (b)

Figure 8–15   Deformation in one direction is often prevented and a plane-strain condition established by the restraint given either by (a) die elements or (b) nondeforming parts of the workpiece adjacent to the deformation zone. (*From J.A. Schey, as Fig. 8–14, p. 13.*)

6. A special condition is reached when all three principal stresses are equal in magnitude (*hydrostatic stress state*, Sec. 4-1-5). An inspection of the yield criteria [Eqs. (8-12) and (8-13)] will show that superimposition of a hydrostatic stress simply shifts all principal stresses by the same amount, thus there is no change in the yield criterion and flow still begins at σ_f.

It will be noted that, for certain stress states, von Mises predicts a critical stress that is 15% higher than the uniaxial flow stress. Not all materials obey the von Mises criterion but to be on the safe side, we will always use $1.15\sigma_f(= 2k)$ as the flow stress in plane strain.

8-2-2   The Relevant Flow Stress

In all calculations, the stress sufficient to maintain plastic deformation, σ_f, must be taken for the temperature, strain, and strain rate prevailing in the process. It cannot be sufficiently emphasized that our interest is not just in initiating but also in *maintaining* plastic flow. Thus, the yield strength found in many handbooks is of little use; the flow stress traverses the true stress–true strain curve (Fig. 8–1*a* or 8–11*a*) within the strain limits defined by the condition of the starting material and the end strain.

1. In cold working it can assumed that the power law, Eq. (8-4), holds and, whenever available, the K and n values should be used (a selection is given in Tables 8–2 and 8–3).

2. For hot working, the flow stress can be calculated from the power function, Eq. (8-11), with the appropriate C and m values (Tables 8–2 and 8–3). If these values are not available for various strains, one has to assume that the flow stress remains constant throughout deformation (as in the curve for a strain rate of 1 s^{-1} in Fig. 8–11*a*; in Tables 8–2 and 8–3 the values are given for a strain of $\varepsilon = 0.5$). If no C and m data are available, one is obliged to make a compression test. It is quite inadmissible to use hot-strength values determined in conventional, slow tension tests, because they often represent only a fraction of the true flow stress prevailing at the much higher (typically, 1–1000 s^{-1}) strain rates attained in deformation processes. Extrapolation from low strain rates to high strain rates is hazardous because m may also change with strain rate (see Fig. 8–12).

It should be noted that the constants used in flow stress calculations are also a function of the starting condition of the material. Data given in Tables 8–2 and 8–3 are representative values for annealed material. Every effort was made to use reliable data, and the two C and m values entered for copper show the worst of the extreme variations occasionally found in published data.

8-2-3   Effects of Friction

In most deformation the workpiece is brought in contact with a tool or die; there-fore, friction between the two contacting bodies is unavoidable. With few excep-tions, our aim will be to reduce friction by the application of a lubricant.

We already examined friction as it is encountered in machinery elements (e.g., in bearings) in Sec. 4-9-2. We described friction by a coefficient of friction μ; because of its importance, we reproduce here Eq. (4-18):

$$\mu = \frac{F}{P} = \frac{\tau_i}{p} \qquad \text{(4-18)}$$

When the interface pressure p is low relative to the flow stress σ_f of the contacting materials (as it would be in a bearing), Eq. (4-18) holds: With increas-ing pressure p the interface shear stress τ_i increases linearly (Fig. 8–16a), and μ could assume any constant value. In deformation processes one of the contacting materials (the workpiece) deforms and in doing so slides against the harder sur-face (the tool or die). A frictional stress τ_i is again generated, but this time there is a limit to μ, because the material will choose a deformation pattern that *minimizes the energy of deformation*. If friction is high, interface shear stress τ_i will reach, in the limit, the shear flow stress τ_f of the workpiece material (Fig. 8–16a). At this point the workpiece refuses to slide on the tool surface; instead, it deforms by shearing inside the body (Fig. 8–16b). Since $\tau_f = 0.5\sigma_f$ (Fig. 8–14b), it is often said that the maximum value of $\mu = 0.5$. This is true only when $p = \sigma_f$; at higher p, the maximum value of μ is lower (Fig. 8–16b). In general, it is more accurate to say that the coefficient of friction becomes meaningless when $\tau_i = \tau_f$, since there is no relative sliding at the interface. This is described as *sticking friction*, even though the workpiece does not actually stick to the die surface.

Because of the conceptual difficulties introduced by the coefficient of friction, it is often preferable to use the actual value of τ_i, especially when interface pressures are very high. Alternatively, τ_i can be denoted as a fraction of the shear flow stress

$$\tau_i = m^* \tau_f = m^* \frac{\sigma_f}{2} \qquad \left(\text{or} = m^* \frac{\sigma_f}{\sqrt{3}} \right) \qquad \text{(8-14)}$$

where m^* is the *frictional shear factor* [the literature uses m but, because of possible confusion with m in Eq. (8-11), we add the asterisk]. For a perfect lubricant, $m^* = 0$; for sticking friction, $m^* = 1$.

We shall use both descriptions of friction in Chap. 9. Friction always increases pressures and forces and it could easily limit the attainable deformation, therefore, in most instances every effort is made to reduce friction with a suitable lubricant.

Table 8-2 Manufacturing properties of steels and copper-based alloys*
(Annealed condition)

Designation and Composition, %	Liquidus/Solidus, °C	Hot Working				Cold Working							
		Usual Temp., °C	Flow Stress,† MPa			Work-ability¶	Flow stress,‡ MPa		$\sigma_{0.2}$, MPa	TS, MPa	Elonga-tion, %	q RA, %	Annealing Temp.,§ °C
			at °C	C	m		K	n					
Steels:													
1008 (0.08C), sheet		<1250	1000	100	0.1	A	600	0.25	180	320	40	70	850–900 (F)
1015 (0.15C), bar		<1250	800	150	0.1	A	620	0.18	300	450	35	70	850–900 (F)
			1000	120	0.1								
			1200	50	0.17								
1045 (0.45C)		<1150	800	180	0.07	A	950	0.12	410	700	22	45	790–870 (F)
			1000	120	0.13								
~8620 (0.2C, 1Mn, 0.4Ni, 0.5Cr, 0.4Mo)			1000	120	0.1	A			350	620	30	60	
D2 tool steel (1.5C, 12Cr, 1Mo)		900–1080	1000	190	0.13	B	1300	0.3					880 (F)
H13 tool steel (0.4C, 5Cr 1.5Mo, 1V)			1000	80	0.26	B							
302 SS (18Cr, 9Ni) (austenitic)	1420/1400	930–1200	1000	170	0.1	B	1300	0.3	250	600	55	65	1010–1120 (Q)
410 SS (13Cr) (martensitic)	1530/1480	870–1150	1000	140	0.08	C	960	0.1	280	520	30	65	650–800
Copper-Base Alloys:													
Cu (99.94%)	1083/1065	750–950	600	130 (48)	0.06 (0.17)	A	450	0.33	70	220	50	78	375–650
			900	41	0.2								
Cartridge brass (30Zn)	955/915	725–850	600	100	0.24	A	500	0.41	100	310	65	75	425–750
			800	48	0.15								
Muntz metal (40Zn)	905/900	625–800	600	38	0.3	A	800	0.5	120	380	45	70	425–600
			800	20	0.24								
Leaded brass (1Pb, 39Zn)	900/855	625–800	600	58	0.14	A	800	0.33	130	340	50	55	425–600
			800	14	0.20								
Phosphor bronze (5Sn)	1050/950		700	160	0.35	C	720	0.46	150	340	57		480–675
Aluminum bronze (5Al)	1060/1050	815–870				A			170	400	65		425–750

*Compiled from various sources; most flow stress data from T. Altan and F. W. Boulger, *Trans. ASME, Ser. B, J. Eng. Ind.* **95**:1009 (1973).

†Hot-working flow stress is for a strain of $\epsilon = 0.5$. To convert to 1000 psi, divide calculated stresses by 7.

‡Cold-working flow stress is for moderate strain rates, around $\epsilon = 1\ s^{-1}$. To convert to 1000 psi, divide stresses by 7.

§Furnace cooling is indicated by F, quenching by Q.

¶Relative ratings, with A the best, corresponding to absence of cracking in hot rolling and forging.

Table 8-3 Manufacturing properties of various nonferrous alloys[a] (Annealed condition, except 6061-T6))

Designation and Composition, %	Liquidus/Solidus, °C	Usual Temp., °C (Hot Working)	Flow Stress,[b] MPa at °C	C	m	Workability[f]	Flow stress[c] MPa K	n	$\sigma_{0.2}$, MPa	TS,[d] MPa	Elongation,[d] %	q RA, %	Annealing Temp.,[e] °C
Light Metals:													
1100 Al (99%)	657/643	250–550	300 / 500	60 / 14	0.08 / 0.22	A	140	0.25	35	90	35		340
~3003 Al (1Mn)	649/648	290–540	400	35	0.13	A		0.15	40	110	30		370
~2017 Al (3.5Cu, 0.5Mg, 0.5Mn)	635/510	260–480	400 / 500	90 / 36	0.12 / 0.12	B	380		70	180	20		415 (F)
5052 Al(2.5Mg)	650/590	260–510	480	35	0.13	A	210	0.13	90	190	25		340
6061-0(1Mg, 0.6Si, 0.3Cu)	652/582	300–550	400 / 500	50 / 37	0.16 / 0.17	A	220	0.16	55	125	25	65	415 (F)
6061-T6	NA[g]	NA	NA	NA	NA	NA	450	0.03	275	310	8	45	
~7075 Al(6Zn, 2Mg, 1Cu)	640/475	260–455	450	40	0.13	B	400	0.17	100	230	16		415
Low-Melting Metals:													
Sn (99.8%)	232	100–200	100	10	0.1	A				15	45	100	150
Pb (99.7%)	327	20–200	75	260	0.1	A				12	35	100	20–200
Zn (0.08% Pb)	417	120–275	225	40	0.1	A				130/170	65/50		100
High-Temperature Alloys:													
Ni (99.4Ni + Co)	1446/1435	650–1250				A			140	440	45	65	650–760
Hastelloy X (47Ni, 9Mo, 22Cr, 18Fe, 1.5Co, 0.6W)	1290	980–1200	1150~	140	0.2	C			360	770	42		1175
Ti (99%)	1660	750–1000	600 / 900	200 / 38	0.11 / 0.25	C / A			480	620	20		590–730
Ti-6Al-4V	1660/1600	790–1000	600 / 900	550 / 140	0.08 / 0.4	C / A			900	950	12		700–825
Zirconium	1852	600–1000	900	50	0.25	A			210	340	35		500–800
Uranium (99.8%)	1132	~700	700	110	0.1	A			190	380	4	10	

[a] Empty spaces indicate unavailability of data. Compiled from various sources; most flow stress data from T. Altan and F. W. Boulger, *Trans. ASME, Ser. B, J. Eng. Ind.* **95**:1009 (1973).

[b] Hot-working flow stress is for a strain of $\epsilon = 0.5$. To convert to 1000 psi, divide calculated stresses by 7.

[c] Cold-working flow stress is for moderate strain rates, around $\dot{\epsilon} = 1\ s^{-1}$. To convert to 1000 psi, divide stresses by 7.

[d] Where two values are given the first is longitudinal, the second transverse.

[e] Furnace cooling is indicated by F.

[f] Relative ratings, with A the best, corresponding to absence of cracking in hot rolling and forging.

[g] NA Not applicable to the -T6 temper.

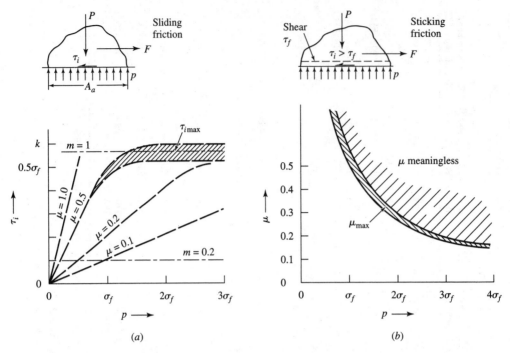

Figure 8–16 (a) The interface shear stress can never exceed the shear flow stress of the material and (b) the maximum possible coefficient of friction decreases when interface pressures exceed the flow stress of the material. (*From J.A. Schey, as Fig. 8–12, p. 15.*)

Example 8-17 | Sticking friction sets in where $\mu p = \tau_f$. Calculate the values of μ for various interface pressures p.

It is convenient to express p as a multiple of σ_f. If the Tresca yield criterion is used, $\tau_f = 0.5\sigma_f$ and, by definition, Eq. (4-18),

$$\mu_{max} = \frac{\tau_f}{p} = \frac{\tau_f}{x\sigma_f} = \frac{\sigma_f}{2x\sigma_f}$$

hence, when

$x =$	1	2	4	8
$\mu_{max} =$	0.5	0.25	0.125	0.062

The points are plotted in Fig. 8–16b. The von Mises criterion gives $\tau_f = 0.577\sigma_f$ and thus slightly higher μ_{max} values.

Example 8-18 | From Eqs. (4-18) and (8-14), $\tau_i = \mu p = m^*\tau_f$. If p is again expressed as an x multiple of σ_f, calculate the equivalent μ and m^* values.

When $p =$	σ_f	$2\sigma_f$	$4\sigma_f$	$8\sigma_f$
then $m^* =$	2μ	4μ	8μ	16μ

There is thus no simple relationship between the two parameters used to describe friction. Furthermore, the above-calculated relations fail to hold when partial sticking sets in.

8-2-4   Lubrication

We already mentioned that lubricant is applied to reduce (or control) friction. A good lubricant accomplishes much more: It separates die and workpiece surfaces and thus prevents adhesion with its undesirable side effects of tool pickup, workpiece damage, and die wear; it reduces die wear due to abrasion and other mechanisms; it controls the surface finish of the part produced; and it cools the system in cold working and helps to prevent heat loss (or removes heat at a controlled rate) in hot working. The lubricant must not be toxic or allergenic, it must be easy to apply and remove, and residues must not interfere with subsequent operations or cause corrosion.

The most frequently used lubricants are listed in Table 8–4 (the basic types are described in Sec. 4-9-2). In a very general sense, lubricants are chosen for "duty." A heavy-duty lubricant film survives sliding under high pressure, substantial expansion of the surface, and higher temperature; it also reduces friction and wear. For lightest duties, a synthetic coolant (a true solution of a chemical in water) may suffice. Emulsions (dispersions of an oily phase in water) and oil-based lubricants can be formulated for a wide range of duties by compounding with boundary (in Table 8–4, FA and FO) and EP additives. For the most severe cold forming, lubricant breakdown can be prevented only by applying a conversion coating (on steel, a Zn-phosphate layer) on the surface. The structure of this coating assures retention of the superimposed lubricant (often a soap that is reactive with the coating), and allows extension of the surface without losing continuity of the lubricant/coating film. Because of their water base, synthetics or emulsions are essential if cooling is important in cold or hot working. Layer-lattice compounds such as graphite or MoS_2 survive high temperatures but concerns for a cleaner plant environment have prompted a search for alternatives.

To minimize cost and impact on the environment, lubricants are often applied in recirculating systems which incorporate tanks (which may hold as little as 10 L and as much as 400 000 L fluid), pumps to develop the necessary pressure, nozzles that deliver the fluid to strategic points, and collecting troughs and return piping. Vital components are *filters* which may be a simple wire mesh or may have sophisticated construction to deal with fines generated in the process, take out undesirable compounds formed in metal/lubricant interactions and, if possible, separate contaminant oil (*tramp oil*) picked up from the equipment. The condition of the lubricant is monitored and appropriate makeup is made to keep concentration and composition within specification. Some lubricants, especially emulsions, are subject to attack by organisms and biocides are added to prevent rancidity and foul smell.

Even with the best care, there is a point beyond which a lubricant cannot be maintained and must be replaced with a fresh batch. The spent lubricant is collected by specialized recycling companies that return the lubricant to its original condition, reformulate it for another application, or make it suitable for disposal by combustion. Some lubricants are difficult to recycle or contain hazardous ingredients and then disposal becomes extremely expensive. For all these reasons, the choice of lubricant cannot be an afterthought; the lubricant must be considered as an integral part of the system from the very beginning.

Table 8-4   Typical lubricants* and friction coefficients in plastic deformation

Workpiece Material	Working	Forging		Extrusion†	Wire Drawing		Rolling		Sheet Metalworking	
		Lubricant	μ	Lubricant	Lubricant	μ	Lubricant	μ	Lubricant	μ
Sn, Pb, Zn alloys		FO–MO	0.05	FO or soap	FO	0.05	FA–MO or MO–EM	0.05 / 0.1	FO–MO	0.05
Mg alloys	Hot or warm	GR and/or MoS₂	0.1–0.2	None			MO–FA–EM	0.2	GR in MO or dry soap	0.1–02
Al alloys	Hot	GR or MoS₂	0.1–0.2	None			MO–FA–EM	0.2		
	Cold	FA–MO or dry soap	0.1 / 0.1	Lanolin or soap on PH	FA–MO–EM, FA–MO	0.1 / 0.03	1–5% FA in MO(1–3)	0.03	FO, lanolin, or FA–MO–EM	0.05–0.1
Cu alloys	Hot	GR	0.1–0.2	None (or GR)			MO–EM	0.2		
	Cold	Dry soap, wax, or tallow	0.1	Dry soap or wax or tallow	FO–soap–EM, MO	0.1 / 0.03	MO–EM	0.1	FO–soap–EM or FO–soap	0.05–0.1
Steels	Hot	GR	0.1–0.2	GL (100–300), GR			None or GR–EM	ST‡ / 0.2	GR	0.2
	Cold	EP–MO or soap on PH	0.1 / 0.05	Soap on PH	Dry soap or soap on PH	0.05 / 0.03	10% FO–EM	0.05	EP–MO, EM, soap, or polymer	0.05–0.1
Stainless steel, Ni and alloys	Hot	GR	0.1–0.2	GL (100–300)			None	ST‡	GR	0.2
	Cold	CL–MO or soap on PH	0.1 / 0.05	CL–MO or soap on PH	Soap on PH or CL–MO	0.03 / 0.05	FO–CL–EM or CL–MO	0.1 / 0.05	CL–MO, soap, or polymer	0.1
Ti alloys	Hot	GL or GR	0.2	GL (100–300)					GR, GL,	0.2
	Cold	Soap or MO	0.1	Soap on PH	Polymer	0.1	MO	0.1	Soap, or polymer	0.1

*Some more frequently used lubricants (hyphenation indicates that several components are used in the lubricant):
   CL = chlorinated paraffin.
   EM = emulsion; the listed lubricating ingredients are finely distributed in water.
   EP = "extreme-pressure" compounds (containing S, Cl, and P).
   FA = fatty acids and alcohols, e.g., oleic acid, stearic acid, stearyl alcohol.
   FO = fatty oils, e.g., palm oil and synthetic palm oil.
   GL = glass (viscosity at working temperature in units of poise).
   GR = graphite; usually in a water-base carrier fluid.
   MO = mineral oil (viscosity in parentheses, in units of centipoise at 40°C).
   PH = phosphate (or similar) surface conversion, providing keying of lubricant.
†Friction coefficients are misleading for extrusion and are therefore not quoted here.
‡The symbol ST indicates sticking friction.
SOURCE: Data extracted from J.A. Schey: *Tribology in Metalworking: Friction, Lubrication, and Wear*, American Society for Metals, Metals Park, Ohio, 1983.

It is often necessary to remove lubricant residues. Organic residues are removed by degreasing. In *solvent degreasing* chlorinated solvents had been extensively used but, because several of these deplete the ozone layer or are potential carcinogens, other methods—such as *aqueous degreasing* with alkaline chemicals, followed by water rinse—have gained prominence. Where solvent degreasing is unavoidable, closed systems or systems with total recovery are used so that no solvent escapes. This applies also to *vapor degreasing* in a chamber saturated with the heated vapor of the solvent; the vapor condenses on the colder parts and oily residues are washed off. Lubricants must be chosen to be compatible with the intended degreasing process. Efforts are also made to minimize the quantity of lubricant applied and to keep the amount of oil mist (volatile organic compounds, VOC) to minimum. Legislated limits are set and gradually tightened for emissions into air and water. Disposal methods are also regulated; therefore, all these factors must be taken into account in selecting a lubricant.

8-2-5   Inhomogeneous Deformation

There is an important source of high pressures and forces which has nothing to do with interface friction, and, therefore, is not affected by lubrication. It can be best understood from the example of *indentation*. Inspection of Fig. 8–17a suggests that a small tool cannot possibly deform the entire bulk of a large (semi-infinite) workpiece. Indeed, experiments show that when the tool penetrates, highly *inhomogeneous material flow* takes place.

The mechanism is shown in Fig. 8–17a: a part of the workpiece (1) immediately under the indentor remains immobile relative to the indenter and moves with it as though it were an extension of the indenter itself. This rigid wedge then pushes two triangular wedges (2) aside, which in turn push up two outer wedges (3), thereby forming humps corresponding to the volume displaced by the indenter. The rest of the workpiece (4) is only elastically loaded. The difficulty of moving the material purely locally—*against the restraint given by the surrounding elastic material*—raises the required interface pressure. Compared to homogeneous deformation, the forging force now has to perform extra work, referred to as *redundant work* (as it was in Example 8-11).

In some processes a workpiece of finite thickness is deformed simultaneously from two sides (Fig. 8–17b). Inhomogeneity of deformation then depends on how far the two deformation zones are separated, and this is most usefully expressed by the h/L ratio, that is, the ratio of height to contact length. It is found from both theory and experiment that when $h/L > 8.7$, the two deformation zones are entirely separated; the material between these zones is only elastically deformed and exerts the same restraining effect as though it were of infinite thickness. At lower h/L ratios the two wedges cooperate (Fig. 8–17b) and the pressure drops. As might be expected, at a ratio of $h/L = 1$ the two deformation zones fully cooperate (Fig. 8–17c) and the material flows at a minimum pressure. If the h/L ratio were to diminish further, deformation would be homogeneous, but friction would now increase die pressures.

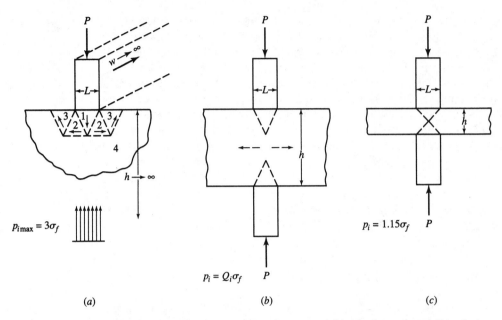

Figure 8–17   Deformation is (a) highly inhomogeneous when a semi-infinite body is indented; (b) at high h/L values, deformation is still inhomogeneous; (c) only at $h/L = 1$ is homogeneity approached.

Inspection of Fig. 8–17b indicates that the two wedges penetrating from top and bottom tend to tear the workpiece apart; in other words, inhomogeneous deformation generates *secondary tensile stresses* (i.e., stresses that are not externally imposed but are generated by the process of deformation itself). Several consequences are possible:

1. Internal fracture may occur in the workpiece during deformation.

2. A residual stress pattern (internal stresses) may be set up that may cause subsequent deformation (warping) of the workpiece, particularly on heating.

3. Surface residual tensile stresses can combine with other effects to cause delayed failures (e.g., stress-corrosion cracking in the presence of a corrosive medium).

In general, therefore, the aim of process development is to make deformation as homogeneous as possible, unless internal fracture is intentionally induced (Sec. 9-7-4). If harmful residual stresses remain, a stress-relief heat treatment is given (Sec. 6-4-1).

We have seen that compressive residual stresses concentrated in a thin surface layer greatly improve the fatigue resistance of the workpiece in service (Sec. 4-5). Highly inhomogeneous compressive deformation is then purposely applied. The surface compressive stresses are balanced by internal tensile stresses, spread over such a large cross-sectional area that their level is harmless.

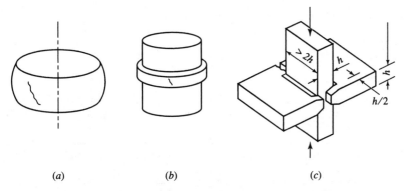

| (a) | (b) | (c) |

Figure 8–18   Workability may be evaluated by (a) compression with sticking friction, (b) upsetting a collared specimen, or (c) partial-width indentation.

8-2-6   Bulk Workability

Once we have determined that a process is feasible from the point of view of pressures and forces, we will want to make sure that the workpiece will survive deformation without fracture. A material of given ductility may fare very differently in various processes, depending on the conditions imposed on it. Therefore, our main concern is not simply ductility, but a more complex property called *workability* in bulk metalworking operations. We saw in Sec. 4-1-4 that ductile fracture is induced by triaxial tensile stresses and we learnt in Sec. 4-1-5 that imposition of hydrostatic pressure delays fracture. Thus, workability has two components:

1. The *basic ductility* of the material allows it to deform to some extent, without fracture, even in the presence of tensile stresses. Therefore, reduction in area measured in the tension test [Eq. (4-10)] is a useful (but not universally applicable) measure of basic ductility; it is essentially a measure of resistance to void formation. Other possible measures are the number of turns to fracture in a torsion test, or the reduction in height in compression (upsetting) tests designed to generate high secondary tensile stresses (Sec. 4-3). Upsetting with sticking friction at the end face causes severe barreling and thus surface cracking in a material of low ductility (Fig. 8–18a). Tensile stresses are higher on a collared specimen (Fig. 8–18b) or in a partial-width indentation test (Fig. 8–18c).

It is generally recognized that ductility is greatly dependent on strain rate in superplastic materials. Less well established is, however, dependence in conventional hot working. The partial-width indentation test is particularly suitable for exploring the ductility of coarse-grained materials at relevant strain rates. In one study, specimens were cut from semicontinuously cast 7075 alloy billets, partially homogenized for 10 h at 470°C. Ductility was expressed as

Example 8-19

reduction of area in the rib of test specimens (Fig. 8–18*c*).

Strain rate in rib, s^{-1}	1	10
Reduction of area at 400°C	75	58
Temperature at which reduction dropped to zero, °C	460	440

Thus, time-sensitive restoration processes affect not only ductility but also the onset of hot shortness. [Data from D. Duly, J.G. Lenard, and J.A. Schey, *J. Mater. Working Technol.*, **75**:143–151 (1998).]

2. The stress state induced by the process modifies ductility. If the process maintains compressive stresses in all parts of the deforming workpiece (if hydrostatic pressure prevails), cavity formation cannot begin and ductile fracture does not set in. (At very heavy deformations, ductility of the material may be exhausted and then brittle, shear-type fracture may occur.) If, however, the process allows tensile stresses to develop, cavity formation can begin and will, finally, lead to fracture.

At what point this fracture should occur is predicted by *workability criteria*, none of which has proven to be universally applicable. The simple yet useful criterion of Cockroft and Latham[2] states that, for a given metal, the work done by the highest local tensile stress must reach a critical value. It follows from this that if the development of secondary tensile stresses is suppressed, deformation can be taken much further, just as fracture in the tension test is delayed by hydrostatic pressure. Therefore, one of the important aims of process design is to increase the *hydrostatic pressure component* $\sigma_H = (\sigma_1 + \sigma_2 + \sigma_3)/3$.

8-2-7   Sheet Formability

We saw in Sec. 8-2-6 that, in bulk deformation processes, limits of deformation are set by the workability of the material. Survival in sheet metalworking is linked to *formability*, which is also a complex property, and must now be related to failure definitions relevant to sheet products:

1. The first objection may arise when a stretched sheet becomes *grainy* in appearance (*orange peel*). This is a reflection of the polycrystalline structure of metals: Individual grains oriented in different crystallographic directions deform to slightly differing degrees. Roughening of the surface has no bearing on the structural integrity of the part. If grainy appearance is esthetically objectionable, a finer-grained material will produce graininess on such a small scale as to be invisible to the naked eye.

2. We saw in Sec. 8-1-2 that in some materials discontinuous yielding results in the formation of surface bands (*Lüders lines, stretcher-strain marks*, or *worms*, as called in the shop). Stretcher-strain marks are harmless but may be

---

[2]M.G. Cockroft and D.J. Latham, *J. Inst. Met.*, **96**:33–39 (1968).

objectionable on exposed surfaces. Once the entire surface is covered, they are no longer distinguishable.

3. Appearance suffers and the functional properties of a part may be affected when *localized necking* occurs. Even though the part is not fractured, its load-bearing capacity may be reduced, although in some configurations the part will remain completely functional. In general, materials are chosen to optimize factors that delay the onset of necking (a large uniform elongation, corresponding to a high n value, Sec. 8-1-1) or help to spread out an incipient neck (a high m value, Sec. 8-1-6, or transformations, Sec. 8-1-7).

4. Once the neck has localized, further deformation occurs by local thinning until, finally, *fracture* sets in. We saw in Sec. 8-1-6 that postnecking strain is a function of the m value. In cold working even a slight increase in m (say, from 0 to 0.05) is helpful; in hot working a high m allows substantial postnecking deformation while maintaining a reasonably uniform thickness (at $m = 1$, the sheet would thin out completely uniformly). A higher reduction of area allows the sheet thickness to reduce further without fracture, but the load-bearing capacity of the part may be lost if local thinning is too severe.

In summary, a highly formable sheet metal has high uniform elongation (or n value), large postnecking strain (or high m value) and may undergo a transformation. In industrial practice, a high total elongation in the tension test [Eq. (4-9b)] has long been regarded as a desirable attribute; inspection of Fig. 8–2a shows that this view translates into a combination of high n and high m, and is thus fundamentally correct. For a given material, ductility decreases with increasing hardness, therefore it is common practice to specify hardness in addition to or instead of elongation. Adequate ductility is a necessary but not sufficient criterion; in addition, a desirable material shows no Lüders bands and has a favorable anisotropy. Just as in bulk deformation, the limits of deformation depend also on the stress state generated in the process (Chap. 10).

8-3 WROUGHT ALLOYS

In all deformation processes, the workpiece shape is formed by displacing material from unwanted locations into positions required by the part shape. This demands that the material should be able to sustain plastic deformation without fracture. We saw that bulk workability and sheet formability are affected by the process itself. Thus, in a general way, we can only say that alloys suitable for deformation processing (traditionally called *wrought metals*) must possess a minimum ductility commensurate with the contemplated process.

This requirement is amply satisfied by all pure metals that have a sufficient number of slip systems (Sec. 6-3-1) and also by most solid-solution alloys of the same metals. Two-phase and multiphase (Sec. 6-3-2) materials are deformable if they meet certain minimum requirements. There must be no liquid or brittle phase on the grain boundaries or across several grains (thus, gray iron, white cast iron, or a hypereutectic Al-Si alloy cannot be cold worked). Excessive amounts of brittle constituent are not permissible even in a ductile matrix, especially if the brittle constituent is also coarse or lamellar. If the matrix is less ductile, it becomes even more important that the material should be free of other weakening features such as inclusions, voids, or grain-boundary contaminants.

Steels represent the largest segment of wrought products (Table 8–1) and, in line with the system adopted in Sec. 7-3 for casting alloys, ferrous materials will be discussed first, followed by nonferrous materials. Relevant properties of selected alloys are given in Tables 8–2 and 8–3.

8-3-1   Carbon Steels

We already discussed, in Sec. 7-3-1, steels as casting alloys. We mentioned that steels (as well as other metals) are normally deoxidized in order to avoid gas porosity. This, however, need not be the case for castings destined for metalworking, and several classes of steel can be distinguished according to deoxidation practice.

1. The so-called *rimmed steels* are not deoxidized. Carbon reacts in the melt with oxygen to form carbon monoxide according to the reaction $2C + O_2 = 2CO$. Since CO is a reducing gas, the large blowholes formed have clean surfaces and, at the high temperatures and pressures prevailing in hot working, weld without trace. An advantage of the large number of gas bubbles formed during solidification is that piping is virtually eliminated. The blowholes are prominent at some distance below the ingot surface and help to move contaminants toward the center, imparting a strong normal segregation pattern (Fig. 7–4a) which persists through all processing steps. The ingot surface (*rim*) is particularly clean and low in carbon. The clean surface is an advantage in many applications, and sheet up to 0.25% C is often produced in this form.

2. Gas evolution is suppressed to some extent when a *cap* (a metal plug) is placed on the ingot (*capped steels*), which thus retains some of the surface cleanliness but achieves greater structural homogeneity than in a rimmed steel. *Semikilled* steels are partially deoxidized and are suitable for applications where great structural uniformity is not required, as in many steels used for construction purposes.

3. The most demanding applications call for *killed steel* in which gas reaction is prevented (*killed*) by the addition of aluminum, silicon, etc. Segregation is virtually absent, properties are uniform throughout, and grain size can be controlled in the finished product. However, proper feeding must be assured to prevent piping.

A further distinction can be made according to carbon content. *Low-carbon steels* (below 0.15% C) contain too little carbon to benefit from hardening and are used in the hot-worked or, for maximum ductility, in the annealed condition, primarily in the form of sheet and wire. Because of their special importance for sheet metalworking, they will be further discussed in Sec. 10-1-1. Steels of less than 0.25% C (often referred to as *mild steels*) have somewhat higher strength but are still easy to weld and are widely employed for structural purposes as hot-rolled bars, sections, and plate. *Medium-carbon steels* (0.25–0.55% C) are often heat-treated (quenched and tempered) after manufacturing by hot or cold metalworking. *High-carbon steels* (0.55–1.0% C) find applications as springs and wear-resistant parts.

Steels that are to be cold worked are usually annealed, and those of higher carbon contents are spheroidized to ensure maximum ductility. A special combination of properties is obtained when the surface of a wrought low-carbon steel part is carburized (Sec. 6-4-4). After heat treatment, a part such as a gear or shaft will have a hard, wear-resistant surface and a tough core.

8-3-2 Alloy Steels

For many applications, carbon steels cannot provide the required combination of properties and then the more expensive alloy steels will be specified.

High-strength low-alloy (HSLA) steels   These rely on very small quantities of Ti, V, or Nb coupled with thermomechanical processing (Sec. 8-1-7) for developing high strength and toughness (see Example 8-16). Carbonitride precipitates inhibit grain growth in the austenite and thus refine the ferrite formed on cooling from the controlled hot working temperature. The combination of grain refinement and precipitation hardening results in high (350–560 MPa) yield strengths. Ductile manganese sulfide inclusions tend to roll out into stringers and the transverse impact properties of the steel suffer. The addition of Zr or Ti reduces the plasticity of inclusions and prevents their spreading, thus removing the harmful effects of inclusions.

Low-alloy steels   Relatively small amounts of alloying elements allow heat treatment of thicker sections (Sec. 6-4-3).

High-alloy steels   Higher alloying element concentrations, in combination with higher carbon content, raise the hardness and hot hardness of tool and die steels by introducing temperature-resistant carbides (such as WC, VC, and chromium carbides). They are most readily worked in the annealed condition, even though the higher carbide content increases forming forces and die wear and reduces ductility. Therefore, these steels are usually hot worked, since in the austenitic temperature range their flow strength is not much higher than that of carbon steels.

Stainless steels   Their corrosion resistance makes them valuable in many applications. Most of them can be hot worked if proper precautions are taken. Those containing both nickel and chromium (*austenitic stainless steels*) are among the most cold-formable materials because of their high strain-hardening rate.

8-3-3 Nonferrous Materials

As in Sec. 7-3-2, nonferrous materials will be discussed in order of increasing melting point, not according to their relative importance.

Tin Alloys   The low strength of tin makes it unsuitable as a structural material, except for foil and collapsible tubes (but then it is used for its corrosion resistance). Of the tin alloys, modern pewter is readily deformed, mostly into decorative products.

Lead Alloys   Lead has low strength but its corrosion resistance encouraged widespread use in the form of sheet, tube, and cable sheathing. It can be strengthened by a number of elements (As, Sn, Bi, Te, and Cu). It also serves as a sound, vibration, and radiation absorber. A Pb–Ca–Sn alloy is used, in the form of expanded sheet, for electrical storage batteries. The toxicity of lead has led to restrictions in many applications.

Zinc Alloys   Pure zinc has use as the material for drawn battery cans, corrugated roofing, and weather stripping (usually with 1% Cu for the last two applications). Because of its hexagonal structure, it is cold worked above 20°C.

A eutectoid transformation in the zinc-aluminum system allows commercial production of extremely fine-grained material that exhibits superplasticity (Sec. 8-1-6). Binary alloys with 22% Al and further-alloyed variants can be deformed at elevated temperatures almost like plastics and attain substantial strength at room temperature. They have been used for prototype work and instrument cabinets where considerable detail of design is to be reproduced.

Magnesium Alloys   The hexagonal structure of magnesium makes it rather brittle at room temperature, but it is worked readily at only slightly elevated temperatures, typically above 220°C. Such low temperatures create no tool or lubrication problems and yield the benefit of great ease of forming. Both solid-solution alloying and precipitation hardening are exploited to obtain material of greater strength. Its low density combines with high strength to give high strength-to-weight ratios, desirable for aerospace and automotive applications.

Aluminum Alloys   The fastest-growing segment of the metalworking industry has been the working of aluminum alloys. An fcc material, aluminum is readily deformable at all temperatures. With the aid of solid-solution and precipitation-hardening mechanisms, materials of great strength can be produced with an often very high strength-to-weight ratio. Aluminum alloys have been the main constructional material for aircraft and are beginning to make larger inroads into the construction of land vehicles as bumpers, wheels, and some body components (including entire space frames). Corrosion resistance and light weight make them attractive for a great many household, food industry, container, marine, and chemical plant applications. Equivalent electrical conductivity may be obtained at a cost often below that of copper, and large quantities are used in high-voltage power lines, busbars, and motor windings.

The metallurgical condition is called temper and is designated by a letter, followed by numbers. Most alloys are formed in the annealed (O) condition. Non-heat-treatable alloys acquire useful strength through cold working (H1 condition) although at the expense of ductility; a second digit describes the degree of hardening (e.g., H12 = quarter-hard; H14 = half-hard; H18 = hard). The H2 temper (strain-hardened and partially annealed) gives higher ductility for a given strength (Fig. 8–9). Heat-treatable alloys may be worked in the annealed condition, then subjected to solution treatment followed by natural aging (T4) or artificial aging (T6 condition). Even greater strength may be obtained by cold working a solution heat-treated material, since on subsequent natural aging (T3) or artificial aging (T8) the precipitates become extremely fine and well distributed (Sec. 8-1-7 and Example 8-15).

High-purity aluminum is an excellent conductor. Commercial purity (1100) aluminum is extensively used for foils, cooking utensils, etc. The solid-solution manganese alloy (3000 series) has higher strength and still adequate ductility, and is a general-purpose material for sheet metal articles. The solid-solution Al–Mg alloys (5000 series) have excellent corrosion resistance and are hence suitable for automotive trim and marine applications. The 2000 series Al–Cu alloys acquire great strength and reasonable ductility in the age-hardened condition. Together with the 7000 series Al–Zn–Mg alloys, they are the primary materials of aircraft construction, although the lighter aluminum-lithium alloys can offer yet higher strength-to-weight ratios (each percent Li reduces density by 3% and increases the elastic modulus by 5%, but special processing is needed to minimize anisotropy). Some alloys, such as Al–Li and Al–Cu–Zr, are superplastic. Recently, scandium additions have pushed the recrystallization temperature to 600°C, resulting in very small grain size and high strength.

Copper-Based Alloys   Copper is one of the most ductile materials, and it has the highest electrical conductivity after silver. Its high thermal conductivity and easy joining by soldering and brazing methods make it the main constructional material for electrical wiring and household water systems. Its strength can be increased without great loss of electrical conductivity by small amounts of Ag, Ca, or Be.

The solid-solution alloys with zinc (brasses) are the most widely used. As its name implies, *cartridge brass* (an α brass) is extremely ductile and is suitable for the heaviest deformations, whereas the $\alpha + \beta$ brasses (such as *Muntz metal*) are less ductile but are tolerant of contaminants, and their good machinability can be further improved by the addition of lead. Lead can be added also to α brasses but then they are not hot-workable because the lead is located on grain boundaries.

Tin bronzes are normally deoxidized with phosphorus, which forms a low-melting ternary eutectic. They are thus hot-short unless homogenized prior to hot working. Aluminum bronzes are readily hot worked, as are the nickel alloys (*cupronickel*) and ternary alloys (such as *nickel silver*, a Cu–Ni–Zn alloy).

The warm glow of copper and the infinite gradations of yellows of brass and bronze have appealed to humans over millennia; their esthetic appeal is often enhanced by corrosion products (*patina*).

Nickel-Based Alloys   Nickel in its pure form is readily deformable, at both elevated and room temperatures. Some of its alloys, particularly those with copper, present no manufacturing problem. Nickel-based superalloys are heavily alloyed with both solid-solution and precipitation-hardening elements to give high creep strength at elevated temperatures. This makes them difficult to work because the hot-working temperature range is very narrow and close to the solidus. Sophisticated melting and pouring techniques are used to exclude contaminants and gases, and a thorough knowledge of the metallurgy of the alloys and of processing technologies is needed to prevent cracking during hot working.

High-Temperature Alloys Hexagonal titanium, stable at room temperature, has reasonable ductility but requires frequent process anneals. The bcc form (over 880°C) is most ductile. For control of finished properties, alloys are often worked just below the transformation temperature, but at higher strain rates they have relatively high strength. Therefore, isothermal forging is frequently employed; since cooling is of no concern, the high strain-rate sensitivity and ductility of the material can be exploited by working at very low strain rates and correspondingly low stresses.

Because of their corrosion resistance, titanium and its alloys—in the forms of tubing and sheet—are extensively used in chemical applications. Heat-treated titanium alloys of high strength-to-weight ratios have become indispensable for critical aircraft components, including the compressor stages of jet engines.

The refractory metal alloys (molybdenum, tungsten, and niobium) readily form a volatile oxide at high temperatures, hence they must be processed in vacuum or protective atmosphere. Tungsten is used extensively in the form of wire in incandescent lamps. Recent developments in refractory metal alloys were spurred by space-age technology demands for materials that function at very high temperatures. An offshoot of this progress is the molybdenum-base die material TZM which is often used in the cold-worked condition; some effects of cold work are retained to about 1000°C.

8-4 SUMMARY

In this chapter we review concepts applicable to all plastic deformation processes.

1. Flow stress is the true stress required to maintain plastic deformation. In cold working, the flow stress increases and ductility decreases because of strain hardening; flow stress is frequently a power-law function of strain. Discontinuous yielding and anisotropy of properties may modify the behavior.

2. Cold working may be exploited to make a strong but less ductile product. Original properties may be regained by applying elevated-temperature restoration processes (recovery and recrystallization) which again may be controlled to give a more desirable set of properties.

3. Strain hardening and restoration processes occur simultaneously in hot working. Since restoration processes take time, flow stress becomes a function of strain rate and temperature. Hot working is eminently suitable for destroying a cast structure. Combination of deformation and

transformations may produce materials with exceptional strength and ductility.

4. Ductility, as observed in the tension test, reflects material properties: Uniform (prenecking) strain increases with strain hardening while postnecking strain increases with strain-rate sensitivity.

5. In practical processes the stress state is typically triaxial and the stresses required for plastic deformation may be found from yield criteria. To compute pressures and forces prevailing in a particular process, it will be necessary to take into account the effects of friction on the die–workpiece interface and the pressure increase due to inhomogeneity of deformation.

6. Ductility in tension may be regarded a material property but is affected by the prevailing stress state, leading to the concepts of bulk workability and sheet formability.

PROBLEMS 8A

8A-1 Define engineering stress and strain in tension (in words or formulas).

8A-2 Define natural (true) stress and strain in tension.

8A-3 Draw a sketch of a cylindrical tension test specimen (*a*) before loading, (*b*) after loading but before necking, and (*c*) after necking. With arrows, indicate the stresses acting in all three cases.

8A-4 Write the power-law equation for strain hardening; define the constants.

8A-5 (*a*) Draw a typical tensile test curve for a ductile material. Identify the (*b*) prenecking and (*c*) postnecking range. Indicate the material property most significant in each range.

8A-6 (*a*) Draw a typical tensile test curve for a material showing yield-point elongation (YPE). (*b*) State what causes the phenomenon. The specimen is unloaded after plastic deformation beyond YPE; draw curves showing (*c*) immediate reloading after unloading and (*d*) retesting after several months.

8A-7 State what significant problem yield-point elongation may cause in (*a*) bulk deformation and (*b*) sheet metalworking.

8A-8 Define *r* for a sheet metal with the aid of a sketch.

8A-9 Draw a diagram showing the change in YS, TS, and el. as a function of cold work.

8A-10 An alloy has been cold-worked to YS = 330 MPa, TS = 350 MPa, and el. = 2%. Draw a diagram showing how these values change as a function of time upon holding the alloy above a temperature of $0.5T_m$.

8A-11 State the power law applicable to hot working. Define the constants.

8A-12 Define superplasticity (*a*) in descriptive terms and (*b*) by reference to the relevant power law. (*c*) State the critical material condition for superplasticity.

8A-13 (*a*) Draw a typical TTT diagram for steel. (*b*) Show major phases. (*c*) Define ausforming by drawing the appropriate line on the diagram.

8A-14 (*a*) Draw a diagram showing the Tresca and von Mises yield criteria for plane stress. Identify the points corresponding to (*b*) uniaxial tension, (c) uniaxial compression, (d) plane-strain compression, (e) balanced biaxial tension, and (f) pure sheer.

8A-15 Recovery and recrystallization are two processes to change the properties of strain-hardened materials. (a) Explain the difference between the two in terms of the operative atomic mechanism. (b) Draw two diagrams, side by side, to show the effects of recovery and recrystallization on yield strength, tensile strength, and total elongation.

8A-16 Draw a three-dimensional diagram showing the dependence of grain size on prior strain and annealing temperature in the course of annealing a strain-hardened material.

8A-17 (a) Give a brief definition of mechanical fibering and state the sources for it. (b) In a simple graph, show its effect on mechanical properties such as strength, ductility, and fatigue strength.

PROBLEMS 8B

8B-1 A part is to be made of a new, heat-resistant alloy steel. It is proposed to estimate the forming forces from the hot tensile-strength data provided by the steel company. (a) Do you agree? (b) Justify. (c) Suggest the appropriate course of action.

8B-2 In a discussion it is claimed that strain-rate sensitivity can be ignored in cold working. (a) Do you agree? (b) If not, what benefit or harm would you expect from low and high strain-rate sensitivity?

8B-3 Draw a sketch of the side view of a cast slab of 100-mm thickness. The slab has centerline porosity. Show tool sets that would (a) open up the pores and (b) close the pores. Justify.

8B-4 (a) Draw a sketch of the gage portion of a sheet metal tensile specimen of w width, h thickness, and l gage length. (b) Draw the specimen after some plastic deformation. (c) Define the r value with reference to the dimensions marked in (a) and (b).

8B-5 A sheet-metal part had been superplastically formed. It distorted badly in service at elevated temperature. (a) Identify the cause of the problem and (b) suggest a remedy.

8B-6 An aluminum company claims that the material it supplied is superplastic, yet it fractured after small deformation in a fast mechanical press. To identify the problem, (a) define what can be expected from a superplastic material and (b) suggest the characteristics of a process that takes advantage of superplasticity.

8B-7 It is suggested that all-round hydrostatic pressure increases YS, TS, and el. To explain your view, (a) draw a typical engineering stress-strain curve for a ductile metal. (b) Identify the axes. (c) Mark YS, TS, and total elongation. (d) Show the effect of hydrostatic pressure. (e) Justify.

8B-8 (a) Draw a diagram to show the grain size obtained on annealing a cold-worked material as a function of temperature and prior cold work. (b) Suggest a process or process sequence to produce a very fine-grained material.

8B-9 A wrought metal is often characterized by fibering. (a) What is the origin of fibering? (b) It is often claimed that such a metal is always better in fatigue than a casting. Subject the claim to a critique.

8B-10 A student wrote "Mechanical fibering results in anisotropy of properties and in a high r value." Subject this statement to a critique.

8B-11 A student wrote "A material free of fibering is isotropic." Subject this statement to a critique.

8B-12 A student answered the question "What is a texture?" by writing "It is the anisotropy of properties." Subject the answer to a critique.

8B-13   In Example 6-13 we saw that annealed brass sheet can be purchased with different grain sizes. Suggest a suitable processing sequence for very fine grain size.

8B-14   (a) Give a brief definition of texture and state the source(s) for it, (b) state its possible effect on mechanical properties such as strength and ductility, and (c) state the consequence on plastic deformation.

8B-15   In a discussion it is stated that the survival of a material in a bulk deformation process can be predicted from its tensile elongation. (a) Do you agree? (b) If not, give a concise explanation of your view.

PROBLEMS 8C

8C-1   Low-carbon steel is known to have a slight but not negligible strain-rate sensitivity at room temperature. To determine the m value, specimens of $w = 12.68$ mm width, $h = 3.23$ mm thickness, and $l = 50$ mm gage length were tested with a crosshead speed of 6 mm/min until a load of 10660 N was reached. At this point speed was instantaneously increased tenfold to 60 mm/min and load increased to 11096N. Calculate the m value. (*Hint:* Since at the moment of speed change the cross-sectional area is constant, force can be substituted for stress).

8C-2   To gain a feel for the effect of strain hardening on flow stress [Eq. (8-4)], (a) calculate the value of ε^n for $\varepsilon = 0.05, 0.1, 0.2, 0.3, 0.5, 0.7, 1.0, 1.2, 1.5,$ and 2.0 and $n = 0, 0.05, 0.1, 0.2, 0.3, 0.4, 0.5,$ and 0.6. (b) Plot the family of n curves as a function of ε.

8C-3   For conversational purposes, engineering strain is widely used. Derive formulas for converting natural strain to engineering tensile and compressive strain.

8C-4   To gain a feel for the effect of strain-rate sensitivity on flow stress [Eq. (8-11)], (a) calculate the value of $\dot{\varepsilon}^m$ for $\dot{\varepsilon} = 1, 10, 100$ and 1000 s^{-1} and $m = 0, 0.05, 0.1, 0.2, 0.3, 0.4, 0.5,$ and 0.6. (b) Plot the family of m curves as a function of $\dot{\varepsilon}$.

8C-5   An aluminum-alloy sheet metal specimen of $w = 10$ mm, $h = 1.5$ mm, and $l = 50$ mm dimension is tested in tension. After 25% elongation, the width is $w_1 = 9.23$ mm. Calculate the r value.

Further Reading

Altan, T., S.I. Oh, and H.C. Gegel: *Metal Forming—Fundamentals and Applications*, 1983.

Avitzur, B.: *Metal-Forming Processes*, Wiley-Interscience, 1981.

Avitzur, B.: *Handbook of Metalforming Processes*, Wiley-Interscience, 1983.

Backofen, W.A.: *Deformation Processing*, Addison-Wesley, 1972.

Blazynski, T.Z. (ed.): *Plasticity and Modern Metal Forming Technology*, Elsevier, 1989.

Boyer, H.E.: *Atlas of Stress-Strain Curves*, American Society for Metals, 1986.

Byers, J. P.: *Metalworking Fluids*, Dekker, 1994.

Dieter, G.E.: *Mechanical Metallurgy*, 3d ed., McGraw-Hill, 1986.

Dieter, G.E. (ed.): *Workability Testing Techniques*, American Society for Metals, 1984.

Frost, H.J., and M.F. Ashby: *Deformation-Mechanism Maps*, Pergamon, 1982.

Ghosh, S.K., and M. Predeleanu (eds.): *Materials Processing Defects*, Elsevier, 1995.

Hosford, W.F., and R.M. Caddell: *Metal Forming Mechanics and Metallurgy*, 2d ed., Prentice Hall, 1993.

Kobayashi, S., S.I. Oh, and T. Altan: *Metal Forming and the Finite-Element Method*, Oxford University Press, 1989.

Mielnik, E.: *Metalworking Science and Engineering*, McGraw-Hill, 1991.

Nachtman, E., and S. Kalpakjian: *Lubricants and Lubrication in Metalworking Operations,* Dekker, 1985.

Prasad, Y.V.R.K., and S. Sasidhara (eds.): *Hot Working Guide: A Compendium of Processing Maps,* ASM International, 1997.

Rowe, G.W.: *Elements of Metalworking Theory,* Arnold, London, 1979.

Schey, J.A.: *Tribology in Metalworking: Friction, Lubrication and Wear,* American Society for Metals, 1983.

Wagoner, R., and J-L. Chenot: *Fundamentals of Metalforming Analysis,* Wiley, 1996.

The aircraft parts seen in the foreground are closed-die forged on a giant 350-GN hydraulic press; as the dies close, flames flare up from the lubricant applied to the die and workpiece. (*Courtesy Wyman-Gordon Company, North Grafton, Massachusetts.*)

chapter

9

Bulk Deformation Processes

We are now ready to explore applications of basic principles to processes in which the bulk of the workpiece is deformed, including:

Open-die forging with simple tools and complex process sequences

Impression- and closed-die forging with complex tools and simple process sequences

Extrusion of net-shape and near-net-shape parts

Extrusion and drawing of wire, tube, and shaped sections

Rolling of flat plate, sheet, strip, and foil and of shaped products

Estimation of die pressure, force, and power requirement

Design of parts for ease of processing

Before embarking on a discussion of processes, we need to explore the application of principles from Chap. 8 to the specific circumstances of bulk deformation. We will then be ready to look at actual processes, with particular emphasis on those that produce end articles. Bulk deformation processes are among processes for which useful computations can be made on the basis of simple analytical methods, and these computations will be introduced in a form suitable for spreadsheet solutions. Derivations of the underlying equations will be found in the Instructor's Manual.

9-1 CLASSIFICATION

Processes may be classified according to several points of view, all of which may be valid under certain conditions.

9-1-1   Temperature of Deformation

We saw that material properties are a function of temperature, with $0.5T_m$ as a rough dividing line between hot and cold behavior. In practice, distinction is made relative to room temperature.

Hot Working   In everyday usage, the term simply refers to the working of preheated material. Usual hot-working temperatures are given in Tables 8–2 and 8–3. Since for most technical materials (except tin and lead), $0.5T_m$ is above room temperature, the everyday definition is correct also from a materials point of view.

Hot working offers several advantages: Flow stresses are low, hence forces and power requirements are relatively low, and even very large workpieces can be deformed with equipment of reasonable size. Ductility is high, hence large deformations (in excess of 99% reduction) can be taken (usually in a succession of passes) and complex part shapes can be generated. The cast structure can be destroyed (Sec. 8-1-7), in general, by a deformation equivalent to 75% reduction in height or area, although reductions of 90% (10:1 ratio) may be needed if highest properties are to be attained.

There are also disadvantages. It takes energy to heat the workpiece to the elevated temperature. Most materials oxidize and the oxides of some metals (e.g., scale on steel) can impair surface finish. Variations in finishing temperatures can lead to fairly wide dimensional tolerances and also to a less well-defined set of properties in the as-hot-worked condition.

Hot working may be carried out by:

1. *Nonisothermal forming.* The deforming tool must be several times stronger than the workpiece, and this usually means that the tool must be kept much colder. It is then necessary to minimize *contact time* to prevent excessive *cooling*. Cooling of the surface layers of the workpiece has several disadvantages: material flow is retarded; cooling of thin sections limits the minimum wall thickness attainable; die pressure and deformation force increase. Cooling affects the product too, because variable cooling introduces variations in properties. Furthermore, periodic contact with the hot workpiece exposes the tooling to thermal cycling which leads to thermal fatigue (Sec. 4-5).

2. *Isothermal forming.* Some of the above problems disappear when the tool is at the same temperature as the workpiece. Contact time ceases to be a problem since there is no cooling; however, it is more difficult to find an appropriate tool material and lubricant.

3. *Controlled hot working.* Usually conducted nonisothermally, controlled hot working is used to impart desirable properties (Sec. 8-1-7).

Cold Working   In the everyday sense, the term refers to working at room temperature, although the work of deformation can raise temperatures to 100–200°C.

Cold working usually follows hot working, and scale and other surface films are normally removed by chemical etching (*pickling*) or shot blasting.

Cold working has several advantages. In the absence of cooling and oxidation, tighter tolerances and better surface finish can be obtained and thinner walls are also possible. The final properties of the workpiece can be closely controlled, and, if desired, the high strength obtained during cold working can be retained or, if high ductility is needed, grain size can be controlled to advantage in annealing (Sec. 8-1-5). Lubrication is, in general, somewhat easier.

There are also drawbacks. For most technological materials, room temperature is below $0.5T_m$; therefore, flow stresses are high and hence tool pressures, deformation forces, and power requirements are high too. The ductility of many materials is also limited, thus limiting the complexity of shapes that can be readily produced.

Warm Working   This combines some of the advantages of hot and cold working, especially in the warm working of steel (typically between 650 and 700°C). Temperatures are low enough to avoid scaling, thus ensuring a good surface finish, yet they are high enough to reduce flow stress and thus allow the forming of parts that would generate excessive die pressures in cold working. The elevated temperature results in substantial strain-rate sensitivity, and the flow stress [Eq. (8-11)] remains low only if strain rates are kept low.

9-1-2   Purpose of Deformation

A further useful distinction may be made according to the purpose of deformation.

Primary Processes   These aim at destroying the cast structure by successive deformation steps. The resulting semifabricated product is destined for further shaping or forming (as shown in Fig. 5–2). Primary processes are usually conducted hot and on a large scale, in specially constructed plants.

Secondary Processes   These take the products of some primary process and further transform them into a finished part; they are at the focus of our discussion. Secondary processes include variants of the bulk deformation processes (Fig. 9–1) and all of the sheet-metalworking processes to be discussed in Chap. 10.

9-1-3   Analysis

From the point of view of understanding and analyzing bulk deformation processes, it is useful to make a different distinction.

Steady-State Processes   In these, all elements of the workpiece are subjected successively to the same mode of deformation. Thus, once the situation is analyzed

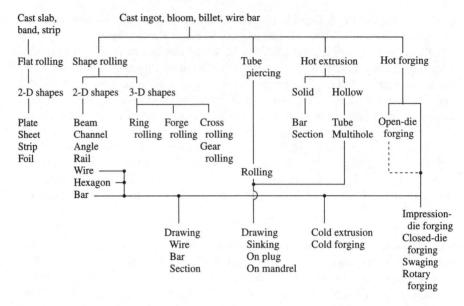

Figure 9–1   Bulk deformation processes convert cast material into semifabricated or finished products, often through a sequence of operations. (*Adapted from J.A. Schey, ASM Handbook, vol. 20, Materials Selection and Design, ASM International, 1997, p. 691. With permission.*)

for the deformation zone, the analysis remains valid for the duration of the process. Drawing of a sheet in plane strain can be taken as the generic case (Fig. 9–2a). The workpiece strain-hardens or suffers other changes as it passes through the deformation zone, and, to simplify computations, a *mean flow stress* σ_{fm} is used. For cold working, this is found by integrating Eq. (8-4) between the limits of strains:

$$\sigma_{fm} = \frac{K}{\varepsilon_2 - \varepsilon_1} \left[\frac{\varepsilon_2^{n+1} - \varepsilon_1^{n+1}}{n+1} \right] \qquad \textbf{(9-1a)}$$

For an annealed material ($\varepsilon_1 = 0$), this reduces to

$$\sigma_{fm} = \frac{K}{\varepsilon} \left[\frac{\varepsilon^{n+1}}{n+1} \right] \qquad \textbf{(9-1b)}$$

Alternatively, the flow stress curve is plotted and the mean is found by visual averaging (Fig. 9–2b). For hot working, a mean strain rate specific for the process is computed and the flow stress is taken from Eq. (8-11).

Non-Steady-State Processes   In processes such as compression, the geometry of the part changes continually (Fig. 9–3a) and the analysis must be repeated for various points in time, from the starting condition to the end of the stroke. Thus, the *instantaneous flow stress* σ_f at the point of interest must be taken (Fig. 9–3b),

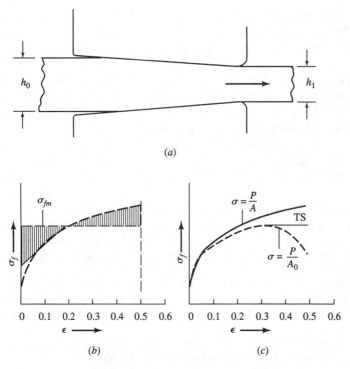

Figure 9–2   In steady-state cold-working processes, (a) the material is subjected to strain hardening during its passage through the die, and (b) the relevant flow stress is the mean flow stress from the true stress–true strain curve; (c) if this is not available, the tensile strength may give a reasonable approximation.

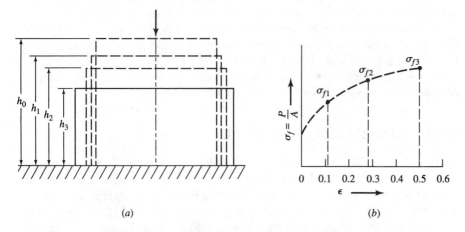

Figure 9–3   In non-steady-state processes, (a) the material is deformed in a succession of incremental deformations and (b) the flow stress is taken at the strain of interest.

from Eq. (8-4) for cold working and from Eq. (8-11) for hot working. It often happens that we are concerned only with the maximum force, developed at the end of deformation, and then the flow stress corresponding to the final strain is used.

A problem arises when the K and n values are not known. If equipment is available, a compression test (Sec. 4-3) can be conducted rapidly. Otherwise, the only guide could be the TS, from Tables 8–2 and 8–3 or other source. Paradoxically, the basically nonsensical method of calculating the TS (Sec. 4-1-3) happens to give a reasonable approximation of the mean true flow stress σ_{fm}, at least for strains on the order of the necking strain (Fig. 9–2c). Since at the point of necking $n = \varepsilon_u$, some reasonable correction can be made for smaller or larger strains.

Example 9-1

In Example 4-4 we found TS = 351 MPa for an annealed Cu-20Ni alloy. In Example 8-1 we determined $K = 760$ MPa and $n = 0.45$. If we had not made the computations of Example 8-1 (or if only the TS of a material were available from the literature), would the TS give any guidance on flow stress?

Applying Eq. (8-4), $\sigma_f = 351 = 760\varepsilon^{0.45}$ and $\varepsilon = \exp(\frac{-0.7725}{0.45}) = 0.18$. Since $\varepsilon = \ln(l/l_0) = 0.18$, $l/l_0 = \exp 0.18 = 1.197$, and the corresponding tensile strain is $e_t = (l - l_0) = (1.197 - 1.0)/1.0 = 19.7\%$. To convert to compressive strain, regard $l = h_0$ and $l_0 = h$; then $e_c = (h_0 - h)/h_0 = (1.197 - 1.0)/1.197 = 16.5\%$. Thus, the TS is a reasonable estimate of σ_f for a small compressive deformation, but would be too low if this heavily strain-hardening material were to be worked to a higher strain.

9-2   OPEN-DIE FORGING

Forging processes are among the most important manufacturing techniques. As shown in Fig. 9–1, three broad groups can be distinguished: *open-die forging* allows free deformation of at least some workpiece surfaces; deformation is much more constrained in *impression-die forging* and is fully constrained in *closed-die forging*. Because at least one of the workpiece surfaces deforms freely, open-die forging processes produce workpieces of lesser accuracy than impression- or closed-die forging; however, tooling is usually simple, relatively inexpensive, and allows the production of a large variety of shapes.

9-2-1   Axial Upsetting of a Cylinder

In the *axial upsetting* of a cylinder, a workpiece of cylindrical shape is placed between two flat parallel dies (*platens*) and is reduced in height by a press or hammer force applied to the platens. Upsetting is a very versatile process, practiced hot or cold. The end products range from huge, 150-Mg or larger steel rotors for power-generation stations to minute components. Frequently, a head is upset at the end of a part, in special-purpose mechanized (automated) machines, producing vast numbers of nails, screws, bolts, pins, and similar components.

Frictionless Upsetting   Let us assume that, by the application of some very good lubricant, we succeed in reducing friction to virtually zero. If we divide the cylinder into many small elements, each element now deforms equally; in other words, deformation is *homogeneous*. The cylinder becomes shorter and, to preserve constancy of volume [Eq. (4-2)], it assumes a larger diameter, but still remains a true cylinder (Fig. 9–4a). Because upsetting is a non-steady-state process, a complete analysis requires computation of variables at several points during the press stroke. In computations of a repetitive kind, it is best to follow a set sequence of operations, as shown here step by step. These steps, adapted to specific processes, will be followed in this chapter throughout.

Step 1: Find the volume of the part. In this case,

$$V = (d_0^2 \pi/4) \, h_0 \qquad\qquad \textbf{(9-2a)}$$

Step 2: In practice, only one of the final dimensions is defined. From constancy of volume, the final end-face area A_1 and diameter d_1 can be found

$$A_1 = A_0 \frac{h_0}{h_1} = \frac{V}{h_1} \qquad\qquad \textbf{(9-2b)}$$

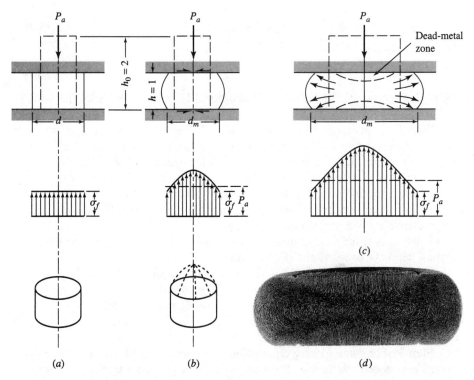

Figure 9–4   Interface pressures are (a) equal to the flow stress in frictionless compression but (b) friction generates a friction hill which (c) is larger for a large d/h ratio. (d) Sticking at the end face, caused by high friction or chilling, leads to folding over of the sides.

and

$$d_1 = \sqrt{\frac{4A_1}{\pi}} \tag{9-2c}$$

Step 3: The compressive engineering strain is needed only for conversational purposes. It is usually computed from the height change [Eq. (4-16)] but, because the volume remains constant, end-face areas can be used equally well:

$$e_c = \frac{A_1 - A_0}{A_1} = \frac{h_0 - h_1}{h_0} \tag{9-3}$$

Step 4: For purposes of computing the flow stress in cold working, true strain is

$$\varepsilon = \ln \frac{h_0}{h_1} \tag{8-5b}$$

Step 5: For hot working, the strain rate is also needed

$$\dot{\varepsilon} = \frac{v}{h} \tag{8-10}$$

Step 6: We are now ready to calculate the relevant flow stress. In cold working

$$\sigma_f = K\varepsilon^n \tag{8-4}$$

In hot working

$$\sigma_f = C\dot{\varepsilon}^m \tag{8-11}$$

(Note that strain rate must always be expressed in units of reciprocal seconds, s^{-1}.)

Step 7: To compute die pressure (also called *interface pressure*, p_a, where the subscript refers to axial symmetry), we need to check the effects of (*a*) stress state, (*b*) friction, and (*c*) inhomogeneity of deformation (Sec. 8-2).

a. The stress state is uniaxial, hence the flow stress is σ_f.

b. Since friction is absent, there is no increase in pressure.

c. The platen overlaps the workpiece, hence no indentation effect is possible and we need not worry about any h/L ratio.

Thus p_a is simply the uniaxial flow stress σ_f. This is the pressure that the tooling will have to withstand (Fig. 9–4a).

Step 8: The press force P_a is interface pressure multiplied by the *contact area* [Eq. (9-2b)]—the *area over which the pressure acts*). If we take the final point, where forces are highest, we find the size of press needed:

$$P_a = p_a A_1 \tag{9-4}$$

Step 9: For some forging equipment, it is necessary to know also the total energy expended in deforming the workpiece. This can be obtained by repeating the computations for the press force P_a at various points (progressively diminishing h) in the stroke. Thus, the force–displacement curve (Fig. 9–5) is defined. Force rises because the area A_1 increases rapidly. The area under the curve has

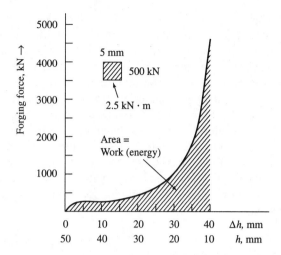

Figure 9–5   Upsetting forces rise steeply with progressing deformation; the area under the curve represents work.

the dimensions of work (work = force × distance). Thus, the work or energy E_a to be delivered by the press or hammer can be obtained by graphical or numerical integration of this area.

Step 10: The energy absorbed by the workpiece is converted into heat. In the absence of cooling, the adiabatic temperature rise ΔT is

$$\Delta T = \frac{E_a}{V\rho\,c} \qquad \text{(9-5)}$$

where V is volume, ρ is density, and c is specific heat (more correctly, heat content per unit volume) of the workpiece. In practice, deformation takes place in finite time, and some of the heat is lost through conduction into the dies and by radiation and convection into the surrounding atmosphere. Therefore, the actual temperature rise is less but can, nevertheless, be significant. In hot working it may raise the temperature above the solidus to cause hot-shortness, and in cold working it may result in lubrication breakdown.

Upsetting with Sliding Friction   In practice it is highly unlikely that zero friction can be attained even with the best lubricant (Table 8–4). Deformation of the cylinder requires that its end faces should slide on the tool surfaces, thus a measurable frictional shear stress τ_i is always present. This shear stress opposes the free expansion of the end faces, with two consequences (Fig. 9–4b):

1. The cylinder assumes a *barrel* shape. We may ignore this in calculating the new diameter (Step 2) simply by taking a mean diameter d_m from constancy of volume [Eq. (9-2c)].

2. In order to overcome the frictional stress, a higher and higher normal pressure must be exerted as we move toward the center of the cylinder. At the free edge, the pressure equals σ_f and rises from here like a hill. With higher friction (from Sec. 8-2-3, expressed as a coefficient of friction μ or frictional shear factor m^*), the *friction hill* will be steeper. This is taken into account in Step 7b, in computing the average interface pressure p_a. A comparison of Fig. 9–4b and 9–4c shows that, for the same magnitude of friction, a cylinder of the same height but of larger diameter gives rise to a taller friction hill and, therefore, higher p_a. Thus, the d/h ratio or *shape factor*—which characterizes the squatness of the cylinder—determines, together with friction, the degree of pressure intensification. The average interface pressure p_a is then conveniently expressed as a multiple of the uniaxial flow stress σ_f. The *pressure-multiplying factor* Q_a (where the subscript signifies axial symmetry) must take into account the effects of both friction (μ or m^*) and workpiece geometry (d/h ratio). From theory, if m^* is used,

$$p_a = \sigma_f Q_a = \sigma_f \left(1 + \frac{m^*}{3\sqrt{3}} \frac{d_1}{h_1} \right) \tag{9-6}$$

Alternatively, if μ is used, Q_a can be taken from Fig. 9–6 and

$$p_a = \sigma_f Q_a \tag{9-7}$$

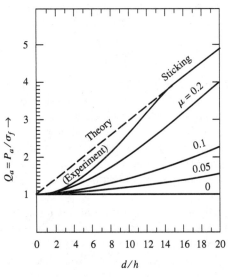

Figure 9–6   Average pressures in upsetting a cylinder increase with increasing friction and squatness of the cylinder. [After J.A. Schey, T.R. Venner, and S.L. Takomana, J. Mech. Work. Tech. **6**:23–33 (1982). With permission of Elsevier Science Publishers.]

The maximum stress $p_{a\ max}$ is of importance for the die material. It is most simply computed with the use of m^*

$$p_{a\ max} = \sigma_f \left(1 + \frac{m^*}{\sqrt{3}} \frac{d_1}{h_1} \right) \qquad \text{(9-8)}$$

Upsetting with Sticking Friction   In the extreme case, when the platen surface is rough and no lubricant is used, the interface shear stress τ_i may reach or exceed the shear flow stress τ_f of the workpiece material (Sec. 8-2-3) and movement of the end face is totally arrested. All deformation now takes place by internal shear in the cylinder; material adjacent to the platens does not move (*dead-metal zones form*) and the sides of the cylinder fold over (Fig. 9–4*d*). In nonisothermal forging, cooling of the end faces aggravates the situation, as shown by the flow lines in the specimen of Fig. 9–4*d*.

This is an unusual case of inhomogeneous deformation in that interface pressure becomes lower. Because the outer fibers of the cylinder are deformed by shearing superimposed on compression, the compressive stress needed is reduced (see Fig. 8–14*b*) and interface pressure remains low. The pressure-multiplying factor remains close to unity as long as $d/h < 2$. Simple theory cannot cope with this complexity and the limiting values of the pressure-multiplying factor given in Fig. 9–6 have been determined experimentally. Finite-element analysis gives similar results.

An AISI 1045 steel billet of $d_0 = 50$ mm and $h_0 = 50$ mm is cold-upset to a height of $h_1 = 10$ mm, on a hydraulic press operating at $v = 80$ mm/s. The lubricant is mineral oil with EP additives. Compute the press force and energy expenditure.

Example 9-2

Flow stress is from Table 8–2, friction from Table 8–4. To obtain the press force it would be sufficient to calculate for the final height only, however, the force is needed at several points of the press stroke if energy is to be determined too. It is best to set up a spreadsheet. The result is:

A	B	C	D	E	F	G	H	I	J	K	L	M	N	O	P
Cold	1045 steel					d0 =	50	mm	h0 =	50	mm	V =	98 175	mm^3	
Flow stress		K =		950	MPa	n =	0.12								
Point	mu	v	h	d1	A1	ec	epsilon	eps dot	sigma f	d/h	Qa	pa	pa	Pa	Pa
No		mm/s	mm	mm	mm^2			1/s	N/mm^2			N/mm^2	kpsi	kN	tonf
				Eq. (9-2c)	Eq. (9-2b)	Eq. (9-3)	Eq. (8-5b)		Eq. (8-10)	Eq. (8-4)		Fig. 9-6	Eq. (9-7)		Eq. (9-4)
0	0.1	80	50	50.0	1963	0	0	1.60	0	1	1.00	0	0	0	0
1	0.1	80	40	55.9	2454	0.20	0.223	2.00	794	1.4	1.10	873	127	2142	241
2	0.1	80	30	64.5	3272	0.40	0.511	2.67	876	2.2	1.25	1096	159	3585	403
3	0.1	80	20	79.1	4909	0.60	0.916	4.00	940	4.0	1.30	1222	177	5999	674
4	0.1	80	15	91.3	6545	0.70	1.204	5.33	971	6.1	1.40	1360	197	8901	1001
5	0.1	80	10	111.8	9817	0.80	1.609	8.00	1006	11.2	1.80	1810	263	17774	1998

Note the rapid rise in force as the height diminishes.

Example 9-3 | The billet of Example 9-2 is now hot-upset at 1000°C without a lubricant. Recompute the press force.

A	B	C	D	E	F	G	H	I	J	K	L	M	N
Hot	1045 st	T=	1000	C		d0=	50	mm	h0=		50	mm	V=	98 175	mm^3
Flow	stress			C=	120	MPa	m=	0.13							
Point	mu	v	h	d1	A1	ec	eps	eps dot	sigma f	d/h	Qa	pa	Pa
No		mm/s	mm	mm	mm^2			1/s	N/mm^2			N/mm^2	kN
				Eq. (9-2c)	Eq. (9-2b)	Eq. (9-3)	Eq. (8-5b)	Eq. (8-10)	Eq. (8-11)		Fig. 9-6	Eq. (9-7)	Eq. (9-4)
0	st	80	50	50	1963	0	0	1.60	127.6	1	1.00	128	250
1	st	80	40	55.90	2454	0.2	0.223	2.00	131.3	1.40	1.10	144	355
2	st	80	30	64.55	3272	0.4	0.511	2.67	136.3	2.15	1.15	157	513
3	st	80	20	79.06	4909	0.6	0.916	4.00	143.7	3.95	1.40	201	988
4	st	80	15	91.29	6545	0.7	1.204	5.33	149.2	6.09	1.80	269	1757
5	st	80	10	111.80	9817	0.8	1.609	8.00	157.2	11.18	3.00	472	4631

Results are plotted in Fig. 9–5. Note the large drop in pressure and force relative to Example 9-2. To obtain the energy requirement, the area under the force–displacement curve is integrated. One square corresponds to $(500 \text{ kN})(5 \text{ mm}) = 2500 \text{ N·m}$; the total area is about 14.5 squares or 36 250 N·m (= 26 700 lb·ft or 320 000 lb·in).

Note that relative to cold forging, press force dropped by some 74%. Normally, a lubricant (with $\mu = 0.1$ from Table 8–4) would be used. Repeating the computation, press force would now be 2470 kN, i.e., one-seventh of the cold upsetting force.

9-2-2   Forging of Rectangular Workpieces

Two fundamentally distinct cases are to be distinguished: forging with *overhanging platens* and forging an *overhanging workpiece*.

Upsetting with Overhanging Platens   When a rectangular slab is upset between two platens that are larger than the workpiece in all directions (Fig. 9–7a), the situation resembles that of the axial upsetting of a cylinder, at least as far as deformation in the *narrower* cross section is concerned. Overall, however, the situation is different, especially if one of the dimensions of the slab is much greater. *The material will always flow in the direction of least resistance.* Frictional resistance is proportional to the distance over which sliding takes place. Therefore, flow in the longer direction (which we shall call, for purposes of analysis, the *width* direction w), is much restricted and the condition of plane strain (Fig. 8–15a) is approximated. *Most material flow takes place in the short direction*, and for purposes of analysis we shall call this the *contact length L* between workpiece and

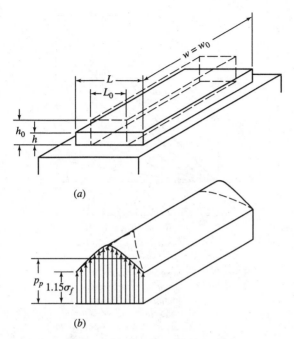

(a)

(b)

Figure 9–7   In upsetting a rectangular workpiece, (a) material flows in the direction of least resistance, marked L; (b) the friction hill now has a ridge shape.

tool surface; evidently, in upsetting with overhanging anvils, *the cross-sectional area remains constant while L increases as compression progresses* (Fig. 9–7a).

Since the two platens are parallel, material flows away from the centerline where no material flow takes place. The centerline is, therefore, called the *neutral line* (the line that divides the flow directions).

In the process of computation, Steps 1–6 are the same as in upsetting a cylinder. However, there are differences in Step 7:

a. First, the material begins to flow only on reaching the plane-strain flow stress $1.15\sigma_f$ (points 4 in Fig. 8–14b).

b. Second, the friction hill resembled a single-pole tent in axial upsetting (Fig. 9–4b) but it is now more like a mountain ridge (Fig. 9–7b). The cross section of the friction hill can still be computed by analogy to axial upsetting, as long as it is clearly understood that the friction hill is defined by that dimension of the workpiece which is measured *in the direction of major material flow*, i.e., the contact length L. Then, the friction hill will be higher for any given μ or m^* and for a larger shape factor (L/h ratio). The average pressure p_p (where the subscript refers to plane strain) is

$$p_p = 1.15\sigma_f \left(1 + \frac{m^*}{4}\frac{L}{h}\right) \qquad \textbf{(9-9)}$$

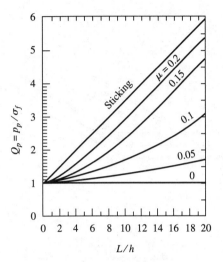

Figure 9–8   Average pressures in upsetting a rectangular slab increase with friction and L/h ratio. (*After J.F.W. Bishop, Quart. J. Mech. Appl. Math.* **9**:236–246 (1956). With permission of Pergamon Press.)

When friction is expressed as μ, the pressure-multiplying factor Q_p is taken from Fig. 9–8 and the average pressure is

$$p_p = 1.15\sigma_f Q_p \tag{9-10}$$

By analogy to upsetting a cylinder [Eq. (9-8)], the peak of the friction hill will be

$$p_{p\ \text{max}} = 1.15\sigma_f \left(1 + \frac{m^* L}{2 h} \right) \tag{9-11}$$

This peak develops at the neutral line.

In Step 8, the force P_p at any one point in the press or hammer stroke is again obtained by multiplying p_p by the *contact area* A_1 over which this pressure acts

$$P_p = p_p A_1 = p_p L w \tag{9-12}$$

Rectangular workpieces are frequently forged, and elements of more complex forgings can often be regarded as rectangular ones.

Example 9-4

A 302 stainless steel pin is to be produced from a square wire. One end is flattened, and the center is pinched. Compute the die pressure and force, assuming that no lubricant is used.

The flattening operation may be considered as upsetting a rectangular workpiece. Because of friction, the 100-mm length of the pin increases very little during flattening, and this

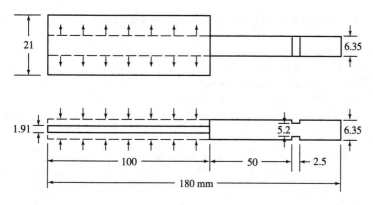

Figure Ex. 9-4

dimension must be regarded as the width w during plane-strain compression (Fig. 9–7a). Most of the material gets displaced in the width of the pin; in terms of analysis, this becomes L ($= h_0 L_0 / L_1 = (6.35)(6.35)/1.91 = 21.1$ mm). Again, a spreadsheet speeds up the task

A	B	C	D	E	F	G	H	I	J	K	L	M	N	O
Cold		302	SS											
Flow	stress			K=	1300	MPa	n=	0.3						
						Cross-section							Contact area	
Pass	mu	v	h	L	w	A (cr)	ec	eps	sigma f	L/h	Qp	pp	A (cont)	Pp
No		mm/s	mm	mm	mm	mm^2			N/mm^2			N/mm^2	mm^2	kN
							Eq. (9-3)	Eq. (8-5b)	Eq. (8-4)	Fig. 9-8	Eq. (9-10)			Eq. (9-12)
0	st	80	6.35	6.35	100	40.3				1		0	645	0
1	st	80	1.91	21.1	100	40.3	0.7	1.2039	1374.4	11.1	3.8	6006	2117	12713

Note the large force required for deforming this relatively small part.

To reduce die pressure and forging force, a compounded mineral oil lubricant is applied. Find its effect on total force.   **Example 9-5**

A repeat of computation with $Q_p = 1.8$ results in $p_p = 2845$ N/mm^2 and $P_p = 6022$ kN. Note the large size of press needed for this seemingly minor operation, even with the application of a lubricant. Die pressures and forces could be reduced by flattening in two steps, with an intermediate anneal.

Forging an Overhanging Workpiece   A very different situation exists when the platen is narrow (and is usually called an *anvil*). Since the forged part now overhangs the anvil, we cannot expect the entire mass of the workpiece to be deformed, and deformation can become inhomogeneous even within the work zone. To judge the degree of inhomogeneity, we must return to Fig. 8–17. When the workpiece is wide, deformation is again in plane strain (Fig. 8–15b). *Major flow occurs in the direction of the short dimension of the anvil*, hence this now becomes L for purposes of analysis. Three distinct possibilities exist:

1. When $h/L > 8.7$ (Fig. 8–17b), the situation is the same as in indenting a semi-infinite body (Fig. 8–17a). It can be shown that the pressure required for indentation $p_{i\,\max}$ is approximately 3 times the uniaxial flow stress σ_f of the material

$$p_{i\,\max} = \sigma_f Q_{i\,\max} = 3\sigma_f \qquad \text{(9-13)}$$

Example 9-6	It will be recognized that while physically the situation shown in Fig. 8–17 appears very different from a hardness test (Fig. 4–12), the strain state is actually very similar. In hardness testing the specimen is, for all intents and purposes, infinite in the width, length, and thickness directions; thus, the indentor has to push the material out, as in Fig. 8–17a. Therefore, the *indentation hardness* of a material is approximately 3 times its uniaxial (compressive) flow strength. Since the highly localized deformation causes rather severe strain hardening, the indentation hardness is 3 times the mean flow stress σ_{fm} prevailing in the shear zones and, for reasons mentioned in Sec. 9-1-3, the TS is a good approximation of this mean value. It is for this reason that the indentation hardness is often taken as $3 \times \text{TS}$ (remember that hardness is quoted in kg/mm^2). For a strain-hardening material, better agreement is obtained when hardness is taken as 3 times the flow stress at 7% cold work.

2. When $8.7 > h/L > 1$, the two deformation zones gradually interact, requiring less and less force to maintain plastic deformation (Fig. 8–17b). Therefore, the pressure-multiplying factor also diminishes, and can be taken from Fig. 9–9. The indentation pressure is

$$p_i = 1.15\sigma_f Q_i \qquad \text{(9-14)}$$

It should be remembered that penetration of the two wedges sets up secondary tensile stresses which, at $h/L > 2$ can lead to internal fracture (*centerburst*) in a less-ductile material.

3. At a ratio of $h/L = 1$ the two deformation zones fully cooperate (Fig. 8–17c) and the material flows at a minimum pressure (at $1.15\sigma_f$).

4. When $h/L < 1$ (or, more conveniently, $L/h > 1$), friction becomes significant and the pressure-multiplying factor must be obtained according to Eq. (9-9) or (9-10). The friction hill is chopped on its sides and the pressure-multiplying factor drops when $w/L < 8$; it drops to Q_a when $w/L = 1$.

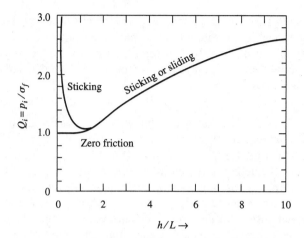

Figure 9–9   Pressures needed to indent a workpiece increase with h/L but are independent of friction. (*After R. Hill, The Mathematical Theory of Plasticity, Clarendon Press, Oxford, 1950. With permission.*)

The pin in Example 9-4 is pinched at the center. Consider the geometry of the operation: the resemblance to Fig. 8–17b is obvious; $L = 2.5$ mm, $w = 6.35$ mm.

Example 9-7

Step 4: $\varepsilon = \ln (6.25/5.2) = 0.2$.

Step 6: $\sigma_f = 1300(0.2)^{0.3} = 802$ N/mm^2.

Step 7c: $h/L = 5.2/2.5 = 2$, thus deformation is indeed inhomogeneous. From Fig. 9–9, $Q_i = 1.3$ for the end of stroke. Thus $p_i = (1.15)(802)(1.3) = 1200$ N/mm^2. Note that, because of the inhomogeneity of deformation, there is uncertainty about the proper value of strain and flow stress. However, the error is usually within practically permissible limits.

---

5. Inhomogeneous deformation is intentionally induced in *shot peening*. Many overlapping indentations are made with high-velocity shot, causing localized compressive deformation of the surface. Since the bulk of the workpiece is not affected, there are two possible consequences:

a. If all surfaces of the part are peened, balanced compressive residual stresses are set up and fatigue life is increased (Sec. 4-7, Fig. 4–18).

b. If only one surface is peened, unbalanced residual stresses cause curvature (Fig. 4–19). Well-controlled *peen forming* is suitable for correcting the shape of large containers and rocket-motor cases and for developing the shape of aircraft wing surfaces.

9-2-3   Open-Die Forging

In addition to upsetting and indentation, open-die forging employs various other processes, all of which can be analyzed by analogy to the processes discussed in

Secs. 9-2-1 and 9-2-2. A great variety of shapes can be produced with relatively simple dies, although often through a complex sequence of deformation steps: *The simplicity of tooling is gained at the expense of complex process control.* Parts of substantial degree of complexity can be forged by a planned sequence of open-die forging steps.

Cogging   The surface area A_1 of a rectangular workpiece can be very large, resulting in an impractically high total force; therefore, it is customary to deform only one part of a large workpiece at a time. Properly sequenced individual *bites* gradually reduce the height of the entire length of the workpiece by the process of *cogging* or *drawing out* (Fig. 9–10). Successive bites must be spaced close enough to produce an even surface, but too short a bite ($b < h_0/3$) will just fold the material down instead of deforming the entire cross section.

Drawing out is sometimes used as a substitute for rolling when quantities are small or the material is prone to hot cracking. The part is often held in mechanical arms (*manipulators*), the motions of which must be closely coordinated with anvil movements (Fig. 9–11); hence, computer control is widespread.

Computation of stresses and forces follows the principles described in Sec. 9-2-2. Contact length L is again measured in the direction of major material flow and is thus equal to the bite (Fig. 9–10b). To obtain an appropriate pressure-multiplying factor, the h/L ratio must be found. When its value is greater than unity, inhomogeneous deformation prevails and the interface pressure is found from Fig. 9–9; when its value is below unity, friction predominates and Fig. 9–8

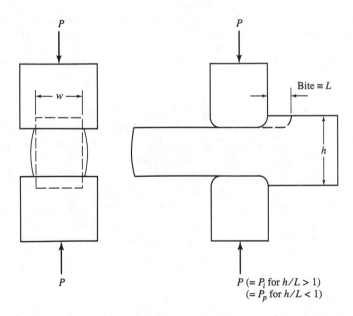

Figure 9–10   Bars may be reduced in height by a sequence of strokes in the process of cogging or drawing out.

Figure 9–11   An alloy steel ingot is held by a manipulator for drawing out on a hydraulic press. (*Courtesy Atlas Specialty Steels, Welland, Ontario.*)

[or Eq. (9-9)] should be used. Plane-strain conditions are approximated only when $w/L > 10$. For narrower pieces the multiplying factor is smaller.

Fullering and Edging   Many parts have thick and thin sections and it is then necessary to redistribute material. Forging between flat anvils is inefficient because some material moves in the width direction (*spreads*) and, when L/h is large, pressures and forces are high. Forging with *inclined surfaces* solves these problems because there is a pressure component acting in the direction of material flow (Fig. 9–12). This has two effects. First, it counteracts frictional retardation (when $\tan \alpha = \mu$, the effect of friction is neutralized) and thus lowers the die pressure. Second, it moves the material perpendicular to the direction of load application. The effect can be exploited to move material away from the center (*fullering*, Fig. 9–12*a*) or toward the center (*edging*, Fig. 9–12*b*). Repeated strokes, with the workpiece rotated around its axis between strokes, allows substantial material redistribution.

Ring Upsetting   When a ring is compressed between flat platens with zero friction, it expands as though it were a solid cylinder. Friction resists expansion, hence the hole expands less and, at higher friction, it becomes actually smaller. Therefore, the *ring-compression test* has become a favorite method of lubricant evaluation. Rings of OD:ID:Height = 6:3:2 ratio are normally used. Lesser contraction of the ID indicates a better lubricant for upsetting operations. Approximate values of μ and m^* may be obtained from the curves of Fig. 9–13.

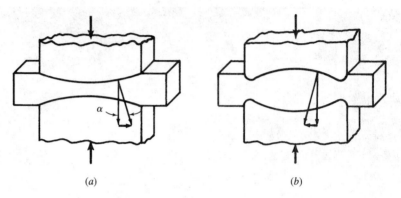

Figure 9–12   Material may be moved (*a*) away or (*b*) toward the center by inclined die surfaces, in the processes of fullering and edging, respectively.

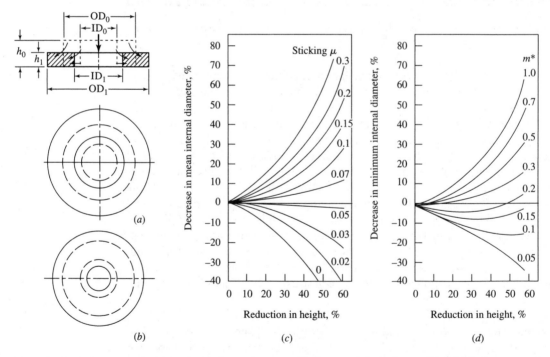

Figure 9–13   In ring upsetting (*a*) with zero friction the ring expands as a solid cylinder would but (*b*) the expansion of the inner diameter is reduced and even reversed when friction prevents free expansion. (*c*) The coefficient of friction may be found from experimental calibration curves and (*d*) the interface shear factor may be derived from theory. [(*c*) After A.T. Male and M.G. Cockroft, *J. Inst. Metals* **93**:38–46 (1964–1965); with permission of The Metals Society. (*d*) After G.D. Lahoti, V. Nagpal, and T. Altan, Trans. ASME, *J. Eng. Ind.* **100**:413–420 (1978). With permission of American Society of Mechanical Engineers.]

Aluminum alloy rings of 30.0-mm OD, 15.0-mm ID, and 10.0-mm height (6:3:2 ratio) were compressed at a press speed of 50 mm/s. The lubricant was stearic acid (a solid at room temperature; may be deposited from an organic solvent or melted above 60°C) or mineral spirits (a paint thinner, sometimes used as a very light lubricant). Rings were reduced to a height of 5 mm; the internal diameter was 15.5 mm with stearic acid and 10.5 mm with the mineral spirits.	**Example 9-8**

The diameter of 15.5 mm corresponds to $(15 - 15.5)/15 = -3\%$ *decrease* in internal diameter. In Fig. 9–13*a*, a horizontal line is drawn at this value; a vertical line is drawn at $(10 - 5)/10 = 50\%$ reduction in height; the two lines intersect at $\mu = 0.05$ for stearic acid. Repeating for $(15 - 10.5)/15 = 30\%$ *decrease* in internal diameter, $\mu = 0.18$ for mineral spirits.

Piercing   Impressions or holes are made in a workpiece by piercing. Several variants of the process are used.

1. For *piercing in a container*, the workpiece is supported at its base and around its sides (Fig. 9–14*a*). Therefore, the workpiece behaves as a semi-infinite body, and the punch pressure is at least $3\sigma_f$ [Eq. (9-13)]. When the punch penetrates to significant depths into a strain-hardening material, pressure rises as high as $4\sigma_{fm}$ to $5\sigma_{fm}$. The material displaced by the punch flows back in a direction opposite to punch movement. Friction on the punch and container surfaces should be minimized; otherwise, the piercing pressure will further rise. Thousands of millions of previously upset bolt heads are indented to make socket-head cap screws and screws with variously shaped recesses.

2. Pressures are much reduced when the bar does not fill out the container and a hole is pierced with *radial expansion* (Fig. 9–14*b*). Buckling is a problem unless the shape of the part provides support.

3. When the workpiece is *unconstrained* (Fig. 9–14*c*), the deformation pattern depends on the ratio of workpiece diameter d_0 to punch diameter D_p. When $d_0/D_p > 3$, the workpiece behaves as a semi-infinite body and Eq. (9-13) applies. At lower d_0/D_p ratios (Fig. 9–14*c*) complex deformation takes place and pressures drop roughly linearly to reach the value of the uniaxial flow stress at $d_0/D_p = 1$

$$p_{\text{pierce}} = \sigma_f \frac{d_0}{D_p} \qquad \textbf{(9-15)}$$

A cylindrical workpiece can be pierced with two punches from the opposing ends to prepare a through-hole; the remaining web is removed in a separate operation.

The most frequent application of piercing is to the indentation of the heads of screws and bolts. Since this is done mostly cold and in a container, pressures on the indenting tool can become excessive. Another limitation is imposed by cracking, resulting from either secondary tensile stresses set up by the expansion of an unrestrained head, or from exhaustion of ductility in the prior heading operation.

$$P_i = p_i A_p = 3\sigma_f A_p$$

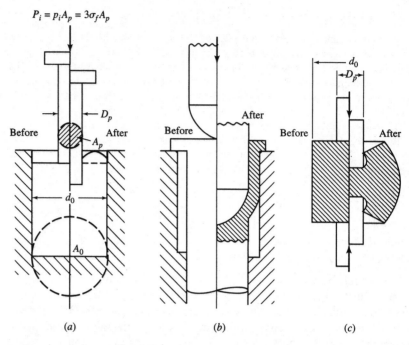

Figure 9–14   Billets may be pierced (a) in a container with reverse flow or (b) radial expansion, or (c) unconstrained with two opposing punches.

Example 9-9

An M10 hexagon socket-head cap screw is made of 1045 steel by upsetting the head of 16-mm diameter from 10-mm diameter rod. The hexagonal cavity of 8-mm across-flat dimension ($a = 4$ mm) is indented. Find the indentation force.

Indentation takes place on the already strain-hardened material of the head.

Upsetting strain $\varepsilon = \ln(A_1/A_0) = \ln(16^2/10^2) = 0.94$

Flow stress $\sigma_f = 950(0.94^{0.12}) = 943$ N/mm^2

Indentation pressure $p_i = 3 \times 943 = 2830$ N/mm^2 (!)

Indented area $= 3.464a^2 = (3.464)(4^2) = 55.4$ mm^2

Indentation force $P_i = 157$ kN

Note the very high pressure acting on the indenter punch.

9-2-4   Process Capabilities and Design Aspects

Open-die forging is one of the few processes capable of producing parts of a very broad size range (Table 9–1). Force limitations can be overcome with incremental deformation, and die pressure becomes critical only in indentation or piercing.

Table 9–1   General characteristics of bulk deformation processes

Characteristics	Deformation Process							
	Hot Forging			Hot Extrusion	Cold Forging, Extrusion	Shape Drawing	Shape Rolling	Transverse Rolling
	Open Die	Impression	Isothermal					
Part								
Wrought alloys	All	All	All	All	All	All	All	All
Shape*	R0-3; B; T1; F0; Sp6	R; B; S; T1, 2, 4; (T6, 7); Sp	As impression	All 0	As hot	All 0	All 0	R1,2,7; T1,2; Sp
Mass, kg	0.1–200 000	0.1–100	0.1–100	50–1000	0.01–50	10–1000	10–1000	0.001–10
Min. section, mm	5	3	1	(0.2) 1	(0.005) 1	0.1	0.5	1
Surface detail	E	C	B	B–C	A–B	A	A–B	A–C
Cost								
Equipment[†]	A–D	A–B	A	A–B	A–C	B–D	A–C	A–C
Die[†]	E	B–C	A	A–D	A–B	C–D	A–C	A–C
Labor[†]	A	B–D	A–C	B–C	C–E	C–E	C–E	C–E
Finishing[†]	A	B–C	C–D	C–D	E	E	E	D–E
Production								
Operator skill[†]	A	B–C	B–C	C–E	C–E	D–E	B–D	B–C
Lead time[†]	Hours	Weeks	Weeks	Days–weeks	Weeks	Days	Weeks	Weeks
Rate (pieces/h)	1–50	10–300	5–20	10–100	100–10 000	0.2–30 m/s	0.5–10 m/s	100–1000
Min. quantity	1	100–1000	100–1000	1–10	100 000	1000 m	50 000 m	1000–10 000

*From Fig. 3–1.
[†]Comparative ratings, with A indicating the highest value of the variable, E the lowest (e.g., cold extrusion produces excellent surface detail, involves high to moderate equipment cost, high die cost, low labor cost, very low finishing cost, and medium to low operator skill. It can be used for high production rates and requires a minimum quantity of 100 000 to justify the cost of the die).

Dimensions and Tolerances   To allow for out-of-roundness and bow, generous machining allowances are applied to hot open-die forgings (e.g., 10 mm on the diameter of a forging of 200 mm diameter and 1500 mm length). Accuracy is greatly improved with computer-controlled forging, and machining allowances and tolerances are usually settled in agreement with the manufacturer. Tolerances are much tighter for parts produced in upsetting dies (Fig. 9–15c and d) and can be very tight in cold forging (Fig. 3–22).

Shape Defects   Parts must be *free of defects* attributable to faulty material flow. Several limits must be observed in production:

1. A very slender cylinder may *buckle* instead of upsetting uniformly. Therefore, it is advisable to limit the h_0/d_0 ratio to 2 when friction is high (Fig. 9–15a).

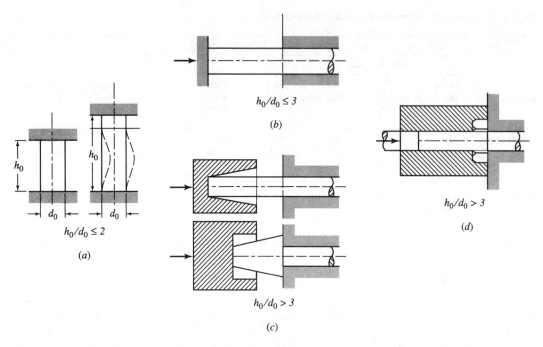

Figure 9–15   The initial height-to-diameter ratio is (a) limited by buckling and may be increased by (b) upsetting the end of a gripped bar, (c) limiting the deflection in conical cavities, or (d) expanding the bar into the cavity of a cold-header tool.

When friction is very low, h_0/d_0 should be less than 1.5 to prevent skewing of the billet. The same limits apply in upsetting rectangular billets, with the narrower dimension taken as d_0.

2. When upsetting is conducted in a *heading* operation, only the end of the workpiece is upset. The longer part of the workpiece—firmly clamped in die halves—becomes fixed, and the increased resistance to buckling allows somewhat greater free lengths (Fig. 9–15b).

3. An even longer length can be upset when deflection of the workpiece is limited in *progressive upsetting* into conical and cylindrical shapes (Fig. 9–15c).

4. In so-called *cold headers* and *horizontal upsetters*, the long overhanging part of the wire or bar is supported in the bore of a die and the head is progressively formed by a punch that spreads material into the space available in this die (Fig. 9–15d). Because the workpiece is guided on both ends, buckling is suppressed and larger heads can be formed in a single stroke.

Fracture   A second group of defects involves actual fracture of the workpiece. If deformation is truly homogeneous (Fig. 9–4a), most ductile materials can take relatively large strain in upsetting before their ductility is exhausted and fracture occurs by shearing at 45° to the application of the compressive stress (Fig. 9–16a).

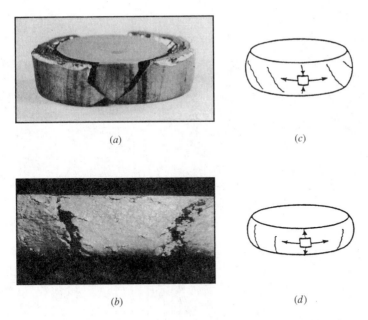

Figure 9–16   Fracture may occur by (a) exhausting ductility in cold working or by (b) intergranular fracture in hot working. (c), (d) The direction of cracks depends on the relative magnitudes of secondary tensile stresses generated by bulging.

In most cases, friction leads to barreling (Figs. 9–4b to 9–4d). It is readily seen that material in the bulge is not directly compressed; instead, it is deformed indirectly, by the radial pushing action of the centrally located material. This expanding action creates circumferential as well axial secondary tensile stresses on the free (barreled) surface and may cause *cracking* (Fig. 9–16b). The direction of cracks depends on the relative magnitudes of the secondary tensile stresses (Fig. 9–16c and d). Similar barreling and fracture occur also in the upsetting of a slab. Since barreling is the primary culprit, improved lubrication (which reduces friction and thus barreling) alleviates the problem.

It is quite common that one has to accept a limited deformation in a single stroke. Reheating in hot working and process anneals in cold working restore ductility and allow further deformation. Mathematical modeling and expert systems are helpful in the concurrent design of part shape.

9-3   IMPRESSION-DIE AND CLOSED-DIE FORGING

More complex shapes cannot be formed with great accuracy by open-die forging techniques. Specially prepared dies are required that contain the negative shape

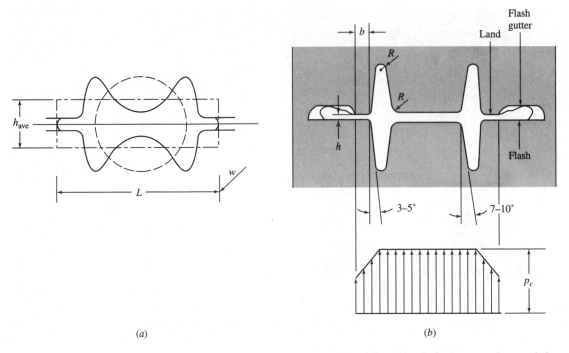

Figure 9–17   Parts with thin webs and tall ribs may be forged by (a) blocking followed by (b) finishing in a die provided with a flash gutter.

of the forging to be produced: *The process is simplified to a sequence of simple compression strokes at the expense of a complex die shape.*

9-3-1   Impression-Die Forging

In one variant of the process (Fig. 9–17) the shape is obtained by filling out the die cavity defined by the upper and lower die halves. Excess material is allowed to escape into the *flash*; since the die is not fully closed, it is properly called an *impression die*. The term *closed-die* is, nevertheless, sometimes applied, and the term *drop forging* has been used to denote forging conducted on a hammer; however, this distinction has no particular technical merit.

Material flow   The first concern is that the material must completely fill the die without defects of material flow, such as could occur when parts of the workpiece material are pinched, folded down, or sheared through. Therefore, the shape of the component must be redesigned to promote smooth material flow (in this, the process is similar to shape casting, Chap. 7).

1. A *parting line* is chosen with proper consideration of the fiber structure of the finished forging. Fibers (Sec. 8-1-7) should follow the contour of the forging

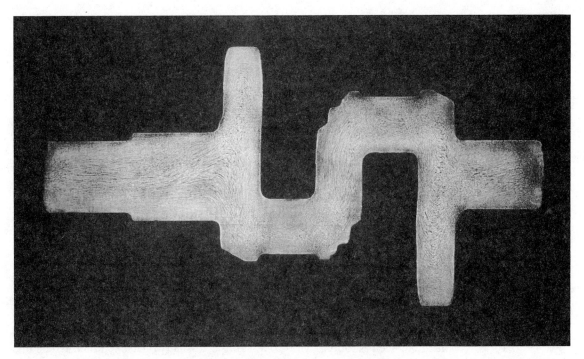

Figure 9–18   Grain flow (flow lines) in a forged steel workpiece can be revealed by macroetching. (*Courtesy Forging Industry Association, Cleveland, Ohio.*)

as far as possible (Fig. 9–18), because this ensures greatest toughness, fatigue strength, and ductility in the finished product. At the parting line the fibers are unavoidably cut through when the flash is trimmed, therefore, the parting line is best placed where minimum stresses arise in the service of the forging (Fig. 9–19).

2. *Fillets* in the part must be given appropriate radii to facilitate smooth material flow, and *corners* must be radiused (Fig. 9–19b) to prevent stress concentrations that would reduce die life.

3. The walls of the die cavity are given sufficient draft to allow removal of the forging.

4. Complex shapes, zero draft, and undercuts are possible when the die is constructed of several movable pieces (*segmented dies*).

Forging Sequence   A complex shape cannot be filled without defects (and excessive die wear) simply by forging the starting bar into the finishing die cavity. Various intermediate steps become necessary:

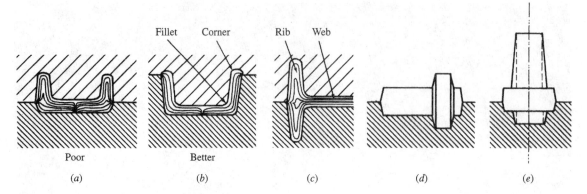

Fillet   Corner   Rib   Web

Poor   Better

(a)   (b)   (c)   (d)   (e)

Figure 9–19   Parting lines and draft angles must be chosen to give sound material flow and allow removal of the forging from the die.

1. The first aim is to distribute the material correctly, so that little change in cross-sectional area occurs in the finishing die. To this end, *free-forging* (open-die forging) operations may be performed, on specially shaped surfaces in the die blocks (Fig. 9–20), or on separate forging equipment, or even by other preparation methods such as rolling. These operations are usually related to fullering, edging, and upsetting.

2. The preform may be brought closer to the final configuration by forging in a *blocker die*, which ensures proper distribution of material but does not give

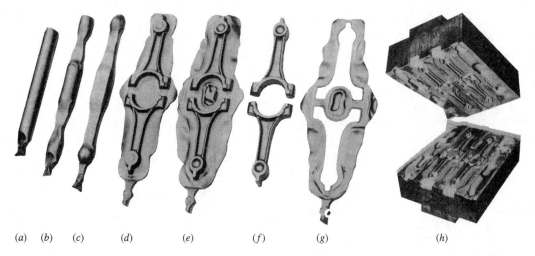

(a)   (b)   (c)   (d)   (e)   (f)   (g)   (h)

Figure 9–20   Hammer forging two connecting rods: (a) bar stock; after (b) fullering, (c) "rolling," (d) blocking, (e) finishing, (f) trimming; (g) the flash; and (h) the forging dies. (*Courtesy Forging Industry Association, Cleveland, Ohio.*)

the final shape (Fig. 9–17a). Excess material is allowed to run out between the flat die surfaces and this *flash* is sometimes removed (*trimmed*) prior to further forging.

3. Final shape is imparted in the *finishing die*. The excess material is again allowed to escape into a flash, which must now be thin to aid die filling and produce close tolerances. As a general rule, flash thickness $h = 0.015(A)^{0.5}$ (mm), where A is the projected area of the forging (mm^2). A thin flash running out between parallel die surfaces would lead to very large L/h ratios and thus high die pressures (Fig. 9–8). Therefore, L is reduced by cutting a *flash gutter* (Fig. 9–17b); this allows free flow of the flash and limits the minimum flash thickness to only a small width, the *flash land* (generally, the land is $3h$ to $5h$ wide). The flash is trimmed either hot or cold, in a separate die resembling a blanking die (Sec. 10-3).

The art of *die design* aims at determining the minimum number of steps that lead from the starting material (usually a round or rectangular bar) to the finished shape. Good die designers judge die filling from basic principles and long experience but modeling can be of great help. Plasticine or wax specimens can be deformed in transparent plastic dies; this is a powerful means of physical modeling and, if properly interpreted, the results are relevant. Tremendous advances have been made in numerical techniques that allow mathematical modeling of material flow and a number of commercial programs are now available. The expense involved in designing many dies can be minimized by the adoption of group technology for similarly shaped parts (Sec. 3-1-2).

Die Pressures and Forces There is no simple yet satisfactory method of computing impression-die pressures and forces, partly because the strain rate varies tremendously in various parts of the workpiece. A very approximate estimate may be obtained by analogy to forging simple shapes, by dividing the forging into parts (cylinders, slabs, etc.) that can be separately analyzed. Alternatively, the entire forging is considered as a simplified shape (Fig. 9–17):

Step 2: Calculate the *average height* from the volume V and the total projected area A_t of the workpiece (complete with the flash-land area)

$$h_{ave} = \frac{V}{A_t} \tag{9-16}$$

Step 4: The *average strain* is found from

$$\varepsilon_{ave} = \ln \frac{h_0}{h_{ave}} \tag{9-17}$$

Step 5: The *average strain rate* is

$$\dot{\varepsilon}_{ave} = \frac{v}{h_{ave}} \tag{9-18}$$

Step 7: *Average die pressure* is found by multiplying the flow stress by a factor Q_c that allows for shape complexity. Its value is taken from Table 9–2;

Table 9-2   Multiplying factors for estimating forces Q_c and energy requirements Q_{fe} in impression-die forging

Forging shape	Q_c	Q_{fe}
Simple, no flash	3–5	2.0–2.5
With flash	5–8	3
Complex (tall ribs, thin webs), with flash	8–12	4

note that the peak pressure develops very rapidly as details of the die are filled. For squat forgings, a crosscheck should always be made against Eqs. (9-7) and (9-10). (As a rule of thumb, die pressures are usually kept to 350 MPa in forging aluminum alloys and to below 700 MPa in forging steels.)

Step 8: The required forging force is

$$P = \sigma_f Q_c A_t \tag{9-19}$$

Step 9: The energy requirement may be estimated with the aid of a multiplying factor Q_{fe} from Table 9–2

$$E = \sigma_f Q_{fe} V \varepsilon_{\text{ave}} \tag{9-20}$$

More sophisticated calculations require larger computational effort because the optimum die configuration can be determined only by iteration. Local die pressures are determined and, if die pressures are found to be too high, the shape is changed and the computations are repeated. Computer programs are available that perform these tasks as well as the modeling of material flow.

One of the most difficult tasks is to determine the relevant flow stress, especially at high strain rates prevailing in hammers. It is found that, because of reduced cooling, hammer forces are only some 25% higher than press-forging forces.

Elastic deflections of the die can reach a significant proportion of tolerance limits in forgings such as the airfoil section of fan or turbine blades. To forge close-tolerance parts, the pressure distribution is computed and the die cavity is designed to compensate for elastic deflections of the die.

Example 9-10

A small connecting rod is forged of AISI 1020 steel at 1200°C. Calculate the press force for forging on a mechanical press which travels at 250 mm/s when the die contacts the workpiece. The volume of the connecting rod is calculated at 28 680 mm^3, and 20% of the starting material is expected to go into flash. In the finishing die the projected area is 3500 mm^2 exclusive of the area of the flash land. The flash land width is 7.6 mm all around the 300-mm circumference, adding 300(7.6) = 2280 mm^2 to the projected area. Thus, $A_t = (3500) + (2280) = 5780$ mm^2.

Step 2: $h_{\text{ave}} = (28\,680)/(0.8 \times 5780) = 6.2$ mm.

Step 5: $\dot{\varepsilon}_{\text{ave}} = 250/6.2 = 40$ s^{-1}

Step 6: In Table 8–2, data are available only for 1015 steel. However, in the austenitic temperature range the carbon content makes little difference (compare 1015 and 1045 steels). For 1200°C, $C = 50$ MPa and $m = 0.17$. Hence $\sigma_f = 50(40^{0.17}) = 93.6$ N/mm^2 (note that in Tables 8–2 and 8–3 C and m are given for $\varepsilon = 0.5$, thus they represent appropriate mean values for a forging such as this).

Step 7: The ribs and webs were approximately 3 mm thick, making the part intermediate in complexity (some experience is needed for this judgment), and $Q_c = 8$.

Step 8: $P = 96.3(8)(5780)/1000 = 4328$ kN $= 486$ tonf. (The data for this example were taken from the course material on *Basic Principles of Forging Die Design* of the Forging Industry Association, Cleveland, Ohio. The forging force was actually measured at Battelle Columbus Laboratories and was found to be 430 tonf).

Forging Practices   The hot-forging sequence shown in Fig. 9–20 is typical of hammer forging. Short contact times and repeated blows in the same cavity allow the forging of parts with thin ribs and webs and intricate details. In press forging the workpiece enters each cavity only once; more preforming cavities may be required and die design is more critical. A typical sequence of hot forging on a special-purpose press (*horizontal upsetter*) is shown in Fig. 9–21.

Cooling—and thus chilling of the part and forces—in hot forging are reduced with heated dies. In nonisothermal forging, dies are preheated to around 200°C; this serves also to reduce thermal shock on the die. In *isothermal forging* (with the die at the workpiece temperature) very slow forging speeds are permissible and complex, thin-walled parts can be forged at low pressures into shapes that have very small or zero draft and require little or no machining (*near-net-shape* and *net-shape forging*, Fig. 9–22). The low forging temperature of aluminum alloys allows isothermal forging in steel dies. Titanium alloys require superalloy or TZM dies; superplastic superalloys may be formed in TZM dies. To prevent oxidation of the molybdenum-alloy dies, special presses with evacuated workspaces are built. Lubrication is vital in isothermal and most nonisothermal forging.

Compressor blades of jet engines are somewhat similar in shape to the casting in Fig. 7–16 but are much larger. Some are made by isothermal forging of a Ti alloy such as Ti-6Al-4V.[1] Assume that for a given blade the airfoil may be regarded as the equivalent of a 120-mm wide, 6-mm thick, and 350-mm long slab. Obtain the relevant flow stress at 900°C for a hydraulic press speed of (*a*) 250 mm/s and (*b*) 2.5 mm/s.

Example 9-11

From Table 8–3, $C = 140$ MPa, $n = 0.4$.

a. $\dot{\varepsilon} = 250/6 = 41.7$ s^{-1}; $\sigma_f = 140(41.4^{0.4}) = 622$ MPa.

b. $\dot{\varepsilon} = 2.5/6 = 0.417$ s^{-1}; $\sigma_f = 98.7$ MPa.

Note the powerful effect of strain rate when m is high.

---

[1] T. Watmough and J.A. Schey, U.S. Patent No. 3,635,068.

Gripper dies   Heading tools (punches)

Figure 9–21   A typical hot-upsetting sequence showing the development of the workpiece from the bar and the associated tooling. (*Courtesy National Machinery Co., Tiffin, Ohio.*)

Figure 9–22   Precision aluminum alloy forgings are made with thin webs and ribs and little or no draft. (*Courtesy Aluminum Precision Products, Inc., Santa Ana, California.*)

Die pressures are high in cold forging and deformation is usually distributed to several cavities (Fig. 9–23). Lubrication is crucial for success, partly to reduce die pressures and partly to prevent die pickup (adhesion) and subsequent scoring of workpieces.

Precision Forging   The term is applied to processes that aim at producing net-shape or near-net-shape parts. Cold forging and most warm forging and isothermal forging fall into this category. When applied to nonisothermal hot forging, the term indicates practices that produce much tighter tolerances, better surface finish, smaller draft angles, and better shape definition than conventional practices.

9-3-2   Closed-Die Forging

In true *closed-die forging* the workpiece is completely trapped in the die and no flash is generated. Economy of forging is thus increased, but die design and process variables must be very carefully controlled. At the end of the stroke the cavity is completely filled with an incompressible solid, and die pressures rise very steeply; this becomes a critical factor in setting up the equipment (Sec. 9-5). Forces are computed as in impression-die forging. A dramatic reduction in forces is attained in thixoforming, with the material in the semisolid state (Sec. 7-5-6).

A special case of closed-die forging is *coining*, in which a three-dimensional surface detail is imparted to a preform. The largest application is, of course, to the minting of coins, but coining is useful for improving the dimensional accuracy, surface finish, or detail of other parts too. The forging pressure is at least $p_i = 3\sigma_f$, but filling of fine details calls for pressures of $5\sigma_f$ or even $6\sigma_f$.

Figure 9–23   A typical cold-upsetting sequence showing the development and transfer of the part in six forming stations at the rate of 300 parts/min. (*Courtesy National Machinery Co., Tiffin, Ohio.*)

9-3-3   Forge Rolling and Rotary Swaging

These are two of the more specialized forging processes.

Forge rolling performs an impression-die forging operation, but this time the die-half contours are machined into the surfaces of two rolls. Reciprocating roll motion is suitable for the rolling of short pieces while unidirectional rotation is used in high-production lines. Forge rolling often replaces open-die forging for preforming but is also suited for finishing more or less flat forgings such as cutlery and scissors.

A special form of hammers is the *rotary swager*. The workpiece is usually stationary, while the hammer itself rotates. The construction resembles that of a roller bearing (Fig. 9–24a): The anvils are free to move in a slot of the rotating shaft and are thus hurled against the rollers, which in turn knock them back. A rapid sequence of blows is obtained and the workpiece, fed axially, is reduced in diameter by a drawing-out process. While, strictly speaking, swaging should be regarded (and sometimes is used) as an open-die forging process, it is capable of producing exceptionally smooth surfaces to close tolerances. The process can be employed for pointing, assembling a bar and collar, or shaping the internal contour of a tube on a mandrel (Fig. 9–24b).

In the impression-die forging of shapes with large d/h ratios, high die pressures are reduced by replacing the top die with an *orbiting tool* that makes contact over only part of the surface. Several proprietary designs exploit this principle.

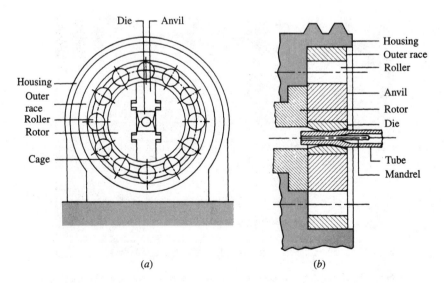

(a)   (b)

Figure 9–24   A rotary swager reduces (a) solid or (b) hollow workpieces with rapid blows.

9-3-4   Process Capabilities and Design Aspects

Impression- and closed-die forging are extremely versatile but shape limitations arise from the need to release the forging from the die (Table 9–1). Design aims to facilitate sound material flow at minimum die pressure.

Dimensions and Tolerances   The part is designed to require minimum machining; machining allowances are typically 1.5 mm in conventional forging and under 0.5 mm in precision forging. Tolerances are shown in Fig. 3–22. Maximum dimensions are governed by the capacity of available equipment. Minimum dimensions depend greatly on die temperature and on the material forged. Minimum web thickness is governed by die pressure; even with the best die design and lubrication, there is a maximum d/h or L/h ratio that can be achieved. Some general guidance is given in Fig. 9–25. Thinner webs are possible if shaped to drive out the material (Fig. 9–26a), as it does in fullering (Fig. 9–12a). Whether the web is flat or tapered, proper preforming is important, otherwise excess material pushed out from the web creates a *lap* (*push through*, Fig. 9–26b) in the rib.

Shape Features   Apart from general shape limitations (Table 9–1), design of the part can aid forging.

1. Draft angles (Fig. 9–26a) depend on material and forging method. Smaller angles are permissible when ejectors are built into the die, and internal draft can be reduced when the temperature difference between forging and die is small. Very general guidance is given below (angles in degrees):

	Blocker Type		Conventional		Precision	
	Outside	Inside	Outside	Inside	Outside	Inside
Aluminum alloy	7	7	5	5	1	1
Steel, Ti alloy	7	9	5	7	2–4	3–5
Heat-resistant alloys	7	10	7	10	5	7

2. Corner and fillet radii on the forging (Fig. 9–26a) depend on the size of the forging and the height of ribs. Filling of very tight corner radii requires excessive die pressure; tight fillet radii lead to defects such as laps. Some typical values are given below (dimensions in mm):

	Conventional		Precision	
	Corner	Fillet	Corner	Fillet
Aluminum	3–5	6–15	1.5–3	3–9
Steel	3–12	10–16		

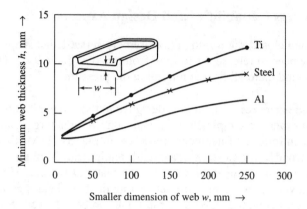

Figure 9–25   Die pressure is a function of d/h or L/h values; therefore, minimum web thickness in conventional forging depends on the smaller dimension of the web.

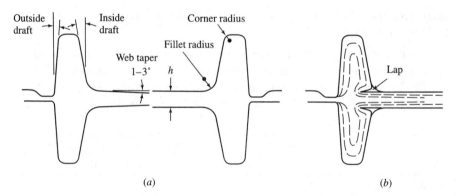

Figure 9–26   Rib-and-web configuration can be forged if (a) appropriate draft and radii are applied and (b) preforming is done to prevent push-through of material from the web.

3. Greater shape complexity is permissible if the die is made in more than two parts. Thus, a *horizontal upsetter* has, in addition to the main ram (similar to the moving crosshead of a press), an auxiliary movement that closes a split die (Figs. 9–21 and 9–27*a*). Thus, shapes that are undercut relative to the ram movement can be forged. Some presses have three or four rams, so that parts such as valve bodies can be forged (Fig. 9–27*b*).

Defects   We already indicated that folds and laps must be avoided. Furthermore, deformation must be as homogeneous as possible to prevent the generation of internal defects. This condition is usually satisfied in impression-die forging, but high h/L ratios may develop in swaging; then the center may open up (*centerburst*

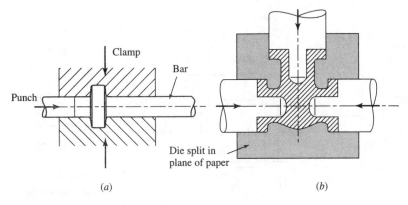

Figure 9–27   Split dies permit release of (*a*) an upset collar and (*b*) a multiram forging.

defect).   Manufactured components of complex shape are often produced by a combination of forging and extrusion; therefore, extrusion processes will be discussed next.

9-4 EXTRUSION

As suggested by the Latin root (*extrudere*: to thrust out), in the extrusion process the workpiece is pushed against the deforming die while it is being supported in a *container* against uncontrolled deformation. Since the workpiece is in compression, the process offers the possibility of heavy deformations coupled with a wide choice of extruded cross sections.

9-4-1   The Extrusion Process

To initiate extrusion, a (usually) cylindrical billet is loaded inside a container and is pushed against a die held in place by a firm support. The press force is applied to the punch and, after the billet has upset to fill out the container, the product emerges through the die (Fig. 9–28). Initially, deformation is non-steady-state but once the product has emerged, steady-state conditions prevail until close to the end of extrusion when continuous material flow is again disturbed.

Types of Extrusion   Two basically different processes are possible:

1. In *direct* or *forward extrusion* the product emerges in the same direction as the movement of the punch (also called *ram*, Fig. 9–28a). The really important

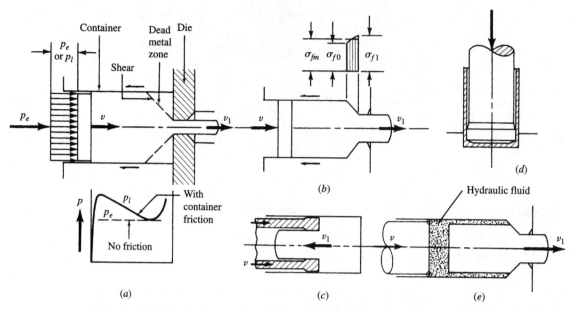

Figure 9–28   Extrusion processes: (*a*) forward or direct without lubrication (and the associated extrusion-pressure/stroke curve); (*b*) forward with full lubrication; (*c*) reverse (or indirect or backward); (*d*) reverse can (impact); (*e*) hydrostatic extrusion.

point is that, for extrusion to take place, the *billet must be moved against frictional resistance* on the container wall (frictional shear stress is shown by half-arrows).

2. In *indirect* (*reverse* or *back*) *extrusion* the product travels against the movement of the punch (Fig. 9–28*c* and *d*). Most important, the billet is at rest in the container; thus, *container friction plays no role.* By definition, piercing in a container (Fig. 9–14*a*) may be regarded as a case of back extrusion.

Lubrication   Further distinctions may be made according to whether a lubricant is used.

1. The material always seeks a flow pattern that results in minimum energy expenditure. When extrusion is carried out without a lubricant and with a die of flat face (180° die opening), the material cannot follow the very sharp directional changes that would be imposed on it. Instead, the corner between the die face and container is filled out by a stationary *dead-metal zone*, and material flow takes place by shearing along the surface of this zone (*unlubricated extrusion*, Fig. 9–28*a*). Thus, the extruded product acquires a completely freshly formed surface.

2. Alternatively, a very effective lubricant is applied to ensure complete sliding on the die face and along the container wall (*lubricated extrusion*). Accordingly, the die is now provided with a conical entrance zone that, ideally, corresponds in shape to the flow pattern of minimum energy (Fig. 9–28b).

3. In a variant of the process, the billet is extruded by pressurizing a liquid medium inside a closed container (*hydrostatic extrusion*, Fig. 9–28e). This reduces friction on the container wall but does not fundamentally change the stress state inside the deforming workpiece; reduced die friction (and thus lower hydrostatic pressure) can even increase the tendency to internal crack formation. However, the absence of container friction permits extrusion of very long billets or even wires, and large reductions can be taken. The process is used for special purposes such as the cold extrusion of copper tubes and of composite copper-aluminum billets to produce copper-clad conductor wires and bars.

Extruded Product   The movement of the punch must be stopped before the conical die entry is touched or, in unlubricated extrusion, before material from the dead-metal zone is moved, since this would create internal defects. Two basic methods of operation are possible:

1. When the purpose of extrusion is to produce a long bar or tube of uniform cross section (*extrusion of semifabricated products*), the remnant (*butt*) remaining in the container is scrap which is removed by taking it out with the die. After the butt is cut off, the extrusion can be extricated from the die, and the die is returned for inspection, conditioning, and reuse.

2. When the purpose is *extrusion of finished components*, the butt forms an integral head of the component. The extrusion is ejected by pushing it back through the extrusion die and lifting it out from the container. Since ejector actuation can be mechanically synchronized with punch movement, high production rates are achieved, provided, of course, that the extruded stem is strong enough to take the ejection force.

A further important distinction can be made according to the temperature of deformation.

9-4-2   Hot Extrusion

While hot deformation is often typical of primary processes, the hot extrusion of shapes offers such a wide scope for custom design that this process can justifiably be regarded as a secondary manufacturing technique. Shapes are usually classified into three groups according to their complexity (Fig. 9–29):

1. *Solid shapes* are produced by extruding through a suitably shaped stationary die.

2. *Hollow products* necessitate the use of a die insert that forms the cavity in the extruded product. This insert may be a *mandrel* fixed to the punch (Fig. 9–30a) or moving inside the punch (Fig. 9–30b), or a *bridge* (*spider*) section attached to the die (Fig. 9–30c). The last method is permissible only if material flow can be

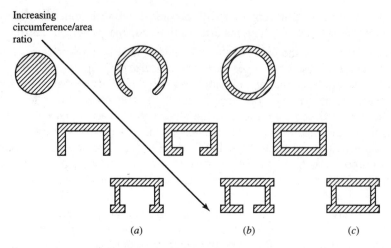

Figure 9–29   In the extrusion of (*a*) solid, (*b*) semihollow, and (*c*) hollow configurations, process difficulty increases with increasing circumference-to-area ratio.

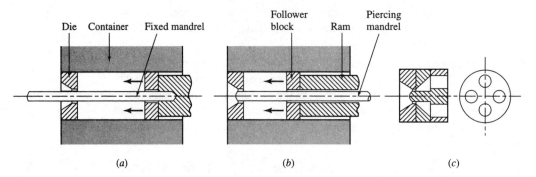

Figure 9–30   Hollow products may be extruded with (a) fixed or (b) piercing mandrels or with (c) bridge- or spider-type dies. [*After J.A. Schey, in Techniques of Metals Research, R.F. Bunshah (ed.), vol. 1, pt. 3, Interscience, 1968, p. 1494. With permission.*]

divided and then reunited prior to leaving the die, with complete pressure welding of the separated streams. This is practicable only in the unlubricated hot extrusion of aluminum and lead; even a trace of lubricant would prevent rewelding.

3. *Semihollow products* appear to be solid sections, but their shape makes the use of a single-piece die impracticable. The die tongue forming the internal shape is connected to the external contour by such a small cross section that it would break off; therefore techniques similar to the extrusion of hollow sections must be used.

Aluminum alloys are extruded isothermally, without a lubricant, and with flat dies made of hot-working die steels. Shearing along the dead-metal zone gives all-new, bright surfaces. Copper and brass are extruded mostly unlubricated, nonisothermally. Cooling on the colder container and die limits the complexity and thinness of shapes. This is true also of the hot extrusion of steel, conducted mostly with a glass lubricant that envelopes the billet and melts in a controlled manner to form a die approach of optimum shape; sometimes shorter lengths and thinner sections are produced with graphitic lubricants. The dies are often coated with a ceramic (e.g., partially stabilized zirconia) for protection.

The starting material is often a cast billet. Extrusion ratio should be at least 4:1 to ensure adequate working, but it may rise to 400:1 in the softer alloys.

9-4-3 Cold Extrusion

The purpose of cold extrusion is mostly to produce a finished part, and the residue (butt) in the container becomes an integral part of the finished product (e.g., in the forward extrusion of a bolt shank or an automobile half-axle, or in the back extrusion of a toothpaste tube).

The low flow strength of tin and lead facilitated their early cold extrusion for collapsible tubes (often called *impact extrusion*, Fig. 9–28d). The low melting point of these metals means that they are really extruded in the warm or hot working regime. True cold extrusion takes place with aluminum and lubrication becomes critical (Table 8–4); nevertheless, severe deformation is possible. Only smaller extrusion ratios are permissible with copper and brass, and the cold extrusion of steel would be quite impossible without a lubricant that withstands very high pressures while also following the extension of the surface. The most successful approach converts the steel surface into a zinc-iron phosphate (*phosphate coating*); this porous surface, integrally joined to the metal surface, is then impregnated with a suitable lubricant, usually a soap (Table 8–4). Steels of higher carbon content can be extruded after a spheroidizing anneal. Strain hardening offers a valuable increase in strength. If a workpiece is to be strain-hardened uniformly, the stem is extruded and the butt subsequently upset. More complex parts, such as stepped hollow shafts, may require several extrusion or forging operations, and the part may have to be process annealed and relubricated. Cold extrusion has made great inroads in the automotive and general equipment industries, for parts previously made by machining.

A component resembling a bolt used to be made by machining and is now to be made by plastic deformation. It is desired to retain the benefits of cold working, and the part is to have equal strain hardening in the head and body sections. What process should be used? | **Example 9-12**

Equal strain hardening can be obtained only if the head is upset and the body is extruded (or otherwise reduced). A bar of intermediate diameter d_0 must be chosen so that the upsetting strain $\ln(A_{head}/A_0)$ is equal to the extrusion strain $\ln(A_0/A_{body})$.

$$A_{head} = 20^2\pi/4 = 314 \text{ mm}^2 \qquad A_{body} = 10^2\pi/4 = 78.8 \text{ mm}^2$$

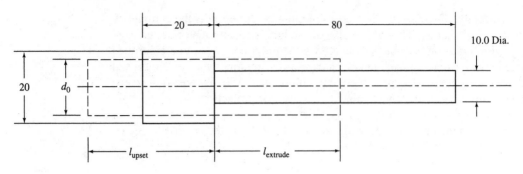

Figure Ex. 9-12

for equal strain, $\ln (314/A_0) = \ln (A_0/78.5)$.

$$A_0^2 = (314)(78.5) = 24\,650 \quad \text{and} \quad A_0 = 157 \text{ mm}^2$$

Initial diameter, from Eq. (9-2c), $d_0 = [4(157)/\pi]^{0.5} = 14.14$ mm.

9-4-4 Extrusion Force

The routine of computations is similar to that followed in forging (Sec. 9-2-1), but there are some significant differences in detail.

Step 2: The dimensions of interest are the cross-sectional areas of the billet A_0 and extrusion A_1. The diameters d_0 and d_1 will also be needed. When the extruded section is not a round bar, an *equivalent diameter* can be calculated from the area A_1 [Eq. (9-2c)].

Step 3: The engineering strain may be computed as reduction of area

$$e_e = \frac{A_0 - A_1}{A_0} \tag{9-21}$$

However, at large reductions, a better feel is obtained from the *extrusion ratio*

$$R_e = \frac{A_0}{A_1} \tag{9-22}$$

Step 4: Strain is simply the natural logarithm of R_e

$$\varepsilon = \ln R_e = \ln \frac{A_0}{A_1} \tag{9-23}$$

Step 5: Strain rate is important in hot working, and a *mean strain rate* may be computed from

$$\dot{\varepsilon}_m = \frac{6v d_0^2 \tan \alpha}{d_0^3 - d_1^3} \varepsilon \tag{9-24}$$

where v is ram velocity. The half angle α is the cone angle of the die entry or, in unlubricated extrusion with a dead-metal zone, it may be taken as 45° (unless experiments or mathematical modeling show it to be different).

Step 6: In cold working the workpiece material strain-hardens during its passage through the die, and a *mean flow stress* σ_{fm} must be obtained as was shown in Fig. 9–2 and Eq. (9-1). In hot working, Eq. (8-11) gives a mean flow stress when the mean strain rate [Eq. (9-24)] is used.

Step 7: Deformation is inhomogeneous and extra work is needed to deform the material to final shape. This *redundant work* is proportional to ε and, for approximate calculations, the *extrusion pressure* p_e may be found from the following formula

$$p_e = \sigma_{fm} Q_e \qquad \text{(9-25a)}$$

where

$$Q_e = 0.8 + 1.2\varepsilon \qquad \text{(9-25b)}$$

Step 8: The total extrusion force P_e acting on the billet is

$$P_e = p_e A_0 \qquad \text{(9-26)}$$

A 6061 Al alloy billet is extruded at 500°C without lubrication on a hydraulic press at a ram speed of 0.5 m/s. Find the basic pressure for extruding a bar of 15-mm diameter, assuming that the half angle of the dead-metal zone is 15, 30, 45, or 60°.

Example 9-13

For repetitive calculations, set up a spreadsheet. The result is:

A	B	C	D	E	F	G	H	I	J	K	L	M	N	O
Hot	6061	Al		T=	500	C								
Flow stress		C=	37	MPa	m=	0.17								
	alpha	v	d0	A0	d1	A1	ee	Re	eps	eps dot	sigma f	Qe	pe	Pe
	deg	mm/s	mm	mm^2	mm	mm^2				1/s	N/mm^2		N/mm^2	kN
							Eq. (9-21)	Eq. (9-22)	Eq. (9-23)	Eq. (9-24)	Eq. (8-11)	Eq. (9-25b)	Eq. (9-25a)	Eq. (9-26)
0		500	200	31416	15	176.7								
1	15	500	200	31416	15	176.7	0.994	177.8	5.18	20.8	62.0	7.02	435.0	13667
1	30	500	200	31416	15	176.7	0.994	177.8	5.18	44.9	70.6	7.02	495.7	15572
1	45	500	200	31416	15	176.7	0.994	177.8	5.18	77.7	77.6	7.02	544.2	17096
1	60	500	200	31416	15	176.7	0.994	177.8	5.18	134.7	85.1	7.02	597.4	18769

Note that a change of angle within reasonable limits (say, 45–60°) has a relatively small effect on forces.

A word of warning is in order here. We already observed (Sec. 9-4-1) that back extrusion of a can is similar to piercing in a container (Fig. 9–14a). The extrusion force, Eq. (9-26), is based on the pressure p_e acting over the base area A_0; at low reductions, however, the force may really be given by the piercing force. This is obtained by multiplying the punch area $A_p = A_0 - A_1$ by the punch (indenter) pressure p_i which, as discussed in Sec. 9-2-3 under "Piercing," can never be less than $3\sigma_f$ [Eq. (9-13)] and is more likely $4\sigma_f$ to $5\sigma_f$. It is advisable, therefore, to calculate the extrusion force from both Eq. (9-26) and from the punch force P_i

$$P_i = p_i A_p = p_i (A_0 - A_1) \qquad (9\text{-}27)$$

and take the *smaller* of the two values. It does not matter whether the indenting punch is solid as in Fig. 9–14a or hollow as in Fig. 9–28c.

Container Friction   In direct extrusion the billet is pushed forward against the frictional resistance developed on the container wall. Correspondingly, the extrusion pressure is higher at the beginning of the stroke when a long length rubs against the container wall (Fig. 9–28a). At high extrusion ratios interface pressures can be very high and the use of a coefficient of friction could be misleading (Sec. 8-2-3). Therefore, it is better to estimate the shear strength of the interface τ_i and add the corresponding pressure to the basic extrusion pressure so as to obtain the ram (punch) pressure p_l at any point in the stroke

$$p_l = p_e + 4\frac{\tau_i l}{d_0} \qquad (9\text{-}28)$$

where l is the length of the billet at the point in the stroke considered, measured from the end of the stroke. Data for τ_i are scarce but an upper limit is given by sticking when $\tau_i = \tau_f$ or $0.5\sigma_f$. With a truly effective lubricant the pressure will drop toward the basic pressure, Eq. (9-25a).

Example 9-14

An 1100 Al can (container) of 50-mm OD and 48-mm ID is to be produced by the cold back extrusion (Fig. 9–28d) of $d_0 = 50$-mm-diameter annealed slugs. Lanolin is used as a lubricant. Calculate the force during steady-state extrusion.

We again set up a spreadsheet, noting that the extrusion is a hollow tube. (Note that the spreadsheet allows for the use of a previously strain-hardened material by providing for ε_{total}).

A	B	C	D	E	F	G	H	I	J	K	L	M	N
Cold		1100	Al										
Flow	stress	K=	140	MPa	n=		0.25						
												Base	
Step	mu	d0	di	A	ec	Re	epsilon	eps	sigma fm	Qe	pe	A0	Pe
No.		mm	mm	mm^2				total	N/mm^2		N/mm^2	mm^2	kN
					Eq. (9-21)	Eq. (9-22)	Eq. (9-23)		Eq. (9-1)	Eq. (9-25b)	Eq. (9-25a)		Eq. (9-26)
0		50	0	1963.5				0					
1	0.1	50	48	153.9	0.92	12.76	2.546	2.55	141.5	3.86	545.4	1963.5	1070.9

With the given geometry and lubricant, wall friction can be ignored.

Check the punch force, for indentation of a strain-hardening material:

$$p_i = 141.5(5) = 707.5 \text{ MPa}$$

$$P_i = (1963.5\text{–}153.9)(10^{-3})(707.5) = 1280 \text{ kN}$$

Thus the value computed for extrusion is lower and will be sufficient to perform the operation.

9-4-5   Process Capabilities and Design Aspects

Extrusion is unsurpassed in its capacity to produce long semifabricated products of complex cross section. Since the dies are relatively inexpensive (Table 9–1), custom design is justifiable for many applications. It is not unusual that custom aluminum extrusions are designed for windows and door frames of larger public buildings, or for computers, medical equipment, conveyor systems, robotics, and many other applications. Cold extrusion and extrusion combined with forging have become important mass-production methods for net-shape and near-net-shape products, such as automotive transmission, steering, and fastener components.

Dimensions and Tolerances   There are dimensional limits: A circle scribed around the section must in general be smaller than container diameters of the available presses. Tolerances can be very tight in cold extrusion and are necessarily larger in hot extrusion (Fig. 3–22). The minimum wall thickness given in Table 9–1 refers to moderate-difficulty materials such as high-strength aluminum alloys or brass. Twice these values are typical of more difficult materials (steels or superalloys) and the very small values given in parentheses are possible for low-strength aluminum alloys (1100 or 3003).

Shape   There is little difficulty in extruding sections with uniform wall thickness; sections of unequal wall thickness would curve on emerging from the die. The reason is that extrusion pressure increases with extrusion ratio [Eq. (9-25)]. A common pressure prevails in the container; since for a given extrusion pressure the rate of extrusion decreases with increasing reduction, the material exits at a lower speed in the thin leg of the extrusion (Fig. 9–31a). To speed up the narrow leg (or restrict flow in the thicker leg), frictional surfaces (*bearings*) are made shorter for the narrow leg (Fig. 9–31b).

Increasing complexity is often expressed as a perimeter/weight or perimeter/cross-sectional area *shape factor*; the higher its value, the more skill is required to produce the part. Computer-aided die design is possible, often with expert programs that capture the knowledge of experienced die designers. Location of die openings, bearing lengths, and die deflections are readily obtained and die-tryout time and costs are minimized. With proper die design, very complex shapes can be produced (Fig. 9–32).

Defects   Even though the material is kept in overall compression and thus the hydrostatic component of stresses is high, the extrusion process is not free of problems.

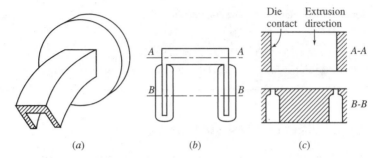

(a)   *(b)*   *(c)*

Figure 9–31   To prevent curving of the emerging extrusion, flow rates are
                     equalized by increasing the die-contact length to retard flow in
                     thicker sections.

Figure 9–32   Hot extrusion of aluminum is capable of producing a large variety of
                     often very complex shapes, including multihole sections, gearing, and
                     tubing, with thin walls and significant wall-thickness variations.
                     (Courtesy Almag Aluminum Inc., Brampton, Ontario.)

1. Deformation tends to be inhomogeneous, especially at extrusion ratios
below 4. Inhomogeneity (Sec. 8-2-5) is, in general, a function of the h/L (mean
height h over compressed length L) ratio. The h/L ratio in Fig. 8–17 refers to
a rectangular slab whereas extrusion is usually conducted with axial symmetry.
Nevertheless, the same principles apply (Fig. 9–33), except that the mean diameter
$(d_0 + d_1)/2$ is now substituted for h.

As before, deformation is inhomogeneous when the h/L ratio is large (Fig.
9–33b); in other words, when the extrusion ratio is small and the die half angle α is

large. As in all inhomogeneous deformation, there is redundant work necessary to create internal deformation that does not show up in the external shape. Because deformation is now concentrated in the outer zones, they are directly elongated, whereas the center of the extrusion is not directly deformed but is dragged along by the surface material. This generates secondary tensile stresses in the core which may ultimately suffer a characteristic *arrowhead fracture* (also described as *centerburst defect*, Fig. 9–33b). The danger is greatest at $h/L \geq 2$. The situation can be remedied by lowering the h/L ratio, which implies either a smaller die half-angle α or a heavier reduction and thus smaller h and larger L (Fig. 9–33c). With a component of fixed geometry, neither of these remedies may be allowable and the only hope is then the use of a more ductile material. Centerburst defects are particularly troublesome when they occur only periodically, affecting the integrity of an unknown number of parts. In very special instances the compressive stress state is maintained, even at critically low extrusion ratios, by extruding into a pressurized space, a process usually described as *extrusion against back pressure* (not to be confused with hydrostatic extrusion).

2. In hot extrusion, the heat generated during extrusion may cause the work-piece temperature to rise [Eq. (9-5)] above the solidus temperature of the material. Hot-shortness then leads to the appearance of circumferential surface cracks (*speed cracking*) which can be eliminated by slowing down the press, thus reducing the strain rate and the rate of heat generation (but also losing output).

3. When the extrusion stroke is taken too far, inhomogeneous material flow leads to the generation of a concentric *pipe* (Fig. 9–33d).

4. When lubricant traces are present in unlubricated extrusion or lubrication breaks down in lubricated extrusion so that a partial dead-metal zone forms, lubricant trapped at the boundary of the dead-metal zone extrudes into the product to form *subsurface defects*. On subsequent heating, gases cause blistering at these locations.

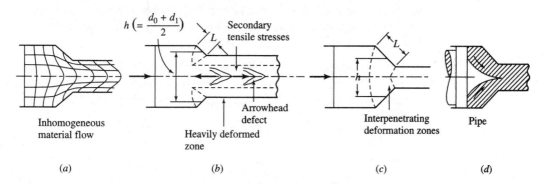

Figure 9–33   Inhomogeneous deformation in extrusion results in (a) acceleration of the center and, if the material has limited ductility, it can lead to (b) internal (arrowhead) defects at high h/L ratios. (c) Development of hydrostatic pressure assures a sound product at low h/L ratios. (d) A pipe forms if extrusion is taken too far.

9-5   FORGING AND EXTRUSION EQUIPMENT

Forging and extrusion are closely related processes. Sometimes they are difficult to distinguish (e.g., piercing in a container versus back extrusion); at other times a distinctly forging-type process is combined with extrusion (e.g., in making a bolt by extruding the shank and then upsetting the head). They also share many types of tooling and equipment.

9-5-1   Tools and Dies

Bulk deformation processes are characterized by high interface pressures and, in hot working, also by high temperatures. *Tool and die materials* are selected and manufactured with the greatest care. In general, ductility is sacrificed in cold-working dies but a compromise between hardness and ductility must be struck for hot-working dies that are exposed also to thermal shock (Table 9–3).

In computing forging and extrusion pressures, the relevant flow stress was σ_f because the workpiece material had to deform. In contrast, interface pressures must be kept low enough not to cause any permanent deformation of the die. Therefore, depending on loading mode, pressures must not exceed a safe fraction (or, in indentation, multiple) of the yield strength $\sigma_{0.2}$ of the die material. From the HRC values given in Table 9–3, the tensile strength can be estimated as follows:

HRC	TS, MPa	TS, kpsi
30	960	140
40	1250	185
50	1700	250
60	2400	350

Allowing for some safety, 80% of the above values can be taken as $\sigma_{0.2}$. The allowable stress depends on the relative configurations of the tool and workpiece, and we can estimate permissible die pressures by *regarding the tool as a workpiece*, the deformation of which must be prevented.

1. Long punch (Fig. 9–34a). Just as a cylindrical billet will buckle when the h/d ratio is too large, so will a punch. For very long punches, the Euler formula is relevant; for shorter ones—more typical of metalworking—the Johnson formula is suitable:

$$p \leq \sigma_{0.2} \left[1 - \frac{4\sigma_{0.2}}{\pi^2 E} \left(\frac{L_p}{D_p} \right)^2 \right] \tag{9-29}$$

where L_p is punch length, D_p is punch diameter, and E is Youngs's modulus for the punch material (210 GPa for steel, 350 GPa for tungsten carbide).

Table 9-3 Typical die materials for deformation processes*

| Process | Die Material† and Hardness HRC for Working | | | |
	Al, Mg, and Cu Alloys		Steels and Ni Alloys	
Hot forging	6G	32–40	6G	35–45
	H12	48–50	H12	40–56
Hot extrusion	H12	46–50	H12	43–47
Cold extrusion:				
Die	W1, A2	56–58	A2, D2	58–60
	D2	58–60	WC	
Punch	A2, D2	58–60	A2, M2	62–65
Shape drawing	O1	55–65	O1, M2	55–65
	WC		WC	
Cold rolling	O1	55–65	O1, M2	55–65
Blanking	W1	62–66	As for Al, and	
	O1, A2	57–62	M2	60–66
	D2	58–64	WC	
Deep drawing	Zn alloy		As for Al, and	
	W1	60–62	M2	60–65
	O1, A2	57–62	WC	
	D2	58–64		
Press forming	Epoxy/metal powder		As for Al	
	Zn alloy			
	Mild steel, cast iron			
	O1, A2, D2			

*Compiled from *ASM Handbook*, vol. 2, ASM International, Materials Park, Ohio, 1991.

†Die materials mentioned first are for lighter duties, shorter runs.

Tool steel compositions, percent:
  6G (pre-hardened die steel): 0.5C-0.8Mn-0.25Si-1Cr-0.25Mo-0.1V
  H12 (hot-working die steel): 0.35C-5Cr-1.5Mo-1.5W-0.4V
  W1 (water-hardening steel): 0.6-1.4C
  O1 (oil-hardening steel): 0.9C-1Mn-0.5Cr
  A2 (air-hardening steel): 1C-5Cr-1Mo
  D2 (cold-working die steel): 1.5C-12Cr-1Mo
  M2 (Mo high-speed steel): 0.85C-4Cr-5Mo-6.25W-2V
  WC (tungsten carbide)

2. Short punch (Fig. 9–34b). This is equivalent to the axial upsetting of a cylinder, and $p = \sigma_{0.2}$. Steel punches are limited to approximately 1200 MPa in simple compression. Some cobalt-bonded WC punches operate at pressures up to 3300 MPa.

3. Flat platen (Fig. 9–34c). When a flat die is larger than the workpiece, the workpiece becomes, in effect, a punch. Therefore, by analogy to piercing (Sec. 9-2-3)

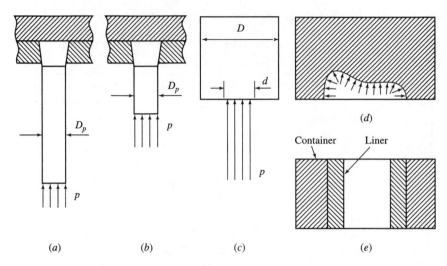

Figure 9–34   Tools and dies fail by various mechanisms: (a) long punches by buckling; (b) short punches by upsetting; (c) flat platens by indentation; (d) die or (e) container cavities by internal pressure.

When the platen $D > 3d$, $p = 3\sigma_{0.2}$ [Eq. (9-13)]

When the platen is smaller, $p = \sigma_{0.2}(D/d)$ [Eq. (9-15)]

4. Cavity (Fig. 9–34d). This is a much more severe case than the flat die, because the workpiece develops an internal pressure which can burst the die. This is also true of extrusion containers. The design of containers is a specialized subject; as a very rough guide, $p = \sigma_{0.2}/2$ when $D \geq 3d$, thus a single-piece container made of high-strength die steel can take up to 1000-MPa pressure. The inner part of the container (*liner*) may be shrunk (Fig. 9–34e) into a larger outer shrink ring (*container*) or it may be wrapped with steel band or wire under high tension. Thus, the internal surface of the container is in compression and can stand up to 1700-MPa internal pressure. Special constructions permit pressures up to 2700 MPa.

Dies are finished to a specified surface roughness; this may be a controlled, random roughness for hot working with solid lubricants, and usually a highly polished finish for cold working with liquid or soap-type lubricants. Many dies are now surface-treated for improved wear resistance (Chap. 19).

All highly stressed tooling must be surrounded by heavy shielding because a fractured die part becomes a potentially deadly projectile.

Example 9-15 | In Example 9-5 we found a die pressure of 2845 N/mm^2 for the lubricated flattening of a stainless steel pin. Is such a high pressure permissible for a tool-steel die?

The pressure exceeds the yield strength of the best steels. However, if the die is made at least $3(21.1) = 64$ mm wide, the nonloaded part of the die will give support (Fig. 9–34c) and a tool of HRC 60 will be safe. After forging several thousand parts, an indentation may gradually develop.

In Example 9-7 we found a pressure of 1200 N/mm^2 acting on the tool used for pinching the pin. Is this pressure permissible?

    If the tool is made in the form of a short punch (or a short extension on a longer but also wider punch), it will be just safe (because it is loaded as a short punch, Fig. 9–34b).

Example 9-16

9-5-2   Hammers

Hammers are energy-limited impact devices in which a mass (the *ram*) is accelerated by gravity and/or compressed air, gas, steam, or hydraulic fluid (Fig. 9–35a). For a ram mass M and impact velocity v the hammer energy E_h is

$$E_h = \frac{Mv^2}{2} = \frac{Wv^2}{2g} \qquad \textbf{(9-30)}$$

where W is weight of the ram and g is gravitational acceleration. The striking velocity v increases with the stroke (drop height) H_d and acceleration ξ

$$v^2 = 2\xi H_d = 2H_d \left(g + \frac{A p_m}{M} \right) \qquad \textbf{(9-31)}$$

where A is the cross-sectional area of the driving piston, p_m is the mean indicated pressure of the pressurized medium, and M is the accelerated mass. Hammers are

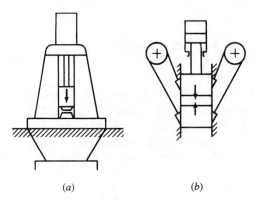

(a)   (b)

Figure 9–35   The deformation force and energy may be delivered by impact devices such as (a) hammers or (b) counterblow hammers. [After J.A. Schey, as Fig. 9-30, p. 1474. With permission.]

available in a large range of sizes (Table 9–4) and with increasingly sophisticated controls for metering the energy per blow.

Table 9–4   Characteristics of hammers and presses*

Equipment Type	Ram mass, kg (or energy)	Force,† kN	Speed, m/s	Strokes/ min	Stroke, m	Bed area, m × m	Mechanical efficiency
Hammers							
Mechanical	30–5 000		4–5	350–35	0.1–1.6	0.1 × 0.1 to 0.4 × 0.6	0.2–0.5
Steam and air	75–17 000		3–8	300–20	0.5–1.2	0.3 × 0.4 to 1.2 × 1.8	0.05–0.5
Counterblow	(5–1000 kN·m)		3–5	60–7		1.8 × 5	0.2–0.7
HERF	(15–750 kN·m)		8–20	< 2			0.2–0.6
Hydraulic presses							
Forging		100–80 000 (800 000)	< 0.5	30–5	0.3–1	0.5 × 0.5 to 3.5 × 8	0.1–0.7
Sheet metal working		10–40 000	< 0.5	130–220	0.1–1	0.2 × 0.2 to 2 × 6	0.5–0.7
Extrusion		1000–50 000 (200 000)	< 0.5	2	0.8–5	0.06 to 0.6 dia. container	0.5–0.7
Mechanical presses							
Forging		10–80 000	< 0.5	130–10	0.1–1	0.2 × 0.2 to 2 × 3	0.2–0.7
Sheet metal working		10–40 000	< 1	180–10	0.1–0.8	0.2 × 0.2 to 2 × 6	0.3–0.7
Horizontal upsetter		500–30 000 (25–230 mm dia.)	< 1	90–15	0.05–0.4	0.2 × 0.2 to 0.8 × 1.2	0.2–0.7
Screw		100–80 000	< 1	35–6	0.2–0.8	0.2 × 0.3 to 0.8 × 1	0.2–0.7

*From various sources, including A. Geleji, *Forge Equipment, Rolling Mills and Accessories*, Akademiai Kiado, Budapest, 1967.
†Divide number by ~10 to get tons force. Numbers in parentheses indicate the largest sizes, available in only a few places in the world.

The energy of impact is absorbed mostly by the energy E required for deforming the workpiece [Fig. 9–5 and Eq. (9-20)]. Some energy is, however, transmitted to the die, the foundation, the ground, and also the hammer components, setting up shock waves in the ground and air. Ground vibration and noise are objectionable and reduce the efficiency of forging. The total hammer energy E_h to be delivered is

$$E_h = \frac{E}{\eta}$$

(9-32a)

where E is the energy required for forging (Fig. 9–5) and the efficiency is

$$\eta = 0.9 \left[1 - \left(\frac{P}{10^3 Mg} \right) \right]$$

(9-32b)

if P is in newtons and M is ram mass in kilograms. Ground shock is avoided in *counterblow hammers* (Fig. 9–35*b*).

High impact velocities and short contact times minimize cooling; therefore, hammers are used mostly for open-die forging and for impression-die forging of intricate shapes. Except for counterblow and *high-energy-rate forging* (HERF) hammers (counterblow hammers driven by gas pressure), the forging is produced by several blows in any one die cavity; therefore, the total energy requirement [Eq. (9-20)] can be delivered by a relatively small hammer. Hammer forging does require, however, considerable operator skill and is less suitable for materials of high strain-rate sensitivity.

9-5-3 Presses

Presses are powered mechanically or hydraulically.

Hydraulic presses (Fig. 9–36*a*) stall out when their load limit is reached and can be used with dies that make contact (*kiss*) at the end of the stroke. Hydraulic presses are particularly suitable for isothermal forging where very low strain rates are required.

Mechanical presses are of various constructions (two examples are shown in Fig. 9–36*b* and *c*). They have a preset stroke and develop an infinite force at the end of the stroke. Therefore, in true closed-die forging the die must allow escape of excess material or the die gap must be set with extreme care. In setting up the press, elastic extension of the frame must be taken into account, as discussed in Sec. 4-1-2. Spring constants of presses seldom exceed 4 MN/mm; thus, a press exposed to a 20-MN force must be set, in the unloaded condition, 5 mm closer than the desired final dimension under load. Because of the lower speeds and

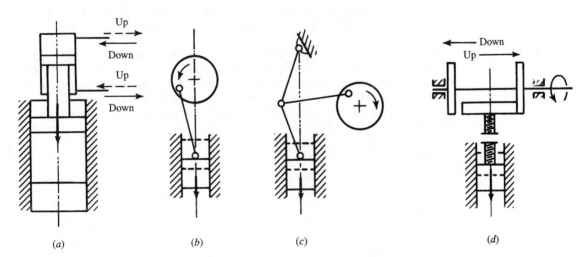

(a) (b) (c) (d)

Figure 9–36 Presses may be (*a*) force-limited hydraulic presses or stroke-limited mechanical presses such as (*b*) crank, (*c*) knuckle-joint, or (*d*) screw presses. [*After J.A. Schey, as Fig. 9–30, pp. 1475–1476. With permission.*]

longer contact times, workpieces must be preformed carefully if complex parts are to be made by hot press forging.

Screw presses (Fig. 9–36d) slow down as the stored energy is exhausted in the blow, hence they have characteristics between mechanical presses and hammers.

As a rule of thumb, a hammer equipped with a 1-Mg ram can do the work of a 10-MN press (ram of 1 ton and 1000-tonf press), because it delivers the total energy required in several blows. Information on typical equipment is given in Table 9–4.

Multiram presses and associated manipulators can be computer-controlled to form the nucleus of a flexible forging system.

Horizontal upsetters for hot and cold working and *cold-headers* for cold working compose a special class of presses. Both start with straight lengths of bar or wire. For cold-headers the material is fed with indexing pinch rollers, sometimes through a multiroll straightener or even a draw die that delivers bar of tight tolerances. The end of the bar or wire is deformed in successive steps ranging from simple upsetting to the most complex combined forging-extrusion operation. Auxiliary movements are synchronized with the main ram movement and are used to open and close clamping dies, actuate auxiliary punches and shearing dies, and transfer the workpiece from one die cavity to another. The workpiece material is cut off the bar or wire either at the beginning or end of the sequence, and either one workpiece may go through the die sequence at a time or a workpiece may reside in each die during each stroke. Examples of hot- and cold-upsetting sequences were given in Figs. 9–21 and 9–23 and one of cold extrusion is shown in Fig. 9–37. The construction and mechanization of these machines is often very ingenious and their production rates are difficult to match with other techniques.

Example 9-17

Estimate the size of press or hammer needed for making the part of Example 9-3.

The press size is given by the maximum load; note that this relatively small part requires a 4600-kN (520-tonf) press because of the large d/h ratio. Cooling would still further increase the force requirement.

Estimation of the hammer size is more difficult because the speed ranges from a high value at impact to zero at the end of the stroke. However, this particular steel is not very strain-rate sensitive ($m = 0.13$) and the high rate of deformation will actually increase the temperature of the workpiece, so that the computed energy requirement will not be too far off.

If the striking velocity from Eq. (9-31) is, say, 6 m/s, then the energy available is, from Eq. (9-30):

Ram Mass, kg	Energy, N·m
500	9 000
1000	18 000
1500	27 000
2000	36 000
4000	72 000

To make the part in one blow, even a 2000-kg hammer would be just sufficient. Perhaps more economically, a 1000-kg hammer could be used to deliver several blows.

Figure 9–37   A typical cold-forging sequence in a seven-station cold former, producing hose connectors by combined forward and back extrusion and forging, at the rate of 160 per minute. (*Courtesy National Machinery Co., Tiffin, Ohio.*)

9-6   DRAWING

Long components of uniform cross section can be produced not only by extrusion but also by *drawing*. Instead of being pushed, the material is now *pointed* (its end reduced, usually by swaging) and then pulled through a stationary die of gradually decreasing cross section. Most *wire* is of circular cross section, but square, rectangular, and shaped wires (*sections*) are also drawn. In addition to direct applications such as electrical wiring, wire is the starting material for many products including wire-frame structures (ranging from coat hangers to shopping carts), nails, screws and bolts, rivets, wire fencing, etc.

Seamless tubes are made by a variety of hot-working techniques but below a minimum size they must be further reduced cold. One of the options is to draw them, and such cold-drawn tubes perform important functions in hydraulic systems of vehicles, aeroplanes, ships, industrial machinery, and water distribution systems, and in such applications as hypodermic needles.

9-6-1   The Drawing Process

The material is deformed in compression, but the deformation force is now supplied by pulling the deformed end of the wire (Fig. 9–38a). Therefore, it is often said that the deformation mode is that of *indirect compression*.

Stationary *draw dies* are made of tool steel, cemented WC, or, for smaller diameters, diamond. The wire is pointed, usually by swaging (Fig. 9–24), fed through the die, and is drawn on a drum (*bull block*, Fig. 9–38c). We shall see (Sec. 9-6-3) that breakage of the wire sets a limit to the attainable reduction, and several successive reductions are usually taken. The task is speeded up with *multiple draw machines* in which the wire is passed through several dies (Fig. 9–38d). Constancy of volume again applies and individual drums are driven at increasing speeds to compensate for reduction in area. The speed of individual motors is controlled to give the appropriate tension for drawing; in older machines, drums are overdriven so that the wire slips on the drum and friction develops the draw force. The die may be replaced with two, three, or four *idling rollers*, all with their axes in a common plane. A *Turk's head* contains four rollers with adjustable positions.

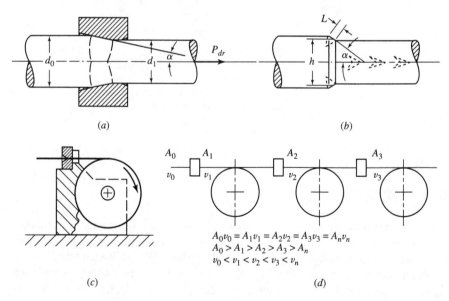

(a)   (b)

(c)   (d)

$$A_0 v_0 = A_1 v_1 = A_2 v_2 = A_3 v_3 = A_n v_n$$
$$A_0 > A_1 > A_2 > A_3 > A_n$$
$$v_0 < v_1 < v_2 < v_3 < v_n$$

Figure 9–38   (a) Deformation in wire drawing takes place under the indirect compression developed by a conical die. (b) High h/L ratios can lead to centerburst in materials of limited ductility. The wire may be drawn on (c) a bull block or, for higher productivity, (d) on a multiple-die wire-drawing machine.

Seamless tubes are sometimes drawn simply through draw dies, either to reduce their diameter (*sinking*, Fig. 9–39a) or to change their shape (say, from round to square). If their wall thickness is to be reduced, an internal die is also needed, which may be of three kinds: a short, conical *plug* held by a long bar from the far end (Fig. 9–39b); a plug shaped so as to stay in the deformation zone (*floating plug*, Fig. 9–39c); or a full-length *bar* of tool steel (Fig. 9–39d).

Drawing processes are very productive because speeds up to 50 m/s are possible on thin wire. Much slower speeds, on the order of around 1 m/s are common in drawing heavier bars, in straight lengths on *draw benches*. Sections that cannot be bent around the drum of a bull block are also drawn in straight lengths, on draw benches at low speeds and, because of the batch-type operation, reduced production rates.

9-6-2   Forces

With a few special exceptions, wire and tube are drawn cold. Initial steps in computing the draw force follow the routine of extrusion computations (Sec. 9-4-4). Here too, engineering strain [Step 3, Eq. (9-21)] is used for conversational purposes, but natural strain [Step 4, Eq. (9-23)] is relevant for computations.

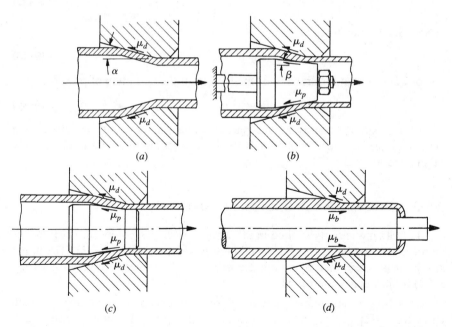

(a)   (b)

(c)   (d)

Figure 9–39   Seamless tubes are drawn: (a) by sinking, (b) on a plug, (c) with a floating plug, (d) on a bar. Half-arrows indicate frictional stresses. (J.A. Schey, *Tribology in Metalworking: Friction, Lubrication and Wear*, ASM International, 1983, p. 353. With permission.)

Drawing is a steady-state process, therefore, σ_{fm} [Eq. (9-1)] is again needed in Step 6.

In Step 7, the effects of die friction and inhomogeneity of deformation must be considered:

$$\sigma_{\text{exit}} = \sigma_{fm} Q_{fr} \phi \, \varepsilon \qquad \textbf{(9-33)}$$

where

$$Q_{fr} = (1 + \mu \cot \alpha) \qquad \textbf{(9-34a)}$$

In this, μ is the coefficient of friction between workpiece and die and α is the half angle of the draw die (Fig. 9–38a). For reasons discussed in Secs. 8-2-5 and 9-4-5, inhomogeneity of deformation requires extra (redundant) work which is a function of the h/L ratio. For drawing wire of circular cross section, h is taken as the mean diameter, L is the length of contact zone (Fig. 9–38b), and the inhomogeneity (redundant work) factor is

$$\phi = 0.88 + 0.12 \frac{h}{L} \qquad (\geq 1) \qquad \textbf{(9-34b)}$$

For deformation in plane strain (e.g., in shaping a rectangular cross section), the factor is

$$\phi = 0.8 + 0.12 \frac{h}{L} \qquad (\geq 1) \qquad \textbf{(9-34c)}$$

In Step 8, the draw force is

$$P_{\text{dr}} = \sigma_{\text{exit}} A_1 \qquad \textbf{(9-35)}$$

The power required for drawing is obtained from the definition of power

$$\text{Power} = P_{\text{dr}} v \qquad \textbf{(9-36)}$$

At high speeds, energy input is large and heating is significant; hence, the lubricant must be able to cool effectively. Fine wires are often drawn completely submerged in a bath.

9-6-3   Process Capabilities and Design Aspects

1. Drawing is fairly limited in the variety of shapes produced, but it makes up for this in the low cost of dies and high productivity (Table 9–1). The upper size limit is determined by equipment capacity. Tolerances are very tight (Fig. 3–22) if die wear is controlled.

2. A limit to attainable reduction is set by breakage of the drawn product. Draw force [Eq. (9-35)] must not exceed the strength of the drawn wire, which can be calculated from yield strength $\sigma_{0.2}$ (if this is not known, it can be taken as 80% of the flow stress at exit from the die):

$$P_{\text{dr}} < \sigma_{0.2} A_1 \qquad \text{or} \qquad P_{\text{dr}} < 0.8 \left(K \varepsilon_{\text{total}}^n \right) A_1 \qquad \textbf{(9-37)}$$

The maximum reduction is typically below 50% (calculated as *reduction of area*, and not as reduction of diameter). Frequent breaking of the wire would severely limit productivity since the end of the wire must be reduced (*pointed*) again so that it can be rethreaded. It is more profitable to limit reductions to below 30% per die (usually to 20% per die in multidie drawing). As seen from Eq. (9–34a), friction increases the draw stress and limits reduction; therefore good lubricating practices are essential (Table 8–4).

3. A second limitation arises from possible nonuniformity of deformation. Just as in extrusion (Sec. 9-4-5), the depth of the compression zone may not be sufficient to ensure homogeneous deformation. This is again governed by the h/L ratio: When $h/L > 2$, secondary tensile stresses can lead to the typical arrowhead (centerburst) defect in less-ductile materials (Fig. 9–38b), especially now that the axial stress is tensile.

4. Secondary tensile stresses arise also when deformation is limited to one part of a section. This will be discussed in more detail for the rolling of shapes (Sec. 9-7-2). Suffice it to say here that cracking of drawn sections may occur when some part of the cross section is not directly subjected to deformation.

Example 9-18

A shaped wire is drawn from annealed, 3-mm-diameter 302 stainless steel wire. The cross-sectional area of the shape is 5.0 mm². A commercial oil-base lubricant is used (from Table 8–4, $\mu = 0.05$), the dies have 12° included angle, and drawing speed is 2 m/s. Compute the draw force and power requirement. Set up a spreadsheet; for more general use, allow for multiple draws. To obtain correction for the inhomogeneity of deformation, the shaped section may be approximated by a circular cross section of equivalent diameter, Eq. (9-2c): $d_1 = 2.52$ mm. From the geometry of a conical die,

$$L = (d_0 - d_1)/2 \sin \alpha = 2.28 \text{ mm.}$$

A	B	C	D	E	F	G	H	I	J	K	L	M	N	O	P	Q	R	S
Cold		302	SS															
Flow	stress	K=	1300	MPa	n=	0.3												
Pass	mu	alpha	v	d	A	ec	eps	eps	sigma fm	h	L	h/L	phi	Qfr	exit	Pdr	safe P	power
No.		deg	m/s	mm	mm^2		pass	total	N/mm^2	mm	mm				N/mm^2	N	N	kW
						Eq. (9-21)	Eq. (9-23)		Eq. (9-1)				Eq. (9-34b)	Eq. (9-34a)	Eq. (9-33)	Eq. (9-35)	Eq. (9-37)	Eq. (9-36)
0				3	7.069			0										
1	0.05	6	2	2.52	5.000	0.293	0.346	0.34	727.5	2.7	2.28	1.2	1.0252	1.476	381.1	1905	3783	3.81

It is always necessary to check whether the draw is feasible. From columns Q and R, the draw is entirely feasible.

Example 9-19 | **S**ome steel-belted radial automobile tires are made with 0.8% C steel wire, patented and heavily cold drawn to 0.25-mm diameter. Surrounded by the ferrite matrix, the very fine carbide platelets are capable of deformation but ductility of the wire is still limited. After coating the wire with brass for rubber adhesion, a final draw is given before weaving the belt. Some tires had suffered early failure on the test track. Wire recovered from the tires showed occasional cup-and-cone type fracture. The problem was blamed on the steel since it was believed that no inhomogeneity could exist in drawing such fine wire. Inquiry showed that final reduction was from 0.27 to 0.25 mm diameter, in dies with 6° half angle. Check whether deformation was inhomogeneous.

$$h = (0.27 + 0.25)/2 = 0.26 \text{ mm}; \qquad L = (0.27 - 0.25)/2 \sin \alpha = 0.0957; \qquad h/L = 2.72$$

thus deformation was inhomogeneous. *Remedy:* Increase reduction or reduce die angle.

9-7   ROLLING

Of all bulk deformation processes, *rolling* occupies the most important position. Over 90% of all materials that are ever deformed are subjected to rolling (see Table 8–1).

9-7-1   Flat Rolling

The process of reducing the thickness of a slab to produce a thinner and longer but only slightly wider product is commonly referred to as *flat rolling*. It is the most important primary deformation process.

The Rolling Process   The process of flat rolling looks deceptively simple (Fig. 9–40*a*). Two driven rolls of cylindrical shape (*work rolls*) reduce the flat workpiece to a thinner gage. The rolls are supported in housings, and the roll gap can be adjusted by mechanical or hydraulic means (Fig. 9–40*b*). Elastic deflection of the rolls would create gage and shape problems; therefore, two support rolls (*backup rolls*) are incorporated into a mill housing of a *four-high mill* (Fig. 9–41*a*) and 18 backup rolls into a *Sendzimir mill* (Fig. 9–41*b*). Almost always, successive reductions are taken. When only one mill is used, it is reversed between passes. After several reductions, the strip becomes very long and is wound up on coilers under tension. For higher productivity, several mills are placed in line (*tandem mills* or *continuous mills*, Fig. 9–41*c*). Since constancy of volume is preserved, the product is speeded up from stand to stand in proportion of the area reduction.

The finished product must have a uniform thickness in length and width, a flat shape, a controlled and uniform surface finish, and reproducible mechanical properties. Satisfying these requirements taxes the ingenuity of the production engineer, equipment designer, control specialist, and theoretician, and makes an apparently simple process into one of the most complex ones (consider holding a tolerance of ± 0.002 mm on kilometers of 0.04-mm-thick strip while rolling at a

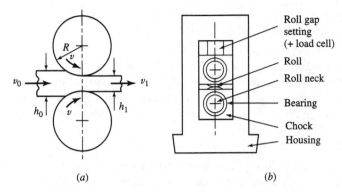

Figure 9–40   Rolling is a steady-state process that (a) reduces the thickness of the workpiece (b) in rolling mills of considerable stiffness.

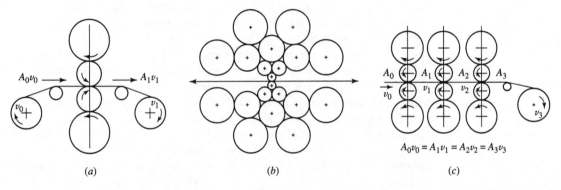

Figure 9–41   Roll force and roll flattening are reduced and deflection of small work rolls is controlled by backup rolls in (a) four-high and (b) Sendzimir mills. Successive reductions on (c) tandem mills greatly increase productivity.

speed of 40 m/s!). More or less complete process models—including those that predict metallurgical structure and properties in hot rolling—take a long time to run on computers and studies on the effects of process parameters are run off line. On-line controls are based on simplified theory and may incorporate empirical, experience-based models. Rolling at high speeds, coupled with a high degree of closed-loop automation, provides high-quality, close-tolerance starting material for various secondary sheet-metalworking processes at low cost.

Flat-Rolled Products   The terminology of rolling makes distinctions between:

1. *Hot-rolled plate*, over 6 mm thick and 1800–5000 mm wide, is rolled on large reversing two- or four-high mills from cast ingots or thick slabs (Sec. 7-5-2) in mass up to 150 Mg (160 tons). Heavy deformation assures that the cast structure is destroyed and defects are healed. Surface finish is relatively rough and dimensional tolerances are not very tight (Fig. 3–22), nevertheless, plate is an important starting material in ship building, boiler making, high-rise and industrial construction, and the manufacture of pipes and miscellaneous welded machine structures.

2. *Hot-rolled sheet* or *band*, of typically 0.8–6 mm thickness and up to 2300 mm width, is rolled on tandem mills. The starting material is often continuously cast slab or band. Long lengths of sheet issuing from the rolling mill are coiled up; coils weigh up to 30 Mg (33 tons). Hot-rolled sheet is an important starting material for the cold pressing of structural parts of vehicles, heavy equipment, and machinery, and also for making welded tubes.

3. *Cold-rolled sheet* (*strip*) is made by further rolling a hot-rolled band on reversing four-high mills or on tandem mills, often at high speeds (up to 30 m/s for steel). The band may be annealed and its surface is cleaned (or pickled, to remove residual scale from steel). Thus, the cold-rolled sheet has thinner gage, better surface finish, and tighter tolerances (Fig. 3–22). It too is wound into coils which may be *slit* into narrower widths or *cut* into shorter lengths, or both, depending on the handling facilities of the secondary manufacturing plants. Standard surface finishes and tolerances are provided at no extra cost; however, controlled or exceptionally smooth finish or tight tolerances can also be produced at an often quite slight premium. The cost, of course, goes up as the gage decreases, especially if the thinner gage necessitates extra passes through a single-stand or tandem mill. Large quantities of steel are rolled to around 0.7 mm for automotive and appliance bodies and down to 0.15 mm for food and beverage containers (cans). Copper is rolled to various gages for roofing, containers, cooking vessels, and down to 0.04 mm for radiator fin stock. Aluminum alloy sheet, of around 1.0-mm thickness, is extensively used in aircraft fuselages, automotive components, and in trailer construction. Aluminum foil is rolled down to 8-μm gage at speeds up to 60 m/s and used in large quantities for packaging. Foils of thicknesses down to 3 μm are produced on special mills (among them, Sendzimir mills) in all materials.

9-7-2   Shape Rolling

The rolling of shapes has a long history, beginning with the rolling of channels of lead for stained-glass windows (Table 1–1). The largest industrial application is now to the hot rolling of *structural shapes*, such as wide-flange beams, U and L channels, and rails. This is a primary deformation process practiced in special-purpose mills. Basically the same techniques can, however, be applied also to the cold rolling of shapes to tight tolerances and excellent surface finish, and these specialized secondary manufacturing processes are gaining popularity as alternatives to drawing and machining.

The starting material for *cold shape rolling* is a wire of square, rectangular, or circular cross section, and the finished shape is approached through a number of *passes* (rolling through shaped rolls) that gradually distribute the material in the desired fashion. The crucial issue is always that of avoiding nonuniform elongation. As seen from the simple example of Fig. 9–42*a*, parts of the cross section that are directly compressed elongate as required to maintain constant

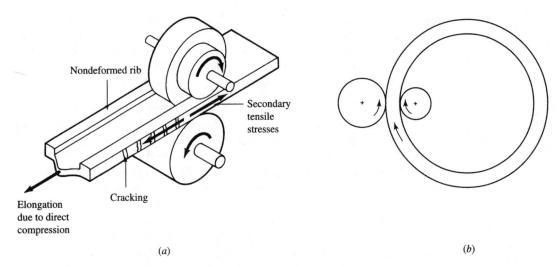

Nondeformed rib

Secondary
tensile
stresses

Cracking

Elongation
due to direct
compression

(a)

(b)

Figure 9–42   (a) Nonuniform elongation in the rolling of shapes can lead to cracking due to secondary tensile stresses. (b) Rings rolling results in an increase in ring diameter while often very complex cross sections are developed.

volume, while parts not subject to direct compression elongate only because of their physical attachment to the deforming portion. Elongation in these noncompressed portions generates secondary tensile stresses which, as remarked before, easily lead to crack formation. Therefore, roll pass design aims at equalizing reductions in all parts of the cross section. This aim can be attained by moving the material sideways, especially in early passes and, if necessary, by the use of vertical rolls that compress the section from its sides. Several roll stands may be placed in *tandem* (in line) and it is then customary to alternate the axes of rolls from vertical to horizontal.

9-7-3 Ring Rolling

Seamless rings are important constructional elements, ranging from the steel tires of railway car wheels to rotating rings of jet engines and races of ball bearings.

The starting material for *ring rolling* is a pierced billet. After making a hole by any suitable technique, the thick-walled ring is rolled out by reducing its thickness and increasing its diameter (Fig. 9–42b). Larger rings are rolled hot in specialized factories but smaller rings, especially those of small cross-sectional area, are frequently rolled cold. In addition to simple rectangular profiles, rings of a fairly complex cross-sectional profile can be rolled.

9-7-4 Transverse Rolling

When a workpiece is placed between two counterrotating rolls with its axis parallel to the roll axes, it suffers plastic deformation (essentially, localized compression) during its rotation between the rolls. The consequences of this deformation depend on the shape and angular alignment of the rolls and, as in all compression (Sec. 8-2-5), on the h/L ratio. The height h is now the workpiece diameter, and L is the length of contact with the roll (equivalent to L of an indenter in plane strain, Fig. 8–17b). Several purposes may be accomplished:

1. When $h/L > 1$, deformation is inhomogeneous and the plastic zones penetrating from the point of contact literally try to wedge the workpiece apart; in other words, high secondary tensile stresses are generated in the center of the workpiece. This is exploited in making thick-walled tubes by *rotary tube-piercing* methods (Fig. 9–43). The rolls have a barrel shape (a) and a mandrel or plug (e) placed against the center of the billet (d) helps in opening up and smoothing out the internal surface. A third roll (b) restrains the billet. Angular misalignment of the deforming rolls (skewing) forces the tube to progress in a helical path; thus, its whole length is pierced through. Such tube-piercing methods are practiced in specialized plants equipped for hot working.

2. The secondary deformation processes based on the same principle have the roll axes aligned and the workpiece rotates in the same plane (*transverse rolling*, Fig. 9–44). If the rolls are shaped so as to avoid the generation of large internal tensile stresses, a sound workpiece of axial symmetry is formed. For example, a dumbbell shape can serve either as a finished part or as a preform for the further forging of, say, a connecting rod or a double-ended wrench. There are a number of other rotary forging/rolling processes with specialized applications.

3. The rolls may be shaped to roll a thread on the workpiece. Large threads are rolled hot, but most *thread-rolling* operations are conducted cold, most often in thread-rolling machines equipped with so-called *flat dies* (Fig. 9–45). One of the dies is stationary, the other reciprocates; at an appropriate point of the stroke, a workpiece (typically, a cold-headed or extruded screw blank) is dropped into the gap, grabbed by the moving die, and rotated against the stationary die, thus the screw-thread profile is gradually developed. Rolled threads have a continuous

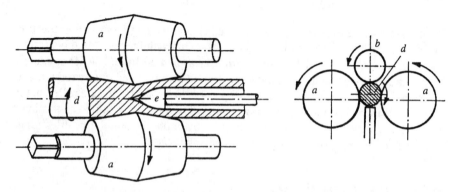

Figure 9–43   Inhomogeneous deformation in rotary tube piercing helps in opening up the center of a billet to make a thick-walled tube.

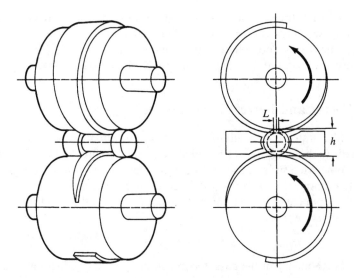

Figure 9–44   Axially symmetric workpieces may be cross-rolled but the h/L ratio must be kept low to avoid opening up the center. [J. Holub, *Machinery* (London), **102**:131 (Jan. 16, 1963). With permission.]

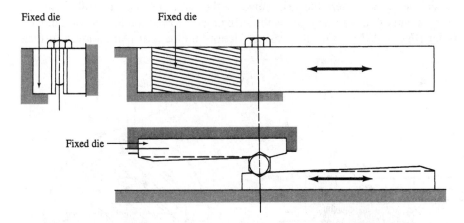

Figure 9–45   Strong threads are rolled at high rates in reciprocating flat dies.

grain flow and are, therefore, more fatigue-resistant than threads cut on a lathe. The productivity of the process is high. Even large, slow machines roll 60 screws per minute while smaller screws are produced at rates of 500 per minute. In machines containing several die pairs, production rates of 2000 parts/min are achieved. The good quality and high productivity of thread rolling has eliminated thread cutting as a competitive process for most mass-production purposes.

Large internal threads could be made by rolling but, apart from cutting, a more practical way is *cold form tapping*. The tool looks like a screw, except that its diameter changes periodically within the screw envelope, so that the protruding portions displace material from the roots into the threads (Fig. 9–46).

9-7-5   Forces and Power Requirements

Rolling is, like drawing, a steady-state process. The situation is, however, more complicated. In drawing, friction could be zero and the wire could still be drawn; in rolling some friction is needed to draw the workpiece into the roll gap. The horizontal component of the friction force must be greater than the opposing horizontal component of the roll force at the point of entry (Fig. 9–47a); thus, the *angle of acceptance* α is

$$\tan \alpha \leq \mu \qquad \textbf{(9-38a)}$$

and, from the geometry of the pass, the maximum possible reduction in the pass is

$$\Delta h_{\max} = (h_0 - h_1)_{max} = \mu^2 R \qquad \textbf{(9-38b)}$$

(A heavier reduction can be taken by pushing the workpiece into the roll gap.)

Once the workpiece (strip) enters the roll gap, there can be only one point (or, rather, one plane) where the strip moves at the same speed as the roll; from the point of entry to this *neutral plane* the strip moves slower and to the exit it moves faster (Fig. 9–47b). With decreasing friction, the neutral plane moves further

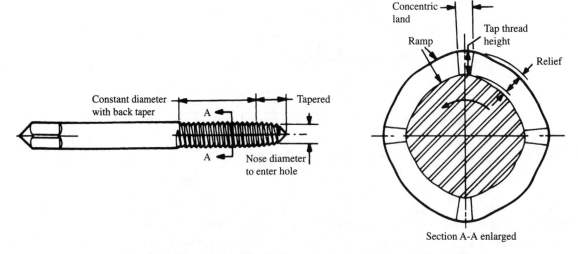

Figure 9–46   A cold form tap forms threads in a hole by displacing rather than removing material. (*From The Tool and Manufacturing Engineers Handbook, 4th ed., vol. 1, p. 12.92. With permission of the Society of Manufacturing Engineers, Dearborn, Michigan.*)

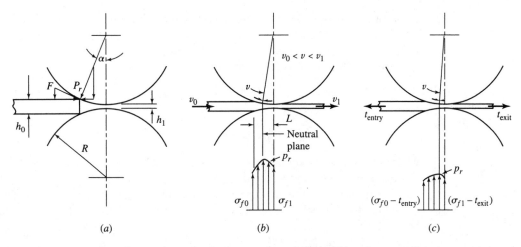

Figure 9–47   In flat rolling (a) the angle of acceptance limits the attainable reduction, and (b) the neutral plane developed in steady-state conditions (c) can be shifted and perssures reduced by the application of tensions.

forward towards the exit and the speed differential between roll and exiting strip (*forward slip*) decreases (with very low friction, skidding occurs). As in plane strain compression (Fig. 9–7), friction results in a rise in interface pressure, but the peak is now not in the middle of the contact zone but at the neutral plane (Fig. 9–47*b*).

Rolling lends itself to analysis and many rolling theories are available. For our purpose, an acceptable estimate of rolling forces can be obtained by analogy to a continuous forging (cogging) process. A comparison of Fig. 9–47*b* with Fig. 9–7 will show that the *projected length of the arc of contact* between roll and workpiece may be regarded as L of the forging tool, because major material flow takes place in the length direction of the slab or strip. The length of contact may be calculated from

$$L = \sqrt{R\,(h_0 - h_1)} \qquad \textbf{(9-39)}$$

where R is the roll radius. Computations follow those for forging an overhanging rectangular workpiece (Sec. 9-2-2), with some differences:

Step 5: $\dot{\varepsilon}$ must now be taken as the average strain rate

$$\dot{\varepsilon} = \frac{v}{L}\,\ln\frac{h_0}{h_1} \qquad \textbf{(9-40)}$$

Step 6: σ_{fm} [Eq. (9-1)] is now needed because rolling is a steady-state process. In hot working, Eq. (8-11) automatically provides a mean flow stress because a mean strain rate was calculated in Eq. (9-40).

Step 7: To find interface pressure, it is first necessary to check for homogeneity of deformation by computing the h/L ratio. When $h/L > 1$, inhomogeneity of deformation predominates and the pressure-multiplying factor Q_i is found from Fig. 9–9. When $h/L < 1$, friction effects are overriding and the pressure-intensification factor Q_p is found from Fig. 9–8 or Eq. (9-9).

Step 8: Roll force is obtained from

$$P_r = (1.15)\ \sigma_{fm} Q_i Lw \qquad (9\text{-}41a)$$

or

$$P_r = (1.15)\ \sigma_{fm} Q_p Lw \qquad (9\text{-}41b)$$

where w is the width of the strip and Lw is area of the contact surface.

Step 9: The torque required to rotate the rolls can be obtained by assuming that the roll force acts in the middle of the arc of contact, thus the moment arm is $L/2$ (Fig. 9–47b). Since there are two rolls to be driven, the total torque M_r will be

$$M_r = \frac{2P_r L}{2} = P_r L \qquad (9\text{-}42)$$

The power requirement is readily calculated in units of watts from

$$\text{Power} = P_r L \frac{2\pi N}{60} = P_r L \frac{v}{R} \qquad (9\text{-}43a)$$

where P_r is roll force in newtons, L and R are in meters, v is in meters per second, and N is in revolutions per minute. To obtain the power requirement in units of horsepower, take

$$\text{Power} = P_r L \frac{2\pi N}{33\ 000} \qquad (9\text{-}43b)$$

where L is in feet and P_r is in pounds.

9-7-6   Process Capabilities and Design Aspects

The rolling process is remarkably forgiving if quality demands are not high, but requires a substantial knowledge and sophisticated control if the product is critical in any respect. There are a number of process limitations:

1. When thin sections are rolled from hard materials, elastic deformation of the rolls may limit the attainable minimum thickness. *Flattening* of the rolls can be minimized with a good lubricant, small roll diameter, and a roll made of a material with a high elastic modulus, such as cemented WC. Roll force is reduced if *tensions* are imposed with the aid of coilers or in a tandem mill (Fig. 9–41). We saw in Fig. 8–14 that a compressive stress (σ_1) required for deformation can be reduced by applying a tensile stress in the other (σ_2) direction. Thus, in Eq. (9-41), we can replace σ_{fm} with ($\sigma_{fm} - t$), where $t = (t_{\text{exit}} + t_{\text{entry}})/2$. The effect is most powerful (Fig. 9–47c).

2. Under the imposed forces, *roll bending* occurs as with any centrally loaded beam, supported at two ends. This makes the roll gap larger in the middle, thus the workpiece is reduced less and is elongated to a lesser degree in the middle, while the edges elongate more and become wavy. Compensation is possible by *cambering* (*crowning*) the rolls (grinding them with a slight barrel shape) and by using backup rolls (Fig. 9–41). Substantial heat is generated in the rolling process and, if the lubricant/coolant is not fully effective, an increased roll diameter in the roll middle (*thermal camber*) results in a wavy middle on the rolled strip.

3. Under the imposed roll force, the entire rolling mill stretches, as discussed in Sec. 4-1-2. The spring constant of mills (the *mill elastic constant*) is usually under 5 MN/mm; hence, the roll gap can open up several millimeters (even if a very thin strip is rolled!). To compensate for elastic deflection, the rolls must be set closer by the amount required by the roll force. Any variations of roll force during rolling must be compensated for by manual or automatic control of the roll gap setting.

4. Inhomogeneous deformation, whether from a large h/L ratio or the absence of direct compression, is always harmful. There is one instance, however, when inhomogeneity is purposely induced. In *roller burnishing* the surface of a thick workpiece is superficially rolled. The deformation zone is very shallow and, in the absence of bulk plastic flow, the material of the surface is put in compression (Fig. 4–18), making the part more resistant to fatigue (as in rolling the journal radii on crankshafts or in finish rolling gears).

An AISI 1015 steel slab of $h_0 = 300$ mm thickness and $w_0 = 1000$ mm width is hot rolled at 1000°C on a mill with rolls of 600-mm diameter. The presence of scale reduces friction to $\mu = 0.3$. A reduction of 27 mm is taken. Roll speed is 1.2 m/s. Calculate roll force and power requirement. The spreadsheet checks the maximum bite (allowable reduction, identified as max Δh).

Example 9-20

A	B	C	D	E	F	G	H	I	J	K	L	M	N	O	P
Hot	1015	st				Roll	R=	300	mm	v=	1200	mm/s	w=	1000	mm
Flow	stress	C=	120	MPa	m=	0.1									
				max									contact		
Pass	mu	h	A	ec	delta h	eps dot	sigma f	h(ave)	L	L/h	Q	p	A	P	power
No.		mm	mm^2		mm		N/mm^2	mm	mm			N/mm^2	mm^2	kN	kW
				Eq. (9-3)	Eq. (9-38b)	Eq. (9-40)	Eq. (8-11)		Eq. (9-39)		Fig. 9-9	Eq. (9-10)		Eq. 9-41)	Eq. (9-43)
0		300	300000												
1	0.3	273	273000	0.09	27	1.2574	122.78	286.5	90.0	0.31	1.50	212	90000	19062	6862

The maximum allowable reduction matches the desired 27 mm. Because L/h is 0.3, Q_i is taken from Fig. 9–9. Note the large roll force and power requirement.

Example 9-21 | After hot rolling, the material of Example 9-20 is cold rolled on a mill of roll diameter 400 mm at a speed of 700 m/min. Compute the force and power requirement for rolling from 1.0 mm to 0.6 mm, if a lubricant reduces the coefficient of friction to 0.05. (The spreadsheet allows for successive reductions.)

A	B	C	D	E	F	G	H	I	J	K	L	M	N	O	P	Q
Cold	1015	st				Roll	R=	200	mm		w=	1000	mm			
Flow	stress	K=	620	MPa	n=	0.18										
					max									contact		
Pass	mu	h	A	ec	delta h	eps	eps	sigma fm	h(ave)	L	L/h	Q	p	A	P	power
No.		mm	mm^2		mm	pass	total	N/mm^2	mm	mm			N/mm^2	mm^2	kN	kW
				Eq. (9-3)	Eq. (9-38b)	Eq. (8-5)		Eq. (9-1)		Eq. (9-39)		Fig. 9-8	Eq. (9-10)		Eq. (9-41)	Eq. (9-43)
0		1	1000			0	0									
1	0.05	0.6	600	0.40	0.500	0.51	0.51	465.6	0.80	8.94	11.2	1.35	723	8944	6465	3373

Note that $\Delta h_{max} > \Delta h$. Note also that even though the reduction is miniscule compared to Example 9-20, roll force is quite high because of the higher flow stress in cold rolling. Power is high because of the higher rolling speed.

Example 9-22 | Assume now that the strip is rolled on a Sendzimir mill with work rolls of 40-mm diameter, under otherwise identical conditions.

For roll radius 20 mm, L drops to 2.83 mm, $Q_p = 1.1$, and roll force drops significantly to 1666 kN. Power is 2749 W, a drop of only 18.5%, because the work of deformation remains unchanged.

Example 9-23 | The flange of Example 7-9 can be made by forging (Table 9–1); not shown, but obviously also suitable, is ring rolling. For forging, it would be attractive to part at the base of the flange (as in Fig. Ex. 7-9b) because of the low cost of a flat lower die. However, location of the preform is not defined, and it is better to forge in the upside-down position. For release from the die, draft angle must be applied on all vertical surfaces and the center of the ring is now filled out

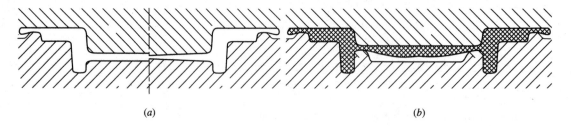

(a)   (b)

Figure Ex. 9-23

by a web. Considerable preforging (blocking) is necessary to prevent push-through of material (Fig. 9–26b). Minimum web thickness from Fig. 9–25 leads to poor material utilization. Die pressures are also high. Relief is obtained by tapering the center of the punch (a) or with an internal flash gutter (b). Ring rolling can yield a near-net-shape ring, with finishing cuts needed only on SID and SFT surfaces. The flange is, however, wide relative to the collar, and the ID is small. A ring of larger diameter would be more practical; even then, skillful preforming is necessary.

The parts of Example 7-10 are routinely made by deformation processes. The straight section (Fig. Ex. 7-10a) can be cut to length from a hot extrusion or a drawn or rolled shape (Table 9–1). The choice of process will hinge on dimensions, material, and tolerances.

Example 9-24

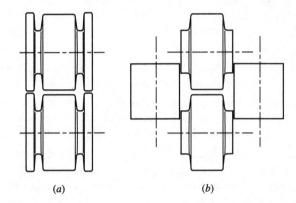

(a)   (b)

Figure Ex. 9-24

Extrusion die design is simple if web and flange thicknesses are equal. If WT is reduced, the faster-emerging flanges would drag the web material along and could cause periodic fracture; therefore, the die must be designed to slow down the flange (or accelerate the web).

Development of the shape in shape drawing requires considerable skill but tight tolerances can be held. Shape rolling is routinely done, on a large scale, in specialized structural-section mills. The section is rolled from rectangular or shaped continuous-cast ingots through a series of roll cavities (*calibers*). Because the web is much thinner than the height of the flange, unequal deformation occurs and roll pass design must minimize secondary tensile stresses (Fig. 9–42a). The parting line is through the middle of the web, and draft and fillet radii must be applied (a). Outside draft can be eliminated by finishing on a universal mill which has both horizontal- and vertical-axis rolls (b).

The presence of a cross rib (Fig. Ex. 7-10b) severely limits the options; indeed, only impression-die or closed-die forging is suitable (Table 9–1). There are several options for placing the parting line, but in all instances draft must be applied to the flanges and a generous fillet radius RF must be given. Minimum web thickness is again dictated by die pressure and thus the L/h ratio (Fig. 9–25); considerable preforging is needed to prevent flow through (Fig. 9–26b).

9-8 SUMMARY

Bulk deformation processes have retained their importance over thousands of years of technological development. They not only provide the starting material for subsequent sheet metalworking, wire and tube bending, and most welding applications, but also ensure the availability of finished components of great structural integrity. The products include hot-forged parts, from turbine blades and gear blanks to garden hoes; cold-forged parts, from nails, screws, and rivets to finished gears; cold-extruded parts, from automotive half-axles and sparkplug bodies to toothpaste tubes; hot-extruded construction sections and valve bodies; and hot- and cold-rolled rings and sections for all purposes. In the design of components and in the control of processes several factors must be considered:

1. It is almost always possible to find a bulk deformation process that can compete with other manufacturing methods, except for parts of greatest shape complexity (Table 9–1).

2. Cold working offers products of increased strength, good tolerances and surface finish, and thin walls, but usually at the expense of lesser ductility and higher flow stress, die pressure, and deforming forces.

3. Hot working offers lower (but strain-rate dependent) flow stresses, die pressures, and forces, but at the expense of extra energy consumption for preheating, and poorer tolerances and surface finish of the product.

4. Die pressures are determined by the flow stress of the material, modified by the effects of stress state (as expressed by the yield criterion), friction, and inhomogeneity of deformation. The aim of process control is to minimize pressures and forces by lubrication and the modification of process geometry.

5. Survival of the workpiece material is a function of workability which encompasses the effects of the hydrostatic pressure developed in the process, superimposed on the basic ductility of the material. The aim of process development is usually that of increasing the hydrostatic pressure component (except for tube-piercing operations).

6. The pressure developed by the process must be accommodated by tools and dies made of appropriate materials, in configurations designed to give maximum resistance to plastic yielding. Elastic deformation of tooling and machinery must be compensated if the shape of parts is to be kept within close tolerances.

7. Mathematical process models and experts systems allow optimization of part shape for the process and for in-service properties.

8. High temperatures, highly stressed tooling, rapid operation, and noise and vibration all present special dangers for the worker, and personal protective outfits, proper guarding of the work space, prevention of accidental operation and, in many cases, total enclosure are necessary.

PROBLEMS 9A

9A-1 Rank (low, medium, high), in a tabular form, the following attributes of cold, warm, and hot working: strain-rate sensitivity, flow stress, die pressure, dimensional tolerances, surface finish, ease of lubrication, effectiveness in healing casting defects.

9A-2 Define hot working (*a*) in the everyday sense and (*b*) in a fundamental sense.

9A-3 (*a*) Draw a cylinder of $h/d = 1$ ratio. Superimpose on it the outlines of the cylinder upset to $0.5h$ with (*b*) zero friction and (*c*) sticking friction. (*d*) In a separate sketch, show the deformation zones within the body of (*c*).

9A-4 (*a*) Draw a typical true-stress–true-strain curve for a strain-hardening material. Choose a strain and mark the appropriate flow stress for (*b*) steady-state and (*c*) non-steady-state deformation.

9A-5 Draw a sketch showing the essential features of (*a*) forward and (*b*) reverse extrusion of a bar, showing the directions of punch movement and material flow. (*c*) Use half arrows to indicate the direction in which friction acts.

9A-6 Draw two sketches showing the essential difference between (*a*) impression-die and (*b*) closed-die forging. (The part can be a simple cylinder.)

9A-7 (*a*) Draw an H section. (*b*) Superimpose the outlines of the die used for impression-die forging. (*c*) Point out the essential features of the as-forged H shape. (*d*) Identify the flash gutter of the die and state its purpose.

9A-8 Draw sketches showing material flow in (*a*) unlubricated and (*b*) lubricated hot forward extrusion. Point out the essential features.

9A-9 Make a sketch showing drawing of a tube on a plug. With half arrows, indicate the frictional stresses.

9A-10 Draw two sketches of flat rolling with rolls of constant diameter, (*a*) one to show inhomogeneous deformation and (*b*) one to show homogeneous deformation. (*c*) Identify the essential difference.

PROBLEMS 9B

9B-1 A billet of 50-mm diameter and 150-mm length is designed to be upset in the axial direction to a height of 75 mm. Make sketches (*a*) of the starting cylinder and (*b*) and (*c*) of potential problems you anticipate in upsetting.

9B-2 (*a*) Draw the plan view of a slab of 100-mm length and 20-mm width (the thickness of 10 mm is not visible in this view). (*b*) Superimpose the outline of the slab after upsetting with sliding friction. Indicate the direction of sliding. (*Hint:* for a given constant coefficient of friction, friction force is proportional to sliding distance).

9B-3 Repeat Problem 9B-2 for a (*a*) square block and (*b*) axially deformed cylinder.

9B-4 A cylindrical part is required to have 50% cold work in it. It has been successfully made with an aspect ratio of $d/h = 2$. As a result of a design change, it is now proposed to make it with the same height but with an aspect ratio of 4. (*a*) Draw sketches to show the two cases. Evaluate the consequences of the change in terms of (*b*) die pressures and (*c*) forging forces (make qualitative arguments, assuming upsetting with sliding friction).

9B-5 Continuing Prob. 9B-4, (*a*) make a sketch to show the change in the design of the part that would reduce die pressures in forging the disk with $d/h = 4$. (*b*) Justify.

9B-6 A gear blank has been successfully hot forged on a mechanical press in a closed die. At the beginning of a new production run, the die burst. What is the likely cause?

9B-7   A billet is placed in a container of 50-mm diameter and a concentric cavity is created with a 25-mm diameter punch. (*a*) Make a sketch of the process and (*b*) state what analysis (or analyses) should be applied.

9B-8   An AISI 1045 steel billet is to be cold-extruded from a 25-mm diameter to a diameter of 16 mm. Suggest (*a*) a suitable lubricant and (*b*) the metallurgical condition of the billet, and (*c*) design in principle a die shape which will avoid problems. Justify each choice.

9B-9   Assume the part of Example 9-12 is made by upsetting an annealed bar of 10-mm diameter. (*a*) Make a sketch of the longitudinal cross section of the bolt. (*b*) Indicate the grain-size variation one should expect, if the bolt is annealed after cold heading. (*c*) Point out the weakest cross section, and (*d*) indicate the outline of the design for a method of production that would avoid this weakness.

9B-10   The shank of the part in Example 9-12 is to be extruded from d_0 diameter to 10-mm diameter. (*a*) In principle, what problems do you expect in creating the shape as specified? (*b*) What changes in design or process would you recommend?

9B-11   Draw sketches showing centerburst defect in (*a*) extrusion and (*b*) bar drawing. (*c*) Indicate the geometry that is likely to induce the defect. (*d*) If the die geometries are identical, which process is more likely to result in the defect? Justify.

9B-12   It is suggested that an H section be extruded. The web is twice as thick at the flanges. (*a*) Make a sketch of the section. (*b*) Do you anticipate any problems in extrusion? (*c*) If you do, what action should be taken in designing the die?

9B-13   Inspect Fig. 9–39 and note the half-arrows indicating the friction stresses. (*a*) For each case, state whether friction increases the drawing stresses. (*b*) Deduce which of the

four processes allows the greatest reduction to be taken. Justify.

9B-14   (*a*) Draw a dumbbell shape. (*b*), (*c*), and (*d*) Make sketches of three possible deformation processes for making the part.

9B-15   Suggest three ways of making a thick-walled tube. Make simple sketches to show the essence of the proposed deformation methods.

9B-16   A flat bar is to be rolled but the rolls refuse to bite (the bar is not drawn into the roll gap). State at least two measures that could be taken.

9B-17   The flange of Fig. Ex. 9-23*a* is to be hot forged of 1045 steel at 1000°C. With the aid of sketches, discuss (*a*) what problem to expect if the part were forged from a flat disk with a diameter equal to the flange diameter, (*b*) the shape of a proposed design for the preform, and (*c*) the design of a die suitable for making such a preform. (*d*) Suggest an alternative starting material.

9B-18   The part shown in Fig. Ex. 9-12 is to be made of a low-alloy steel. It will be heat-treated to HRC40 and will be subjected to bending fatigue in service. State the steps involved in heat treatment. Processes proposed for making the part include: (1) hot forging with the parting line through the axis; (2) hot forging a head on a bar of 10-mm diameter; (3) hot extrusion of a 20-mm-diameter billet; (4) combined cold forging and extrusion, as in Ex. 9-12; (5) cold upsetting a 10-mm-diameter bar; and (6) cold extrusion of a 20-mm-diameter billet. As a part of your design, evaluate each proposal with reference to (*a*) the benefit for service, (*b*) the change in part shape required, and (*c*) potential production problems. (*d*) Recommend a preferred method.

9B-19   A tractor component is to be made of 1045 steel bar by cold forging with heavy deformation. To establish the forces to be expected, a good measure of the flow strength of the material is needed. Your materials-testing laboratory proposes to perform standard tension tests. (a) Explain why this might not give sufficient information. (b) Suggest a more appropriate method. (c) Identify the precautions to be taken if relevant data are to be generated.

PROBLEMS 9C

9C-1   As a part of evaluating the economies of your design, calculate how much material is saved by making the part described in Example 9-12 by plastic deformation instead of machining.

9C-2   It is proposed that the end of a $d_0 = 6$ mm cartridge brass bar be cold upset over a length of $h_0 = 5$ mm to form a flat head of $h_1 = 0.8$ mm height (thickness). To assess feasibility of design, calculate the upsetting pressures and forces, (a) first assuming that a good lubricant reduces friction to $\mu = 0.1$, and then (b) for rough, unlubricated dies (as would apply if the lubricant had broken down or the lubricant supply had failed). (c) Check whether the operation can be performed without buckling. (d) If the answer is yes, suggest a suitable die material.

9C-3   In preparation of forging a large gear blank, a 1045 steel billet of 200-mm diameter and 400-mm height is upset at 1000°C to a 100-mm thick "pancake." A graphite lubricant reduces friction to $\mu = 0.2$. As a part of your process design, (a) make a sketch of the operation. Calculate the (b) average die pressure and (c) the force required to forge the part if a hydraulic press with 3 m/min speed is used. Express the end result also in USCS units.

9C-4   Analytical methods of determining pressures in upsetting are often based on the assumption that pressure rises exponentially from the edge of the slab: $p_x = 2k \exp(2\mu x/h)$, where x is the distance from the edge. Integration gives the average die pressure as in Eq. (9-10) with $Q_p = (h/\mu L)[\exp(\mu L/h) - 1]$. (a) Compute Q_p over the range of $1 < L/h < 40$ for $\mu = 0.05, 0.1$, and 0.2. (b) Plot the results and compare with Fig. 9–8. (c) Explain any discrepancies.

9C-5   Many researchers prefer to use the interface shear factor m^* for computing pressures and forces in metalworking. Explore the possible differences by returning to the pin forged in Example 9-4. Recalculate the average die pressure and forging force. For unlubricated upsetting, $m^* = 1$; for the lubricated case, estimate m^* from μ, as shown in Example 8-18.

9C-6   For Example 9-12, calculate the length of bar of d_0 diameter needed to (a) upset the head and (b) extrude the shank. If the part is made of 1045 steel and is lubricated with a phosphate-soap lubricant, compute the (c) upsetting force and (d) extrusion force (ignoring container friction). (e) Show what design changes have to be made to the shape of the part.

9C-7   The flange of Example 9-23 is forged of 1045 steel on a hydraulic press at 1000°C. Press speed is 70 mm/s; a graphited lubricant is used. Take the dimensions of the ring from Fig. Ex. 7-9b. The flashland is 6 mm wide. To evaluate the relative merits of alternative designs, estimate the forging force for the configuration of (a) Fig. Ex. 9-23a (5-mm-thick flat web) and (b) Fig. Ex. 9-23b (internal flash).

9C-8   A $d_0 = 50$ mm and $h_0 = 75$ mm billet of a 2017 Al alloy is to be compressed to an $h_1 = 20$ mm height in a hydraulic press (ram velocity 100 mm/s) at 500°C between unlubricated anvils. (a) For the end of the press stroke, calculate interface pressure and press force. (b) What increase in stress and force would occur if the workpiece were to cool to 400°C?

9C-9   We computed in Example 9-3 that an energy of 36 250 N·m was needed to hot forge the billet. In Example 9-17 we found that a 2000-kg hammer would be just sufficient to deliver this energy. Assume now that a 1000-kg hammer is the largest available. From Example 9-17 we know that it will deliver 18-kN·m energy. The plant proposes to forge in three blows. As part of your process designs, make approximate calculations to see whether this is feasible. (*Hint:* In Fig. 9–5, divide the area under the force–displacement curve into 3 unequal areas, remembering that the initial, softer blow is more efficient and can thus deliver more energy. Start by assuming a height of 20 mm in the first blow; compute the energy required; if it is less than that delivered by the hammer, proceed to the second blow to 12.5 mm, and then to the third blow taking the billet to the final 10mm. You may have to iterate to find a reasonable solution.)

9C-10   It is proposed to make a tube-shaped part of 1045 steel by cold piercing a $d_0 = 30$ mm billet with a $D_p = 20$ mm diameter punch. The length of the part (the depth of the hole) is 80 mm; for constructional reasons, the punch is 120-mm long. Check several aspects of the design intent: (*a*) Specify the condition of the billet material. (*b*) Calculate the punch pressure. (*c*) Suggest a suitable punch material. (*d*) Check the punch for possible failure modes. (*e*) Specify a lubricant suitable for this task.

9C-11   In extruding the part in Prob. 9B-10, through a die of $\alpha = 45°$ half-angle, many components are found to have centerburst defects. (*a*) Make a sketch (to scale) to find an explanation and (*b*) suggest ways of getting out of trouble by changes in process or part design. (Assume that the composition of the material cannot be changed.)

9C-12   Commercial-purity (1100) Al billets of 250-mm diameter are extruded at 500°C, with a ram speed of 0.6 m/min, into bars of 125-, 50-, 25-, and 12.5-mm diameter. Assuming a dead-metal zone of 45° and ignoring friction, (*a*) compute the basic extrusion pressure and extrusion force for the four bar diameters, (*b*) determine the speed at which the extrusions emerge, (*c*) find a ram speed that will give an extrusion force of 4000 kN for a bar of 12.5-mm diameter. (*d*) Make a judgment whether the press is large enough for the economical extrusion of bars of this size.

9C-13   The H section shown in Fig. Ex. 7-10*a* is to be extruded of 6061 Al alloy. The section is 50 mm wide, 50 mm high, and thickness in the flange and web is 3 mm. (*a*) Make a sketch of the section; make any design changes necessary, in your judgment, for ease of extrusion. (*b*) Obtain the minimum extrusion pressure and force for the unlubricated extrusion of a 150-mm-diameter billet at 500°C with the extrusion emerging at a speed of 1 m/s. (*c*) Obtain the maximum extrusion pressure for a billet of 450-mm length.

9C-14   The extrusion of Prob. 9C-13 is now to be made in 7075 Al alloy at 450°C. The extrusion emerges with severe cracks transverse to the extrusion direction. (*a*) Identify the cause of the problem. (*b*) On the basis of data given in Example 8-19, define the extrusion conditions that would make extrusion safe.

9C-15   A flat, fluted shape of $h = 1$ mm average thickness and $w = 10$ mm width is to be made by cold drawing. It is proposed to start from flat, annealed Cu-5Sn bronze wire of 10-mm by 2-mm cross section, so as to gain the advantage of strain hardening in the product. The flat is drawn at 1 m/s in dies with $\alpha = 7°$ half-angle, and, to keep the part cool, an emulsion is chosen as the lubricant. Check if the process as designed is feasible by calculating (*a*) the relevant flow stress, (*b*) the drawing force and power, and (*c*) the force sustained by the drawn shape. (*d*) If the draw is not feasible, suggest a way of making the required end product.

9C-16 Continue Prob. 9C-15. Recalculate on the assumption that the shape is made in two passes, with 30% reduction in the first pass and no annealing after the first pass.

9C-17 Continue Prob. 9C-16. Draw in four passes, with 20% reduction in the first three passes and a fourth pass to the final thickness of h = 1 mm. (*a*) Find exit stress and (*b*) draw force. (*c*) Check whether the draws are feasible. (*d*) Check whether centerburst defect may occur. (*e*) If there is danger of centerburst, distribute reductions evenly so that the same strain is imposed in all four passes.

9C-18 Continue Example 9-18. (*a*) Explore the effect of changing the half-angle from 1° to 20° in 2° increments, for $\mu = 0.02$, 0.05, and 0.1. (*b*) Plot Q_{dr} versus half angle for the three μ values. (*c*) For each μ, find the minimum exit stress.

9C-19 A small, shallow U channel of Cu-5Sn bronze is cold-rolled. The shape is shallow enough to regard it as a $w = 20$ mm wide, $h = 1.5$ mm thick strip of rectangular cross section. According to a preliminary process design, a 40% reduction in height is taken in a single pass, on a mill with 150-mm-diameter rolls, at $v = 0.8$ m/s speed, with a mineral-oil lubricant ($\mu = 0.07$). (*a*) Check whether the reduction is feasible; if not, take two reductions, then calculate (*b*) the roll force and (*c*) power requirement.

9C-20 The rolling mill of Example 9-21 has a spring constant of 4000 kN/mm. If the rolls are set to kiss at zero load, what will be the gap between them at the roll load developed in the pass?

9C-21 Commercial-purity (1100 Al) aluminum is routinely rolled, in several passes but without annealing, to a total reduction of over 98%. Find (*a*) the uniform strain (ε_u), total elongation (e_f), and reduction in area (*q*) for this material. (*b*) Compare these to the rolling reduction, and (*c*) explain the reasons for the difference.

9C-22 A 70/30 brass sheet is rolled to 06 temper; by definition, this is obtained by 50% reduction. Calculate the TS expected and compare with the value given in Example 8–7. (*Hint:* Since engineering and true stress are not much different in a heavily strain-hardened material, TS can be taken as the flow stress of the material after cold working. Strain in the tension test should be added to the rolling strain.)

9C-23 A 2017 aluminum alloy slab of 200 mm thickness and 800 mm width is hot rolled, at 500°C and 100 m/min, on a mill equipped with 600-mm-diameter work rolls, using an emulsion lubricant that gives $\mu = 0.2$. In a preliminary process design, it is proposed that a reduction of 30 mm be taken in the first pass. (*a*) Make a sketch of the process, to scale. (*b*) Check whether the reduction is feasible; if not, calculate the permissible reduction. (*c*) Calculate the roll force and (*d*) net power requirement for the allowable reduction. State whether there is any chance of developing (*e*) internal defects or (*f*) edge cracking; state why.

9C-24 In the final hot-rolling pass on the slab of Prob. 9C-23, the reduction is from 5.0 mm to 3.5 mm. Temperature has dropped to 400°C. Recalculate the roll force and power requirement.

9C-25 To reduce cooling, rolling speed is increased to 10 m/s in Prob. 9C-24. Find its effect on strain rate, flow stress, roll force, and power.

9C-26 The strip of Prob. 9C-24 is cold rolled, annealed at a gage of 2.0 mm, and then finished by cold rolling to deliver a product with controlled strain hardening. The tandem mill has 3 stands with 300-mm-diameter work rolls. Reductions are from 2.0 to 1.5 to 1.0 to 0.7 mm. Speed is 120 m/min in the first stand and increases in successive stands in proportion to the length increase of the strip. An oil-based lubricant gives a coefficient of friction of 0.05. Set up a spreadsheet to compute for each stand: (*a*) speed, (*b*) flow stress (take into account the progressive strain hardening of the material), (*c*) roll force, (*d*) net power requirement. (*e*) Convert answers into USCS units. (*f*) State whether there is any danger of internal defects. Justify.

9C-27   Explore the effect of roll diameter on roll force and power for the last pass (1.0 to 0.7 mm) of Prob. 9C-26 in rolling on a two-high ($D = 800$ mm), four-high ($D = 300$ mm), and Sendzimir ($D = 30$ mm) mill.

9C-28   A strip is cold-rolled on a tandem mill.

Reductions are in four stands, from 0.6 to 0.45 to 0.3 to 0.2 to 0.14 mm. (*a*) Calculate the engineering and natural strain for each pass. (*b*) Calculate the strain in a single pass from 0.6 to 0.14 mm. (*c*) Add up the engineering strains from (*a*) and compare to the strain calculated in (*b*); repeat for natural strains. (*d*) Are engineering strains additive? (*e*) Are natural strains additive?

9C-29   Assume that in Example 9-11 the maximum allowable die pressure is 70 MPa, and that friction and geometry gives a pressure-multiplying factor of $Q_p = 4$. In designing the process, what press speed is allowable? (Note that the long length of the airfoil section results in deformation in plane strain.)

9C-30   Derive Eqs. (9-38*a*) and (9-38*b*).

9C-31   Continuing Prob. 9C-27, compute roll force if a back tension equal to one-half the entry flow stress and a front tension equal to one-half the exit flow stress are imposed.

FURTHER READING (see also Chap. 8)

ASM Handbook, vol. 14, *Forming and Forging*, ASM International, 1988.

Wick, C., J.T. Benedict, and R.F. Veilleux (eds.): *Tool and Manufacturing Engineers Handbook*, 4th ed., vol. 2: *Forming*, Society of Manufacturing Engineers, 1984.

The Aluminum Extrusion Manual, Aluminum Extruders Council.

Byrer, T.G. (ed.): *Forging Handbook*, Forging Industry Association, 1985.

Davis, J.R. (ed.): *Tool Materials*, ASM International, 1995.

Ginsburg, V.: *High-Quality Steel Rolling*, Dekker, 1993.

Geleji, A.: *Forge Equipment, Rolling Mills and Accessories*, Akademiai Kiado, Budapest, 1967.

Hoffmann, E.G. (ed.): *Fundamentals of Tool Design*, 2d ed., Society of Manufacturing Engineers, 1984.

Lange, K. (ed.): *Handbook of Metal Forming*, McGraw-Hill, 1985 (now published by Society of Manufacturing Engineers).

Laue, K., and H. Stenger: *Extrusion—Processes, Machinery, Tooling*, American Society for Metals, 1981.

Lenard, J.G., M. Pietrzyk, and L. Cser: *Mathematical and Physical Simulation of the Properties of Hot Rolled Products*, Elsevier, 1999.

Magad, E.L., and J.M. Amos: *Tool Materials Management*, Chapman and Hall, 1995.

Open Die Forging Institute: *Open Die Forging Manual*, 3d ed., Forging Industry Association, Cleveland, Ohio, 1982.

Pietrzyk, M., and J.G. Lenard: *Thermal-Mechanical Modelling of the Flat Rolling Process*, Springer, 1991.

Giant transfer press is 24 m long, 10 m wide, and 11.4 m high. Two slides exert a total of 30 MN (3300 tonf) force. Sliding bolsters visible in the foreground allow complete change of the five 2- by 3-m dies in only 5 minutes. Sheet-metal parts are moved from die to die by electronically controlled transfer bars. (*Courtesy Verson Corporation, Chicago, Illinois*).

10

Sheet-Metalworking Processes

*In this chapter we will learn how to convert sheet into an endless variety
of products, in processes such as:*

Cutting parts for further working or immediate assembly

Bending on presses or roll-forming lines

Stretch forming of shapes and its limitations

Deep drawing by pulling material into a die

Combined stretch-drawing to make complex shapes for auto bodies

Because of the low cost of mass-produced sheet of high quality, sheet metalworking has gained an
outstanding position among manufacturing processes. Originally, the semifabricated starting material
was sheet, rolled and supplied in limited sizes. Since the appearance of continuous tandem rolling mills,
sheet has really been produced in coils of wide strip. Coils may be cut up, either in the rolling mill or
in service centers, for easier handling in the facilities of the secondary manufacturer; however, there is
an increasing trend to ship entire coils (sometimes slit into narrower widths) which are then fed into the
presses and press lines of the manufacturer.

10-1 SHEET MATERIALS

All wrought alloys (Sec. 8-3) are suitable for sheet-metalworking applications.
The critical properties are, however, somewhat different from those discussed
for bulk deformation, partly because deformation now occurs mostly in tension
rather than compression, and partly because many sheet parts are large and highly
visible, making appearance a major concern.

10-1-1   Steels

Much steel is used in the hot-rolled condition for automotive wheel rims, axle cases, chassis parts, compressed gas cylinders, etc., but much steel sheet is cold-rolled. Table 10–1 shows the properties of the most frequently used sheet steels.

Table 10–1   Typical properties of steel sheets*

Steel	YS, MPa	TS, MPa	el., %	n Value	r Value
Hot rolled					
Commercial quality	195–280	315–390	28–36		
Drawing quality (DQ)	180–250	295–365	34–42		
DQSK (special killed)	180–270	295–380	36–44		
Cold rolled					
Commercial quality	200–250	280–335	36–41		
DQ	175–225	280–320	38–43		
DQSK	160–200	270–320	40–45		
Interstitial-free (IF)	150	320	46	0.22	
Cold rolled, ASTM A109					
No. 1, hard		550–690	–		
No. 2, half hard		380–520	4–16		
No. 3, quarter hard		310–450	13–27		
No. 4, skin rolled		290–370	24–40		
No. 5, dead soft		260–340	33–45		
Cold rolled, SAE J2329†					
Grade 1	N/A	N/A	N/A	N/A	N/A
Grade 2	140–260	270	34	0.16	N/A
Grade 3	140–205	270	38	0.18	1.5
Grade 4	140–185	270	40	0.20	1.6
Grade 5	110–170	270	42	0.22	1.7
Dual-phase steel, hot rolled	440	630	29		
Cold rolled	570–740	730–1050	18–13		

*From various sources.
†Proposed.

Low-Carbon Steels   Among carbon steels (Sec. 8-3-1) low-carbon steels with up to 0.15% C are used in largest quantities. Capped steels are suitable for structural parts and tubing but not for deep drawing. For the latter purpose, rimmed or killed steels are needed. Maximum ductility is found in annealed material.

1. *Rimmed steel.* The high ductility and relatively low cost of rimmed steels has made the commercial-quality and drawing-quality steels the favorites for less critical applications. The very-low-carbon surface is an advantage in enameling. Grain size is controlled by heavy (50–70%) cold rolling followed by annealing (Fig. 8–10). However, the presence of carbon and nitrogen results in yield-point elongation (Fig. 10–1a) and objectionable stretcher-strain marks. Therefore, the strip is usually given a *temper pass*, i.e., a very light rolling reduction, on the

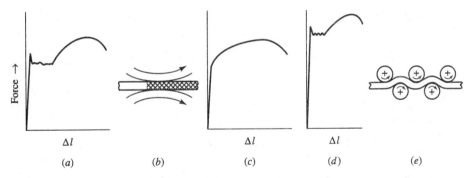

Figure 10-1   The undesirable effects of (a) yield-point elongation (b) may be masked by temper rolling. (c) Yield-point elongation is then absent and Lüders' bands are not visible but (d) strain aging leads to their return. (e) Lüders' lines may then be masked again by roller leveling.

order of 1% or less (Fig. 10–1b). This is less than the yield-point elongation and thus barely affects ductility, but produces very finely spaced Lüders' bands so that on subsequent tensile deformation there is no yield point (Fig. 10–c) and no visible bands appear. However, if the material is stored prior to drawing, strain aging takes place within a few weeks or months (depending on composition and storage temperature), ductility is reduced, and yield-point elongation returns (Fig. 10–1d). *Roller leveling* (Fig. 10–1e) bends the strip repeatedly and helps to disguise Lüders' bands by a mechanism similar to temper rolling, but ductility suffers.

2. *Killed steel.* Drawing-quality special-killed (DQSK) steel has uniform properties and high n and r values, and is specified when severe draws or stretching (such as the oil pan of an auto engine) are to be made or when storage is unavoidable. Killed steel also contains carbon and nitrogen but the nitrogen is combined with aluminum into a compound, and only carbon remains in a form that would allow condensation onto dislocations. Annealed sheet again shows yield-point elongation (Fig. 10–1a) but temper rolling (Fig. 10–1b) permanently eliminates the yield point (Fig. 10–1c); carbon does not diffuse to dislocation sites unless the steel is heated to 120°C. These steels can be processed to high r values.

3. *Interstitial-free steel.* Carbon content is reduced to very low levels and nitrogen is tied down by alloying with small amounts of Nb or Ti. There is no yield-point elongation, the YS is low, and the n value is high.

Since material properties such as YS, TS, el., K, n, and r are relevant for forming performance and mathematical modeling, new steel specifications are based on these values. (Source: J.R. Fekete, *SAE Paper 970715.*)

High-Strength Steels   We already mentioned in Sec. 1-3 the pressures that resulted in reducing the weight of automobiles. Thinner gage must be balanced by higher strength, hence the quest for reduced mass prompted the development of stronger materials. All strengthening mechanisms are utilized:

1. *Cold-rolled sheet.* Strain hardening is the least expensive strengthening mechanism and, if possible, the sheet metalworking process is directed so that the finished part should have considerable cold work imparted to it. If the starting sheet itself is strain-hardened, the remaining ductility (Fig. 8–7 and 8–8*a*) may be too small for all but the lightest deformation.

2. *Partially annealed sheet.* Greater ductility combined with reasonable strength is given by heavy cold rolling followed by recovery anneal (Fig. 8–9).

3. *Annealed sheet.* Grain refinement is a powerful strengthening mechanism (Fig. 6–18). Heavy cold rolling followed by recrystallization (Fig. 8–10) can be applied to all materials.

4. *Solution-hardened steel.* Steels solid-solution hardened with Mn, P, or Si strain-harden more rapidly.

5. *Bake-hardenable steels.* These are usually temper rolled; after forming, they undergo rapid aging during paint baking, resulting in a 30–40 MPa gain in strength due to the condensation of a C atmosphere. Gains from strain aging can be maximized by increasing the nitrogen content.

6. *Dual-phase steels.* Higher carbon contents, needed for quenching and tempering heat treatment (Sec. 6-4-3), are useful in spring steels but would reduce the ductility of deep-drawing steel too much. However, low-carbon steels with 1.4% Mn can be annealed to produce a structure consisting of ferrite strengthened by dispersed martensite. Such dual-phase steels have a low yield strength, which is an advantage when springback is objectionable; at the same time, rapid strain hardening during working imparts a high strength (up to TS = 1000 MPa) in the formed product.

7. *High-strength low-alloy (HSLA) steels.* HSLA steels are increasingly used in vehicles and other structures. An important difference to observe is that their high $\sigma_{0.2}/E$ ratio results in large springback.

Coated Steels   Much sheet metal is formed with preapplied coatings that improve the service properties or appearance of finished parts.

1. *Tinplate.* Tin-coated steel sheet is corrosion-resistant as long as the tin layer is free of scratches, and the nontoxicity of Sn makes tinplate suitable for food containers.

2. *Galvanized sheet.* A zinc coating protects steel by the preferred (sacrificial) corrosion of zinc, thus even a damaged coating protects. In the past used chiefly for roofing, ductwork, and similar low-tech applications, galvanized sheet has become the principal material for auto body and appliance construction (see its growth in Table 8–1). The zinc is applied by passing the strip through a Zn melt (*hot-dip galvanizing*) or by electrodeposition (*electrogalvanizing*). Heating a hot-dip zinc layer to around 500°C for a controlled time converts it into a Fe–Zn alloy (*galvannealing*). There are a number of other coating systems (e.g., Zn–Al coatings) that perform similar functions. The frictional characteristics of coated sheets are substantially different from those of bare steel and lubricants must be chosen to minimize pickup as well as friction. Some sheets are further treated (e.g., galvannealed sheet is prephosphated) to improve their response.

3. *Terne plate.* Lead-coated sheet resists corrosion in some media for which tin or zinc offer no protection but, because of the toxicity of Pb, terne plate is limited to nonfood applications and is increasingly restricted even in those.

4. *Aluminum-coated sheet.* An Al–Fe alloy formed at elevated temperatures protects from corrosion by hot gases, thus the sheet is suitable for heat exchangers, automotive exhaust systems, grill parts, etc.

5. *Prepainted sheet.* Coatings of paints as well as thicker polymeric films (plastics, such as vinyls) offer both protection and a pleasing finish. If the sheet had been properly pretreated and is formed with care, the coatings remain adhered to the surface. The need for finish painting the part is eliminated, the quality of the coatings is often superior to paint finishes applied after forming, and economies can be realized too.

Stainless Steels   High strain-hardening capability and thus excellent formability, combined with corrosion resistance, make austenitic steels the choice for food processing equipment and other heavily deformed products such as kitchen sinks, heat exchangers, and chemical processing equipment. The less-expensive ferritic and martensitic steels are used when the lesser formability and corrosion resistance is acceptable.

10-1-2   Nonferrous Metals

We already mentioned that copper, and especially brasses, are among the most formable materials, and alloying and grain-size control are routinely used. With aluminum alloys the challenge is serrated yielding (especially in the 5000 series Al–Mg alloys) which leads to objectionable surface marking. Precipitation-hardening alloys (6000 series) are free of this problem. They also benefit from thermomechanical processing: When they are worked in the naturally aged (T4) condition, further strengthening takes place during the paint-bake cycle. Aluminum sheet is the main constructional material of subsonic aircraft and has application in automobiles (primarily deck lids and hoods).

10-1-3   Surface Topography

For many applications, the roughness characteristics of sheets are critical. Although it may seem that sheets should be as smooth as possible, in reality tightly controlled upper and lower roughness limits are needed for two reasons: First, roughness helps to hold paint and thus affects the surface appearance of the finished product, but asperities of a too-rough sheet would show through the paint film. Second, a very smooth surface would not entrap enough lubricant; sheet-to-die contact would result in cold welding and tool pickup, and pressed parts would have to be rejected because of scoring of the surface. However, on a very rough surface there would be too few asperities on which the pressworking pressure could be distributed, and controlled flow of the metal into the die could not be achieved. A random orientation of surface features is preferable for appearance and is essential for assuring equal draw-in in all directions. Typical roughness is 1–1.5 μm R_a with 3–6 peaks/mm. The most widespread technique for producing such surfaces is to finish-roll with randomly *shot-blasted rolls* (Fig. 3–23*b*). More recently, rolls are *textured*: Craters are formed in a predetermined pattern by laser beam or electrical discharge.

10-2   CLASSIFICATION

The emphasis in this chapter will be on sheet forming, but some of the basic processes are applicable also to wire, bar, rolled or extruded sections, and tube (Fig. 10–2). Most operations are carried out cold; heating is necessary only for special purposes (Sec. 10-9). More often than not, production of a sheet-metal part involves more than one operation and Fig. 10–2 gives simplified flow paths.

10-3   SHEARING

Irrespective of the size of the part to be produced, the first step involves cutting the sheet or strip into appropriate shapes by the process of *shearing*. The terminology is fairly descriptive: Cutting a sheet along a straight line is simply called *shearing*. Cutting a long strip into narrower widths between rotary blades is referred to as *slitting* (Fig. 10–3*a*), often conducted in service centers where full-width coils from the rolling mill are slit for shipment to sheet-metalworking plants. A contoured part (whether it be circular or more complex in shape) is cut between a punch and die in a press, and the process is called *blanking* (Fig. 10–3*b*). The same process is also used to remove unwanted parts of a sheet, but then one refers to *punching* or *piercing* (Fig. 10–3*b*) a hole, of circular or any other shape. Individual parts are made by *cutoff* (Fig. 10–3*c*) or *parting* (Fig. 10–3*d*). Cutting out a part of the sheet edge is called *notching*, and a partially cut hole, with no material removed, is made by *lancing* (Fig. 10–3*e*). A contoured part may be cut by repeated small cuts in the process of *nibbling*. Drawn products are finished by *trimming* off excess material (Fig. 10–3*f*).

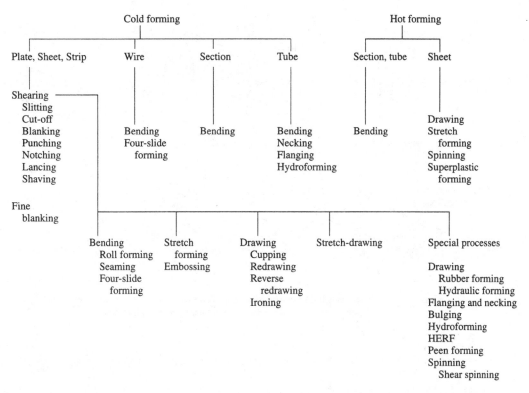

Figure 10–2   General classification of sheet-metalworking processes. (*Adapted from J.A. Schey, ASM Handbook, vol. 20, Materials Selection and Design, ASM International, 1997, p. 691. With permission.*)

10-3-1   The Shearing Process

The process of separating adjacent parts of a sheet through controlled fracture cannot be described as either purely plastic deformation or as machining. The sheet is placed between two edges of the shearing tools—in the instance of blanking, a *punch* and a *die* (Fig. 10–4). The events taking place during the stroke of the press can be followed by recording punch force as a function of stroke (Fig. 10–5) and by inspecting the cut surfaces.

On penetration of the tool edges, the sheet is first pushed into the die, and plastic deformation results in a rounding of the edge of the blank (*roll-over*). Then the blank is pushed into the die by extrusion-like plastic deformation, indicated by the parallel, *burnished zone* on the blank, and characterized by steadily increasing forces. After some critical deformation, cracks are generated at a slight angle to the cutting direction, first usually at the die edge. When these cracks meet, shearing is complete and the cutting force drops (Fig. 10–5) even though the cutting edges had moved only partly through the thickness of the sheet (Fig. 10–4*a*). The *fracture surface* is not perfectly perpendicular to the sheet surface

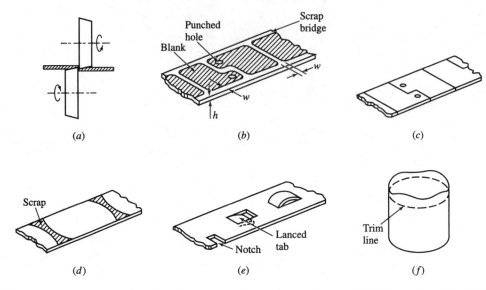

(a)   *(b)*   *(c)*

(d)   *(e)*   *(f)*

Figure 10–3   Processes based on shearing: (a) slitting with rotary knives, (b) blanking and punching, (c) cutoff, (d) parting, (e) notching and lancing, and (f) trimming.

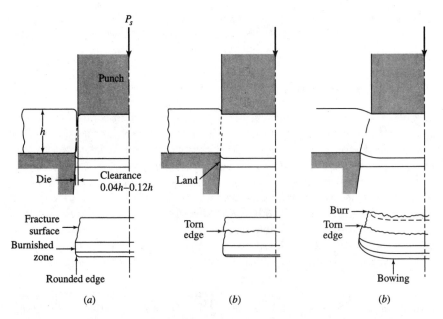

(a)   *(b)*   *(b)*

Figure 10–4   Sheared parts of acceptable finish are produced (a) when blanking is done with optimum clearance. (b) The skirt of torn edge produced with a small clearance and (c) the burr produced with excessive clearance are undesirable.

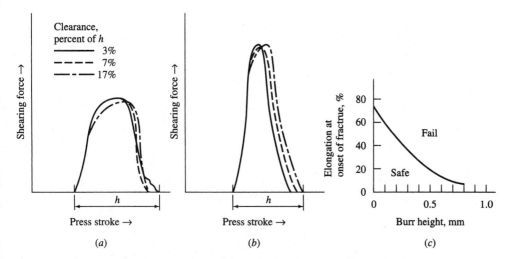

Figure 10–5   Total fracture sets in after the cutting edges penetrate (*a*) more than half the sheet thickness when shearing soft materials but (*b*) earlier on hard materials. (*c*) Burr produced in blanking reduces the elongation attainable in subsequent tensile deformation. [*Part (c) after S. P. Keeler, Machinery **74**:101 (1968).*]

and exhibits some roughness; nevertheless, the finish is acceptable for many applications. The part would hang up in the die and must be pushed by the punch beyond the parallel *die land* (indicated in Fig. 10–4*b*).

The quality of the cut surface is greatly influenced by the *clearance* between the two shearing edges. With a very tight clearance, the cracks—originating from the tool edges—miss each other and the cut is then completed by a secondary tearing process, producing a jagged edge roughly midway in the sheet thickness (Fig. 10–4*b*). Excessive clearance allows extensive plastic deformation, separation is delayed, and a long fin (*burr*) is pulled out at the upper edge (Fig. 10–4*c*).

In the course of shearing thousands of parts, the tool edges wear, become rounded, and burr forms even with an optimum clearance. The jagged edge of the burr with its sharp roots acts as a stress concentrator (Sec. 4-1-6); the harmful effect is evident in the reduced elongation measured in the tension test (Fig. 10–5*c*). Burr initiates fracture during subsequent forming or in the service of the part. Therefore, proper choice of clearance and regular tool maintenance are vital aspects of the process. A small clearance leads to more rapid tool wear, therefore, greatest economy is obtained when the clearance is chosen as large as permissible for the given application. From experience, clearance is taken between 4 and 12% of sheet thickness (a smaller clearance goes with a more ductile material).

10-3-2   Forces

One can readily estimate the size of press required for conventional shearing. Since deformation is concentrated in a very narrow zone with severe strain hard-

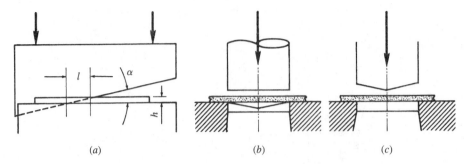

(a)   (b)   (c)

Figure 10–6   Shearing forces can be reduced by giving a rake or shear to (a) the blades in a guillotine, (b) the die in blanking, or (c) the punch in piercing.

ening, the maximum force can be obtained from an empirically determined shearing stress multiplied by the cross section to be cut. Shearing stress decreases with increasing clearance, but average values may be found in handbooks or may be taken as a C_1 fraction of the TS. Thus, the *shearing force P_s* is

$$P_s = C_1\,(\text{TS})\ hl = C_1 K \left(\frac{n}{e}\right)^n hl \tag{10-1}$$

where h is sheet thickness, l is length of cut, and C_1 is 0.85 for ductile materials and 0.65 for less ductile ones (or 0.7 on the average). The TS of most materials is known (as in Tables 8–2 and 8–3). [If only the K and n values are available, the TS may be approximated by substituting $\text{TS} = K\,(n/e)^n$ where e is the base of the natural logarithm.]

When the shearing edges are parallel, l is the entire length of the contour cut. This can lead to very high forces, which can then be reduced by placing the two shearing edges at an angle to each other (at a *shear* or *rake*, as in a guillotine, Fig. 10–6a), thus only the instantaneously sheared length l needs to be considered. In blanking, the *scrap bridge* can be allowed to bend and the rake is on the die (Fig. 10–6b). In punching, the punched-out *scrap* can be bent, and the rake is on the punch (Fig. 10–6c).

The *shearing energy E_s* to be delivered by the press is equal to the area under the force–displacement curve (Fig. 10–5a and b). An approximate value can be obtained from

$$E_s = C_2 P_s h \tag{10-2}$$

where $C_2 = 0.5$ for soft materials (Fig. 10–5a) and 0.35 for hard materials (Fig. 10–5b).

Example 10-1 | Circular blanks of $d_0 = 200$-mm diameter are to be cut from $h = 3$-mm thick, annealed 5052 aluminum alloy. What press force and energy are needed?

From Table 8–3, TS = 190 MPa.
From Eq. (10-1), $P_s = 0.85(190)(3)(200\pi) = 304$ kN.
From Eq. (10-2), $E_s = 0.5(304)(3) = 457$ N·m.

Example 10-2

Mild steel plate of 5 mm thickness and 2 m width is cut in the width direction. Estimate the shearing force for cutting (*a*) with parallel blades and (*b*) in a guillotine with 6° shear.

From Table 8–2, for 1015 steel, TS = 450 MPa.

(*a*) The length to be cut $l = 2$ m; $P_S = 0.85(450)(0.005)(2) = 3830$ kN.

(*b*) From the geometry of the operation (Fig. 10–6*a*), $l = h/\tan\alpha = 5/0.105 = 47.6$ mm. Hence $P_S = 0.85(450)(5)(47.6) = 91$ kN. This value is approximate but shows the large benefit to be gained.

10-3-3   Improving the Quality of Cut

There is great demand for processes that produce clean-cut edges, perpendicular to the sheet surface and of a surface finish sufficiently smooth to allow immediate use of the parts, e.g., as gears in lightly loaded machinery and close-tolerance, contacting members in instruments. Several approaches are possible; in most of them, a *counterpunch* cooperates with the main punch and, as an additional benefit, eliminates curvature of the part.

1. We saw that fracture can be delayed by the imposition of a high hydrostatic pressure (Fig. 4–7). This principle is exploited in *precision blanking* or *fine blanking* (Fig. 10–7*a*). A specially shaped blankholder (*V ring, impingement ring*) is pressed into the part just prior to beginning the cut; thus, the deformation zone is kept in compression and the whole thickness is plastically sheared.

2. A high hydrostatic pressure is also maintained on *shearing with a negative clearance*, and the part is actually pushed (extruded) through the cutting die (Fig. 10–7*b*).

3. In two-sided shearing (*counterblanking*) the sheet is clamped between two dies (Fig. 10–7*c*). The punches penetrate in one direction until cracks are initiated, and then the cut is completed in the other direction.

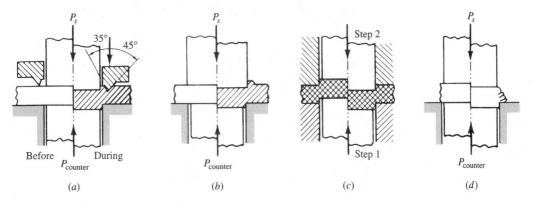

Figure 10–7   Parts with finished edges can be produced by (*a*) precision blanking, (*b*) negative-clearance blanking, (*c*) counterblanking, or (*d*) shaving a previously sheared part.

4. A conventionally blanked part may be *finish-shaved* in a die set with tight clearances (Fig. 10–7*d*). This is equivalent to cutting with a tool of zero rake angle (Sec. 16-1-1).

5. The quality of cut greatly improves in *high-speed cutting* when cutting velocities exceed the dislocation propagation velocity in the metal. This requires very high speeds, on the order of 30 m/s.

10-3-4   Processes

The punch and die are made of tool steel or, for longest runs, of sintered WC (Table 9–3). The scrap bridge (*skeleton* or, in punching, the part) would bind onto the punch (or stick in the die) and must be stripped with fixed, spring-supported, or cam-driven stripper plates (Fig. 10–8) or with a plastic (usually polyurethane) foam pad. Process choice is governed primarily by product characteristics and quantity:

1. Holes of standard sizes and shapes can be cut on general-purpose *punch presses* with interchangeable tooling. Die change is speeded up with the rotating

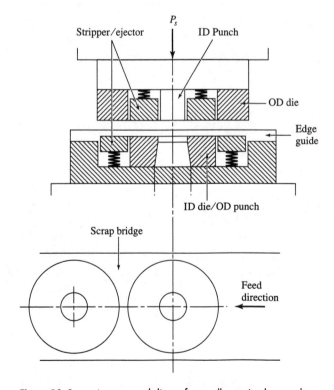

Figure 10–8   A compound die performs all cuts simultaneously.

tool changers of *turret presses*. Numerically controlled presses (*CNC punching machines*), equipped with an *x-y* table and tool magazines, allow rapid and accurate location of the sheet and selection of the punch and die, and thus permit low-cost, flexible production of small and medium quantities. Larger holes can be made by repeated cutting with the same punch or by nibbling.

2. Complex geometries can be created in *compound dies* in which several cutting edges work simultaneously (Fig. 10–8).

3. In *progressive dies* several punching and blanking operations are sequentially performed with die elements fastened to common die plates, while the strip is fed in exact increments (*indexed*). Often, blanking and punching are among many processing steps taken in progressive dies producing complex parts (Fig. 10–9). Productivity is high, limited only by the rate of feeding material into the press and by the rate of stroking the press (some presses operate at several hundred strokes per minute). Multiple punches are used when many parts or holes are to be produced, as in the blanking of circles for can making or the punching of holes

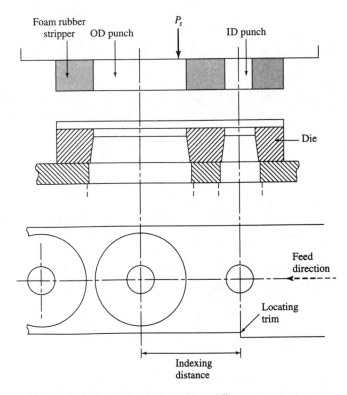

Figure 10–9   A progressive die performs different cuts in successive stations (die elements for cutting the locating trim are not shown).

for perforated metals. *Lamination* dies are used to blank sheet for transformers and motors.

4. Highest production rates are obtained in *roll piercing*, with the die and punch located on the surfaces of rolls.

5. For smaller quantities of, say, a few hundred pieces, the die cost can be lowered if a greater scrap loss is tolerable. In *rubber pad blanking* the die is simply a steel plate cut to size, and the cutting action occurs by pressing the sheet around this die with a rubber cushion (Fig. 10–10*a*). The overhanging part of the sheet is bent down and clamped against the base plate by the cushion, and tearing occurs around the edges of the die plate.

6. For short runs, an inexpensive solution is the *steel rule die*. In this, the "die" is made of bevel-edge strips of high-carbon or tool steel, tightly pressed into slots in plywood and backed by a steel plate (Fig. 10–10*b*). The punch (*die plate, template*) is made of steel when cutting metal.

7. *Punching cells* contain one or two punching presses or punching centers and some kind of transfer mechanism. Sheet loading, tool change, part transfer between machines, and unloading of the part are coordinated by flexible automation.

8. The basic shearing process can be adapted for wire, bar, sections, and tubing in preparation for further work. Quality of cut is improved by providing support for the cut piece and also by applying compressive stress (hydrostatic pressure) during the cut.

9. There has been rapid development of cutting methods based on welding techniques (laser beam, electron beam, plasma torch, electric arc, or oxyfuel

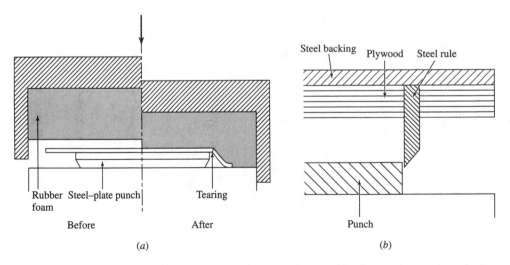

Figure 10–10   Low-cost blanking and punching is possible with a (a) rubber foam cushion or (b) steel-rule die.

cutting, Sec. 18-7-3) and waterjet cutting. Combined with x-y tables (and sometimes turret presses), such computer-controlled *cutting centers* become extremely versatile and productive.

10-4   BENDING

Many parts are further shaped by *bending* in one or several places.

10-4-1   The Bending Process

Characteristic of this process is stretching (tensile elongation) imposed on the outer surface and compression on the inner surface (Fig. 10–11). For a given sheet thickness h, tensile and compressive strains increase with decreasing forming radius R_b (i.e., with *decreasing R_b/h ratio*). For the part to retain its shape, the R_b/h ratio must be small enough to bring much of the sheet cross section into the state of plastic flow. There is, as in elastic bending (Sec. 4-1-7), only one line (the *neutral line*) which retains its original length.

When bending with relatively generous radii, the neutral line is in the center. When bending around tight radii, the neutral line shifts toward the compressive side, the centerline is elongated, and constancy of volume is preserved by the thinning of the sheet. The increased length of the centerline is usually taken into account for bends of $R_b < 2h$ by assuming that the neutral line is located at one-third of the sheet thickness. When the sheet is relatively narrow ($w/h < 8$), there is also a contraction in width w.

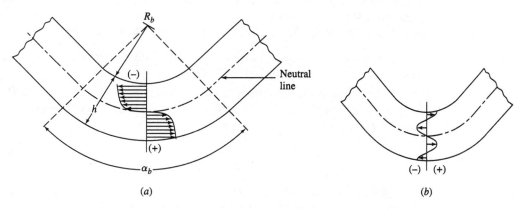

Figure 10–11   In the course of bending (a) the entire stress-strain curve is traversed; (b) elastic stresses result in springback and retention of a residual stress pattern.

Example 10-3 | The part shown is to be made of 3-mm-thick sheet. Find the length of strip.

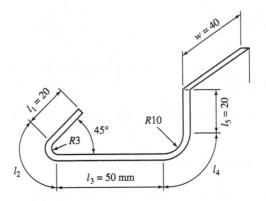

Figure Ex. 10-3

In bending to $R_b = 10$-mm radius, $R_d/h = 3.3$, hence the neutral plane will be in the center of the sheet; since the bend is over a 90° angle, $l_4 = 2\pi(R_b + 0.5h)90/360 = 2\pi(10 + 1.5)90/360 = 18.06$ mm.

In bending to $R_b = 3$ mm, $R_b/h = 1$, hence the neutral line is at $0.33h$; for a bend of $180 - 45 = 135°$ angle, $l_2 = 2\pi[3 + (0.33)(3)]135/360 = 9.4$ mm.

Thus the total starting length is $l = l_1 + l_2 + l_3 + l_4 + l_5 = 20 + 9.4 + 50 + 18.06 + 20 = 117.5$ mm. (If we had ignored the shift of the neutral line, l would have been 118.7 mm.)

10-4-2   Bending Limits

Corresponding to the limits discussed in Sec. 8-2-7, the *minimum bend radius* (the smallest permissible die radius R_b or, more generally, the *minimum radius-to-thickness ratio R_b/h*), can be defined according to several criteria.

1. *Orange peel* may be esthetically undesirable but is not a defect since it can be remedied by choosing a finer-grain material.

2. *Localized necking* causes a structural weakening of the bent part. Necking occurs when elongation in the outer fiber, e_t, exceeds the uniform elongation of the material e_u in the tension test

$$e_t = \frac{1}{(2R_b/h) + 1} \le e_u \qquad \textbf{(10-3a)}$$

For materials that obey the power law of strain hardening, Eq. (8-4), $\varepsilon_u = n$ and the engineering uniform strain e_u may be obtained from

$$e_u = (\exp n) - 1 \qquad \textbf{(10-3b)}$$

The relationship holds best for steels; for most other materials, the actual e_u measured in the tension test should be used. Because the strain is redistributed to adjacent zones during bending, a somewhat higher strain is usually permissible. A burr acts as a stress raiser and, if on the outer (tensile) surface, leads to much earlier fracture. Therefore, if possible at all, *the burr is oriented toward the punch.*

3. *Fracture* represents an absolute limit. This is directly related to the reduction in area q measured in the tension test [Eq. (4-10), and Tables 8–2 and 8–3)]. The minimum permissible bend radius may be estimated for less-ductile materials from the following formula

$$R_b = h \left(\frac{1}{2q} \right) - 1 \quad \text{for } q < 0.2 \quad \text{(10-4a)}$$

and for ductile materials, because of the shift of the neutral radius in tight bends, from

$$R_b = h \frac{(1-q)^2}{2q - q^2} \quad \text{for } q > 0.2 \quad \text{(10-4b)}$$

A material of $q > 0.5$ can usually be bent on itself (zero bend radius).

4. *Crushing* on the inside surface may occur in bending to very tight radii.

Anisotropy, of any origin, affects bending. We have seen that mechanical fibering (Sec. 8-1-7) results in greater ductility in the rolling direction, and it is usually more favorable to bend sheet with the bend line oriented across the rolling direction. A textured material of low r value thins down easily (Fig. 8–6a) and thus can be bent around tighter radii than a material of high r value.

Example 10-4

The part of Example 10-3 was originally made of annealed cartridge brass. It is now proposed that, as a weight-saving measure, it should be made of 5052-H34 aluminum alloy. Is there any problem to be expected?

The relevant properties for the two materials, from Table 8–2 and *Metals Handbook Desk Edition*, p. 6.33 and 6.35, are:

	Brass	5052-H34	5052-0	6061-T4
YS, MPa	100	215	90	145
TS, MPa	310	260	195	240
el., %	65	10	25	22
q, %	75			

The yield strength is perfectly adequate. The tightest bend is $R_b/h = 3/3 = 1$; from Eq. (10-3), uniform elongation of $1/(2 + 1) = 33\%$ would be desirable. Since the total elongation of 5052-H34 is only 10%, uniform elongation must be even less and the material will fail in the bend. A soft 5052-0 sheet with 25% total elongation would perhaps survive because of strain redistribution, but the YS is slightly low. Aluminum alloy 6061-T4 would do better, although the bend radius may have to be relaxed.

10-4-3   Stresses and Springback

The stress state is extremely complex in bending. The complete tensile and compressive stress–strain curves of the material are traversed on the tensile and compressive sides of the bend, respectively. This means that around the neutral plane the stresses must be elastic. When the forming tool is retracted, the moment developed by the elastic components of the stress causes *springback*, and a residual stress pattern, shown in Fig. 10–11*b*, develops. Springback elastically recovers some of the total deformation; thus, it increases both the angle and radius of the bent part (Fig. 10–12). The elastic zone is more extensive for a relatively gentle bend (large R_b/h ratio) and for a material with a high ratio of yield strength $\sigma_{0.2}$ to elastic modulus E; therefore springback also changes according to the approximate formula

$$\frac{R_b}{R_f} = 1 - 3 \left(\frac{R_b}{h} \frac{\sigma_{0.2}}{E} \right) + 4 \left(\frac{R_b}{h} \frac{\sigma_{0.2}}{E} \right)^3 \qquad \textbf{(10-5)}$$

where R_b is the radius of the bending die and R_f is the radius obtained after the forming pressure is released.

Since the length of the neutral line does not change, the angle after springback, α_f, can be obtained (in radians) from

$$\alpha_f \left(R_f + \frac{h}{2} \right) = \alpha_b \left(R_b + \frac{h}{2} \right) \qquad \textbf{(10-6)}$$

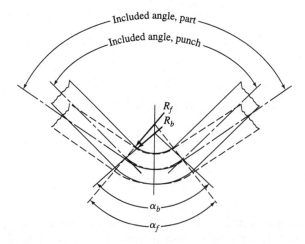

Figure 10–12   Dimensions used to characterize springback.

Springback establishes a new force equilibrium with a residual stress distribution typified by a compressive stress on the outer and tensile stress on the inner surface (Fig. 10–11*b*).

Several techniques are used to combat springback.

1. If springback for a given material is known and if the material is of uniform quality and thickness, compensation for springback is possible by *overbending* (Fig. 10–13*a* and *b*). This is the most basic form of bending, in which no compressive pressure is imposed in the thickness direction of the sheet (*air bending*).

2. The elastic zone can be eliminated at the end of the stroke by one of two means. First, the two ends of the sheet may be clamped before the punch bottoms out, so that the end of the stroke involves *stretching* of the part, causing tensile yielding in the entire sheet thickness. In the second method the punch nose is shaped to *indent* the sheet, so that plastic compression takes place throughout the thickness (Fig. 10–13*c*).

3. If a *counterpunch* is used with a controlled pressure, compressive stresses are maintained in the bend zone during the entire process (Fig. 10–13*d*). Since this also has the effect of imposing a hydrostatic pressure on the bend zone, bending beyond the limits given by Eq. (10-4) is possible.

4. Less-ductile materials may have to be bent at some elevated temperature; because the yield strength is lower, springback is also less.

Bending Force   A simple estimate of the *bending force* in free bending to 90° may be obtained from

$$P_b = \frac{wh^2 \, (\text{TS})}{W_b} \tag{10-7}$$

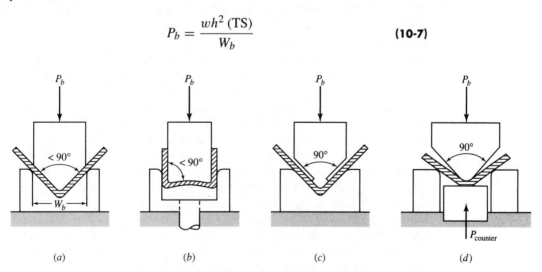

Figure 10–13   Springback may be neutralized or eliminated by: (*a*), (*b*) overbending, (*c*) plastic deformation at the end of the stroke, and (*d*) subjecting the bend zone to compression during bending. [*Part (d) after V. Cupka, T. Nakagawa, and H. Tyamoto, CIRP,* **22**:73–74 (1973).]

where W_b is the width of the die opening (Fig. 10–13a) and w is the width of the strip (the length of the line over which bending takes place).

Example 10-5

How much is the springback in making the 90° bend of the part of Example 10-3 if the workpiece material is (a) annealed cartridge brass or (b) 6061-T4 aluminum alloy. From Table 5–2, E(brass) $= 140$ GPa, E(Al) $= 70$ GPa. $R_b/h = 3.3$, and $\sigma_{0.2}$ is taken from Example 10-4.

   Thus, for the brass, $R_b/R_f = 1 - 0.0071 + 0.0 = 0.9929$; for the Al alloy, $R_b/R_f = 1 - 0.0205 + 0 = 0.9795$. Springback is negligible with the brass but not with the aluminum alloy.

Example 10-6

Estimate the force required for making the 90° bend in the part of Example 10-3, assuming that the workpiece material is brass.

   From Table 8–3, TS $= 310$ MPa. The minimum die opening must accommodate l_4 plus some straight length (say, twice 10 mm). Thus $W_b = 18 + 20 = 38$ mm. The width of the part is $w = 40$ mm; $h = 3$ mm. From Eq. (10-7), $P_b = (40)(3)^2(310)/38 = 2937$ N. It is usual to allow some 20% more, thus the force is 3.5 kN.

10-4-4   Bending Methods

The equipment used for bending depends on the size, mostly length, of the bent part.

   1. *Mechanical presses* can bend short lengths at high rates in dies, as in Fig. 10–13.

   2. *Press brakes* are special presses with very long beds. In these, simple tooling is used to form complex shapes by repeated bending of a long sheet (Figs. 10–14 and 10–16a). Tooling costs are reduced when a polyurethane foam slab replaces the female die. The advantage of press brakes is that a great variety of parts can be produced with a limited number of tools. Elastic deflection of the tooling (opening up in the middle under the press force) would result in variations of bend angle, therefore, the tool must be made to bend closer in the center. This is achieved by shimming or, in modern press brakes, by mechanical or hydraulic flexing, often on the basis of computed bending forces. In conjunction with mechanized sheet feeders, the press lends itself to computer control, including a back gage for sensing sheet position. Compensation for springback is made with the help of empirical tables or Eq. (10-6). More sophisticated control schemes take material properties and gage variations into account. In one scheme, the angle of bend is measured in the first stroke; the force is then released to obtain springback, and a compensating second stroke is made. In other schemes, the elastic–plastic stress–strain curve is derived from information obtained by force and displacement transducers, and a control algorithm computes the required overbend.

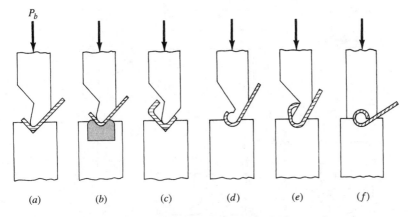

Figure 10–14   Press-brake forming of (a) a 90° angle; (b) the same but with a polyurethane female die; (c) a U channel; and (d)–(f) a bead.

3. *Wiping* is an alternative method for bending along a straight line (Figs. 10–15a and 10–16b). To estimate the bending force, W_b may be taken as $(2R+h)$.

4. *Three-roll benders* impart a uniform but adjustable curvature to sheet, plate, or section by forming with rolls arranged in a pyramidal fashion (Fig. 10–15b). This is an important preparation step for making large rings and welded-plate structures.

5. *Roll forming* is a highly productive continuous production method. Bending is now done progressively, by passing the strip between contoured, driven

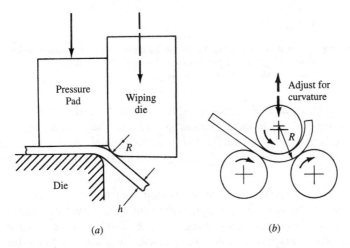

Figure 10–15   Sheet may be bent also with (a) a wiping die or (b) bending rolls (pyramidal rolls).

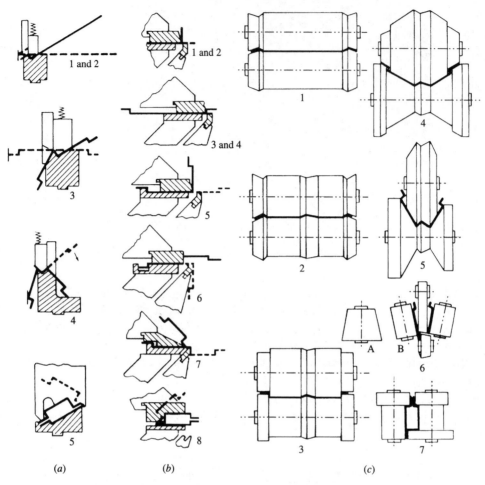

(a)   *(b)*   *(c)*

Figure 10–16   Complex profiles such as door frames may be formed by a sequence of operations on (a) press brakes, (b) wiping dies, or (c) by profile rolling. *(From G. Oehler, Biegen, Carl Hanser Verlag, Munich and Vienna, 1963. With permission.)*

rolls placed in tandem. A typical product is corrugated sheet. For many other shapes, idling rollers are used to press the sides of the partially formed shape. Thus, tubes for subsequent welding, sections that replace hot-rolled or extruded sections, as well as complex shapes (such as door frames) can be formed (Fig. 10–16*c*).

6. Bending of sections and tubes is an important manufacturing activity. The problem in free bending (e.g., in pyramidal benders) is usually that of distortion and buckling of more complex shapes. Better results are obtained when the profile or tube is wrapped around a *form block.* Conformance to the form block is ensured by *winding under tension,* by passing a *wiper roll* or *wiper block*— hinged at the center of the radius of curvature—around the section or tube (Fig.

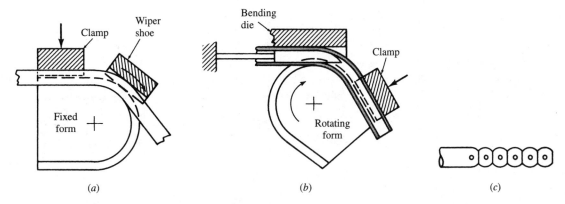

Figure 10–17   Tubes and sections may be formed by (*a*) compression bending or (*b*) draw bending, sometimes with the use of a (*c*) linked-ball mandrel.

10–17*a*), or by a *rotating form block* (Fig. 10–17*b*). To prevent the collapse of tubes in bending over tight radii, several methods are available. The inside is supported with sand, a low-melting-point metal or, more economically, by a mandrel made up of individual sections (Fig. 10–17*c*), or the tube is drawn over a fixed mandrel (Fig. 10–17*b*). CNC bending machines can be programmed to make tubes with several bends in different orientations, as are required for auto and aircraft hydraulic systems and automobile exhausts.

10-5   STRETCH FORMING

Enormous quantities of sheet metal are formed into more or less deep, container-like components of a great variety of shapes. In contrast to most bent parts, they are characterized by curvatures in two directions (they are 3-D shapes). They can be produced by stretch forming, deep drawing, or their combinations.

10-5-1   Stretch Forming Processes

In pure *stretch forming* the sheet is completely clamped on its circumference and the shape is developed entirely at the expense of sheet thickness. Physically this can be achieved in a variety of ways:

1. The sheet may be clamped with a multitude of fixed or swiveling clamps (Fig. 10–18*a*). The advantage is that only one die (*male die* or *form punch*) is needed, but productivity is low; hence, such stretch forming is most suitable for low-volume production as is typical of the aircraft industry. Very large parts

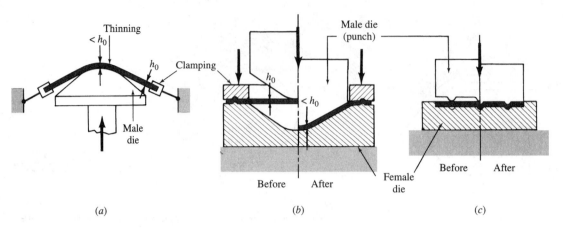

Figure 10–18   The shape is developed entirely at the expense of wall thickness in stretch forming with (a) clamps or (b) lock-bead dies and in (c) embossing.

(fuselage skins, wing skins, boat hulls) can be formed. Springback can be substantial in forming gently curved shapes, and then forming at elevated temperatures (sometimes by allowing creep or superplastic deformation over the die) is helpful. Rolled and extruded sections may also be stretch formed.

2. For mass production, such as is typical of the automotive and appliance industries, the blank is clamped with an independently movable *blankholder* which clamps the sheet with the aid of *locking draw beads* (Fig. 10–18*b*); the punch cooperates with the *female die* to define the shape. One part is finished for each press stroke; thus, productivity is high but die costs are higher too.

3. In the process of *embossing* (Fig. 10–18*c*) the restraint is given by the sheet itself, through the multiple contact points with the die.

10-5-2   Stretch Formability

The first limit is reached in stretching when a localized neck becomes visible, and the ultimate limit is given by subsequent fracture. The *formability limit* is a technological property and the limit strain depends on the material, the strain state, and friction on the punch surface.

The influencing factors are clearly shown when a clamped sheet is stretched by a hemispherical punch (Fig. 10–19*a*). Localized strain variations (the *strain distribution*) can be revealed simply by applying a grid of small (typically, 2–6-mm diameter) circles (or a *square/circle grid*) onto the sheet surface, usually by electrolytic etching or a photoresist technique. In the course of straining, thinning of the material is accompanied by a growth of the circles, as required by constancy of volume [Eq. (4-2)]. When deformation is the same in all directions,

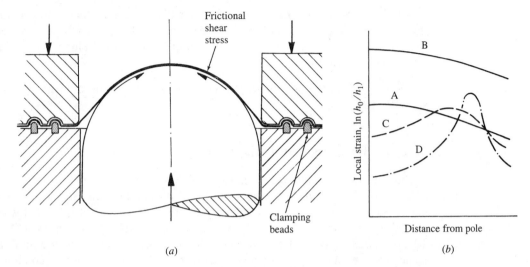

Frictional
shear
stress

Clamping
beads

Local strain, $\ln(h_0/h_1)$

Distance from pole

B

A

C

D

(a)   (b)

Figure 10–19   (a) Friction on the punch surface opposes thinning and leads to (b) changes in strain distribution from pole to flange. (*Adapted from J.A. Schey, Tribology in Metalworking: Friction, Lubrication and Wear, ASM International, 1983, p. 520. With permission.*)

as it would be on blowing up a balloon (*balanced biaxial strain*), the circle expands into a circle of larger diameter. When deformation is different in different directions, the circle distorts into an ellipse: the major axis gives the major strain and, perpendicular to it, the minor axis gives the minor strain.

When a sheet of a given material is stretched over the hemispherical punch, strain distribution depends on a number of factors:

1. In the total absence of friction (which could in reality be achieved only in bulging with hydraulic pressure), the sheet thins out gradually toward the apex where fracture finally occurs (line A in Fig. 10–19b). Thinning is more uniformly distributed with a material of high n value, and a deeper dome can be obtained before necking becomes localized (line B). It will be recalled that in the tension test necking occurs at $\varepsilon_u = n$ (Sec. 8-1-1). In balanced biaxial tension the presence of the transverse strain prevents the formation of a localized neck (*diffuse necking*). Straining can continue until a local neck develops at or close to the apex, at some point where there is some inhomogeneity in the material or the sheet was originally thinner. In stretching over a punch, the depth of stretch never reaches that obtained in frictionless, hydraulic bulging but, as long as friction is very low, failure still occurs at the apex.

2. Friction on the punch surface hinders free thinning at the apex, and with increasing friction the position of maximum strain moves toward the die radius (lines C and D in Fig. 10–19b). Strain becomes more localized and fracture occurs in plane strain, close to or at the punch-to-sheet contact.

3. Under otherwise identical conditions, thicker sheet gives a deeper stretch, because bending superimposed on stretching improves ductility.

4. The depth of stretch increases with all material variables that delay necking (high n value, transformations) or increase the postnecking strain (high m value). Indeed, a good empirical correlation is found between total elongation in the tension test and limiting dome height.

Strain distribution is important because it determines the properties of the stretched part.

10-5-3   Forming Limit Diagram

A comparison of various materials is possible with the aid of the *forming limit diagram* (FLD). Gridded sheet-metal strips of different widths are tested with a very good lubricant (e.g., oiled polyethylene film) on the punch. A sheet wide enough to be clamped all around gives the balanced biaxial tension point (Fig. 10–20). As the strip width diminishes, the minor strain decreases too until, at some characteristic strip width, it becomes zero. By definition, this is a condition of plane strain (Fig. 8–15). The FLD is usually constructed for localized necking (another curve could be constructed for fracture). The FLD moves higher for thicker sheet and is, obviously, lower for a material of lower ductility (lower n value).

The FLD is a system characteristic and the FLDs of two materials can be compared only if determined under identical conditions. With decreasing circle size, the FLD moves higher and changes its shape, because more of a small circle falls into the necked zone where strain is high. The quality of production lots of sheet can be more quickly checked by conducting *limiting dome height* (LDH) tests. Samples of different widths are again tested and the minimum height obtained at plane strain is quoted in millimeters.

The FLD was introduced in the 1960s and quickly became an important tool in diagnosing production problems. When parts are found to fail in production, gridded sheets are placed into the production die and are stretched. Distortion of the circles is measured (sometimes with the aid of an instrument called an *optical grid analyzer*). The circle nearest to the fracture line gives the strain ratio at the critical point and defines, say, point A in Fig. 10–20. Several remedies, some of them not intuitively evident, may then be explored to bring strains within allowable limits:

1. *Increase the minor strain* by clamping more firmly in that direction.

2. If fracture occurred away from the apex, *improve lubrication* to redistribute strains (as in Fig. 10–19b);

3. If all else fails, the part will have to be *redesigned* to reduce the major strain, or some material must be allowed to flow into the die, changing the process into combined stretch-drawing (Sec. 10-7).

The FLD of Fig. 10–20 is typical of steel and some aluminum alloys. Some other materials, such as austenitic stainless steels and brass, show no improvement

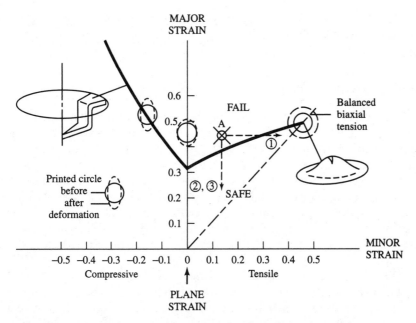

MAJOR
STRAIN

FAIL

Balanced
biaxial
tension

Printed circle
before ——
after ——
deformation

SAFE

Compressive

Tensile

MINOR
STRAIN

PLANE
STRAIN

Figure 10–20   The forming limit diagram typical of low-carbon steel gives the
permissible deformations at various strain ratios. (*Left-hand side
typically after G.M. Goodwin, SAE Paper 680092, 1968; right-hand
side after S.P. Keeler, SAE Paper 650535, 1965.*)

in the forming limit with increasing biaxiality of strains. Nevertheless, some of
them have a high n value (Tables 8–2 and 8–3) and are then eminently suitable
for stretching.

Example 10-7

Complex shapes such as automotive hub caps are often made of low-carbon steel (chromium
plated), stainless steel, or 5052 Al alloy. If the caps are to be made by stretching, which of
these alloys allows the deepest stretch?

A first approximation can be obtained by comparing the n values. From Tables 8–2 and
8–3, austenitic stainless steels are best ($n = 0.3$), followed by low-carbon steel (0.25), 5052 Al
(0.13), and martensitic stainless steel (0.1). If deep details are to be made in aluminum alloy
or martensitic stainless steel, draw-in of the metal must be encouraged (see Sec. 10-7).

10-6   DEEP DRAWING

The difference between stretching and deep drawing is substantial: In the former,
the blank is clamped and depth attained at the expense of sheet thickness; in

the latter, the blank is allowed—and even encouraged—to draw into the die, and thickness is nominally unchanged.

10-6-1   Drawing Processes

In the simplest case of pure *deep drawing* or *cupping*, a circular blank of diameter d_0 is converted into a flat-bottomed cup by drawing it through a draw die with the aid of a punch of diameter D_p (Fig. 10–21). Both the die and punch must have well-rounded edges (*die* and *punch radii*), otherwise the blank might be sheared. The finished cup is *stripped* from the punch—for example, by machining a slight recess (a ledge) into the underside of the draw die. After the cup has been pushed through the die, its top edge springs out because of springback, gets caught in the ledge on the return stroke of the punch, and the ledge strips the cup. A central hole is often provided in the punch to prevent development of a vacuum and thus aid stripping.

The stress state prevailing in the part *during* drawing is shown in Fig. 10–22a about halfway through the draw. The base is in balanced biaxial tension; the side wall is in plane-strain tension because the punch does not allow circumferential contraction; material in the transition between wall and flange is subjected to bending and rebending (straightening out); and the flange is in radial tension and circumferential compression, because the circumference of the blank is reduced while it is forced to conform to the smaller diameter of the die opening.

The circumferential compressive stresses cause the blank to thicken, and the punch-to-die clearance is usually some 10% larger than the sheet thickness to accommodate this *thickening* without the need for reducing (*ironing*) the wall. Compression can also lead to *wrinkling* (equivalent to buckling in upsetting, Fig. 9–15a) in the flange. In practice, two methods of operation are feasible:

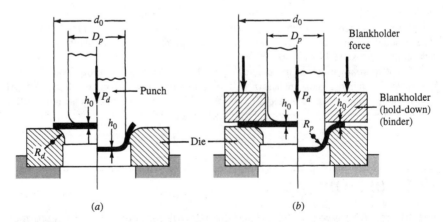

(a)   *(b)*

Figure 10–21   Containers may be formed by drawing (a) without or (b) with a blankholder.

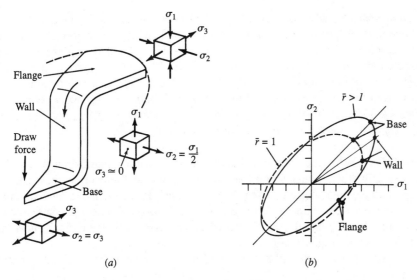

Figure 10–22   (a) The stress state varies greatly in different parts of a partly drawn cup. (b) A material of higher r value benefits from strengthening the base and wall.

Drawing without a Blankholder   Wrinkling can be avoided when the sheet is sufficiently stiff (Fig. 10–21a). This is always the case for very shallow draws, when the drawing ratio $d_0/D_p < 1.2$. Blanks that are thick relative to blank diameter allow higher drawing ratios (Fig. 10–23); wrinkling depends also on the die profile, which determines the rate of circumferential compression. Most favorable is the tractrix die (Fig. 10–23).

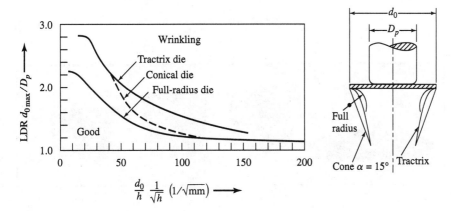

Figure 10–23   The limiting draw ratio in cupping low-carbon steel without a blankholder is a function of die geometry and of blank diameter–to–sheet thickness ratio. (After G.S.A. Shawki, Werkstattstechnik, **53**:12–16, 1963. With permission of Springer Verlag, New York.)

Drawing with a Blankholder   When the blank is relatively thin, and the draw ratio is beyond the limits indicated in Fig. 10–23, the flange must be constrained with a blankholder (also called *hold-down*, Fig. 10–21*b*). The blankholder must exert sufficient pressure to prevent wrinkling (Fig. 10–24*b*), but excessive *blankholder pressure* would restrict free movement of the material in the draw ring and cause fracture in the partly formed cup wall (Fig. 10–24*d*). To produce a sound cup (Fig. 10–24*c*), the blankholder pressure may be taken, as a first approximation, as 1.5% of the yield strength ($\sigma_{0.2}$) of the material.

Draw Force   When the optimum blankholder pressure is applied, draw force rises as the partly drawn flange strain-hardens; as the flange diameter decreases, force drops until the thickened edge of the blank is ironed (Fig. 10–25, line A). Excessive pressure causes early fracture (line B and Fig. 10–24*d*). Too low pressure allows wrinkling (line C) and, if the wrinkles cannot be ironed out, the cup fails near the end of the draw (Fig. 10–24*b*).

A very approximate estimate of the *drawing force* may be obtained from the formula

$$P_d = \pi D_p h \ (\text{TS}) \left(\frac{d_0}{D_p} - 0.7 \right) \tag{10-8}$$

10-6-2   Limiting Draw Ratio

When the draw force exceeds the force that the cup wall can support, the partly drawn cup fractures (Fig. 10–24*d*). There is thus a limit to the attainable deformation, expressed as *reduction* $(d_0 - D_p)/d_0$ or, more commonly, as *drawing ratio* d_0/D_p. The maximum diameter of the circle that can be drawn under ideal conditions is expressed as the *limiting draw ratio* (LDR)

$$\text{LDR} = \frac{d_{0 \ (\text{max})}}{D_p} \tag{10-9}$$

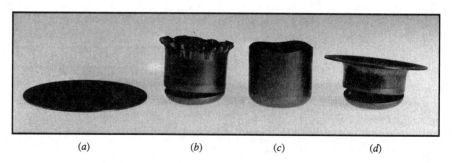

(*a*)	(*b*)	(*c*)	(*d*)

Figure 10–24   Deep drawing of low-carbon steel cups from (*a*) a round blank, with (*b*) insufficient, (*c*) optimum, and (*d*) excessive blankholder pressure. Note in (*c*) the typical earing due to planar anisotropy. (*From J.A. Schey, as Fig. 10–19, p. 527.*)

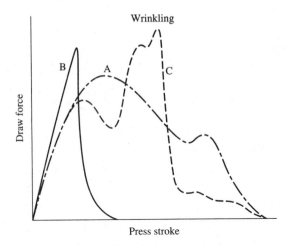

Figure 10–25   Draw force curves typical of drawing with optimum (line A), excessive (line B), and insufficient (line C) blankholder pressure.

We have seen that the draw force is composed of the forces required to compress the sheet in the flange circumferentially, overcome friction between blank and blankholder and die surfaces, bend and unbend the sheet around the draw radius, and overcome friction around the draw radius. Therefore, the LDR is not a material constant but a system property that depends on all variables that affect the draw force and the strength of the cup wall.

1. A high n strengthens the cup wall but also increases the draw force, hence it is fairly neutral; a slight improvement in LDR is often found with higher n because of a later development of the force maximum.

2. A high m strengthens an incipient neck in the wall while barely affecting the draw force, thus it is slightly positive in its effect.

3. The most powerful material variable is the r value. We saw in Fig. 8–6b that a material of high r value resists thinning while voluntarily reducing its width. This helps the blank to conform to the reduced diameter of the cup and is thus a positive factor. Furthermore, a high r value causes the yield ellipse (Fig. 8–14) to expand in the balanced biaxial direction (Fig. 10–22b). The partly drawn cup wall is in plane-strain tension, in which a high r-value material is stronger, whereas the flange is subjected to combined tension and compression, in which it is slightly weaker than an isotropic one. The combined result is that the LDR increases with increasing r (or more precisely, $\bar{r}$) value (Fig. 10–26). The effect is more powerful than it appears from Fig. 10–26, because an LDR of 2.0 gives a cup of approximately $0.8D_p$ depth, whereas an LDR of 3.0 gives a depth over $2.0D_p$.

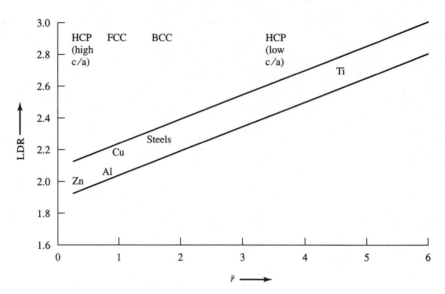

Figure 10–26   High normal anisotropy is a powerful factor in increasing the limiting draw ratio.

4. Tight punch and part radii impose severe bending strain and thus increase the draw force yet without affecting the strength of the wall; therefore, they decrease the LDR. However, very large radii would leave much of the blank unsupported and *puckering* (wrinkling between punch and die) could occur. Hence radii are optimized, usually within the limits of $R > 4h$ for thick (> 5 mm) and $R > 8h$ for thin (< 1 mm) sheet.

5. Friction between blankholder, die, and flange surfaces adds to the draw force and is thus harmful. Contact pressures are below σ_f and, therefore, Eq. (4-18) holds. The friction stress can be reduced by reducing the normal stress (the blankholder pressure), but this is limited by wrinkling. Therefore, a good lubricant must be applied that reduces μ and thus the friction force.

6. In drawing relatively thin sheet, of d_0/h ratios over 50, the frictional force becomes a larger part of the total drawing force; hence, the LDR drops with increasing d_0/h ratio.

7. Friction on the punch is helpful because it transfers the draw force from the cup to the punch. Thus a rough punch, or a blank that is lubricated only over the flange area, gives a higher LDR.

There is still no international standard for LDR determination, and only data obtained under identical conditions are comparable.

The LDR does not necessarily give the usable cup depth. A material with planar anisotropy (Sec. 8-1-3) shows different properties (Fig. 8–5b) in the rolling, transverse, and 45° directions ($r_0 \neq r_{90} \neq r_{45}$). This leads to *earing*, a periodic variation of cup height (Fig. 10–24c); the ears reflect the crystal symmetry and come in pairs (4, 6, or 8). The flange thickens less in the higher-r direction; thus, ears form in these directions.

A low-carbon steel blank of 200-mm diameter and 2-mm thickness is to be drawn into a cylindrical cup of 100-mm internal diameter. The bottom radius is 5 mm. (*a*) Check whether the draw is feasible and, if it is, (*b*) estimate the press force. | **Example 10-8**

(a) From Fig. 10–28, LDR = 2.4, hence the draw is feasible.

(b) From Table 8–2, for 1008 steel, TS = 320 MPa. From Eq. (10-8),

$$P_d = \pi(100)(2)(320[(200/100) - 0.7] = 261 \text{ kN}$$

Drawing-quality aluminum-killed steel of $r = 1.7$ has an LDR of 2.4 (Fig. 10–28). A cylindrical cup is drawn from sheet of $h = 2$-mm thickness with a punch of diameter $D_p = 100$ mm and nose radius $R_p = 5$ mm. Find (*a*) the maximum blank diameter $d_{0(\text{max})}$, (*b*) the depth of cup, assuming a constant wall thickness of 2 mm, and (*c*) the height-to-diameter ratio. | **Example 10-9**

(*a*) From Eq. (10-9), $= \text{LDR} \times D_p = 2.4(100) = 240$ mm.

(*b*) Since wall thickness remains unchanged and the cup is relatively thin, constant volume equals constant area, and we can equate the area of the starting blank with the area in midplane of the cup. Area of the radius = length of quarter circle × length swept by center of gravity (for a quarter circle, the center of gravity is at 0.6*R*). Thus,

$$\text{Area of blank} = (\text{area of base}) + (\text{radius area}) + (\text{wall area})$$
$$240^2 \pi/4 = [(100 - 10)^2 \pi/4] + [(6\pi/2)(97.2\pi)] \times [(100 + 2)\pi h]$$

wall height $h = 112.3$mm; internal depth of cup $= 112.3 + 5 = 117.3$ mm

(*c*) $h/D_p = 117.3/100 = 1.17$.

Repeat the calculation for LDR = 2; note the effect of LDR on height-to-diameter ratio. | **Example 10-10**

(*a*) $d_{0(\text{max})} = 200$ mm.

(*b*) internal depth of cup $= 69 + 5 = 74$ mm.

(*c*) $h/D_p = 0.74$.

There is usually some reduction in wall thickness, hence the oft-quoted ratio of $h/D_p = 0.8$. The effect is quite marked: For an LDR increase of 20%, cup height increased by 58%.

10-6-3   Further Drawing

Cups of a depth greater than permitted by the LDR can be made by further forming after initial cupping.

1. *Redrawing* (Fig. 10–27*a*) leaves the wall thickness essentially unchanged.

2. *Ironing* (Fig. 10–27*b*) leaves the inner diameter virtually unchanged and achieves greater depth by reducing the wall thickness. It will be recognized that ironing is similar to drawing a tube on a bar (Fig. 9–39*d*).

3. A basic phenomenon, not mentioned hitherto, is that a cold-worked material exhibits greater ductility when the deformation direction is reversed in successive operations (*strain softening*); this is exploited in the *reverse redrawing* of cups (Fig. 10–27*c*).

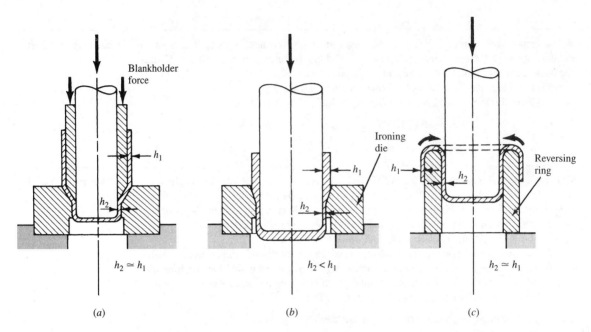

Figure 10–27   Cups are further deformed by (a) redrawing, (b) ironing, or (c) reverse redrawing.

Redrawing is extensively used for food containers, fountain-pen caps, oil-filter housings, shock-absorber pistons, etc. Ironing is used in the mass production of drawn-and-ironed beverage cans and ammunition cartridges.

There is, of course, wide opportunity, but often combined with greater difficulty, to change the basic shape of the drawn part. In drawing square or rectangular containers the degree of difficulty increases with increasing part depth–to–corner radius ratio; earing in the corners is helpful. A punch with a curved or hemispherical end imposes a different, combined deformation state, to be discussed next.

10-7   STRETCH-DRAWING

In many practical applications, most notably in the production of automotive body and chassis parts, the drawing process is neither pure stretching nor pure drawing. The sheet is not entirely clamped (therefore it is not pure stretching), neither is it allowed to draw in entirely freely (thus it is not pure drawing). Instead, the complex shapes are developed by controlling the draw-in of the sheet, retarding it where necessary with *draw beads* inserted into the die and blankholder surfaces (Figs. 10–18*b* and 10–19*a*). To prevent die pickup and regulate draw-in, a lubricant is applied and the roughness and directionality of sheet surface is specified (sheet with random finish is widely used). In some cases, blankholder pressure is varied in a programmed manner during the stroke or the blankholder is subjected to a pulsating load. In the most advanced systems the blankholder is actuated by several independently programmable hydraulic cylinders so that the retarding force can be locally controlled.

The shape of the part is often represented by a "sculptured" surface, one that can be described only with cubic patches or point-by-point in spatial coordinates. The application of CAD/CAM to such shapes has greatly reduced the time and effort involved in the design and analysis of parts and in programming CNC machine tools for making the dies. Curvatures can be gentle and nonsymmetrical, resulting in problems of springback and distortion after release from the die, especially with materials of high $\sigma_{0.2}/E$ ratio, and computer modeling can help in defining a die shape that compensates for this. In other instances, forming is taken close to the limits allowed by the material, and fracture could easily occur in the absence of tight controls. Computer modeling can allow exploration of the effect of process variables.

Forming Limits   In the last few years, there has been a remarkably swift acceptance of formability concepts for production control purposes. The forming limit diagram is useful for analyzing the causes of failures, as discussed in connection with Fig. 10–20. Ellipses next to the fracture location give the critical position on the FLD. Then various corrective measures can be taken: The minor strain may be increased (line 1 in Fig. 10–20) by increasing the restraint of the sheet in that direction (by inserting a draw bead or increasing the number of draw beads); the major strain may be reduced (vertical arrow in Fig. 10–20) by reducing the depth of stretch or by allowing more material to draw in (by reducing the number of or completely eliminating draw beads); localized thinning in a deep part of the drawing can be reduced by increasing friction on that part of the male tool.

Shape Analysis   The overall severity of the operation is better judged by *shape analysis*, which takes the contributions of both stretching and drawing into account. For this, combined stretch-draw charts (*forming lines*) are determined in the laboratory. For one end point, the LDR is determined. The other end point is found in pure stretching, by pressing a steel ball into a clamped sheet until a localized neck is observed. The ball is of 20-mm diameter for sheet of 1.5-mm and lesser thickness, and of 50-mm diameter for up to 3.5-mm sheet thickness. The *stretching limit* SL is the height of stretch h_s divided by the diameter of the die D_s. The stretch-draw limit is obtained by connecting these two end points (Fig. 10–28).

The really important application is, of course, to the prediction of success or failure before the expensive tools are built, so that modifications can be made in time. Progress has been made in predicting the relative contributions of draw and stretch; analyses of new part designs are often made by reference to similar parts for which experience exists. In the most advanced applications, the database established by CAD is used for preliminary analysis. This is a rapidly developing field which can already boast some successes and should, ultimately, allow true concurrent engineering before a production die is finalized. Until then, it is still

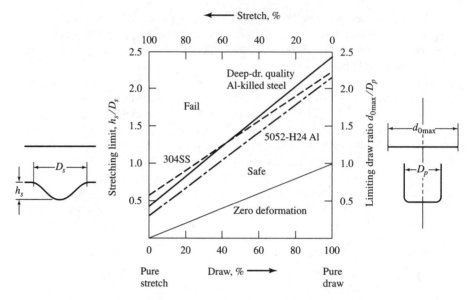

Figure 10–28   Combined stretch-draw limit diagram used for judging the severity of combined operations. [*After A.S. Kasper, Metal Progress, **99**(5):57–60 (1971). With permission of ASM International.*]

necessary to do *die tryout*, using dies made of a less expensive material, usually a zinc alloy which can be cast to shape. The magnitude of challenge can be sensed from Fig. 10–29, showing a part of complex shape. It is important to note that springback can reach significant proportions and modeling is helpful in computing the necessary correction with the die shape.

Dent Resistance   An important consideration is the dent resistance of larger panels. When a local indenter such as a sphere is applied, the panel first flexes and then a dent (an indentation) forms. Events depend in a complex manner on material properties, mode of load application, and the thickness and shape of the panel. In static loading, a flat panel will simply deflect; a gently curving panel may indent or snap through (reverse its curvature), while sharply curved panels support the load through membrane compression. In dynamic loading (as in hailstorms), stiffer panels are unable to absorb the impact energy by elastic deflection; contact forces become large and denting results. Thus, sharply curved panels—which perform well under static loading—are more susceptible to dynamic denting.

Tailored Blanks   In a recent development, the blank is welded from several sheets of different thicknesses or different compositions, so that strength is provided where needed. Such *tailored blanks* (Fig. 10–30) allow reduction in the mass and number of parts.

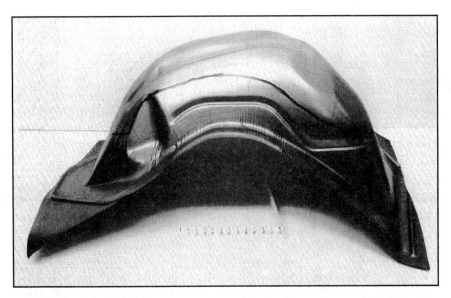

Figure 10–29   A severely formed part (an automotive wheel housing) made in a single operation by a combination of stretching and drawing. Note the etched circle grid, the draw beads, and the line marking the boundary between the stretched nose and the drawn-in sides. (*Courtesy A.S. Kasper, Chrysler Corporation, Detroit.*)

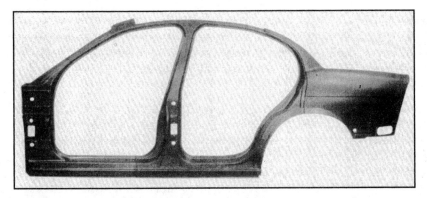

Figure 10–30   The body side outer panel of the UltraLight Steel Auto Body was produced from a blank in which sheets of different gage and mechanical properties were joined by laser welding, so as to provide strength where needed, yet without extra weight. (*Courtesy American Iron and Steel Institute, Southfield, Michigan.*)

Example 10-11

It is estimated that even for cars of current design, 3 to 5 welded blanks will be used per vehicle. At 50–60 million vehicles per year, the worldwide demand will grow into hundreds of millions of blanks.

10-8 PRESS FORMING

In industrial usage, the term *press forming* or *pressworking* serves to describe all sheet-metalworking operations performed on power presses with the use of permanent (steel) dies. It incorporates all steps required to complete a part of any complexity, whether they be blanking, punching, bending, drawing, stretching, ironing, redrawing, embossing, flanging, trimming, and so forth. The die sets used depend on production quantities, required production rates, and the number of operations necessary to complete the part.

Single-Operation Dies   Minimum complexity is gained if each operation is performed separately, in individual dies and presses. Die costs still add up and labor and handling costs can be high. Nevertheless, this is the most usual option when total production quantities are insufficient to justify more complex dies, or when the part is very large. This latter situation prevails in the production of automotive body parts. For increased productivity, presses are lined up behind each other, and the part is moved from press to press with mechanical arms, programmable robots, or specialized transfer mechanisms. In-process inventory is reduced and greater flexibility is secured if *quick-die-change* schemes are adopted [even very large dies are changed in minutes, allowing several die changes in each shift for just-in-time (JIT) production].

Compound Dies   Two or more operations performed in a single die (Fig. 10–8) assure greatest accuracy, but dies are limited to relatively simple processes such as blanking, punching, flanging, perhaps combined with bending or a single draw. Special (telescoping) dies are made for multiple draws.

Multistation Dies   Many parts are of a geometry that cannot be directly formed, either because the depth-to-diameter ratio is too large or because the shape has steps, conical portions, etc., requiring several successive draws for which a compound die is often inadequate. The part can still be made in a single press if all die elements needed to complete the part are incorporated within one die set, so that one finished part is obtained for each press stroke. Coil stock is fed at preset increments, and parts are transferred by one of two techniques:

1. *Progressive dies* are fed with strip; the blank is only partially cut, remaining attached with connecting tabs to the remnant of the strip, and this skeleton is used to move the part through the forming stages, with the final separation reserved for the last stage (Figs. 10–9 and 10–31).

2. *Transfer dies* are constructed on the same principle, but the blank is cut out first and the scrap bridge is chopped up and disposed of. The blank is moved through successive stages of the die with indexing transfer mechanisms, usually in a straight line, but sometimes along a circular path.

Figure 10–31   A typical example of progressive die work: forming of two seat-frame parts at a time, by a sequence of blanking, flanging, piercing, flattening of flange, and, in the final stage, cutting off and bending. (*Courtesy General Seating Products Division, Lear-Siegler Industries Ltd., Kitchener, Ontario.*)

Presses for both progressive and transfer dies have to be large enough to accommodate all die stages on the press bed and to provide the force for all simultaneous operations. Very high die costs are counterbalanced in mass production by low labor costs and high production rates. In some special cases, very large parts (the entire side wall panel of an automobile) are formed in a die progression from blanks up to 4200 × 2000-mm size. Such huge transfer presses were introduced in the mid-1990s, with a total press force of up to 80 MN. All stations, with separate slides, are driven from a common shaft, and the part is transferred with an indexing mechanism.

Four-Slide and Multislide Machines   These special-purpose machines were originally developed for complex wire-bending operations but are now increasingly used also for sheet metalworking. A great variety of shapes can be produced at high rates.

10-9   SPECIAL PROCESSES

There are many processes that defy simple classification yet share some features of processes already discussed.

Drawing   Specialized drawing processes are designed to give greater depth of draw, more complex shapes, lower die costs, or a combination of any of these features.

   1. *Rubber forming* replaces the die with a rubber cushion (Fig. 10–32) and the sheet is draped over the punch (often made of a resin or a zinc alloy). There is no need for more expensive, mating steel dies, and the presence of compressive stresses and friction on the punch surface help to obtain deeper draws and shapes that are otherwise difficult to make (e.g., conical parts).

   2. *Hydroforming* replaces the rubber cushion with a fluid contained by a rubber diaphragm (Fig. 10–33a). Hydraulic pressure is programmed throughout the stroke, often with CNC, to press the sheet onto the punch and thus obtain parts of great depth and complexity. Alternatively, the die space is closed by seals and the sheet is directly deformed with the fluid, thus all friction is eliminated and the part strain-hardens uniformly (Fig. 10–19b, line B). The prebulged sheet can then be formed around the die (Fig. 10–33b)

Flanging, Seaming, and Necking   Some complex forms of bending—combined with elements of compressive forming and/or stretching—are encountered in working the edges of blanks, holes, tubes, and drawn parts.

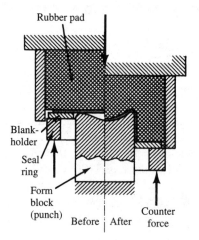

Figure 10–32   Deep parts may be drawn with relatively inexpensive tooling using a rubber pad.

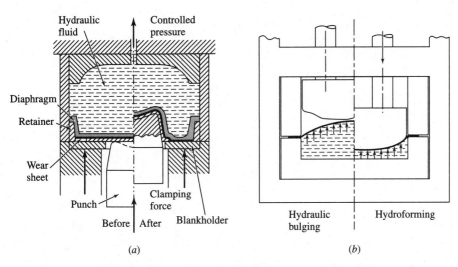

Figure 10–33   Hydraulic forming methods offer capability for (a) deep draws and (b) uniform strain hardening by prebulging a sheet.

1. *Flanging of a blank* (and *shrink flanging* a sheet) puts the outer edge in compression (Fig. 10–34a). It is similar to a shallow deep-drawing operation and sets no great demand on ductility but wrinkling may occur.

2. *Flanging of a hole* and *stretch flanging* of a sheet (Fig. 10–34b) impose severe tensile strains on the edge. If burr is present on the cut edge or if the sheet material contains inclusions or other defects, splitting occurs at a much lower strain than would be expected from the tensile elongation measured in the absence of a burr (for the effect of burr on ductility, see Fig. 10–5c). In critical cases, deburring, shaving, and even reaming of the hole may become necessary.

3. Severe tensile strain is imposed also in the *expansion* or *flanging* the ends of a tube or drawn cup (Fig. 10–34c). In contrast, the *necking* of a tube or cup (Fig. 10–34d) imposes compressive stresses; the reduction that can be taken in a single operation is limited only by the axial collapse of the tube or by the formation of internal wrinkles. Necking is an important step in making cartridge cases and pressurized-gas cylinders.

4. *Seaming* is an important assembly process. A previously flanged part is joined to another part by continuing deformation (Fig. 10–35a and b), as in joining (*hemming*) the inner and outer trunk lid, hood, or door of automobiles. Examples of flanging a sheet and flanging the end of a tube (container) are encountered in forming double seams for sealing food and beverage cans (Fig. 10–35c and d).

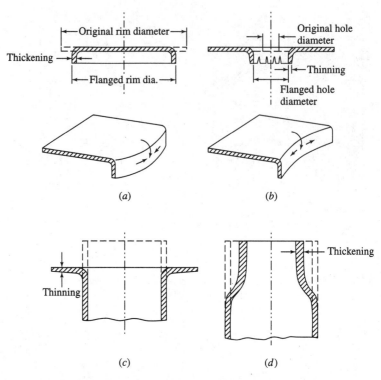

Figure 10–34   Deformation is (a) compressive in flanging a disk or shrink flanging a plate but (b) tensile in flanging a hole or stretch flanging a plate (cracks shown are a consequence of excessive tensile strain). Strain is (c) tensile in flanging a tube but (d) compressive in necking.

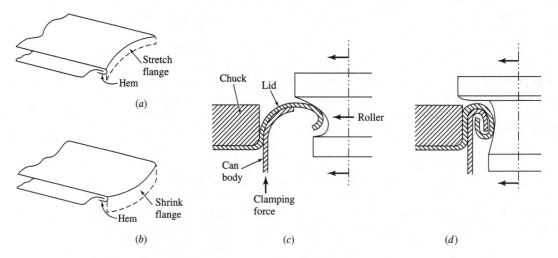

Figure 10–35   Two sheet-metal parts are joined by hemming; on curved edges stretch (a) or shrink (b) flanges are folded over. Lids are attached to can bodies by double lock seams formed in two operations (c) and (d).

Example 10-12

Beverage cans are examples of developments for a mass market. In Sec. 5-4-1 we mentioned that in North America alone some 110 billion (110×10^9) aluminum cans are produced annually. The can has to withstand internal pressure, support the load when stacked, resist corrosion, and do all this at minimum cost. Years of development have resulted in designs and manufacturing methods that meet these goals. The can is a sophisticated pressure vessel with a domed base and a thin but very strong wall (reducing the wall by only 0.0025 mm saves about 136 g/1000 cans, and this saves a 15 000 Mg/year aluminum refinery). To attain the necessary strength, the starting material is heavily strain-hardened by cold rolling. Circles of typically 140-mm diameter are blanked from 0.28-mm thick 3004-H19 strip (YS = 280 MPa, TS = 300 MPa, el. = 5%). Such hard material will not take very high draw ratios; hence, in the first draw with blankholder (Fig. 10–21b), a cup of 85–92-mm diameter is made. The cup is then transferred to the "body maker" where it is redrawn (Fig. 10–27a) to 66-mm diameter and—in the same stroke—the wall is reduced to 0.10-mm thickness in three ironing rings (Fig. 10–27b), with reductions of 20–25% in the first and second rings and 40% in the third ring. Body makers operate at some 400 strokes/min. Recent trends are for even higher rates, necessitating shorter strokes, so that only two ironing rings with 40% reduction each are used. After trimming (Fig. 10–3 f) the body is washed, decorated on the outside, baked, coated on the inside with a polymer, baked again, necked (Fig. 10–34d) to about 55-mm diameter, flanged (Fig. 10–34c), inspected, and shipped to the filling plant where the lid is attached (Fig. 10–35). The necked configuration allows a smaller and thus cheaper and stronger lid. (*Data courtesy of G.L. Smith, Alcoa.*)

Bulging   Tensile deformation is typical of the *bulging* of tubes, containers, and similar products, using rubber (polyurethane foam) plugs or hydraulic pressure (Fig. 10–36a). The technique also represents the first step in making metal bellows (Fig. 10–36b); the prebulged tube forms the bellows when axially compressed.

Hydroforming   In an increasingly important group of processes, seamless or welded tubes are further deformed by high-pressure fluid. The processes are based on the recognition that very large deformations are possible if axial compressive stresses are applied simultaneously with the expanding pressure. The tube, constrained in a split die, is compressed between two punches while a pressurized fluid is applied internally. Originally the process was applied to parts such as copper T fittings (Fig. 10–36c) and a deep bulge, necessary for the T shape,

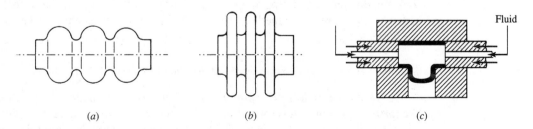

(a)   (b)   (c)

Figure 10–36   Tube expansion can be (a) the first step in (b) making bellows or (c) it can be used to make complex structural components.

was formed without danger of fracture. Recently the technique has become a mass-production process, primarily for auto construction, replacing welding in fabrication of assemblies. The tube, bent to the general shape of the part, is placed into a split die and is expanded against the die cavity. Contact with the cavity wall arrests thinning at that point, and further shape development takes place at the expense of localized reduction in thickness. Better thickness distribution is obtained if expansion is first conducted at low pressure while the end plugs are pressed in, so as to make the material flow in a combined compressive-tensile stress state (point 5 in Fig. 8–14*b*). Pressure is then raised to fill details. If required, holes can also be punched at this point.

Example 10-13

For maximum strength at lowest mass, the roof panel of the ULSAB (Example 5-1) was hydroformed (as in Fig. 10–33*b*) to benefit from uniform stain hardening. To provide a load path for structural performance and crash energy management, the side roof rail was hydroformed from a tube.

High-Energy-Rate Forming (HERF)   These processes use a single (male or female) die. There is no press; the energy required for deformation is derived from various sources. In *explosive forming*, an explosive mat is placed over the sheet; in *electromagnetic forming*, a magnetic field is applied by discharging a capacitor bank through a coil surrounding the part; in *electrohydraulic forming* a pressure shock is created in water by discharging a capacitor bank through a spark gap or a through a wire which evaporates. Pressure application is sudden but the rate at which the material deforms is usually not much higher than in a fast mechanical press. Of the many possible applications, drawing in of necks and internal expansion of tubular and container-like parts is frequently encountered. The latter serves as an alternative to expansion with a rubber plug or hydraulic fluid, and can be used for field repair of condensers and similar tube/header structures.

Peen Forming   We saw in Fig. 4–19*b* that unbalanced internal stresses cause distortion of the part. As shown in Sec. 9-2-2, the principle is exploited in *peen forming* by the judicious shot peening of one of the surfaces. With steel shot of 2–6-mm diameter impacting at speeds of 60 m/s, gently curved surfaces such as aircraft wing skins are shaped and shape defects in products such as rocket cases are corrected.

Spinning   Parts of axial symmetry are produced by several variants:

1. In the basic form of *spinning*, a circular blank is held against a male die (*form*) which in turn is rotated by some mechanism similar to a lathe spindle. Shaped tools are pressed, by hand, tracer mechanism, or under NC control against the blank, so that the metal is gradually laid up against the surface of the form (Fig. 10–37a). Wall thickness remains more or less unchanged.

2. In the process of *shear spinning* (also called *power spinning*, *flow turning*, or *spin forging*), the diameter of the workpiece remains constant and the shape is developed by thinning the wall (Fig. 10–37b). The maximum reduction obtainable is limited by the ductility of the material and correlates well with reduction of area in the tension test. At $q > 50$, a reduction of 80% can be achieved. Very large thick-walled shapes are spun hot.

3. *Tube spinning* is a form of power spinning in which the wall thickness of a tube or vessel is reduced (Fig. 10–37c).

Hot Working   Materials are worked at elevated temperatures for one of three reasons:

1. Thick plate, sections, and tubing are heated to reduce forming forces.

2. Some metals can be deformed only at elevated temperatures, thus, beryllium is formed at 540–820°C, magnesium alloys at 150–400°C, and titanium and its alloys are heated to 480–790°C unless deformation is very slight.

3. *Superplastic forming* allows the manufacture of complex shapes by techniques borrowed from the thermoforming of plastics (Sec. 14-4-2). Alternatively, the part is formed isothermally between heated dies. Practical applications include forming of aluminum alloys (such as fine-grained 7475 alloy: 5.6Zn-2Mg-1.5Cu-0.2Cr, at 520°C) and most titanium alloys (particularly, Ti-6Al-4V which is superplastic at 840–870°C even without special preparation), primarily for aircraft applications and also for prototyping.

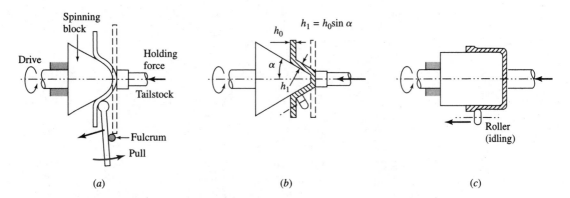

Figure 10–37   Parts of axial symmetry are made by (a) spinning, (b) shear spinning, and (c) tube spinning.

10-10   SHEET-METALWORKING DIES AND EQUIPMENT

Tool materials are chosen mostly on the basis of the expected size of the production run. Blanking tools are subjected to severe wear and are made from the various cold-working die steels (Table 9–3). Bending and drawing dies are made of similar materials, although cast iron and even hard zinc alloys or plastics are suitable for short production runs or softer workpiece materials.

In contrast to bulk deformation, die pressures seldom limit sheet-metalworking processes. The problem is, more likely, that of finding an economical die material and die-making method. Surface coating (Chap. 19) of tools exposed to severe wear is gaining in popularity, and lubricants are always chosen to give control of the process as well as reduce die wear.

Dies—their design, manufacture, maintenance, and modification—represent a substantial part of production costs. CAD/CAM techniques and computer modeling of processes minimize the design and tryout effort and allow faster response at lower cost, especially for the design of progressive dies and dies of complex (sculptured) configurations.

Apart from special-purpose equipment, most press forming makes use of mechanically driven and, increasingly, hydraulic presses. Suitable clutches permit operation of mechanical presses in single strokes (initiated by the operator) or continuously, at rates of 30–600 strokes per minute. The principle of construction is similar to that of presses used in bulk deformation (Fig. 9–36 and Table 9–4) but special features and, for the same rated tonnage, much larger beds make them more adaptable to the working of sheet metal.

Smaller presses often have inclinable press frames which facilitate removal of the stamped part by gravity. Larger presses may have two or even three independently movable rams, one moving inside the other. Such double- and triple-acting presses provide built-in facilities for blank holding or clamping and for ejection, and allow more complex operations. Spring-, air-, or hydraulic-powered *cushions* provide blankholder pressure on single-acting presses and add flexibility to the operation. Mechanical or robotic feeding and part removal speed up production. Die change and alignment are time-consuming but can be greatly speeded up by quick-die-change techniques, moving prealigned dies in and out of the press through side or front openings in the press frame.

10-11   PROCESS CAPABILITIES AND DESIGN ASPECTS

Sheet metalworking processes are highly versatile, but some shape imitations must be recognized (Table 10–2). Dimensions span a very wide range, from miniature electronic components to 4-m-long auto body outer side pressings and 25-m-long creep-formed or peen-formed wing skins for the Boeing 747 aircraft. Tolerances can be very tight and many processes make net-shape parts. The design of parts must take specific limitations into account.

Table 10–2   General characteristics of sheet-metalworking processes

Characteristics	Forming process						
	Blanking	**Bending**	**Stretching**	**Deep Drawing**	**Stretch-Drawing**	**Rubber Forming**	**Spinning**
Part							
Sheet metal	All	All	All	All	All	All	All
Shape*	F0-2, T7	R3; B3; S0, 3, 7; SS; T3; F3, 6	F4; S7	T4; F4, 7	F4; S7	As blanking, bending, drawing	T1, 2, 4, 6; F4, 5
Max. thickness, mm	10	100	2	5	2	2	25
Cost[†]							
Equipment	B–D	C–E	B–C	A–C	A–C	A–C	B–D
Die	C–E	B–E	A–C	A–B	A–C	C–D	B–D
Labor	C–E	B–E	B–E	C–E	B–E	A–D	B–C
Finishing	D–E	D–E	C–E	D–E	C–E	B–D	D–E
Production							
Operator skill[†]	D–E	B–E	B–E	D–E	B–E	C–E	A–C
Lead time	Days	Hours–days	Days–months	Weeks–months	Days–months	Days	Days
Rates (piece/h)	10^2–10^5	10–10^4	10–10^2	10–10^4	10–10^4	10–10^2	10–10^2
Min. quantity	10^2–10^4	1–10^4	10–10^5	10^3–10^5	10^3–10^5	10–10^2	1–10^2

*From Fig. 3–1.
[†]Comparative ratings, with A indicating the highest value of the variable, E the lowest. For example, deep drawing involves high equipment cost, high die cost, low to very low labor cost, low to very low finishing cost, and low to very low operator skill. It can be used for medium to high production rates and requires a minimum quantity of 1000 to 100 000 to justify the cost of die.

In blanking, the scrap bridge (Fig. 10–3b) represents material loss. The minimum width of the bridge is limited by the danger of pulling the bridge material into the die clearance and is typically $w = 2h$ (Fig. 10–3b) but can be reduced to $w = h$ with high blankholder pressure and thicker, stiffer sheet. Hole diameters can be seldom less than sheet thickness and must be up to $2h$ in harder materials. The shape of parts should allow economical nesting (Fig. 10–3b) or even scrapless cutoff (Fig. 10–3c). Material utilization can be optimized by proper *layout and nesting* of parts, an art which is considerably aided by computer programs. Productivity is further increased and material losses cut with multirow blanking from wider strip.

The minimum radius of bent parts is chosen to avoid fracture and, if appearance or finish strength requires, necking (Sec. 10-4-2). The maximum radius is reached when there is no plastic deformation. Springback increases with increasing R_b/h ratio (Sec. 10-4-3) and process design must provide for compensation. If bending is combined with stretching, very large R_b/h ratios are possible, provided that friction is low enough to assure sliding over the die.

Stretch-formed parts suffer thinning by the nature of the process. This is actually of benefit when the material strain-hardens, and thinning is an important means of increasing the dent resistance of a part. Sliding on the punch surface and thus strain hardening can be encouraged by avoiding locking features in part design and by applying a low-friction lubricant in the process. Deeper stretch can be attained with a part design that imposes transverse stresses.

The simplest deep-drawn shapes (flat-bottom cylindrical cups) with a bottom radius of $5h$–$10h$ are most favorable. Thinner walls are easily obtained by ironing. Stepped cups can be readily drawn by successive redrawing. Conical shapes are more difficult; a stepped cup can be made into a cone but die lines will show. Alternative processes such as hydroforming and spinning should be considered. More complex shapes require several operations, but the expense may be justified if a multipiece assembly can be replaced with a single part.

Parts of rectangular or irregular shape can be drawn or stretch-drawn. In general, tight die corners and deep local details make manufacturing difficult but not impossible, as exemplified by auto engine oil pans, spare tire wells in the trunk, and many other pressings of an auto body. Minor (and functionally insignificant) adjustments to the shape of the part (typically, more generous radii) often provide the most economical relief from production problems. Modeling of metal flow is already at the point where initial judgment of the feasibility of producing a part can be made.

The scope of shapes can be further expanded if conventional limitations are relaxed. A good example is the multicompartment dinner tray in which wrinkling and folding is not only permitted but even encouraged. This provides the necessary stiffness while also facilitating deep draws that would far exceed the stretchability of the hard aluminum alloy sheet.

Example 10-14 | Consider making the flange of Example 7-9 (Fig. Ex. 7–9b) by sheet metalworking. From Table 10–2, possible approaches are: bending a section into a ring and welding; flanging a sheet; flanging a tube; or spin-flanging a tube.

1. Bending a 20-mm × 20-mm × 5-mm section into a ring of 60-mm ID entails severe deformation and, while it can be done, it is not practical.

2. Flanging a sheet requires cutting a circular blank with a hole and stretch flanging the collar (a). Assume that all deformation is by stretching the collar. The hole to be punched is: $(\mathrm{ID} - 2\mathrm{FH} + 2\mathrm{FT}) = 60 - 40 + 10 = 30$ mm. Tensile strain is $e_t = [(60\pi - 30\pi)/30\pi]100 = 100\%$. This obviously exceeds the cold-deformation capability of all metals.

3. Flanging the end of a tube (b). Assume that no material flows in the axial direction; thus, the dx width of the outer fiber remains unchanged and t reduces in proportion to the increase in circumference. Tensile strain is:

$$e_t = [(100\pi - 60\pi)/60\pi]100 = 67\%$$

This again exceeds the uniform elongation of all materials.

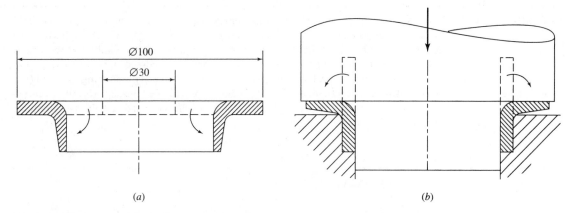

(a)   (b)

Figure Ex. 10-14

4. Make a flange by spinning the end of a tube. The limiting factor is now reduction of area q rather than tensile elongation. From (b), volume of the outer fiber remains constant: $60\pi(5) = 100\pi(t)$; hence, $t = 3$ mm. Compressive strain is:

$$e_c = [(5 - 3)/5]100 = 40\%$$

well within the capacity of most metals (because spinning imposes substantial compressive stresses).

Example 10-15

In general, sheet metalworking is not attractive for a small flange of great wall thickness. The situation changes if the flange diameter–to–wall thickness ratio increases. Take now OD = 140, ID = 100, FH = 20, and CD = 104 (FT = 2 mm).

1. For flanging, ignoring the RF radius, the hole to be punched is (ID − 2FH + 2FT) = 64 mm. Stretch is:

$$e_t = [(100-64)/64]100 = 56\%$$

Again, tensile strain is excessive.

2. A possibility, not obvious from Table 10–2, is to partially draw a cup and then pierce out the bottom. Part geometry changes because the draw die must be radiused with $R_F = 6$ mm or preferably 10 mm and the punch must have a radius (say, $R_p = 5$ mm). Ignoring these radii, and assuming no change in sheet thickness, the surface area of the part is:

$$[(140^2 - 100^2)\pi/4] + (100\pi)(25 - 4) + 100^2\pi/4 = 21991 \text{ mm}^2$$

from this, the blank diameter is:

$$d_0 = [(21991)4/\pi]^{0.5} = 167.3 \text{ mm}$$

and draw ratio = 1167.3/100 = 1.67, well within allowable range. Upon piercing, the collar will have a sharp edge. If this is inadmissible, one can trim across the collar (Fig. 10–3f). If the OD of the flange is critical, a larger blank must be used so that the outer contour can be trimmed. This is especially important if earing would cause irregular flange width.

Example 10-16 The ULSAB project (Example 5-1) illustrates many aspects of designing for sheet metalworking. More than 90% of the body structure consists of high-strength (YS 210–550 MPa) and ultra-high-strength (YS > 550 MPa) steels, and die design had to allow for the high springback of these materials. Almost half of the body mass consists of tailored blanks of varying sheet thickness and strength, and die design had to allow for this too. Hydroforming was used (Example 10-13) where conventional forming could not provide the requisite properties. Process simulation was confirmed by actual production with soft dies (except for tubular hydroforming, for which hard dies were needed).

10-12 SUMMARY

Well over one-half the total metal production ends up in the form of sheet-metal parts. More often than not, parts are joined into larger assemblies. The variety of products is immense, from aircraft skins and automobile bodies to appliance shells, from construction girders and truck frames to furniture legs, from supertankers and bathtubs to beverage cans, and from wheel rims and fan blades to watch bands.

1. Because most deformation is the result of an imposed tensile stress, and because many parts are highly visible in service, tensile ductility, yield point phenomena, and anisotropy of plastic deformation become important.

2. Shearing (blanking, punching) does not result in a perfectly smooth and perpendicular cut, but acceptable quality can be obtained with the proper die clearance, provided that bending of the sheet is prevented. An increased hydrostatic pressure changes the shearing process to resemble extrusion with a resulting smooth "cut" edge. Most applications are to mass production but economical production in small lots is made possible by CNC machines.

3. Bending, hole flanging, spinning, and stretch forming are limited by the onset of necking or by fracture; the former is related to uniform elongation (and thus the n value); the latter to resistance to triaxial tension (and thus to reduction in area q in the tension test). Localization of the neck may be delayed by changing from uniaxial to biaxial tension and by controlling friction on the stretch-forming punch. If a complex shape is genuinely required, it can be produced in a sequence of operations. The suitability of materials for a given task may be judged from forming limit diagrams.

4. Deep drawing (cupping) is limited primarily by the r value, although the LDR is also a function of process variables such as friction and process geometry. Subsequent redrawing or ironing permit the production of parts of large depth-to-diameter ratio, thin wall combined with thick bottom, zero corner radius, and tapered or stepped shape.

5. Combination of several processes is possible, and the variety of shapes produced by combined stretching and drawing is almost unlimited. Prediction of success is possible by shape analysis and by computer-aided

process modeling. Parts may have varying cross sections (e.g., a neck or bulge on a cup-shaped part) and transverse features (e.g., holes pierced into the side of vessels).

6. In addition to the usual precautions, special care must be taken in guarding the workspace of presses and to protect against noise.

PROBLEMS 10A

10A-1   Make three sketches of blanking with (a) optimum, (b) insufficient, (c) excessive clearance. Under each, show the edges of sheared blanks (identify the salient features of the cut surface). (d) State whether the optimum clearance is larger for an annealed strip than a cold-rolled, hard-temper strip of the same material.

10A-2   Draw sketches of and name at least two methods for improving the quality of cut in the process of blanking (not subsequent to it).

10A-3   (a) State the two mechanical properties of materials that are most important in influencing springback. (b) State whether springback increases or decreases with these properties.

10A-4   State whether for a given sheet thickness springback is greater for a larger or smaller bending radius.

10A-5   In brief descriptions, suggest at least three methods of making bends with exactly 90° angle.

10A-6   (a) Draw a sketch of stretching a sheet over a hemispherical punch. In an accompanying diagram, show the strain distribution with (b) low friction and (c) high friction.

10A-7   (a) Draw a forming limit diagram typical of low-carbon steel and some aluminum alloys. Identify the axes. Mark the points corresponding to (b) balanced biaxial tension and (c) plane strain.

10A-8   (a) Draw a typical tensile-test curve for a low-carbon steel with yield-point elongation. (b) State the source of the phenomenon. (c) State at least one method for eliminating it temporarily. (d) Define strain aging. (e) Superimpose on curve (a) the typical tensile-test curve of a strain-aged material.

10A-9   Define, with the aid of a sketch of a tension-test specimen, the r value.

10A-10   What does LDR stand for? Define it with the aid of a sketch.

10A-11   Draw a diagram showing the dependence of the LDR on r value. Indicate the range of typical r values for fcc, bcc, and hcp metals.

10A-12   State the consequences of using (a) too low and (b) too high blankholder pressure in deep drawing.

10A-13   (a) Draw a sketch of a deep-drawing operation, showing a partially drawn cup. (b) Define LDR. (c) State the condition for fracture. (d) Show the effects of YS, TS, el., n, m, q, r, blankholder friction, and punch friction on the LDR [tabulate the effects: use + sign to show when a quantity increases the LDR, − sign when it decreases the LDR, and 0 when the effect is very small]. (e) On the basis of (d), state what combination of variables would give the largest LDR.

PROBLEMS 10B

10B-1   The edges of shearing punches and dies gradually become rounded in service because of wear. (a) Explain, with the aid of a sketch, the consequences in terms of the process geometry. (b) In a companion sketch, explain the changes one should expect in the quality of cut.

10B-2 Small spur gears are to be made; the gear-tooth surfaces must be parallel to the axis of the gear and must have a smooth finish. Blanking from a sheet is considered. (a) Is this a feasible proposition? (b) If the answer is yes, make a sketch of a process (identifying the die elements) that assures the required quality. (c) Explain why the process works (if relevant, with the aid of another sketch).

10B-3 A hard steel plate is blanked on a mechanical press. In each cut, the press suddenly "snaps," with a loud bang. (a) Find an explanation for the phenomenon. (b) Suggest a way of minimizing it (the material cannot be changed).

10B-4 The Quality Control Department rejected a bent part because its surface shows orange peel; production wants to use it. To mediate the situation, (a) explain, with the aid of a sketch, the source of the effect and (b) pose a question, the answer to which will settle the issue.

10B-5 What sheet properties would you specify for (a) bending without orange peel; (b) bending to zero radius; (c) greatest resistance to permanent deformation in service. Justify the choices.

10B-6 A sheet-metal part is made by bending 1015 steel of 5-mm thickness on a sharp edge (zero radius). Many parts fail in service and, for increased strength, it is now proposed to change to 1045 steel. Make a quick engineering judgment whether this will be feasible.

10B-7 You are asked to explain whether plastic anisotropy has an effect in bending. (a) First draw a sketch to define the r value. Then assume that a bar of 20-mm width and 10-mm thickness is bent over a die radius of 10 mm. Make sketches

to show qualitatively the changes in width and thickness one should expect for materials of (b) high and (c) low r value.

10B-8 Consider the deformation forced upon the material during bending around a sharp radius. State what (a) n, (b) m, (c) q, and (d) r values are desirable to avoid excessive thinning or fracture. Answer simply "high" or "low," and justify each in a brief sentence.

10B-9 A lever-type component is made by bending a blanked preform. In the middle of a large production run, it is noted that a number of parts fracture, partially or fully, during bending. (a) Suggest the most likely cause, assuming that all blanks are sheared from the same batch of material, and suggest remedies in the (b) blanking and (c) bending operation.

10B-10 A small lever-type precision component of a camera must be bent to exactly 90° angle and the angle must always be the same. (a) State what variations in strip material can be expected in a production batch. (b) It is suggested that the desired 90° angle will always be assured by over-bending to a smaller angle. Do you agree? Why? (c) If not, sketch one possible process that will always deliver exactly 90°.

10B-11 HSLA sheet can show fibering. (a) Define HSLA. (b) Define fibering. (c) Will fibering affect bending? (d) If it does, which direction is more favorable for bending?

10B-12 An automotive pressing fails in production. The part is formed by almost pure stretching, using drawbeads in the dies. (a) What would you do to analyze the problem? (b) What is the likely strain state at the point of fracture (use forming limit diagram). (c) Indicate in the FLD two possible remedies, keeping the shape of the pressing unchanged. (d) If none of this works, what else could be attempted?

10B-13 Stretcher-strain marks are visible on slightly stretched surfaces of pressed steel parts. (*a*) State the cause of this phenomenon. (*b*) Identify the type of steel used. (*c*) Is the rolling mill or the press shop at fault? (*d*) What could the rolling mill do to eliminate the problem (without changing the composition of the steel)? (*e*) Could the press shop be responsible for the problem? (*f*) Is there anything the press shop could do to save the situation? (Make sketches whenever they help to clarify your answers.)

10B-14 The laboratory ran a check on two steel coils and concludes that (*a*) the coil that shows yield point elongation is rimming steel and that (*b*) the other coil, which does not show it, is killed steel. List all possible interpretations for the two cases.

10B-15 (*a*) Draw a typical engineering stress–strain curve showing yield-point elongation (YPE). Mark the axes. Explain what effect YPE has in (*b*) blanking, (*c*) bending, (*d*) light stretch forming, (*e*) severe stretch forming, and (*f*) deep drawing.

10B-16 In stretch forming a part, premature fracture sets in near the punch contact line. Suggest a remedy; justify with a sketch.

10B-17 A part fails in the course of deep drawing. (*a*) Fracture occurs toward the end of draw; identify a likely source of the problem and suggest a possible remedy. (*b*) Fracture occurs earlier; identify a likely source of the problem and suggest as many possible remedies as you can.

10B-18 Obtain samples of containers (beverage cans, food cans, sardine tins, etc.). With tin snips, cut them open, measure the wall, base, and lid thickness, and make informed judgments on the likely methods of manufacture. (*Caution:* Thin sheet has razor-sharp edges and must be handled with utmost caution, using protective gloves.)

10B-19 It is suggested that, for highest reduction in deep drawing, the punch and die radii should be as large as possible. Subject this

suggestion to a critique, using a sketch to support your argument.

10B-20 It is suggested that, to avoid earing in deep drawing, the sheet should be free of (*a*) mechanical fibering, (*b*) planar anisotropy, and (*c*) normal anisotropy. Subject each statement to a critique, justifying your answers.

10B-21 An assortment of drawn cups, made of various materials, shows the following features: (*a*) one ear, (*b*) two ears, (*c*) four ears, (*d*) six ears, (*e*) eight ears. Make a reasoned judgment of the most likely cause of the problem in each case.

10B-22 Someone asserts that a blankholder is always needed in deep drawing. (*a*) Do you agree? (*b*) If not, state (qualitatively) the defining conditions.

10B-23 To explain the significance of the *r* value, draw roughly to scale the cross section of a 12×6-mm tension test specimen (*a*) before deformation and after 50% deformation if the *r* value is (*b*) zero, (*c*) unity, and (*d*) infinity. (*e*) In a separate sketch, show the effect of a high *r* value on the yield surface in plane stress. (*f*) With the aid of a sketch of a segment of a partially drawn cup, explain the effect on the LDR.

10B-24 Some two-piece food cans are made of tinplate. (*a*) Define tinplate. Assuming that the can has a height-to-diameter ratio of 2:1 and the base and wall are of the same thickness, (*b*) suggest a process or process sequence for making the body (answer by drawing sketches of the key features of the dies used).

10B-25 A typical two-piece aluminum soft-drink can has much thinner wall than base. Suggest a likely production sequence (use sketches of the tooling, with relative wall thickness dimensions indicated).

10B-26 An 1100-Al sheet of $\frac{1}{4}$-hard temper (H12, Sec. 8-3-3) is drawn into cylindrical cups. The cups show earing and are of insufficient height when trimmed. It is suggested that annealing the sheet will eliminate the problem. (*a*) Define the causes of earing and (*b*) state whether annealing will always solve the problem.

10B-27   In early work on drawing beer-barrel halves of an aluminum alloy, prototype production was successful. For full-scale production, the tooling was carefully polished. All parts failed. Tests showed that there was no change in the material. Offer an explanation.

PROBLEMS 10C

10C-1   A hemispherical bowl is to be made from 302 stainless steel of 1.5-mm thickness. In the first step, a circular blank of 600-mm diameter is cut. (*a*) Calculate the force required. (*b*) Show in a sketch how this force could be reduced by appropriately shaping the die or punch (the blank must remain flat!).

10C-2   The blank of Prob. 10C-1 is now made into a bowl by stretch forming a hemisphere of 400-mm diameter. (*a*) Make a sketch of a suitable die set. (*b*) From the information available, make an estimate whether fracture will occur. (*c*) If the answer is yes, suggest at least one process to make the part.

10C-3   Circular blanks (slugs) of 1100 Al, diameter $d = 25$ mm and thickness $h = 3$ mm, are to be mass-produced as the starting material for the extrusion (impact extrusion) of collapsible tubes. The available presses are of 800-kN capacity and can take maximum 300-mm-wide strip. Economy of material utilization increases if more rows are cut from the strip width. Calculate (*a*) the force required for blanking a single slug and (*b*) the maximum number of slugs that can be blanked simultaneously with the available press. (*c*) Design the optimum layout for the slugs and the width of the strip if the web (remaining material between cuts and at edges) is approximately h.

10C-4   Blanks are to be made of 70/30 brass for the part of Fig. Ex. 10-14*a*. Calculate the forces required for (*a*) blanking and (*b*) punching the hole.

10C-5   Calculate the material utilization factor in blanking the rings for the part of Fig. Ex. 10-14*a* in (*a*) single-row, (*b*) double-row, and (*c*) triple-row configuration. (Take scrap bridge dimensions from Sec. 10-11). (*d*) Recalculate on the assumption that the 30-mm diameter punchouts can be utilized.

10C-6   Derive Eqs. (10-3*a*) and (10-3*b*) from basic definitions.

10C-7   A 302 SS sheet of 1-mm thickness is bent around radii of 2, 10, 50, 100, and 250 mm. Calculate the approximate values of radii after springback.

10C-8   From simple elastic bending theory, derive the force necessary for wipe bending a sheet of h thickness and w width. To allow for the effects of plastic deformation, double the end result.

10C-9   The roof of automobiles could be made of (*a*) DQSK steel, (*b*) HSLA steel ($\sigma_{0.2} = 310$ MPa), or (*c*) an aluminum alloy (similar to 6061-T6). This will influence die design; make a quantitative judgment on which of these will give the greatest springback after forming.

10C-10   Continuing problem 10C-9, design the die radius to obtain a finished radius of 100 mm with the three sheets, all of 0.75 mm thickness [ignore the cubic term in Eq. (10-5)].

10C-11   A bent sheet metal part is made of 6-mm-thick 410 stainless steel. The bend radius is 1 mm. All parts show necking and some have even fractured. (*a*) Draw a sketch of a bending process, including the stress distribution in the part. (*b*) State what properties influence necking and (*c*) fracture. (*d*) In the course of a redesign, it is now proposed that the part be made of 1008 carbon steel, subsequently coated for corrosion resistance. Will this steel show less necking and fracture? Make a quantitative judgment (no calculations are needed).

10C-12 A customer wishes to order 200 dishes of diameter $d = 1000$ mm and depth = 250 mm, of 5052-0 Al, in the shape of a spherical segment. As a first step, a conceptual design of processes is made. (*a*) The first thought is to make the dishes by deep drawing. Make a sketch and indicate the main problem. Consider also the economy of the process. (*b*) Next it is proposed to make the dish by pure stretch forming. Determine feasibility. (*c*) Suggest a possible alternative process.

10C-13 Parabolic reflectors are made of an Al–Mg alloy (similar to 5052) sheet by pure stretching. (*a*) Make a sketch of the process, clearly marking the relative wall thickness changes and the die elements. (*b*) Some of the stretched parts show fracture away from the apex. Suggest a remedy (process must still be pure stretching). (*c*) If the reflector still cannot be made, would it help to change the design to 302 or 410 stainless steel? Why?

10C-14 A flat-bottomed cup has been successfully produced of drawing-quality low-carbon steel sheet. For greater corrosion resistance, the design is now changed to make the cup of either (*a*) 5052 Al alloy or (*b*) pure titanium. Give an explanation of the changes to be expected in drawing behavior.

10C-15 A cylindrical container (a cooking pot) of 200-mm OD, 160-mm depth, 2-mm wall, and 5-mm bottom thickness is to be produced from 5052-H24 Al alloy by first drawing a container (cup) and then reducing the wall thickness. Check the feasibility of design. (*a*) Assuming that the wall thickness of the container in the first draw is the same as the starting sheet thickness, calculate the diameter of the starting blank; (*b*) calculate the blanking force; (*c*) check whether the first-draw container can be made in a single draw. If it cannot be made, make sketches of the suggested process sequence. (*d*) Select the punch diameter for the first draw and calculate the drawing force. (*e*) Suggest two methods of making the finished pots, illustrating with sketches.

10C-16 Show that, for a material obeying the power law of strain hardening, $TS = K(n/e)^n$.

10C-17 In Example 10-15 we indicated that a flange could be formed by partially drawing a blank of 167.3-mm diameter into a cup of 100-mm diameter. (*a*) Make a sketch of the partially drawn part. (*b*) You are asked to predict the required press size if the material is 70/30 brass. Is there enough information for this in the present textbook? (*c*) If not, can you make a judgment of an upper bound for press force? (*d*) If the answer is yes, make the calculation.

10C-18 Using basic definitions, convert LDR = 2.4 into percent reduction in diameter.

10C-19 In Example 10-14, part 3, we had a tensile strain of 67%; in part 4 we had a compressive strain of 40% for essentially the same deformation. Reconcile the apparent contradiction (*a*) from basic definitions of engineering strains and (*b*) from natural strains. (*Hint:* Make sketches to show the dimensions of the outer circumferential fiber.)

FURTHER READING (see also Chaps. 8 and 9)

Fundamentals of Tool Design, 4th ed., Society of Manufacturing Engineers, 1998.
Progressive Dies, Society of Manufacturing Engineers, 1994.

Benson, S.D: *Press Brake Technology*, Society of Manufacturing Engineers, 1997.

Davis, J.R. (ed.): *Tool Materials*, ASM International, 1995.

Dinda, S., K.F. James, S.P. Keeler, and P.A. Stine: *How to Use Circle Grid Analysis for Die Tryout*, American Society for Metals, 1981.

Eary, D.F., and E.A. Read: *Techniques of Pressworking Sheet Metal*, 2d ed., Prentice Hall, 1974.

Iliescu, C.: *Cold-Pressing Technology*, Elsevier, 1990.

Lascoe, O.D.: *Handbook of Fabrication Processes*, ASM International, 1988.

Marciniak, Z., and J.L. Duncan: *The Mechanics of Sheet Metal Forming*, Edward Arnold, 1992.

Pearce, R.: *Sheet Metal Forming*, Adam Hilger, 1991.

Schuler Incorporated: *Metal Forming Handbook*, Springer, 1998.

Smith, D.A. (ed.), *Die Design Handbook*, 3d ed., Society of Manufacturing Engineers, 1990.

Suchy, I.: *Handbook of Die Design*, McGraw-Hill, 1997.

Wagoner, R.H., K.S. Chan, and S.P. Keeler (eds.): *Forming Limit Diagrams*, The Minerals, Metals and Materials Society, 1989.

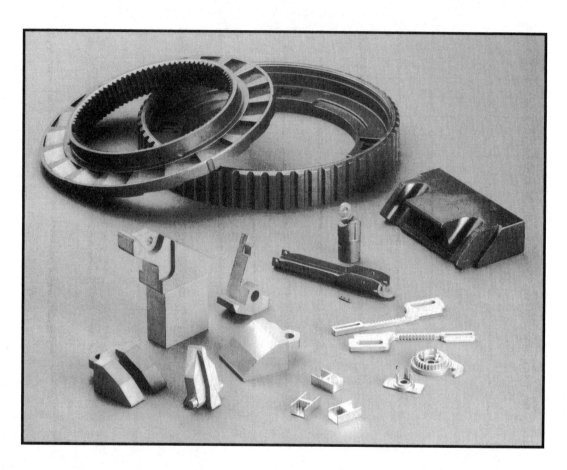

Powder metallurgy is capable of providing complex parts to high precision (foreground: stainless steel mortise lock parts, aluminum optical sensor holders with 0.63-mm wall, injection-molded latch door gear and racks; center: injection-molded ring and clevis, padlock body; background: notch plate and pocket plate for one-way auto clutch). (*Courtesy Metal Powder Industries Federation, Princeton, New Jersey.*)

chapter

11

Powder Metallurgy

In this chapter we explore a shortcut to making net-shape or near-net-shape parts of metals and other materials. We will find out:

How to make alloy powders of unique structure

How powder can be processed into shapes of adequate green strength for handling

Why high-temperature sintering is necessary to create permanent bonds

How a permanently lubricated bearing can be made

Why superior tool steels can be made by this technique

How the filament of an incandescent light bulb is made

As indicated in Fig. 5–2, manufactured components or articles of consumption may be directly produced by bringing a powder of the starting material into the desired end shape. The process was first applied to metals that could not be melted by the technology of the day (platinum around 1800, tungsten a hundred years later). Rapid developments in the twentieth century led to an explosive growth of applications. There are a number of incentives to use the process: Net-shape structural parts of fairly complex shape can be economically produced; materials of unique properties, such as controlled porosity, can be made; parts may really be composites, such as self-lubricating bearings impregnated with oil, brake pads with embedded ceramic fibers, or brushes for electric motors combining copper with graphite. Sometimes powder metallurgy is the route to wrought materials, such as tool steels of superior properties or tungsten wire for incandescent light bulb filaments. The essential feature is that the bond between particles is produced without total melting, although in some instances localized melting may occur.

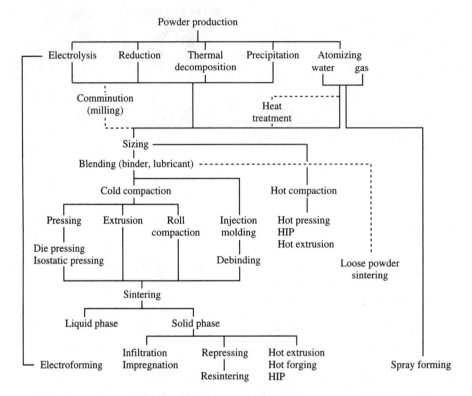

Figure 11–1   General sequence of producing powder metallurgy parts. (*Adapted from J.A. Schey, ASM Handbook, vol. 20, Materials Selection and Design, ASM International, 1997, p. 694. With permission.*)

11-1   CLASSIFICATION

A general scheme of processing is given in Fig. 11–1. Powder is produced by one of many techniques, subjected to a number of preparatory steps, and consolidated to give it shape and temporary strength until sintering establishes metallurgical bonds. An alternative route develops the shape and strength by hot consolidation. Consolidation on the atomic scale takes place in electroforming.

11-2   THE POWDER

The processing steps involved in making, characterizing, and treating the powder have decisive influence on the quality of the end product.

11-2-1   Powder Production

Powders may be made by a number of techniques.

Winning   The metal is obtained from its compound.

1. *Reduction* of an oxide by carbon or hydrogen (Fe, Cu, Co, Mo, W) often results in a porous cake (hence, for example, the name *sponge iron*) which is ground up by techniques similar to those used for ceramics (Sec. 12-4-1).

2. *Thermal decomposition* of a compound [such as the carbonyl $Ni(CO)_4$] yields spiky particles.

3. *Electrolysis* is directed so as to produce a highly uneven, often dendritic deposit which is then broken up (Fe, Cu, Be).

4. *Precipitation* of the metal from an aqueous solution is possible by *cementation* (precipitation with a less-noble metal, e.g., Cu with Fe) or by reduction with hydrogen (e.g., Ni).

Deposition   Precipitation of solid material from the vapor phase yields an extremely fine powder (Zn).

Atomization   This is the dominant process for prealloyed materials. Intensive research has made close control of the process possible. With direct observation of particle formation, some closed-loop control has already been achieved.

1. *Water atomization* accounts for the largest quantities. Melt emerging from a nozzle is broken up with water jets (Fig. 11–2a). It is used for low-alloy steels, stainless steel, Cu and Ni alloys, and Sn. The size and shape of particles are readily changed by controlling the process parameters (the median particle size is inversely proportional to jet pressure), but the powder is always oxidized.

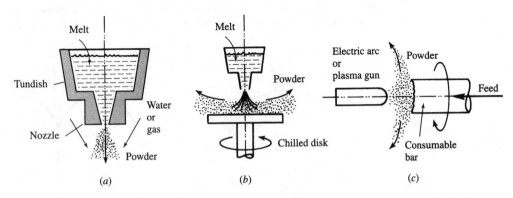

Figure 11–2   Metal and alloy powders may be made by (a) water or gas atomization, (b) centrifugal atomization, or (c) rotating electrode process.

2. *Gas atomization* which (Fig. 11–2*a*) produces spherical powders. When oxidation is allowed or the oxide can later be reduced, air is suitable (Al, Cu, Sn), but inert gas is preferred for superalloys, stainless steels, tool steels, and Ti alloys.

3. *Centrifugal atomization* is based on directing a melt stream at a chilled, rotating disk (Fig. 11–2*b*). In *rotating-electrode* processes the alloy to be atomized is in the form of a fast-rotating (15 000 r/min) electrode which is gradually melted by an electric arc or helium plasma arc (Fig. 11–2*c*). Material melted is immediately hurled off; the spherical particles solidify without touching any surface and remain very clean.

Fiber Production   In *melt extraction* a water-cooled, notched rotating disk is brought in contact with the melt surface (Fig. 11–3*a*). In *melt spinning* the melt is directed onto a cooled, fast-rotating disk (Fig. 11–3*b*) to form a ribbon of 20–100 μm thickness. A ribbon is formed also in *roller quenching* (Fig. 11–3*c*) between two chilled rolls.

Mechanical Powder Production   Some metal powders (notably beryllium) are produced by *machining* a cast, coarse-grained billet and comminuting the swarf by ball milling and impact grinding. Some powders of ductile titanium alloys are also made from castings: hydrogen gas is introduced to form brittle hydrides which can be milled into a powder; ductility is then restored by driving off the hydrogen.

Structure of Powders   In all atomizing and fiber (ribbon) processes the cooling rate is much higher than in conventional solidification, on the order of 100°C/s in gas atomization, 1000°C/s in water atomization, and over 10^6°C/s in centrifugal and rotating-electrode atomization (hence the term *rapid solidification technology*, RST). Consequently, the equilibrium conditions described in Sec. 6-1 are never achieved, and unusual and, in many respects, highly desirable metallurgical structures are formed. At cooling rates of 1000°C/s, secondary dendrite arm spacing is very small (on the order of 1 μm) and intermetallic particles are

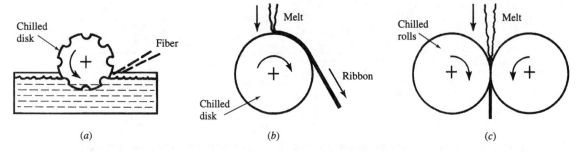

Figure 11–3   Very high cooling rates are obtained in (*a*) melt extraction, (*b*) melt spinning, and (*c*) roller quenching.

finely distributed. At ultrahigh cooling rates (over 10^6 °C/s), supersaturated solid solutions are retained, some crystalline phases are suppressed and, in the limit, there is no time even for the arrangement of atoms into a lattice, and the solid (especially the eutectics) remains amorphous. *Amorphous metals (metallic glasses)* are made by melt spinning of approximately 40-µm-thick ribbons. Some metallic glasses devitrify on heating to 350–500°C and become brittle, but others can be transformed into a microcrystalline structure of high strength (up to 1250 MPa TS) and reasonable ductility. They have already found applications as (Nd–Fe–B) magnetic materials, reinforcing elements in ceramics, and brazing alloys.

11-2-2   Characterization of Powders

Properties that affect both the consolidation of powders and the properties of the final product are routinely determined.

Morphology   Particle shape, size, and size distribution are important variables.

1. *Particle shape* is an important factor in determining the processing characteristics and is the subject of ISO Standard 3252. Formal methods of morphological analysis may be used; in a less quantitative sense, it is usual to speak of spheroidal, nodular (slightly elongated, roundish), irregular, angular, lammellar (plate-like), acicular (needle-like), and dendritic particles. Thin lammellar particles agglomerate into flaky powders; long, thin needles into fibrous powders; and spherical or irregular powders into granular powders. Some powders are porous, while others are more or less complete hollow spheres or other shapes.

2. *Particle size* should be neither too large nor too small. Too large particles may not display the desired structure which is often the reason for choosing the powder route, and they may not allow the development of high densities. Too small particles may be difficult to handle and tend to agglomerate; furthermore, their large surface area–to–volume ratio may introduce large quantities of undesirable adsorbed substances and oxides.

3. *Particle size distribution* is analyzed by passing the powder through a series of sieves of gradually diminishing hole size (increasing number of holes per unit area). The fraction of particles passing a certain sieve is given in percentages (usually weight %). The sieve size is quoted as mesh number (for mesh numbers 50 and higher, the particle diameter, in millimeters, is 15 divided by mesh number). Sieve analysis is usually conducted dry, but vacuum or wet sieving is necessary for powders below 325 mesh (45 µm). Techniques based on laser diffraction, light intensity fluctuation, electrical pulses, or sedimentation are suitable for analyzing wide size distributions. Optical and scanning electron microscopy can be used for both size and shape analyses.

Physical Properties   The powder possesses a number of properties that are of importance to further processing:

1. *Specific surface area* (area/unit mass, in units of m^2/g or, for metals, often cm^2/g) is determined by physical adsorption of gas or chemical adsorption of a dye. It indicates the surface available for bonding and also the area on which adsorbed films or contaminants may be present.

2. *True density* (also called *theoretical density*) is the mass per unit volume of solid, and is a material property. The *apparent density* or *mass per unit volume* (g/cm^3) is a very important value because it defines the actual volume filled out by the loose powder. It is often expressed as a percentage of the fully dense material (as a percentage of true density). *Tap density* is obtained by tapping or vibrating the receptacle, and is a measure of compaction achievable without pressure. Both apparent and tap densities depend on particle shape and distribution as well as interparticle friction.

3. *Flow properties* are given by the *flow rate* (the time required for a measured quantity of powder to flow out of a standard funnel) and *angle of repose* (the base angle of a cone of powder resting on a circular plate).

4. *Compressibility* as a term describes the change in green density with increasing compacting pressure. It is usually given as the density at some specified pressure or, in a graphical or tabular form, at several pressures.

11-2-3 Powder Preparation

Most powders are subjected to various preparatory steps.

Classification This is the process of separation into fractions by particle size. Milling is sometimes necessary to break up agglomerates, flatten (flake) the particles, or modify their properties by strain hardening. Excessively large particles are removed and, if required, size fractions are separated by *screening* the entire production lot. Fine particles may be separated by screening a slurry. Settling from a liquid solution (*elutriation*) or classification of dry powders in an air stream within a *cyclone* are also useful for separating fine powders. Superfine particles can be removed by *electrostatic separation.*

Powder Conditioning Some metals, such as iron, have oxides that are readily reduced by a suitable atmosphere during sintering. Others, such as titanium, dissolve their own oxide and are thus reasonably suitable for powder processing. Aluminum oxides contribute to dispersion strengthening. Still other alloys are covered with a thin but very tenacious and persistent oxide film that greatly impairs the properties of the finished part, and these materials (typically those containing chromium and, in general, the high-temperature superalloys) must be treated by special techniques to keep oxygen content very low. Some powders, such as water-atomized high-speed steel powders, are annealed and deoxidized in a single operation, making them more compressible and easier to sinter. Contaminants that segregate on the surface are bound to create not only consolidation and sintering problems but will also greatly detract from the service properties of the material. Any nonwetted remnants of a surface film at grain boundaries may act as crack initiators (Sec. 6-3-4).

Finely distributed metal powders can be hazardous and must be treated with special care. Some (such as beryllium and lead) are *toxic*, others (such as zirconium, magnesium, aluminum) present danger of explosion, many others are *pyrophoric* (ignite spontaneously on air) below some critical particle size. A *thermite reaction*, in which an oxide (such as iron oxide) is reduced by another, more reactive metal (such as aluminum) is also possible, and proceeds at high temperatures.

11-2-4 Blending

A single powder may not fulfill all requirements of production or service properties, and powders are then *blended*. Blending may serve several purposes: Uniformity of size distribution is ensured in a large lot; response to imposed stresses (*rheology*) is controlled for improved handling; density of the compacted body is adjusted; and composition or service properties are changed. Blending must always be thorough, with each particle uniformly coated and with the various constituents uniformly dispersed. Ball milling is often employed.

1. Blending a coarser fraction with a finer fraction (Fig. 11–4*a*) ensures that interstices between large particles will be filled out. Thus, tap densities over 65% of theoretical density can be obtained with powder of favorable (spherical or nodular) shape.

2. *Metallic alloys* can be made by blending different elemental powders (for some steels, bronzes, and Al and Ti alloys). The alloy is formed in the course of sintering; the driving force is the chemical potential gradient due to concentration differences.

3. In the intense milling of two otherwise immiscible metals, severe strain hardening, fracture, and cold welding result in *mechanical alloying*; milling of two phases (such as a metal and an oxide) can be used to produce dispersion-hardened materials. Metal–nonmetal composites (such as WC bonded by Co) are also made.

4. *Lubricant additives* reduce friction between particles and between particles and mold wall.

5. *Binders* (such as wax or thermoplastic polymers) are added to powder that would otherwise fail to develop adequate green strength.

6. *Sintering aids* are added to accelerate densification on heating.

7. When the powder is to be made into a slurry, *deflocculants* and other chemicals are added to impart favorable rheological characteristics.

8. *Spray drying* ensures uniform distribution of constituents in fine powders and allows forming free-flowing powder from fines that otherwise would not flow.

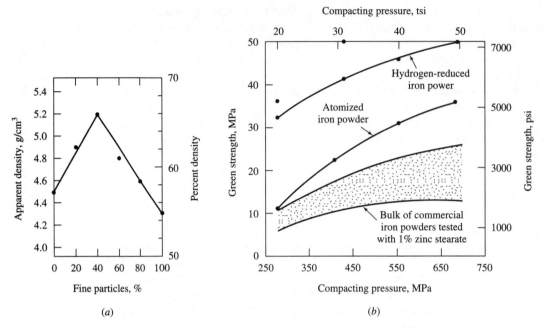

Figure 11–4   In compacting powders, (a) optimum fill density is obtained by a mixture of fine and coarse powders, (b) green strength increases with compacting pressure and is greater for more irregularly shaped powder. [(a) *After H.H. Hausner, in Handbook of Metal Powders, A.R. Poster (ed.), Reinhold, 1966. (b) From Metals Handbook, 9th ed., vol. 7, ASM International, 1984, p. 289. With permission.*]

The ingredients are made—with water or an organic liquid—into a slurry, which is then *atomized* by air or a gas. The liquid is evaporated from the droplets in their flight and the coarser, roughly spherical powder thus formed is collected. It is the principal method of preparing oxide-dispersion-strengthened alloys and cemented carbides. Less favorable shapes are obtained when the powder is *granulated*: A damp mix is forced through orifices in a plate or holes in a screen.

11-3   POWDER CONSOLIDATION

The method chosen for *consolidation*, i.e., for bringing the particulate material into the required shape, depends on the particulate and on the intended density of the product.

11-3-1   Cold Compaction

Dry powders, which may be coated with a lubricant or dry binder, are *compacted* by the application of pressure to form a so-called *green body* (a body without permanent bonding). For a given particle-size distribution, the density of the compact increases with applied pressure (Fig. 11–4b). Density is a function also of particle shape: A spherical powder compacts to a higher density than an irregularly shaped one. In the course of pressing, air is expelled from between particles, and particles slide against each other and against the surface of the compacting tool. At higher pressures, the applied force is concentrated at contact points between particles; the high local pressures cause local deformation or fracture, and the compact acquires some *green strength*, usually expressed as rupture strength, measured on transverse bend specimens (Fig. 4–9).

There are several sources of green strength. Sliding combined with pressure promotes adhesion (and even cold welding with some powders); therefore strength increases with increasing pressure (Fig. 11–4b). Another source of green strength is mechanical interlocking, especially with particles of irregular shape, therefore green strength is less for spherical powders even though they pack more densely. If neither mechanism is available, bonding agents are added which evaporate in the course of sintering. Of course, green strength is lower when the powder is coated with a lubricant. Various compacting techniques are available:

Die Pressing   This finds widest application for net-shape (or near-net-shape) parts. If the part shape is fairly simple and a die can be made of steel, high applied pressures are permissible and, if the particulate can deform plastically, densities in excess of 90% of theoretical density can be achieved.

1. The effectiveness of pressing with a *single-acting punch* is limited because particulate material does not transmit pressures as a continuous solid would and wall friction also opposes compaction. The pressure tapers off rapidly and density diminishes away from the punch (Fig. 11–5a), limiting the attainable depth-to-diameter ratio. For an applied pressure of p_0, the pressure at l depth in the body is

$$p_l = p_a \exp\left(\frac{-\mu k A_{\mathrm{fr}}}{A_0}\right) \tag{11-1}$$

where μ is wall friction, A_{fr} is the frictional surface area, A_0 the compacted area, and k is the factor expressing the ratio of radial stress to axial stress and is, for an elastic solid

$$k = \frac{p_r}{p_a} = \frac{v}{1 - v} \tag{11-2}$$

where v is Poisson's ratio.

2. The situation improves with a *floating container* (Fig. 11–5b) which is moved against the stationary punch by the friction force between powder and container.

3. Good results are obtained in special presses with two *counteracting punches* advancing from the two ends of the die cavity (Fig. 11–5c).

4. When the thickness of the part changes considerably from point to point, green density is equalized by constructing dies with *multiple punches* guided

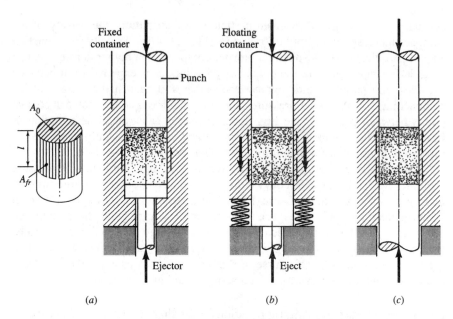

Figure 11–5   The density of a green body depends on the method of compaction: (a) it is higher under the punch when compacting is done with a single punch in a fixed container; better uniformity is obtained with (b) a single punch and floating container or (c) with two counteracting punches.

within each other (Fig. 11–6), so that the same degree of compaction can be applied everywhere. Clearances between moving parts must be kept very small (below 25 μm) to prevent entry of particles.

Dies are usually built of high-strength tool steel or, for larger production runs and severe abrasive conditions, of cemented tungsten carbide. Allowable die pressures are as in cold forging and extrusion (Sec. 9-5-1), although moderate pressures (of 100 MPa) are usual, rising to 500 and even 900 MPa in some cases. Dies can be quickly and automatically filled by gravity, with the excess powder simply wiped away.

Cold Isostatic Pressing   The powder is filled by vibration into a deformable (reusable rubber) mold. Hydrostatic (omnidirectional) pressure is applied by means of a hydraulic fluid inside a pressure vessel (Fig. 11–7, page 452), assuring uniform density. The fluid is usually water. Shape limitations are few in the *wet-bag method* (Fig. 11–7a). Shorter cycle times are achieved in the *dry-bag method* (Fig. 11–7b) because fluid is introduced only into the space between the fixed die and the elastomeric mold, but shapes are somewhat more limited. In many instances, neither lubricants nor binders are needed. Pressures of 300 MPa are usual and up to 550 MPa can be achieved.

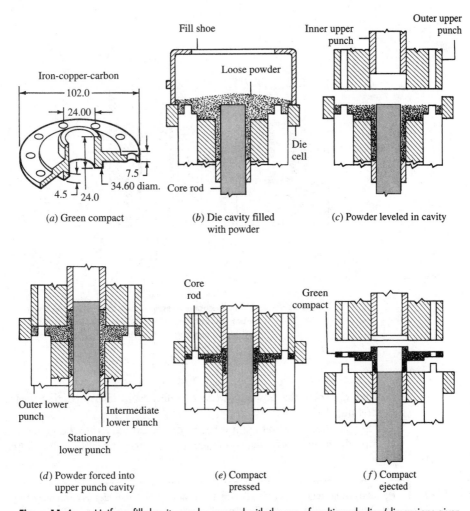

Figure 11–6   Uniform fill density can be assured with the use of multipunch dies (dimensions given in inches). (*From Metals Handbook, 8th ed., vol. 4, ASM International, 1969, p. 461. With permission.*)

Rolling   Compaction by rolling, followed by sintering and perhaps rerolling, is an important method for making Ni strip and is used also for cladding a solid base metal. *Impact compaction* on fast hammers or with the aid of explosive charges yields parts of high density.

Gravity   Often assisted by *vibration*, gravity filling of a mold gives a compact of low density and little strength. Only with careful handling can it be converted into a porous sintered product; more likely, the mold itself will be heated to at least initiate sintering or to set the dry binder. In the process known as *loose*

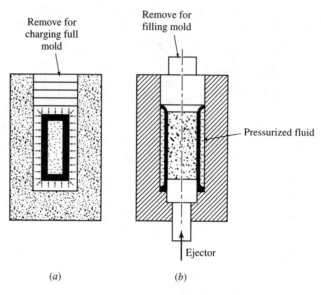

Figure 11–7   Omnidirectional pressure helps to produce parts with more uniform density distribution in isostatic pressing by the (*a*) wet-bag method and, somewhat less so, by the (*b*) dry-bag method.

powder sintering the powder is finish sintered in the compacting mold to make porous filters.

Shape elements that cannot be produced directly may be generated by machining the green body; warm pressing gives more strength to the delicate body.

Example 11-1 | Perfectly spherical nickel powder of 0.1-mm particle diameter is compacted by vibration. (*a*) What percentage of the theoretical density can be achieved? (*b*) Will this increase or decrease if the particle diameter is uniformly increased to 0.2 mm?

(*a*) In the ideal case, closest packing—equivalent to a fcc structure—will be achieved. From Fig. 6–2*a*, there are (6 half) + (8 one-eighth) = 4 spheres of R radius in each unit cell of a side. The body diagonal is $4R = a(2^{0.5})$. The packing factor is

$$\frac{\text{Volume of spheres}}{\text{Volume of cube}} = \frac{4(4\pi/3)R^3}{a^2} = \frac{4(4\pi/3)R^3}{(4R/\sqrt{2})^3} = \frac{\pi}{3\sqrt{2}} = 0.74$$

This is also the ratio of powder to solid volume; thus, 74% of the theoretical density is obtained. (In practice, a density of less than 67% of theoretical density is more likely, see Fig. 11–4*a*.)

(*b*) The particle radius R dropped out, thus no change in fill density will occur.

Example 11-2 | The green compact shown in Fig. 11–6 has a hub of 24-mm height and a flange of 7.5-mm thickness. Calculate the position of the lower punches for filling the die cavity (Fig. 11–6*b*) if the density of loose powder is 38% and the density of the green compact is 78% of theoretical density.

The mass of loose powder after leveling (Fig. 11–6c) must be equal to the mass of green compact. Powder cannot be counted upon to flow from hub to flange or vice versa; therefore, the same mass of powder remains in the flange (and in the hub) before and after pressing. Since mass is (volume) × (relative density), we may write

$$V_l(\text{rel. dens. loose}) = V_g(\text{rel. dens. green})$$

where V_l is the loose volume and V_g the volume of the green compact. Since no powder flows from punch to hub and vice versa, we may write

$$V_l = Ah_l \quad \text{and} \quad V_g = Ah_g$$

where h_l is the height of loose powder column and h_g is the thickness of the green compact. Consequently,

$$h_l = h_g \left(\frac{\text{rel. dens. green}}{\text{rel. dens. loose}} \right)$$

For the hub, $h_l = 24(0.78/0.38) = 49.3$ mm; for the flange, $h_l = 7.5(0.78/0.38) = 15.4$ mm.

A Ni-powder cylinder is to be made by uniaxial pressing (Fig. 11–5a). Calculate the maximum allowable length l if the pressure drop is limited to two-thirds of the original value. No lubricant is used ($\mu = 0.5$) and $\nu = 0.31$.

Example 11-3

From Eq. (11-2), $k = (0.31)/(1 - 0.31) = 0.45$. $A_0 = d^2\pi/4$; $A_{\text{fr}} = d\pi \cdot l$.

From Eq. (11-1)

$$\frac{p_l}{p_0} = \frac{1}{3} = \exp\left(\frac{-0.5(0.45)\,d\pi l}{d^2\pi/4} \right) = \exp\left(\frac{-0.9l}{d} \right)$$

hence $l = 1.22d$. Note that friction between particles will further reduce pressure; therefore, a lubricant would be used.

11-3-2   Injection Molding

When the proportion of the binder or other liquid is high enough to allow the relative displacement of individual particles within a liquid matrix, the mixture acquires rheological properties suitable for processing by *plastic forming* techniques. The mixtures usually behave as Bingham solids or non-Newtonian (pseudoplastic) bodies (Fig. 7–5) and can be processed by techniques familiar from die casting (Sec. 7-5-6) and used also in polymer processing (Sec. 14-3-3).

Injection molding of fine (< 20 μm) powder is feasible for fairly thin-walled (0.5–5 mm) parts. The powder is combined with 25–45% wax or thermoplastic polymer and injected at 135–200°C, at pressures of 100–140 MPa, into molds on standard injection-molding machines. Cooling in the mold fixes the shape. Shrinkage on sintering is large, but complex shapes, including transverse holes, can be produced in all materials, including refractory metal alloys. The polymer has to be removed by heating (*debinding*) in a protective atmosphere; this may

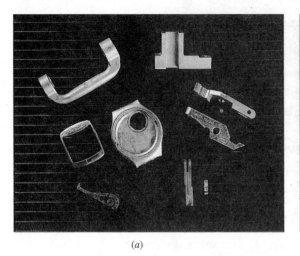
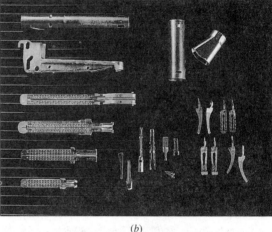

(a)   (b)

Figure 11–8   High-precision parts of complex configuration are made by precision powder injection molding. Note in (a) the part with transverse hole. The smaller parts in (b) are for minimally invasive surgery; the smallest is only 6 mm long. (*Courtesy Thermat Precision Technology, Corry, Pennsylvania.*)

take a long time and softening causes some loss of shape. Shrinkage is large, up to 20% linearly. Better shape retention is assured by an acetal binder which can be chemically decomposed at low temperature. Further significant advances are represented by *precision powder injection molding* (PPIM).[1] The powder is coated with a thin film of thermosetting polymer in the B stage where it is thermoplastic (see Sec. 13-3). The coated powder is mixed with a water-soluble binder and injection-molded. The part is passed through a water bath to remove the binder. Polymerization is completed thermally, chemically, or by ultraviolet light, and the part is sintered. Since the powder is bonded with a thermoset, no softening takes place; at the sintering temperature the polymer decomposes and escapes as gas. Thus, parts of high precision can be made in a wide range of sizes (Fig. 11–8). By molding around plastic cores, internal reentrant shapes (U2 and 4 in Fig. 3–1) are possible. Shrinkage is only 10%.

11-4   SINTERING AND FINISHING

The compact is given its strength by sintering and, if necessary, subsequent processing.

---

[1] A proprietary process of Thermat Precision Technology Inc.

11-4-1   Sintering

The green compact is heated to attain the required final properties. In the course of heating, several changes take place.

Drying   At low temperatures, liquid constituents are driven off. The required residence time increases with increasing wall thickness; fast heating would cause sudden vaporization and could result in the disintegration of the compact. Vacuum accelerates drying. If organic binders are to be burnt off, sufficient oxygen must be available for their combustion.

Sintering   At higher temperatures (above $0.5T_m$ but, more typically, at hot-working temperatures, about 0.7–$0.9T_m$), sintering takes place. The compact contains particles of the material in close proximity to each other. The energy of the system will decrease by reducing the total surface area—in other words, the driving force for sintering is the surface energy which is reduced by joining adjacent particles (Fig. 11–9). Several mechanisms come into play, of which evaporation and condensation are usually much less important than solid diffusion. To begin with, interatomic bonds are established between adjacent surfaces, and necks grow by the movement of atoms from the surface and bulk of particles toward the necks. Plastic or viscous flow may also take place and this, together with more massive diffusion, reduces the size of pores (Fig. 11–9b). Thus, the volume shrinks and density increases (Fig. 11–9a). To achieve the same sintered density, shrinkage is greater for lower green densities. If phase changes take place during heating, shrinkage may be negligible or even growth may occur. It is usually necessary to determine the shrinkage experimentally: If process steps are closely controlled, shrinkage is reproducible and finished parts can be held to close tolerances.

At this stage, strength increases markedly and a ductile material exhibits an increase in ductility (Fig. 11–9a). However, fatigue properties are likely to be inferior as long as there is any porosity left. Further heating does not necessarily improve the situation because grain boundaries begin to migrate, some grains are consumed, average grain size increases with a consequent drop in strength, and pores left behind (inside the new grains) become stable. These pores, together with any porosity resulting from entrapped gases, impair properties.

Liquid-Phase Sintering   The process is accelerated when one of the constituents melts and envelopes the higher-melting constituent. A liquid that wets the solid particles exerts capillary pressure which physically moves and presses particles together for better densification. The phases formed can be determined from the relevant phase diagrams; best bonding is achieved when there is mutual solubility. A liquid phase forms at the surface of particles when elemental powders are mixed and transient low-melting phases form before diffusion creates a homogeneous alloy (as is the case with Al alloys, brasses and bronzes). Density can reach theoretical density.

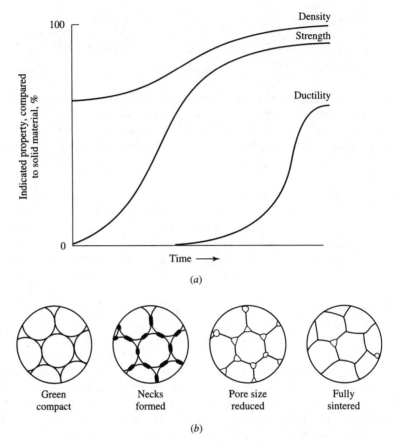

Figure 11–9   In the course of sintering (a) the compact acquires permanent strength while the volume shrinks (density increases) as a result of (b) elimination of most pores between particles.

Shrinkage   It is usually necessary to conduct tests to determine the exact value of shrinkage but a good estimate can be obtained by considering that shrinkage is uniform in all directions and that there is no change in mass (unless volatile constituents are present in significant quantities).

$$\text{Mass} = \text{constant} = (\rho_{\text{green}})(V_{\text{green}}) = (\rho_{\text{sintered}})(V_{\text{sintered}}) \quad \textbf{(11-3a)}$$

$$\text{Volumetric shrinkage} = \frac{V_{\text{sintered}}}{V_{\text{green}}} = \frac{\rho_{\text{green}}}{\rho_{\text{sintered}}} \quad \textbf{(11-3b)}$$

$$\text{Linear shrinkage} = \left(\frac{\rho_{\text{green}}}{\rho_{\text{sintered}}} \right)^{1/3} \quad \textbf{(11-3c)}$$

The green compact shown in Fig. 11–6a is sintered to 96% theoretical density. The composition is Fe-8Cu-2C. Calculate (a) the theoretical density and (b) shrinkage of the 24-mm dimension.

Example 11-4

(a) The densities of constituent elements are Fe: 7.87; Cu: 8.96; C(graphite): 2.25 g/cm^3. The volume of 100-g powder is

$$(90/7.87) + (8/8.96) + (2/2.25) = 13.21 \text{ cm}^3$$

Theoretical density is $100/13.21 = 7.566$ g/cm^3.

(b) In Example 11-2 we had the green compact density $\rho_{green} = 0.78$. The sintered density is $\rho_{sintered} = 0.96$. From Eq. (11-3c),

$$\text{Linear shrinkage} = (0.78/0.96)^{1/3} = 0.933.$$

Thus, the 24-mm dimension shrinks to $(0.933)(24) = 22.4$ mm.

A cube of 25-mm sides and 95% theoretical density is to be made by injection molding. The binder is 40% by volume and is completely removed during debinding and sintering. Calculate (a) the volumetric shrinkage and (b) the dimensions of the green body.

Example 11-5

(a) The volume changes, but the mass of powder remains constant

$$\text{Volume injected} = (0.6 \text{ powder}) + (0.4 \text{ binder})$$

$$\text{Volume sintered} = (0.6/0.95) = 0.6316$$

$$\text{Volumetric shrinkage} = 1 - 0.6316 = 36.84\%$$

(b) Linear shrinkage $= (0.6316)^{1/3} = 0.858.$

Side of injection-molded cube $= 25/0.858 = 29.14$ mm.

Sintering Furnaces   They are of the batch or continuous type. Continuous furnaces have a preheat (drying or burn-off) zone, a high-heat (sintering) zone, and a cooling zone. Except for powders of high fill density, covered with a protective oxide (such as aluminum), all sintering is done in an atmosphere chosen to provide a neutral to nonoxidizing or reducing environment. Among the extensively used gases, nitrogen is neutral. Sintering in a vacuum furnace also provides a neutral environment but at high temperatures it favors the deoxidation of many metals. Hydrogen is a very effective reducing agent but must be handled with caution to avoid explosions. Frequently used gases are: nitrogen with 10% hydrogen plus methane; dissociated ammonia; and partially combusted (exothermic or endothermic) hydrocarbon gas. In sintering steel, the carbon content is also controlled and, in some instances, steels are carburized in a CO-containing atmosphere.

11-4-2   Finishing

The porosity of a fully sintered part is still significant (4–15%), depending on powder characteristics, compacting pressure, and sintering temperature and time. Density is often kept intentionally low to preserve interconnected porosity for bearings, filters, acoustic barriers, and battery electrodes, or when components are to be infiltrated. Powder metallurgy offers unique opportunities for tailoring properties to needs: By pressing different sections of a part to different densities, strength and porosity can be locally adjusted. The residual porosity makes sintered compacts rougher than the compacting die. Impact and fatigue properties are lower than in a wrought material but may be improved. Heat treatment is one of the possibilities; other processes are unique.

1. *Cold restriking* (coining or sizing) of the sintered compact increases its density and improves dimensional tolerances. Further densification and improvement of strength can be achieved by *resintering* the repressed compact (Fig. 11–10). Heat treatment is often performed.

2. *Impregnation* of a sintered compact of interconnected porosity is possible by immersion in heated oil. Capillary action distributes the oil; application of vacuum aids the process.

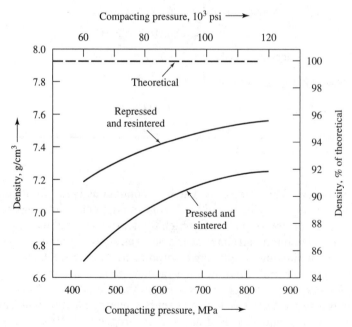

Figure 11–10   The density of an electrolytic powder-iron compact can be increased by repressing at 700 MPa and resintering at 1120°C for 1 hour. (*From Metals Handbook, 8th ed., vol. 4, ASM International, 1969, p. 455. With permission.*)

3. *Infiltration* is impregnation with a metal (e.g., Cu for ferrous parts), carried out by immersion in the molten metal or by placing the infiltrant metal in the form of a sheet above or below the compact in a furnace; again, capillary action fills the pores.

11-5   HOT COMPACTION

All particulate processing sequences described hitherto involve consolidation followed by sintering; advantages are often gained by combining the two into one simultaneous operation. Sufficient pressure is applied, at the sintering temperature, to bring the particles together and thus accelerate sintering. If possible, individual grains are deformed to ensure greater compliance; most importantly, grains are moved relative to each other, so as to break through surface films that would slow down or prevent diffusion. Under such conditions porosity may be completely eliminated. Since a particulate body has no strength whatsoever prior to pressing, the applied pressures must be all compressive—and preferably omnidirectional—so that the generation of secondary tensile stresses is avoided.

Hot Pressing   This is feasible in heated graphite or ceramic dies but suffers from the difficulty of transmitting pressure uniformly to all parts of the compact (Fig. 11–5). In *spark sintering* a high alternating current generates electric discharges during the early phase of consolidation, thus activating the surface of particles.

Hot Isostatic Pressing (HIP)   This process has gained wide acceptance. In the basic form of HIP, the powder is encased (*encapsulated*) in a deformable metal can or shroud which is evacuated and then placed inside a furnace, which in turn is enclosed in a cold-wall high-pressure chamber (similar to Fig. 7–27). The chamber is pressurized with an inert gas up to 300 MPa or, more typically, to 100 MPa. Furnace temperatures range from 480 to 2000°C; more typically to 1200°C. In a variant of HIP, the powder is placed in a glass mold; in yet another variant, into a ceramic shell mold. Parts presintered to 92–95% density (chiefly, tool bits and similar small components) have no interconnected porosity and are treated without encapsulation to remove residual porosity. The high-pressure gas can be replaced with a ceramic powder or a softer metal, and then hot pressing on a conventional press yields results similar to HIP.

Hot Rolling and Extrusion   For these, provision must be made to prevent undesirable reactions with the surrounding atmosphere. Therefore, the powder is sometimes encased in a can made of a metal that resists the high temperatures. Solid bars are rolled or extruded and hollow tubes are also extruded. Their properties can be superior to material produced by the conventional route because of the unusual structure of powders.

Figure 11-11   On a connecting rod made by
powder forging, the end cap was
separated by fracture splitting,
assuring perfect fit of the two halves.
(*Courtesy Ford Motor Company,
Dearborn, Michigan.*)

Hot Forging of Powder Preform   The process can yield parts of superior properties, including toughness, and can make shapes otherwise too complex to attain. For example, connecting rods can be forged to finish dimensions (Fig. 11–11) and rotating parts of turbines can be made.

Spray Deposition   In this direct-forming method, atomized particles are immediately deposited on a mandrel, thus no sintering is needed. Prime applications are to superalloy rings and tubes which are subsequently rolled.

Example 11-6	Connecting rods ("conrods") for internal combustion engines have traditionally been hot-forged, often with a separate cap (as in Fig. 9–20). Powder metallurgy has emerged as a strong competitor. The preform is cold-pressed, sintered in a reducing atmosphere to 75–85% of theoretical density, and directly transferred to a forging press where it is finish-forged in a closed die, to a much closer weight tolerance than conventional forgings, and with both pin and crank holes fully formed. The product is also a rare example of introducing a crack intentionally: Sharp notches are formed into the crank end and the cap is separated by "fracture splitting" (imposing internal expanding force). The granular fracture surface assures perfect fit and prevents lateral movement. [Source: D.R. Bankovic and D.A. Jager, *Adv. Mater. Proc.*, 1994(8):21–23.]
Example 11-7	Tungsten, a bcc metal, is very brittle in the recrystallized form at room temperature, yet filaments for incandescent lamps must be ductile. To satisfy this requirement, a still-used process was developed in 1909 by W.D. Coolidge at General Electric Company. Tungsten powder, produced by reducing WO_3 with H_2, is cold-pressed, self-resistance heated to 2500°C, sintered to 90% theoretical density, and hot-worked above 1500°C to achieve full density (at this point, the ductile-to-brittle transition temperature is still 300°C). Further deformation is at gradually

lower temperatures. Since recrystallization temperature is 1700°C, this is technically in the cold-working regime; grains elongate and a marked texture develops with the [110] direction aligned in the axis of the wire. The ductile-to-brittle transition temperature gradually drops, so that the wire is ductile at room temperature and has extraordinary strength (TS = 5000 MPa). (Source: J.P. Wittenauer, T.G. Nieh, and J. Wadsworth, *Adv. Mater. Proc.*, 1992(9):28–37.)

11-6   POWDER-METALLURGY PRODUCTS

Shipments of powder metallurgy products (Table 11–1) amount to only about 2% of the total weight in most metal categories, but their value and industrial significance are much greater, partly because of special applications.

Table 11–1   Production of powder-metallurgy parts (North America)*

Metal	Shipments, 1000 Mg[†]	
	1972	**1997**
Iron and steel	139.0	350.0
Copper and alloys	24.3	22.0
Aluminum	81.5	40.0
Molybdenum	1.6	2.3
Tungsten	4.6	0.6
Tungsten carbide		5.7
Nickel	4.3	10.4
Tin		0.9

*Data from Metal Powder Industries Federation, Princeton, N.J.
[†]Mg = 1000 kg = metric tonne = 2200 lb.

1. Structural parts are competitive with conventionally produced parts because only as much material is used as is needed for the finished part. Even though the starting material may be more expensive, the savings in intermediate processing steps and scrap losses often more than compensate for this, particularly on parts of complex shape. This is especially true when net shapes or near-net shapes are produced.

The largest quantities are made of iron powder, often mixed with 4–6% copper and 1% graphite for greater strength; of porous iron infiltrated with copper; and, increasingly, of atomized steel. They find applications in the automotive, off-road equipment, appliance, and business-machine industries as gears, transmission and pump parts, bearings, and fasteners. Smaller but increasing quantities are made of copper and especially aluminum alloys which are preferred in business machines because of their low mass.

Of great importance are structural components for aircraft, jet-engine, and rocket-motor applications, such as superalloy turbine disks, titanium-alloy bulkheads and fuselage components, and rocket nozzles made of tungsten or molybdenum infiltrated with copper. In superalloys, the powder metallurgy approach avoids the problems of alloy segregation, carbide clustering, and residual cast structures.

2. *Bearings* can be made that combine the load bearing and wear resistance of one component with the lubricating function of another. Examples are oil-impregnated iron or bronze "permanently lubricated" bearings, plastic-filled bearings, lead-filled iron bearings, and bearings pressed with graphite (strictly speaking, all these should be regarded as composites).

3. A small but important application is for *surgical implants* (as in Fig. 1–5). A long-established application is filling of teeth with dental amalgams. They represent room-temperature transient liquid-phase sintering, in which an Ag–Sn alloy is amalgamated with Hg; the mercury is used up in the reaction.

4. Some metals can be produced only by the powder metallurgy route. Beryllium is hot vacuum pressed. Tungsten is sintered and hot forged in preparation for wire drawing into incandescent-lamp filaments; doping with small quantities of alloying elements (e.g., 0.5% Ni) accelerates sintering.

5. In the electrical industry, *contacts* must be good conductors while also resisting wear. Tungsten or molybdenum with 25–50% silver or copper, or tungsten carbide with 35–55% silver fulfill the requirements. *Brushes* consist of graphite bonded with 20–97% copper or silver.

6. Magnetic applications include *magnetically soft materials* such as Fe, Fe-3Si, Fe-50Ni, which, because of their mechanical softness, are difficult to machine but are easily formed into final shape by powder metallurgy. Among *permanent magnets*, Alnico (Fe–Al–Ni–Co) magnets can be sintered instead of cast to final shape. Only powder metallurgy is suitable for the extremely powerful, elongated single-domain magnets which consist of $R \cdot Co_5$ (where R is a rare earth such as Sm) particles. Compaction takes place while the powder is oriented in a magnetic field.

7. In *dispersion-strengthened alloys* a high hot strength is secured by the presence of finely distributed, stable oxides which prevent grain-boundary migration and slip. Often, the oxide is produced by internal oxidation after the powder compact of the constituent metals has been made [e.g., jet engine components from thoria-dispersed (TD) nickel with 2% ThO_2, copper welding electrodes with dispersed alumina, and sintered aluminum powder (SAP) with up to 14% Al_2O_3].

8. Increasing quantities of *tool steels* are made by powder metallurgy. Such high-speed steel tools have a much finer carbide distribution, and the carbide content can be raised beyond the limits encountered in conventionally made steels; hence, tool life increases too.

9. Tool, die, and wear-resistant materials of great importance are the *cemented carbides*. Tungsten carbide (WC) powders are milled with cobalt, so that each particle is coated with the metal. After pressing, liquid-phase sintering establishes full density. Sometimes the final shape is given by grinding a presintered compact, which is then finish-sintered. Powders of 1–5 µm particle size are sintered for tools and dies. With the cobalt content increasing from 3 to 15%, hardness decreases but ductility increases; die components subjected to bending stresses often contain up to 30% cobalt. Further improvements, at least for steel-cutting purposes, are obtained by replacing some of the WC with TiC. For wear components the particle size is < 10 µm, although submicron powders are increasingly used for higher properties.

10. Cemented carbides belong to the broader class of *cermets* (ceramic-metal composites) which, by definition, include heavy-duty metallic friction materials such as Cu–Sn or Cu–Zn alloys with embedded SiO_2, Al_2O_3, etc.; nuclear fuel elements with UO_2 dispersed in aluminum or stainless steel; and metal–nonmetal composites such as electrical contacts and brushes that contain graphite, and metal-bonded diamond (typically with copper and 15–20% tin). Most often the term is used in conjunction with cutting tools. Newer members of the family are TiC bonded with a Ni–Mo alloy or with 50–60% tool-steel binder. The latter has the advantage that the part can be machined after sintering; strength is imparted by final heat treatment. Some typical applications are noted in Table 9–3 and in Sec. 9-5-1.

11-7 PROCESS CAPABILITIES AND DESIGN ASPECTS

Powder metallurgy is capable of producing near-net-shape or net-shape parts; hence its principal application is to small- to medium-size parts of complex shape such as gears, cams, and levers. General capabilities of the technique are shown in Table 11–2. The materials indicated are those most widely used, although others can also be processed. Shape limitations can be readily deduced from comparisons with other techniques previously discussed. The limitations have essentially two main sources: First, the particulate material must be able to fill the mold or die cavity, and second, the completed compact must be of a shape that can be released from the mold or die.

In conventional powder metallurgy, size is limited by press capacity and by a need to allow escape of binder and lubricant from the inside of the body. Tolerances are governed by reproducibility of shrinkage. Thickness in the axial (pressing) direction is limited by the decay of pressure in a fixed container (Fig. 11–5a) to typically $2D$ and with counteracting punches to $4D$ (Fig. 11–12a). With a single punch, steps must be limited (Fig. 11–12b) to avoid density variations. Multisleeve punches allow much greater differences (Fig. 11–12c).

For longer die life, part design should provide radii for die elements (Fig. 11–12c). Feather-edged punches wear rapidly and should be changed to present a flat face (Fig. 11–12d). Sleeves should be minimum 1 mm thick (Fig. 11–12c). Internal cavities can be of various shapes but require draft (Fig. 11–12e). Sharp

Table 11-2   General characteristics of powder metallurgy processes

Characteristics	Process				
	Conventional	HIP	Injection Molding	Precision Injection Molding	Preform Forging
Part					
Metal*	All	All (super, SS)	All (steel, SS)	All	steel, super
Shape†	Not S3, T2, 3, 5, 6, F3, 5, U	Not T5, F5	Not T5, F5, U1, 4	Not T5, F5, U2, 4	Not S3, T2, 3, 5, 6, F3, 5, U
Surface detail‡	B	B–C	B	A	A
Mass, kg	0.01–5 (30)	0.1–10 not encapsulated 10–7000 encapsulated	0.01–0.2	0.005–0.2	0.1–3
Min. section, mm	1.5		1	0.1	3
Min. core diam., mm	4–6		1	0.2	5
Tolerance, ±%	0.1	2	0.3	0.1	0.25
Cost‡					
Equipment	B–C	A	A–B	A–B	A–B
Die	B–D	B–C	A–B	A–B	A–B
Labor	D–E	C–D	D–E	D–E	D–E
Finishing	C–E	B–D	C–E	D–E	D–E
Production					
Operator skill‡	D–E	D–E	D–E	D–E	D–E
Lead time	Weeks	Weeks	Weeks	Weeks	Weeks–months
Rates (piece/h)	100–1000	5–20	100–2000	100–2000	200–2000
Min. quantity	1000–50 000	1–100	10 000	10 000	100 000

*Most frequently used metals: super = superalloy, SS = stainless steel.

†From Fig. 3–1.

‡Comparative ratings, with A indicating the highest value of the variable, E the lowest. For example, conventional PM produces good surface detail, involves medium to high equipment cost, moderate to high die cost, low labor cost, medium to low finishing cost, and low operator skill. It can be used for medium to high production rates and requires a minimum quantity of 1000 to 50 000 parts to justify the die cost.

corners and feather edges should be avoided on the punch too (Fig. 11–12*f*). Through-holes should be of minimum 4–5-mm diameter to prevent premature core rod failure (Fig. 11–12*c*). Large wall thickness differences can cause problems with shrinkage stresses (Fig. 11–12*g*). Narrow spaces cannot be filled even under pressure; therefore, wall thickness should be minimum 1 mm (or, preferably, 1.5 mm, Fig. 11–12*h*).

Transverse holes, undercuts, and reentrant shapes are not permissible. Despite these limitations, hard tooling is the most suitable for mass production. Dimensions can be well controlled; shape complexity in the plan view can be substantial (as in gears), and production rates are high. It is possible to join several separately molded pieces and assure the virtual disappearance of the joint during sintering.

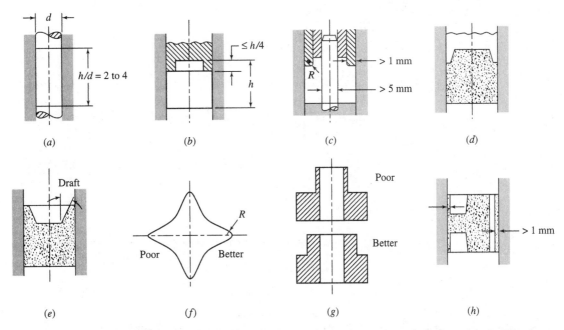

Figure 11-12   Design features of powder-metallurgy parts: (a) length to diameter ratio, (b) stepped end, (c) multisleeve die, (d) elimination of feather edge on punch, (e) draft on punch, (f) avoiding sharp edges on punch contour, (g) avoiding unequal wall thickness, (h) minimum wall thickness. (*Adapted from Powder Metallurgy Design Manual, 2d ed., Metal Powder Industries Federation, Princeton, New Jersey, 1995. With permission.*)

Greater freedom is afforded by flexible isostatic compacting molds which permit undercuts or reverse tapers, but not transverse holes.

Most limitation are overcome by metal injection molding which is subject to the same rules as die casting (Sec. 7-8-2). Uniform wall thickness is preferred, and transitions to a different wall thickness should be tapered or at least radiused (Fig. 11–13a). Localization of mass results in shrinkage (*sink, draw-in*) and can be avoided as in casting (Fig. 11–13b).

Example 11-8

Explore the possibility of making the flange of Example 7-9, part (b) by powder metallurgy.

The height-to-thickness ratio of the collar is $20/5 = 4$, acceptable, and wall thickness is well in excess of the minimum. To provide for the much longer stroke needed for the collar, the die set will have a separately operated sleeve that also acts as an ejector. Assume: fill density $= 35\%$; green density $= 72\%$. Thus, fill height $= 0.72/0.35 = 2.06$ (green height). The depth of flange cavity must be $5(2.06) = 10.3$ mm and of the collar cavity $20(2.06) = 41.2$ mm.

This ignores shrinkage. If sintering increases density to 95%, volumetric shrinkage is, from Eq. (11-3b): $= 0.72/0.95 = 0.758 (24.2\%)$ and linear shrinkage, from Eq. (11-3b) $= 0.9117$. Therefore, increase all dimensions by $1/0.9117 = 1.0968$ or 9.68%.

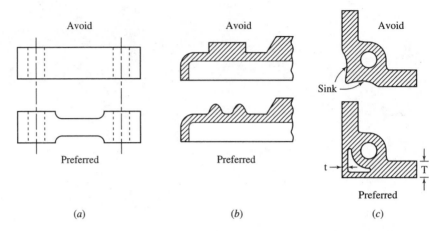

Figure 11-13   Design features of injection-molded powder-metallurgy parts: (*a*) minimizing wall thickness differences, (*b*) breaking up a heavy section, (*c*) avoiding large mass by coring. (*As Fig. 11–12.*)

Example 11-9 | The part of Example 7-10 is to be produced by powder metallurgy. The same considerations apply as in Example 11-8, with the difference that movement of the die elements becomes more complicated. After filling, the lower center punch must retract together with the upper center punch and compaction can begin only after they have reached their proper positions.

11-8   ELECTROFORMING

The smallest particulate is the atom (or, for a compound, the molecule). Components may be produced through controlled deposition of atoms on a surface; one speaks of *plating* and *coating* (Chap. 19) when the deposit is to stay in place (as in the chromium plating of a car bumper) and of *forming* when the deposit, stripped from the form (variously called *matrix, mandrel, die,* etc.), serves as a component.

In the process of *electroforming*, a plate or slab of the metal (the *anode*) is immersed into an aqueous solution of a salt of the same metal (the *electrolyte*), and is connected to the positive terminal of a low-voltage, high-current dc power supply (Fig. 11–14). An electrically conductive mold (matrix or mandrel) of the desired shape is immersed at some distance from the anode and is connected to the negative terminal (and thus becomes the *cathode*). Metal atoms are removed as positive ions from the anode, transported through the electrolyte toward the cathode (and are, therefore, called *cations*), and deposited on the cathode as neutral atoms.

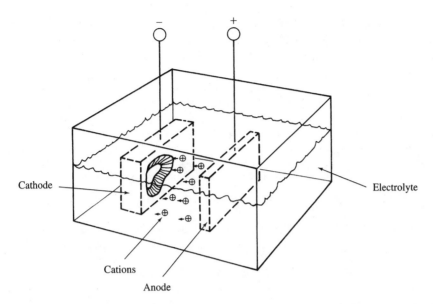

Cathode

Electrolyte

Cations

Anode

Figure 11-14   Complex shapes may be reproduced with great detail and accuracy by electroforming.

It takes 96 500 coulomb ($=$ ampere·second) to remove 1 mol of monovalent metal (*Faraday constant*). The metal transfer rate is then

$$W_e = \frac{j}{96\,500} \frac{M}{Z} \eta \tag{11-4}$$

where W_e is g/s·m^2, j the current density (A/m^2); M the gram atomic weight (g/mol), Z the valence (charge/ion) and η the efficiency (typically around 0.9).

The composition, temperature, and circulation of the electrolyte and the current density need careful control. Once a deposit of sufficient thickness is obtained (and this may take hours or days), it is stripped from the matrix.

Permanent matrixes may be made of metal, or of glass or rigid plastic with a metallized surface (e.g., metallized by a chemical deposition technique). Adhesion is minimized by passivating the metal matrix surface and by the application of a thin coating of a parting compound. A slight taper is allowed to facilitate stripping.

Expendable matrixes are made of a metal (aluminum or zinc) that can be chemically dissolved, or of a low-melting alloy (such as eutectic Sn–Zn alloy), wax, or plastic that can be melted out. Since the finished part is not stripped, great freedom in shape complexity is gained (comparable to investment casting, Sec. 7-5-5).

The atom-by-atom deposit reproduces the matrix surface with the greatest accuracy, and this, together with the attainable shape complexity, defines the economical application range of the process to finished parts (such as waveguides, bellows, venturi tubes, reflectors, seamless screen cylinders for textile printing, filters, and typing wheels) and dies (for stamping of high-fidelity records and for plastic molding in general). Internal stresses can be severe, and there is an art to producing sound parts.

Example 11-10

A highly decorated vase is made by electroforming into a conductive die (matrix). The total surface area is 0.2 m^2; deposition proceeds at 6 V and 80 A, from a copper sulfate ($CuSO_4$) solution. How long will it take to attain a wall thickness of 0.5 mm (500 μm)?

Current density $j = 80/0.2 = 400$ A/m^2; $M = 63.54$ g/mol; $Z = 2$. From Eq. 11-4

$$W_e = (400)(63.54)(0.9)/(96\,500)(2) = 0.1185 \text{ g/s·m}^2$$

Since the density of copper is 8.96 g/cm^3 = 8.96 Mg/m^3, the thickness of the layer is $0.1185/8.96(10^6) = 0.013$ μm/s. Thus, time $= 500/0.0132 = 37\,800$ s $= 630$ min $= 10.5$ h.

11-9 SUMMARY

Particulate matter, ranging in size from atoms to coarse powder, has been consolidated into usable products from the earliest times. The technique, first applied to ceramics, is suitable also for metals. Powder metallurgy processes comprise several steps critical to success:

1. Elemental metal powder may be obtained from oxides or other compounds by reduction. Increasing quantities of pure metal and, particularly, alloy powder are produced by atomization of melts. High cooling rates in rapid solidification technology offer a number of benefits: The size and spacing of features such as dendrite arms and second-phase particles are reduced; supersaturated solid solutions may be retained; in the limit, amorphous (glassy) metals may be produced.

2. The powder is comminuted when necessary, classified according to size and shape, cleaned, and blended to impart the required composition, fill density, and rheological properties that allow easy handling.

3. A green body is produced by a variety of processes generally classified as cold consolidation (die pressing, isostatic pressing, injection molding, rolling, or extrusion).

4. Permanent bonds are established by sintering at high temperatures, thus developing strength while the volume shrinks. Remaining voids or grain-boundary defects impair fatigue and impact properties, and one of the aims of advanced techniques is improving the fracture toughness of finished parts. Liquid-phase sintering is applicable to cermets and partial liquid-phase sintering to aluminum alloys made from elemental powders.

5. Great improvement in properties is attainable by hot consolidation, i.e., the application of pressure at the sintering temperature. Particles are brought into close proximity and are moved relative to each other, thus promoting adhesion between them, and full density is attained.

6. In addition to fully densified parts of improved properties, such as jet engine components and metal-cutting tools, parts of great shape complexity and close tolerances, such as gears, may be produced. Powder injection molding avoids many of the shape limitations of conventional processes. Powder metallurgy is often the most practical method for producing composites of unusual properties and parts with controlled porosity.

7. Powder metallurgy processes and equipment present the same hazards as foundry and metalworking operations, and similar protective measures must be taken.

PROBLEMS 11A

11A-1 (*a*) In powder metallurgy, does the term "atomization" refer to the production of particles of atomic size? (*b*) Make a simple sketch of the water atomization process.

11A-2 (*a*) Define RST and (*b*) indicate the most advanced (unusual) product that can result from this technique.

11A-3 Make sketches of dies suitable for compacting cylinders with height-to-diameter ratios of (*a*) 0.5 and (*b*) 2. Indicate the movement of die elements and the direction of friction.

11A-4 Make a sketch of a die that assures equal density in an axially symmetrical part resembling a flange. Show the direction of die-element movement.

11A-5 Draw a diagram showing changes of (*a*) density, (*b*) strength, and (*c*) ductility as a function of time in the course of sintering (plot properties as fractions of the properties of wrought products). (*d*) State the temperature of sintering (in generic terms).

11A-6 A machine component (such as a spur gear) is made of a low-alloy (AISI 8620) steel by the powder metallurgy technique. List the sequence of steps in the course of production.

11A-7 State the mechanisms that give strength to a green compact made by cold pressing.

11A-8 Make sketches of isostatic pressing with (*a*) wet bag and (*b*) dry bag methods.

11A-9 (*a*) Define HIP. (*b*) Make a sketch of the process, identifying the principal elements.

11A-10 List the steps in making a connecting rod by powder forging.

11A-11 Briefly list the essential steps in powder injection molding.

11A-12 Define spray forming.

11A-13 (*a*) Define the term *cermet* and (*b*) give two examples.

11A-14 State the sequence of operations in making a permanently lubricated bearing. (State also the critical condition for achieving the aim.)

PROBLEMS 11B

11B-1 Make sketches to show (*a*) acicular, (*b*) spheroidal, (*c*) nodular, (*d*) lamellar, (*e*) dendritic, and (*f*) irregular powder shapes. (*g*) State which of these is likely to give highest fill density.

11B-2   State whether a uniformly fine or uniformly coarse spherical powder will give highest fill density. Justify your choice.

11B-3   State whether, in general, brittle or ductile particles can be consolidated to a higher density. Justify your choice.

11B-4   State whether, in general, soft or hard particles can be consolidated to a higher density. Justify your choice.

11B-5   Is it possible to comminute a highly ductile material such as pure aluminum?

11B-6   Explain why (a) porosity impairs the mechanical properties of powder-metallurgy parts, and why (b) it impairs tensile and impact properties more than (c) compressive strength or hardness.

11B-7   A small spur gear is needed for an appliance. Suggest at least three materials from which it could be made by powder metallurgy, and identify the benefits for each.

11B-8   (a) Is it physically possible to design a die that allows the part of Problem 7B-10 to be made by powder metallurgy techniques? (b) If it is, is it likely to be technically and economically attractive?

11B-9   To explain the difference between a hot-forged and powder-forged part, make a simplified sketch of the longitudinal cross section of a connecting rod made by (a) impression-die forging (before removal of the flash) and (b) powder forging.

11B-10   The design of a cylindrical part incorporates a transverse hole. (a) Make a sketch to visualize the part. (b) Exercise critical judgment to evaluate whether it can be made by any powder metallurgy technique.

11B-11   (a) State the sources of strength in a green metal-powder body which contains no binder. (b) State whether normal pressure alone or normal pressure combined with sliding is more effective in developing green strength. Justify.

11B-12   In powder metallurgy, why are oxide films (a) disastrous for the ductility of superalloy products but (b) less harmful for titanium-alloy products?

11B-13   Explain the differences to be expected in the structures of a high speed tool steel produced by conventional ingot metallurgy and powder metallurgy

11B-14   It is customary to speak of tungsten carbide tooling. (a) Define what it is and (b) list the major steps in producing it.

11B-15   A steel part must have 100% density. (a) Explain whether this can be achieved by sintering alone. (b) Briefly define two other suitable P/M processes.

11B-16   (a) Suggest at least two methods by which the mechanical properties of an already sintered powder metallurgy part can be improved. Justify. (b) Suggest a method by which a part of improved mechanical properties could be made directly (without sintering first).

11B-17   Sketch a die suitable for making a cup-shaped part by powder metallurgy.

11B-18   Conceptual designs include each of the four shapes in group U of Fig. 3–1. Make a judgment whether they can be made by any powder metallurgy technique. Identify the appropriate technique(s) for each.

11B-19   Powder metallurgy is one of the routes to producing: (a) steel gears, (b) tungsten wire, (c) HSS cutting tools, (d) "permanently lubricated" bearings, (e) elongated single-domain magnets. State, for each case, the principal reason for choosing this technique in preference to other techniques.

11B-20   The turbine disk of a jet engine is to be made of a nickel-base superalloy by powder metallurgy techniques. State, in one sketch each, the essential steps of the process; specify the pressing operation that gives the most uniform density.

PROBLEMS 11C

11C-1 Continue Example 11-3 to determine what maximum thickness (length) is allowed in the design of the part if stearic acid is added ($\mu = 0.1$).

11C-2 What is the pressure in the midlength of an $l/d = 4$ cylinder pressed on a double-acting press (Fig. 11–5c) if compaction takes place (a) unlubricated ($\mu = 0.5$) and (b) lubricated ($\mu = 0.1$). Poissons's ratio is 0.31.

11C-3 An Al-5Mg alloy is prone to coring. (a) With the aid of a sketch of the relevant portion of the phase diagram, explain why this should be so. (b) Consider whether coring could be eliminated more rapidly in a casting or in a powder-metallurgy part.

11C-4 A compacted body of iron powder is sintered into a 20-mm-diameter, 45-mm-tall cylinder which, upon weighing, is found to have a mass of 98 g. Calculate (a) the apparent density, (b) the percentage of theoretical density, and (c) the void volume (porosity) in percent.

11C-5 Assuming that the cylinder of Problem 11C-4 is sintered until full theoretical density is obtained, and shrinkage is uniform in all directions, calculate the dimensions of the cylinder.

11C-6 For cold pressing the part of Example 11-8, (a) design the die (show which parts move), (b) calculate the press size if the cold-compaction pressure is 200 MPa, and (c) calculate the force the ejector must develop to assure equal compaction in the collar. (d) If, for higher density and greater dimensional accuracy, the part is sintered and then repressed, can the same die be used?

11C-7 For the part of Example 11-9, design the die. Show the position of die elements for filling and before and after compaction. For simplicity, assume a fill density of 33% and compact density of 66%. (*Hint*: Study the die movements in Fig. 11–6.)

11C-8 A cylinder of $d_0/h_0 = 1$ is compacted to 82% theoretical density by cold pressing an atomized alloy steel powder. Full density and high strength are to be obtained by hot upsetting the cylinder to one-third of its original height. (a) What diameter should one expect, approximately, after upsetting? (b) Should one anticipate cracking in upsetting? If yes, where and why? (Illustrate with a sketch.) (c) If cracking is a danger, how could it be prevented or minimized? (d) Could cracking be prevented in a die of the final diameter?

FURTHER READING

ASM Handbook, vol. 7, *Powder Metallurgy*, ASM International, 1998.

Atkinson, H.V., and B.A. Rickinson: *Hot Isostatic Pressing*, Adam Hilger, 1991.

Dowson, G.: *Powder Metallurgy: The Process and Its Products*, Adam Hilger, 1990.

German, R.M.: *Sintering Technology and Practice*, Wiley, 1996.

German, R.M.: *Powder Metallurgy Science*, Metal Powder Industries Federation, 1985.

German, R.M., and A. Bose: *Injection Molding of Metals and Ceramics*, Metal Powder Industries Federation, 1997.

Hausner, H.H., and M.K. Mal: *Handbook of Powder Metallurgy*, Chemical Publishing Co., New York, 1982.

Karlsson, L. (ed.): *Modeling in Welding, Hot Powder Forming and Casting*, ASM International, 1997.

Kuhn, H.A., and B.L. Ferguson, *Powder Forging*, Metal Powder Industries Federation, 1990.

Lawley, A.: *Atomization: The Production of Metal Powders*, Metal Powder Industries Federation, 1992.

Lenel, F.V.: *Powder Metallurgy: Principles and Applications*, Metal Powder Industries Federation, 1980.

Liebermann, H.H. (ed.): *Rapidly Solidified Alloys*, Dekker, 1993.

Powder Metallurgy Design Manual, 2d ed., Metal Powder Industries Federation, 1995.

Advances in ceramics processing have resulted in the development of ceramic machine elements. The low density of silicon nitride balls allows hybrid roller bearings to operate at higher speeds; all-ceramic bearings can operate in hostile environments. (*Courtesy Norton Advanced Ceramics, Worcester, Massachusetts.*)

chapter

12

Processing of Ceramics

After reviewing the basic properties of ceramics, we will explore:

The manufacture of technical ceramics

Consolidation and sintering

Measures to mitigate the brittle nature of ceramics

The many applications of technical ceramics

Glasses and their manufacture

Ceramics, even though brittle, have been indispensable in human development as building materials, containers, cooking vessels, and, generally, corrosion-resistant materials. Indeed, without ceramics, much evidence of the life and art of our ancestors would be missing. Ceramics have lost none of their importance, and new members of the family have only expanded their field of application. High-technology ceramics of exceptional electrical and magnetic properties have made the microelectronics revolution possible; others have opened up the possibility of producing structural components of hitherto unattainable temperature and wear resistance. Many ceramics processes are of great antiquity, yet others are at the cutting edge of development (Table 1–1).

In the narrower definition, *ceramics* are compounds of metallic and nonmetallic elements. This leaves out such materials as diamond, SiC, and Si_3N_4, and a broader definition regards ceramics as everything that is not a metal or organic material and is subjected to high temperature during manufacture or use. In addition to a vast variety of naturally occurring silicates and oxides, the definition includes manufactured materials, sometimes of similar composition but of greater purity, and at other times carbides, nitrides, and other compounds not found in nature. By definition, glass is a ceramic and will be discussed in this chapter.

12-1   CHARACTERISTICS OF CERAMICS

Before we can embark on a discussion of processes, we need to examine the unique characteristics of ceramics; this will be a review for those who have had an introductory course in materials science.

12-1-1   Bonding and Structure

In metals the atoms were mobile because bonding was by a cloud of electrons; in ceramics, mobility is sharply limited by different bonds.

Nature of Ceramic Bonds   Several forms of bonding may play a role:

1. *Covalent bonds* are formed by electrons shared between adjacent atoms. These are very strong bonds. The directional nature of bonds often leads to the formation of a spatial framework in which atoms are not necessarily closely packed. The high bond strength reflects in a high melting point, strength, and hardness coupled with brittleness; thermal expansion is often low, electrical resistance high. Carbon in the form of diamond is a purely covalently bonded material (Fig. 12–1a).

2. *Ionic bonds* form when one atom gives up one or more electrons to complete the outer electron shell of another atom or atoms. Electrical charge balance is maintained, but the electron donor is now deficient in electrons and thus be-

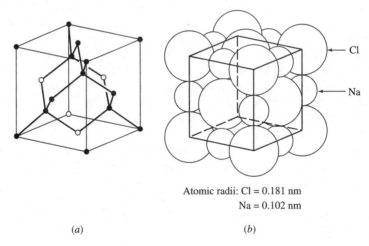

Atomic radii: Cl = 0.181 nm
Na = 0.102 nm

(a)   (b)

Figure 12–1   The two basic bond types in ceramics are (a) covalent bonding, as in diamond in which each carbon atom shares electrons with four adjacent atoms and (b) ionic bonding, as in NaCl in which sodium and chlorine ions alternate regularly.

comes a positive charge center (*positive ion* or, because it is attracted to the negative terminal, also called *cation*), whereas the recipient atom has an excess negative charge (*negative ion* or *anion*). Attraction between the opposing charges (coulombic attraction) provides the bond strength.

An example is common salt, NaCl, in which the charge of each Na^+ iron is balanced by the charge of the Cl^- ion. A spatial network is again formed, but this time it is closely packed. For example, the NaCl lattice can be visualized as interpenetrating cubic Na and Cl lattices (Fig. 12–1b). Electrical conductivity is low at low temperatures but increases at higher temperatures when ions move and carry the electric charge (*ionic conductivity*). In such a regular structure dislocations can propagate somewhat as they do in metals; hence, even though the structures are brittle at low temperature, some ductility is evident at high temperatures and when hydrostatic pressure is applied. Bond strength and melting point increase with increasing charge. Thus, NaCl is relatively soft; MgO (double charge) is harder, Al_2O_3 is harder yet, and SiC (in which there are four bonds) is hardest.

3. *Dual bonds* exist in many ceramics in which electrons tend to concentrate toward atomic centers, giving the bonds some degree of covalent character even in ionic compounds. The bond becomes more covalent in character with a decreasing difference in *electronegativity* between the atomic species (electronegativity is a measure of the affinity of atoms to attract electrons). The degree of covalent character can be estimated for different ceramics: $MgO = 0.25$; $SiO_2 = 0.5$; $Si_3N_4 = 0.7$; $SiC > 0.9$; $C = 1$. More complex ceramics may incorporate both bond types: in gypsum ($CaSO_4$) the S is bonded covalently to O but the SO_4^{2-} group is ionically bonded to Ca^{2+}.

4. *Secondary bonds* are extremely important when ceramics form layered structures of platelets. Within the platelets the strong, predominantly covalent bonds ensure great strength; between platelets only secondary bonds act and the strength of these bonds can be influenced by the addition of molecules of a gas or liquid. Such molecules adsorb on the surfaces of platelets, allowing easy movement and thus facilitating processing.

5. *Multicomponent ceramics* are made up of two or more simple ceramics such as oxides and can be regarded as the ceramic counterparts of metal alloys. The resulting phases are shown in equilibrium diagrams which reveal features similar to those found in the phase diagrams of metals (Sec. 6-1). When ceramics form separate phases, one speaks of *multiphase* ceramics. Thus, a ceramic may be toughened by one phase dispersed in the matrix formed by the other phase.

Crystallinity   In the stable, lowest-energy form, most ceramics are *crystalline*. Atoms occupy defined lattice sites over long distances. As the temperature or pressure changes, different crystalline structures, *polymorphs*, may become more stable. Because there is a change in crystal structure, *polymorphic transformations* are accompanied by a volume change. The magnitude of this change is usually larger than that caused by allotropic transformations in metals, and can lead to crazing, fracture, or total destruction of the part. In a *displacive transformation* the bonds are retained but distorted to allow a new crystal structure to form, somewhat like in the austenite–martensite transformation in steel (Sec. 6-4-3). In a *reconstructive transformation* the new structure is formed by breaking existing bonds; this takes more driving energy and the transformation may be suppressed by rapid cooling.

Some ceramics are noncrystalline (*amorphous*) and are then called *glasses*. A glass forms when a normally crystalline ceramic is heated above its melting point and then cooled so fast that crystallization is suppressed. The bonds are the same as in a crystalline ceramic, but the long-range lattice arrangement is missing in this *glassy state* (also called, from the Latin, *vitreous state*). Whereas extremely fast cooling rates are needed to make metallic glass (Sec. 11-2-1), ceramic glasses are formed at industrially practical cooling rates. If such a glass is held at elevated temperatures for a long period of time, the more stable crystalline form is again obtained (the glass crystallizes or, as it is often said, it is *devitrified*). A more detailed discussion of glasses will be given in Sec. 12-5. Noncrystalline solids can also be formed by chemical reaction; the resulting *gels* are colloidal structures related to the everyday jelly but of substantial strength.

In general, the properties of amorphous ceramics and some ceramics crystallizing in the cubic form are isotropic, whereas the properties of those crystallizing in more complex forms can be highly anisotropic.

12-1-2   Properties of Ceramics

Ceramics are used as engineering materials because of their high strength, hardness, hot strength, corrosion resistance, and desirable electric, magnetic, and optical properties.

Mechanical Properties   Mechanical properties are measured by techniques described in Chap. 4. A common feature of ceramics is that there are minute cracks in most of them. Thus, they are brittle and their mechanical properties are subject to random variations (they are *probabilistic* in nature, in contrast to most metals which are *deterministic*). This means that a sufficient number of tests must be carried out to obtain a meaningful measure of the scatter to be expected. Evaluation is often based on statistical treatment and design must take these realities into consideration. A condensed treatment of the topic will be found in *Engineered Materials Handbook Desk Edition*, ASM International, 1995.

1. Tension testing is difficult. Extreme care must be taken not to introduce surface defects during specimen preparation and to avoid eccentric loading in the test. Bending tests are easier to carry out to determine the modulus of rupture (Sec. 4-1-7) but specimen preparation is equally critical. The four-point bending test is favored because the more uniform stress distribution makes the discovery of cracks more likely (see Example 4-6). The large stress concentrations introduced by sharp cracks make ceramics vulnerable to fatigue failure, and ceramic components are increasingly designed by the fracture-mechanics approach.

2. Cracks are less harmful in compression (Sec. 4-3) and the compressive strength of ceramics is generally several times higher than their tensile strength. Perfect alignment of platens is critical if bending stresses are to be avoided in the compression test. In the absence of plastic deformation, elastic deformation is terminated with catastrophic disintegration when a critical density of defects is reached. Hardness tests are also used.

3. Ceramics retain their strength and hardness to high temperatures, and those destined for elevated-temperature applications are subjected to high-temperature creep tests of long duration, since creep rates are extremely low. Only close to their melting point do some ceramics exhibit significant plasticity.

Because of the critical role of cracks, ceramic components are extensively tested by various NDT methods (Sec. 4-8).

We saw in Example 4-6 that Si_3N_4 gave a rupture strength of 930 MPa in a three-point bending test but only 725 MPa in a four-point bending test. The uniform stress distribution between the loading points in the four-point test (Fig. 4–9*b*) brought to light imperfections that were missed in the three-point test. A tension test on the same material gave a maximum load of 11.3 kN on a specimen of 3.2 mm × 6.4 mm cross section. From Eq. (4-8), TS = 11 300/[(3.2)(6.4)] = 552 MPa. Thus, the tension test is the most critical and is most likely to detect defects. However, tensile test results are greatly affected by bending that may be inadvertently introduced by the slightest misalignment in the test apparatus; therefore, the four-point bending test is preferred.

Example 12-1

Improving Mechanical Properties   Several techniques are available:

1. *Reduce particle size.* Tensile properties and toughness improve with decreasing particle size because flaws are generally of the size of the constituent grains; therefore, ceramics of *submicron* particle size are often used in high-technology applications. Yet higher properties are obtained with *nanoscale* (< 0.1 μm) particles.

2. *Retard the propagation of large cracks.* Very small cracks may be unavoidable, but damage can be limited by incorporating into the structure features that retard crack propagation. Such features are more effective if they are smaller and more closely spaced. Three approaches are possible:

a. Incorporate particles that suffer *phase transformation.* A prime example is *partially stabilized zirconia* (PSZ). Zirconia has a tetragonal lattice at high temperatures, and on cooling transforms to a monoclinic form with a disastrous

(3.25%) volume change. However, by adding small amounts of Y_2O_3, MgO, or CaO, it can be fired to have a predominantly cubic structure. Within this stable cubic matrix, small islands of the metastable tetragonal phase are included. The pressure exerted by a crack propagating through the material causes transformation of these islands into the monoclinic form; some of the energy is absorbed in the transformation and, most importantly, expansion of the transformation product puts the matrix in compression.

b. Incorporate fine *fibers* (such as graphite or silicon carbide fibers of less than 50-μm diameter); the crack follows the weak interface and is arrested.

c. Create intentionally weak interfaces that produce a multitude of crack-retarding *microcracks* ahead of the major crack. Propagation of cracks may be arrested also by incorporating particles of differing thermal expansion, so that very fine cracking is induced on cooling; even though these cracks reduce static strength, they prevent the propagation of large cracks.

3. *Induce compressive residual stresses.* We have seen that strength properties can be improved by imparting compressive residual stresses to the surface of the component (Sec. 4-7). In ceramics this can be achieved by a number of techniques: quenching (Sec. 12-5-2); replacing some surface ions with larger ions (ion exchange or ion stuffing, Sec. 12-5-2); forming a low-expansion surface layer at high temperatures which is then put into compression on cooling (as in coating Al_2O_3 parts with an Al_2O_3–Cr_2O_3 coating); formulating the ceramic so that a displacive polymorphic transformation results in the expansion of a surface layer; and grinding conducted under conditions that lead to surface deformation.

4. *Reduce creep.* The strength and hardness of crystalline ceramics remains high close to the melting point. They are, however, subject to creep. Creep in single crystals can occur only by dislocation movement, which can be blocked by precipitate particles. In polycrystalline ceramics creep involves diffusion and grain-boundary sliding, both of which are made easier when porosity is present. Hence, *minimization of porosity* is one of the aims of manufacturing processes. However, if densification was aided by the formation of a glass at particle boundaries, viscous creep of this glass will allow sliding of grains and accelerates the creep of the entire body.

Physical Properties   Since the covalently bonded ceramics are not closely packed, they can accommodate increasing atomic vibrational amplitudes without a change in macrodimensions; thus, their thermal expansion is lower than that of metals. Some polycrystalline ceramics such as lithium aluminum silicate ($LiAlSi_2O_6$) have zero expansion and can be heated or cooled rapidly without damage.

The electrical properties (Sec. 4-9-3) of ceramics range from conductors (graphite) through semiconductors (SiC) to insulators (Al_2O_3). Many ceramics also have a high dielectric strength; thus, they withstand high electrical fields without breaking down, and have allowed the miniaturization of capacitors. Some ceramics exhibit *piezoelectricity*: A crystal subjected to mechanical loading generates a potential difference and can be used as a force transducer; in the reverse mode, a potential difference applied to the crystal causes a dimensional change which can be exploited in ultrasonic transducers and force generators. The bulk resistivity of *piezoresistive* materials changes significantly upon the imposition of a stress. Yet other ceramics are *pyroelectric*: they develop a voltage in response to a temperature difference. Ceramics also exhibit the complete range of magnetic properties (Sec. 4-9-4). Without ceramics, the solid-state electronics revolution would have been impossible.

Ceramics can be formulated to provide the full range of *optical properties* (Sec. 4-9-6). Single crystals of ionically bonded ceramics are usually transparent whereas covalently bonded ceramics may range from transparent to opaque. Grain boundaries and defects such as pores and cracks that create internal reflecting surfaces reduce transparency; only isotropic ceramics are transparent in the polycrystalline form. By appropriate additions, a selective wavelength of the visual spectrum can be absorbed, giving ceramics the widest range of colors. The index of refraction can also be controlled; this affects the change in direction when a light beam enters a solid (at point B in Fig. 3–17a) and is most important in applications such as lenses or decorative "crystal" glass.

Light tubes, video-display terminals, and color television rely on *phosphorescence*: Ceramic phosphors emit light of a characteristic wavelength when stimulated by an electric discharge or electron beam. Of rapidly increasing industrial importance are lasers (Sec. 17-5-2), some of which utilize a single-crystal rod made of a ceramic.

Chemical Properties   A great advantage of ceramics is that they are often resistant to chemical attack by gases, liquids, and even high-temperature melts. Combined with their remarkable high-temperature strength, this makes them suitable for such applications as temperature-resistant furnace linings (*refractories*), insulators, and even mechanical components such as turbine disks, turbine blades, and various components of internal combustion engines.

12-2 CLASSIFICATION OF CERAMICS PROCESSES

We already mentioned that some ceramics occur in nature and others are manufactured. Natural raw materials have been dominant for thousands of years and are the starting material for what are now described as *traditional ceramics*. The high demand placed by engineering applications led to the development of modern ceramics: Some natural ceramics are replaced by manufactured (synthetic) versions of controlled purity, grain size, and porosity, and ceramics are made that do not occur in nature at all. Advanced forms of these ceramics are usually called *advanced* or *high-technology ceramics*, or, in Japan, *fine ceramics*. Processing routes are similar for all (Fig. 12–2).

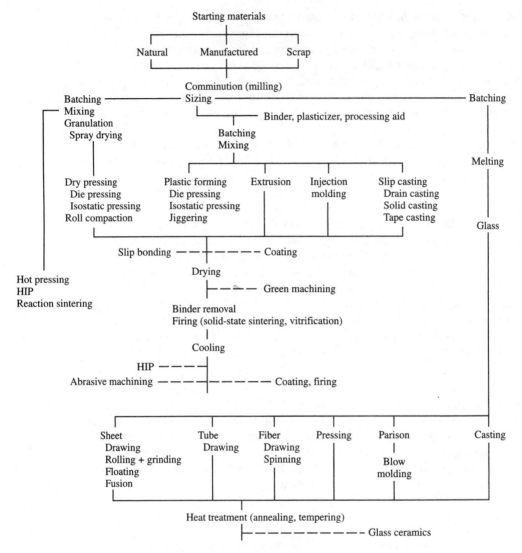

Figure 12–2   Processing sequences in ceramics manufacturing. (*Adapted from J.A. Schey, ASM Handbook, vol. 20, Materials Selection and Design, ASM International, 1997, p. 698. With permission.*)

12-3   CERAMIC MATERIALS

The typical process sequence for ceramics (Fig. 12–2) shows that starting materials include a significant proportion of scrap. Process scrap has always been recycled but the rising cost of disposal has led to increased use of postconsumer scrap, especially in glass making.

12-3-1   Natural Ceramics

Natural ceramics are mined, in open-pit mines whenever possible. After comminution (reduction in size), undesirable components are removed by screening, magnetic separation, filtering, or flotation. In *flotation* the particulate mass is suspended in water, and a frothing agent is added which preferentially attaches itself to one or the other of the mineral species, causing it to rise to the surface. Thus, either the desirable mineral or the unwanted species (*gangue*) can be separated economically. The most frequently used natural ceramics are:

1. *Silica* (SiO_2) is abundant in nature. It forms a high-viscosity melt at 1726°C. On cooling it crystallizes and undergoes several polymorphic transformations. Since Si is tetravalent, it forms a tetrahedron with four oxygen atoms (Fig. 12–3). The tetrahedrons then join into a spatial network, with each O atom attached to two Si atoms, resulting in the ratio SiO_2. The hexagonal form is called *quartz*. The large single crystals found in nature or grown in manufacturing plants are valuable because they exhibit piezoelectricity and, ground to exact thickness, are used to control the frequency of oscillators.

2. *Silicates* are obtained when other atoms or oxides are introduced into the silica framework. A tremendous variety exists. Some silicates form chains or fibrous crystals (*asbestos* family). In others the SiO_4 tetrahedra join into sheets, with the negative charge of the top oxygen atoms remaining available for bonding with other cations, giving rise to the vast number of *layer silicates*, including talcs, micas, and clays. In some sheet minerals such as mica cleavage occurs over long distances on the same plane; prior to the discovery of manufactured dielectrics,

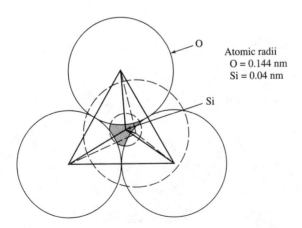

Atomic radii
O = 0.144 nm
Si = 0.04 nm

Figure 12–3   Tetravalent silicon forms a SiO_4 tetrahedron with oxygen; the remaining oxygen valences are available for the formation of a spatial silica network, compounds, or glasses.

mica sheet was used extensively. Now mica is more often comminuted and then bonded with glass to make high-precision insulators. Further ions or oxides may also be introduced into the spatial network of SiO_4. For example, stuffing the framework with Na or Ca ions leads to *feldspars*. Glasses are three-dimensional networks in which crystallinity is lost.

3. *Clay minerals* represent the most important family of natural ceramics. They can generally be described as hydrated aluminosilicates of a layer structure. Each layer crystal consists of several sheets, as shown on the example of the unit cell of kaolinite (Fig. 12–4). Each unit cell is in electrical charge balance, and the layer crystals are held together only by relatively weak van der Waals forces between the surface sheets of O^{--} and OH^- ions. Therefore dry clay is brittle and crumbly. The weak polarization of the surface is sufficient to adsorb water (*physically adsorbed water*), which facilitates sliding of the thin (approximately 50-nm-thick) plates relative to each other, making the clay plastic.

12-3-2   Manufactured Ceramics

Here we discuss only the general approaches to making the starting materials; specific ceramics—including graphite and diamond—will be discussed in Sec. 12-4-6.

Solid-phase synthesis relies on high-temperature reactions. An example of a solid-solid reaction is the mixing of a Si or SiO_2 aerosol with an aerosol of C to produce very fine *silicon carbide* (SiC) powder. Large quantities of SiC are made by passing an electric current through a long mound of coke surrounded by

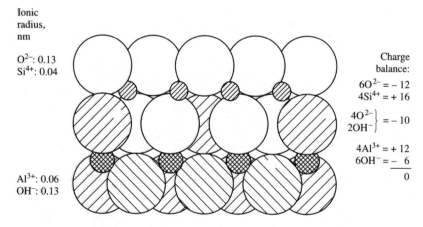

Ionic radius, nm

O^{2-}: 0.13
Si^{4+}: 0.04

Al^{3+}: 0.06
OH^-: 0.13

Charge balance:

$6O^{2-} = -12$
$4Si^{4+} = +16$

$\left.\begin{array}{r}4O^{2-}\\2OH^-\end{array}\right\} = -10$

$4Al^{3+} = +12$
$6OH^- = -\ 6$

$\overline{0}$

Figure 12–4   One constituent of clay is kaolinite, the chemical formula of which $2(OH)_4Al_2Si_2O_5$ does not reveal that layers of O^{2-}, OH^-, Al^{3+}, and Si^{4+} combine to form a layer structure which is in electrical charge balance.

high-purity sand. The purest reaction product, found in the core of the mound, is suitable for electrical applications such as high-temperature heating elements; the adjacent, less pure layer is suitable for abrasives. An example of a solid-gas reaction is the high-temperature reaction of silicon with nitrogen gas to produce *silicon nitride* (Si_3N_4), a ceramic that does not occur in nature.

Chemical synthesis is the method of producing many of the advanced ceramics of great purity and homogeneity. *Liquid-phase synthesis* produces hydroxides, carbonates, or oxalates of high purity. From these, the oxide is obtained by *calcination*, a high-temperature endothermic decomposition. Particles can be very fine (of submicron size) and have large specific surface area ($100 \text{ m}^2/\text{g}$ and more). Coarser particles are produced at higher temperatures. The largest-scale application is to *alumina* (aluminum oxide, Al_2O_3). It occurs in nature as corundum or as single-crystal gems (ruby, colored by Cr ions, and sapphire), but most industrial alumina powder is produced by thermal reduction of aluminum hydroxide. By controlling the process and cooling rates, crystallization and properties of the alumina can be adjusted to yield products ranging from relatively soft to hard. *Magnesia* (magnesium oxide, MgO) also occurs naturally, but for industrial use it is made from carbonate or hydroxide.

Vapor-phase synthesis by PVD and CVD (Secs. 19-6 and 19-7) is assuming greater industrial significance. Many varieties of oxides, carbides, nitrides, borides, and more complex ceramics are made.

Hydraulic cements contain calcium silicates and calcium aluminates. The cement is finely powdered and mixed with water, whereupon hydration takes place. The water reacts to form, over a period of days, a partially crystalline ceramic of substantial strength. The main use of cements is in *concrete*, a composite of *aggregate* (sand and gravel) and cement. An even more complex composite is formed when concrete is made more resistant to tensile stresses by the incorporation of steel reinforcing bars and even metal, polymer, or ceramic fibers.

12-4   PROCESSING OF PARTICULATE CERAMICS

The basic processing steps are the same for ceramics and metals but—primarily because of differences in the type of bonding—there are also differences in the roles various mechanisms play.

12-4-1   Preparation of Powders

Several processing steps precede consolidation.

Comminution   Most natural ceramics are mined in coarse lumps and most manufactured ceramics made at high temperatures are in a coarse mass, therefore, some process of *comminution* (reduction in size) is frequently necessary.

1. *Crushing* in jaw, gyratory, cone, or roll (Fig. 12–5*a*) crushers is suitable for earlier stages of preparation.

2. *Impacting* of particles between two hard bodies results in finer powder. The hard bodies may have a variety of shapes but are always made of a metal or ceramic chosen so as to avoid contamination. In a *ball mill* (Fig. 12–5*b*), impact energy is provided by balls dropping in a partially filled, horizontal-axis rotating drum. Higher energy is imparted to the balls and milling is accelerated by vibration (*vibratory mill*, Fig. 12–5*c*) or by the action of horizontal arms attached to a vertical, rotating shaft inside a vertical-axis drum (*attrition mill*, Fig. 12–5*d*). A similar action is exerted by rods placed in a horizontal-axis rotating drum (*rod mill*, Fig. 12–5*e*) and by rotating blades called hammers (*hammer mill*, Fig. 12–5*f*).

3. *Impact milling* yields finer powder by hurling particles against a hard, stationary surface, either by an air or other gas jet or in a slurry, usually an aqueous suspension (Fig. 12–5*g*). Alternatively, particles entrained in two opposing fluid streams are comminuted by impact on each other (*fluid-energy milling*, Fig. 12–5*h*).

A powder is termed granular when particle size is larger than 44 μm (325-mesh sieve). When particle size has diminished to a certain value (typically, below 1 μm), secondary bonding forces (van der Waals forces) lead to *agglomeration*, i.e., the formation of larger clusters of particles. Uncontrolled agglomeration is undesirable and can be prevented by appropriate additives. In

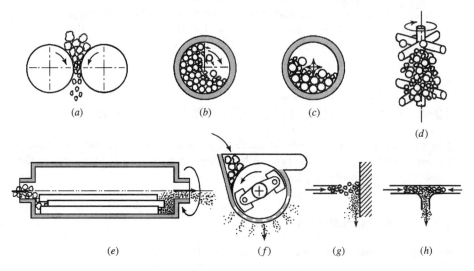

Figure 12–5   Particulate material may be comminuted (reduced to a smaller size) by many techniques, some of which are (*a*) roll crushing, (*b*) ball milling, (*c*) vibratory ball milling, (*d*) attrition milling, (*e*) rod milling, (*f*) hammer milling, (*g*) impact milling, and (*h*) fluid energy milling.

wet milling there are chemicals that impart an electric charge to the surface of particles so that particles repel each other. Thus *flocculation* (formation of loosely bonded, woolly masses) is prevented by these *deflocculants*.

Sizing   Particle size distribution, shape, specific surface area, and physical properties are determined by the methods described for metal powders (Sec. 11-2-2). Natural and many manufactured ceramics are angular and size fractions are readily separated by *screening*. Fine particles may again be separated by classification techniques described in Sec. 11-2-3.

Granulation   Powders produced by reaction in the vapor phase or from chemical precursors are very fine and it may be necessary to form hard *aggregates* for easier processing. Spray drying and granulation (Sec. 11-2-3) are often used. A related process is *freeze drying* in which a water solution of salt is atomized, the droplets rapidly frozen, the water *sublimated* (evaporated without melting), and the dry powder thus obtained is calcined to decompose the crystalline salts and obtain the ceramic in the form of dry, pure powder. Because of the short diffusion paths in the fine droplets, such particles are homogeneous and particle size can be well controlled.

Batching and Mixing   Many ceramics are blended with other ceramics, binders, lubricants, and other processing aids. *Binders* are substances that give temporary strength to the green compact and must be *plasticized* to make consolidation possible. Inorganic binders (clays, silicates, phosphates, etc.) are usually plasticized with water. Organic binders (polymers, waxes, gums, starches, etc.) are plasticized with low-molecular-weight organics or, if compatible, with water. *Plasticizers* are added during comminution, fine screening, classification, or in a separate mixing process. By controlling the proportions of various constituents, the right consistency is obtained for subsequent forming. Intimate mixing is essential. When the powder is to be made into a slurry, deflocculants and other chemicals are added to ensure favorable rheological characteristics. Pressing properties and shrinkage can be controlled also by adding comminuted presintered ceramics (*grogs*) to the particulate body.

12-4-2   Consolidation of Ceramic Powders

In addition to techniques described in Sec. 11-3, there are processes particularly well suited to ceramics.

Dry Pressing   Spray-dried ceramics of closely controlled particle size (typically, 20–200 μm) are cold-consolidated with techniques familiar from powder metallurgy. *Cold pressing* into metal dies (Figs. 11–5 and 11–6) requires pressures of 20–300 MPa, tight (< 50 μm) die clearances and, because of the abrasive character of ceramics, fairly expensive dies, but it allows mass production of parts to close (typically, ± 1%) tolerances. Thus, tens of millions of spark plug insulators, ceramic capacitor dielectrics, circuit substrates, and enclosures are pressed, sometimes to a thickness below 1 mm. Lubricants and binders are used as required. *Isostatic pressing* by the wet-bag method (Fig. 11–7*a*) is slow, labor-intensive but suitable for low-quantity and prototype production at pressures up to 500 MPa. Tolerances are wide (±3%) and the part usually requires green machining. The rubber mold stays in the pressure vessel in the dry-bag method (Fig. 11–7*b*); thus, production rates are high and the process can be automated. With a metal mandrel, hollow products can be made. Pressures are typically 200 MPa. *Roll compaction* is practiced for thick-film substrates.

Wet Pressing   With more liquid added, the ceramic behaves as a pseudoplastic or Bingham body (Fig. 7–5*b*, lines *C* and *D*); it is deformable but supports its own weight. Thus, it can be processed by techniques used for metals or polymers.

1. *Extrusion* with a piston (Fig. 9–28) is practiced but screw extrusion (see Sec. 14-3-2) ensures better mixing of the constituents and allows continuous operation. Dies are tapered (Fig. 9–28*b*) and the die angle is chosen to avoid centerburst defect (Fig. 9–33). Since divided streams can be reunited, hollow tubes and multihole filters may also be extruded with bridge-type (spider) dies (Fig. 9–30*c*). The higher water or organic carrier content results in a larger shrinkage and less tight (±2%) tolerances.

2. *Injection molding* at pressures of 200 MPa is gaining in importance for high-technology ceramics. *Compression molding* and *transfer molding* are related to the similar polymer processing techniques (see Sec. 14-3-4).

3. *Jiggering* is a mechanized form of throwing on the wheel. Hollow shapes, plates, etc. are formed by pressing the plastic mass against a vertical-axis, rotating mold with appropriately shaped, profiled templates or rollers. In this respect, the process resembles spinning (Sec. 10-9).

Casting   When sufficient liquid is added to allow viscous flow (Fig. 7–5*b*, line *A*), casting becomes possible. The suspension is called a *slip* or *slurry*. A suspension of smaller (less than 20 μm) particles exhibits non-Newtonian viscous flow. Deflocculants and dispersants, combined with acidity (pH) control are used to prevent flocculation of fine (typically < 5 μm) particles. It is usually found that slips of given solid concentration have a minimum viscosity at some specific pH value. Slips made of very fine (< 1 μm) powder tend to dilatancy (Fig. 7–5*b*, line *B*) and are more difficult to handle. The high liquid content makes for large shrinkage and tolerances are fairly wide, except when the part is of simple shape, as in tape casting ceramic substrates for electronic circuits (see below).

Techniques familiar from casting (Sec. 7-5) may be applied. The difference is that the mold is usually porous, so that fluid from the slurry is absorbed by capillary action, leaving behind a powder compact. To make the mold, water is added to plaster of paris ($CaSO_4 \cdot \frac{1}{2}H_2O$); a hydration reaction takes place which results in the precipitation of gypsum ($CaSO_4 \cdot 2H_2O$) in the form of acicular crystals, arranged in a randomly oriented network. Excess (entrapped) water is driven off by drying the mold. The mold is now a solid structure which has pores below 1 μm in diameter. Mold-release agents are usually applied to the mold surface, and the slurry is then poured.

1. To produce a *solid casting*, the mold cavity must be topped up with slip until it is filled with a powder body. Vacuum aids mold filling and speeds up the process of liquid extraction. Centrifuging (Fig. 7–26) also helps filling.

2. *Drain casting*, also referred to simply as slip casting, is the most frequently used variant. It is a relative of slush casting (Sec. 7-5-6): The mold is filled with slip and, after sufficient time has elapsed to form a dewatered particulate shell, the mold is tipped to pour out the excess slip, leaving behind a hollow product.

3. *Tape casting* is an important variant of slurry casting. Thin (0.025 to 1.5 mm thick) tapes are formed, usually with organic binders, by several techniques: The slurry may be cast onto a thin, moving plastic film while controlling the tape thickness with a blade (*doctor-blade process*). The slurry may be cast onto a paper carrier which is subsequently burnt off (*paper-tape process*). The green tape obtained by any of these processes is flexible enough to be rolled up; it can be blanked, mechanically scored or marked with a laser beam prior to firing, so that it can be made into small components such as substrates for thin-film devices (see Sec. 20-4-1).

12-4-3   Drying and Green Machining

Free water is often reduced by *holding* at room temperature. An initial, low-temperature heating (*drying*) stage is most important when *physically adsorbed water* (also called *mechanical water*) must travel long distances to reach the surface. Moisture trapped in the center would blow up the compact. Slip-cast and extruded parts shrink some 3–12% (no shrinkage of dry-pressed parts). Organic binders are driven off (with accompanying shrinkage) and burnt to prevent discoloration of the ceramic.

Even though the strength of green compacts is low, they can be machined, usually in vertical-axis lathes, if held in appropriate fixtures. Since drying shrinkage is allowed to take place prior to machining, fairly complex shapes can be created to close tolerances, as in the manufacture of high-tension electric insulators made of porcelain.

12-4-4   Sintering

During sintering (*firing*) of ceramics, several events take place as the temperature is increased:

1. Water contained in the form of *water of crystallization* (also called *chemically bound water*) is removed at relatively low temperatures (between 350 and

600°C). In clay-type ceramics *dehydroxylation* (breakdown of hydroxy groups) also takes place. In some instances, salts are not calcined prior to compaction and then conversion to oxides must take place during heating to sintering temperatures. For all these reasons, the rate of heating is slow and temperature may have to be held for some considerable period of time.

2. At yet higher temperatures, *sintering* begins:

a. In single-component ceramics (such as oxides, borides, and carbides) diffusional growth of necks (Fig. 11–8) dominates. On extended heating, grain growth occurs just as in metals. Polymorphic transformations may take place, but the dimensional change is accommodated in the partially sintered ceramic.

b. In the many ceramics formulated of more than one component, reactions between adjacent particles take place. Solid-phase diffusion can lead to the formation of solid solutions and other phases, as dictated by the phase diagram. Various transformations may also take place.

c. Most important, liquid phases form at higher temperatures, aiding densification but also increasing the danger of grain growth. The relative quantities of liquids can be estimated from phase diagrams. Just as minor elements that form low-melting eutectics in metals lead to hot shortness, minor contaminants that form low-melting phases in ceramics impair the hot strength.

d. In some systems a liquid is present but is then used up in further reactions; such *reactive liquid sintering* often yields products of very good high-temperature properties because the glassy phase is absent.

e. The very fine particle size of high-technology ceramics makes for short diffusion paths and allows sintering at significantly lower temperatures, thus avoiding grain growth that would reduce their strength.

f. In *reaction sintering* a reactive gas is formed or is introduced to change the chemistry (e.g., Si powder is reacted with nitrogen to make reaction-bonded Si_3N_4). Porosity is high but dimensional stability is good.

In most cases there is significant (35–45%) shrinkage during sintering.

3. After sintering is completed, the ceramic is *cooled* to room temperature at closely controlled rates. In ceramics containing a glassy phase, cooling rates determine the degree of crystallization. Ceramics of high thermal expansion could fracture on sudden cooling. Polymorphic transformations may also occur and the accompanying volume change results in microcracking.

The need for slow heating and cooling makes for long cycle times (days or even weeks), even though sintering itself is quite rapid. Furnaces (*kilns*) are potential sources of air pollutants. The process is conducted to generate minimum dust, and undesirable (or controlled) gases and flue gases are passed through effective dust-precipitation and gas treatment equipment to minimize hazardous air pollutants (HAPs).

12-4-5   Hot Compaction

Excellent properties are obtained by consolidating at high temperature.

Hot Pressing   Fine (< 0.1 μm) powder can be hot pressed at pressures of 7–70 MPa, using graphite dies supported by ceramics. Temperatures can be kept lower than in static sintering because the simultaneous application of pressure aids densification. This, together with the short high-temperature exposure, allows the production of fine-grained (*microcrystalline*) ceramics of high strength. Thus, the technique finds wide application for high-technology ceramic structural components, but shapes are severely limited by the decay of pressure with distance.

HIP   Many shape limitations are eliminated by applying a gas pressure of 70–200 MPa to encapsulated ceramics. Sometimes the part is consolidated in a molten glass or metal envelope (*rapid omnidirectional compaction*). HIP at pressures of only 0.1–10 MPa is used to fully densify ceramics presintered to 92–96% density; for this, no encapsulation is needed.

Superplastic Forming   Nanocrystalline ceramics exhibit superplasticity at high temperatures where grains can slide against each other, allowing precision parts to be formed.

Most ceramics are sintered (fired) to finish dimensions. However, in some critical applications the surface of the sintered body is ground with a yet harder ceramic to improve surface finish, dimensional tolerances, or impart a more complex shape.

A block of graphite (density: 1.9 g/cm³) has a length of 36.34 mm, width of 24.28 mm, and height of 12.70 mm. Its dry weight is 18.878 g. When weighed in diethyl phthalate (density: 1.120 g/cm³), its weight is 6.919 g. It is then again weighed (saturated with the fluid), and is found to weigh 19.235 g. Calculate the (*a*) true volume, (*b*) total (bulk) volume, (*c*) open, closed, and total porosity, and (*d*) bulk and apparent density.

| | **Example 12-2** |

(*a*) True volume = dry weight/density = $18.878/1.9 = 9.936$ cm³.

(*b*) Total volume, from geometry: $(36.34)(24.68)(12.7) = 11.390$ cm³. Often the part is of irregular shape and then the Archimedes principle may be used:

Fluid displaced by the fluid-saturated block = $19.235 - 6.919 = 12.316$ g.

Since the density of the fluid is 1.120 g/cm³, the block displaced $12.316/1.120 = 10.996$ cm³ fluid. This is the total (bulk) volume, including closed porosity.

(*c*) Total porosity = (total vol. − true vol.)/total vol. = $(11.390 - 9.936)/11.390 = 0.12766$ or 12.766%. Open (or apparent) porosity = vol. of fluid absorbed/total vol. = (saturated weight − dry weight)/(fluid density)(total vol.) = $(19.235 - 18.878)/(1.120)(11.390) = 0.027985$ or 2.80%. Closed porosity = total porosity − open porosity = $0.12766 - 0.027985 = 0.099675$ or 9.968%.

(*d*) Bulk density = mass/total vol. = 18.878 g/11.390 cm³ = 1.657 g/cm³ (since both the graphite and fluid are weighed in the same gravitational field, weight is a direct measure of mass).

Apparent density = mass/(total vol. − open pore vol.) = $(18.878$ g$)/(11.390$ cm³$)(1 - 0.027985) = 1.705$ g/cm³.

Check the calculation: Density = mass/(total vol. − open pore vol. − closed pore vol.) = $18.878/(11.390)(1 - 0.027985 - 0.099675) = 1.9$ g/cm³.

Check the accuracy of measurements: Bulk volume = true vol. + closed porosity = 9.936 + (11.390)(0.099675) = 11.0713 cm^3, in good agreement with the 10.996 cm^3 found in (*b*).

12-4-6   Applications

Ceramics are used in all phases of our life. The emphasis here will be on technical applications.

Clay-Based Ceramics   Some products are made from natural clays. Clays of high (60–80%) SiO$_2$ and low (5–20%) Al$_2$O$_3$ content are pressed into bricks and tiles and fired at 900–1000°C to a porous but reasonably strong condition.

All other clay-based ceramics are made from mixes of controlled composition. Because three components, quartz (flint), clay, and feldspar (aluminosilicates of K, Na, and Ca) are used, one speaks of *triaxial bodies*. Feldspars reduce the firing temperature by increasing the ratio of eutectic melt. The proportions depend on the field of application (Fig. 12–6). *Earthenware* is fired only once, at 1150–1280°C, to a slightly porous body. *Stoneware* is fired at 1200–1300°C to dense bodies. Vitrified *whiteware* for bathroom fixtures is usually slip-cast and coated with a glaze prior to firing at 1260°C. Dry-pressed floor tiles and lathe-turned electrical porcelains are also glazed and then fired at 1290°C. Most other whiteware, such as semivitrified tableware and hard porcelain, is fired twice.

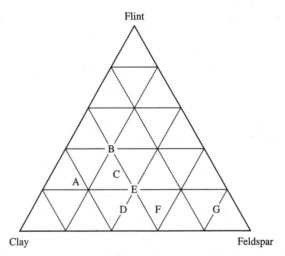

Figure 12–6   Traditional ceramics are often triaxial bodies. A: wall tile; B: semivitreous whiteware; C: hard porcelain; D: vitreous whiteware; E: electrical porcelain; F: floor tile; G: dental porcelain.

The first or *biscuit firing* liquid-sinters the body, making it translucent but at the expense of lower strength and electrical resistivity; after cooling down, the glaze is applied and is then melted in the second or *glost firing*.

Refractories   Like refractory metals, refractory ceramics are noted for their resistance to high temperatures. They are formulated to resist specific melt and atmospheric conditions. Thus, they may be acid (based on SiO_2), neutral (Al_2O_3; mullite $3Al_2O_3 \cdot 2SiO_2$; chromite $FeO \cdot Cr_2O_3$), or basic (magnesite MgO; dolomite Ca–Mg–O). Larger grog particles are usually embedded in a finer ceramic matrix to make bricks. Mortar and furnace hearths produced in situ are made of refractory granules bonded with a cement. Some refractories are melted and cast into shape. Refractory powders and porous blocks serve as thermal insulation in high-temperature applications.

Oxide Ceramics   Fine-grained ceramics of a single oxide can have high strength. Most widespread is *alumina*, Al_2O_3 (melting point 2054°C) which is sintered into cutting tool bits, sparkplug insulators, high-temperature tubes, melting crucibles, wear components, and substrates for electronic circuits and resistors. Small additions of MgO remain concentrated on grain boundaries, facilitating densification and preventing grain growth. Finest-grain material is obtained by hot pressing. *Zirconia*, ZrO_2 (melting point 2710°C) is more heat-resistant but, as already indicated, it suffers a polymorphic transformation with a catastrophic volume change. A stable cubic solid solution is obtained by adding 5–15% Y_2O_3 or CaO, and such *stabilized zirconia* is useful to 2400°C as a furnace lining and, above 1000°C, as a heating element. Partially stabilized zirconia is used as a die for the hot extrusion of metals.

Complex Oxide Ceramics   Many of the most important manufactured ceramics consist of carefully controlled combinations of several oxides.

1. In the $MgO–Al_2O_3–SiO_2$ system there are several compositions suitable for electrical and electronic applications. For example, *steatites* are used as insulators in high-frequency circuits.

2. *Ferrites* are generally composed of a metal oxide MeO (where Me can be any bivalent metal) and Fe_2O_3. To avoid confusion with ferrite in irons and steels (Sec. 6-2-1), the term *ferrospinel* is also used. They fall into two major groups:

a. The structure of $MeFe_2O_4$ ferrites (where Me is Ni, Mn, Mg, Zn, Cu, or Co) is cubic, the same as that of the mineral spinel ($MgAl_2O_4$ or $MgO \cdot Al_2O_3$). They have a low magnetic hysteresis combined with high electrical resistance, hence losses due to eddy currents are low. They make excellent cores for high-frequency applications in radios, television, and recording heads. In powder form, they can be deposited on an insulating (plastic) substrate to provide magnetic recording mediums. Rare-earth garnet ferrites, deposited on a nonmagnetic substrate, serve as bubble memory.

b. The more complex ferrites, especially those of Ba, Sr, and Pb, are hexagonal in structure. They combine high resistivity with high coercive force and are thus excellent low-cost magnets for loudspeakers, small motors (as used also in automobiles) and, added to polymers, as magnetic elastomeric seals (as used on refrigerator doors).

3. *High-temperature superconductors* are based on copper oxides such as $(Bi,Pb)_2Sr_2Ca_2Cu_3O_{10}$. They have applications as thin films or silver-encased wire (Sec. 15-5) which is capable of carrying high current densities at liquid-nitrogen (77 K) temperature.

4. *Titanates* contain TiO_2 as one of the constituents. Most significant is $BeTiO_3$ which has a high dielectric constant, making it suitable for capacitors. It also exhibits *ferroelectricity* (spontaneous alignment of electric dipoles) and, because of anisotropy of properties, piezoelectricity.

Carbides, Nitrides, Borides, and Silicides   These ceramics are noted for their high hardness (see Table 16–3)

Carbides have the highest melting point of all substances; an 80TaC-20HfC ceramic has a melting point of 4050°C. Silicon carbide (SiC) is difficult to sinter but solid SiC bodies such as high-temperature resistance-heating elements, rocket nozzles, and sandblast nozzles can be obtained by pressure sintering or reactive sintering. Melting crucibles are made with a clay bond. The powder is one of the most important abrasives for grinding (see Sec. 16-8-3). The extremely hard B_4C is used as a grinding grit and, in a sintered form, for wear-resistant parts and body armor. Other carbides are important as coatings (Chap. 19) and in cemented carbides.

Nitrides have only slightly lower melting points than carbides. One form of boron nitride (BN) is hexagonal (also called *white graphite*). It can be used as a high-temperature lubricant; it is also a good insulator and can be processed into large bodies. The cubic form (CBN) has a structure similar to that of diamond and is, after diamond, the hardest material, suitable for metal-cutting tools.

Silicon nitride (Si_3N_4) has good thermal conductivity, low expansion, and high hot strength, making it the prime candidate for ceramic engine components, turbine disks, and rocket nozzles. It can be processed by hot pressing, reaction bonding, vapor deposition, and injection molding. The *oxynitrides* (trade name: Sialon, from Si–Al–O–N) have better oxidation resistance and are used as cutting tools and welding pins.

Borides (TiB_2, ZrB_2, CrB, and CrB_2) have high melting points, strength, and oxidation resistance, and are used as turbine blades, rocket nozzles, and combustion chamber liners.

Molybdenum disilicide ($MoSi_2$) has high oxidation resistance and serves as a heating element.

Carbon   Carbon can be amorphous (lampblack), but the industrially most important forms are crystalline.

Graphite, of hexagonal structure, occurs in nature. Adsorption of volatile fluids or gases reduces the bond strength in the *c* direction (Fig. 6–2*c*), allowing slip along the basal plane, therefore graphite is a good solid lubricant up to 1000°C (although it begins to oxidize at 500°C). Mixed with clay, it forms the "lead" of pencils. The technically important solid bodies are made of coke, formed into final shape with a pitch or resin binder, and converted to graphite above 2500°C. It is a good electrical conductor, it has low heat expansion, and resists high temperatures; hence, it is used for heating elements, electrodes, EDM electrodes (Sec. 17-4), compacting dies, and crucibles. High-purity graphite is used for moderators and reflectors in nuclear power plants.

Graphite fibers are produced by the conversion of a polymer fiber, such as polyacrylonitrile. Holding the fiber under tension at 2500°C, approximately 10 μm-diameter fibers of oriented graphite are obtained with strengths between 2.0 and 3.5 GPa (the elastic modulus of the weaker fiber is higher, 400 GPa, compared to the 200-GPa modulus of the stronger fiber, see Table 15–1).

Diamond is totally covalently bonded in a cubic structure (Fig. 12–1*a*), is an insulator, and is the hardest known material. Natural diamonds are used in wear-resistant applications such as wire-drawing dies, cutting tools, and grinding wheels. Small but relatively defect-free crystals of diamond can be made from carbon at high pressures and temperatures; manufactured diamond outperforms natural diamond in many applications. It can be sintered to give polycrystalline bodies (*megadiamonds*) or 0.5–1.5-mm-thick layers on a tougher substrate such as cemented carbide.

Fullerenes are the newest members of the carbon family (see Sec. 20-5-2). The soccer-ball shaped C_{60} has been most extensively investigated.

12-4-7   Process Capabilities and Design Aspects

The choice of ceramic processes (Table 12–1) is greatly influenced by the shape of the part.

Parts for dry pressing are designed by the rules given for conventional powder-metallurgy parts (Sec. 11-7 and Fig. 11–12). Tolerances are good, in the ±1% range. The next-greatest freedom is afforded by flexible isostatic compacting molds which permit undercuts or reverse tapers. The depth of undercuts is limited by the need to remove the part from the elastomeric mold; tolerances are poorer (±3%) and transverse holes are not feasible. Hot pressing is limited to simple shapes by the difficulty of manipulating dies at very high temperatures. Plastic forming (jiggering, pressing) are limited only by the need of releasing the part from the rigid mold. By the nature of the process, roll compaction and extrusion are suitable only for two-dimensional shapes.

Injection molding is subject to the rules of die casting (Sec. 7-8-2). Slip casting is the most versatile and any shapes (including hollow and undercut shapes) can be formed provided that they can be released from the mold. It is possible to join several separately molded pieces and assure the virtual disappearance of the

Table 12–1   General characteristics of ceramics manufacturing processes*

Characteristics	Process							
	Dry Pressing	Isostatic Pressing	Hot Pressing	Roll Compaction	Extrusion	Plastic Forming	Slip Casting	Injection Molding
Part								
Shape[†]	Not S3, T2, 3, 5, 6, F3, 5 U	Not T3, 5, F5	R0, B0, F0, 4	R0, B0, S0, F0	All 0	Not T3, 5, 6, F5	All	Not T5, F5, U2, 4
Surface detail[‡]	A–B	B–D	C	A	A	A–B	A–B	A–B
Mass, kg	0.05–30	0.1–100 wet bag 0.1–50 dry bag	0.01–100	not limited	not limited	0.02–30	0.05–200	0.02–3
Min. section, mm	1	1	1	0.02	0.2	1.5	0.2	1
Tolerance[‡]	A–B	B–D	B–C	A–B	C	C	D–E	A–B
Cost[‡]								
Equipment	A–C	A–B	A	A	A	C–E	C–E	A–B
Die	A–B	C–E	B–C	A–B	A–C	C–E	B–C	A–B
Labor	C–E	A–C wet bag C–E dry bag	B–D	D–E	D–E	B–E	C–D	D–E
Finishing	C–E	B–D	B–D	D–E	D–E	C–E	B–D	C–E
Production								
Operator skill[‡]	D–E	B–C wet bag C–E dry bag	C–E	D–E	D–E	D–E	D–E	D–E
Lead time	Weeks–months	Weeks	Weeks–months	Weeks–months	Weeks	Weeks	Weeks	Weeks–months
Rates (piece/h)	1000–10 000	1–10 wet bag 50–1000 dry bag	10–100	continuous	continuous	10–100	1–100	100–1000
Min. quantity	1000–50 000	1–10	10–1000			1000–50 000	1–100	10 000

*Adapted from J. A. Schey, *ASM Handbook*, vol. 20, *Materials Selection and Design*, ASM International, 1997, p. 697. With permission.

[†]From Fig. 3–1.

[‡]Comparative ratings, with A indicating the highest value of the variable, E the lowest. For example, dry pressing produces good surface detail and tolerances, involves medium to high equipment cost, high to very high die cost, medium to low labor cost, medium to low finishing cost, and low operator skill. It can be used for medium to high production rates and requires a minimum quantity of 1000 to 50 000 parts to justify the die cost.

joint during sintering. One needs to think only of the complex yet low-cost mass-produced figurines made in porcelain and other ceramics to realize the potential of the process. Tolerances are relatively poor ($\pm 3\%$).

Size limitations stem partly from the capacity of usual equipment and partly from processing difficulties: removal of binders from the green body and prevention of the development of dangerous stresses during firing and cooling in thick-walled parts. Process choice is affected also by achievable tolerances (Table 12–1).

12-5 GLASSES

We already observed that glasses are, by definition, ceramics, which differ from other ceramics in that they are produced by the melt processing route. Accordingly, the starting materials are typical of ceramics, whereas the processing techniques are closer to that of thermoplastic polymers (Chap. 14). An elementary knowledge of structure is essential for an understanding of glass processing technologies.

12-5-1 Structure and Properties of Glasses

The basic building block of glass is the silica (SiO_4) tetrahedron (Fig. 12–3). In the noncrystalline, glassy form (*fused silica*) the free valences join in a rather loose, three-dimensional, covalently bonded network (Fig. 12–7*a*). Since each oxygen ion is shared by two silicon ions, silica glass is really a $(SiO_2)_n$ polymer. A few other oxides (such as B_2O_3 and P_2O_5) are also *network formers* on their own, yet others (such as Al_2O_3) enter into the SiO_2 network. Another group of oxides, the *network modifiers*, depolymerize the network by breaking up the O–Si–O bonds; their oxygen attaches itself to a free Si bond, while the metal cation is distributed randomly, its charge balanced by the negative charge of the dangling oxygen ions (Fig. 12–7*b*), maintaining overall charge balance. *Intermediate oxides* (MgO, BeO, TiO_2) may enter into the network or depoly-

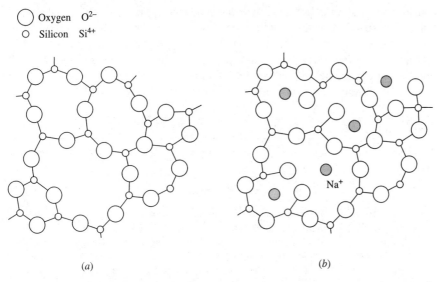

⬭ Oxygen O^{2-}
○ Silicon Si^{4+}

(a)   (b)

Figure 12–7 Simplified two-dimensional representation of (*a*) fully polymerized silica network and (*b*) partially depolymerized glass (fourth bonds are outside the plane of the illustration).

merize it. Substantial fundamental knowledge exists that allows the formulation of glasses for special purposes. In general, Al_2O_3 increases hardness and reduces thermal expansion, whereas PbO reduces hardness and increases the refractive index. It should be noted that some elements (S, Se, Te), nonoxide compounds (e.g., As_2S_3), and organic compounds (e.g., abietic acid) also form glasses.

On cooling from the melt state, the spatial network of Fig. 12–7 forms without a sudden change in specific volume (as shown in Fig. 13–5a for amorphous polymers) and the glass can be regarded as a *metastable, undercooled liquid.* Because of the gradually reducing amplitude of thermal vibrations, the specific volume decreases gradually with dropping temperature until, at some *critical temperature* T_g, the glass assumes properties typical of a solid; it is now regarded as being in a nonequilibrium, glassy (vitreous) state. The *glass-transition temperature* T_g depends on cooling rate and is higher for fast cooling. Hence, as in polymer technology, one speaks of a transformation temperature range.

Mechanical Properties   The distinction is important from a mechanical point of view:

1. Above T_g, the undercooled liquid exhibits Newtonian viscous flow behavior (Fig. 7–5b, line A). Therefore, glass cannot be used as a structural material above T_g; a Newtonian fluid deforms under even the slightest stress (Eq. 7-1). At the same time, viscous flow is extremely desirable for manufacturing processes because the fluid can be exposed to tension without the danger of localized necking (Sec. 8-1-6).

2. Below T_g, the glass is an elastic-brittle solid. By agreement, the glass is considered to have reached a solid form when viscosity rises to $10^{13.5}$ P (the unit of poise is still used extensively, and is sometimes expressed in the SI unit of dPa·s). The strength calculated from bond strength is never reached in practice because scratches and cracks act as stress raisers in the brittle state (Eq. 4-11b). The presence of cracks makes glass subject to *static fatigue*: under an imposed tensile load, fracture may occur suddenly after some considerable time has elapsed.

Example 12-3

Prove that a fiber of a Newtonian fluid such as glass thins down in proportion to the applied force.

$$\sigma = P/A = k\eta\dot{\varepsilon}$$

where k is a proportionality constant. By definition, $\varepsilon = dl/l = -dA/A$ [integration of this leads to Eq. (8-3)]. Strain rate, by definition, $\dot{\varepsilon} = d\varepsilon/dt = (-dA/A)/dt = -\dot{A}/A$.

Then $P = k\eta\dot{\varepsilon}A = -k\eta\dot{A}$ or $\dot{A} = -P/k\eta$. Thus, for a given viscosity, the fiber will thin down more rapidly if a larger force is applied.

Chemical and Physical Properties   Glasses are often chosen for their resistance to corrosion by liquids or gases. This does not, however, mean that all glasses are corrosion-resistant even under mild conditions. Indeed, the main source of surface cracks is *atmospheric corrosion*. Water vapor present in air attaches itself to the glass surface; hydrogen ions replace monovalent cations (chiefly Na^-) by a stress-corrosion mechanism, creating cracks of very small tip radii. The attack is rapid; hence, freshly drawn glass fiber quickly loses the very high strength typical of defect-free glass. Corrosion can be reduced by replacing monovalent alkali metals by calcium. High strength is retained if freshly drawn (or fire-polished) fiber is coated with a polymer that is impervious to water. Corrosion by water is exploited in waterglass, a fully depolymerized $Na_2O \cdot xSiO_2$ glass which is dissolved in water; when treated with CO_2, a strong gel forms which serves as a binder for sand molds (Sec. 7-5-4) and grinding wheels (Sec. 16-8-4). High-lead glass is attacked by acids; therefore, there are now limits set to leachable lead in glassware, such as pitchers and tumblers.

Glasses are electric insulators at low temperature but become ionic conductors in the melt regime, allowing electric heating of melts.

Optical properties are most important. Amorphous glass is transparent and may be colored by appropriate oxide or metal additions. *Photosensitive* eyeglasses are made from glass that contains AgCl. When this glass is energized by ultraviolet rays, Ag^+ ions form and impart a deeper color to the glass.

Glass Ceramics   We mentioned that the glassy state is metastable. Therefore, all glasses can be converted, upon heating for a prolonged period of time, into a crystalline form. For some applications, glasses are formulated to avoid unwanted crystallization. In contrast, other glasses are formulated for controlled conversion into the crystalline form. Such fully dense *glass ceramics* are usually based on the $Li_2O–Al_2O_3–SiO_2$ system. Nucleating agents (metals such as Cu, Ag, Au, Pt, Pd, or oxides such as TiO_2) are added to promote the formation of many small crystals. Very low thermal expansion combined with high strength makes these glass ceramics suitable for cookware (the Pyroceram of Corning Glass Works) and many industrial applications.

Glass-ceramic construction materials are made from inexpensive natural and waste materials which crystallize spontaneously (e.g., fused basalt and blast-furnace slag for floor tiles, building cladding, and paving blocks).

12-5-2   Manufacturing Processes

Complete melting of a ceramic charge is a slow process, therefore, components of the charge are finely comminuted, blended, and then spread on top of a molten bath held in a batch or, more frequently, continuous melting furnace. Scrap (*cullet*) helps the melting process and this is one of the reasons for recycling postconsumer glass. Furnaces are usually heated by gas, although auxiliary electric heating is used because it creates intensive convection.

A typical charge for a window glass would be made up of quartz (sand, of 0.1–0.6-mm particle size), limestone ($CaCO_3$), and soda ash (Na_2CO_3). During melting the carbonates decompose and react with SiO_2. Gas evolution helps to homogenize the melt, but bubbles would remain. In the final or *fining* (refining) stage bubbles are allowed to rise, a process which is accelerated by additions such as As_2O_3. Further additions are made to control the color. For example, the green-blue of FeO can be changed to the much lighter yellow of Fe_2O_3, which then can be further masked by adding oxides that give a complementary color. Molten glass is highly corrosive to refractory furnace linings; therefore, linings are chosen for resistance to attack and so as not to introduce harmful components into the glass. The temperature of the bath is controlled, often in a *forehearth* (an extension of the melting furnace), to impart the optimum viscosity for the subsequent forming process. The glass is kept here agitated, by electric current or mechanical stirrers, to maintain uniformity.

Glass compositions are chosen for specific applications (Table 12–2). The composition also determines the viscosity–temperature relationship (Fig. 12–8).

Table 12–2   Manufacturing properties of some glasses*

	Corning Glass Works Code Number and Type						
Property	7940 Fused Silica†	E-glass	7740 Borosilicate	1720 Aluminosilicate	0080 Soda-Lime-Silicate	8871 Potash-Lead	8830 Soda-Borosilicate
Composition, weight %							
SiO_2	99.9	54	81	62	73	42	65
B_2O_3		10	13	5			23
Al_2O_3		14	2	17	1		5
Na_2O			4	1	17	2	7
K_2O						6	
Li_2O						1	
CaO		17.5		8	5		
MgO		4.5		7	4		
PbO						49	
Viscosity, P‡ at °C	956	507	610	667	473	350	460
$10^{14.5}$ (strain point)							
10^{13} (annealing point)	1084	657	560	712	514	385	501
$10^{7.6}$ (softening point)	1580	846	821	915	695	525	708
10^4 (working point)			1252	1202	1005	785	1042
Coefficient of linear expansion $\times 10^{-7}/°C$	5.5	60	33	42	92	102	49.5
Typical uses	High temperature aerospace windows	Fiber	Chemical, baking ware	Ignition tube	Container, sheet, plate	Art glass, optics, capacitors	Sealing glass for Kovar

*Data compiled from D.C. Boyd and D.A. Thompson, *Glass,* in *Kirk-Othmer Encyclopedia of Chemical Technology,* 3d ed., vol. 11, Wiley, New York, 1980, pp. 807–880.

†Produced by vapor deposition.

‡Multiply poise by 0.1 to get $N \cdot s/m^2$.

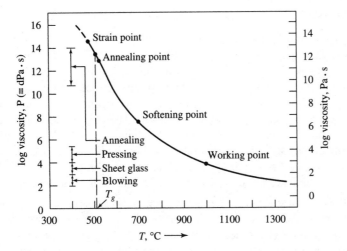

Figure 12–8   Glasses soften gradually; their viscosity depends on composition.

Viscosity η decreases with increasing temperature T according to

$$\log_{10} \eta = C + \frac{B}{T} \qquad \textbf{(12-1)}$$

where C and B are constants, and T is absolute temperature. The lowest temperature at which forming is still practical is signified by the Littleton softening point at which a standard fiber extends under its own weight.

The 0080 glass of Table 12–2 is occasionally used as a lubricant for hot extrusion (Sec. 9-4-2). What is its viscosity at 1200°C?

Example 12-4

From Table 12–2, viscosity is $10^{7.6}$ P at 695°C and 10^4 P at 1005°C. From Eq. (12-1):

$$7.6 = C + \frac{B}{695 + 273} \quad \text{and} \quad 4.0 = C + \frac{B}{1005 + 273}$$

Solving the two equations simultaneously we obtain $B = 14\,366$ and $C = -7.24$. Thus, at 1200°C,

$$\log_{10} \eta = -7.24 + \frac{14\,366}{1200 + 273} = 2.51$$

Thus, viscosity is $10^{2.51}$ P at 1200°C, which agrees reasonably with Table 8–4, where a viscosity of 100–300 P at the working temperature is indicated.

Forming Processes

1. Flat glass used to be made by drawing or rolling from the forehearth. These processes have largely been replaced by the *float glass* process in which the glass flows (at a typical viscosity of 10^4 P) onto the surface of a molten tin bath in a controlled atmosphere (Fig. 12–9a). The bottom surface is atomically smooth and the top surface is smoothed by surface-tension effects. Surface tension holds the glass to 67-mm thickness. Width is up to 4 m. The glass ribbon can be reduced down to 1.5-mm thickness by stretching with rollers. The sheet leaves the furnace at about 600°C and can then be supported on rollers without marking. Heavy plate glass used to be made by casting a thick plate and polishing it mechanically; now all is made by the float process. The process is limited to soda-lime glass.

2. All glass varieties can be made by the *fused glass* process in which two streams of viscous (2×10^5 P) glass, overflowing from a trough, are united (Fig. 12–9b). Thickness is controlled by the speed of pulling and may range from 0.4 to 10 mm. The surface is not touched by any tool and is smoothed by surface tension.

3. *Glass tube* is made by flowing glass onto a hollow, rotating mandrel through which air is blown (Fig. 12–10a); the gradually stiffening tube is mechanically drawn to thinner dimensions.

4. A similar principle is used in making continuous *glass fiber* of 3–20-μm diameter for insulating fabric and reinforcing fiber for plastics. A glass of high electrical and corrosion resistance (hence called E glass) is melted in (or transferred from the forehearth to) a platinum tundish called *bushing* in which there are 200 to 400 nozzles (Fig. 12–10b). Glass flows out at a rate q determined by nozzle dimensions (radius r and length l), the kinematic viscosity v (dynamic viscosity η divided by density ρ) of the melt, and the hydrostatic pressure generated by the melt of height h

$$q = \frac{khr^4}{vl}$$

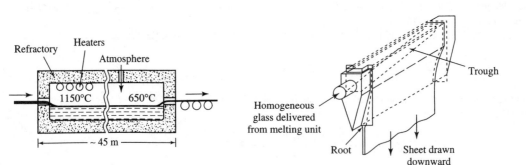

(a)   (b)

Figure 12–9   Sheet glass is produced by the (a) float glass or (b) fused glass process. (*Part (b) courtesy Corning Inc., Corning, New York.*)

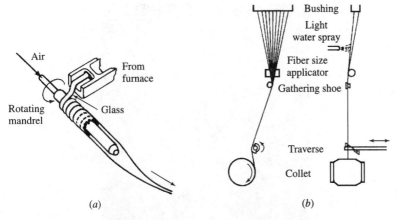

(a)   *(b)*

Figure 12–10   Methods of making (a) tube and (b) fiber by continuous methods
[(a) adapted from P.C. Boyd and D.A. Thompson, in Kirk-Othmer
Encyclopedia of Chemical Technology, 3rd ed., Wiley, 1980, vol. 11,
p. 864; (b) K.L. Loewenstein, The Manufacturing Technology of
Continuous Glass Fibers, 2nd ed., Elsevier, 1983, p. 29. With
permission.]

where k is a constant. The emerging fibers are cooled and subjected to mechanical *attenuation* (stretching) by winding the take-up spool at a higher speed
(50–60 m/s). The fiber is coated with an organic *size* (such as starch in oil)
which allows processing with minimum damage. Fibers may be chopped, or
short fibers (*staple*) directly made by attenuating the fiber with compressed air
or steam which breaks up the fibers while also thinning them down. *Glass wool*
of 20–30-μm-diameter fibers is spun (ejected) from rotating heads; it is often
immediately matted to form insulating blankets.

5. A special class of fibers is used as *light guides* in fiber optics and as
fiber-optical wave guides for long-distance transmission of digital signals sent by
pulsed lasers or photodiodes. Attenuation (losses) must be very low; hence, all
effort is made to ensure unimpeded passage of light while preventing the escape
of light from the fiber. The former aim is achieved by melting extremely pure raw
materials (Fig. 12–11) or by forming the fiber by vapor deposition. The second
aim is satisfied by surrounding the core glass (often, germanium-doped silica)
with an envelope of lower refractory index (pure silica), so that reflection takes
place at the interface between the two. Optical fiber is a prime candidate for
manufacture in space.

6. Individual articles may be made by *pressing* a measured quantity of glass
(*gob*) into steel or cast iron molds. The process is related to closed-die forging
(Sec. 9-3-2).

7. Articles with thinner walls and reentrant shapes are often needed. For
these, the gob is dropped into a mold (or glass is sucked into the mold by vacuum)
and a preform (*parison*) is formed with a puff of air or by pressing with a punch
(Fig. 12–12). After transfer to a second, split mold, the part is allowed to reheat

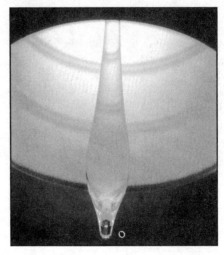

Figure 12–11   Pure silica blank, to be drawn into a fine strand of optical fiber. (*Courtesy Corning Incorporated, Corning, New York.*)

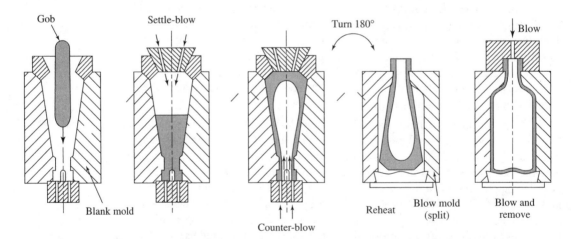

Figure 12–12   A bottle may be made by forming a gob of glass into a parison which is then blow-molded to final shape.

and is then *blow-molded* to final shape. Newtonian viscous flow ensures that the wall should thin out uniformly. To maintain a good surface finish, the mold is coated with a mineral oil, an emulsion, or a wax-sawdust mix that is then converted into carbon. Prior to blowing, the carbon layer is slightly dampened; the steam generated during blowing separates the paste-mold from the glass and gives a smooth finish. Light bulbs are produced on multistage blowing units at the rate of 2000 per minute. Hand blowing is now limited to artistic work.

Finishing Operations   Some glasses (especially fused silica) have a low thermal expansion and can be cooled rapidly. However, in most glasses rapid cooling sets up residual stresses which could cause explosive disintegration and must be relieved by *annealing* in a furnace called *lehr*. The temperature is chosen to allow stress relief in a reasonable time without losing the shape of the article. Stress relief occurs (stress is reduced to 2.5 MPa) in 4 h at the *lower annealing temperature or strain point* and in 15 min at the *upper annealing temperature* or *annealing point* (Fig. 12–8). To avoid reintroduction of stresses, the part must be cooled slowly from the annealing-point to the strain-point temperature.

Glass that has not been in contact with any tool exhibits an extremely smooth surface (*natural fire finish*); therefore, cut edges are *fire polished* by a pencil torch.

The strength of typical glass is around 70 MPa. Resistance to tensile stresses may be increased by inducing compressive stresses in the surface and thus hold cracks in compression. This can be achieved by subjecting the finished glass article (ovenware, containers, eyeglasses, etc.) to thermal or chemical treatment.

1. *Tempering* is a thermal process, shown in Fig. 12–13 for the example of a plate. The plate is heated above T_g, then its surfaces are quenched with an air blast, causing the surfaces to contract and stiffen. At this point, the center is still soft and follows the contraction of the surface layers. On further cooling, the center cools too and in doing so contracts; since the surfaces are now stiff, they cannot follow the contraction of the center and are put in compression. Compressive surface stresses on the order of 140 MPa or more are balanced by internal tensile stresses; they are harmless because there are no cracks in the center.

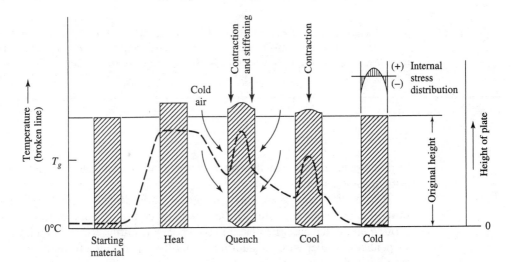

Figure 12–13   Glass is toughened by the sequence of operations shown; high compressive surface residual stresses make it resistant to tensile loading. The variation of height across the thickness of the plate is shown schematically for each processing step. The broken line shows the corresponding temperature distribution.

2. *Ion exchange* introduces favorable compressive stresses in a much thinner surface layer. When a soda–lime glass is immersed, below T_g (at about 400°C), into molten KNO_3, Na ions are replaced by the larger K ions (or Li ions may be replaced by Na ions in a $NaNO_3$ bath). Alternatively, Na or K ions are exchanged for smaller Li ions at temperatures above T_g; the surface has a lower coefficient of expansion and, since it shrinks less, is subjected to compressive stresses by the bulk of the glass.

Example 12-5 | **A** glass container is heated in an oven to 350°C and is plunged into boiling (100°C) water. Assuming that the surface stress should not exceed the tensile stress (70 MPa) of the glass, determine which glasses in Table 12–2 can be safely subjected to the above treatment.

In the simplest case, the stress state is uniaxial and, from Eq. (4-5),

$$\sigma = E e_t = E \alpha \, \Delta T$$

where E is Young's modulus (typically $E = 70\ 000$ MPa) and α is the coefficient of linear expansion (Table 12–2).

Glass	α, 10^{-7} per °C	σ, MPa
Fused silica	5.5	9.6
Borosilicate	33	57.8
Aluminosilicate	42	73.5
Soda–lime	92	161
Potash–lead	102	178.5

Sample calculation for fused silica: $\sigma = 7(10^4)(5.5)(10^{-7})(350 - 100) = 9.6$ MPa.

The advantage of a low-expansion glass is obvious. A slightly more sophisticated treatment takes into account the fact that the stress state on the surface of a plate or container is usually biaxial.

12-5-3   Coatings

Ceramic surface layers are deposited on metal and ceramic components for both esthetic and technical reasons.

Glazes and *enamels* are glassy or partially crystalline coatings applied in the form of an aqueous slip (slurry). For optimum control of composition, the glass is often premelted and granulated in water. The resulting *frit* is comminuted and suspended in water with the addition of antiflocculants etc. To retain the slip on vertical surfaces, its properties are adjusted by the addition of clay or organic binders to impart pseudoplastic or Bingham behavior (Fig. 7–5*b*).

The composition is chosen so that the glaze or enamel should have a slightly lower thermal expansion coefficient than the part and will thus be put into compression on cooling. Lead reduces firing temperatures but must be avoided in all food-related applications because of its toxicity. Coatings range from transparent to completely opaque, and a wide range of colors and special effects may be obtained.

Vitreous enamels are applied to low-carbon steel, cast iron, or aluminum articles for improved corrosion resistance. Iron carbide reacts with enamels to form CO_2; therefore, better quality and adhesion is obtained on steel of very low carbon content (hence the preference for rimmed steel with its segregation, Fig. 7–4*a*, and for special decarbonized steel of 0.03% C). Adhesion on cast iron is purely mechanical; on aluminum it is aided by first forming a surface oxide.

Dental crowns and bridges are usually made from a gold or palladium alloy as lost-wax castings. For esthetic reasons, they are covered with several layers of porcelain, chosen for controlled translucency, color, and matching thermal expansion. Adhesion to the noble metals is poor; therefore, the alloys contain indium or tin which allow the development of an oxide to which the porcelain bonds.

Example 12-6

Glazes are applied to ceramics to make them impermeable. Some glazes are made up of raw materials, but better control is obtained with frits.

12-5-4   Process Capabilities and Design Aspects

The gradual softening of glass makes it one of the most versatile materials in terms of shape. The artisan creates products of high complexity with only a few tools, often by joining several pieces, if desired, of different colors.

Industrial production is much more limited by the need of using metal molds (dies) which have to be of a shape to release the part, if necessary, by opening a split die. In contrast to all other materials considered heretofore, glass allows cavities to be formed without a die, simply by blowing, and this allows many parts of reentrant internal shape (F5 and U4 in Fig. 3–1) to be formed.

Size limitations are few. Wall thickness can be vanishingly small (witness the glass Christmas-tree ornaments) or huge (as for telescope mirrors), but cooling rates and thus production rates drop precipitously with increasing thickness. Stress relief anneal also becomes imperative.

12-6   SUMMARY

Ceramics play an exceptionally wide role in human life. Our emphasis is on technical ceramics, distinguished by their tightly controlled and often unusual mechanical properties, resistance to wear or high temperature, and controlled optical, electrical, or magnetic properties. They may originate in natural oxides, silicates, and clay minerals, or they are manufactured in the form of materials of exceptional purity or of a composition not found in nature, such as carbides or nitrides.

1. The starting material is subjected to preparatory steps, including comminution, classification and, when necessary, agglomeration to impart the desired rheological properties when blended and mixed with a binder.

2. Consolidation is by processes familiar from powder metallurgy, including dry pressing, isostatic pressing, and injection molding, and also by screw extrusion. Ceramics may also be made into slurries that exhibit viscous flow and can thus be treated by casting processes, including slip casting into plaster of paris molds which allow greatest shape complexity.

3. The binder is removed in drying and early stages of sintering, and permanent bonds are established at high temperatures by diffusion mechanisms. Partial melting (vitrification) increases density but may affect hot strength.

4. Glass is amorphous and softens gradually, allowing processing based on viscous flow. Glasses too have entered the high-technology age, tailor-made for specific applications and subjected to special treatments. The metastable supercooled liquid can be transformed into the stable crystalline form and such glass ceramics have full density.

The processing of ceramics invariably involves high temperatures, and protective measures similar to those taken in foundries and metalworking plants must be taken.

PROBLEMS 12A

12A-1   Make a distinction between ceramics, glasses, and glass ceramics.

12A-2   Briefly describe the steps involved in making several copies of a porcelain statuette.

12A-3   Indicate which statements are true of ceramics: (*a*) A green body is so named because of its color. (*b*) Glass ceramics are ceramics that have an amorphous structure. (*c*) Devitrification means crystallization of a glass. (*d*) A vitreous substance is amorphous. (*e*) A vitreous substance is usually transparent.

12A-4   The term *polymorphism* means which of the following: (*a*) a single crystal charac-terized by many facets; (*b*) the substance undergoes an allotropic transformation; (*c*) the substance can possess different crystal structures, depending on temperature and pressure; (*d*) the shape of individual grains varies within a polycrystalline body.

12A-5   (*a*) State whether ceramics have the same tensile and compressive strength. (*b*) Justify.

12A-6   (*a*) State what tests are usually performed to determine the tensile strength of ceramics. (*b*) State which of these tests is likely to give the lowest value and why.

12A-7   Define (*a*) jiggering, (*b*) slip casting, (*c*) drain casting, and (*d*) tape casting.

12A-8   Define (*a*) earthenware, (*b*) stoneware, (*c*) porcelain, and (*d*) refractory.

12A-9   In a series of sketches and with brief sentences, show the steps involved in tempering a glass plate.

12A-10   Use simple sketches to define the main features of making (*a*) sheet glass, (*b*) float glass and (*c*) parison.

12A-11   Define glass transition temperature (state the properties above and below this temperature).

12A-12   Define the term *rheology*.

PROBLEMS 12B

12B-1   Explain why the mechanical properties of ceramics improve with diminishing crystal size.

12B-2   Suggest at least two methods for improving the impact properties of ceramics. Justify.

12B-3   (*a*) Define the term *liquid-phase sintering*. Give examples of (*b*) powder-metallurgy and (*c*) ceramic products made by the technique.

12B-4   A ceramic part is made for high-temperature service. It is argued that creep properties improve with higher density, therefore, liquid-phase sintering should be used. Do you agree and why?

12B-5   Suggest appropriate forming techniques for making a (*a*) common brick with insulating holes; (*b*) face brick with decorative surface configuration; (*c*) alumina tube; (*d*) glass tube; (*e*) toilet bowl; (*f*) field tile; (*g*) porcelain figurine; (*h*) porcelain dinner plate; (*i*) porcelain electrical insulator for high-tension power lines.

12B-6   Many processes used for the consolidation of ceramic powders have their counterparts in processes used for making metal components. For the following processes, indicate the counterparts, highlighting differences in the mechanisms that allow their application: (*a*) dry pressing, (*b*) wet pressing, (*c*) jiggering, (d) injection molding, and (*e*) slip casting.

12B-7   The term *ferrite* may be applied to which of the following: (*a*) an iron of FCC structure; (*b*) a ferrospinel; (*c*) a ceramic of desirable magnetic properties; (*d*) one of the constituents of pearlite; (*e*) a very low-carbon iron-carbon alloy at room temperature; (*f*) an iron of exceptional hardness.

12B-8   Explain the similarities and differences in (*a*) composition, (*b*) structure, and (*c*) mechanical properties of carbon black, graphite, and diamond.

12B-9   (*a*) State why glass is weak in tension. (*b*) Suggest two methods by which the tensile (or bending) strength of glass can be increased; briefly explain why they should work.

12B-10   Explain why a glass bottle can be blown to a uniform wall thickness, without the danger of fracture.

12B-11   Define the terms (*a*) *vitrification* and (*b*) *devitrification* as applied to ceramics. (*c*) State whether equivalent phenomena can be found in metals. Give examples for each.

12B-12   State what would be the best characterization of window glass from the point of view of (*a*) structure, (*b*) composition, and rheology at (*c*) elevated temperature and (*d*) room temperature.

12B-13   Explain why substantial deformation is possible in (*a*) metals and (*b*) glasses but not in (*c*) ceramics.

12B-14   Many glass objects of great age are exhibited in museums. Some show opaque areas. Why?

12B-15   Bathtubs are often made of enameled iron sheet. The bathtub is normally at room temperature, but is subjected to sudden heating by hot water. Should the thermal expansion of the enamel be greater or lower than that of iron?

12B-16 A student wrote "A glass can be distinguished from a ceramic by its transparency." Do you agree?

12B-17 Freshly drawn glass is very strong. (*a*) Explain why it loses much of its strength and (*b*) suggest a way of maintaining high strength.

12B-18 Which is easier to recycle: (*a*) ceramic or (*b*) glass ceramic. Justify.

PROBLEMS 12C

12C-1 An electrical insulator block of $10 \times 20 \times 150$ mm dimensions is slip-cast. Immediately after removal from the mold it weighs 48 g. After drying, the weight is 35 g, and the length has shrunk to 130 mm, with proportional shrinkage in the thickness and width directions. Calculate (*a*) the weight loss, percent; (*b*) the coefficient of linear shrinkage; (*c*) the dry dimensions and volume.

12C-2 Recalculate the stresses generated by the quenching described in Example 12-5.

Consider that in balanced biaxial tension $\sigma_1 = \sigma_2$ and $\sigma_3 = 0$. Apply the generalized Hooke's law (ν is Poisson's ratio).

$$e_1 = \frac{1}{E}\left[\sigma_1 - \nu(\sigma_2 - \sigma_3)\right] = \frac{1}{E}\sigma_1(1 - \nu)$$

$$\sigma_1 = \frac{eE}{1 - \nu} = \frac{E}{1 - \nu}\alpha\Delta T$$

12C-3 (*a*) State what rheological models best describe the behavior of a glass below and above the fictive point. (*b*) Write the governing equations for tensile loading. (*c*) Sketch the corresponding spring/dashpot models. (*d*) Show why a glass rod thins out uniformly if stretched above T_g (use a physical argument based on the behavior of an incipient neck).

12C-4 A square alumina tool bit of $a = 20$-mm sides and $h = 10$-mm thickness is pressed at a pressure of $p = 100$ MPa. Poisson's ratio is 0.23. Determine (*a*) the press force, and the pressure prevailing at the bottom of the pressing if a single punch is used in a fixed container for pressing (*b*) dry ($\mu = 0.5$) and (*c*) with a lubricant ($\mu = 0.1$).

FURTHER READING

ASM Engineered Materials Handbook, vol. 4, *Ceramics and Glasses*, ASM International, 1991.
ASM Engineered Materials Handbook: Desk Edition, ASM International, 1995.

Ceramics

Barsoum, M.: *Fundamentals of Ceramics*, McGraw-Hill, 1997.
Brook, R.J. (ed.): *Concise Encyclopedia of Advanced Ceramic Materials*, Pergamon, 1991.
Cheremisinoff, N.P.: *Handbook of Ceramics and Composites*, 3 vols., Dekker, 1991.
Chiang, Y-M.: *Physical Ceramics; Principles for Ceramics Science and Engineering*, Wiley, 1995.
Cranmer, D.C., and D.W. Richerson: *Mechanical Testing Methodology for Ceramic Design and Reliability*, Dekker, 1998.
Jones, J.T. and M.F. Berard: *Ceramics—Industrial Processing and Testing*, 2d ed., Iowa State University Press, 1993.

Kingery, W.D., H.K. Bowen, and D.R. Uhlmann: *Introduction to Ceramics*, 2d ed., Wiley, 1976.

Lu, H.Y.: *Introduction to Ceramic Science*, Dekker, 1996.

McHale, A.: *Phase Diagrams and Ceramic Processes*, Chapman and Hall, 1998.

Musikant, S.: *What Every Engineer Should Know About Ceramics*, Dekker, 1991.

Norton, F.H.: *Elements of Ceramics*, 2d ed., Addison-Wesley, 1974.

Phillips, G.C.: *A Concise Introduction to Ceramics*, Van Nostrand Rheinhold, 1991.

Rahaman, M.N.: *Ceramic Processing Technology and Sintering*, Dekker, 1995.

Reed, J.S.: *Principles of Ceramics Processing*, 2d ed., Wiley, 1995.

Richterson, D.W.: *Modern Ceramic Engineering*, 2d ed., Dekker, 1992.

Schwartz, M.M.: *Handbook of Structural Ceramics*, McGraw-Hill, 1992.

Terpstra, R.A., P.P.A.C. Pex, and A.H. DeVries (eds.): *Ceramic Processing*, Chapman and Hall, 1995.

Vincenzini, P.: *Fundamentals of Ceramic Engineering*, Elsevier, 1991.

Nanoceramics, Institute of Materials, 1993.

Glasses

Doremus, R.H.: *Glass Science*, 2d ed., Wiley, 1994.

Hlavac, J.: *The Technology of Glass and Ceramics*, Elsevier, 1983.

Lewis, M.H. (ed.): *Glasses and Glass-Ceramics*, Chapman and Hall, 1989.

Loewenstein, K.L.: *The Manufacturing Technology of Continuous Glass Fibers*, 2d ed., Elsevier, 1983.

Tooley, F.V. (ed.): *The Handbook of Glass Manufacture*, Ashlee Publishing Co., 1984.

Zarzycki, J.: *Glasses and the Vitreous State*, Cambridge University Press, 1991.

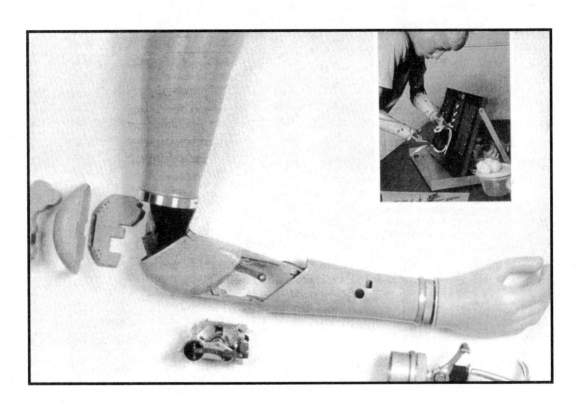

Polymers are compounded with various additives so that the resulting plastics become suitable for a vast variety of applications. Plastics can be formed into parts of highly complex shapes, such as the artificial arm shown here. (*Courtesy Canadian Plastics Industry Association, Mississauga, Ont.*)

13

Polymers and Plastics

Plastics are based on polymers and, before embarking on the study of their manufacture, we need to review the answers to some basic questions, such as:

Why some polymers are deformable while others are completely brittle

Why applications are so greatly limited by temperature

What allows a rubber to undergo repeated elastic deformation without failure

What mechanisms can be exploited for manufacturing

Why some plastic products can be used in place of metals

A *polymer* is, as indicated by the Greek roots *poly* (many) and *meros* (part), any substance made up of many (usually, thousands) of repeating units, building blocks, called *mers*. Most polymers are based on a carbon backbone and are thus organic materials. There are many natural polymers and, after concrete, wood is still the most widely used structural material. Our concern here, however, is with *synthetic polymers*; when in a form ready for further working, they are also called *resins*. They are seldom used in their neat form; most often, they are compounded with various additives and the resulting material is usually referred to as a *plastic* (again, from the Greek *plastikos*, derived from *plassein*: to form, to mold). Frequently, the terms polymer, resin, and plastic are used interchangeably.

From relatively recent beginnings (Table 1–1), the growth of the plastics industry has been phenomenal. As shown in Table 13–1, the sale of plastics—on a volume basis—has grown to more than double the steel production in the United States. The growth trend is expected to continue, partly because new uses are still being found and partly because improved plastics can substitute for other materials. Initially, most plastics were used in applications where their low density, high corrosion resistance, electrical insulation, and ease of manufacturing into complex shapes presented advantages and where mechanical strength was of secondary importance. These applications still consume the largest quantities. A more recent and technically most important trend has been the emergence of structural polymers (Table 5–4) which can be made into load-bearing components and structures, at least for applications in which temperatures are only moderately high, typically, below 150–250°C.

Table 13-1   Sales of plastics* (United States) 1000 Mg†

Plastic	1972	1997
ABS	388	630
Acrylic	208	270
Epoxy	78	
Nylon	67	665
Phenolic	652	
Polycarbonate	25	75
Polyester	416	765
Polyethylene, low density	2 372	7 050
High density	1 026	6 000
PET		1 700
Polypropylene	767	5 850
Polystyrene	1 240	2 900
Polyurethane	460	2 160
PVC and vinyls	2 326	6 350
Urea and melamine formaldehyde	411	

*Compiled from *Modern Plastics*, January 1973 and January 1997. For some plastics data are not available in a given year.
†Mg = 1000 kg = metric tonne = 2200 lb.

Many of the principles discussed hitherto apply also to polymers, nevertheless, there are sufficient differences to justify a review of polymer structure and properties, with reference to concepts previously explored for metals.

13-1   POLYMERIZATION REACTIONS

Polymers, like metals, are produced from raw materials in specialized plants and are provided to manufacturing facilities in forms suitable for processing into finished articles (Fig. 5–3). Primary manufacturing will be discussed here only to the degree necessary for an understanding of manufacturing properties.

There are many ways of classifying polymers, one of which is the technology used for making them. Details are beyond the scope of this book; it will be sufficient to note that macromolecules may be obtained by one of two processing techniques (Fig. 13–1):

1. Chain polymerization. Carbon is a tetravalent element and carbon chains can be formed with single, double, or triple bonds between adjacent carbon atoms. The starting material for chain polymerization is usually a *monomer* in which there

HYDROCARBONS

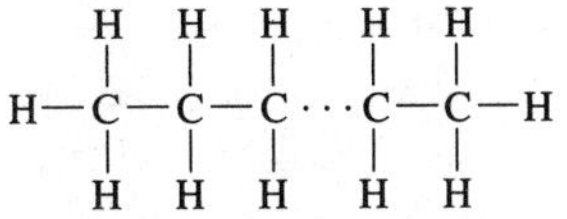

Paraffin (aliphatic) 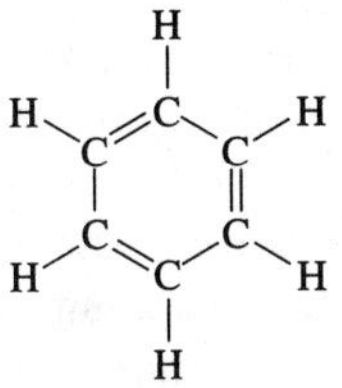   Benzene (aromatic) 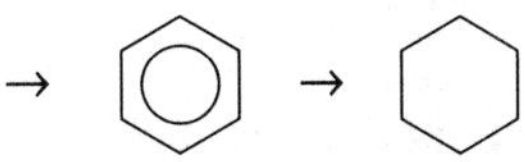   symbolic presentation
(hydrogen not shown)

CHAIN-REACTION POLYMERS

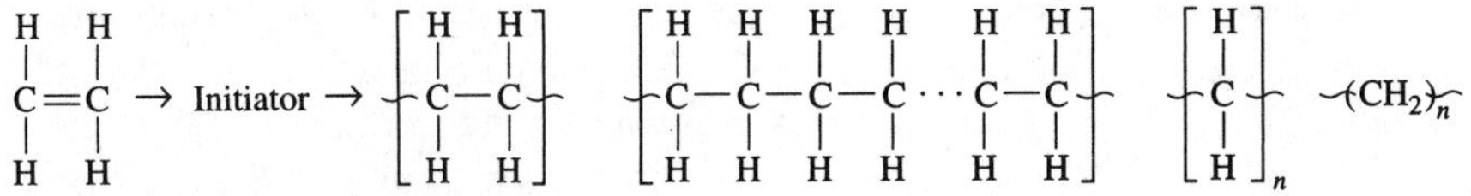

Ethylene   repeat   Polymer: PE (Polyethylene)   alternative
monomer   unit   presentation

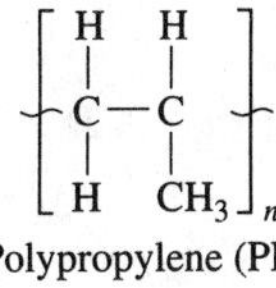

Polypropylene (PP)   Polyvinyl-   Polyvinylidene-
                        chloride (PVC)   chloride

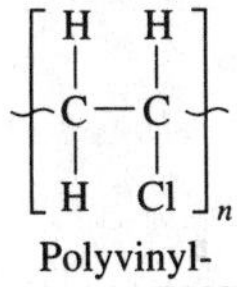

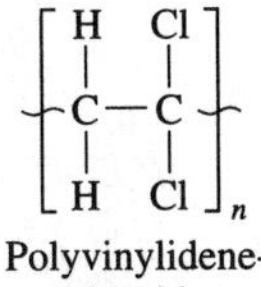

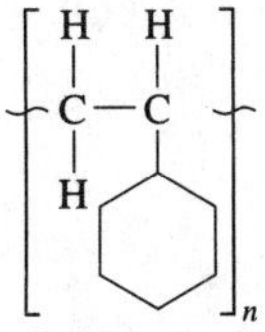

Polystyrene   Polyvinyl-   Polytetraflouroethylene
                        flouride   (PTFE)

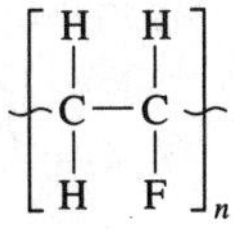

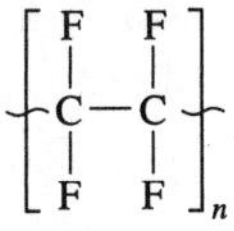

STEP-REACTION POLYMERS

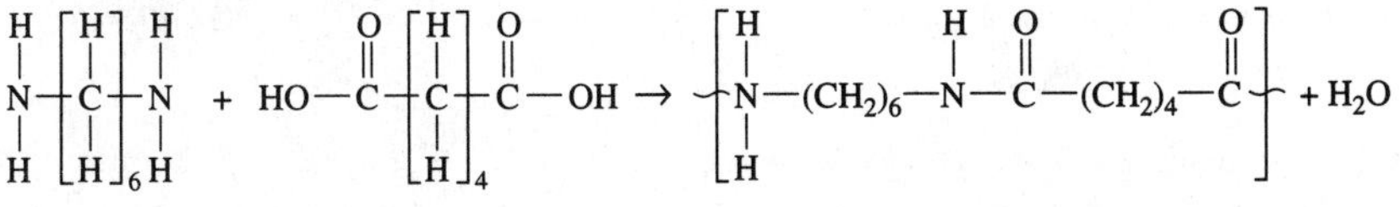

Hexamethylene   +   Adipic acid   Polyamide (nylon-6,6)   + Water
diamine

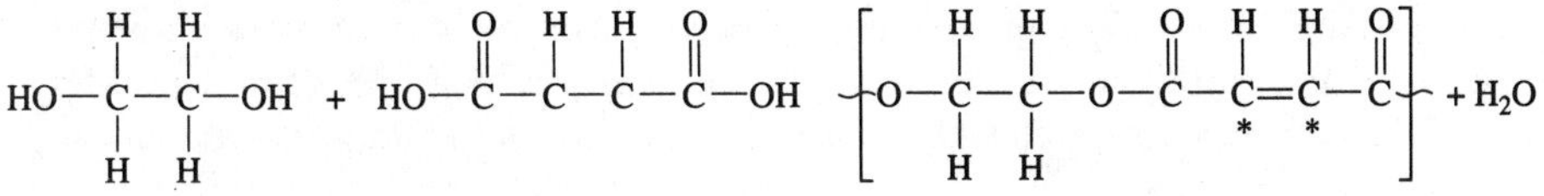

Ethylene glycol   +   Maleic acid   Linear unsaturated polyester   + Water
(alcohol)   (* possible site for crosslinking)

Figure 13-1   Thermoplastic polymers are made up of essentially linear molecules, formed by
chain-reaction (addition) or step-reaction (condensation) polymerization. The presence of
double bonds in the polymer, as in unsaturated polyesters, makes cross-linking possible.

is a double bond that can be opened up with the aid of a compound called *initiator* (organic or inorganic substance or catalyst, which is not consumed in the reaction), whereupon polymerization occurs simultaneously in the entire batch in a few seconds. Conducted at some elevated temperature and under pressure, the process is also called *addition polymerization*. The most frequently occurring structures are hydrocarbons (Fig. 13–1) in which carbon and hydrogen may form straight chains (*aliphatic hydrocarbons*) or benzene rings (*aromatic hydrocarbons*). Other polymer molecules may contain also N, O, S, P, or Si (in the backbone or side chains) while the monovalent Cl, F, or Br may replace H. (Some of these can be recycled by high-temperature cracking.)

2. Step-reaction polymers. In the majority of these processes, two dissimilar monomers are joined into short groups which gradually grow; a byproduct of low molecular weight is often also released (water in the example of nylon-6,6, Fig. 13–1), and then it is customary to speak of a *condensation reaction*. (Unless cross-linked, some of these polymers can be recycled by depolymerization: at higher temperatures and in the presence of a reactive agent they can be "unzipped" to regenerate the monomer.)

Either way, the polymer chemist can control the average length of the molecules by terminating the reaction. Thus, the *molecular weight* (the average weight, in grams, of 6.02×10^{23} molecules) or *degree of polymerization* (the number of mers in the average molecule) can be controlled. For example, the length of molecules may range from some 700 repeat units in low-density polyethylene (LDPE) to 170 000 repeat units in ultrahigh-molecular-weight polyethylene (UHMWPE).

13-2   LINEAR (THERMOPLASTIC) POLYMERS

It will be noted that all examples shown in Fig. 13–1 result in the formation of more or less straight chains; hence, these polymers are called *linear polymers*.

13-2-1   Structure of Linear Polymers

In metals the basic unit was the atom or, at the most, the unit cell of an intermetallic compound, and these units readily conformed to long-range order to give a crystalline structure. The great length of polymer molecules combines with other spatial features to make for a much greater variety of possible structures. Here we review them with emphasis on the significance of structure to manufacturing and service properties. The molecules of a linear polymer are not simple straight chains for the following reasons:

1. Even the simplest chain, that of polyethylene (PE), is not straight. The C—C bond is at a fixed bond angle of 109.5°. The spacing between C atoms is 0.154 nm along the bond but only 0.126 nm in a straight line. Thus, a PE molecule of 2000 carbon atoms would have a length of 252 nm (or 0.25 μm) when fully stretched. However, the single bond between the carbon atoms allows rotation around the bond; thus, the molecule can become randomly coiled and twisted, and the actual average end-to-end distance is typically only 18 nm. Such a molecule will not readily fit into a long-range ordered structure and the polymer will be *amorphous*.

2. The chains of some polymers, such as HDPE, are smooth when stretched out (Fig. 13–2*a*) whereas others, such as polypropylene (PP), have pendant groups (in this case, —CH₃) in certain positions. The ordering (in Greek: *taktika*) of these groups determines whether the polymer is *isotactic* (with all groups on one side of the backbone, Fig. 13–2*b*), *syndiotactic* (alternating on the two sides, Fig. 13–2*c*), or *atactic* (randomly arranged, Fig. 13–2*d*). The pendant group (a benzene ring) is still larger in polystyrene (Fig. 13–1). Tight packing of molecules is obviously more difficult and flexibility is reduced if the pendant groups are large and randomly oriented. Thus, atactic PP is amorphous with poor mechanical properties, whereas isotactic PP can be made highly crystalline and is then widely used even in engineering applications.

3. Even simple molecules such as LDPE may not be truly straight chains but have side branches which further increase the difficulty of close packing and ordering. Only some polymers, such as linear LDPE (LLDPE) or HDPE polymerized in the presence of special catalysts and polytetrafluoroethylene (PTFE), are free of branching.

4. In aromatic polymers the presence of the benzene ring offers the possibility of creating a backbone of a double strand with regular cross-links, resembling a ladder (*ladder polymer*). Because two bonds must be broken before a lower molecular weight product is formed, such structures can be highly temperature-resistant.

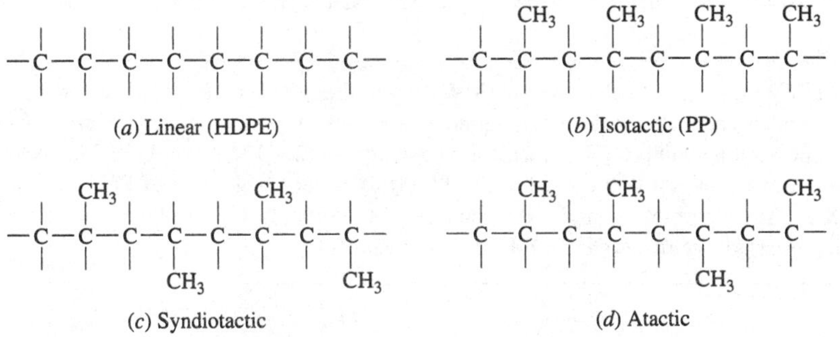

(a) Linear (HDPE)

(b) Isotactic (PP)

(c) Syndiotactic

(d) Atactic

Figure 13–2   The type of backbone and the ordering (spatial arrangement) of pendant groups around the backbone determine many properties of linear polymers. See text for significance of various arrangements. (For simplicity of presentation, hydrogen atoms are not shown.)

5. The examples shown in Fig. 13–1 are made up of one kind of repeating unit and are called *homopolymers*, even if the repeating unit is made up of two precursor molecules.

6. It is possible to polymerize two types of monomers (generally, A and B) to obtain *copolymers* (more exactly, *binary copolymers*) somewhat analogous to solid-solution alloys. In copolymers each repeating unit is capable of forming a polymer on its own, as in an ethylene–propylene copolymer. (A three-component polymer is a *ternary copolymer* or a *terpolymer*, an example of which is ABS, made of acrylonitrile, butadiene, and styrene monomers).

The repeat units may occur in a *random* (AAABBABAABBA); *alternating* (ABABABAB), or *block* (AAAAABBBBBAAAABBBBBBAAAA) sequence, or one species may branch off in a *graft polymer*

AAAAAAAAAAAAAAAAAAAAAAAAA

B	B
B	B
B	B
B	B
B	B

7. Two miscible polymers or a polymer and a monomer may be combined into a homogeneous *polymer alloy*. There is a single T_g and properties often exceed the properties of the constituents.

8. A further possibility is to have two incompatible polymers (which do not enter into a joint chain) mixed with one another, with one serving as the matrix. These are called *polymer blends* (*polyblends*, but sometimes also called *polymer alloys*) and can be regarded as the polymeric counterparts of two-phase metal alloys. There are two distinct T_g values. Desirable properties of the constituent polymers are combined but not necessarily exceeded, as will be seen later in the example of thermoplastic elastomers. When the two polymer networks interpenetrate (*interpenetrating network polymers*), properties can greatly improve.

Example 13-1 | An UHMWPE has a molecular weight (MW) of 4 million. If this refers to the number average of molecules, calculate the degree of polymerization and the length of the stretched chain.

The building unit is C_2H_4, of a molecular weight of $(2 \times 12) + (4 \times 1) = 28$. The degree of polymerization is $4\,000\,000/28 = 143\,000$ (this is very large; that of PTFE is typically 30 000). The length of the molecule would be $(143\,000)(1.26) = 18\,000$ nm $= 18$ μm. Of course, we shall see that it will be folded, as in Fig. 13–4b.

13-2-2 Sources of Strength

An engineering part consists of many macromolecules. Bonding *within* each molecule is provided by the electrons shared between adjacent atoms (*covalent bonding*). Bond energy (the energy required to break one mole, i.e., 6.02×10^{23}

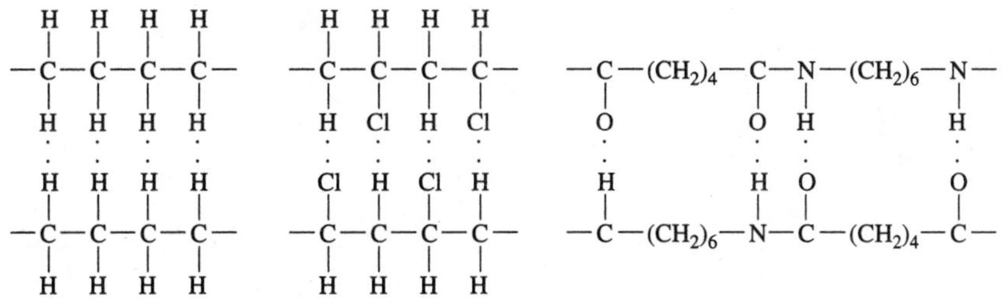

Figure 13-3 Polymers owe their strength to secondary bonds between molecules: (a) weak van der Waals forces between nonpolar molecules, (b) dipole bonds between polar molecules, and (c) the strong hydrogen bond between H and O, N, or F.

bonds) is on the order of 350–830 kJ/mol, making the molecule itself very strong. However, this tells us little about the actual strength of a polymeric part; to appreciate the strength or lack of strength of certain polymers, we must look into forces that hold the multitude of molecules together. Entanglement accounts for some of the strength, but the predominant source of strength is the presence of *secondary bonds*. These are of several kinds:

1. At the least, there are always *van der Waals forces* present, even though they are very weak (2–8 kJ/mol) (Fig. 13–3a).

2. When atoms share electrons in covalent bonds, the atom that loses electrons appears to have a positive charge and vice versa; thus, a permanent dipole is set up (the molecule has a polar character). Polar molecules, such as those containing free Cl, F, or O valences, make stronger (6–13 kJ/mol) *dipole bonds* (as in polyvinylchloride, PVC, Fig. 13–3b).

3. The *hydrogen bond*, established between hydrogen and O, N, or F, is a special case of dipole bonds. Bond energy is high (13–30 kJ/mol), as in nylon-6,6 (Fig. 13–3c).

The number of secondary bonds increases with chain length, giving increasing strength to the body. Thus, secondary bonds are sources of strength. At the same time, at higher temperatures—where thermal excitation is significant and secondary bonds are easily broken and reformed—they allow molecules to move relative to each other. The ease of movement depends on the number of secondary bonds present; very long-chain plastics, such as UHMWPE and PTFE, may char before ever reaching the moldable state.

High-density polyethylene is produced in a wide range of molecular weights to satisfy various needs.

Example 13-2

	Molecular weight
Commodity grade (injection molding) resin	80 000
Blow-molding resin	120 000
High-performance resin	200 000–500 000
UHMWPE	4 000 000

13-2-3   Crystalline and Amorphous Polymers

When linear polymers are heated to some high temperature (but not so high as to break primary bonds), one can visualize a mass of polymer molecules as a bowl of spaghetti. Since no long-range order exists, the polymer is amorphous (Fig. 13–4a). The comparison to spaghetti is, however, incomplete. Under the influence of the elevated temperature, the molecules are in constant motion and the free volume (the space within the polymer) is large. Upon cooling, one of two events may take place.

Crystallization   If the molecule is of relatively simple shape, with chemical regularity along the chain, and conditions are favorable, some long-range order may develop: The polymer begins to crystallize upon cooling below the melting point T_m. As in metals (Sec. 6-1-7), the process begins by nucleation followed by growth. This involves the repeated folding of chains into thin, approximately 10-nm-thick *lamellae* with ordered (Fig. 13–4b) or random (Fig. 13–4c) reentry. The crystals grow in the two lateral dimensions and several lamellae growing from a common nucleus form a *spherulite* (Fig. 13–4d). Adjacent regions are held together by *tie molecules*. Perfect crystallinity—such as is found in metals and ceramics—is never achieved; there is always some amorphous material between the spherulites. Thus, the term *crystalline polymer* refers to a structure in

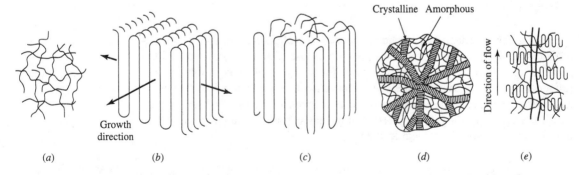

Figure 13–4   Linear molecules may be arranged (a) randomly (amorphous polymers) or in thin, crystalline lamellae with (b) ordered or (c) random reentry. Lamellae growing from a common nucleus form (d) spherulites in an amorphous matrix. Lamellae are (e) oriented in crystals that grow during deformation.

which crystalline regions predominate. During the formation of the crystalline zones, there is a corresponding drop in specific volume (Fig. 13–5a), but it is neither as steep nor as complete as in metals (Fig. 6–1d). Mechanical properties, such as elastic modulus, are also greatly affected by crystallinity (Fig. 13–5b). If the polymer is deformed while it is cooling below T_m, polymer chains become oriented and crystals form with the folding in the direction of flow (*row-nucleated structure*, Fig. 13–4e).

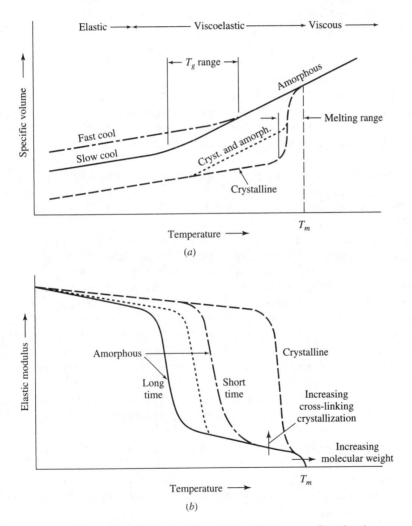

Figure 13–5   Structural changes in thermoplastic polymers are reflected in changes in specific volume (a) and in time- and temperature-dependent mechanical properties, exemplified here by the elastic modulus (b).

The degree of crystallinity depends on many factors. It is greater for polymers formed by mers of regular shape, without side branches or pendant groups. If there are pendant groups (Fig. 13–2), the isotactic form will more readily crystallize than the atactic form. For the same reasons, copolymers in general, and random and graft copolymers in particular, will not crystallize. Since it takes time for molecules to fold, slower cooling promotes crystallization. For rapid crystallization from the melt, the mold temperature is set to $(T_m + T_g)/2$. Above T_g, crystallization may continue for a prolonged period of time. This causes after-shrinkage, as in nylon-6,6 which has a T_g below room temperature. Heating at elevated temperature results in rapid completion of crystallization.

Cooling rate affects also the size of spherulites. As in metals, the rate of growth reaches its maximum at a higher temperature than the rate of nucleation, hence spherulites become coarser at slower cooling rates. Nucleation rate can be increased and thus the size of spherulites reduced by seeding, for example, with very fine silica (heterogeneous nucleation).

Mechanical alignment induced by directional deformation such as drawing or extrusion also contributes to crystallization, and the pronounced directionality of structure (texture, Sec. 8-1-3) results in a directionality (anisotropy) of properties, with higher strength in the length direction of molecules.

Amorphous Polymers   If the structure and process conditions are unfavorable for crystallization, the polymer continues to cool while remaining amorphous. Specific volume drops at the rate typical of the molten state (Fig. 13–5a); even though the excitation of molecules is reduced, they can still move relative to each other to reduce free volume and, in some instances, further secondary bonds may form. Freedom of movement is lost at some typical temperature, called—as in glasses (Sec. 12-5-1)—the *fictive point* or *glass-transition temperature*, T_g. No further bonds are established, and the specific volume changes at a much lower rate, as a result of the reduced thermal motion of molecules fixed in space. The situation is somewhat analogous to metals in that T_g is often around $0.5T_m$ (these are homologous temperatures, Sec. 4-6). We saw that in metals the onset of hot behavior can be shifted to higher temperatures by alloying. Similarly, T_g can also be shifted substantially. It is around $0.66T_m$ in typical homopolymers, it can drop to $0.25T_m$ in block copolymers, and a value of $0.9T_m$ may be reached in random copolymers. In contrast to metals, the open molecular structure results in a substantial increase of T_g with increasing pressure.

Liquid-Crystal Polymers   A small but important class of thermoplastic polymers is unique in that the rod-like molecules line up parallel in the liquid state. Upon solidification, the molecules draw closer to each other but retain their orientation; hence, there is practically no shrinkage in the longitudinal direction and the dimensions are highly stable.

13-2-4   Rheology of Linear Polymers

The *rheology* of polymers (from the Greek *rheos* = flow) deals with their response to stresses. This response is a function of structure and temperature.

Viscous flow   Above T_m, molecules can move, slide relative to each other, and low-molecular-weight polymers may exhibit Newtonian viscous flow (Fig. 7–5b, repeated here as Fig. 13–6a. This is subject to Eq. (7-1), repeated here because of its importance:

$$\tau = \eta \frac{dv}{dh} = \eta \dot{\gamma} \qquad \textbf{(13-1)}$$

Viscosity increases with increasing molecular weight because of the greater number of secondary bonds available along a longer chain (for many high-molecular-weight polymers, viscosity is proportional to the 3.4 power of the weight-average molecular weight).

Viscosity is also a function of molecular structure: entanglement, less open structure (fewer side chains), and restricted molecular segment rotation (lesser chain flexibility) all contribute to higher viscosity, whereas a wide molecular-weight distribution (which signifies the presence of shorter chains) leads to a lower viscosity.

Viscosity increases with decreasing temperature, because of the drop in free volume and the lessened mobility of molecules. As with thermally activated processes in general, viscosity changes exponentially with inverse temperature

$$\eta = A \exp \left(\frac{-E}{RT} \right) \qquad \textbf{(13-2)}$$

where A is a material constant, E is activation energy, R the universal gas constant, and T temperature (K).

The change in viscosity with temperature can be expressed by considering the change in free volume; this leads to the Williams–Landel–Ferry (WLF) equation.

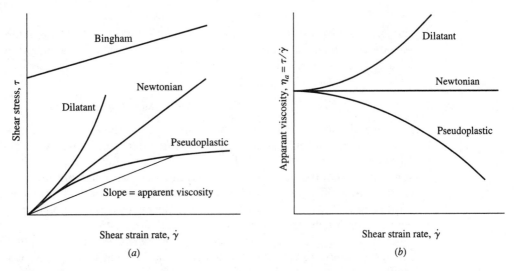

(a)   *(b)*

Figure 13–6   Many plastics show pseudoplastic flow (a) and the apparent viscosity declines with increasing shear strain rate (b).

With experimentally determined constants, it takes the following form:

$$\log_{10}\left(\frac{\eta_T}{\eta_{T_g}}\right) = \frac{-17.44(T - T_g)}{51.6 + (T - T_g)} \tag{13-3}$$

where T is the temperature of interest (K), and η_T and η_{T_g} are the viscosity at T and T_g, respectively. The importance of the glass-transition temperature is obvious.

Molecules uncoil when subjected to shearing; uncoiling reduces entanglement and hence viscosity. Thus, most polymers exhibit non-Newtonian flow: The shear stress needed for deformation is not a linear function of shear strain rate (Fig. 13–6a). Stated differently, viscosity is not constant but varies with strain rate (Fig. 13–6b). Many polymers are pseudoplastic (Fig. 13–6a) and obey a power law

$$\tau = \eta_a \dot\gamma^m \tag{13-4}$$

where $m < 1$. The similarity to Eq. (8-11) (flow stress in hot working a metal) is evident, except that Eq. (13-4) is written in terms of shear stress τ and shear strain rate $\dot\gamma$. Relative to metals, the strain-rate sensitivity exponent m of a pseudoplastic polymer is negative and reaches values close to unity at temperatures above T_m.

For computational purposes it is convenient to express an *apparent viscosity* η_a as the shear stress at a given strain rate (Fig. 13–6a)

$$\eta_a = \frac{\tau}{\dot\gamma} \tag{13-5}$$

This it is not a constant but very much a function of shear strain rate (Fig. 13–7a), temperature (Fig. 13–7b) and, not shown, of pressure. Since viscosity represents resistance to the sliding of molecules,

$$\eta \propto \frac{\text{entanglement}}{\text{free volume}} \tag{13-6}$$

and since pressure reduces while temperature T (above T_g) increases the free volume, the combined effects of temperature, pressure, molecular weight, and—for a given molecular weight—ordering, can be expressed qualitatively as:

$$\eta \propto \frac{\text{pressure}}{(T - T_g)} \frac{\text{molecular weight}}{\text{long-chain side branches}} \tag{13-7}$$

It is generally found that a pressure increase of 100 MPa is equivalent to a temperature drop of 30–50°C. Many polymers behave like Bingham substances (Fig. 13–6a) and begin to flow in a viscous manner only after a certain initial shear stress has been imposed; this initial stress increases with pressure. Some others are thixotropic: Their viscosity decreases with time at a constant strain rate.

For practical processing purposes, the flow properties of polymers under realistic conditions need to be known. Thus, shear stress is determined as a function of temperature and shear rate in *torque rheometers*. For some polymers, a *melt flow index* is given which is simply the mass of plastic (in grams) extruded in 10 min through a standard orifice (ASTM D1238) at a pressure of 300 kPa at

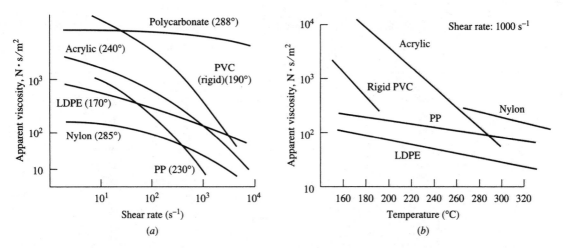

Figure 13-7   The apparent viscosity of polymers is highly sensitive to (a) shear strain rate and (b) temperature. (*From D.H. Morton-Jones: Polymer Processing, Chapman & Hall, p. 40, Fig. 2.5 and p. 41, Fig. 2.6. With permission.*)

a specified temperature (190°C for PE and 230°C for PP). Thus, the melt index is inversely related to viscosity and is suitable for comparing different grades of the same type of polymer.

Upon imposing a load, a viscous material deforms at a rate dictated by its viscosity. When the shear force is removed, the material remains in its deformed shape, hence it can be modeled by a damper or dashpot in which a piston is displaced against the resistance exerted by the shearing of an oil (Fig. 13–8a). It will be noted from Eq. (13-1) that deformation will occur, no matter how slowly, even under the slightest imposed load. Therefore, *the limiting temperature for structural use is below T_m for highly crystalline polymers and well below T_g for amorphous polymers.*

Elastic deformation   When the temperature drops below T_m for a highly crystalline polymer or well below T_g for an amorphous polymer, there is no chain mobility (except on an exceedingly long time scale). Upon imposing a load, the polymer deforms purely elastically, and the original dimensions are regained immediately upon removing the load. The polymer behaves like a spring (Fig. 13–8b). By analogy to Eq. (4-5), the shear stress τ is

$$\tau = G\gamma \tag{13-8}$$

where G is the *shear modulus* and γ is shear strain. This relates to E according to

$$G = \frac{E}{2(1 + \nu)} \tag{13-9}$$

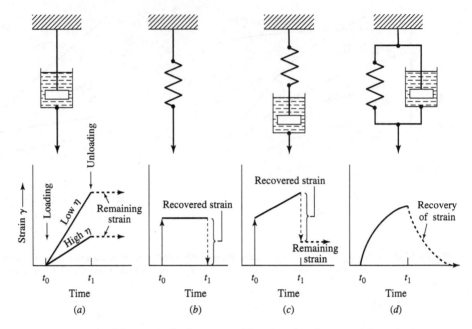

Figure 13–8 The deformation of polymers may follow the rules of (*a*) viscous flow, (*b*) elastic deformation, or viscoelastic flow according to the (*c*) Maxwell or (*d*) Voigt (Kelvin) model. (*Note*: Load is applied at time t_0 and is removed at time t_1.)

where ν is Poisson's ratio and ranges from 0.25 for stiff polymers to 0.5 for flexible ones.

Because elastic deformation in a polymer is accompanied by stretching of the relatively weak intermolecular bonds, much larger elastic strain is possible than in metals. If, however, the stress exceeds a critical value (the tensile strength), the part breaks in a brittle manner (Fig. 4–4*a*), hence one speaks of *glassy elastic* or brittle behavior. Such a polymer is useful as an engineering constructional material because it will keep its shape under imposed loads, but its brittleness (lack of toughness, low notched impact strength) is a drawback in many applications. Obviously, polymers can be processed in this temperature range only by machining.

Viscoelastic Flow From below T_g to above T_m, high-molecular-weight amorphous polymers exhibit *viscoelasticity*. The increased mobility of molecules allows some deformation to take place—in addition to the relative sliding of molecules against each other—by: uncoiling and stretching of molecules, rotation of molecular segments around single bonds (such as a C—C bond), and the cooperative movement of molecular segments. Thus, upon removing the load, some of the strain is recovered as elastic deformation (the polymer has *memory*), but part of it may remain as viscous flow.

Viscoelastic behavior can be described by various (and often very complex) models. Among the simplest is the *Maxwell element* (a damper and spring in series, Fig. 13–8c), in which the initial elastic (spring) deformation is followed by viscous flow, and the elastic component is regained immediately upon unloading. Deformation occurs at the rate

$$\frac{d\gamma}{dt} = \dot{\gamma} = \frac{d\tau}{dt}\left(\frac{1}{G}\right) + \frac{1}{\eta}\tau \qquad \textbf{(13-10)}$$

Alternatively, the polymer may be described by a *Voigt element* (a damper and spring in parallel, Fig. 13–8d), in which elastic deformation is damped by viscous flow in the dashpot and is only gradually recovered after unloading. The active stress is

$$\tau = G\gamma + \eta\frac{d\gamma}{dt} \qquad \textbf{(13-11)}$$

Realistic models of polymers usually require several Maxwell and Voigt elements acting in series and/or parallel. This explains why the modulus in Fig. 13–5b is a function not only of temperature but also of time: Upon fast loading, there is less time for viscous flow, and the modulus is higher.

Heat-shrinkable sleeves are sold in the expanded state (i.e., they were expanded at some elevated temperature and cooled to room temperature where their shape is stable). Upon heating, they shrink. For maximum shrinkage, is it better to have a behavior characterized by a Maxwell element or Voigt element?

In a Maxwell element (Fig. 13–8c) the deformation corresponding to viscous flow is irreversible. In a Voigt element, the spring is in parallel with the dashpot and, if the viscosity of the fluid in the dashpot is decreased (by heating the polymer), the spring will pull it back. Thus, the Voigt element is more suitable.

Example 13-3

Because of the viscous flow component, polymers can be processed in the viscoelastic temperature regime. The behavior of amorphous polymers above T_g is termed *rubbery*. It is important that these polymers be cooled well below T_g prior to release from the mold, so that the new molecular arrangement is frozen in. This means, however, that the part must not be allowed to heat above T_g in use because distortion, due to recovery of the elastic strain, will follow. In partially crystalline polymers deformation results in a fragmentation of crystalline regions, stretching of the tie molecules, and ordering of the crystal fragments (Fig. 13–9); this inhibits deformation above T_g (*leathery* behavior). Therefore, substantially crystalline polymers (with at least 50% crystalline phase) can be used above their T_g (such is the case for PE and PP).

From the manufacturing point of view, the viscous and viscoelastic behaviors are of utmost importance. In tension, very large total elongation is possible:

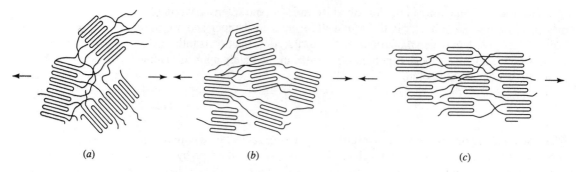

Figure 13-9   Deformation of a partially crystalline polymer results in a break-up and orientation of lamellae.

Even though a neck may appear after only 2–15% uniform elongation, the neck is resistant to fracture because of strain-rate effects and strengthening due to the alignment of molecules (Sec. 8-1-6 and Fig. 4–16). Successive necks may form, and total elongation of several hundreds or even thousands of percent may be attained before fracture occurs.

Thermoplastic Polymers   It is evident from the above discussion that the behavior of linear polymers is greatly affected by temperature. When an amorphous polymer is heated above T_g (or a crystalline above T_m), it becomes deformable, and it will retain its shape upon cooling below that temperature. The heating and cooling sequence can, in principle, be repeated; hence, it is customary to speak of *thermoplastic* polymers. It is possible to form such a polymer into a semifabricated shape (pellet, bar, tube, sheet, or film), which is then cooled and shipped to the secondary manufacturer who reheats and forms it into the final shape. Clean scrap can be recycled by adding it to virgin polymer, at least to some limited extent (although some degradation may occur).

Cross-linking of already formed thermoplastic polymers is possible by the application of high-energy radiation such as ultraviolet (UV) light, electron beams, γ-rays, x-rays, and particle beams. Cross-linking by radiation is exploited to increase strength, as of wire and cable insulation, or to endow the polymer with a capacity for large shape changes, as in heat-shrinking sleeves and films. Excessive exposure leads to concurrent chain degradation, oxidation, and gas formation, and then one speaks of *radiation degradation* of the polymer.

Example 13-4 | Heat-shrinkable PE tubing is made by cross-linking an extruded tube, followed by heating, expansion, and cooling. When it is reheated, it returns to the original cross-linked diameter.

13-3   CROSS-LINKED (THERMOSETTING) POLYMERS

In the linear polymers discussed hitherto, covalent bonds existed only *within* molecules and molecules were held to each other by secondary bonds that could be broken and reformed during deformation. When covalent bonds are established *between* molecules, they are firmly bonded to each other, and breaking the bond would result in failure.

Thus, *thermosetting polymers* or *thermosets* are so named because once polymerization is completed (the polymer is *cured*, e.g., by the application of heat or a catalyst) no further deformation is possible. The cross-link density is so high that no significant elastic deformation is possible either; the polymer is brittle but has a relatively high elastic modulus and temperature resistance. The covalent bonds form a *spatial network*, as in phenol formaldehyde (Fig. 13–10) and thermosetting polyester resins, permanently fixing the shape of the part. Prolonged heating of a polymer cured at a relatively low temperature causes further cross-linking; a fully cross-linked polymer is unaffected. Once a thermosetting polymer is cured, the only possible change is destruction of the bonds by overheating (Fig. 13–11) which results in the formation of a carbonaceous mass (*char*). Scrap cannot be recycled, except as ground-up filler.

It should be noted that it is not necessary to complete polymerization and cross-linking in a single step to the final stage.

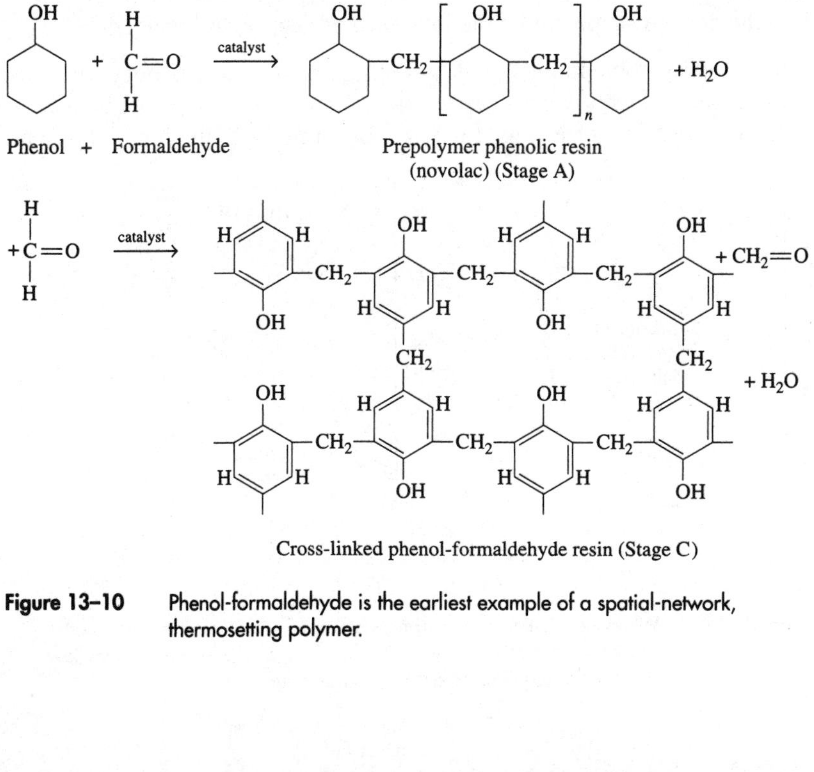

Phenol + Formaldehyde

Prepolymer phenolic resin
(novolac) (Stage A)

Cross-linked phenol-formaldehyde resin (Stage C)

Figure 13–10   Phenol-formaldehyde is the earliest example of a spatial-network, thermosetting polymer.

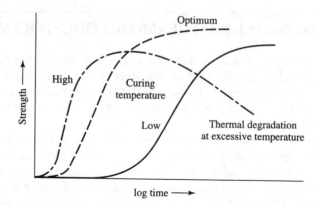

Figure 13–11   Cross-linking and polymerization are accelerated by temperature, but excessively high temperatures lead to the destruction of the polymer.

1. With some thermosets, it is possible to polymerize the precursor materials partially into a linear, mostly monomeric *A-stage* resin.

2. After adding fillers, colorants, and some curing agents (also called, incorrectly, catalysts), a partially polymerized *oligomer* or *prepolymer* (*B-stage* resin) is formed which is supplied to the secondary manufacturer. This resin is stable at room temperature and may be thermoplastic for a limited time at elevated temperature.

3. Final polymerization and cross-linking (*C-stage*) then occurs after molding into the desired shape, under the influence of heat or curing agents.

The viscosity of monomers or of B-stage resins is generally not unduly high, but it increases suddenly when—under the influence of curing agents or heat—many large molecules form (Fig. 13–12*a*). Because cross-linking and polymer-

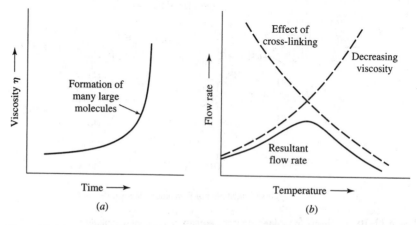

Figure 13–12   When thermosetting prepolymers are molded, (*a*) viscosity increases suddenly when many large molecules form, and (*b*) the maximum flow rate is achieved at some optimum temperature.

ization occur simultaneously with the flow of the material into the mold, the rheology of thermosets becomes complicated, and maximum flow rates are attained at some intermediate temperature (Fig. 13–12b).

The essential difference between thermoplastic and thermosetting polymers is, then, that *the thermosetting polymer is heated to stabilize its shape*, whereas *the thermoplastic polymer is heated to make it moldable and must be cooled to fix its shape*.

13-4   ELASTOMERS

Elastomers are a special class of polymers. When subjected to a tensile load, they are capable of extensions of minimum 200% and often over 500%. Upon removing the load, all deformation is regained: The elastomer behaves as a spring. Elastomers are basically of two types.

Thermoset Elastomers   These are made of amorphous linear polymers, used above their T_g. The repeating unit must contain a double bond which can be opened up to establish *cross-links*. Once cross-linked, these elastomers cannot be further shaped: The entire structure is a single giant molecule. Properties depend on the repeating unit and the degree of cross-linking.

The —CH_3 group in polyisoprene (Fig. 13–13a) presents greater resistance to deformation and, when the main chains are cross-linked (*vulcanized*) with sulfur, highly desirable properties are obtained. (The accidental discovery of cross-linking *natural rubber* with sulfur by Charles Goodyear in 1839 laid the foundations of the rubber industry.) When only a few cross-links are formed, say, between every 500–1000 repeat units (as in rubber gloves), the coiled molecules not only unwind, but can also move relative to each other as far as the cross-links permit (Fig. 13–13b). An increasing degree of cross-linking (say, between every 20 or 50 mers) makes the rubber harder and suitable, for example, for automotive tires.

Since there is no —CH_3 group in polybutadiene, a *synthetic butadiene rubber* has lower strength and tear resistance. The addition of styrene results in a copolymer of superior properties which can also be cross-linked (Fig. 13–13c), as can be polyurethane (Fig. 13–13d).

Example 13-5

Calculate the amount of sulfur required to establish a cross-link between every 10 mers of polyisoprene. Assume that only one sulfur atom is involved in each cross-link.

Each mer of isoprene (C_5H_8) has a molecular weight of $(5 \times 12) + 8 = 68$ g. Sulfur has an atomic weight of 32 g. Since a link is to be made between every 10 mers, the weight fraction of sulfur is

$$\frac{32/10}{(32/10) + 68} = \frac{3.2}{71.2} = 0.045$$

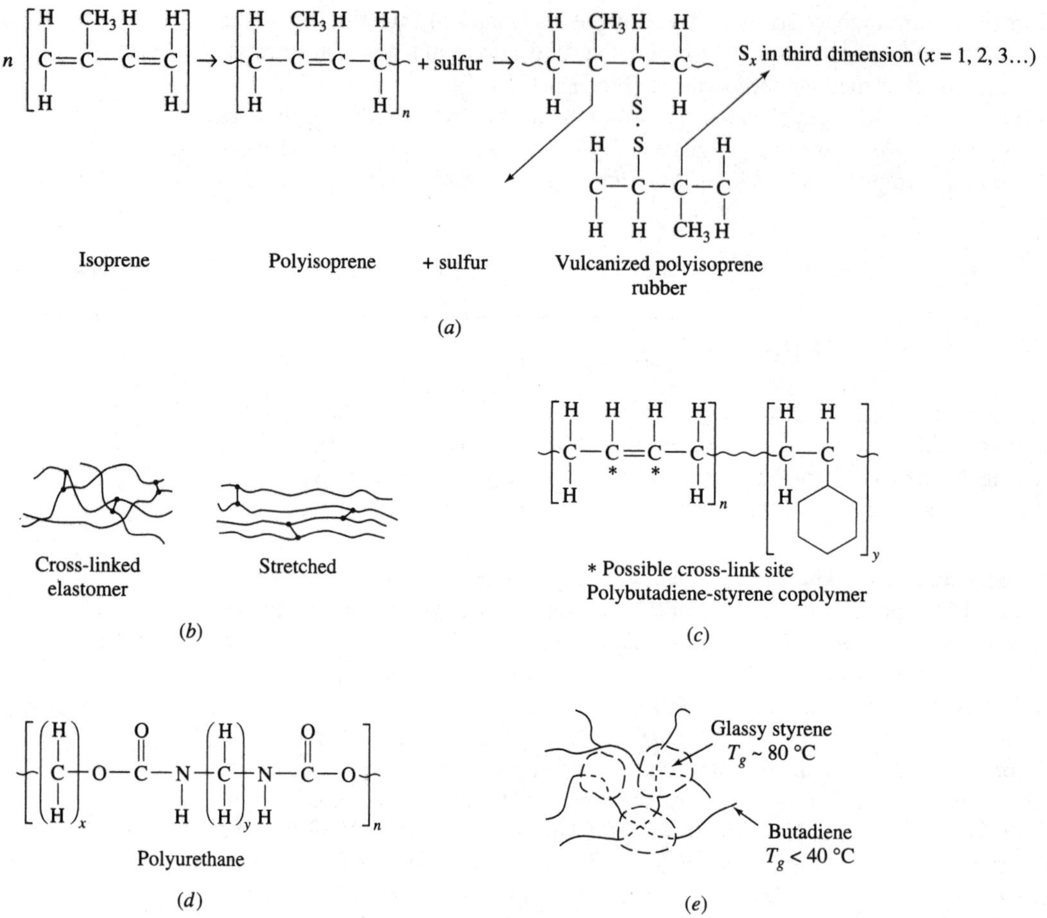

Isoprene   Polyisoprene   + sulfur   Vulcanized polyisoprene
                                                                    rubber

(a)

Cross-linked   Stretched
elastomer

(b)

* Possible cross-link site
Polybutadiene-styrene copolymer

(c)

Polyurethane

(d)

Glassy styrene
$T_g \sim 80\ °C$

Butadiene
$T_g < 40\ °C$

(e)

Figure 13–13   Elastomers are formed by (a) establishing cross-links between linear molecules thus (b) preventing permanent flow. (c) Glassy polymers copolymerized with a rubbery polymer and (d) polyurethanes may be cross-linked to form an elastomeric, spatial (thermosetting) network. In thermoplastic elastomers, glassy regions in an amorphous matrix act as anchor points and prevent viscous flow (e).

or 4.5% of the vulcanized rubber. As shown in Fig. 13–13a, sulfur itself forms chains and, for this reason, a higher proportion of sulfur will be needed.

Thermoplastic Elastomers   More recent is the development of thermoplastic elastomers. They owe their spring-like properties not to covalent cross-links but to the *presence of glassy regions in an amorphous matrix*. This is the case with styrene–butadiene block copolymers (Fig. 13–13c): The styrene segments congregate to form polystyrene-like glassy regions of high T_g which act as anchor points—similar to cross-links—in the polybutadiene-like rubbery regions (Fig. 13–13*e*). Copolymers of polypropylene and polyethylene exhibit similar properties. Most importantly for processing and recycling, these elastomers can be repeatedly heated and cooled, and scrap can be recycled. They cannot, however, be used above T_g of the glassy phase.

13-5   ADDITIVES AND FILLERS

The wide range of properties available in various polymers can be further modified by the addition of agents which fall into two groups. Those that enter the molecular structure are usually termed *additives*, whereas those that form a clearly defined second phase are termed *fillers*. Intimate mixing of additives and fillers is a critical step in processing.

13-5-1   Additives

Additives are agents designed to change properties.

1. *Antioxidants* are added particularly to chain-reaction polymers such as PP; they also act as heat and UV light stabilizers. Specific stabilizers must be added to PVC and other chlorine-containing polymers because the release of HCl initiates a destructive chain reaction.

2. *Flame retardants* are important because all carbon-base polymers support combustion. In general, control is exerted for various purposes: increase the temperature at which a flame is supported; slow down the rate of flame propagation; generate an atmosphere that does not support combustion; eliminate or reduce the emission of noxious fumes. A whole battery of tests is available for evaluating the flammability of plastics, and codes and regulations have been gradually tightened.

Plastics can never be fireproof in the sense of ceramics or most metals, but they do show some large differences in behavior. Some plastics (e.g., those containing chlorine) are more flame-resistant than others, although then they may release toxic gases. Polymers having lower-strength side groups will not support a flame but are reduced to a solid, carbonaceous body (they carbonize or *char*). In other polymers, breakage of primary bonds results in the formation of liquid or gaseous compounds of lower molecular weight which then ignite.

3. *Plasticizers* make an otherwise rigid thermoplastic polymer flexible. By plasticizing nitrocellulose with camphor, Parkes in England and the Hyatt brothers in the United States introduced, in the 1860s, celluloid, the first moldable plastic. Today, the major application is to PVC. A plasticizer is usually a fluid of high molecular weight (over 300), the molecules of which are interspersed between the polymer molecules, loosening the structure, thus depressing T_g and allowing greater flexibility.

4. *Solvents* are organic substances of short chain length. Their molecules enter between polymer molecules, breaking the secondary bonds. Thus, they are undesirable in service, and polymers are selected to resist specific solvents. Cross-linking prevents dissolution; at the most, the solvent molecules cause swelling. Highly cross-linked polymers are completely resistant to many solvents.

Solvents can, however, serve a useful purpose in manufacturing by allowing processing in the fluid state; subsequently they must be removed by evaporation. Some solvents, such as methylene chloride, attack amorphous thermoplastic polymers such as polystyrene, PMMA, polycarbonate, and ABS, and this is useful for solvent welding. Crystalline polymers such as PP and PTFE are much more resistant.

5. Many cast or molded plastics are transparent, as are PMMA and poly-carbonates (PC); others are translucent (PE, PP). Yet others are opaque or may have a natural color. Organic *dyes* (which are soluble in solvents) and organic or inorganic *pigments* (finely divided insoluble substances such as oxides) are added to impart a desired color. For example, titanium dioxide is an excellent white pigment; iron oxides give yellow, brown, or red color; carbon black is not only a pigment but also UV-light absorbent. Finely divided calcium carbonate dilutes (extends) the color and is used in large quantities as a low-cost filler.

6. *Lubricants* and *flow promoters* are added primarily to PVC, PS, and ABS to facilitate flow in molds and dies. Mold release agents may be added to the polymer or to the mold surface.

13-5-2   Fillers

Fillers may be incorporated to improve mechanical properties and then they are often called *reinforcing agents*. At other times, their main purpose is to reduce material cost, reduce shrinkage and thermal expansion coefficient, increase the electric or thermal conductivity of the plastic, or ease processing, and then they are usually called *extenders* (or simply, fillers).

Since many fillers are structurally or chemically significantly different from the basic polymer, their effect on mechanical properties can be discussed with reference to Fig. 6–15 which deals with the properties of two-phase structures.

1. Fillers that are not wetted and thus produce a weak interface act as stress raisers, reducing the strength and toughness of the polymer (Fig. 6–15, case 4 and 5).

2. The effectiveness of fillers wetted by the polymer depends on their shape, size, and distribution.

a. More or less equiaxed fillers of larger size simply serve to increase the bulk, with little effect on properties (Fig. 6–15, case 3a). However, when finely distributed, they change the character of the polymer because the bond length between particles is reduced, increasing the stiffness of the structure (Fig. 6–15, case 3b). Particulate fillers include fine (between 0.5 and 20 μm) minerals such as calcium carbonate, clay, talc, quartz, diatomaceous earth (siliceous fossils of algae); metal oxides, such as alumina; metallic powders; carbon black; coarser (70–500-μm) natural organic fillers such as wood flour; and synthetic polymers.

b. A more powerful effect is obtained when the filler is platelike (flake), because short bonds developed over a larger surface area (as in Fig. 6–15, case 2d) are stronger. A typical example is mica. During processing, flakes often become aligned and the structure acquires directional properties.

c. The greatest strengthening effect is obtained when the second phase is a stronger fiber, of higher elastic modulus, and of at least 50:1 length-to-diameter ratio. The strength of the plastic is increased in tension because stresses are transferred—through the interfacial bond—to the fiber. Strength is also increased in compression because elastic buckling of the fiber is prevented by the surrounding polymer. Further strengthening is derived from the shorter bond length between fibers. If the fibers become oriented, markedly directional properties can be obtained. Impact strength is increased by the energy needed to break the fibers, debond fibers from the matrix, and pull fibers from the matrix. Fibrous fillers include cellulosics, asbestos, glass, carbon (graphite), boron, metals, and polymers of higher strength and/or greater toughness. Polymers reinforced with continuous fibers (filaments) are regarded as composites and will be discussed in Sec. 15-3. Toughness of brittle plastics is often improved by incorporating a separate elastomeric phase into the brittle matrix.

Some fillers are immediately wetted by the plastic, whereas other polymers such as PE, PTFE, and silicones need special surface treatments. In many other instances special *coupling agents* are needed; for example, silanes are applied to fiberglass or mineral fibers.

13-6   SERVICE PROPERTIES OF PLASTICS

The vast numbers of available plastics exhibit a wide range of characteristics, making them suitable for a very broad range of applications.

13-6-1   Mechanical Properties

The properties of plastics are determined in tests similar to those described in Chap. 4. Prior to testing, the specimens are brought into a standard condition (temperature and moisture) according to ASTM D618.

Strength   Standard tension tests are performed on molded or machined specimens (ASTM D638) and films (ASTM D882). The typical stress–strain curve may show purely brittle behavior for thermosets and for thermoplastics well below T_g (Fig. 13–14a). Above T_g a neck forms after the maximum stress has been reached, and the force (and thus engineering stress) drops (Fig. 13–14b). The neck spreads over the length of the sample at a roughly constant force, and the force rises only when molecules are stretched prior to fracture. (In terms of true stress,

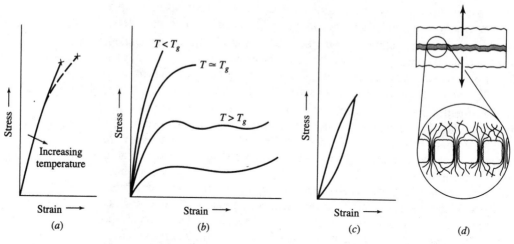

Figure 13–14   The mechanical properties of polymers are greatly affected by increasing temperatures: (*a*) The elastic modulus of thermoplastics drops and (*b*) strength decreases while elongation increases. (*c*) A hysteresis loss is often encountered on repeated loading. (*d*) Crazing results when internal voids form under the influence of tensile stresses but microfibrils continue to support load.

the stress–strain curve rises continuously.) For reasons explained in Sec. 8-1-6, postnecking elongation increases with increasing m and thus with temperature above T_g. Increasing strain rate has the same effect as lower temperature, and a plastic that is capable of substantial deformation at low strain rates may suffer brittle fracture at high strain rates.

In most engineering applications, dimensional stability is a prime requirement. Even the slightest permanent deformation is undesirable; therefore, the useful loading range is limited to the elastic regime. The initial slope of the force-displacement recording again gives the elastic modulus E [Eq. (4-6)], also called the *tensile modulus*. It is some two orders of magnitude lower than in metals (inspect data in Tables 5–2 and 5–4) and is not well defined, because there is a slight deviation from linearity even at very low strains. Furthermore, in most polymers the modulus drops with increasing temperature (Fig. 13–5*b*). Many thermoplastics show some strain-rate sensitivity even below T_g, and the tensile modulus is then also a function of loading rate. The strength of some plastics, such as nylon-6,6 is markedly reduced by moisture.

When deformation is taken past the linear deflection regime, deformation may still be largely elastic but a different curve is followed upon unloading (Fig. 13–14*c*). The area of the *hysteresis loop* represents work which is transformed into heat. On repeated loading this may lead to overheating and fatigue.

Tensile stresses imposed on glassy and semicrystalline thermoplastics may lead to localization of strain in bands perpendicular to the loading direction. Internal voids formed in these bands (Fig. 13–14*d*) make them visible as *crazing*.

Although these bands may appear as fine cracks, oriented molecules drawn out from the bulk (*microfibrils*) continue to support the load. Ultimately, failure may result even at loads much below the tensile strength.

The stiffness of plastics is often determined in a three-point bending test (Fig. 4–9a). The maximum stress at fracture is the *flexural strength*; for tough plastics, the maximum fiber stress at a strain of 5% is taken. The *flexural modulus* is the initial slope of the bending force–strain curve and is usually somewhat higher than the tensile modulus.

When loaded in compression, localized deformation in *shear bands* leads to a temporary drop in flow stress; upon continued deformation, the stress rises. Compression and indentation hardness testing are used mostly for processing checks on specific kinds of plastics. For applications such as gaskets, springs, and force fits, stress relaxation tests are made to obtain the decay of stress under constant-strain conditions (see below).

Toughness   Impact tests provide a quick evaluation of the toughness of plastics. The Charpy test (Fig. 4–10) is suitable, but the *Izod test* (ASTM D256)—in which the notched specimen is clamped at one end—is often preferred. Because of the high strain rates and imposed bending, impact toughness may differ considerably from toughness determined in the tension test, and an *impact tension test* (ASTM D1822) often correlates better with field performance. In thermoplastic polymers, impact strength is meaningful only below T_g. Toughness can be increased with moderate crystallinity, decreasing spherulite size, and, in block copolymers, by incorporating a component which has a T_g much below the service temperature (creating, essentially, an elastomer).

For sheet and film, shear strength and puncture strength are often important. For applications involving fatigue, an endurance limit (Fig. 4–14a) is determined.

Elevated-Temperature Properties   The great temperature and time sensitivity of polymers makes creep deformation significant, and creep is the primary agent in losing dimensional stability. It is also responsible for stress relaxation: When a part is subjected to loading, the initial stress gradually decays. The decay can be followed by a Maxwell model (Fig. 13–8c). The initial strain imposed at $t = 0$ time induces a stress τ_0. Since the strain is held constant, $d\gamma/dt = 0$ and Eq. (13-10) becomes

$$0 = \frac{d\tau}{dt}\left(\frac{1}{G}\right) + \frac{1}{\eta}\tau \qquad \textbf{(13-10')}$$

Solving it, the stress at any point in time is

$$\tau = \tau_0 \exp\left(\frac{-Gt}{\eta}\right) \qquad \textbf{(13-12)}$$

If we define η/G as the relaxation time t_r, we obtain

$$\tau = \tau_0 \exp\left(\frac{-t}{t_r}\right) \qquad \textbf{(13-13)}$$

Thus, relaxation time is the time required for the stress to decay to $1/e$ of its initial value. The equation can be written in a similar form for tensile or compressive stresses.

Example 13-6 | **A** polymer washer is used as a gasket. The polymer was tested in a gasket relaxometer by compressing between two hollow cylinders. The force was measured with a load cell. For a given initial compressive strain, the stress was 10 MPa, which decayed to 8 MPa in 5000 h. Calculate the remaining stress at one year.

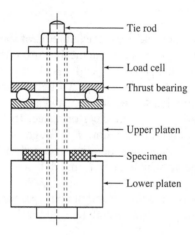

Figure Ex. 13-6

First, find t_r. Eq. (13-13), with compressive stresses

$$8 = 10 \exp\left(\frac{-5000}{t_r}\right)$$

and $t_r = 22\,407$ h. One year $= 365 \times 24 = 8760$ h and

$$\sigma = 10 \exp\left(\frac{-8760}{22\,407}\right) = 6.764 \text{MPa}$$

For longer service, the gasket needs to be periodically retightened. Note that the calculation assumes that the polymer follows predictions for longer times than in the original experiments. This is usually justified for times not exceeding 10 times the original test period.

Creep is a very complex function of temperature, the magnitude and state of stress, and solvents (and for hygroscopic materials, humidity) in the environment.

Design is therefore often based on pseudo-elastic principles: It is assumed that the maximum temperature and load operate throughout, and creep-stress data are substituted into classical elastic stress-analysis equations.

The upper temperature limit of application is often defined as the *heat-deflection temperature* (ASTM D648). A bar of 12.7-mm width, 127-mm length, and 3–13-mm thickness is loaded in a three-point bending test (Fig. 4–9a), on a span of 102 mm. The load is chosen to give a maximum stress [Eq. (4-12a)] of 1820 kPa (264 psi) for very rigid polymers and 455 kPa (66 psi) for flexible ones. The bar is heated, under load, at a rate of 2°C/min; the heat-distortion temperature is reached when the midspan deflection is 0.25 mm. The test is, however, a poor predictor of creep at lower temperatures.

For highly crystalline thermoplastics, the temperature at which a standard flattened needle penetrates the plastic is taken as the *Vicat softening* point (ASTM D1525).

Gradual changes at elevated temperatures lead to loss of properties and the resistance to *thermal aging* is often specified with reference to Underwriters Laboratories (Northbrook, Illinois) Standard UL 746B.

Many components are subjected to fatigue and design must then take this into account. Design with plastics is beyond the scope of the present treatment and references given at the end of Chap. 14 should be consulted.

Residual Stresses   Manufacturing processes often induce residual stresses in plastic components. This is of particular concern for thermoplastics in which differential cooling rates lead to varying degrees of crystallization and thus shrinkage. Furthermore, the flow of the plastic may result in varying unwinding of molecules in different sections of a part. Surface and center layers of thicker sections have different thermal histories because of poor heat conductivity, and this can induce further differences in structure. In both thermoplastics and thermosets, the shape of the mold may prevent uniform contraction of all sections, and residual stresses are then again set up.

If creep deformation is possible, residual stresses gradually diminish with consequent distortion (Fig. 4–19b) and loss of dimensional stability. In polymers subject to solvent attack, stresses lead to the development of cracks; a multitude of fine cracks is referred to as *crazing*. Very fine cracks appear as a clouding of the surface.

Because linear polymer molecules are uncoiled and stretched during manufacture, a part heated to above T_g may return to the original shape. This is exploited in heat-shrinkable plastics but is undesirable in structural parts.

13-6-2   Physical and Chemical Properties

Density is important in many applications. Amorphous hydrocarbons have a density close to that of water (0.86–1.05 g/cm^3; e.g., 0.91 for LDPE and 0.96 for HDPE). Chlorinated polymers are heavier (PVC 1.4, polyvinylidene chloride 1.7) and the crystalline PTFE is heavier yet (2.2) (see also Table 5–4).

The tribological properties of polymers are generally good. Many of them give a very low coefficient of friction either because of low adhesion (as do fabric-reinforced phenolic bearings in the presence of aqueous lubricants) or because a transfer film forms which then ensures low dry friction (nylon, PTFE). Polymers may have intrinsically high wear resistance in particular applications (as does UHMWPE) or their wear resistance may be increased with cross-linking or with fillers that increase hardness or also impart lubricating properties (as does molybdenum disulfide in nylon).

Polymers are, in general, poor heat conductors, although conductivity can be somewhat improved by filling with metal powders (Table 4–1). Their thermal expansion is large (Table 4–1), and this can create problems when assemblies contain both plastics and metals.

The resistance of polymers to weathering—involving moisture, ozone, UV light, and temperature changes—can be improved by the incorporation of opaque fillers and UV absorbers. Otherwise, darkening, crazing, and embrittlement may occur.

Amorphous polymers tend to be transparent since there are no internal reflecting boundaries in them. Crystalline polymers are translucent or opaque except when crystals are very small or few or the polymer is in the form of a very thin film.

Polymers have many desirable electrical properties: high resistivity, dielectric strength, arc resistance, and dielectric constant, as well as a small dissipation factor (i.e., little heat is generated in them when placed in an electric field). Thus, they enjoy wide application for insulation and, in general, as substrates. If necessary, they can be made conductive by adding metal powder (6–8% if on bead boundaries, 35–40% if randomly distributed). A special class of polymers is intrinsically conductive and exhibits electroluminescence and piezoelectricity. Very high static charges may build up in plastics as a result of friction against some other surface, and antistatic additives are designed to dissipate the charge by electrical conduction into the polymer or by radiation to the surrounding atmosphere.

13-7   PLASTICS

As indicated at the beginning of this chapter, we make a distinction between neat polymers and their compounded versions, plastics. There is an ever-growing variety of plastics available and, for detailed data, handbooks listed at the end of this chapter should be consulted. An outstanding attribute of plastics is the relative ease with which they can be manufactured into parts, and "general-purpose" plastics (also termed *commodity plastics*) are often employed, simply because of their cost-effectiveness, for products that do not set very high demands. However, general-purpose plastics also have many engineering applications, and are often reinforced with appropriate fillers for higher strength in load-bearing components. The term *engineering plastics* is loosely applied to describe polymers that are too expensive for general applications but offer high strength, elastic modulus, toughness, temperature and chemical resistance, desirable electrical properties, and good surface finish, so that they can often be used in structural applications, replacing metals. Many of them can be further strengthened by fiber reinforcement. Therefore, they are finding increasing application in instrument and machinery housings, electrical connectors, gears, cams, impellers, valves, and, in general, structural components.

13-7-1   Thermoplastics

The number of thermoplastic polymers has increased greatly in the last four decades. Classification is possible from various viewpoints; here, thermoplastics will be discussed in groups defined by their structure. The properties of some frequently used thermoplastics are given in Table 13–2.

1. *Polyolefins.* These are based on straight-chain alkenes (i.e., hydrocarbons with double bonds between the carbon atoms, such as ethylene in Fig. 13–1).

a. *Polyethylenes* (PE, Fig. 13–1). These are very versatile plastics with good chemical and electrical properties but need antioxidants and UV stabilizers. The shorter-chain, branched LDPE (density: 0.910–0.925 g/cm^3) finds its main use in packaging, although it is losing ground against linear LDPE (LLDPE). The linear HDPE (0.941–0.965 g/cm^3) is suitable for containers and pipes. UHMWPE (with a molecular weight of minimum 3 million) is a true engineering thermoplastic. It has extremely high abrasion resistance and impact toughness; hence, it is used in many industries for wear surfaces.

b. *Polypropylene* (PP, Fig. 13–1) is used in the isotactic form (Fig. 13–2*b*). It has good stiffness, stress-crack resistance, and a higher temperature resistance than PE. Appliance parts and containers are typical uses. Large amounts are recycled from the break-up of lead-acid battery cases.

c. *Ionomers* contain inorganic compounds which give ionic bonding between chains (cross-linking), while still allowing plastic molding processes at elevated temperatures. They are transparent, very tough, and are used typically for golf-ball coatings, hammer heads, and other high-impact applications.

2. *Vinyls* are based on ethylene monomers in which one H is substituted.

a. *Polyvinylchloride* (PVC) (Fig. 13–1) is a low-cost, easily processed plastic of good water resistance and strength-to-weight ratio. It needs stabilizers to prevent decomposition at high processing temperatures and in sunlight. Rigid PVC is hard and relatively brittle but finds extensive use in buildings as window frames, gutters, pipe, and also for wire insulation and bottles. Flexible PVC, made by adding some 30–80% plasticizer, finds extremely broad applications, e.g., as garden hose, boots, tubing, packaging, and raincoats. When burnt, PVC generates HCl and is self-extinguishing.

Table 13-2   Manufacturing properties of selected thermoplastic polymers*

Type	T_m, °C	T_g, °C	Heat Deflection Temperature, °C — 1850 kPa	450 kPa	Compression Molding† Temp., °C	Injection Molding Temp., °C	Shrinkage, %	Tensile properties — Strength, MPa	Elongation, %	Flexural Modulus, GPa	Izod Impact,‡ J/25 mm
ABS	175	110–125	95	100	175–260	190–260	0.4–0.9	30–55	5–25	2.5	4–16
Acetal copolymer	175		110	155	170–205	195–230	2	60	40–75	2.8	2
With 25% glass			160	165	170–205	190–250	0.4–1.8	130	3	8	2.5
Acrylic (PMMA)		90–105	80	85	150–220	160–260	0.2–0.8	50–80	2–10	3	0.7
Cellulose acetate	230		55	65	125–220	165–255	0.3–1.0	20–65	6–70	0.8–2	1–8
PTFE	327			120			3–6	15–35	200–400	0.6	4
Fluorinated ethylenepropylene	275			70	320–400	330–405	3–6	20	300	0.6	No break
PE-TFE copolymer	270		70	104	300–330	300–350	3–4	45	100–400	1.4	No break
With 25% glass	270		210	265	300–330	300–350	0.2–3.0	85	8	6.7	12
Nylon-6,6	265	50	75	245	300–330	270–330	0.8–1.5	85	60–300	1.3	2.5
With 30% glass	265	50	255	260	300–330	270–300	0.4–0.6	155	5–7	6	4
Polycarbonate		150	130	135	270–300	250–345	0.5–0.7	55–70	100–130	2.4	20
Polyester											
PBT	250		65	150		225–275	1.5–2.0	55	50–300	2.7	1.5
PET	250	70	40		230–380	280–315	2.0–2.5	50–70	50–300	3.0	0.7
Polyethylene											
LD	110	–120		40	E135–230	150–230	1.5–5.0	8–30	100–650	0.3	No break
HD	135	–110		80	E170–275	170–260	1.4–4.0	20–35	10–1200	1.2	0.5–5
UHMW	130		45	80	200–260		4	40	420–500	1.0	No break
Polyimide		310–365	340	115	330–365			120	10	3.5	2
Polypropylene	168	–18	55	100	E205–260	205–290	1–2.5	30–40	100–600	1.5	0.5–1
Polystyrene		100	95	100	150–205	175–260	0.4–0.7	35–55	1–2	3.2	0.5
high-impact			80	80	E190–260	175–275	0.4–0.7	20–45	20–65	2.0	1.5–4
PVC, rigid	200	95–105	60	62	140–205	150–210	0.2–0.6	35–65	40–80	3.0	1.5–30
Flexible		75–105			140–180	160–195	1.0–5.0	10–25	200–450	0.02–1	Varies
Styrene–butadiene block copolymer		75–105			120–160	150–220	0.1–0.5	4–20	300–1000		No break

*Compiled from *Modern Plastics Encyclopedia*, McGraw-Hill, New York, 1995. Values given are approximate.

†E = extrusion.

‡Izod impact values J/25-mm notch, measured on 3.2 mm-thick specimen (divide by 1.35 to obtain ft · lb/in notch).

b. *Vinylidene chloride* contains two substituted Cl atoms per mer (Fig. 13–1). It is chemically inert, has low permeability to liquids and gases, and is relatively resistant to combustion. It is used for seat covers, pipes, and valves. The copolymer with vinyl chloride (Saran™) makes a packaging film.

3. *Fluorocarbons.* Replacing H with F creates chemically inert polymers (Fig. 13–1). The fully substituted polytetrafluoroethylene (PTFE) is crystalline and has great heat resistance, extreme chemical inertness, and inherently low friction. It is nonflammable but releases toxic gases when it decomposes on over-heating. It does not melt and must be produced by metal technologies (sintering). Billets are skived (shaved) to make films. The fluorinated ethylene propylene (FEP) copolymer is moldable.

4. *Styrenes.* These are obtained by substituting a benzene ring for H.

a. *Polystyrenes* (Fig. 13–1) are amorphous, transparent, low-cost but brittle polymers. Elastomers may be included to improve impact resistance (*high-impact polystyrene*). Apart from solid forms, polystyrene is employed also in the form of expandable beads. It finds wide use in packaging, trays, and houseware.

b. *Styrene acrylonitrile* (SAN) random copolymer (Fig. 13–15a) is also transparent but has better stiffness. It is used in appliances and for glazing.

c. *ABS* is a copolymer of butadiene rubber grafted with SAN. The very finely distributed rubber-like regions (of 0.1–1.0-μm diameter) impart low-temperature ductility and toughness while maintaining the high hardness and rigidity of the glassy styrene. It is an engineering plastic, the properties of which can be varied by changing the ratios of the monomers and by the method of manufacture. It needs UV stabilization. It is extensively used for business machines, electrical hand tool housings, camper tops, and pipes.

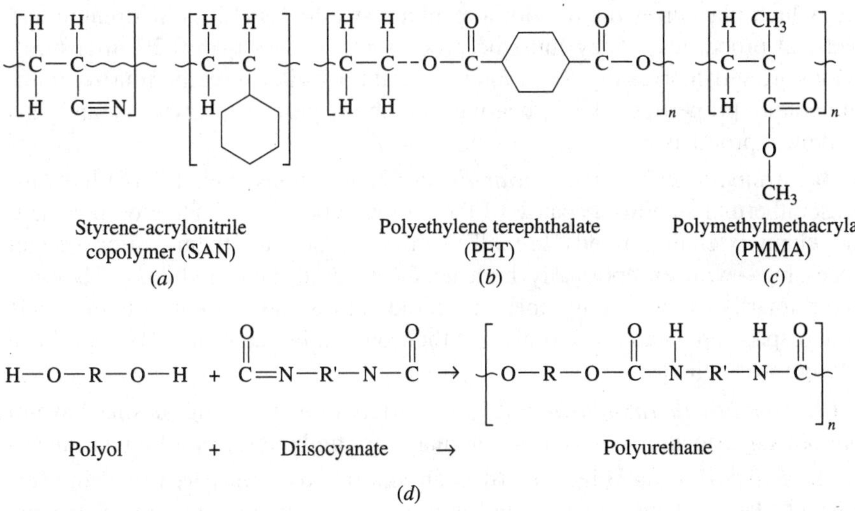

Styrene-acrylonitrile copolymer (SAN)

(a)

Polyethylene terephthalate (PET)

(b)

Polymethylmethacrylate (PMMA)

(c)

$$H-O-R-O-H \;+\; C=N-R'-N-C \;\rightarrow\; \left[O-R-O-C-N-R'-N-C\right]_n$$

Polyol + Diisocyanate → Polyurethane

(d)

Figure 13–15   The chemical formulas of some important linear thermoplastic polymers.

5. *Polyesters.* Esters are reaction products between an organic acid and an alcohol (Fig. 13–1); thus, ester groups are in the main chain.

a. *Thermoplastic polyesters. Polyethylene terephthalate* (PET, Fig. 13–15*b*) is a clear, tough, orientable polymer of higher temperature resistance. It finds wide use in containers for carbonated beverages and liquors and is frequently recycled as a postconsumer plastic. *Polybutylene terephthalate* (PBT) is semicrystalline. After reinforcement, it has good creep resistance, making it suitable for auto and appliance bodies and electrical parts.

b. The aromatic ester *polyarylate* is an engineering plastic of high creep resistance.

6. *Acrylics* have side-chain ester groups. PMMA (Fig. 13–15*c*) is transparent, brittle, and notch-sensitive. High-impact grades contain a plasticizer or copolymer. It is polymerized in sheet form from a solution of monomer and prepolymer, or the polymer is treated by thermoplastic techniques. It is extensively used for glazing, skylights, fixtures, lenses (including large Fresnel lenses).

7. *Cellulosics.* Cellulose is a natural polymer from cotton fiber or wood pulp (after removal of lignin). It does not melt. Chemically modified (acetate, etc.), it becomes a tough thermoplastic for containers, tool handles, tape cases.

8. *Polyurethanes* (Fig. 13–15*d*) formed from the reaction of a polyol (an alcohol with at least two —OH groups) and an isocyanate (with at least two —NCO groups) are extremely versatile. Thus, they can be linear thermoplastic polymers which find use as wire coatings, fascia boards, and other components.

9. *Polyamides* are formed by the condensation of an amine and a carboxylic acid.

a. *Nylons* (Fig. 13–1), the first engineering polymers, are tough, crystalline polymers. Modified grades are amorphous and transparent. They absorb moisture, which plasticizes them, with a resultant swelling and loss in strength and electrical properties. They must be processed dry (less than 0.2% moisture). Nylons are self-lubricating. Glass and mineral fiber fillers greatly improve high-temperature properties. Gears, bearings, autobody parts, caps, wheels, and fans are typical products.

b. *Aromatic polyamides* (*aramids*, aromatic nylons, Fig. 13–16) have increased thermal stability because of the aromatic backbone. They do not melt. The chain-extending bonds are all parallel, endowing them—after special processing—with exceptionally high tensile strength and modulus. They are used primarily as reinforcing fibers, as in radial tires and in composite materials for aerospace applications, but also on their own in bulletproof vests, as cables, or tough work gloves.

10. *Aromatic thermoplastics.* A great variety of engineering plastics have in common an aromatic backbone which makes for higher temperature resistance.

a. *Polycarbonate* (Fig. 13–16) is an amorphous, transparent and impact-resistant plastic of high clarity, and is used in auto lamps, instrument panels, optical and electrical applications (including compact disks), and returnable milk bottles.

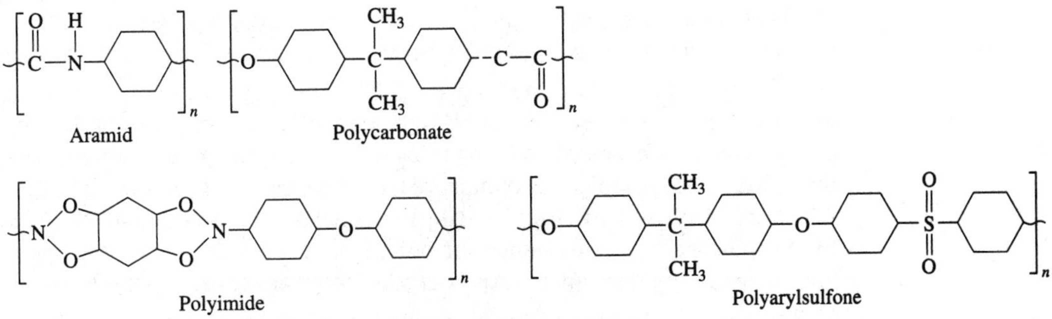

Figure 13-16 The benzene ring in aromatic linear polymers makes these polymers stronger and more resistant to temperature.

b. *Polyimide* (Fig. 13–16) has a semiladder structure, giving it high temperature resistance and toughness. Among the copolymers, polyamide–imide is an amorphous, tough, opaque polymer with $T_g > 275°C$, of many engineering uses. Polyether–imide is transparent and UV-resistant. It is highly microwave transparent and can thus be used, for example, for microwave cookware.

c. *Sulfones* include polysulfone, polyarylsulfone (Fig. 13–16), and polyether sulfone.

d. *Phenylene-based resins* such as polyphenylene oxide (PPO) also find uses in high-temperature applications.

e. *Polyaryletherketones* have similar properties. Polyetheretherketone (PEEK) has high impact strength.

11. *Polyacetals* are made by the polymerization of formaldehyde, thus they have a —CH_2O— backbone (Fig. 13–17). These are highly crystalline, opaque engineering plastics, often glass filled. Copolymers contain randomly distributed C—C bonds and are highly creep resistant. They are used for plumbing, sinks, fittings, containers, and engineering parts.

12. *Silicones* have a siloxane bond (Si—O—Si) in the backbone (Fig. 13–17) which makes them usable over a wide temperature range. They are water

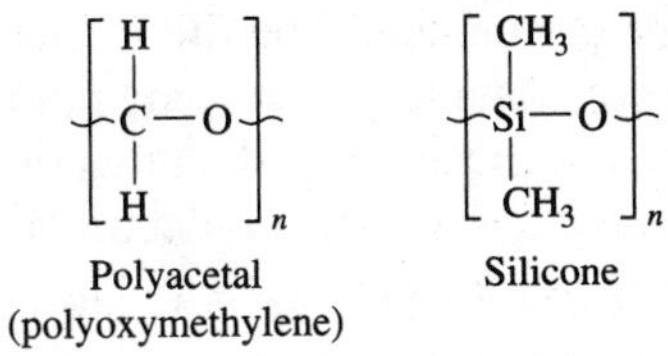

Polyacetal
(polyoxymethylene)

Silicone

Figure 13-17 The backbone of the molecules of some polymers contain oxygen; in silicones, carbon is replaced with Si.

repellent, weather resistant, and have excellent electrical properties. Many are used as rubbers, whereas rigid resins are used for encapsulation and molding.

13. *Copolymers and alloys* represent a rapidly growing field, because new plastics with tailored properties can be made without the expense of developing new polymers. We have already mentioned ABS and polybutene–styrene copolymers. Many alloys are made to improve the properties of the matrix; for example, ABS second-phase particles of 1–10-μm diameter may be added to PVC, PS, or PC to improve notch resistance, or polyphenylene oxide to nylon to improve high-temperature resistance. Liquid-crystal polymers are also polyblends.

14. *Biodegradable polymers* have found niche applications such as sutures that are absorbed by the body. By definition, they have no engineering application and none have yet been found to be free of problems.

13-7-2   Thermosets

Thermosets offer, in general, greater dimensional stability than thermoplastics, but at the expense of greater brittleness. However, these properties can be modified, and in many applications thermosets and thermoplastics are directly competitive. The properties of some frequently used thermosets are given in Table 13–3.

1. *Phenol-formaldehyde* (Fig. 13–10) is the oldest synthetic resin and is still used extensively (the original trademark Bakelite has become a generic term). Dark in color, it is almost invariably filled with wood flour or, for better impact properties, with glass, asbestos, or cotton fibers. Further improvements are obtained by adding butadiene–acrylonitrile copolymer rubber to the A-stage resin. Molding compounds are used for electrical, automotive, and appliance purposes, such as switch gear, ignition parts, and handles. The good adhesion of phenolics to other materials makes them suitable for bonding foundry sand (Sec. 7-5-4), plywood, and grinding wheels (16-8-4).

2. *Amino resins* are so named because they contain the amino (—NH_2) group. Strictly speaking, only *melamine formaldehyde* belongs here, but *urea formaldehyde*, which contains amide (—ONH_2) groups, is often classed with it. They are translucent and can be pigmented. The filler is usually α-cellulose (a high-molecular-weight cellulose), obtained from wood pulp or cotton fiber, for applications such as countertops, wall paneling, dinnerware, circuit breakers, and wall plates. Wood flour is also used as a filler for electrical purposes.

3. *Polyesters*, as discussed in Sec. 13-7-1, are thermoplastic when in the linear form (Fig. 13–1). However, these linear esters can be treated as prepolymers to which colorants, fillers (such as ground limestone), or reinforcing agents (chiefly glass fiber) can be added to make a *premix* which then can be molded and cured by heat. The term *alkyd* (*alc*ohol reacted with *acid*) is, in principle, generic to all polyesters having a carbon–carbon double bond, but it is also used to describe specific types. *Unsaturated polyester plastics* are cured by the

Table 13-3 Manufacturing properties of selected thermosetting polymers*

Type	Deflection Temperature, °C		Molding Temperature, °C		Shrink-age, %	Tensile Properties		Flexural Modulus, GPa	Izod Impact, J/25 mm
	1850 kPa†	450 kPa†	Compression	Injection		Strength, MPa	Elonga-tion, %		
Epoxy, casting	45–290				0.1–1	30–90	3–6		0.3–1.3
Molding, glass-filled		120–260	150–160		0.02	120–180	4	18–28	0.4–0.7
Melamine-formaldehyde (cellulose-filled)	175–200		150–190	90–170	0.5–1.5	35–90	0.5–1	8	0.3–0.6
Phenol-formaldehyde, casting	115				1–1.2	36–60	1.5–2		0.3–0.6
Wood-flour-filled	150–190		145–195	165–205	0.4–0.9	35–65	0.4–0.8	7–8	0.3–0.8
Polyester, glass-filled, SMC	190–260		130–175		0.1–0.4	55–175	3	7–15	10–30
BMC	160–175		155–195	150–190	0.05–0.4			7–15	5–18
Polyurethane, casting	Varies		85–120		2	1–70	100–1000	0.07–0.7	30–No break
Urea-formaldehyde, cellulose-filled	125–145		135–175	145–160	0.6–1.4	40–90	0.5–1	9–11	0.3–0.5

*Compiled from Modern Plastics Encyclopedia, McGraw-Hill, New York, 1994.
†Divide by 7 to get psi.

addition of catalysts (more correctly, initiators), which can be chosen to cause cross-linking polymerization with added styrene at room or elevated temperature. Major applications are pipes, boats, tanks, walls (bathrooms), auto hoods, deck lids, quarter panels, helmets, structural beams, and even oil-well sucker rods. No by-product is formed; therefore, unsaturated polyesters are used extensively in bulk- and sheet-molding compounds. Vinyl esters have better toughness at a slightly higher cost.

4. *Epoxy resins* are available as room-temperature-curing, two-component systems (an intermediate resin and a hardener reactant, often misnamed a catalyst) or as elevated-temperature-curing, single-component resins. Cross-linking proceeds with minimum shrinkage; therefore, they are widely used for encapsulation (potting) of electrical parts, for fiber-reinforced epoxy structures, or—mixed with sand, glass, or marble—as concrete. They are also used in prepregs.

5. *Polyimides* can be cross-linked for high-temperature (up to 315°C) applications.

6. *Polyurethanes* can be cross-linked to make a variety of parts. They are suitable for reaction-injection molding.

7. *Silicones* can be cross-linked to give rigid plastics for applications where water resistance or high electrical resistance are needed.

13-7-3   Elastomers

Natural rubber is primarily polyisoprene. In its natural form it is a sticky substance because the molecules slide readily. Cross-linking with sulfur (Fig. 13–13a) produces a rubber which, because of its relatively low hysteresis loss, is used in automobile tires. Carbon black increases wear resistance.

Most synthetic rubbers are random copolymers. For example, butyl rubber is a copolymer of isobutene and a few isoprene units, with the isoprene providing the sites for cross-linking. The high hysteresis loss of this rubber makes it suitable for vibration-damping engine mounts. There are also styrene–butadiene and ethylene–propylene rubbers.

For all applications, the T_g of the rubber must be below the lowest service temperature. Silicone rubber serves down to −90°C, most others to −50 to −60°C.

We already discussed the styrene-butadiene copolymer thermoplastic (Fig. 13–13e) and the polyurethane elastomers (Fig. 13–13d). Polyurethanes can be elastomeric thermosets. They form tough and abrasion-resistant coatings. Large quantities are formed into foams, ranging from soft (seating, bedding), semiflexible (dashboard padding), rigid (insulation), to high-density (structural) foams. They are castable and can be made of a prepolymer or in one-shot processes using the monomers, as in reaction-injection molding (Sec. 14-3-4). Reinforced with glass or other fibers, they can serve as automotive door panels, trunk lids, and fenders. Fluoroelastomers find application in high-temperature, corrosive environments in engines and power plants. Silicone elastomers are usually filled with silica.

13-8 SUMMARY

Polymers have been indispensable to humans for millenia. Wood was and still is one of the most important structural materials. Wood and wood products and animal and vegetable fibers were the only organic polymers available until fairly recently. Manufactured organic polymers, beginning with celluloid and later Bakelite, were initially regarded with the suspicion accorded to substitute materials. However, the phenomenal development of the polymer industry has changed our technological outlook in the last 60 years. Polymers are seldom used in their neat form; more often, they are blended with colorants, fillers, and other additives. The product is then more properly referred to as a plastic. From containers to packaging, from household utensils to toys, from electrical and electronic components to fabrics, and from gears to aircraft structures, plastics have moved in to take markets from traditional materials.

Much of this success must be attributed to basic research that has given a better understanding of properties and processes, and to the translation of this understanding into practice. By the manipulation of the molecular structure and processing sequence, an unprecedented variety of products of exceptional properties can be obtained. From the manufacturing point of view, there are two major, fundamentally different classes of polymers: thermoplastic and thermosetting substances.

1. Thermoplastic polymers are capable of viscous or viscoelastic flow when heated above some critical temperature. This is the melting point T_m in crystalline (and therefore opaque) polymers, and the glass-transition temperature T_g in amorphous (and transparent) polymers. Thus, these plastics can be formed into the desired shape upon heating, and the shape is fixed by cooling. The process can be repeated and recycling is feasible.

2. Thermosetting plastics must be brought into the final shape before polymerization and cross-linking takes place under the influence of heat or catalysts. Once polymerization is complete, the part is one single, giant, spatial-network molecule, and the process is irreversible. The potential for recycling is limited.

3. Elastomers can be cross-linked amorphous thermoplastics of a glass-transition temperature well below room temperature. These are related to thermosetting resins in that cross-linking must take place during or immediately after shaping. Thermoplastic elastomers are now also available.

PROBLEMS 13A

13A-1 Make simple sketches of molecular arrangements for (*a*) amorphous, (*b*) crystalline, (*c*) liquid-crystal, and (*d*) cross-linked polymers.

13A-2 One property characterizing the flow of thermoplastic polymers is viscosity. (*a*) Define viscosity, with the aid of a sketch and the relevant equation. (*b*) Define apparent viscosity. (*c*) Make a graph showing the typical variation of apparent viscosity for a shear-thinning polymer as a function of strain rate.

13A-3 Many oils are paraffins and are formed of the same units as polyethylene (Fig. 13–1), yet they are liquids at room temperature. Explain the reason.

13A-4 Explain the difference between a pseudoplastic and shear-thinning polymer. Define for each the characteristic response to shear.

13A-5 State the major difference in molecular structure between (a) thermoplastic and (b) thermosetting polymers (use sketches, if desired).

13A-6 State in what condition (a) thermoplastic and (b) thermosetting polymers can be plastically deformed.

13A-7 Describe the changes in molecular arrangement occurring upon cooling from melt temperatures for an (a) amorphous, (b) partially crystalline, and (c) liquid-crystal thermoplastic polymer (use sketches, if desired).

13A-8 Draw spring–dashpot models of (a) viscous, (b) elastic, and (c) two viscoelastic materials. (d) Which of these will show permanent deformation upon unloading?

13A-9 Make a sketch showing the change of specific volume as a function of temperature for a thermoplastic polymer, drawing (a) one line for an amorphous and (b) one line for a partially crystalline polymer. (c) Identify the critical temperatures. (d) Name the typical rheological behavior in each temperature regime.

13A-10 Draw a diagram showing the change in elastic modulus as a function of temperature for an amorphous polymer at (a) short and (b) long loading time.

13A-11 Make a sketch of spring-dashpot models with the two (a) in parallel and (b) in series. Underneath make sketches to show the strain developed in response to loading at time t_0 and removal of the load at time t_1.

PROBLEMS 13B

13B-1 Define the term crystal for (a) metals and (b) polymers (use sketches, if desired).

13B-2 Draw a coordinate system with shear strain rate as the abscissa and shear stress as the ordinate. Draw lines corresponding to (a) Newtonian flow, (b) pseudoplastic flow, (c) dilatant flow, and (d) Bingham behavior. Which of these is desirable for (e) filling a mold of complex shape under pressure, and for (f) paint to be applied to a vertical surface. Justify.

13B-3 (a) With reference to characteristic temperatures, list the regimes of mechanical response of amorphous polymers. (b) State in which regime is the dimensionally most stable product obtained; justify.

13B-4 For thermoplastic polymers, indicate what effects an increase in the following variables have on apparent viscosity: (a) temperature, (b) molecular weight, (c) pressure, (d) presence of long-chain side branches. Give your reasoning for each choice.

13B-5 Explain, briefly and simply, why a part may change its shape in use (use a spring–dashpot model; refer also to changes in molecular arrangement).

13B-6 State the effects of hydrostatic pressure on the (a) strength and (b) ductility of metals and the (c) viscosity of polymer melts.

13B-7 From basic principles, deduce ways of dealing with postconsumer scrap if the material is (a) low-density polyethylene or (b) wood-flour-filled phenolic resin. List as many possibilities as you can.

13B-8 A student wrote: Elastomers are thermosetting polymers. Apply a critique.

13B-9 A student wrote: There is no glass-transition temperature in crystalline thermoplastic polymers. Apply a critique.

13B-10 There is a glass transition temperature in (a) glasses and (b) polymers. Explain if there are differences in terms of response to imposed stresses both above and below T_g.

13B-11   Explain the difference between (a) metal melts and (b) polymer melts in terms of response to imposed stresses.

13B-12   The dimensions of a plastic part gradually change over several weeks. (a) For an explanation, make a sketch of the simplest spring–dashpot model that would give such behavior. (b) Would you expect this to be a thermoplastic or thermosetting polymer? (c) What else could you conclude about the plastic?

13B-13   A colleague claims that thermoplastic polymers are deformable, thermosetting polymers not. Do you agree? Justify.

PROBLEMS 13C

13C-1   The behavior of strain-rate sensitive materials is sometimes expressed by the equation

$$\dot{\varepsilon} = b\sigma^n \quad \text{or} \quad \dot{\gamma} = B\tau^n$$

where b and B are material constants and n is a strain-rate sensitivity exponent (not to be confused with our n value, the strain-hardening exponent). What is the value of n for (a) a Newtonian fluid and (b) an ideal rigid-plastic, non-strain-rate-sensitive material? (c) What is the relation of this n to m in Eq. (13-4)?

13C-2   Which of the polymers listed in Table 13–2 is viscoelastic at room temperature? Justify.

13C-3   A design requires that the gasket of Example 13-6 remain under a stress of 12 MPa after 2 years of service. Calculate the required initial stress.

13C-4   In Example 13-1 we found that UHMWPE had a degree of polymerization of 143 000. For comparison, calculate the degree of polymerization for a blow-molding grade of HDPE from Example 13-2.

13C-5   Because of its low cost, PP is often specified for ambient-temperature applications. (a) Check T_g from Table 13–2. (b) On this basis, would you expect significant creep at room temperature? (c) If the answer is yes, state why the plastic is still usable.

13C-6   Crystallinity depends on, among other factors, molecular structure. From information available in this text, place in order of likely increasing crystallinity the following polymers: LDPE, HDPE, PP, and polystyrene. Justify.

13C-7   A part is designed for service at 200°C. From data in Table 13–2, which polymer would be your first and second choice? Justify.

FURTHER READING

Kirk-Othmer Encyclopedia of Chemical Technology, 4th ed., Wiley-Interscience, 1991–1998 (numerous articles on individual polymers).

Harper, C.: *Handbook of Plastics, Elastomers, and Composites*, 3d ed., McGraw-Hill, 1997.

McCrum, N.G., C.P. Buckley, and C.B. Bucknall: *Principles of Polymer Engineering*, Oxford University Press, 1989.

Progelhof, R.C., and J.L. Throne: *Polymer Engineering Principles: Properties, Tests for Design*, Hanser, 1993.

Rudin, A.: *The Elements of Polymer Science and Engineering*, 2d ed., Academic Press, 1999.

Seymour, R.B.: *Reinforced Plastics: Properties and Applications*, ASM International, 1991.

Ulrich, H.: *Introduction to Industrial Polymers*, Hanser, 1993.

Wollrath, L., and H.G. Haldenwanger: *Plastics in Automotive Engineering*, Hanser, 1994.

Young, R.J., and P. Lovell: *Introduction to Polymers*, Chapman and Hall, 1991.

A top-entry robot removes an automotive bumper molding from a 20-MN, two-platen plastic injection-molding machine. (*Courtesy Husky Injection Molding Systems Ltd., Bolton, Ontario.*)

14

Processing of Plastics

We are now ready to see how the properties of plastics are exploited for making a vast variety of products by various techniques, including:

Casting of solid and hollow bodies of often very complex shapes

Processing of melts under pressure, to make extrusions or molded 3-D parts

Blowing up previously formed preforms into bottles and open containers

Shaping sheet and film into often very large parts, such as boat hulls

Creating parts by reactions in a mold

Producing foams for structural and consumer use

Many processing techniques for polymers have their counterparts in processes for metals and ceramics; hence, we will discuss them with reference to the principles introduced earlier. In general, plastics can be processed at much lower temperatures than metals, and this removes many processing difficulties and also allows some processes which are impracticable for metals. Many plastics can be processed by several techniques, but some techniques may be more suitable than others, and this will be pointed out when applicable.

14-1 CLASSIFICATION

Most processes are suitable for making many products from a large variety of plastics, therefore, classification by process rather than product or material is more relevant (Fig. 14–1). Following the logic adopted in covering metal processes, plastics manufacturing processes will be discussed in order of decreasing

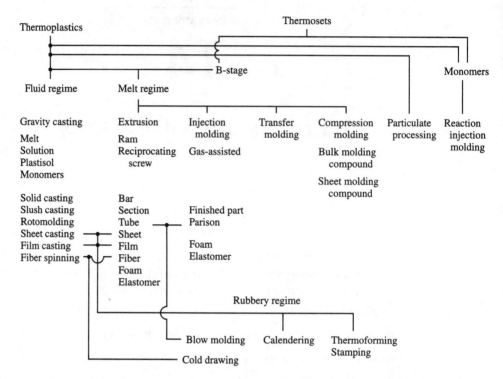

Figure 14–1   Processing sequences for plastics. (*Adapted from J.A. Schey, ASM Handbook, vol. 20, Materials Selection and Design, ASM International, 1997, p. 699. With permission.*)

processing temperature. It will be noted that in many instances a plastic may be subjected to a sequence of processing steps. The terminology is fairly diffuse; therefore, we will follow a classification based on physical principles rather than the names of processes.

14-2 CASTING

The term *casting* will be used here to describe *filling a mold by gravity.* Thus, the material must have a low enough viscosity to flow freely. This can be achieved by several means:

1. Thermoplastics can be heated above T_m (*hot-melt plastics*) and cast into molds. In a variant, high-molecular-weight, highly crystalline, and hence strong nylon parts such as gears and bearings are obtained by melting the monomer, adding the catalyst and activator, and pouring the mix into molds.

2. Liquid resins can be monomers (such as epoxy resins) or short-chain poly-mers (such as A- or B-stage polyester or phenolic thermosets). When the polymer is used to fix a component in place, one speaks of *potting*; when it surrounds the component entirely, of *encapsulation*.

In all instances, absence of moisture is critical and gases either must be removed from the liquid by processing in vacuum or must be kept in solution by the application of pressure during polymerization. Molds may be made of metal, glass, and rigid or flexible plastics. The latter can be peeled off the castings and thus allow the production of complex, undercut shapes. In common with metal casting, shrinkage can present problems, especially with acrylics that shrink greatly during polymerization. Internal voids may result in a collapse of the surface. Rules for designing metal castings (Sec. 7-8-2) can be adapted to plastics.

3. Among thermoplastics, PMMA sheet is produced by pouring catalyzed MMA between glass plates (*cell casting*) or between endless stainless steel belts (*continuous casting*). Polymerization takes place by heating. Processing times are shorter when a partially polymerized "syrup" is cast.

4. Of great importance is the casting of *plastisols*, especially for flexible PVC. A plastisol is a suspension of PVC particles in the plasticizer; it flows as a liquid and can be poured into a heated mold. When heated to around 177°C, the plastic and plasticizer mutually dissolve each other. On cooling the mold below 60°C, a flexible, permanently plasticized product results. Slush casting is used extensively for thin-walled products such as snow boots, gloves, and toys. Thermosets too can be slush-cast by pouring the prepolymer into a heated mold and draining it after a layer is cured.

5. Solutions (*syrups*, such as an acrylic polymer dissolved in an acrylic monomer) and *organisols* (in which the polymer is dissolved in a volatile solvent) may also be cast. Thus, solutions of polymers, especially of PVC, are cast onto a traveling stainless steel belt (*solvent casting* of films). In *wet spinning* fibers are formed by passing the solution through stationary multihole dies (from textile-industry tradition, called *spinnerets*). Because of the short diffusion path, the solvent is easily removed by heating and can be recirculated. Large quantities of cellulose acetate, cellulose triacetate, and polyacrylonitrile fibers are made this way.

6. *Rotational molding*, also called *rotomolding*, is a variant of slush casting and, even though named molding, is properly classified with casting processes. A measured quantity of polymer (liquid or powder) is placed into a thin-walled metal mold, and the mold is heated while rotating around two mutually perpendicular axes (Fig. 14–2). Thermoplastics (such as PE, nylon, or polycarbonate) melt whereas thermosets polymerize and cross-link. The mold is cooled and the part is removed. To increase production rates, three-arm carousels are often used, with one mold each in the load-unload, heat, and cool positions. Since no pressure is involved, the mold is simple. The part is free of molded-in stresses. The process is suitable for making large, relatively thin-walled, hollow (open or closed) parts. Even very large parts (such as 80 000-liter containers) can be made. The technique is suitable also for plastisols.

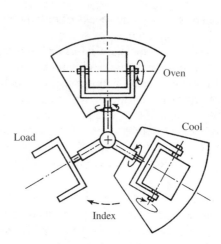

Figure 14–2   Rotomolding produces
hollow products by rotating
the mold around two
mutually perpendicular
axes; three stations speed
up production.

14-3   MELT PROCESSING (MOLDING)

Most plastics are too viscous even at high temperatures to flow under the force of
gravity, and the term *melt processing* refers to techniques in which polymers are
deformed *with the aid of applied pressure.* It is applicable to both thermoplastics
and thermosets. *Extrusion processes* result in a rod, tube, sheet, or film (and are
equivalent, in their end product, to metal extrusion); *molding processes* result in
a finished part (and are equivalent to die casting or hot forging in metals). Even
though these processes are related to some of the previously discussed metal
processes, there are significant differences rooted in the properties of plastics.

14-3-1   Principles of Melt Processing

By definition, the plastic must be capable of viscous flow.

Thermoplastics   A thermoplastic must be heated above T_m for crystalline polymers and well above T_g for amorphous polymers. The shape is fixed by cooling
well below T_g (or, for highly crystalline polymers, below T_m). Relative to metals,
there are two points to observe: First, special methods of producing and convey-
ing the viscous material may be applied and, second, viscoelastic effects may
cause changes in the molded shape.

1. The starting material is usually powdery, granular, cut-up strand, diced sheet or, in the case of recycled material, chopped (reground) and, sometimes, compacted scrap. Conveyance and consolidation by screw is possible. For freedom from gas bubbles, the plastic must be free of water. Heating may be partially external and partially internal (by transforming the work of viscous shearing into heat). Overheating can cause permanent damage. For example, PMMA depolymerizes and monomer gas bubbles form; PVC needs stabilizers; PE and PS are relatively insensitive; some others (such as polyacetal with PVC) may even form explosive mixtures.

2. The substantial volume change on cooling (Fig. 13–5a) reflects the reduction in free volume due to rearrangement of molecules and establishment of secondary bonds. Since these are time-dependent processes, shrinkage increases with slower cooling (higher melt temperature), decreasing pressure, and shortened deformation time (higher strain rate).

3. Fast solidification also means that the orientation of uncoiled molecules will be frozen in. Orientation can be desirable when molecules are aligned in the direction of maximum service stress but may cause distortion in service. Distortion is minimized if the molecules are given time to recoil before freezing. Unfortunately, measures that reduce distortion also increase shrinkage; hence, collapse of the surface is often observed in thicker sections. Furthermore, because thicker parts cool more slowly, molecules have more time to recoil inside the already solidified shell and set up residual stresses.

Thermosets   We saw in Sec. 13-3 that, prior to cross-linking (i.e., in the A or B stage), thermosetting polymers are capable of flowing under pressure. They may be granular, and then they can be treated as powders, or they may become thermoplastic on heating. Hence, processing techniques can be similar to those used for thermoplastics. There is, however, a major difference: Whereas thermoplastics are cooled to fix their shape, thermosets must be held in a heated mold for a long enough time for polymerization and cross-linking to occur. Some polymers can be removed from the mold as soon as their shape is fixed and then full cross-linking is obtained during cooling or holding in a separate oven. At other times, cross-linking begins immediately upon heating; then the prepolymer must be introduced into a cold mold, and the mold must be taken through a cycle of heating and cooling for each part, resulting in a very long cycle time.

14-3-2   Extrusion

Extrusion accounts for the largest production volume, since it is used not only for the production of bar, tube, sheet, and film in thermoplastic materials but also for the thorough mixing of all kinds of plastics and for the production of pellets. Ram extrusion (similar to Fig. 9–28b) is restricted to special cases (such as the extrusion of PTFE); the first major difference relative to metal extrusion is the use of *screw extruders*.

Screw Extruders   In its basic form, the equipment is fairly standard (Fig. 14–3). The polymer is fed through a *hopper* to a *barrel* in which a helical *screw* transports the polymer toward the die end. The screw has three sections: the *feed section* of constant root diameter (constant flight depth) takes in the granules or pellets from the feed hopper and moves them to the *compression section* (*melting section, transition section*) in which flights of gradually reducing cross section compress the softened pellets. Viscous shearing usually generates sufficient heat to bring the polymer to the required temperature; the barrel may be externally heated to compensate for heat losses or the barrel (or screw) may be cooled to prevent overheating. At the end of this section a viscous fluid is delivered to the *metering section*. This, like the feed section, has a constant but smaller free cross section. Here the melt is further heated by shearing at a high rate.

The design of the screw is critical. The pitch angle ϕ is typically 17.5° but may be higher for some plastics. The *compression ratio* (the ratio of free areas at the start and end of the screw, ranging typically from 2:1 to 4:1) and the length (or, more properly, the *length-to-diameter ratio*, ranging typically from 16:1 to 32:1) of the screw are chosen with due regard for the polymer. Heat-sensitive polymers (such as PVC) are extruded with minimum shear whereas polymers with a sharp melting point (such as nylon) call for a long metering and short compression section. For successful operation, temperatures (heating and cooling), back pressure, screw speed, injection rates, etc., must be tightly

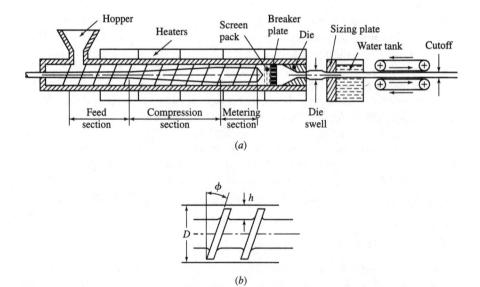

(a)

(b)

Figure 14–3   A thermoplastic resin, loaded into a screw extruder, is compressed, melted, and extruded through a die. Upon exiting from the die, the extrudate is cooled and die swell is reduced by pulling through a sizing plate under controlled tension (*a*). The screw has flights of *h* depth and ϕ pitch angle (*b*).

controlled. Temperature control along the barrel becomes even more critical when a *general-purpose screw* of compromise design is used for a variety of plastics. The screw may be in two sections, allowing decompression at about the middle of the barrel, so that gases can be vented off, and pressure is then built up again.

To prevent any unmelted polymer or entrapped dirt entering the extrusion, a *screen pack* (fine wire screens) is placed into the stream of polymer. It is backed by a sturdy *breaker plate* which has numerous holes of about 3-mm diameter. The screen increases back pressure, thus improving mixing and homogenization, and flow though the breaker plate removes the "turning memory" from the melt. The streams of plastic are then reunited before entering the die; the temperature is high enough to assure complete continuity. Pressures at die entry are 1.5–15 (less frequently, up to 70) MPa.

Twin- and multiple-screw extruders are more suitable for heat-sensitive materials such as rigid PVC, because they rely less on shear and drag to move the material: Intermeshing screws provide positive feed with minimum shearing.

Screw Output We saw in Sec. 7-2-1 that shearing a fluid between two surfaces sets up a shear stress. It is this stress that creates a *drag flow* through the barrel of the extruder (Fig. 14–3a and b) at a rate of q_{dr}

$$q_{dr} = 0.5\pi^2 D^2 N h \sin \phi \cos \phi \qquad \text{(14-1)}$$

which is the maximum possible output for a given extruder. Conveyance of the plastic through the gradually decreasing free cross section and the resistance of the screen pack generate a back pressure which reduces the flow rate by the *back-pressure flow* q_{bp}

$$q_{bp} = \frac{p\pi D h^3 \sin^2 \phi}{12\eta L} \qquad \text{(14-2)}$$

Thus, the output of the extruder is

$$q_e = q_{dr} - q_{bp} \qquad \text{(14-3)}$$

where D = screw (barrel) diameter (m), h = depth of flight channel (m), L = barrel length (m), N = screw revolutions per second, p = head pressure in barrel, ϕ = flight angle (degree), and η = viscosity (N·s/m^2). There is also a small, usually negligible output loss due to leak flow in the gap between screw and barrel. While back pressure reduces flow, it is essential for proper plastication. In the limit, back pressure becomes high enough to reduce output to zero; at this point, $q_{dr} = q_{bp}$ and the maximum pressure is

$$p_{max} = \frac{6\pi D N L \eta \cot \phi}{h^2} \qquad \text{(14-4)}$$

This gives the end point of the so-called *extruder characteristic curve* (Fig. 14–4). For a given extruder, the geometry terms in Eqs. (14-1) and (14-2) are constant,

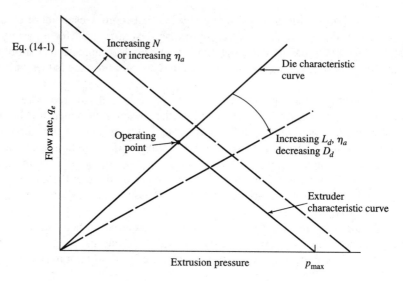

Figure 14–4   The output characteristics of a given extruder and the back pressure developed by the die determine the flow rate for a given set of operating conditions.

and Eq. (14-3) reduces to

$$q_e = \alpha N - \left(\frac{\beta p}{\eta} \right) \tag{14-3'}$$

This form clearly shows that output increases with increasing speed, decreasing pressure, and increasing melt viscosity. This simple treatment ignores, however, the effects of friction and non-Newtonian behavior (Fig. 13–6) which make analytical solutions much less reliable than for metal extrusion, and more complex numerical models, using apparent viscosity at the appropriate temperature and stain rate, are usually needed.

A useful, approximate estimate of a typical single-screw extruder capacity for $L/D = 24$ can be obtained simply from screw diameter (D mm or inch):

$$q_e = C_e D^{\mathrm{scr}} \tag{14-5}$$

where C_e and the exponent scr are empirical constants. The values of constants given in the literature underestimate the output; the values recommended below are in better agreement with outputs given by Levy and Carley [S. Levy and J.F. Carley (eds.), *Plastics Extrusion Technology Handbook*, 2d ed., Industrial Press, 1989, p. 270].

	Usual		**Recommended**	
	C_e	scr	C_e	scr
For output in kg/h	0.006	2.2	0.006	2.3
For output in lb/h	16	2.2	20	2.35

Actual output may differ by ± 20% and is higher with specially designed screws. An alternative approach to capacity estimation is based on the assumption that essentially all heat is obtained from mechanical work (external heat just compensates for heat losses):

$$q_e = \frac{2700 \ (\text{kW})}{C_p \Delta T} \tag{14-6}$$

where C_p = heat capacity (kJ/kg·K) and ΔT is the temperature rise in °C. (The room-temperature heat capacity of unfilled plastics is typically 1.2 kJ/kg·K for polystyrene, 1.7 for nylon and PP, and 2.3 for LDPE and acrylics; heat capacity increases with temperature). This formula is also useful for estimating the temperature rise if the power (kW) is known. Very approximately, the energy requirement is 0.15 kWh/kg.

The low heat conductivity of plastics limits cooling rates, therefore, extrusion speed (the speed of the emerging extrusion) is very sensitive to cross section. It ranges from a few mm/min for a 250-mm-diameter rod to 1000 m/min for wire coating, with 3 m/min typical of many products.

Example 14-1

A screw extruder has D = 75-mm-diameter screw with h = 5-mm-high flights of 17.5° pitch angle in the metering section. The screw rotates at 100 rev/min. The plastic has a density of 1 g/cm³. Calculate (*a*) the output for zero back pressure, (*b*) the output to be expected for typical conditions, (*c*) the approximate power requirement, and (*d*) the temperature rise attainable with PP.

(*a*) From Eq. (14-1)

$$q_{dr} = 0.5 \ \pi^2 (75^2)(100/60)(5)(\sin 17.5)(\cos 17.5) = 66340 \ \text{mm}^3/\text{s} = 239 \ \text{kg/h}$$

This is reduced by the back-pressure flow due to resistance in the screen pack and die.

(*b*) From Eq. (14-5), $q_e = 0.006(75^{2.3}) = 123$ kg/h or approximately one-half the maximum output.

(*c*) With kW = 0.15 kWh/kg, the power requirement is 0.15(123) = 18.45 kW.

(*d*) From Eq. (14-6), $\Delta T = (2700)(18.45)/(123)(1.7) = 238°C$, which brings the plastic into its usual processing temperature range (Table 13–2).

Dies   Flow through the die generates a back pressure which must be taken into account in computing the output. For a simple cylindrical flow channel (Fig. 14–5a), flow rate is given by Poiseuille's equation

$$q_c = \frac{p \pi D_d^4}{128 \eta_a L_l} \tag{14-7}$$

where D_d = die diameter, L_l = length of die land, and η_a = apparent viscosity. Thus, flow rate increases linearly with pressure, leading to the so-called *die characteristic curve* (Fig. 14–4). Note that D_d is to the fourth power, thus die dimensions have an extremely powerful effect on the extruder output (*operating*

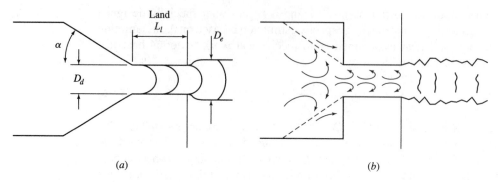

Figure 14–5   (a) A tapered die with a long land makes for orderly flow and minimum die swell. (b) A flat-face die creates a dead-material zone and turbulent flow which leads to melt fracture.

point). Beyond this, details of die design are critical for the production of a good extrusion.

There are several similarities to metal extrusion, one of which is the formation of a dead-material zone (as in Fig. 9–28a) with a flat-face die (*plate die*). Shearing on the boundaries of this zone causes thermal degradation of heat-sensitive plastics such as PVC. Furthermore, turbulent flow can lead to melt fracture (Fig. 14–5b). A tapered entry, as in Fig. 9–28b, but with an included angle of typically 60° (Fig. 14–5a), improves the situation; a streamlined die gives optimum flow. Dies often have a relatively long parallel length (*land*) so as to orient the molecules and control dimensions. In contrast to metals, however, neither the shape nor the dimensions of the extrudate are fixed. On leaving the die, internal stresses are released, molecules recoil, and *die swell* increases the dimensions (Fig. 14–5a). Tapered die and long land minimize but cannot eliminate die swell.

Round Bars.   The diameter increase is of no concern if the extrudate is chopped up for pellets, but it must be corrected for structural extrusions. Two measures are usually taken: first, the bar is passed through a *calibrating die* (*sizing plate*) while cooled with water (or occasionally air) and, second, tension is developed (*pull down*) by a continuous pulling device such as twin belts or caterpillars bearing on the surface of the bar (Fig. 14–3). Pseudoplastic materials (for example, PVC) are easier to work with because they develop sufficient strength upon emerging from the die; others could sag if not cooled and pulled sufficiently. The bar is cut to length or, if permissible, it is coiled up.

Sections.   The corners of a square are subjected to greater drag, reducing flow and resulting in rounded corners (Fig. 14–6a). To obtain sharp corners, extra material must be provided by changing the die shape (Fig. 14–6b) or drag must be reduced by reducing the die land length at the corners. Similar measures are necessary in extruding sections of unequal thickness. As in metal extrusion, thicker parts of the section are retarded by increasing the contact length on the

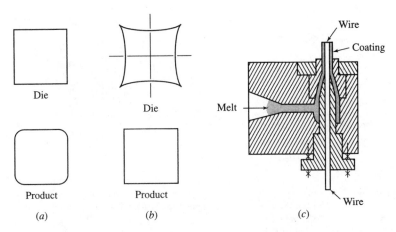

Figure 14–6   Higher drag at the corners of a square-section die result in rounded corners on the extrusion (*a*). The shape of the die can compensate for this (*b*). Flow can be directed around a wire (*c*).

land (see Fig. 9–31); in properly streamlined dies, this results in complex die shapes, often made by EDM (Sec. 17-4).

Pipes and Tubes.   Hollow (tubular) products are readily made with spider-type dies (similar to Fig. 9–30*c*). The highest pressure is obtained just in front of the die land; thus, the separated streams are reunited and a sound product is extruded. A tapered die with a long land minimizes the weld lines (*spider lines*). Wire and cable insulation is applied by feeding the wire through a crosshead die (Fig. 14–6*c*).

Sheet and Film.   A unique feature of polymer extrusion is that it is not necessary to keep the extrusion within the barrel diameter. If the melt is properly distributed, sheets much wider than the barrel can be made. Distribution in width is readily accomplished with a manifold (Fig. 14–7), but the pressure drops off from the center and the edges would be starved. Therefore, material flow must be retarded in the center by making the land longer; this leads to the so-called coat-hanger die configuration (Fig. 14–7*a*). Alternatively (or additionally), flow is locally regulated with adjustable *choke bars* (Fig. 14–7*b*) or by adjusting screws that close the die lips in the center. The emerging sheet is usually guided around a set of two or three highly polished, internally cooled rolls (*chill rolls*) so that it is cooled while its surface is polished. Thin films (< 0.1 mm thick) are similarly extruded and are guided around a chill roll prior to further treatment (*cast-film operation*). Applying a tensile force causes elongation (*draw-down*) and unidirectional alignment of molecules; simultaneous sideways stretch (by *tenterhooks*) gives biaxial orientation. This is really deformation in the viscoelastic regime, a process discussed in greater detail later, in the section dealing with blowing an extruded tube.

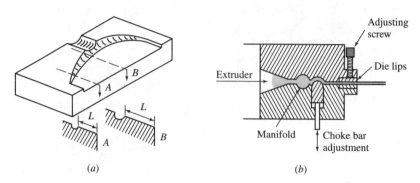

Figure 14-7   Wide sheet and film can be extruded but flow rates across the width must be equalized with a "coat-hanger die" (a) or a choke bar and die lip adjustment (b).

Fibers. In *melt spinning* the thermoplastic melt is extruded from a spinneret (a die with many—even tens of thousands—holes of about 0.1-mm diameter), typically at 230–315°C, in an arrangement similar to that of Fig. 12–10b, except that air is used for cooling the emerging fibers and an extension (*draw-down ratio*) of 2–8 is imposed. Nylon, polyester, and PP fibers are made by this technique. In *wet spinning* a polymer solution is extruded into a chemical bath in which the fiber is formed by mutual diffusion. In *reaction spinning* a prepolymer is extruded into a fluid reactive medium to produce a different (for example, cross-linked) fiber. Fiber diameters are small (typically, between 2 and 40 μm) and are usually expressed in units of *denier* (the weight, in grams, of a 9000-m fiber) or the SI unit *tex* (the mass, in grams, of a 1000-m fiber).

Special Processes   The scope of extrusion can be further widened.

Coextrusion. The possibility of uniting two or more streams of different plastic melts allows coextrusion of sheet, film, profiles, and tubing. For example, in products for food packaging one layer serves as a barrier to oil and water and the other is of food-contact grade, and wire insulation is extruded with colored tracers.

Example 14-2 | Coextruded plastic film is used for the bag in cereal boxes. The sealing joint must be strong enough to withstand handling yet weak enough to allow separation by hand. Therefore, the inner layer has a lower melting point, so that it can be heat crimped without softening the outer layer.

Reactive Extrusion. In this, chemical reactions are intentionally induced during extrusion. First used for the processing of rubber, it has been extended to thermosets and long-chain thermoplastics. Several purposes may be served: production of a high molecular-weight polymer from monomers or monomers and prepolymers; reaction of monomer and polymer to form graft polymers; reaction of polymers to form copolymers; cross-linking; and controlled degradation (cracking) of polymers to modify their rheology. The process is often more economical than processing in solution because there is no solvent to remove.

Calendering. This in-line postprocess is related to continuous strip casting (Sec. 7-5-2) in that the thermoplastic melt is fed from the extruder directly to a multiroll *calender*. The first roll gap (nip) serves as a feeder and spreads the plastic in width; the second roll gap acts as a metering device; and the third one sets the gage of the gradually cooling polymer which is then wound, with controlled stretching, onto a drum (Fig. 14–8); this is related to the hot rolling of metals (Sec. 9-7-1). As in metal rolling, parallelism of the roll gap must be maintained: Temperature distribution is carefully controlled, and the roll camber (crown) is controlled either by roll bending or skewing the central roll (which, in effect, changes the roll gap from center to edge). Calendering is a high-production-rate (typically, 100 m/min) process, producing sheet or film in widths up to 3 m. It induces less shear than direct sheet extrusion and is used mostly for flexible PVC (including the coating of paper and fabric, for tapes, upholstery, rainwear, shower curtains, etc.) and rigid PVC (trays, credit cards, laminations). Some ABS is also formed. Calendering, together with direct extrusion, is the principal method of making elastomeric (rubber) products for further forming into finished parts such as belts and tires.

In-Line Forming. Rolls or twin caterpillars can be used for embossing, vacuum forming (Sec. 14-4-2), corrugation, and other forming of the emerging extrusion.

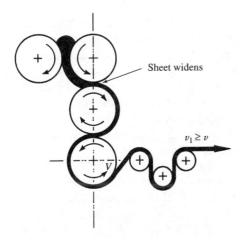

Sheet widens

$v_1 \geq v$

Figure 14–8   Wide sheet and foil can be produced by calendering.

Process Control   Even though the extrusion of plastics is more tolerant than the extrusion of metals, the process is not free of problems. Overcompression in the compression section causes blockage by solid and results in erratic *surging* (a drop-off in pressure), usually when the process is operating at maximum output. Gear pumps placed between the end of the screw and the die stabilize extrusion and can increase production rates. Friction on the die retards the surface layers (Fig. 14–5a); upon exiting from the die, elastic recovery creates tensile stresses on the surface which can result in random roughening (*sharkskin effect*). In the extreme, stick-slip leads to a periodic circumferential opening up of the surface, and the extrusion resembles a bamboo shoot (*bambooing*).

14-3-3   Injection Molding

Injection molding is the most widespread technique for making 3-D configurations. It is utilized for thermoplastic and, more recently, also thermosetting resins. The process (Fig. 14–9) resembles the hot-chamber die casting of metals.

Transport of Plastic   There are two basic ways of transporting the polymer:

1. Machines with hydraulically driven *reciprocating plungers* are capable of developing pressures of 70–180 MPa (Fig. 14–9a). The plastic is heated by external heaters on the barrel and by shearing around the *torpedo* (*spreader*) which also assures uniformity of flow.

2. More frequently, a rotating screw (Fig. 14–9b) is used. To deliver the required amount of molten plastic to the mold, the screw of a *reciprocating screw machine* is supported by a hydraulic ram that is pushed back when the pressure in front of the screw builds up to a preset value and the amount of melt needed for filling the mold is accumulated (Fig. 14–9c). At this point, rotation is stopped; the hydraulic ram pushes the screw forward and thus injects the plastic into the mold while backflow is limited with a nonreturn valve (Fig. 14–9d). Screws are sometimes used also to feed compression- and transfer-molding presses. In some machines the screw is used only for plasticating, i.e., feeding the melt to an injection chamber, from which a separate plunger injects the melt into the mold.

a. Thermoplastics are heated above the melting point (170–320°C) while the mold is held at a lower (typically 90°C) temperature. Injection pressures are around 140 MPa but can rise to 350 MPa for thin-walled products. Typically, 2–6 cycles are completed every minute.

b. For thermosets, the barrel is preheated just sufficiently (to 70–120°C) to ensure plastication. Injection under high pressures (up to 140 MPa) generates enough heat to reach 150–200°C in the sprue. The mold itself is heated to 170–200°C. The process is also used for molding glass-filled bulk molding compounds (BMCs); however, the charge would hang up in the hopper, and the barrel is stuffed.

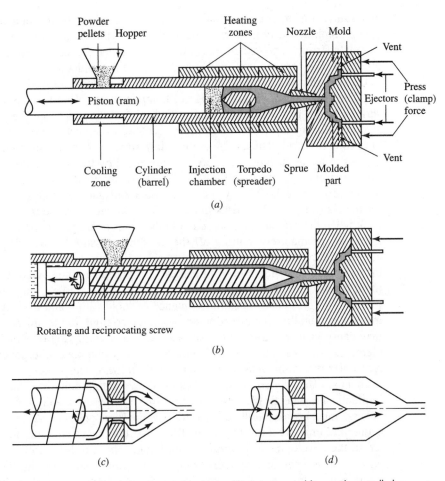

Figure 14-9   High productivity can be obtained by injection molding, with controlled amounts of melt injected by (a) a reciprocating plunger or (b) a reciprocating rotating screw; with the latter, flow is controlled by a nonreturn valve which allows buildup of melt (c) but closes during injection (d). [(a) and (b) adapted from Petrothene Polyolefins: A Processing Guide, U.S. Industrial Chemicals Co., New York, 1971. With permission.)

A freely cooling plastic shrinks by 7–20% from the molding temperature. Plastic melts are, however, compressible, and the high injection pressures serve not only for mold filling but also for stuffing the cavity. Furthermore, the pressure is maintained for the first part of solidification to make up for shrinkage. Flow rates can be very high, and erosion by hard filler particles may become severe.

Dies (Molds)   As in die casting (Fig. 7–25), the die is split to allow removal of the product. It must be kept firmly shut during injection, with the aid of a large hydraulic cylinder, or hydraulically actuated mechanical clamps, or a mechanical clamp combined with a short-stroke hydraulic cylinder. The clamping force is calculated from the projected area of the moldings and the recommended injection pressure. *Ejectors* are provided for removing the molded component, and fine (0.02–0.08 mm × 5 mm) *vents* ensure that no air remains trapped.

The structure of a molding is not homogeneous: With thermoplastics, the surface layers in contact with the mold cool rapidly and retain the orientation acquired during injection, whereas molecules in the slower-cooling inner part have time to recoil. Thus, there are molding stresses in the finished parts.

As in metal casting, the process is governed by the laws of fluid flow (this time, a strain-rate-sensitive fluid) and heat conduction. Therefore, feeding of the mold is critical. The system of runners and gates is similar to that used for metals (Sec. 7-5-3). Gates must not be too large, because melt would flow back when the pressure is released. On the other hand, gates that are too small freeze off prematurely, cutting off the molding pressure before full packing is attained. Nevertheless, small in-gates (*pin gates*) are sometimes used because they increase the shear strain rate, heat the plastic, reduce viscosity, and aid mold filling; they also make it easier to break off the sprue. The number and location of gates determines the sequence of mold filling and the alignment of molecules (and thus the direction of maximum strength in the finished part) and may cause visible marks on the surface. In many configurations, melt streams merge (as in Example 7-4) and failure to attain complete interpenetration of molecules results in weaker *weld lines* (*knit lines*), corresponding to cold shuts in metals. The low strength of plastics allows gating solutions that would be impractical for metals (Fig. 14–10). Multiple cavities are readily accommodated but care must be taken to feed each cavity at the same pressure. As in die casting, economies improve if material in the flow-distribution system is minimized. This led to the development of *sprueless molding*: The nozzle extends to the mold cavity and is heated; a sudden drop in temperature shuts off the flow, while rapid heat-up prevents freeze-up. In other cases, a valve is used to shut off the flow.

Example 14-3 | **A** die cavity of 120 cm³ volume is fed through a single gate of $L = 50$ mm length and $D = 4$ mm diameter. The extruder injects PP of 150 Pa·s apparent viscosity at $\dot{\gamma} = 1000$ s^{-1} under a pressure of 140 MPa. Calculate the filling time.

From Eq. (14-7)

$$q_c = \frac{(140 \times 10^6)\pi(0.004^4)}{(128)(150)(0.05)} = 0.0001173 \text{ m}^3/\text{s} = 117.3 \text{ cm}^3/\text{s}$$

and filling time is $120/117.3 = 1.023$ s. If the gate is reduced to $D = 3$-mm diameter, filling time increases more than three-fold (if the gate diameter is halved, filling time increases 16-fold). The calculation is very approximate because a change in shear strain rate changes the apparent viscosity, but it demonstrates the importance of gate dimension.

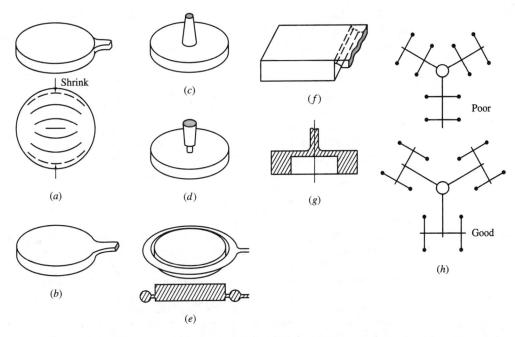

Figure 14–10   Gating is a decisive factor in injection molding. A side gate is simple, but oriented flow results in higher transverse contraction (a); the situation improves with a fan gate (b). Good feeding is given by a sprue gate (c), but the gate is difficult to remove. A pin gate (b) can be broken off, but "jetting" may be objectionable. Ring gates for cylindrical parts (e) and film gates for rectangular parts (f) feed uniformly, as do disk gates for rings (g). Multiple parts can be molded, but the gating system must assure equal pressure to each cavity (h).

The importance of proper gating is shown by the example of molding a plastic hinge. The material is isotactic PP (Fig. 13–2b); the molecule coils into a helix and bends freely as a helical spring would. A durable hinge is obtained if the molecular alignment is perpendicular to the hinge line. For this, melt must flow straight through the die restriction that will form the hinge, a goal that can be achieved by placing the gate as far away from it as possible.   **Example 14-4**

With the aid of computer programs, mold filling can be modeled, just as in metal casting. This greatly reduces or eliminates the trial-and-error approach otherwise needed for finding the optimum gating scheme. Mold filling is the critical factor also in terms of shrinkage and distortion. Shrinkage is different for different plastics but is also greatly affected by part thickness and processing conditions such as temperature, injection pressure, and hold time. Thicker parts solidify last, and shrinkage may cause *sink marks* to develop. *Gas-assisted injection molding* minimizes this by injecting a gas into the partially filled mold. The gas replaces the least viscous melt, forms internal cavities in thicker sections, and aids in filling intricate molds.

Temperature and pressure control is critical. Machine controls have become sophisticated, allowing rapid filling of the runner system, slowing down for the beginning of injection to prevent jetting, speeding up for mold filling, and holding the pressure (or increasing it) during solidification. Commercial programs of ever-increasing sophistication help in designing the mold and process controls with due regard to operating and material variables.

Production rates are greatly increased with *multistation, rotary turntable machines* on which loading, injection, and stripping (and, if appropriate, placing of inserts) take place simultaneously.

Example 14-5

An injection-molding machine of 1-MN (112-tonf) capacity (clamp force) is to be used for making a 125-mm-wide × 250-mm-long × 75-mm-deep box. (*a*) What molding pressure can be sustained? (*b*) How will this pressure change if a box of the same width and length but 150-mm depth is to be made? (*c*) How will the pressure change if the wall thickness of the boxes is doubled?

(*a*) Only the pressure acting perpendicular to the parting plane (parallel to the clamp force action) has to be resisted by the clamp. Thus, only the projected area is of interest: $A = 125 \times 250 = 31\,250$ mm^2. The sustainable pressure is $1\,000\,000/31\,250 = 32$ N/mm^2 (MPa).

(*b*) The projected area does not change with the depth of the box; hence, the pressure remains unchanged.

(*c*) Any change in wall thickness is immaterial, as long as the projected area remains the same. (However, the quantity of plastic to be delivered doubles.)

14-3-4   Other Molding Techniques

Several molding processes work independently of extruders.

Reaction Injection Molding   *Reaction injection molding* (RIM) differs from other processes in that, not the polymer, but the reactants are heated and brought together under high (10–20 MPa) pressure so that they inpinge upon each other as they enter the die through a pressure-reducing chamber (Fig. 14–11). Good mixing results, and the polymer is produced directly in the mold. The primary application is to polyurethane (solid bodies, foams, and elastomers), with some applications to nylon and epoxy. Pressures in the mold are low (300–700 kPa); thus, the mold-closing forces are low too. Dies can be simple, of low-cost construction. Since polymerization takes place in the mold, internal stresses are minimal, and the process is suitable even for large, complex, filled-plastic parts, such as auto-body and appliance components, and cycle times are 1–12 minutes. To avoid directionality of reinforcement, flake glass is preferred.

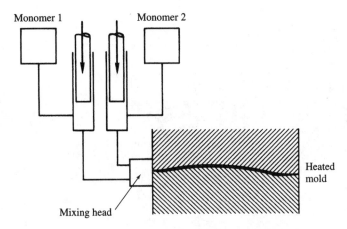

Figure 14–11   Reaction injection molding (RIM) produces parts by rapid impingement mixing and injection of monomers into the die cavity where the polymer forms.

Compression Molding   *Compression molding* is the equivalent of closed-die forging (Sec. 9-3-2). A premeasured quantity of polymer is introduced into the mold. When the plastic is completely trapped, one refers to a *positive mold* (Fig. 14–12a); any variation in polymer quantity results in a variation of part thickness. Closer tolerances can be held if a small flash is allowed to extrude—usually along the punch perimeter—in *semipositive molds*. More plastic is lost in *flash molds*, similar to those used in impression-die forging (Fig. 9–17b).

The molding temperature is chosen from experience (Tables 13–2 and 13–3). The press force is calculated by multiplying the projected area by the empirically determined molding pressure (typically 7–15 MPa for BMC, 15–40 MPa for phenolics, up to 70 MPa for ABS, PMMA, and PS, and 140 MPa for PTFE and UHMWPE).

Although suitable for thermoplastics, the main application of the process is to thermosets. Shortly after closing the mold, it is often opened slightly to vent gases, steam, and air, and then it is closed again until curing takes place (typically, 1–5 min). The advantages of the process are that there is no or little material loss, fillers do not become oriented, and internal stresses are minimal.

Cold molding is a variant related to powder metallurgy: a powder or fillers (often of refractory materials) are mixed with a polymeric binder, compressed in a cold die, and removed for curing in an oven.

Transfer Molding   *Transfer molding* utilizes an extrusion-molding principle (Fig. 14–12b). An excess quantity of the polymer is loaded into the *transfer pot* from which it is pushed through an *orifice* (sprue) into the mold cavity, by the action of a punch, at approximately 35–100 MPa pressure, while the die is held closed by another, higher-capacity ram. Pressures up to 300 MPa may be needed for

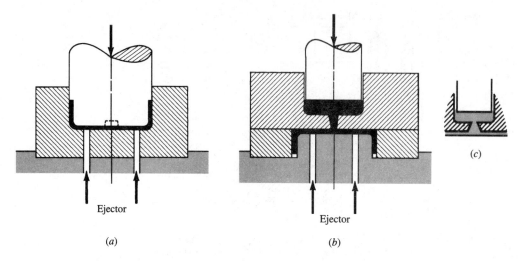

Figure 14–12   The processes of (*a*) compression molding and (*b*) transfer molding are suitable for both thermoplastics and thermosets. If material in the sprue hardens between shots, the sprue is tapered to break away (*c*).

some plastics such as cellulose acetate, rigid PVC, or styrene–butadiene block copolymer. The die is of closed construction and multiple cavities may be used, with appropriate runners and gates to feed them (as in Fig. 7–13). Thus, reasonable production rates are achieved on both thermosetting and thermoplastic compounds.

A great advantage is that the plastic acquires uniform temperature and properties in the transfer pot prior to transfer. The plastic is further heated by shearing through the orifice, viscosity is reduced, and the plastic fills intricate mold details. The technique is favored for making electrical connectors and for the encapsulation of microelectronic devices, because the low-viscosity plastic does not damage delicate wires and inserts.

Example 14-6 | A transfer-molding machine is operated from a 20-MPa hydraulic supply. The transfer cylinder is of diameter 125 mm, the clamp cylinder of diameter 250 mm. If the polymer calls for a transfer pressure of 50 MPa, calculate the diameter of the transfer punch (the diameter of the transfer pot) and the maximum allowable projected cross-sectional area of the molded parts.

The transfer cylinder exerts a force of $P = pA = 20(125^2\pi/4) = 245.4$ kN. At a required pressure of 50 N/mm^2, the punch area is $P/p = 245\,400/50 = 4909$ mm^2. The diameter is 79 mm (say, 80 mm).

The clamp exerts a force of $P = pA = 20(250^2\pi/4) = 981.7$ kN. Since the transfer pressure is acting on the entire projected area, the permissible area is $A = P/p = 981\,700/50 = 19\,635$ mm^2. To prevent the formation of a flash, it is usual to calculate with 1.15 × transfer pressure. Thus, $A = 19\,635/1.15 = 17\,074$ mm^2.

14-3-5   Process Capabilities and Design Aspects

Many of the principles discussed for extrusion (Sec. 9-4-5), rolling (Sec. 9-7), die casting (Sec. 7-5-6), and forging (Sec. 9-3) are also applicable to the processing of plastics, provided that the special characteristics of plastics are taken into account. General attributes of plastics processes are given in Table 14–1.

Table 14–1   General characteristics of polymer processes

Characteristics	Manufacturing process					
			Molding			
	Casting	Extrusion	Injection	Compression	Transfer	Thermoforming
Part						
Material	All	All	All	All	All	Thermoplastics
Preferred		Thermoplastics		Thermosets	Thermosets	
Shape*	All	All Group 0	All (with lost core)	All but T3, 5, 6, F5, and U4	As Compression	T4; F4, 7, and S5
Min. section, mm	(0.2) 4	0.4	0.4	0.8	0.8	1
Cost[†]						
Equipment	D–E	A–B	A–C	B–C	B–C	B–D
Tooling	B–E	A–C	A–C	A–C	A–C	B–C
Labor	A–C	D–E	D–E	C–E	C–E	B–E
Production						
Operator's skill[†]	B–E	D–E	D–E	D–E	D–E	B–E
Lead time	Days	Weeks	Weeks	Weeks	Weeks	Days–weeks
Cycle time		10–60	10–60	20–600	10–300	10–60
Min. quantity	1		1000	100–1000	100–1000	10–1000

*From Fig. 3–1.
[†]Comparative ratings, with A indicating the highest value of the variable, E the lowest. For example, injection molding involves high equipment cost, high die cost, low to very low labor cost, and low to very low operator skill. It takes weeks to have a mold made, and requires a minimum quantity of 1000 to justify the cost of mold.

Extrusion   By its nature, extrusion is limited to 2-D shapes (Group 0 in Fig. 3–1), although wall thickness can be varied by the programmed movement of a tapered mandrel. Within these limitations there is substantial freedom and high productivity. Dimensions are limited only by cooling. Tolerances are wider than in metal extrusion: Wall-thickness tolerances are ±8% and linear (width) tolerances range from ±8% on small (under 3 mm) dimensions to ±1.5% on large (over 100 mm) dimensions. While the extrusion die can be designed to produce nonuniform wall thickness, differential cooling results in bowing, and uniform wall thickness is preferable. Large flat areas may distort and can be stiffened with ribs.

The market for extruded products is large. Profiles are used in buildings and automotive applications; tubes, sheet, and film in a great variety of general applications and as the starting material for further processing; pipes in drainage, waste, and vent lines.

Molding   The low molding temperature of plastics greatly extends the life of die-steel molds and permits the use of aluminum molds for shorter (up to tens of thousands parts) runs.

1. *Shape.* Shape limitations are similar to die casting (Sec. 7-8-2) but, because of the relatively low operating temperatures, molds of considerable complexity can be made. Even rotating cores are feasible for threaded parts. Injecting around an expendable core (*lost-core method or fusible-core molding*) removes many shape limitations, although at the expense of more complex processing.

2. *Dimensions.* The removal of gases (produced by reactions, or entrapped during the compaction of particulate starting materials) must be allowed and encouraged. This sets a practical limit of 100–200 mm to the maximum thickness attainable without gross porosity. The low heat conductivity of plastics (Table 4–1) leads to slow cooling rates and limits the economical thickness typically to below 6 mm (preferably 3 mm). The minimum wall thickness is limited by the difficulty of removing very thin parts from the mold, and also by the high pressures required to fill at a large width-to-thickness ratio. Large wall-thickness differences lead to porosity, internal voids, or sink marks, just as in metal casting (Fig. 7–30), and part design should aim at avoiding them (Fig. 14–13a and b).

3. *Tolerances.* While molds and dies can be made to close tolerances, the great sensitivity of dimensions to processing conditions, postprocessing changes

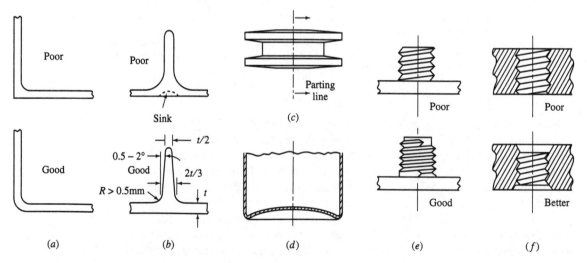

Figure 14–13   Some features of design for injection molding.

(polymerization, crystallization, loss of plasticizer, aging, relief of residual stresses) dictates that tolerances be as wide as permissible for the given application. Dimensional variations are reduced if sufficient time is allowed for the polymer to become rigid prior to removal from the mold. It is often possible to reduce cycle time by removing the part as soon as it cools to a temperature where it holds its shape, so that it can cool to room temperature out of the mold.

4. *Surface finish.* In all molding processes the plastic reproduces the surface finish of the mold, and, once the often very expensive die is made, a smooth or patterned surface can be obtained at no further expense. Wear becomes a significant factor primarily with filled plastics. Ejector marks ruin the surface, and ejectors should be placed where their marks are not visible.

5. *Parting line.* As in die casting and forging, the parting line must be chosen to minimize the complexity of the mold, avoid unnecessary undercuts that would necessitate complex movable inserts and cores, and minimize the cost of removing flash, for example, by allowing flash removal by tumbling. This becomes more difficult when the mold is split through the axis (Fig. 14–13c). Distortion is minimized when the gate is placed so as to give symmetrical mold filling.

6. *Ribs.* Distortion can be minimized by ribbing larger surfaces, but the width of ribs must be kept small to prevent the creation of large hot spots (Fig. 14–13b). Doming is an attractive alternative, especially in cylindrical parts (Fig. 14–13d).

7. *Drafts and radii.* Release from the mold requires a draft of 0.5–2°, and even larger drafts on ribs and bosses. Tight radii can be molded, but generous radii improve material flow, increase die life, and prevent stress concentrations in service. Minimum radii of 1–1.5 mm are recommended. Excessive radii result in hot spots and sinks (Fig. 14–13b).

8. *Holes.* Through-holes are limited only by the strength of the core pin and are usually held below a length-to-diameter ratio of 8. Freely extending core pins are needed for blind holes; therefore, such holes are limited to a depth-to-diameter ratio of 4 for $d > 1.5$ mm and to a ratio of 1 for smaller holes. Threaded holes of 5-mm diameter and over can be molded directly, preferably with a coarse thread, although a more complex mold is needed. Smaller holes are best drilled. Relief must be provided to avoid feather edges and stress raisers (Fig. 14–13e and f). Highly complex shapes requiring nonretractable cores can be formed with fusible cores (Fig. 14–14).

9. *Inserts.* The use of molded-in metal inserts greatly expands the scope of application for plastics and very often eliminates problems in subsequent assembly, although at some expense. Threaded inserts, binding posts, electric terminals, anchor plates, nuts, and other metallic components are molded into plastics by the millions. Some precautions are necessary, however. Gating must be carefully designed to prevent weak weld lines when melt streams flow around the insert. The shape of the metal part must ensure mechanical interlocking with the plastic, for example, by heavy knurling, since there is no adhesion between metals and

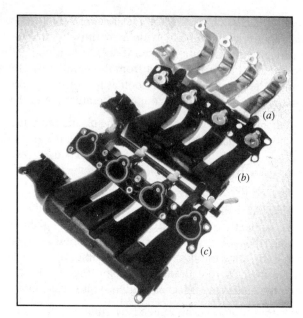

Figure 14–14   Complex internal passages can be formed with very smooth walls by the lost-core method: The core is made of a low-melting metal which is melted out after the glass-fiber-reinforced nylon had been injection molded. (*Courtesy Siemens Automotive, Auburn Hills, Michigan.*)

plastics—at least not without special surface preparation. The thermal expansion of plastics is much larger than that of metals (Table 4–1); this helps to shrink the plastic onto the insert, but could also cause cracking of a brittle plastic. The wall thickness around the insert must therefore be made large enough to sustain the secondary tensile stresses.

Example 14-7   Pins in connectors for computer and other electronics hardware used to be spaced at 2.54-mm (0.1-in) centers. With the increasing drive toward miniaturization, spacing is reduced to 1 mm or even 0.5 mm, with a corresponding reduction in the wall thickness of connectors. The task then is that of filling very thin channels without excessive pressure, so as not to bend the delicate connectors. Liquid-crystal polymers, even though expensive, are often chosen because their structure allows injection at high speed and low apparent viscosity. Tolerances are very tight and temperature resistance is high, allowing soldering at up to 230°C temperature.

Example 14-8   In Example 5-8 on the recycling of automobiles we saw that shredder fluff, consisting mostly of plastics, cannot be effectively recycled. The problem is becoming more severe with the growing use of plastics, driven by economy and the need for weight reduction. There is, therefore, a drive for recycling plastics. The aim is to have at least 25% postconsumer plastic in various parts. This may, occasionally, result in replacement of one plastic by another. For example, a panel topper pad for a full-size car used to be made from a glass-filled plastic which suffered breakage during part shipment and was difficult to recycle. It was replaced by an identical part made of a PC/ABS polymer blend which contains 25% postconsumer recycled ABS. [Source: *Adv. Mater. Proc.*, 1997(5):27.]

Example 14-9

One of the growth areas of engineering plastics is in underhood components for automobiles. Temperatures are high and shapes are often complex. Rigid cores limit the shape of internal cavities to configurations that can be retracted in straight-line or rotary (helical) motions. Much more complex shapes are possible with expendable cores. Thus, in making air intake manifolds (Fig. 14–14) with exhaust air recirculation, a low-pressure die-cast core of a low-melting (< 150°C) Sn–Bi alloy (Fig. 14–14a) is placed into the injection mold. A nylon reinforced with 35% glass fiber is injected; after a few minutes, the part (Fig. 14–14b) is removed from the die and transferred to a melt-out tank where the alloy is drained from the molding. Thus, a part with complex internal passages is produced (Fig. 14–14c). Relative to an aluminum casting, weight is reduced by 60%, and internal surfaces are exceptionally smooth. A similar approach is used for thermosetting phenolic resins: the Sn–Bi alloy is melted out during post curing after removal from the mold. [Source: *Adv. Mater. Proc.*, 1994(1):19.]

14-4   PROCESSING IN THE RUBBERY STATE

The viscoelastic behavior of thermoplastics above T_g allows the further processing of cast, extruded, injection-molded, or calendered semifabricated products. Deformation is again by uncoiling and alignment of molecules, but this time in tension. The molecular alignment that gave pseudoplastic behavior (*reduced stress due to shear thinning*) in shear flow [Eq. (13-4)] now results in positive strain-rate sensitivity [Eq. (8-11)]. Tensile strength (sometimes referred to as *melt strength*) *increases with strain rate* and a high m value allows substantial stretching without the localization of necks. Since the strength of an incipient neck increases more with higher strain rate [Eq. (8-11)], rapid stretching is beneficial. Pressures can be quite low, 0.2–1 MPa, seldom to 2 MPa. Uncoiling and alignment of molecules in the direction of stretching also affects the product: Strength increases in the direction of alignment but strength and, particularly, impact strength are lower perpendicular to the alignment. Biaxial stretching increases strength in all directions, and biaxially oriented products show superior performance in many applications.

14-4-1   Blow Molding

In *blow molding* an extruded tube or preform emerging from the melt process is expanded by internal pressure (usually by hot air).

Blown Film Extrusion   This technique is used to make biaxially oriented thin film by expanding with air a tube emerging from the extruder (Fig. 14–15), typically to a *blow-up ratio* of 3:1, in materials such as PE. Molecular alignment amounts to partial crystallization and results in reduced transparency (*frost line*). Polymers that have a low m value and are slow to crystallize cannot be directly processed. Thus, a PP melt is in-line quenched to retain an amorphous structure, reheated above T_g and is blown; stress-induced crystallization gives small lamellar crystals which do not affect transparency (the process configuration is upside down relative to Fig. 14–15). Bags are made by heat sealing or by pinching off, and very wide (over 6 m) flat films are produced by slitting along the length. Plastics that crystallize rapidly (such as nylon and acetate) cannot be processed.

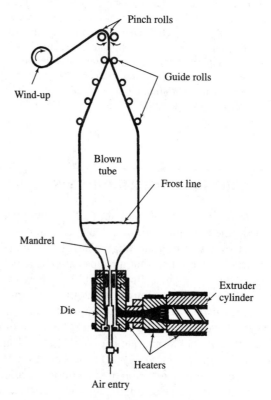

Figure 14–15   Extruded tube is expanded by air into a biaxially oriented film; temperatures are closely controlled to keep the frost line (where the film becomes slightly opaque) at a constant position. (*From D.C. Miles and J.H. Briston: Polymer Technology, Chemical Publishing Co., 1979, p. 551. With permission.*)

Example 14-10

Blown film of 25-μm thickness is extruded for grocery bags. The width of the bags (the lay-flat width of the blown tube) is 50 cm. Optimum properties are obtained by a blow-up ratio (bubble-to-tube diameter ratio) of 2.5:1. On emerging from the extrusion die, the tube is pulled at a faster rate; the draw-down ratio (the ratio of die opening to tube thickness) is 15:1. Calculate the required extrusion die diameter and opening.

The circumference of the bubble equals twice the lay-flat width. Thus $d\pi = 2(500) = 1000$ mm and $d = 318.3$ mm. The blow-up ratio of 2.5 gives a tube diameter of $318.3/2.5 = 127.3$ mm (a die of 125- or 130-mm diameter is likely to be chosen). The tube thickness, before blowing, is $2.5(25) = 62.5$ μm $= 0.0625$ mm. The draw-down ratio of 15:1 gives a die opening of $15(0.0625) = 0.9375$ mm (an opening of 1 mm will do).

Extrusion Blow Molding A continuous tube is extruded, pinched off so that a solid-state weld is produced at the pinch line, and the preform thus produced is then blown (Fig. 14–16), typically to a blow ratio of 1.5–3 and less frequently to 7. This method has the highest productivity. Even very large parts such as 2000-L drums can be molded, and engineering plastics are formed into complex parts such as car heating/cooling ducting and instrument panels. To avoid the need for very high-capacity extruders, melt is gathered in an accumulator from which it can be quickly extruded, with a separate ram, into the die. Localized thinning of parts can be avoided or the wall thickness of parts may be intentionally varied along their length by the programmed movement of a tapered mandrel in

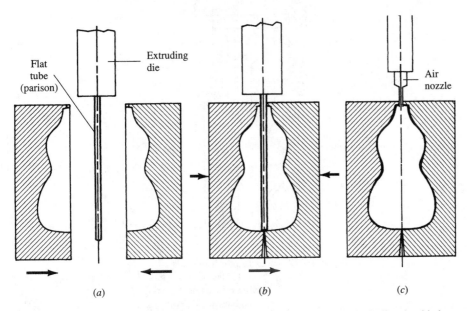

(a)   *(b)*   *(c)*

Figure 14–16   In extrusion blow molding (a) a flat extruded tube is (b) pinched off and welded at the bottom by the closing die and (c) expanded into the mold by air pressure; the bottle is cooled and removed at third and fourth stations (not shown).

the extrusion die orifice (Fig. 14–17). Pressures are relatively low (0.5–0.8 MPa); therefore, thinner-wall, more economical dies can be made of aluminum or steel.

Injection Blow Molding   This is the same process as used for glass bottles (Fig. 12–11) except that the parison is injection-molded—complete with neck—around a hollow core rod, and is then blown through the core rod. Parisons may be allowed to cool, stored, and reheated for blowing. Many PVC, PP, PET, and polycarbonate bottles of smaller (say, up to 1.5 L) sizes are blown on indexing machines.

Stretch Blow Molding   This term is applied to the simultaneous axial and radial expansion of a parison, yielding a biaxially oriented container. The parison is lengthened mechanically by the core rod while it is expanded by air. The process is subject to the rules of blown film extrusion. The extruded or injection-molded parison is first conditioned, either by controlling its temperature immediately after extrusion, or by reheating the cold parison. PET soda bottles are thus biaxially strengthened.

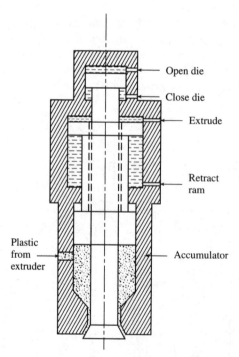

Figure 14–17   Large parts are made by accumulating a large mass of molten plastic and extruding it into a blow mold. Wall thickness is controlled by the programmed movement of a tapered mandrel.

14-4-2   Thermoforming

Thermoforming is a term applied to forming a thermoplastic sheet in the rubbery regime into open, container-like, but often very complex shapes (superplastic forming of sheet metal, Sec. 10-9, was derived from this process).

Thermoforming employs a clamp that grips the sheet around its circumference, a heater to bring the polymer above the glass-transition temperature (usually around 55–90°C), and a die which may be male or female. Conformance to the die shape may be achieved by mechanical means or by air pressure. Since the die is cooler, the polymer is chilled (and stiffened) by die contact, and portions of the workpiece that first touch the die become stiff while still of a thicker gage. Subsequent deformation is limited to the freely deforming portions of the workpiece, and excessive thinning could lead to fracture. Much of process design is aimed at controlling wall-thickness distribution by the planned sequence of operations. Numbers in Fig. 14–18 refer to this sequence.

1. In the simplest (straight) techniques all forming is done with vacuum or pressure. In *vacuum forming* (Fig. 14–18a) the sheet is (1) clamped, (2) heated above T_g, and then (3) vacuum is applied to draw the sheet into intricate recesses of the female die.

2. Alternatively, hot *air pressure* of 100–2000 kPa is applied (Fig. 14–18b) to drive the sheet into the female die cavity (provided, of course, that venting holes are furnished at the underside). The corners of all straight-formed parts

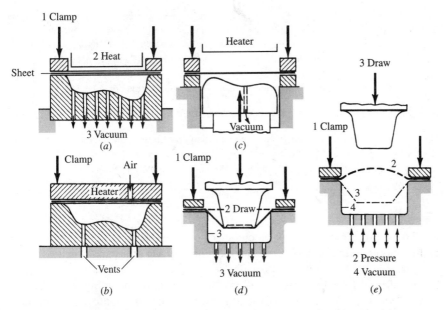

Figure 14–18   Thermoforming processes convert a sheet into complex shapes by (a) vacuum, (b) pressure, (c) drape-vacuum, (d) plug-assist, and (e) pressure-bubble plug-assist methods. Numbers refer to the process sequence.

will be thin, as will be seen from considering the points of first die contact in Fig. 14–18a and b. As in blow molding, high strain rates are beneficial, hence the vacuum or pressure is applied suddenly.

3. The shape can be developed by *drape forming* with a male die (similar to the stretch forming of metals, Fig. 10–18b); reentrant shapes can be formed by providing the punch with holes through which vacuum is drawn, and thus the polymer is pulled into the recessed parts (Fig. 14–18c).

4. Corners can be made thicker by first deforming the sheet with a punch (now called a plug) in *plug-assist forming* (Fig. 14–18d). After (1) clamping and heating the sheet, (2) the cooler punch moves in to stiffen the polymer at the points that will become the corners. Thereafter, (3) the part is finish formed by the application of vacuum.

5. Further wall-thickness control is achieved in *reverse draw forming* (Fig. 14–18e). After (1) clamping and heating the sheet, it is (2) free-formed into a dome by pressure (*pressure bubble*) or vacuum; this preform is then (3) deformed by the plug, and (4) the final shape is achieved by making the sheet conform to the plug, die, or plug and die combination with the application of pressure or vacuum (note the similarity to hydroforming, Fig. 10–33b).

6. Contact with die surfaces is eliminated and dome-shaped parts of optical quality (canopies) can be made by blowing up a clamped sheet (*free forming*, a counterpart of hydraulic bulging of metals).

Example 14-11

A cylindrical container of $d_o = 200$ mm outside diameter and $h_i = 70$ mm inner depth is to be made by vacuum forming. The average wall thickness is 1.5 mm. Calculate the diameter and thickness of the starting blank, assuming that a 20-mm edge (flange) is needed to hold the blank.

The principle of volume constancy applies. The volume of the container is

$$V = [(d_o^2 - d_i^2)\pi/4]h_i + (d_o^2\pi/4)t = [(200^2 - 197^2)\pi/4]70 + (200^2\pi/4)1.5 = 112\,603\text{ mm}^3$$

This is formed from a blank of $d_0 = 200$-mm diameter, thus blank thickness is $112\,603/(200^2\pi/4) = 3.58$ mm. The starting diameter is $200 + 2(20) = 240$ mm. If the edge (flange) has to be trimmed off, material utilization is $200^2/240^2 = 0.695$ or 69.5%.

The starting material may be cut sheet or coiled strip which is fed through multistation continuous web systems (the counterparts of progressive die lines in metalworking). Heat may be provided to the cold input sheet, or the line may be directly fed from an extruder. Many thermoplastics including ABS, PP, PS, PVC, PMMA, and polyesters, filled and unfilled, are formed into varied products such as aircraft canopies, automobile head liners, wheel-well liners, building panels, bathtubs, boat hulls, refrigerator liners, and vast quantities of packaging products. Computer modeling of localized thickness changes has taken much of the trial-and-error experimentation out of part and process design.

14-4-3 Cold Drawing

The term is applied to the continuous stretch-drawing of filaments and fibers, including fibers emerging from a spinneret or ribbons slit from a film. Alignment of linear molecules results in a substantial increase in strength. The process is most important for textile and reinforcing fibers, including PP, polyester, nylon, and aramid fibers.

14-4-4 Matched-Die Forming

The terminology is somewhat diffuse. *Solid-state forming* (also called *solid-phase forming*, and sometimes misnamed as cold forming) describes deformation just above T_g (or, for highly crystalline polymers such as PP, just below T_m) by metalworking techniques. Forging is applied to a limited extent, e.g., for making gears, but most forming is done by sheet forming (also called *mechanical forming*) techniques, including bending, stretching, and drawing, The sheet is heated and pushed by a heated plug into the mold and forced against the cold die (similar to Fig. 10–18b), often in multistage rotary or shuttle-type arrangements that allow the part to cool before ejection. Many food-packaging tubs and containers are formed of PP essentially by stretch forming.

Thermoplastic stamping (or *matched-die forming*) is a term used to describe the deformation of filled thermoplastic polymer sheet at melt temperatures, between cold, mating dies, thus giving reduced cycle times. At lower preheating temperatures, elastic behavior becomes more dominant and more springback is to be expected.

14-4-5 Process Capabilities and Design Aspects

Deformation in the rubbery regime is the counterpart of hot working in metals. The processing temperatures are, however, much lower, and plastics have a much higher m value than metals (except for superplastic metals, which can be worked with plastics processing techniques). All these benefits combine to make blow molding and thermoforming the processes of choice for relatively thin-walled products of often great complexity. The 2-L PET soda bottles are pressure vessels of remarkable performance, and one need to inspect only a refrigerator door liner to see the severe deformations that can be achieved. There are less visible but equally impressive technical applications, such as heavily ribbed gasoline tanks for automobiles, dash boards, double-walled components, fascia panel, and air ducts. Coextruded sheets or tubes can provide tailored properties. Thus, a gasoline tank may have a gasoline-resistant nylon lining and a stronger and cheaper PP outside layer.

Shape and size are limited only by the size and operation of the die. As always, straight parting lines make for less-expensive molds, but complex parting lines may be justifiable for mass production and part consolidation. Undercuts require movable cores but can be made. In all cases, sharp corners cause excessive local thinning and generous radii are mandatory. Blow-molded parts need not be of closed shape; different parts can be made in the two mold halves and then separated.

14-5   PARTICULATE PROCESSING TECHNIQUES

Even though many polymers are processed from particles (powder, pellets, etc.), the techniques described hitherto are not particulate processes in the sense adopted in Chap. 11, since the particles fuse and melt. However, some very high-molecular-weight polymers such as UHMWPE, PTFE, and polyimide decompose before reaching a temperature where viscous flow is possible. Therefore, they must be processed by techniques typical of metals (Sec. 11-3).

The powder is compacted cold, with pressures up to 350 MPa (50 kpsi), and then sintered at 360–380°C to form a solid billet or shaped component. Alternatively, the powder is compression-molded at a somewhat higher temperature (as in the hot pressing of metal powders), or powder is hot extruded in a ram-type press (as in Fig. 9–28b). The extrusion of solids causes alignment of molecules, and extrusion at a high extrusion ratio is one of the methods of making ultra-high-strength polymer fibers.

14-6   CELLULAR OR FOAM PLASTICS

Blowing agents create gas-filled voids, cells in the polymer; thus, one speaks of the production of *cellular (foam) plastics*. Many techniques are available: Air may be introduced by mechanical whipping; gases can be diffused in at elevated pressures; low-boiling liquids, gases (such as CO_2), or chemical compounds that decompose on heating, may be incorporated; hollow glass or hollow or expandable plastic spheres may be added; or the polymer may be molded with an additive that can be subsequently bleached out.

In general, a highly viscous polymer gives closed cells, whereas a less-viscous polymer results in open (interconnecting) porosity. Porosity is uniform throughout products such as flexible or elastomeric foams used in insulation, packaging, cushions, etc. *Structural foams* have a solid skin. For these, the polymer is poured or injected into a cold mold which suppresses expansion of the polymer, forming a solid skin on the surface; in contrast, the slowly cooling core is foamed. The foam core prevents collapse of the surface, and the final product has low density combined with reasonable strength.

Various techniques of manufacture are used:

1. Expanded polystyrene. A volatile hydrocarbon (usually, pentane) is added to polymer beads as a blowing agent. When heated with live steam (steam introduced into the polymer), preexpanded beads are obtained. The beads are stored to allow evaporation of moisture, and are then fed into the appropriate mold. Both mold and beads are heated with live steam to cause final expansion and fusing, followed by cooling. Insulating board is composed of large beads at densities of 15–30 kg/m^3; foam cups are made of small beads at 50–65 kg/m^3.

2. Extruded thermoplastic foams. Most frequently, the gas (pentane, fluorocarbons) is directly introduced into the molten plastic in the extrusion barrel. Expansion takes place as the extruded sheet or profile leaves the die. To control cell size, a fine, dry powder is mixed in the plastic as a nucleating agent; gas comes out of solution on the powder particles. The main application is for trays and packaging, at densities of 30–150 kg/m^3 in PE and PS. PVC extrusions are used as substitutes for wood moldings in the construction industry.

3. Structural foams. All thermoplastics (such as ABS) can be injection-molded, at rapid injection rates, to 60–90% of solid density. There are many applications: computer and appliance housings, material-moving pallets and bins, doors and shutters, and automotive instrument panels.

4. Multicomponent liquid foam processing. In this, chemical compounds are poured or injected into molds to form thermosetting foams in situ. RIM is also applicable. Most widely used is polyurethane in both the open-cell (flexible) and rigid (up to 500-kg/m^3) forms. Load-bearing structures such as furniture frames and doors are also made. Polyester and silicone rubber foams are also produced by this technique.

The density of polystyrene is 1.05 g/cm^3. Calculate the volume of air space in the walls of an expanded polystyrene foam cup of 50 kg/m^3 density. | **Example 14-12**

For ease of conversion, consider that 1 $g/cm^3 = 1000$ kg/m^3 (you may wish to check this). Hence the volume of polymer in 1-m^3 bead is 50 kg/1050 kg/m^3 = 0.0476 m^3 or 4.75%. The volume of air is $100 - 4.76 = 95.24\%$.

14-7   PROCESSING OF ELASTOMERS

We saw in Sec. 13-7-3 that elastomers are either cross-linked thermosets or anchored thermoplastics. We referred to their processing wherever appropriate. It will, nevertheless, be useful to give a brief review with emphasis on differences.

The elasticity of rubbers permits liquid-state processing by dipping (also called *dip casting*, the reverse of slush casting). The form is dipped (if necessary, repeatedly) into a *latex* (an emulsion of rubber in water) or into a solution of rubber. The layer thus built up is dried and cross-linked in an oven. The product, such as a glove, is then removed.

Vulcanized rubber is cross-linked by heat; hence, processing is controlled to prevent premature vulcanization (called *scorching*). First the rubber is mixed with carbon black and processing aids in powerful mixers (especially, so-called Banbury mixers) which break down the rubber into a lower-viscosity mass while raising its temperature to 150°C. The vulcanizing agent, usually sulfur, is added in the second stage where temperatures are kept lower. The product is then extruded, calendered, or injection-, transfer-, or compression-molded the same way as other thermosets. Fibers or fabrics are incorporated during molding. Cross-linking takes place in the process or during subsequent heating.

14-8   PLASTICS-PROCESSING EQUIPMENT

Molds (dies) used in plastics processing can be of relatively light construction if pressures are low, as in thermoforming, rotary molding, and blow molding, and are often made of fully heat-treated 7075 aluminum alloy. Injection-molding and extrusion dies are made of heat-treated steel. Surfaces are usually highly polished (unless given a decorative finish) and, for greater wear resistance, may be chromium plated or built up with a wear-resistant surface coating. Wear is of particular concern when polymers containing highly abrasive fillers move at high velocity over die surfaces, as in injection-molding dies. Temperature control of dies is critical and sophisticated heating-cooling systems are needed.

Plastics processing equipment shares many features with metal-processing equipment; however, the greater sensitivity of plastics to temperature, shear rate, and residence time makes process control more critical. Therefore, substantial effort has been directed at the development of sensors (mold pressure and temperature, ram speed and position) that allow data acquisition and closed-loop or adaptive control of all important process variables. The introduction of programmable logic controllers and microcomputers has made processing much more reproducible. For example, adaptive control is used in the extrusion of sheet and film to equalize thickness across large widths: Sheet thickness is continuously measured at various points across the width, and the flow of polymer is redistributed by trimming the die shape with the aid of die bolts (Fig. 14–7b). The introduction of CAD/CAM techniques has taken much of the trial-and-error experimentation out of process design. Variable-speed electric drives and servo-controlled hydraulics provide opportunity for CNC control of most processes.

14-9   SUMMARY

The availability of a vast variety of plastics makes the task of part and process design more difficult than for metals. The difficulty is further compounded by the different responses of various plastics to service conditions such as static and impact loading at service temperature; the effects of humidity, radiation (including light), and other environmental factors on the long-term deterioration of properties (aging); dimensional stability; wear; the release of undesirable chemical compounds on contact with food or air; and response to electrical fields. An ever-growing body of technical literature and software aids the designer. As in all design, the interaction of the process with the material must be considered early on, with due regard to the quantities to be produced. It is often found that the initial choice of material must be changed in light of constraints imposed by the economics of processing. Hence, the final material choice also fixes the process, and design then proceeds with allowance for the capabilities and limitations of that process. An often decisive advantage of plastics is the possibility of part integration, which reduces assembly and results in lower cost.

Plastics processes are related to casting and plastic deformation processes in metals, with some significant differences attributable to the nature of polymers.

1. Casting under the force of gravity is possible for low-viscosity melts, plastisols, and precursors. Casting into molds, slush casting, and continuous casting are all practiced.

2. Deformation under pressure in the melt (viscous flow) regime is employed for both thermoplastics and thermosets.

   a. Screw extrusion relies chiefly on heat developed in shearing the plastic. Bars, sections, and tubes are extruded in dies related to metalworking dies. The extrusion of wide sheet and film is unique to plastics, but die design follows universal principles.

   b. Injection molding is capable of mass producing parts of great complexity, because plastics exhibit shear thinning and are capable of flowing into narrow channels. The relatively low processing temperatures allow the operation of dies with movable cores. In reaction injection molding the polymer is formed in the mold. Compression and transfer molding further widen the scope of processing, especially for thermosets.

3. The molecular alignment responsible for shear thinning serves to promote high uniform elongation in tensile deformation; hence, preforms and sheet made by extrusion or injection molding can be further formed in blow molding and thermoforming. Biaxial deformation results in products free of directional properties.

4. Many techniques are also suitable for processing foam plastics and elastomers (rubbers).

5. Plastics processing equipment presents the same dangers as metal-casting and metalworking machinery. In addition, many constituents of plastics create a health hazard and release noxious fumes.

PROBLEMS 14A

14A-1 (*a*) Define plastisol. (*b*) Suggest a process in which it would be used.

14A-2 (*a*) Define melt processing. (*b*) Define thermoplastic. (*c*) Define thermosetting. (*d*)

On the basis of these definitions, can a thermoset be melt-processed? Justify your answer.

14A-3 (*a*) Make a sketch of a screw extruder, identifying the principal elements. (*b*) Mark the various zones of the screw and indicate their distinguishing features. (*c*) Show the die and an emerging round product; indicate the difference.

14A-4 Make a sketch to explain the phenomenon of melt fracture.

14A-5 (*a*) Make a sketch of calendering. (*b*) Indicate what metalworking process this resembles.

14A-6 (*a*) Make a simplified sketch of screw injection molding, complete with a simple die. (*b*) Indicate the relative temperatures at the exit of the barrel and in the die for a thermoplastic and for a thermosetting plastic. (*c*) Identify the essential die elements.

14A-7 Draw sketches illustrating the principles of (*a*) compression molding and (*b*) transfer molding. Show the essential die elements, indicating (generically) the temperatures for an amorphous thermoplastic polymer and for the die elements. (*c*) Indicate which of the two processes gives higher material utilization. (*d*) Give the main reason for using the process of lower material utilization. (*e*) Name the metal processing techniques closest to these polymer techniques.

14A-8 Define, generically, the temperatures for (*a*) melt processing and (*b*) processing in the rubbery regime for an amorphous and a crystalline plastic.

PROBLEMS 14B

14B-1 Six identical parts are to be injection molded at once in a die. Make a sketch of a gating system that (*a*) gives uniform flow into each cavity and (*b*) one that does not.

14B-2 Choose a gating solution for injection molding the ring of Example 7-9; justify your choice.

14B-3 Hollow products are made in metals by slush casting and in ceramics by slip casting. (*a*) State the counterpart of these processes in polymer (not plastisol) technology, and give a succinct definition of this process. Give the principal process sequence and the characteristic temperatures (generically) for the polymer and mold if the polymer is (*b*) an amorphous or (*c*) a crystalline thermoplastic or (*d*) a thermosetting polymer.

14B-4 A thermosetting polymer is injection-molded, in the prepolymer stage, into an intricate shape. Some parts of the cavity fail to fill. Make a recommendation whether mold temperature should be raised or lowered for complete filling (to support your argument, draw a graph showing, in principle, the change in flow rate as a function of mold temperature).

14B-5 A rectangular bar is to be extruded in a thermoplastic. (*a*) Superimpose on the rectangle of the die the shape of the extruded product. (*b*) Suggest two possible changes to the die design to obtain the correct cross section.

14B-6 A round thermoplastic bar emerging from an extrusion die has a surface similar to a shark skin. (*a*) Suggest a reason for its appearance (*hint*: tensile stresses in the surface can cause such deformation). (*b*) Suggest a way of eliminating the problem.

14B-7 A colleague argues that a pseudoplastic polymer cannot be blow-formed because of its negative strain-rate sensitivity. Make a reasoned evaluation of the claim.

14B-8 Thermoplastic polymer pipes are extruded over a water-cooled mandrel, hence the inner layers cool first. What residual stress distribution is to be expected? Is it likely to persist over a long period of time?

14B-9 Returnable HDPE milk jugs must be ground up and remanufactured after some 12 cycles to the customer because they shrink. Explain why this should be so.

14B-10 (*a*) Make a sketch of a process by which an open rectangular container of uniform wall thickness can be made of an amorphous thermoplastic polymer sheet. Specify the temperatures of the workpiece and of the die elements (generically). State the sequence of operations. (b) The container made in (a) is observed to lose shape in service. Identify the probable cause. (c) Make a recommendation to eliminate the problem.

14B-11 Collect several different plastic bottles and containers (include, if available, a bottle with a handle). Inspect the outer surfaces for evidences of manufacturing techniques. After sectioning, inspect the inner surfaces and gage the wall thickness variations. From your conclusions, describe the most likely manufacturing process for each container.

14B-12 A square box is made by thermoforming. In straight vacuum forming the edges thin out excessively. To explain your analysis and recommendation, (a) make a sketch of pure vacuum forming, showing die and polymer temperatures; (b) show why the corners and edges should thin out and (c) suggest a production method for reduced thinning.

14B-13 List at least three processes by which a plastic sheet of 2-m width and 0.2-mm thickness can be made.

14B-14 A bottle is to be made by blowing up a preform from 40 mm to 100 mm diameter. The wall of the bottle must be uniform and no local thinning is permissible. Candidate materials are (a) thermoplastic polymer, (b) glass, and (c) metal. State what property is essential for successful manufacture, and then judge whether the three materials possess this property, and if they do, under what conditions.

PROBLEMS 14C

14C-1 A bottle is blown, as in Fig. 14–16, from an extruded HDPE tube of 30-mm OD and 28-mm ID. (a) What will be the finished wall thickness at the waist (diameter 70 mm) and at the bulge (diameter 110 mm)? (b) How could the wall thickness be equalized?

14C-2 Measure the wall thickness of a soft-margarine or cottage cheese container at the rim, side, and bottom (use a ball-point micrometer and exert very small pressure). If available, check two containers of identical diameter but different depth. Show that the depth was developed from a circular blank with a thickness equal to the thickness of the rim.

14C-3 Grocery sacks are often made of 25-μm-thick, blown LDPE (0.920 g/cm^3) film. A typical sack is 60 cm long and 50 cm wide, and the price of resin is \$0.8/kg. It is now suggested that the sack be made of LLDPE (0.922 g/cm^3, \$0.96/kg). Because this material is stronger and stretches less, the gage can be reduced to 18 μm. Calculate whether the proposition is economical.

14C-4 A pressure of 300 MPa is applied to the punch of a transfer molding die. The punch diameter is 50 mm, and the projected area of the mold cavities, including runners and gates, is 3600 mm^2. Calculate the press size and the clamp force required.

14C-5 Check the calculation of Example 14-12 by measuring the dimensions of an expanded polystyrene drinking cup and weighing the cup. Explain the causes of possible discrepancies.

14C-6 A round bar of 15-mm diameter is extruded from a single-screw extruder of 100-mm barrel diameter. The material is LDPE. Calculate (a) the approximate flow rate, (b) speed of emerging extrusion and (c) expected power requirement.

14C-7 The part shown in Fig. Ex. 7-9b is to be made of a wood-flour-filled phenol-formaldehyde resin. (a) From data in Tables 13–3 and 14–1, choose an appropriate process. (b) Taking average values from

Table 13–3 and the relevant text, specify the process conditions. (*c*) Design, in principle, the die, showing the main die elements. (*d*) Determine the size of equipment needed.

FURTHER READING (see also Chapter 13)

ASM Engineered Materials Handbook: Desk Edition, ASM International, 1995.

ASM Engineered Materials Handbook, vol. 2, *Engineering Plastics*, ASM International, 1988.

Mitchell, P. (ed.): *Tool and Manufacturing Engineers Handbook*, vol. 8, *Plastic Part Manufacturing*, Society of Manufacturing Engineers, 1996.

Modern Plastics Encyclopedia, McGraw-Hill, Annual.

Agassant, J.F., P. Avenas, J.Ph. Sergent, and P.J. Carreau: *Polymer Processing: Principles and Modeling*, Hanser, 1991.

Buckleitner, E.V. (ed.): *Dubois and Pribble's Plastics Mold Engineering Handbook*, Chapman & Hall, 1995.

Bryce, D.M.: *Plastic Injection Molding*, SME, 1996

Campbell, P.: *Plastic Component Design*, Industrial Press, 1996.

Chabot, J.F.: *The Development of Plastics Processing Machinery and Methods*, Wiley, 1992.

Chanda, M., and S.K. Roy: *Plastics Technology Handbook*, 2d ed., Dekker, 1992.

Charrier, J.-M.: *Polymeric Materials and Processing*, Hanser, 1991.

Crawford, R.J. (ed.): *Rotational Moulding of Plastics*, Wiley, 1992.

Griskey, R.G.: *Polymer Process Engineering*, Chapman & Hall, 1995.

Lee, N.C. (ed.): *Plastic Blow Molding Handbook*, Van Nostrand Reinhold, 1990.

Levy, S., and J.F. Carley (eds.): *Plastics Extrusion Technology Handbook*, 2d ed., Industrial Press, 1989.

Malloy, R.A.: *Plastic Part Design for Injection Molding*, Hanser, 1994.

Michaeli, W.: *Extrusion Dies for Plastics and Rubber*, 2d ed., Hanser, 1992.

Morton-Jones, D.H.: *Polymer Processing*, Chapman & Hall, 1989.

Muccio, E.A.: *Plastic Part Technology*, ASM International, 1991.

Muccio, E.A.: *Plastics Processing Technology*, ASM International, 1994.

Potter, K.: *Resin Transfer Molding*, Chapman & Hall, 1997.

Pye, G.W.: *Injection Mold Design*, Longman-Wiley, 1989.

Rosato, D.V., D.P. DiMattia, and D.V. Rosato: *Designing with Plastics and Composites: A Handbook*, Van Nostrand Reinhold, 1991.

Rosato, D.V., and D.V. Rosato: *Injection Molding Handbook*, 2d ed., Chapman & Hall, 1995.

Shastri, R.: *Plastics Product Design*, Dekker, 1996.

Strong, A.B.: *Plastics: Materials and Processing*, Prentice Hall, 1996.

Xanthos, M. (ed.): *Reactive Extrusion*, Hanser, 1992.

Plastics have long been used for the body of automobiles but generally as skin panels attached to a load-bearing steel spaceframe. A possible future solution is shown by the Composite Concept Vehicle of Chrysler: the entire body is made from four fiberglass-reinforced PET thermoplastic moldings, saving mass and simplifying assembly onto a steel subframe. (*Courtesy Ticona, Summit, N.J.*)

chapter

15

Composites

Now we are ready to look at methods of combining the benefits of different materials to influence the bulk properties of composites that:

Make plastics competitive with steel

Allow building a single-piece boat hull

Increase the elastic modulus and strength of light metals

Construct wires that carry current without resistance.

The terminology of composites presents some difficulties. Composites are always formed to combine the benefits of two or more materials for improved strength, stiffness, corrosion resistance, electrical properties or simply appearance. By definition, composites are structures made of distinct materials, the identities of which are maintained even after the component is fully formed. Thus, there is always an interface between the two materials, and the properties of this interface will have a decisive influence on the properties of the composite. We saw in Sec. 6-3-2 that this is also the case for two-phase materials and, indeed, two-phase metals, ceramics, and polymeric blends are, in one sense, composites. However, their properties are affected by events on the atomic or molecular scale and, to distinguish them from composites in the generally accepted sense, we have to narrow our definition of composites to *structures formed on the macro scale*. What is "macro" depends on the particular system but is generally on the order of several micrometers.

Frequently, the desirable properties are obtained by applying a surface layer, and then one speaks of a *coating* or a *cladding*. Since this is sometimes done as the very last step in the manufacturing sequence, coatings will be discussed with surface treatments in Chapter 19; here we are concerned with *composites which are formed to influence the properties of the bulk* of a product.

15-1   CLASSIFICATION OF COMPOSITES

The material that surrounds the other component of the system is the matrix, and one speaks of metal-, ceramic-, or polymer-matrix composites. The embedded component may also be a metal, ceramic, or polymer; in terms of its effect, shape is of particular importance.

15-1-1   Particulate Composites

In these the embedded material is in the form of more or less equiaxed particles, and properties are nearly isotropic. This is true also of small lamellar (flake-like) or fibrous particles if randomly oriented. For simplicity, we call them *fillers*, even if they affect properties.

Rule of Mixtures   The mass m_c of the composite body is

$$m_c = m_f + m_m = \rho_f V_f + \rho_m V_m = \rho_c V_c \qquad \textbf{(15-1)}$$

where m is mass, ρ is density, V is volume, and the subscripts c, f, and m refer to the composite, filler, and matrix, respectively. The density of component is then:

$$\rho_c = \frac{\rho_f V_f + \rho_m V_m}{V_c} \qquad \textbf{(15-2)}$$

If we denote $V_f/V_c = f$ as the volume fraction of the filler,

$$\rho_c = f\rho_f + (1 - f)\rho_m \qquad \textbf{(15-3)}$$

This is the rule of mixtures: Each component contributes to the properties of the composite in proportion to its volume fraction.

15-1-2   Fiber Reinforcement

When the stronger or higher-modulus filler is in the form of thin fibers, strongly bonded to the matrix, properties depend on the properties and proportion of the fibers, and on their orientation relative to load application (Fig. 15–1). A minimum (critical) volume fraction of fiber must be present, and the fiber must be continuous or, if chopped, long enough for stresses to be transferred from the matrix to the fiber. An approximate *critical length* l_{cr} can be obtained from the condition that it should take more force to shear the matrix at the fiber boundary (to pull out the fiber) than to break the fiber:

$$l_{cr} = \frac{d(\text{TS})}{2\tau_i} \qquad \textbf{(15-4)}$$

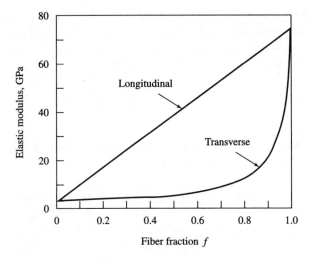

Figure 15–1   The elastic modulus of a unidirectional composite increases rapidly in the fiber direction but little is gained in the transverse direction. (The values are for the glass-fiber-reinforced epoxy of Example 15-1.)

where TS is tensile strength and d is diameter of fiber and τ_i is interfacial bond (shear) strength. In general, when the fiber has a length over $15l_c$, its effect is similar to that of a continuous fiber.

Unidirectional Composites (UDCs)   In these, the fibers are all aligned. When the load is applied parallel to the fibers, fiber and matrix are equally strained (*iso-strain condition*) and the elastic modulus follows the rule of mixtures. Thus, the composite elastic modulus E_c is

$$E_c = fE_f + (1 - f)E_m \qquad \textbf{(15-5)}$$

where f is again the volume fraction of fiber and E_f and E_m are the elastic moduli of the fiber and matrix, respectively. Accordingly, the elastic modulus (and flexural modulus) can be increased by incorporating into the matrix a material of higher modulus, and the increase is directly proportional to the fiber fraction (Fig. 15–1).

Tensile strength is subject to more complex events. Upon loading, a single fiber breaks first, and the consequences of this depend on the properties of the matrix and on local conditions. The aim is to neutralize the effect of local failure; for this, the matrix should have high enough ductility not to propagate a crack and should be able to carry the shear stress on the fiber-matrix interface. Even though a first estimate of strength can be made from the rule of mixtures, actual design is based on statistical treatment of properties.

When the fibers are oriented perpendicular to the load application, more fiber is needed for stiffening, and the transverse properties of oriented composites are poorer. The fibers transfer stresses to the matrix, and fiber and matrix are equally stressed (*isostress condition*). The modulus increases, if only slightly, according to

$$E_c = \frac{E_f E_m}{f E_m + (1 - f) E_f} \tag{15-6}$$

It takes a very large fiber volume to make a significant difference (Fig. 15–1) and the effect on tensile strength is even less.

Biaxial Composites   When the component is exposed to biaxial stresses (as a pressure vessel or a tube would be), the best results are obtained with biaxially oriented (cross-ply) fibers. Random orientation gives useful improvement (Fig. 15–2).

Laminates   An alternative to biaxially oriented fibers is to join unidirectional composites into a laminate of two or more (preferably, an odd number) layers. Plywood is a familiar example. Wood itself is an extremely complex natural unidirectional composite. On the microscopic scale, it is composed of tube-

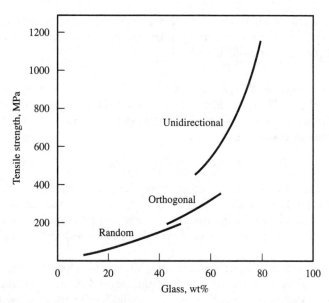

Figure 15–2   Orientation of fibers is an important factor in determining the properties obtained in polyester-glass fiber composites. [*After R. M. Ogorkiewicz, The Engineering Properties of Plastics (Engineering Design Guide No. 17), The Design Council, London, 1977. With permission.*]

like cells, the walls of which are made up of crystalline cellulose $(C_6H_{10}O_5)_n$ chains. Cellulose microfibers of 10–30-nm diameter and 2–4-mm length provide the skeleton which is enclosed in semicrystalline hemicellulose and bonded by a matrix of lignin. Alignment of cells gives the wood its high tensile strength in the axial (parallel) direction, and several layers combined in mutually perpendicular alignment give the superior biaxial properties of plywood. Laminates are often made up for other than load-bearing applications and the layer may then be a non-reinforced material. Bimetallic sheet and coextruded plastic sheet are examples. However, the properties of the bulk are not affected; thus, these products fail to meet one of our criteria of a composite and will be discussed in Sec. 18-12.

The toughness of reinforced structures is a complex question and depends greatly on the mechanism by which the matrix and reinforcement fail. A strong fiber, strongly bonded to a polymer or metal matrix, increases toughness because extra energy is required to break the fiber. We shall see, however, that bond strength must be optimized in ceramic-matrix composites to allow controlled pull-out of fibers. An advantage of composites is that fracture of one fiber does not impair the total structure; failure is often gradual, and repairs may be possible although often expensive.

The improvements can be gaged from data for reinforced plastics in Tables 13–2 and 13–3. The potential benefits relative to metals are shown in Fig. 15–3,

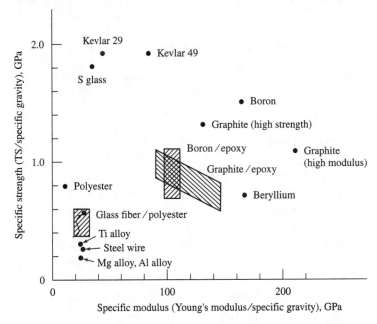

Figure 15-3   For a given weight, composite materials with fibers oriented in the direction of loading offer tremendous gains in strength and stiffness, as shown by their specific strength and modulus (shaded areas). For comparison, typical values are shown for bare fibers and for the highest-strength members of alloy groups.

where specific strength and stiffness are given for various composites. (In principle, strength and elastic modulus should be divided by density but, to keep units in GPa, they are divided here by specific gravity.)

15-2 REINFORCING FIBERS

A good reinforcing fiber has a high elastic modulus, high strength, low density, and is wetted by the matrix. It is often necessary to improve wetting by the application of coupling agents, coatings, or conversion treatment of the fiber surface.

The major types of fibers are shown in Table 15–1. Glass compositions are tailored for high electrical resistance (E glass), strength (S glass), or chemical resistance (C glass). Polymer fibers are primarily suitable for tensile applications. Graphite fibers are made by *pyrolysis* (heating in an inert atmosphere at 800–1200°C) of polymer, pitch, or rayon precursors. Boron and SiC monofilaments are made by chemical vapor deposition (Sec. 19-7) onto a tungsten substrate wire. *Whiskers* are single crystals of diameters on the order of micrometers and a length equal to tens or hundreds of diameters. Since they are free of dislocations, they have very high strength. Single-crystal platelets, less effective but still useful

Table 15–1   Properties of some reinforcing fibers

Fiber	Density, g/cm^3	Young's Modulus, GPa	Tensile Strength, GPa
High-strength steel	7.83	210	2.1
Tungsten	19.3	350	4.2
Beryllium	1.84	300	1.3
E glass	2.48	75	3.5
S glass	2.54	85	4.6
C glass	2.56		3.3
Alumina	3.15	320	2.1
SiC	3.0	400	2.8
Boron	2.6	420	4.0
Graphite			
High modulus	1.9	390	2.1
High strength	1.9	240	4.0
Kevlar* 29	1.44	83	2.8
Kevlar* 49	1.44	130	3.2

*Trademark for DuPont aramids.

for nondirectional toughening, are typically < 1 μm thick and have 10–30-μm diameter.

Fibers are of typically 2–30-μm diameter; the small diameter is helpful because it allows bending around small radii without exceeding the elastic limit [see Eq. (10-3a)]. Fibers are available in *filaments* (or *monofilaments*, i.e., very long, continuous single fibers), *yarn* (twisted bundle of filaments), *roving* (untwisted bundles of parallel-gathered filaments), *tows* (untwisted bundles of thousands of filaments), *woven fabrics* (made from filaments, yarn, or roving, woven at 90° angles to each other), *braids* (fibers woven into a tube instead of a flat fabric), *mats* (continuous fiber deposited in a swirl pattern or chopped fiber in a random pattern), *combination mats* (one ply of woven roving bonded to a ply of chopped-strand mat), *surface mats* (very thin, monofilament fiber mats for better surface appearance), *chopped fibers or roving* (of 3–50-mm length) and, in brittle materials, *milled fibers* (of 0.5–3-mm length). Glass fibers are most frequently used, at least in polymer matrixes, partly because of their low cost. If the cost of a unit weight of E glass is 1, the cost of aramid and carbon fibers is about 10, SiC 30, and boron over 100.

15-3   POLYMER-MATRIX COMPOSITES

By far the largest quantities of composites are based on polymers which otherwise have low strength, stiffness, and creep resistance. The proportion of reinforcement is seldom below 20% and may go as high as 80% in oriented structures. We already saw in Sec. 13-5 that many plastics contain fillers (which may simply increase the bulk but may also affect other properties) and/or reinforcing particles (added with the express purpose of improving mechanical properties). Conventionally, they are called reinforced or filled plastics, even though they may satisfy our definition of composites. Only when the fibers are long (or continuous), is the term composite always applied, and specific processes are then used for their manufacture (Fig. 15–4).

15-3-1   Polymers

The polymer is most often a thermoset. Cross-linkable polyesters are relatively inexpensive, easy to process, and are the dominant resins for intermediate (chopped-fiber) reinforcement. For higher-temperature applications, phenolic resins are suitable but develop a condensation by-product. If higher strength is also needed, epoxy or bismaleimide resins are used which polymerize without a by-product and give pore-free parts. For highest temperature, polyimides are employed. Processing temperatures are low, around the T_g of the prepolymer (20–175°C for epoxies but higher for polyimide). Processing times are long, often of several hours. Recycling is limited to the incorporation of reground material.

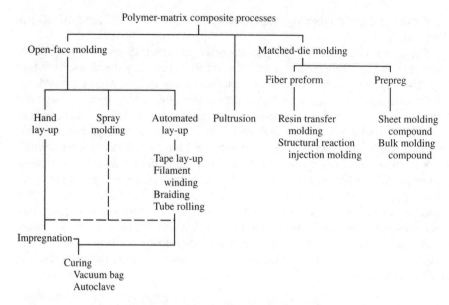

Figure 15–4   Classification of manufacturing processes for polymer-matrix composites. *(Adapted from J.A. Schey, ASM Handbook, vol. 20, Materials Selection and Design, ASM International, 1997, p. 701. With permission.)*

There is an increasing use of thermoplastics, primarily PP, polyesters, polyamides, and polysulfones, especially for reinforcement by short fibers. Unless formed in situ from monomers, their processing temperature is much higher (above T_m) and wetting of the fibers is more difficult than for thermosets. Cycle times are, however, shorter (on the order of minutes), defects can be healed, and full recycling is possible. Fillers are frequently incorporated in both thermoplastic and thermoset matrixes.

15-3-2   Application of Polymers

The polymer (or prepolymer) and reinforcement may be applied separately but large quantities of rovings, mats, etc. are sold also as *prepregs* (preimpregnated with the resin). They are in two basic forms:

Bulk molding compounds (BMC) of a putty-like consistency are made by premixing the polymer with fibers, fillers, colorants, initiators, catalysts, etc.

Sheet molding compounds (SMC) are prepared for large sheet-like parts. First, a mixture of resins (chiefly polyester, with some thermoplastics added for improved impact properties) is deposited onto a PE carrier film (with a doctor blade defining the thickness), then chopped fiber roving of 6–75 mm length is deposited on top of this, finally, another layer of resin mix—applied to a carrier film—is used as a cover. The sandwich is then compacted by rollers and is

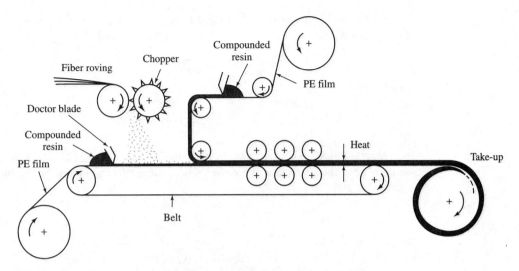

Figure 15–5   Sheet-molding compounds (SMCs) are made in large quantities for subsequent matched-die molding. The PE film is peeled off before molding.

heated to ensure complete wetting of the fibers and to begin thickening (Fig. 15–5). This SMC can be rolled up into coils and stored (if necessary, refrigerated) for some limited time. A similar processing sequence is used for unidirectional tape and for laminated tape composed of several layers, each with a specific orientation. Thermoset prepregs must be used before cross-linking begins and are often refrigerated for storage. Thermoplastic prepregs have the advantage of infinite shelf life. Critical issues are wetting and complete penetration of the reinforcement by the plastic. Lamination between rollers is an effective method. Further treatment of SMCs is similar to sheet metal. Blanks may be cut with steel rule dies (Fig. 10–10b), laser beam, or reciprocating knives used in the textile industry.

A unidirectional composite board is to be made of epoxy resin reinforced with E glass fiber. A longitudinal elastic modulus of 20 GPa is desired. Calculate (*a*) the proportion of glass fiber required by volume and (*b*) by weight, (*c*) the density of the composite, (*d*) the elastic modulus in the transverse direction, and (*e*) how much of the imposed load is carried by the fibers. The modulus of E glass is 75 GPa (Table 15–1) and of an unfilled epoxy is 3.0 GPa (Table 5–4). | **Example 15-1**

(*a*) From Eq. (15-5), $E_c = fE_f + (1 - f)E_m$. Thus $20 = 75f + (1 - f)3.0$ and $f = 0.236$ or 23.6% by volume.

(*b*) The density of epoxy is 1.22 g/cm^3, that of the glass 2.48 g/cm^3. In 100 cm^3 composite, there is $23.6(2.48) = 58.5$ g glass and $(100 - 23.6)(1.22) = 93.2$ g epoxy. Total weight is $58.5 + 93.2 = 151.7$ g. Of this, $58.5/151.7 = 0.3856$ or 38.56 wt% is glass and $93.2/151.7 = 0.6143$ or 61.43 wt% is epoxy.

(c) Density is 151.7 g/100 cm^3 = 1.517 g/cm^3. This could have been obtained directly from the rule of mixtures, Eq. (15-3): $\rho_c = f\rho_f + (1 - f)\rho_m$ and

$$0.236(2.48) + (1 - 0.236)(1.22) = 1.517 \text{ g/cm}^3$$

(d) Eq. (15-6)

$$E_c = \frac{E_f E_m}{f_f E_m + f_m E_f} = \frac{3.0(75)}{(0.236)(3.0) + (0.764)(75)} = 3.88 \text{ GPa}$$

thus, there is a minor increase in modulus.

(e) If the glass represents 23.6% of the volume, it also represent 23.6% of the cross-sectional area. The glass and epoxy are forced to extend together under loading, thus strain is equal in the two. From Eq. (4-5)

$$e = \frac{\sigma}{E} = \frac{P}{AE}$$

thus

$$\frac{P_f}{P_m} = \frac{A_f E_f}{A_m E_m} = \frac{0.236(75)}{0.764(3)} = 7.72$$

in other words, 88.5% of the load is carried by the 23.6% fiber.

15-4   FABRICATION OF POLYMER-MATRIX COMPOSITES

The fabrication method depends largely on the shape and size of the product and on the quantities required.

15-4-1   Open-Mold Processes

Only one mold (male or female) is needed and may be made of any material such as wood, reinforced plastic or, for longer runs, of sheet metal or electroformed nickel. Plaster molds can be broken up; thus, ducting with complex passages can be made. The surface of the part that contacts the mold reproduces the mold surface and can be very smooth.

Shaping   To allow separation of the part, a mold release agent (silicone, polyvinyl alcohol, fluorocarbon or, sometimes, plastic film) is first applied. For best surface quality, an unreinforced surface layer (*gel coat*) may then be deposited. The free surface is much coarser because the high shrinkage of plastics makes the reinforcement stand out.

1. In *hand lay-up* the resin and fiber (or pieces cut from prepreg) are placed manually, air is expelled with squeegees and, if necessary, multiple layers are built up. Hardening is at room temperature but may be speeded up by heating. Void volume is typically 1%. Foam cores may be incorporated (and left in the part) for greater shape complexity. Thus, essentially all shapes in Fig. 3–1 can be produced. The process is slow (deposition rate around 1 kg/h) and labor-intensive, and quality is highly dependent on operator skill.

2. Labor costs are lower in *spray-up*, with a spray gun supplying resin in two converging streams into which roving is chopped (Fig. 15–6). Automation with robots results in highly reproducible production.

3. High productivity (5 kg/h) is obtained with *tape-laying machines* which cut and lay the ply or prepreg under computer control and without tension; thus reentrant shapes may be made. Cost is about half of hand lay-up. All the preceding techniques are extensively used for products such as airframe components, boats, truck bodies, tanks, swimming pools, and ducts.

4. In *filament winding* spherical, cylindrical, and other shapes are produced by winding a continuous filament in a predesigned pattern and under constant tension onto a mandrel. Because of the tension, reentrant shapes cannot be produced. CNC winding machines with several (sometimes 7) degrees of freedom are frequently employed. The filament (or tape, tow, or band) is either precoated with the polymer or is drawn through a polymer bath so that it picks up polymer on its way to the winder. Void volume can be higher (3%). The cost is about half that of tape laying and productivity is high (50 kg/h). Pipes, tanks, and pressure vessels are made this way because the filament can be oriented. The angle in helical (Fig. 15–7a) and polar (Fig. 15–7b) winding is chosen in the direction of maximum stress; hoop winding (Fig. 15–7c) secures maximum circumferential strength. Cylindrical vessels are often made with several layers of different orientations. Carbon-fiber-reinforced rocket motor cases used for the Space Shuttle and other rockets are made this way. (The solid-propellant booster rockets of the Space Shuttle each have four cases of 3.6-m diameter and 9-m length.) More complex shapes can be formed at lower cost by the application of textile weav-

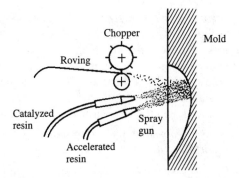

Figure 15–6   Spray-up is a low-cost method of creating medium-strength composite parts.

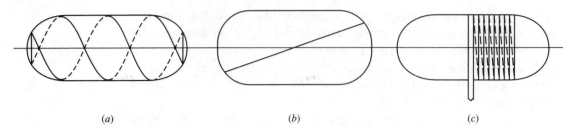

(a)   (b)   (c)

Figure 15–7   Filament winding offers the possibility of orienting the reinforcement in the direction of maximum stress by making (a) helical, (b) polar, or (c) hoop windings.

ing techniques that produce braiding in two- or three-dimensional configurations. Knit fabrics can be draped around complex contours.

Example 15-2	The fuselage of the Premier 1 business jet of Raytheon Aircraft is made in only two pieces of 1.8-m diameter. A 7-axis tow placement system places up to 24 carbon-fiber tows of 3-mm width simultaneously. The tows are fed through a head with segmented rollers which compress the tow into the work area. Several layers are built up and the control program feeds, cuts, and restarts the tows as needed to make window and door openings. The inner wall is first deposited on the mandrel, then honeycomb (Sec. 18-12) is placed by hand, the outer layer of carbon fiber ply is laid next, and finally a fabric layer with metal filaments completes the outer skin. Curing takes place in an autoclave. The carbon-fiber composite, in combination with the honeycomb structure, has a wall thickness of 21 mm compared to the 75 mm typical of aluminum construction; thus, it increases cabin volume substantially. [Source: *Manufacturing Engineering*, 1998(3):48.]

Curing   Even though curing often takes place in air, various techniques are employed for higher quality.

1. In *vacuum-bag molding* several layers are first applied on top of the composite. Among these, the *release film* is a porous, nonsticking film (cloth) removed at the end of the process; the *bleeder* (polyester, fiberglass, or cotton felt) absorbs excess resin; the *barrier* (a porous layer) prevents resin from blocking the vacuum lines. The whole assembly is then covered with a plastic film and, after sealing the edges, a vacuum is drawn; thus, curing takes place under atmospheric pressure, voids are reduced, and the free surface is smoother.

2. Higher pressures are possible in *pressure-bag molding*, in which pressure is applied to the free surface with a tailored rubber bag.

3. Highest properties are obtained with *autoclave molding*. The autoclave is a large (up to 30-m diameter by 50-m length) heated pressure vessel with gas circulation. Because of danger of explosion, nitrogen is preferred over air. As in vacuum-bag molding, the composite laminate is covered by layers designed to absorb excess resin and permit the escape of volatile by-products. For better control of thickness and definition of the free surface, metal plates (*caul plates*) may be placed on top of the assembly. The part, complete with mold, is then *bagged* so that entrapped gases can be removed by vacuum. Curing takes place under increased pressure and temperature (for example, 600 kPa and 175°C for epoxy).

15-4-2   Pultrusion

Fibers are impregnated with a prepolymer, exactly positioned with guides, pre-heated, and pulled through a heated, tapering die where curing takes place. The emerging product is cooled and pulled by oscillating clamps or a caterpillar (Fig. 15–8); small-diameter products are wound up. Two-dimensional shapes, including solid rods, profiles, or hollow tubes, similar to those produced by extrusion, are made, hence the name *pultrusion*. Production rates are around 1 m/min. Applications are to sporting goods (golf club shafts), vehicle drive shafts (because of the high damping capacity), nonconductive ladder rails for electrical service, and structural members for vehicle and aerospace applications.

15-4-3   Matched-Die Molding

For high production rates, close tolerances, and best surface finish, the part is made between two heated *matched dies*.

1. The resin and reinforcement are placed in the lower die and the die is closed in so-called *wet molding*.

2. BMCs are molded essentially by compression molding although, more and more, transfer and injection molding and RIM are also practiced. If the directionality resulting from injection is objectionable, glass flake is used for reinforcement. Several mass-produced automobiles and minivans have plastic skin parts—including side panels, tailgates, roofs, etc.—attached to a welded-steel, precision-machined, drivable space frame.

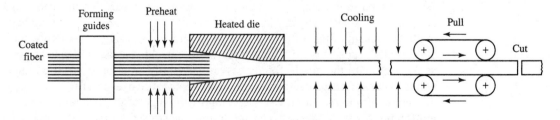

Figure 15–8   Continuous-fiber-reinforced two-dimensional shapes are produced by pultrusion.

Example 15-3 | Persons of limited mobility are helped to walk with long leg braces having a foot plate and long, thin uprights. They are typically made of steel and aluminum alloy. A carbon-fiber-reinforced nylon was chosen to make braces that are two-thirds lighter yet 40% stiffer and twice as strong. A significant advantage is that the thermoplastic allows postforming of the braces for custom fit. Long carbon fiber was cowoven with nylon, with carbon fiber in the warp (the lengthwise direction) and nylon fiber in the fill direction. Fully continuous fiber would break during postforming, therefore, the cowoven cloth was cut into 60-mm-long and 25-mm-wide pieces and laid end to end in overlapping layers. The braces were then formed by compression molding at 250°C and 20 MPa. [Source: *Adv. Mater. Proc.*, 1993(3):47–47.]

3. SMC sheets, cut to size, are formed by techniques similar to metal pressing, except that deformation is affected by the geometry of the reinforcement and pressures are low (7–10 MPa). Cycle time is 2–8 min and several molds are kept in circulation for reasonable production rates. The part is removed from the mold and final curing takes place outside the press, often in fixtures designed to prevent distortion. Structural parts usually need continuous reinforcement and care must be taken to avoid stretching that would result in fiber breakage.

4. Rubber plug (Fig. 10–32) and hydroforming (Fig. 10–33*a*) methods reduce die cost and permit more complex shapes because some of the forming stress is transferred to the rubber. Long sections and similar shapes may be made by roll forming (Fig. 10–16*c*).

5. In a variant of RIM, a fiber form is preplaced in the mold (*structural RIM* or *SRIM*, pronounced s-rim). *Resin transfer molding* (RTM) differs in that the already reacted resin is injected into the mold into which a fiber form has been placed. Both processes allow great complexity and reasonable production rates, hence the interest from the automotive industry. Penetration of the fiber form and wetting of the fibers are critical; therefore, special low-viscosity resins have been developed. Since completion of the part takes place in the mold, good surface quality and dimensional tolerances can be achieved.

15-4-4   Process Capabilities and Design Aspects

We have seen in the sections dealing with various processes that the size range and variety of shapes are very large. Rapid advances in computer modeling have made the design of parts for load-bearing applications possible. Generous radii are essential, especially with long-fiber reinforcement. Surface finish is as good as the surface of the mold. Tolerances are, generally, inferior to metal products but parts consolidation can make composites competitive, especially if matched-die molding is used. In all cases, the high shrinkage of plastics relative to the reinforcing fibers (Table 4–1) can set up internal stresses in the part which then lead to distortion. Angles close up; thus, a 90° angle becomes 87–89° ("negative springback" or "spring-in"). To compensate for this, the mold can be made to a larger angle. Differential shrinkage on the mold is another source of distortion. Most advantageous is to use a mold of the same thermal expansion coefficient. Carbon-fiber composites often have zero shrinkage and the mold is then made of a carbon-fiber composite or, for longer production runs, of Invar (Table 4–1).

Capital costs are highly variable. Investment is very little for hand lay-up and very high for multiaxis computer-controlled winding. Autoclaves are also capital-intensive installations requiring close monitoring and safety precautions. Matched-die molding dies are less expensive than metalworking dies and presses have lower rated capacity.

Most advantages of fiber reinforcement are lost when the bond between matrix and fiber is weak. Fibers pull out and delamination may occur at low stresses. Thus, strict process control is an essential prerequisite. Computer-aided modeling of infiltration helps in anticipating problems and in finding solutions. NDT techniques (primarily, ultrasonic scanning) are used for detecting defects such as improper fiber orientation, lack of bonding, delamination, resin-rich (weaker) and resin-starved (unbonded) regions. Processing of NDT signals by computer imaging and analysis allows 100% inspection of such large and critical parts as the carbon-fiber reinforced wing skin of the Harrier II jumpjet. Defects are difficult or impossible to repair in thermosets and parts have to be rejected at a point where most costs had already been incurred; therefore, efforts are made to develop in-process measurement techniques to reveal problems in deposition and curing. In aerospace applications the cost of quality assurance may equal the cost of production.

15-5   METAL-MATRIX COMPOSITES

The terminology is, again, diffuse. In this chapter we narrowed the definition of composites to structures in which bulk properties are affected. By this definition, many powder-metallurgy products are composites: Copper-infiltrated iron qualifies as a metal–metal composite; cermets (including cemented carbides) are metal-bonded ceramics, and metal bearings infiltrated with PTFE or nylon are metal–polymer composites. Yet, in everyday usage, none of these are regarded as composites; the term *metal-matrix composite* (MMC) is reserved for more recent combinations in which the embedded material is a fiber or ceramic powder. Most of the time, it is incorporated for higher strength but other purposes may be served as well.

The matrix is usually a low-density metal, primarily aluminum or magnesium, less frequently titanium. The reinforcement is most often SiC, Al_2O_3, or carbon, and is incorporated by one of several techniques. Ceramic and metal powders may be mixed and processed directly by powder metallurgy techniques (Chap. 11). For example, composites of TiC in a Ti-6Al-4V matrix are made by cold isostatic pressing, followed by sintering and HIP. Alternatively, the composite is produced in the form of flakes obtained by rapid solidification technology, particularly, melt spinning (Sec. 11-2-3), and is then consolidated. Some composites are processed by casting into ingots (Sec. 7-5-2) which are worked like a metal (Chap. 9).

Many applications are to shape casting in which the composite melt is directly cast into final shape. Particulate reinforcement is pushed ahead of growing dendrites, and control of solidification is important for proper distribution of particulates. Alternatively, a fiber preform is placed into the mold and infiltrated by the metal. In some applications, penetration is aided by pressure or vacuum, as in the squeeze casting of pistons, but in other instances infiltration is instantaneous, as in permanent mold casting. Wetting of fibers is crucial and is assured by various proprietary treatments. Graphite is not wetted by metals, therefore, the fiber is first coated with a metal, metal carbide, or SiO_2 layer; the preform is then readily infiltrated by molten metal. In some applications, as in casting pistons, the main benefits of composites are increased elevated-temperature strength and reduced thermal expansion.

Example 15-4

A 356 Al alloy (7Si-0.3Mg-0.35Zn-0.25Ti), reinforced with SiC particles, can be investment cast by conventional techniques. In the fully heat-treated (T6) condition it gave the following results:

	YS, MPa	TS, MPa	el., %	E, GPa	Thermal exp., μm/m
356	200	255	4	75	21.4
20% SiC	296	317	0.5	96	16.4

The low thermal expansion is of benefit for higher-temperature applications and also for substrates and supports of electronic components; thus, benefits can be had at little expense. [Data from D.O. Kennedy, *Adv. Mater. Proc.*, 1991(6):42–46.]

A special class of MMCs is *in-situ composites*. We have already come across several examples, such as eutectics, rapidly solidified supersaturated alloys, and the hypereutectic Al–Si alloy in which primary silicon crystals are embedded in a eutectic matrix. More generally, the term is applied to all multiphase materials in which *the reinforcing phase is synthesized within the matrix during composite fabrication*. Such composites can be made, for example, by exothermic reactions (hence the term XD = *exothermic dispersion*); the second phase thus formed is very stable at elevated temperatures. A prime example is TiB_2 dispersion in a matrix of titanium aluminide (Ti_3Al and $TiAl$) intermetallic. The material has favorable high-temperature properties at only half the density of the superalloys it may replace.

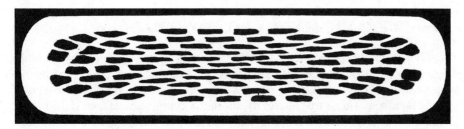

Figure 15–9   A PbBiSrCaCu oxide is encased in silver tubes to form a filament which, when combined into a multifilament wire and cooled below its critical temperature, conducts a hundred times more current than a copper wire of the same cross section, and does so without resistive losses. (*Courtesy American Superconductor, Westborough, Massachusetts.*)

Superconductor wires have numerous continuous filaments of the superconducting material embedded in a metal matrix (Fig. 15–9). Their manufacture presents several technical challenges.

Low-temperature superconductors operate in liquid helium at 4.2 K. The superconducting material is a ductile Nb–Ti alloy or a brittle Nb_3Sn intermetallic in a copper matrix. Round rods of the superconducting material are inserted into copper tubes with a hexagonal outer surface, which are then assembled inside a large tube into extrusion billets of typically 305-mm diameter. Hot extrusion produces an 85-mm diameter bar which is cold drawn, through dozens of draws, to 0.68-mm diameter. By this time, each superconducting filament is reduced to 10-μm diameter. Several of these wires may then be combined and further deformed to obtain wire with thousands of 1-μm filaments, which is then used for such products as magnets of MRI units. (Source: B.A. Zeitlin, *ASM Handbook*, vol. 14, pp. 338–342.)	**Example 15-5**

We saw in Sec. 4-9-3 that ceramics can be high-temperature superconductors. For practical applications, long wires or tapes are needed with sufficient ductility to allow winding on cores. In the oxide-powder-in-tube (OPIT) technique a calcined oxide is placed into a conductive metal (usually Ag) tube which is then reduced by processes such as swaging, drawing or, for tapes, rolling. A multitude of such metal-ceramic composite wires is then united into a conductor wire. A high degree of crystallographic alignment of the oxide is needed for superconductivity at high current densities, and this is obtained by a thermomechanical processing sequence using intermediate or final heat treatments. One promising application is to SMES (superconducting magnetic energy storage) units in which current flowing in a (low-temperature) superconductive coil generates an intense magnetic field to store electricity with very little loss. Connected to a power source, the SMES releases its energy in a few milliseconds, sufficient to maintain voltage during momentary brownouts which could cause massive disruptions in chemical process industries. The efficiency of these units has been increased with OPIT current leads which reduce heat leak. (Source: A.P. Molozemoff et al., *Applied Superconductivity Conference*, Palm Desert, California, 1998.)	**Example 15-6**

15-6   CERAMIC-MATRIX COMPOSITES

In polymer-matrix composites the reinforcement is always stronger and of much higher elastic modulus than the matrix; hence, a significant increase in strength can be achieved by transferring stresses from the matrix to the fiber through a strong interface. This is true also of metal-matrix composites, at least if the metal has a low elastic modulus, as Al, Mg, and even Ti have. Ceramics (except glasses) have high elastic moduli, close to that of potential reinforcing fibers. Hence the purpose of *ceramic-matrix composites* (CMCs) is generally that of increasing toughness.

A monolithic ceramic fails in a completely brittle, catastrophic mode (Fig. 15–10). Reinforcements have little effect on the modulus (they may even reduce it) and failure in the matrix may set in earlier. The difference is that failure is not catastrophic and is retarded by several mechanisms:

Microcracks. We saw in Sec. 12-1-2 that microcracks retard major cracks. Such microcracks are caused by a *slight* mismatch in thermal expansion coefficient (a large mismatch would cause fiber breakage).

Crack deflection. A crack meeting the reinforcement is deflected along the interface where energy is used to effect separation (Fig. 15–10, *A*).

Crack propagation barrier. The reinforcement can force the crack front to bow out and thus increase the stress needed for propagation (Fig. 15–10, *B*; the effect is similar to that of obstacles to dislocation propagation, Sec. 6-3).

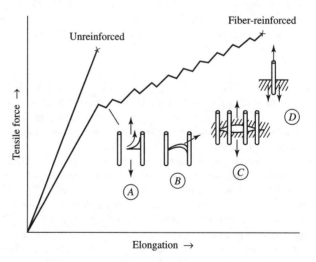

Figure 15–10   Increased toughness in fiber-reinforced ceramics is obtained by (*A*) crack deflection, (*B*) crack bowing, (*C*) fiber bridging, and (*D*) fiber pull-out.

Fiber bridging. If the geometry is favorable, fibers bear the load across a crack, putting the crack in compression (Fig. 15–10, *C*).

Fiber pull-out. Most importantly, energy is used up in pulling a fiber out of the matrix (Fig. 15–10, *D*). For this reason, *interfacial bond strength must not be so high that the fiber would break rather than pull out.*

Even this brief review of toughening mechanisms makes it evident that the processing of CMCs presents many challenges. The reinforcing material is usually an oxide (primarily Al_2O_3), carbide (primarily SiC), or nitride (e.g., Si_3N_4). Typical fiber diameters are 3–30 μm; whiskers are also used. Bond strength to the matrix is controlled (usually degraded) by coatings, such as carbon for temperatures below 400°C and boron nitride for higher temperatures.

Short-fiber composites are consolidated by typical ceramic processing techniques, primarily by HIP. Long fibers may be coated with a slurry of the matrix mixed with an organic binder; after consolidation into the desired shape, the binder is burned out and final consolidation occurs by HIP. Alternatively, the fiber structure is infiltrated with molten ceramic, or the ceramic is formed by reactions of starting materials. Most of these processes are highly specific and expensive, justifiable mostly for high-value aerospace applications.

Glass-Matrix Composites are based on particulates (primarily, Al_2O_3) or fibers (primarily, SiC) dispersed in the glass.

An alkali-borosilicate glass reinforced with SiC fiber has greatly increased strength, toughness, and thermal shock resistance. Therefore, it can be used for supporting hot glass parts in the production of glass tableware. | **Example 15-7**

	E, GPa	Flexural Strength, MPa	Work of Fracture, J/m^2	Thermal Shock Resistance, °C
Glass	63	30–50	120	> 100
40% SiC	110	450	35 000	> 530

[Source: W. Beier and S. Markman, *Adv. Mater. Proc.*, 1997(12):37–40.]

Carbon–Carbon Composites CCCs have low density (1.9 g/cm^3) combined with high heat capacity and thermal shock resistance. They were developed for applications where high heat is suddenly generated, as on nose tips of rockets and brakes of aircraft and racing cars. For the latter, the high coefficient of friction is also beneficial. The matrix is carbon formed by the pyrolysis of a thermosetting resin or tar, often under pressure, or by infiltrating the fiber preform with a carbonaceous gas (*chemical vapor infiltration*). Because carbon oxidizes rapidly above 400°C in air, surface coatings must be applied to the finished part.

15-7 SUMMARY

Composite structures greatly expand the scope of application of all engineering materials by combining the desirable properties of two or more materials. The identities of the constituent materials are retained, the properties of the interface are controlled, and the bulk properties of the composite are affected. The embedded phase may be a particulate or, for controlled oriented properties, a fibrous structure.

1. Plastics particularly benefit from the increased strength, stiffness, toughness, and creep resistance imparted by fibers and other reinforcing fillers, and components and assemblies of exceptional strength-to-weight and stiffness-to-weight ratios can be produced. Higher operating temperatures may also be permissible. The matrix is a thermoset or, increasingly, a thermoplastic. Open-mold processes are capable of producing large parts of complex shape, while matched-die molding offers better dimensional control and higher production rates and can be competitive with metal forming in medium-volume and, for certain parts, even in mass production. Cosmetic repairs are relatively easy but structural repairs can be expensive and not always reliable.

2. Metal-matrix composites are based chiefly on light metals (Al, Mg, less frequently, Ti) and the reinforcements are mostly ceramics or carbon. A special combination is a superconducting ceramic in a metal matrix.

3. Ceramic-matrix composites are different in that the interfacial bond is kept intentionally weaker to allow fiber pull-out as the main toughening mechanism. Carbon–carbon composites serve special situations where heat is generated at high rates.

4. The processing of composites involves safety concerns and health hazards typical of the plastics, metal, and ceramics industries, and appropriate measures must be taken for the protection of the worker and the environment.

PROBLEMS 15A

15A-1 List at least three classes of fibers that can be used for polymer-matrix composites.

15A-2 Define what is meant by critical fiber length.

15A-3 Draw a diagram showing the elastic modulus of a unidirectional continuous-fiber-reinforced composite as a function of fiber proportion. Draw lines for (*a*) longitudinal loading and (*b*) transverse loading. (*c*) Add a line for random chopped fiber of critical length.

15A-4 Define gel coat.

15A-5 Define pultrusion.

15A-6   Define the terms (a) prepreg, (b) bulk molding compound, and (c) sheet molding compound.

15A-7   Suggest three ways of making a simple open boat hull of glass-fiber-reinforced plastic.

15A-8   State the essential steps in hand lay-up of a plastic reinforced with chopped glass fiber (state also, generically, the polymer to be used).

15A-9   (a) What is an autoclave and (b) what is it used for?

15A-10   Define (a) structural RIM and (b) RTM.

15A-11   Define whisker (a) geometry, (b) structure, and (c) possible materials.

15A-12   Is the highest possible fiber-to-matrix bond strength beneficial for (a) polymer-matrix, (b) metal-matrix and (c) ceramic-matrix composites? Justify your answers.

15A-13   Explain the difference between a composite and a clad product, on the basis of definitions adopted in this text.

PROBLEMS 15B

15B-1   Continue Prob. 15A-3 and add a line for a composite with 90° woven fabric, loaded (a) in the fiber direction and (b) 45° to the fiber direction. (c) Explain your answer to (b) by showing the deformation of the fabric under loading.

15B-2   To explain the difference between a filled plastic and a polymer-based composite, formulate a definition for each.

15B-3   In producing a particular component, either a thermoplastic or thermoset prepreg may be used. The customer needs quick delivery of the finished product but cannot tell when. Which prepreg would you chose and why?

15B-4   Reinforcing fibers generally used are made of metal, glass, or polymer. Most of them are drawn or stretched (attenuated) to reduce their diameter. (a) Name the process and state the usual temperature range for drawing each of the three classes of materials. State the structural source of strengthening, if any, in drawing (b) metal, (c) glass, or (d) polymer fibers.

15B-5   A 1000-liter cylindrical tank subject to some internal pressure is made of a fiber-reinforced polymer. The designer made the ends flat. Consider the method of making the tank and suggest design changes if necessary. Clarify with sketches.

15B-6   A polymer is reinforced with 20–50% glass in the form of (a) powder, (b) flakes, (c) randomly oriented, milled fibers of 10:1 aspect ratio, (d) unidirectionally oriented, chopped fibers of 100:1 aspect ratio, and (e) continuous unidirectional fibers. Draw graphs showing your qualitative judgment on the increase in yield strength to be expected relative to the matrix for each composite in the (f) longitudinal and (g) transverse direction. Justify.

15B-7   A unidirectional continuous-fiber-reinforced composite is loaded in the fiber direction. Compared to the fiber, the matrix is very weak. A colleague argues that, because of this, bond strength between fiber and matrix is immaterial. Do you agree? If not, justify.

15B-8   A ski is to be made with a fiberglass-reinforced polymer core. From basic engineering considerations, would you choose (a) longitudinal, (b) transverse, (c) random, (d) longitudinal/transverse or (e) 45° crossply? Justify your choices. (Hint: It is desirable that the ski should resist twisting while allowing longitudinal flexing.)

15B-9   Surface treatment of carbon fibers is an important step in making composites with them. State the primary purpose for (a) metal-matrix and (b) carbon-matrix composites.

PROBLEMS 15C

15C-1   We saw in Example 15-4 that 20% SiC particles improve the properties of 356 Al casting alloy. Calculate the expected improvement in (*a*) thermal expansion (take 4×10^{-6}/K for SiC), and (*b*) elastic modulus. (*c*) Compare to the reported data and comment on the agreement or lack of it.

15C-2 Some cutting tool inserts are made of SiC-reinforced alumina. Using data from this book and considering the loading mode in metal cutting, determine if increased strength or increased elastic modulus is the main incentive. If neither, suggest a reason for making such inserts.

15C-3 Check data from Example 15-7 to see whether the increase in elastic modulus obeys Eq. (15-5) or Eq. (15-6).

15C-4 Derive Eq. (15-4) for critical fiber length. (Remember that the fiber is pulled from two directions!)

15C-5 An epoxy-matrix composite is to be made with E glass fiber of 10-μm diameter. Calculate the critical length if $\tau_i = 30$ MPa.

15C-6 A polyimide-matrix composite is reinforced unidirectionally with 30% high-modulus carbon fiber. What elastic modulus is to be expected in the fiber direction and transverse to it?

15C-7 Calculate and plot the change in longitudinal and transverse elastic modulus for 20% to 80% unidirectional continuous fiber if the ratio of fiber/matrix elastic modulus is 100:1.

15C-8 Repeat problem 15C-7 for a fiber/matrix elastic modulus ratio of 100:20. From the comparison, draw conclusions on the importance of the modulus of the matrix.

15C-9 An aluminum mold is used for making a 500-mm-long, 300-mm-wide carbon-fiber-reinforced epoxy resin composite tray. Reinforcement is unidirectional in the length of the tray. The thermal expansion coefficient of the composite is close to zero in the fiber direction and 19×10^{-6}/K in the transverse direction. In the autoclave, the epoxy solidifies (reaches gel temperature) at 120°C. Calculate the length and width of the mold, if room temperature is 20°C.

FURTHER READING

ASM Engineered Materials Handbook, vol. 1, *Composites*, ASM International, 1987.

Belitskus, D.L.: *Fiber and Whisker Reinforced Ceramics for Structural Applications*, Dekker, 1993.

Buchanan, R.C.: *Ceramic Materials for Electronics*, 2d ed., Dekker, 1991.

Chawla, K.K.: *Composite Materials*, 2d ed., Springer, 1998.

Clyne, T.W., and P.G. Withers: *An Introduction to Metal Matrix Composites*, Cambridge University Press, 1993.

Gutowski, T.G. (ed.): *Advanced Composites Manufacturing*, Wiley, 1997.

Hollaway, L.: *Handbook of Polymer Composites for Engineers*, Woodhead, Cambridge, 1994.

Jang, B.Z.: *Advanced Polymer Composites: Principles and Applications*, ASM International, 1994.

Mallick, P.K. (ed.): *Composites Engineering Handbook*, Dekker, 1997.

Mallick, P.K.: *Fiber-Reinforced Composites: Materials, Manufacturing, and Design*, 2d ed., Dekker, 1993.

Patter, K.: *An Introduction Composite Products: Design, Development and Manufacture*, Chapman and Hall, 1997.

Schwartz, M.M.: *Composite Materials Handbook*, McGraw-Hill, 1992.

Schwartz, M.M.: *Handbook of Composite Ceramics*, McGraw-Hill, 1992.

Schwartz, M.M.: *Joining of Composite Matrix Materials*, ASM International, 1994.

Strong, A.B.: *Fundamentals of Composites Manufacturing*, Society of Manufacturing Engineers, 1989.

Woishnis, W.A.: *Engineering Plastics and Composites*, 2d ed., ASM International, 1993.

Machining processes create a shape by removing material; the chip produced represents a loss but complex shapes can be made to close tolerances and excellent surface finish. (*Courtesy Kennametal, Latrobe, Pennsylvania.*)

chapter

16

Machining

We will now change our focus and, instead of forming or consolidating material, will look at creating shapes by removing material in the form of chips. For this, we will examine:

The idealized process of chip formation in metals

Chip formation in real materials

Factors affecting the quality of the machined part

Cutting tools and advances in tool materials

Methods by which the basic process of chip formation is put to use

Optimizing process conditions

Machining with abrasive processes

Design of parts for machining.

In the processes discussed so far, the shape of the workpiece was obtained by the solidification, plastic deformation, or consolidation of the material. The amount of material lost in scrap was relatively small, and the scrap particles tended to be large enough and relatively easily separated by alloy type, allowing easy and economical recycling. In contrast, machining aims to generate the shape of the workpiece from a solid body—or to improve the tolerances and surface finish of a previously formed workpiece—by removing excess material in the form of chips. Machining is capable of creating geometric configurations, tolerances, and surface finishes often unobtainable by any other technique. However, machining removes material which has already been paid for, in the form of relatively small particles that are more difficult to recycle and can easily become mixed. Therefore, developments often aim at reducing or—if at all possible—eliminating machining, especially in mass production. For these reasons, machining has lost some important markets, yet, at the same time, it has also been developing and growing and, especially with the application of computer numerical control, has captured new markets. With increasing metal removal rates, it has sometimes become more economical to "hog out" a part from a solid extruded o rolled blank rather than machining a forging or casting, especially in low- to medium-volume productior

If absolutely essential, a machining process can be found for any engineering material, even if it ma be only grinding or polishing. Nevertheless, economy demands that a workpiece should be machinabl to a reasonable degree. Before the concept of machinability can be explored, it is necessary to identify basic process, that of metal cutting. *Machining* is a generic term, applied to all material removal, whil *metal cutting* refers to processes in which the excess metal (or alloy) is removed by a harder tool, throug a process of extensive plastic deformation or controlled fracture. We shall see in Chap. 17 that at lea some of the phenomena observed in metal cutting are relevant also to the machining of other materials

16-1   THE METAL-CUTTING PROCESS

The variety of metal-cutting processes is very large; nevertheless, it is possible to idealize the process of chip removal.

16-1-1   Ideal Orthogonal Cutting

As indicated by its name, in *orthogonal cutting* the cutting edge of the tool is straight and perpendicular to the direction of motion (Fig. 16–1a). In the simplest case, the workpiece is rectangular and is of large enough width w for width changes to be neglected (plane strain). Cutting is performed with a tool inclined at a *rake angle* α, measured from the normal of the surface to be machined. To prevent excessive rubbing on the machined surface, the tool is relieved at the back or *flank* by the *clearance angle* θ.

In principle, it makes no difference whether the tool or workpiece is moved. We may visualize a stationary workpiece, with the tool moving at a *cutting speed* v. The tool is set to remove a layer of thickness h. To avoid confusion, this is *not* called the depth of cut, but rather the *undeformed chip thickness* h. In the simplest case, deformation takes place by intense shearing in a plane, the *shear plane*, inclined by the *shear angle* ϕ. The chip thus formed has a thickness h_c. The shear angle ϕ determines the *cutting ratio* r_c

$$r_c = \frac{h}{h_c} = \frac{l_c}{l} \qquad \qquad \textbf{(16-1)}$$

where l is the cut length and l_c is the length of chip. Frequently, the reciprocal value of r_c, called the *chip-compression factor*, is quoted. Both can be obtained from measured chip thickness or, if the chip is ragged and uneven, from the measured length l_c or, if the width of the chip has changed, from the weight of a chip of measured length.

In practice, the chip is always thicker than the undeformed chip thickness and $r_c < 1$. With a decreasing rake angle, ϕ and, thus, r_c also drop and become particularly low when cutting with a tool of negative rake angle (Fig. 16–1b). The

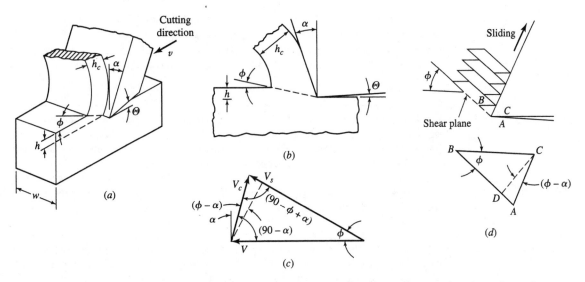

Figure 16–1   In orthogonal cutting the cutting edge is perpendicular to the direction of motion. The rake angle may be (a) positive or (b) negative. (c) Velocities are obtained if, to a first approximation, (d) chip formation is visualized as simple shearing.

value of r_c gives valuable clues regarding the efficiency of the process. From the geometry of the process, the shear angle is defined by

$$\tan \phi = \frac{r_c \cos \alpha}{1 - r_c \sin \alpha} \qquad \textbf{(16-2)}$$

Because of constancy of volume, the chip-thickness ratio can be expressed also from the chip velocity v_c and cutting velocity v (Fig. 16–1c)

$$r_c = \frac{v_c}{v} = \frac{\sin \phi}{\cos(\phi - \alpha)} \qquad \textbf{(16-3)}$$

thus, *with increasing shear angle ϕ, the chip becomes thinner and comes off at a higher speed.*

In the ideal case all shear is concentrated in an infinitely thin shear zone. The shear strain γ is (Fig. 16–1d)

$$\gamma = \frac{AB}{CD} = \frac{AD}{CD} + \frac{DB}{CD} = \tan(\phi - \alpha) + \cot \phi \qquad \textbf{(16-4)}$$

and, expressed as natural strain, drops from 5 to 2 as ϕ increases from $10°$ to $35°$.

For an infinitely thin shear plane, the shear strain rate $\dot{\gamma}$ would reach infinity. In reality, the shear plane has some finite thickness Δy, typically 0.03 mm (0.001 in), and the shear strain rate can be calculated from

$$\dot{\gamma} = \frac{v_s}{\Delta y} = \frac{\cos \alpha}{\cos(\phi - \alpha)} \frac{v}{\Delta y} \qquad \textbf{(16-5)}$$

It can easily reach values of thousands per second.

The magnitude of the shear angle is of fundamental importance. For any given undeformed chip thickness, a small angle means a long shear plane and, therefore, a high cutting force and energy; also, a small angle results in a high shear strain, hence the chip will be heavily strain-hardened.

Example 16-1	AISI 4340 steel, heat-treated to HB 270, is turned with high-speed steel (HSS) tools of $\alpha = 8°$ rake angle. The undeformed chip thickness is 0.3 mm, depth of cut (chip width) is 1.5 mm, and cutting speed is 0.6 m/s. The chip comes off in a continuous helical form; a 1-m-long chip weighs 5.7 g; the chip width is unchanged at 1.5 mm. Calculate the chip-thickness ratio. Volume of chip = weight/density = 5.7/7.9 = 0.72 cm^3. Chip thickness h_c = 0.72/(100)(0.15) = 0.048 cm = 0.48 mm. Chip-thickness ratio, from Eq. (16-1), r_c = 0.3/0.48 = 0.625 (chip compresson ratio =1.6).

Example 16-2	In the course of cutting the steel of Example 16-1 the chip changes to a straight type. A repeat of calculations shows that r_c is now 0.5. What is the new shear angle and chip velocity?
	    Shear angles, from Eq. (16-2): for $r_c = 0.625$,
	$\tan \phi = 0.625(0.9903)/[1 - (0.625)(0.1392)] = 0.678$ and $\phi = 34°$; for $r_c = 0.5$, $\phi = 28°$.
	    Chip velocity changed, according to Eq. (16-3), from $v_c = 0.625(0.6) = 0.375$ m/s to $v_c = 0.5(0.6) = 0.3$ m/s. Thus, the change in chip shape was accompanied by a decreasing chip velocity.

16-1-2   Forces in Cutting

The magnitude of the shear angle ϕ depends on the relative magnitudes of forces acting on the tool face (Fig. 16–2). Three views may be taken of the situation:

1. With the aid of a dynamometer (a device containing several load cells to resolve forces acting in mutually perpendicular directions), *the external forces acting on the tool holder* can be measured. In orthogonal cutting there are two forces: the *cutting force P_c* is exerted in the direction of cutting, parallel to the surface of the workpiece, while a *thrust force P_t*, acting perpendicular to the workpiece surface, is necessary to keep the tool in the cut. (At high positive rake angles, the thrust force is negative, and the tool is pulled into the material.)

2. From the point of view of *forces acting on the tool*, the *resultant force P_R* may be regarded as being composed of the normal force P_n acting perpendicular to the tool face and the friction force F acting along the face. Their magnitude may be calculated from measured forces and the rake angle

$$P_n = P_c \cos \alpha - P_t \sin \alpha \qquad \text{(16-6a)}$$

$$F = P_c \sin \alpha + P_t \cos \alpha \qquad \text{(16-6b)}$$

A simple way of interpreting the results would be in terms of a model that postulates sliding of the chip along the rake face of the tool. The magnitude of

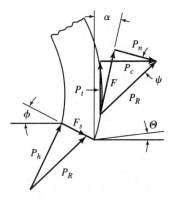

Figure 16–2   Forces acting on the chip, tool, and tool holder can be readily resolved.

the friction force F would then determine a *friction angle* ψ for which

$$\tan \psi = \frac{F}{P_n} \quad (= \mu)$$

(16-7)

Alternatively, from measured forces,

$$\tan \psi = \frac{P_t + P_c \tan \alpha}{P_c - P_t \tan \alpha}$$

(16-8)

This presentation can, however, be misleading because with decreasing friction F often drops less rapidly than P_n and μ may actually rise. Therefore, we call μ the *apparent mean coefficient of friction*.

3. From the point of view of *forces acting on the material*, the resultant force P_R can be resolved into a shear force F_s acting in the plane of shear and a compressive force P_h that exerts a hydrostatic pressure on the material being sheared. We saw in Sec. 4-1-5 that hydrostatic pressure does not affect the flow stress of the material but does delay fracture; thus, in a reasonably ductile material, the chip can form in a continuous fashion even though strains are high. The balance of the two forces depends on the shear angle ϕ. Clearly, with increasingly positive rake angle α, ϕ must increase and the chip must thin out. Friction on the rake face has the opposite effect: it resists the free flow of the chip and reduces ϕ. Thus, in contrast to most other metalworking processes, the geometry of the cutting process depends not only on tool/workpiece geometry but also on the process itself. To find a quantitative relationship, the assumption may be made that the material will choose to shear at an angle that minimizes the required energy.[1] This leads to

$$\phi = 45° - \frac{1}{2}(\psi - \alpha) \qquad \textbf{(16-9a)}$$

Another approach based on upper-bound analysis[2] leads to a qualitatively similar result,

$$\phi = 45° - (\psi - \alpha) \qquad \textbf{(16-9b)}$$

Thus, the shear angle decreases and the shear force (and with it the work of cutting) increases with a decreasing rake angle (by approximately 1.5% for each degree change) and an increasing friction angle. One may conclude that favorable conditions, in terms of energy consumption, could be secured by using large positive rake angles and by minimizing friction along the tool face.

We may now return to the thrust force P_t. It is important for several reasons. First, it causes deflection of the workpiece, and this makes support of long, slender parts necessary; if the part has thin walls, support may be difficult and the thrust force must be minimized. Second, it causes deflection of the tool and tool holder, and this impacts directly on achievable tolerances. Third, it can vary as a function of cutting conditions, and this leads to variable dimensions and also instabilities. While thrust force can be directly measured, it is useful to consider how cutting conditions influence it. From Fig. 16–2

$$P_t = P_R \sin(\psi - \alpha) \qquad \textbf{(16-10a)}$$

$$P_t = P_c \tan(\psi - \alpha) \qquad \textbf{(16-10b)}$$

thus, for a given cutting force P_c (or given resultant force P_R), the thrust force decreases when friction (ψ) decreases or the rake angle (α) becomes more positive. When $\psi < \alpha$, the tool is drawn into the workpiece.

16-1-3 Realistic Orthogonal Cutting

While the above conclusions are qualitatively correct, the model is oversimplified in many ways.

The Cutting Zone   In the idealized view, we regarded metal cutting as proceeding by shear in an infinitely narrow zone (Fig. 16–1d). Reality is different (Fig. 16–3).

1. Most metals strain-harden when deformed. The shear plane broadens into a *shear zone* (usually denoted as *primary shear zone*). The thickness of the shear zone is greater for a more heavily strain-hardening material (larger n) and also for a material of high strain-rate sensitivity (high m). The situation is complicated because the energy expended in shear raises the temperature and reduces the flow

---

[1]H. Ernst and M.E. Merchant, in *Surface Treatment of Metals*, American Society of Metals, New York, 1941, pp. 299–378.
[2]E.H. Lee and B.W. Shaffer, *Trans. ASME J. Appl. Mech.* **73**:405–413, 1951.

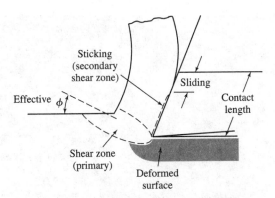

Figure 16–3   Chip formation is quite complex in reality. The shear zone broadens and extends ahead of the cut and a secondary shear zone forms on the rake face of the tool.

stress. In general, with a higher n the shear zone becomes wider and longer, and energy consumption increases. Strain hardening extends ahead of and below the shear zone and the newly formed surface is also strain-hardened.

2. Interface pressure on the tool face is high. We saw in Sec. 8-2-3 (Fig. 8–16) that sliding of the workpiece over the tool surface is arrested (sticking friction sets in) when the product of interface pressure and coefficient of friction exceeds the shear flow stress of the deforming material ($\mu p > \tau_f$). Therefore, the coefficient of friction is really meaningless. Since there is no movement on the tool face, the chip must flow up, in the *secondary shear zone*, over the stationary material found next to the rake face. This intense shearing is a second source of heating. Sliding contact is limited to a short distance (typically, less than 30% of the total contact) where the chip begins to curl away (Fig. 16–3).

3. Around the tool edge, conditions become even more complex. The workpiece material is upset and *plowed* by the tool edge, and rubbing against the freshly formed metal surface (essentially, a slight ironing) creates a third zone of heat input (sometimes termed the *tertiary shear zone*).

Chip Formation   Realistic metal cutting also differs from ideal cutting in the mode of chip formation.

1. In the ideal case, the shear zone is well defined, primary shear can be assumed to take place on closely placed shear planes (Fig. 16–1d) and a *continuous chip* is formed (Fig. 16–4a). This situation is approximated under various process conditions:

a. At moderately low speeds, in the presence of a cutting fluid which finds access to both the rake and flank faces and acts as a lubricant, the chip slides on the rake face. The newly formed surface is smooth (Fig. 16–5b), as is the

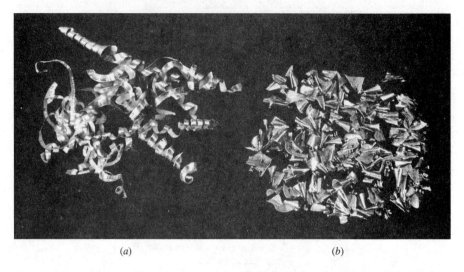

Figure 16–4   Chips are (a) continuous, straight or helical in cutting ductile materials but (b) short, broken up in cutting free-machining metals.

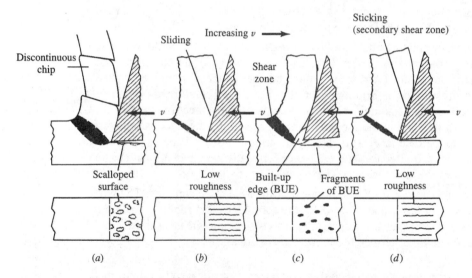

Figure 16–5   The process of chip formation changes with cutting speed. In cutting steel, the chip (a) is discontinuous at speeds below 2 m/min, (b) is continuous and slides on the rake face at 7 m/min, (c) forms with a built-up edge at 20 m/min, and (d) develops a secondary shear zone at 40 m/min. (After M.C. Shaw, in Machinability, Spec. Rep. 94, The Iron and Steel Institute, London, 1967, pp. 1–9. With permission.)

underside of the chip. The chip is still continuous but the inner side is jagged, bearing evidence of chip formation by shearing.

b. At somewhat higher speeds, heat generation causes a rise in temperatures. Friction increases until sliding at the tool face is arrested, and the system seeks to minimize the energy expenditure by finding the optimum process geometry. It will be recalled that in the processes of indentation (Sec. 8-2-5) and extrusion (Sec. 9-4-1) sticking friction led to the formation of dead-metal zones (Figs. 8–17a and 9–28a). In metal cutting too, at some intermediate speed, shearing takes place along a nose of stationary material attached to the tool face. This so-called *built-up edge* (BUE) acts like an extension of the tool (Fig. 16–5c): Shear takes place along the boundary of the BUE; hence, the *effective rake angle* becomes quite large and the energy consumption drops. However, a penalty is paid: Dimensional control is lost and, because the BUE becomes periodically unstable, it leaves occasional lumps of metal and damaging cracks behind, and the surface finish is poor. Under certain conditions a small stable BUE may be maintained; this is desirable because it protects the tool without producing an unacceptably poor surface finish.

c. With increasing speed, the material of the BUE heats up and softens, the BUE gradually disappears or, rather, degenerates into the secondary shear zone (Fig. 16–5d). The speeds at which the transition to BUE formation and secondary shear zone development occur are indicated in Fig. 16–5 for steel. Similar changes take place when cutting other materials; the critical speeds depend on the temperature reached in the shear zone but are also affected by adhesion between tool and workpiece materials. As in metal deformation processes, temperature effects can be normalized by reference to the homologous temperature scale. Examples of continuous chips are given in Fig. 16–6 with and without BUE formation.

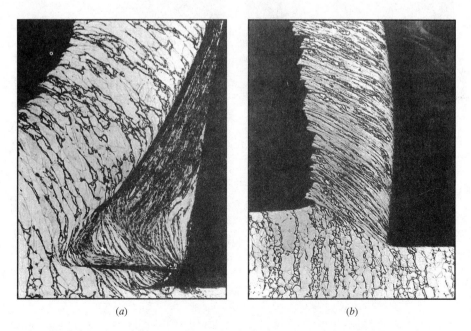

(a)   (b)

Figure 16–6   In cutting 60Cu-40Zn brass, chip formation proceeds with (a) BUE formation at 30 m/min and (b) in a continuous form at 100 m/min. (*Courtesy Dr. P.K. Wright, University of California at Berkeley.*)

2. Under special conditions, the chip may be continuous yet show a periodic change in thickness. A *wavy chip* (Fig. 16–7a) exhibits roughly sinusoidal variations in thickness. Such variations are usually related to *chatter* (vibration) attributable to periodic variations in cutting forces. As in all machines, the imposed forces cause elastic deflections of the workpiece, tool, tool holder, and machine tool (Sec. 4-1-2). Any variations in forces result in a change of undeformed chip thickness and hence in a visible and measurable waviness of the machined surface. Out-of-balance forces accelerate tool wear and can cause breakage, thus impose serious limitations on speed and production rates.

a. In *regenerative chatter* (*self-excited vibration*) the source of vibration is a change in undeformed chip thickness (from waviness produced in a preceding cut by the presence of a hard spot or other irregularity) or a periodic loss of the BUE. A remedy is usually found in changing the process conditions (speed, feed, workpiece support, tool support).

b. Chatter may also originate in *forced vibrations* due to a periodic variation of forces acting within the machine tool (e.g., from a gear box or coupling) or may be transmitted from an external source such as a nearby, vibrating machine tool. Vibration-isolation mountings or moving the offending machine tool eliminate the problem. Interrupted cuts in milling may also set up vibrations, and uneven spacing of teeth is then helpful.

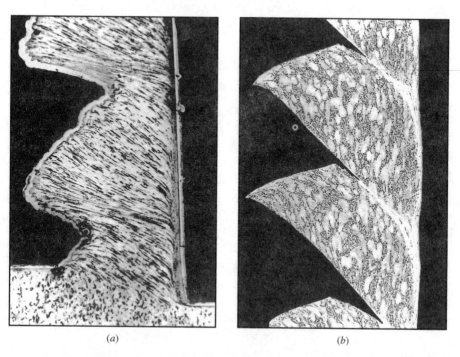

(a)   *(b)*

Figure 16–7   Under some conditions, the chip formed is (a) wavy (AISI 1015 steel, 55 m/min) or (b) segmented (Ti-6Al-4V, 10 m/min) (*Courtesy Dr. R. Komanduri, Oklahoma State University, Stillwater.*)

3. *Segmented chips* show a sawtooth-like waviness. The thick sections are only slightly deformed and are joined by severely sheared, thinner sections (Fig. 16–7b).

a. An extreme form of segmented chip formation is observed in materials of low heat conductivity, such as titanium. The process starts by upsetting ahead of the tool, resulting in *shear localization*. Because heat generated in the shear plane cannot dissipate, the material heats up, weakens, and shears until a chip segment is moved. The process then starts again by upsetting.

b. Segmented chips also form at very high speeds (in *high-speed machining*) when the drop in shear stress due to heating is greater than the increase due to strain hardening. The speed at which this occurs depends on material; in cutting heat-treated steel it is around 1000 m/min. Cutting force drops and much of the heat is removed in the chip. It is important to note, however, that high-speed machining does not necessarily equate to high-performance machining, because speed may be limited by temperature rise and higher material removal rates may be obtained with heavier cuts (Sec. 16-1-8).

4. Under certain conditions *discontinuous chip* forms:

a. When ductile materials are cut at very low speeds, severe strain hardening of the material causes upsetting until sufficient strain is accumulated to initiate shear. Elastic members in the system (e.g., the tool holder) allow sudden acceleration and the complete separation of a chip, to be followed by a new upsetting cycle. Cutting forces fluctuate violently, the new surface is torn (Fig. 16–5a) and wavy. High adhesion and low cutting speeds that generate low homologous temperatures favor such chip formation.

b. Discontinuous chip formation is intentionally introduced in some free-machining alloys by the incorporation of inclusions or second-phase particles that serve as stress raisers and cause total separation of tightly spaced chip fragments (Fig. 16–4b). The second-phase particles or inclusions often reduce the shear strength in both the primary and secondary shear zones; thus, cutting forces are low. Because chip separation is facilitated, the surface finish is good and the tendency to chatter is reduced.

The above discussion was based on the assumption of continuous engagement of the cutting edge, as is typical of operations such as turning and drilling. In other processes, most notably milling, cutting is interrupted, with the edge emerging from the cut after limited engagement. This has advantages in terms of chip disposal but subjects the tool to impact loading and rapid temperature fluctuation.

Removal of Chips   With increasing metal removal rates, large quantities of chips are produced and must be effectively removed. There is little problem with the short chips produced in cutting free-machining materials (Fig. 16–4b). In contrast, the continuous chip formed when machining ductile materials under stable conditions (Fig. 16–4a) is a nuisance: It is difficult to remove; it may clog up the work zone; it may wrap around the workpiece or tool; and it may present danger to the tooling, machine, and operator alike.

A partial remedy is found in *chipbreakers*. Some are designed to impart additional strain to the chip, causing it to break into shorter lengths or at least curl up into tight coils that break frequently. They increase the effective rake angle and thus reduce cutting force, temperature, and wear. At other times, the chip is forced to bend and hit an obstruction such as the tool holder, tool flank, or the workpiece itself. With one end fixed, the chip grows until bending stresses make it snap. Continuous chips formed by ductile materials must be chopped up or dragged out of the work zone. Chip removal is easier if the cut surface is vertical and chips drop by gravity.

A chipbreaker may be incorporated into the tool by giving the rake face a curvature, away from the cutting edge. Such *groove-type* chipbreakers (Fig. 16–8*a*) are typical of inserts; they now incorporate pads and bumps designed to deflect the chip (Fig. 16–8*c*). Alternatively, a separate chip breaker may be attached to the rake face (*obstruction type*, Fig. 16–8*b*). Chip breakers increase the effective rake angle but, if moved too close to the cutting edge, localize heat at the edge and may cause rapid loss of the tool because of overheating. The natural curvature of chips is a function of many process variables; in general, the chip curl radius becomes smaller (the chip is tighter) with increasing *h* and decreasing speed. Correspondingly, chip breakers work most efficiently at specific undeformed chip thicknesses and speeds, and inserts are specially designed for roughing or finishing cuts. Near-net-shape cast or forged parts present unusual challenges in that the part surface may have scale and local hard spots normally removed in roughing; now the depth of cut is small, the chip does not break easily, and inserts must have features that facilitate chip breaking. Computer

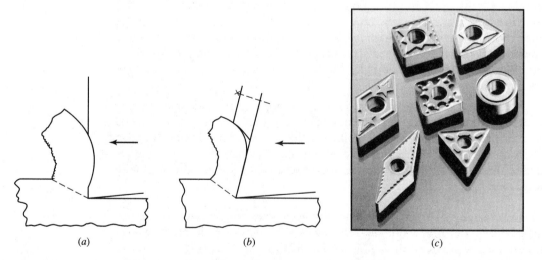

(*a*)   (*b*)   (*c*)

Figure 16–8   Chips are forced into tighter curls and break up when (*a*) groove-type or (*b*) obstruction-type chip breakers are used. (*c*) Complex chip-braking patterns may be incorporated into indexable inserts. [*Part (c) courtesy Kennametal, Latrobe, Pennsylvania.*]

modeling of chip formation has advanced to the point where it is applied to speed up chipbreaker design.

Chatter marks are found on the surface of a workpiece being turned. Identify the cause of chatter. | **Example 16-3**

The wavelength of chatter (the distance between consecutive high spots) can be measured. Cutting speed divided by wavelength gives frequency. (*a*) If the frequency agrees with or is a simple multiple of the rotational frequency of a machine part or of the vibrational frequency of a nearby machine tool, then the vibration is forced. The best remedy is elimination of the source, although changes in cutting conditions are helpful if they increase the cutting forces in the direction of vibration (because heavier loading results in increased stiffness). (*b*) If vibration is not forced, it can often be eliminated by changing the cutting speed, because machine tools have higher stiffness at certain speeds; additionally, conditions may be changed to reduce cutting forces.

Better discrimination is obtained if the test is run at two different speeds. If the frequency of vibration changes and again equals some multiple of spindle speed, the vibration is forced. If the frequency of vibration remains essentially unchanged, it is self-excited.

16-1-4   Oblique Cutting

In most practical cutting processes, the tool edge is set at an *angle of inclination i* (Fig. 16–9*a*). Such *oblique cutting* differs from orthogonal cutting in several respects:

1. The chip curls into a helical rather than spiral shape and is more readily removed from the work zone. It is normally found that the chip flows, at a velocity v_c, at an angle η_c equal to the angle of inclination i (Stabler's rule, Fig. 16–9*b*).

2. The *normal rake angle* α_n is measured in the plane containing the normal to the workpiece surface and the tool velocity v. The *effective rake angle* α_e is measured in the plane containing v and v_c and is larger than α_n

$$\sin \alpha_e = \sin^2 i + \cos^2 i \sin \alpha_n \qquad \text{(16-11)}$$

Thus, the cutting force is lower than with an orthogonal tool of equal rake angle. In general, for equal effective rake angles, an oblique cutting tool is stronger than an orthogonal one.

In many processes the tool edge is not wide enough to take a cut over the whole width of the workpiece and then the surface layer is removed in increments, by feeding the tool across the width of the workpiece. *Feed is the distance between successive engagements of the cutting edge.* In the example shown in Fig. 16–10*a*, the tool is moved in a straight-line path; a cut is taken during forward movement and the tool is lifted out of contact during the return stroke. The feed f

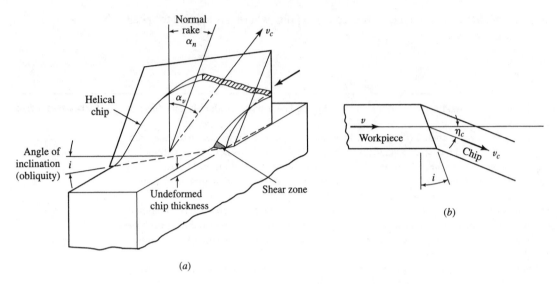

Figure 16–9 In oblique cutting, the chip flows at an angle over the rake face to form a helix.

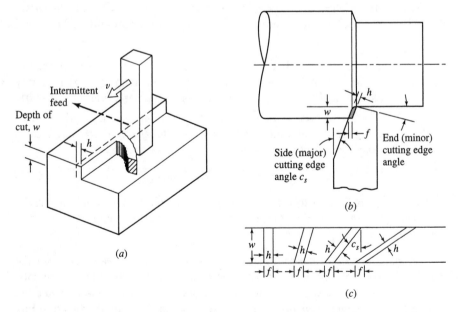

Figure 16–10 To machine a large surface, (a) the tool must be given a feed. For easier chip removal (b) the cutting edge is at an angle, which (c) affects the undeformed chip thickness for a given feed.

is taken before the next forward movement begins and is equal in magnitude to the undeformed chip thickness h.

The tool geometry shown in Fig. 16–10a is not practical, and a number of changes are desirable to allow the chip to flow easily and to prevent damage to the freshly formed surface. These changes are best shown for cutting a cylinder (turning, Fig. 16–10b).

1. The cutting edge is set at an angle in the direction of feed. An increase in this *side cutting-edge* (or *approach* or *lead*) angle C_S results in a smaller h for the same feed ($h = f \cos C_S$, Fig. 16–10c) and increases radial as well as axial forces. Typical angles are 15–30° except for long, slender parts for which zero angle gives less deflection.

2. The end of the tool is relieved, creating *major* and *minor cutting edges*. The two meet at the *corner* or *nose*; the transition between the cutting edges is *radiused* to obtain a smoother finish.

Further changes may be made, for example to impart a positive or negative rake angle, and the major cutting edge may also be inclined.

16-1-5   Forces and Energy Requirements

There are a number of theories that take a realistic view of the cutting process. Such theories are highly valuable in parametric studies, i.e., in exploring the effects of process variables. For the prediction of forces and energy requirements, there is, however, the problem of determining the relevant flow stress. We saw in Sec. 16-1-1 that large shear strains are produced in a narrow shear zone and that strain rates reach very high values. The flow stress of most materials is increased by such high strain rates even at cold-working temperatures; however, counteracting this is the large temperature rise which lowers the flow stress. Therefore, only flow stress values determined at the appropriate—and often unknown—temperatures and strain rates are relevant. Even then, predictions of the shear angle and the shear zone width are needed before a reasonable estimate can be made. Nevertheless, modeling—often by numerical methods—has advanced.

Approximate calculations of forces and power requirements, of sufficient accuracy for all practical purposes, are made from experimentally determined material constants. Three approaches are usual:

1. The cutting force P_c divided by the cross-sectional area of the undeformed chip gives the nominal cutting stress or *specific cutting pressure* p_c

$$p_c = \frac{P_c}{hw} \quad \left(\frac{N}{m^2}\right) \tag{16-12}$$

Note that p_c is not a true stress, even though it has the dimensions of stress.

2. The energy consumed in removing a unit volume of material is called the *specific cutting energy* E_1. Energy (or work) is force P_c multiplied by the distance

l over which the force acts. Since the volume of material removed is $V = hwl$, the specific cutting energy may be written as

$$E_1 = \frac{P_c l}{hwl} \qquad \left(\frac{\text{J}}{\text{m}^3} \text{ or } \frac{\text{N}}{\text{m}^2}\right) \tag{16-13}$$

It will be noted that, *when expressed in consistent units*, the numerical values of p_c and E_1 are the same. Since the purpose of calculation is often that of finding the size of the drive motor, E_1 is usually expressed in units of W·s/m³ or equivalent (Table 16–1). For dull tools, E_1 is increased by 30%.

Table 16–1   Approximate specific energy requirements for cutting* [Multiply by 1.3 for dull tools. Undeformed chip thickness: 1mm (0.040 in).]

Material	Hardness HB	HRC	Specific Energy E_1 W·s/mm³	hp·min/in³
Steels (all)	85–200		2.1	0.8
		35–40	2.4	0.9
		40–50	2.9	1.1
		50–55	3.2	1.4
		55–58	6.0	2.2
Stainless steels	135–275		2.3	0.8
		30–45	2.5	0.9
Cast irons (all)	110–190		1.3	0.5
	190–320		2.4	0.9
Titanium	250–275		2.1	0.8
Superalloys (Ni and Co)	200–360		4.5	1.6
Aluminum alloys	30–150 (500 kg)		0.5	0.2
Magnesium alloys	40–90 (500 kg)		0.3	0.1
Copper		80 HRB	1.8	0.7
Copper alloys		10–80 HRB	1.2	0.5
		80–100 HRB	1.8	0.7
Zinc alloys			0.3	0.1

*Extrapolated from data in *Machining Data Handbook*, 3d ed., Machinability Data Center, Metcut Research Associates, Cincinnati, Ohio, 1980.

3. The *material removal factor* K_1 is the reciprocal of the specific cutting energy

$$K_1 = \frac{1}{E_1} \qquad \left(\frac{\text{m}^3}{\text{W} \cdot \text{s}}\right) \tag{16-14}$$

It is convenient because it gives a feel for the amount of material that can be removed in unit time with a drive of unit power.

The above material constants cannot be immediately used for calculations because they are not truly constants but depend also on process parameters such as undeformed chip thickness, rake angle, and cutting speed. Undeformed chip thickness is the most powerful factor because the total energy requirement is actually a sum of at least two components:

1. Energy expended in the primary shear zone is proportional to undeformed chip thickness, but so is the amount of material removed. This would make p_c, E_1, and K_1 true material constants.

2. There is, however, additional energy needed to provide the flank friction and plowing forces; since this energy is virtually independent of undeformed chip thickness, it accounts for a larger proportion of total energy when h is small. Thus, the energy required to remove a unit volume of material increases with decreasing undeformed chip thickness. Therefore, material constants such as E_1 must be determined for some agreed-upon h such as $h_{\text{ref}} = 1$ mm, and the *adjusted specific cutting energy* E for any other h can then be found from an empirical power law

$$E = E_1 \left(\frac{h}{h_{\text{ref}}} \right)^{-a} = E_1 h^{-a} \qquad \textbf{(16-15)}$$

where a ranges from 0.2 to 0.4 and may be taken as 0.3 for most materials. It should be noted that below an undeformed chip thickness of 0.1 mm, the energy requirement increases even more steeply. This becomes significant in multipoint machining such as drilling and milling, and even more so in abrasive machining.

The power to be developed by the machine tool can then be estimated if the rate of material removal V_t and the efficiency of the machine tool η (usually around 0.7–0.8) are known:

$$\text{Power (W)} = \frac{E V_t}{\eta} \qquad \left(\frac{\text{W} \cdot \text{s}}{\text{mm}^3} \frac{\text{mm}^3}{\text{s}} \right) \qquad \textbf{(16-16a)}$$

or, in conventional units

$$\text{Power (hp)} = \frac{E V_t}{\eta} \qquad \left(\frac{\text{hp} \cdot \text{min}}{\text{in}^3} \frac{\text{in}^3}{\text{min}} \right) \qquad \textbf{(16-16b)}$$

The cutting force P_c to be resisted by the tool holder and the machine tool can be calculated by recalling that power divided by speed gives force. If the cutting speed v is in m/s:

$$P_c = \frac{\text{power (W)}}{v} \qquad \text{(N)} \qquad \textbf{(16-17a)}$$

or, if the cutting speed v is in units of ft/min:

$$P_c = \frac{33\,000\ \text{power (hp)}}{v} \quad \text{(lbf)} \tag{16-17b}$$

Alternatively, the force can be determined from Eq. (16-12). To a first approximation, the thrust force P_t may be taken as one-half of P_c in cutting with zero or low positive rake tools; with increasingly positive rake angle the thrust force diminishes and, in the extreme, the tool is pulled into the workpiece.

Example 16-4

As discussed in Example 16-1, a 4340 steel bar of HB 270 hardness is cut at a speed of 0.6 m/s. The undeformed chip thickness is 0.3 mm and the width of the chip is 1.5 mm. Calculate the power requirement and cutting force.

From Table 16–1, $E_1 = 2.4$ W·s/mm^3. Hence the adjusted specific cutting energy is, from Eq. (16-15), $E = 2.4(0.3)^{-0.3} = 3.44$ W·s/mm^3. The rate of material removal is simply chip cross section multiplied by cutting speed: $V_t = 0.3(1.5)(600) = 270$ mm^3/s. Power, from Eq. (16-16a): $3.44(270)/0.7 = 1328$ W or, for dull tools, $1328(1.3) = 1727$ W ($= 2.32$ hp). Cutting force, from Eq. (16-17a), $1727/0.6 = 2878$ N ($= 647$ lbf).

16-1-6   Temperatures

The energy expended in machining is concentrated in a very small zone. Only a small fraction of it is stored in the workpiece and chip in the form of increased dislocation density, and the vast majority of energy is converted into heat.

Because the cutting zone keeps moving into the workpiece, there is little heating ahead of the tool and, at least at high cutting speeds, most of the heat (over 80%) is carried away by the chip. However, the tool is in continuous contact with the chip and, in the absence of an effective heat-insulating layer, the rake face of the tool heats up. Rubbing on the rake face (or deformation in the secondary shear zone) is also a substantial source of heating. Detailed calculations show that the maximum temperature is developed at the rake face some distance away from the tool nose but before the chip lifts away (Fig. 16–11a). As would be expected, both maximum ($T_{\max}$) and average interface (T_{int}) temperatures increase with increasing cutting speed (Fig. 16–11b). The highest temperature that can be reached is the melting point of the material; for this reason, alloys whose melting point is lower than the softening temperature of the tool (such as the Al and Mg alloys) can, in principle, be cut at any speed, especially with carbide tools.

A rough estimate of temperatures may be obtained by dimensional analysis, assuming that all energy E [Eq. (16-15)] is converted into heat.[3] Then the mean tool face temperature T_T is

$$T_T = E \left(\frac{vh}{k\rho c} \right)^{1/2} \tag{16-18}$$

---

[3]M.C. Shaw, *Metal Cutting Principles*, 3d ed., Oxford University Press, 1984.

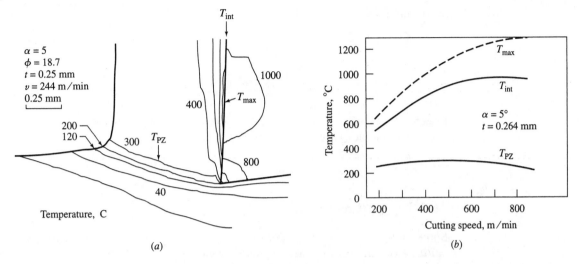

Figure 16–11   (a) Calculated temperature distribution in chip and tool and (b) variation of temperature with cutting speed when cutting AISI 1016 steel with a carbide tool. [After A.O. Tay, M.G. Stevenson, G. DeVahl Davies, and P.L.B. Oxley, Int. J. Mach. Tool Des. Res., **16**:335–349 (1976). Reprinted with permission of Pergamon Press, Oxford.]

where k is heat conductivity, ρ density, and c specific heat (heat content per unit volume) of the workpiece material. Thus, higher temperatures are to be expected in cutting stronger materials (higher E) at higher speeds, especially if the workpiece material is a poor heat conductor, of low density, and low specific heat. Materials such as titanium and superalloys are difficult to machine, whereas aluminum and magnesium are easy. Even though most of the heat is taken away in the chip, some diffuses into the workpiece, and the resulting dimensional change must be compensated for, and heat buildup in the tool is a significant factor limiting cutting speed.

A feel for the effect of material properties on permissible cutting speeds may be obtained from Eq. (16-18) if undeformed chip thickness and temperature rise are kept constant for a given tool (say HSS). Take, for example, the Ni-base superalloy IN-100 of TS = 1000 MPa and the steel 4140 heat treated to TS = 1000 MPa (hardnesses are about equal and around 300 HB).

Example 16-5

	IN-100	4140	Source
E_1 (W·s/mm^3)	4.5	2.4	Table 16–1
k, at 500°C (W/m·K)	17	37	*Metals Handbook*, 9th ed.,
ρ (g/cm^3)	7.75	7.85	vol. 1, pp. 148, 149
c at 500°C (J/kg·K)	480	520	vol. 3, p. 243

Substituting gives

$$\Delta T = \text{const.} = 4.5 \left[\frac{v_{\text{super}}}{(17)(7.75)(480)} \right]^{1/2} = 2.4 \left[\frac{v_{\text{steel}}}{(37)(7.85)(520)} \right]^{1/2}$$

$$v_{\text{steel}} = \frac{3058439}{364262} v_{\text{super}} = 8.4 v_{\text{super}}$$

From Fig. 16–45, $v_{\text{steel}} = 0.45$ m/s; from Fig. 16–46, $v_{\text{super}} = 0.06$ m/s; the ratio is $0.45/0.06$ = 7.5, in reasonable agreement with our calculation.

16-1-7   Cutting Fluids

Some cutting operations are performed *dry*, i.e., without the application of a *cutting fluid* or, as sometimes called, a *coolant*.

Actions of Cutting Fluids   The fluid fulfills basically three major functions:

1. *Lubrication.* Access of the fluid to the rake face is difficult, especially at higher cutting speeds. The fluid does, however, enter the sliding zone, and some fluid may seep in from the sides of the chip too. Effects attributable to lubrication can be frequently observed, especially when contact with the cutting tool is intermittent. In low-speed cutting with sliding friction, rake-face friction is reduced, therefore, the shear angle increases, the chip becomes thinner and curls more tightly, and power consumption drops. At speeds where a BUE forms in the absence of a lubricant, the onset of BUE formation is shifted to higher speeds. At higher speeds, where a sticking zone develops, the length of the sticking zone is reduced. At all speeds, lubricant access to the flank face is possible and rubbing is reduced. The combined effect is that, in general, surface finish also improves.

2. *Cooling.* Because shear is highly concentrated and the shear zone moves extremely rapidly, temperatures in the shear zone are not affected. However, a cutting fluid reduces the temperature of the chip as it leaves the secondary shear zone, and it cools the workpiece. It may also reduce the bulk temperature of the tool. While relationships are by no means straightforward, it is often found that a cutting fluid reduces temperatures sufficiently to allow cutting at higher speeds. In the majority of instances it is, however, essential that the fluid be applied to the cutting zone. In interrupted cuts, such as in milling, the tool is subjected to rapid temperature fluctuations. A coolant is helpful only if it floods the entire cutting zone; otherwise, it subjects the tool to more extreme temperature fluctuations.

3. *Chip removal.* Cutting fluids fulfill an additional and sometimes extremely important function: They flush away chips from the cutting zone and prevent clogging or binding of the tool. Chips are then transported away by conveyors or vacuum.

Cutting Fluids   In selecting cutting fluids, many of the considerations relating to metalworking fluids (Sec. 8-2-4) apply.   Cutting fluids fall into two main categories.

1. *Cutting oils* are based on mineral oils with appropriate additives, and are used mostly at lower speeds and with high-speed steel (HSS) tooling.

2. *Water-base* (*aqueous*) *fluids* may be *emulsions* (oils dispersed in water with the aid of surface-active substances), *semisynthetic fluids* (also called semi-chemical fluids or chemical emulsions, in which large quantities of surface-active agents are used to reduce oil particle size to the point where the fluid becomes translucent or transparent), or *synthetic fluids* (also called chemical fluids, which contain no oil, only water-soluble wetting agents, corrosion inhibitors, and salts).

Because of intimate tool–workpiece contact, high temperatures, and the danger of wear, many cutting fluids contain boundary and EP agents.   Some typical fluids are listed in Table 16–2.

Table 16–2   Commonly used machining fluids*

Process	Tool	Steel (< 275 BHN)	Steel (> 275 BHN)	Stainless Steel Nickel Alloy	Cast Iron	Aluminum Alloy	Magnesium Alloy	Copper Alloy
Turning	HSS	O1, E1, C1	O2, E2, C2	O2, E2, C2	E1, C1	E1, C1, Sp	O1, Sp	E1, C1, Sp
	Carbide	D, E1, C1	D, E1, C1	D, E1, C1	D, E1, C1	D, E1, C1	O1, Sp	E1, C1
Milling	HSS	O1, E1, C1	O2, E2, C2	O2, E2, C2	E1, C1	D, O1, Sp	O1, Sp	E1, C1, Sp
Drilling	Carbide	D, E1, C1	O1, E1, C1	O2, E1, C1	D, E1, C1	D, O1, Sp	O1, Sp	E1, C1
Form turning	HSS	O2, E2, C2	O2, E2, C2	O2, E2, C2	E1, C1	E1, C1, Sp	O1, Sp	E1, C1, Sp
	Carbide	D, E1, C1	E2, C2	O2, E2, C2	D, E1, C1	D, E1, C1	O1, Sp	E1, C1
Gear shaping	HSS	O2, E2, C2	O2, E2, C2	O2, O3	E1, C1	O1, Sp	D, O1, Sp	O1, Sp
Tapping	HSS	O1, E2, C2	O2, E2, C2	O2, O3	E1, C1	D, O1, Sp	D, O1, Sp	O1, Sp
Broaching	HSS	O2, E2, C2	O2, E2, C2	O2, E2, C2	E2, C2	E1, C1, Sp	D, O1, Sp	E1, C1, Sp
	Carbide	O1, E1, C1	O1, E1, C1	O1, E1, C1	D, E1, C1	D, E1, C1	D, O1, Sp	E1, Cl, Sp
Grinding		O1, E1, C1	O2, E1, C1	O2, E2, C2	E1, C1	O1, Sp	O1, Sp	O1, Sp

*From J. A. Schey, *Tribology in Metalworking: Friction, Lubrication and Wear*, American Society for Metals, Metals Park, Ohio, 1983. Code: D—Dry. O1—Mineral oil or synthetic oil. O2—Compounded oil. O3—Heavy-duty compounded oil. E1—Mineral-oil emulsion. E2—Heavy-duty (compounded) emulsion. C1—Chemical fluid or synthetic fluid. C2—Heavy duty (compounded) chemical or synthetic fluid. Sp—Specialty formulated fluid, with boundary and/or E.P. additives.

Application of Cutting Fluids   The method of applying cutting fluids is as important as is their selection.

1. *Manual application.* The application of a fluid from a squirt can or in the form of a paste (for low-speed operations) is commonly practiced even though it is not really acceptable even in job-shop situations. In the absence of cooling, cutting speeds are limited, and it is difficult to keep the machines and plant clean.

2. *Flooding.* Most machine tools are equipped with a recirculating system that incorporates filters. The fluid is applied at a rate of up to 15 L/min for each simultaneously engaged cutting edge. For convenience, the tool is usually flooded from the chip side (Fig. 16–12a), although better cooling is secured by application into the clearance crevice (Fig. 16–12b), especially when the fluid is supplied under a pressure of 300 kPa (40 psi) or more. High-pressure systems apply pressures of 5–35 MPa (0.8–5 kpsi) and, hitting at 350–500 km/h, help to carry away chips, but the entire workspace must be enclosed. A second nozzle may be necessary to clear away the chips in some operations (Fig. 16–12c). Flow rates in drilling are typically 5 L per millimeter of drill diameter. However, fluid access to the cutting edges is limited and chip removal is difficult.

3. *Coolant-fed tooling.* There are drills and other tools available in which holes are provided through the body of the tool so that pressurized fluid can be pumped to the cutting edges, ensuring access of fluid and facilitating chip removal. For inserts, coolant nozzles may be built directly into tool holders, and some inserts have holes through which coolant is delivered to the underside of the chip.

4. *Mist application.* Fluid droplets suspended in air provide effective cooling by evaporation of the fluid, although separate flood cooling of the workpiece may be required. Measures must be taken to limit airborne mist, for example, by the use of demisters.

5. *Treatment of lubricants and chips.* The principles of lubricant care given in Sec. 8-2-4 apply here too. Recirculating systems may be small for individual machine tools or very large, integrated installations for an entire plant. In either case, lubricant quality must be carefully monitored and appropriate makeup of oil or water made. Because disposal is becoming increasingly difficult and costly,

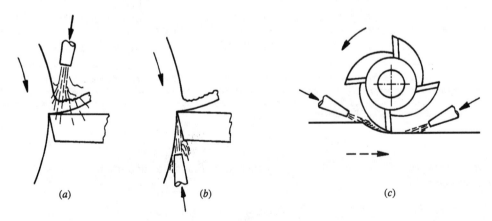

(a)   (b)   (c)

Figure 16–12   Cutting fluids are usually applied (a) to the chip although (b) better cooling is obtained by applying it to the flank face. In milling, (c) an additional jet removes the chip. *(From J.A. Schey: Tribology in Metalworking, ASM International, 1983, p. 633. With permission.)*

fluids are selected that are more readily recycled. In some processes, such as milling and gear hobbing with carbide cutters, dry machining is also growing. Filtration separates chips and fines from the lubricant/coolant but chips are next to worthless if mixed and oily. Chips kept carefully segregated and deoiled in a centrifuge can be returned for recycling.

16-1-8   Tool Life

In deformation processes (Chaps. 9 and 10) tool lives are measured in thousands of parts or in weeks or hours of operation, and concern over wear is often overshadowed by considerations of die pressure or material flow. In contrast, tool wear is the dominant concern in metal cutting. This is not surprising since the tool of relatively small mass is exposed to high pressures and temperatures and often also to shock loading. Tool lives on the order of tens of minutes are common and reach only hours on production lines for mass production. Therefore, the economy of the process is controlled very largely by *tool life*.

Tool Wear   As might be expected, *tool wear* can take several forms (Fig. 16–13), and all wear mechanisms discussed in Sec. 4-9-2 may play a role.

1. *Flank wear.* Intense rubbing of the clearance face of the tool over the freshly formed surface of the workpiece results in the formation of a wear land. The rate of wear can be characterized by interrupting the cut and measuring the average width of the wear land *VB* (Fig. 16-13a). After rapid wear during the first few seconds, wear settles down to a steady-state rate only to accelerate again toward the end of tool life (Fig. 16–13b). Flank wear is due usually to both abrasive and adhesive mechanisms, and is generally undesirable because dimensional control is lost, surface finish deteriorates, and heat generation increases. It is, nevertheless, the normal wear mode.

2. *Notch wear.* A notch or groove of *VN* depth often forms at the depth-of-cut line where the tool rubs against the shoulder of the workpiece (Fig. 16–13a). Abrasion by surface layers is often accelerated by oxidation or other chemical reactions. In the limit, notch wear may lead to total tool failure.

3. *Crater wear.* The high temperatures generated on the rake face (Fig. 16–11a) combine with high shear stresses to create a crater some distance away from the tool edge. Wear is usually quantified by measuring the depth *KT* or the cross-sectional area of the crater perpendicular to the cutting edge. Crater wear progresses linearly under the influence of abrasion, adhesion followed by dragging out tool material, diffusion, or thermal softening and plastic deformation. Crater wear in itself is not damaging; indeed, a stable BUE may develop and the tool then acts as though it had a larger positive rake angle (Fig. 16–14a). Ultimately, however, crater wear leads to catastrophic edge failure; therefore, crater wear is generally avoided.

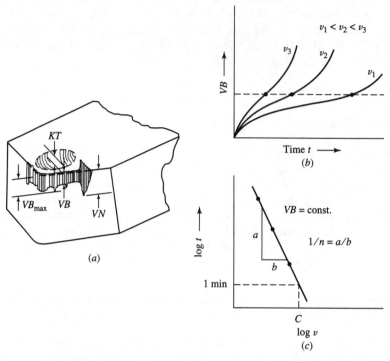

Figure 16–13   Flank and crater wear (*a*) may be characterized by the dimensions shown. (*b*) From the progression of flank wear, (*c*) the tool-life constants C and *n* may be extracted.

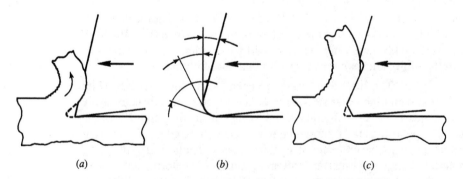

Figure 16–14   The shape of the tool edge has significant effects: (*a*) groove-type chip breakers promote BUE formation, (*b*) rounding of the tool edge by wear results in increasingly negative rake angles toward the flank face; (*c*) a small negative rake ground at the nose encourages the formation of a stable BUE.

4. *Edge rounding.* The major cutting edge may become rounded by abrasion. Cutting then proceeds with an increasingly negative rake angle toward the root of the cut (Fig. 16–14*b*). When the undeformed chip thickness is small, cutting action may cease and all energy may be expended in plastic or elastic deformation of the workpiece. At high cutting speeds (high temperatures) and high tool pressures the tool edge may deform plastically; the nose of HSS tools may be entirely lost. Problems with edge rounding may be minimized, at least when hard tools are used, by grinding a double rake (also called T land, Fig. 16–14*c*) so that cutting proceeds with a stable BUE. Because cutting forces are higher, the machine tool must be very stiff.[4]

5. *Edge chipping.* This may be caused by periodic break-off of the BUE or when a brittle tool is used in interrupted cuts. Surface finish suffers and the tool may finally break.

6. *Edge cracking.* Cyclic mechanical loading leads to cracks parallel to the edge while thermal fatigue causes cracks to form perpendicular to the cutting edge of brittle tools (*comb cracks*).

7. *Catastrophic failure.* Tools made of more brittle materials are subject to sudden failures (breakage). This is a problem of all brittle materials, especially ceramics, in interrupted cuts. Improved tool manufacturing processes, zero or negative rake, and selection of the proper machining conditions all help.

Cutting fluids are designed to extend tool life, although under certain conditions (chemical reactions, or thermal stressing in interrupted cuts) they may shorten it.

Tool Life Criteria   Tool life affects the choice of tool, process conditions, economy of operation, and the possibility of automation and computer control. Unfortunately, no simple definition of tool life is possible: Tool life must be specified with proper regard to the aims of the process. Thus, in finishing operations surface quality and dimensional accuracy are most important; in roughing, greater deterioration of surface quality and dimensional accuracy may be tolerated in exchange for high metal removal rates; an absolute limit is reached when cutting forces increase to high enough values to cause tool fracture.

All these considerations are usually translated into some easily measurable values. Most frequently, flank wear VB or VB_{max} (Fig. 16–13*a*) is specified as the end of useful tool life:

HSS tools, roughing	$VB_{max} = 1.5$ mm
Finishing	$VB = 0.75$ mm
Carbide tools	$VB = 0.4$ mm (or $VB_{max} = 0.7$ mm)
Ceramic tools	$VB_{max} = 0.6$ mm

---

[4]T. Hoshi, in *Cutting Tool Materials*, American Society for Metals, Metals Park, Ohio, 1981, pp. 413–426.

Other criteria include a specified crater wear, total loss of the tool edge or nose, or total (flank and crater) wear volume.

Tool life is usually given as the time (in minutes) it takes to reach the specific wear criterion under specified process conditions (speed, feed, depth of cut), although for mass production on transfer lines tool lives of hours are desirable. For tools such as drills and taps a more practical measure is the number of holes drilled or tapped under specified conditions.

Prediction of Tool Life   Even though various wear mechanisms come into play, gradual wear is produced by temperature-dependent mechanisms (even abrasive wear is accelerated by temperature because the strength and abrasion resistance of the tool drops at high temperatures). We saw that temperatures are greatly affected by cutting speed [Eq. (16-18)], and it is known that gradual wear is a function of rubbing distance which, for a given cutting speed, is proportional to time. It is to be expected then that, for a given tool life criterion such as flank wear, tool life should drop as a function of speed. It was first observed by Taylor[5] that the relation follows a power law

$$vt^n = C \qquad\qquad \text{(16-19)}$$

where v is the cutting speed (m/min or ft/min), t is tool life (min), and C is the cutting speed for a tool life of one minute. Strictly speaking, the equation should be written as

$$vt^n = Ct_{\text{ref}}^n \qquad\qquad \text{(16-19}'\text{)}$$

where $t_{\text{ref}} = 1$ min.

Accordingly, there is a straight-line relation when tool life is plotted against speed on a log-log scale (Fig. 16–13c). Since metal cutting is a complex system, the constants also depend on a number of variables. Nevertheless, C is basically a constant for a given workpiece material whereas the *Taylor exponent n* [not to be confused with the strain-hardening exponent of Eq. (8-4)] is characteristic of the tool material. Its value is typically 0.08–0.1 for HSS, 0.25–0.4 for cemented carbides, 0.4–0.6 for coated carbides, and 0.5–0.7 for ceramic tools. Note, however, that in high-speed machining tool life can be much longer than predicted by the Taylor equation; because high temperatures stabilize the secondary shear zone, the adhering material protects the tool from abrasive and diffusional wear. Only when speed is raised further will the protective layer disappear, followed by a rapid increase in wear. A tool coating that prevents diffusion protects to higher speeds.

A better feel for the importance of the Taylor exponent is gained by rearranging the formula to express tool life:

$$t = \frac{K}{v^{1/n}} \qquad\qquad \text{(16-20)}$$

---

[5]F.W. Taylor, *Trans. ASME,* **28**:31–279, 1907.

It will be noted that for $n = 0.1$, tool life decreases extremely rapidly with the tenth power of speed.

Heat generation is affected by the total heat input (or energy input), which increases with undeformed chip thickness h and chip width (or depth of cut) w. The Taylor formula can be extended to take these into account:

$$t = \frac{K}{v^{1/n_1} f^{1/n_2} w^{1/n_3}} \qquad (16\text{-}21)$$

where, in general, $n_1 < n_2 < n_3$. These exponents are not completely independent of each other and are affected also by tool geometry and process. Typical values for HSS are $n_1 = 0.1$, $n_2 = 0.18$, $n_3 = 0.45$. Therefore, for increased material removal rates (for high-performance machining), it is preferable to increase first the depth of cut, then feed, and only last, speed. Of course, when tool life is limited by catastrophic tool failure, the Taylor equation is useless and must be replaced by a statistical life criterion. Even when the Taylor equation is used, the statistical distribution of tool lives must be taken into account, especially, if the equation is used to program tool changes under unattended automatic (computer) control.

Example 16-6

Find the effect of doubling speed on tool life in machining with HSS, carbide, coated carbide, and ceramic tooling.

Denote the initial speed v_1 and the doubled speed v_2. Then, from Eq. (16-19)

$$v_1 t_1^n = v_2 t_2^n \qquad \text{and} \qquad (t_2/t_1) = (v_1/v_2)^{1/n} = (1/2)^{1/n}$$

	n	t_2/t_1	t_1/t_2
HSS	0.1	0.001	1024
WC	0.3	0.1	10
Coated carbide	0.4	0.177	6
Ceramic	0.5	0.25	4

The change is catastrophic for HSS.

16-1-9   Surface Quality

Machining aims to create a part of a given geometry, to specified dimensions and dimensional tolerances. To permit proper function of the part, the surface finish (Sec. 3-5) is also specified. Beyond these geometrical considerations, it is also important that the surface produced should be free of defects such as cracks, have no harmful residual stresses, and not be subjected to undesirable metallurgical changes. These are particularly important aspects when the part operates in a hostile environment, is subject to fatigue loading, or when its failure could have catastrophic consequences. With the growth of such critical applications, particularly in the aerospace industries, the term *surface quality* has acquired a complex meaning.

Surface Roughness   The surface formed in simple orthogonal (Fig. 16–1a) or oblique (Fig. 16–9a) cutting is, ideally, perfectly smooth (roughness is zero). When a tool of R radius is moved by the feed f between successive cuts (Fig. 16–10), the *ideal transverse roughness* can be calculated approximately by considering the geometry (Fig. 16–15). The peak-to-valley height is

$$R_{max} = \frac{f^2}{8R} \qquad \text{(16-22a)}$$

The arithmetic average for a triangular roughness is $R_a = R_{max}/4$, hence

$$R_a \approx \frac{f^2}{32R} \qquad \text{(16-22b)}$$

The longitudinal roughness will still be zero. Similar relationships can be developed for other processes.

Superimposed on the ideal roughness are features introduced by the chip-forming process itself. This results in a measurable roughness in the longitudinal direction and a modification of surface profile (and hence of roughness values) in the transverse direction. Several features may be observed:

1. In cutting at very low speeds and typically also with all discontinuous chip formation, the surface is scalloped (Fig. 16–5a) and cracks may develop transverse to the cutting direction.

2. In cutting with an unstable BUE, heavily strain-hardened fragments are welded to the surface, covering some 5–10% of it (Fig. 16–5c).

3. When a continuous chip is formed without a BUE, the surface configuration comes close to the ideal one, even though localized wear or chipping of the tool edge gives some roughness increase in the transverse direction (Fig. 16–5b and d).

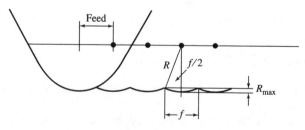

Figure 16–15   An ideal roughness value may be calculated from the geometry defined by the nose radius.

4. Chatter introduces a periodic variation of surface geometry which is readily visible and shows up as waviness on a recorded longitudinal trace (Fig. 3-19).

5. The surface finish changes in the course of cutting and, in general, deteriorates with the progression of wear. Indeed, tool life is sometimes specified as the time for which an acceptable finish is produced.

A bar of free-machining steel (HB 200) is to be finish turned to $R_a = 1.6\mu m$ with a carbide tool. Suggest the appropriate cutting conditions.

| | **Example 16-7** |

We shall see (Fig. 16–45 and Table 16–5) that a suitable feed would be $f = 0.38/2 = 0.19 = 0.2$ mm. From Eq. (16-22b), $R_a = f^2/32(R_a) = 0.2^2/(32)(0.0016) = 0.78$ mm. This is the nose radius for ideal roughness; to allow for some roughening due to chip formation, a tool of minimum 1-mm nose radius should be used.

Surface Integrity   The term *surface integrity* has been introduced to indicate the absence of undesirable features on the surface as well as in the subsurface region of the workpiece.

1. Strain hardening of a surface layer is a natural consequence of chip formation (Fig. 16–3). A residual stress may also be generated which is, most of the time, compressive and is thus beneficial.

2. Cracks formed in low-speed cutting are harmful, as are those sometimes found in cutting with an unstable BUE.

3. Cutting of heat-treatable steels at high speeds can result in heating above the transformation temperature. As the tool leaves the heated zone, the cold mass of the workpiece quenches the surface at a high enough rate for martensite to form. Such transformed surfaces are resistant to attack by common etching agents and are, therefore, referred to as white layers. Since untempered martensite is very hard and brittle, cracks are often formed, if not during machining then in service. The danger is more acute when machining quenched-and-tempered steels. The problem is aggravated when excessive tool wear gives large flank-friction forces. It is present also in high-speed machining with ceramic tools where the negative rake angle results in large normal forces.

4. Unfavorable cutting conditions can leave large residual tensile stresses in the surface.

Some aspects of surface integrity can be evaluated only by destructive techniques under the microscope, particularly, SEM. On the basis of such tests, cutting conditions that ensure good surface integrity can be specified. For the most critical applications, NDT techniques—including x-ray analysis for residual stresses—are employed.

16-2   WORK MATERIAL

The discussion of the metal cutting process in Sec. 16-1 made it clear that the response of metals must depend on the process itself. Thus, machinability is a system property and no general ranking of materials is possible. Nevertheless, it is customary to speak of *machinability* as a material property, and in the most general sense a material is highly machinable when satisfactory parts can be made from it at low cost, with minimum difficulty.

16-2-1   Machinability

A closer definition of machinability requires that quantitative judgments be made. There are several possibilities.

1. A *machinability index* is often quoted, which is an average rating stated in comparison with a reference material: for steels, a free-machining Bessemer steel B1112, very similar to the present AISI 1212 steel; for copper-base alloys, a leaded free-machining brass; and for aluminum alloys, 7075-T6 aluminum. It is based on cutting speed in turning for 60-min tool life. The system can be misleading because the ranking is different for different processes.

2. A more quantitative measure is *tool life* to total failure by chipping or cracking under specified conditions. Specifications are given as the cutting speed for a given tool life in minutes or seconds, or as the volume of material removed for a given tool life criterion.

3. Another measure is *tool wear*. This can be related to the gradual wear of the flank face or development of the crater. It is given as the change in the dimension of the machined part due to wear per unit time for a given cutting speed and feed, or as the time required for a standard flank-land wear to develop. In other cases crater depth is specified.

4. Another quantitative measure is *surface finish* produced at standardized cutting speeds and feeds.

5. Yet other rankings are based on cutting force, power, temperature, or chip formation.

Since machinability is a system property, all parts of the system must be well defined if reproducible and relevant data are to be obtained. The principles of such testing are laid out in ISO Standard 3685-1993, "Tool-life testing with single-point turning tools." Evaluation is based on tool wear. Wear is quoted as a function of time when testing at a single speed, and as tool wear–time curves (or Taylor constants) when testing at several speeds. Full evaluation is time-consuming and expensive. Some shortened tests are also available, although they tend to have limited validity.

16-2-2 Machinable Materials

Since machinability is such a many-faced property, it is influenced by a number of material properties. Good machinability may mean one or more of the following: cutting with minimum energy, minimum tool wear, good surface finish. This means that:

1. A material of low ductility is desired, so that chip separation occurs after minimum shearing and the chip breaks up easily. This is exactly the opposite of what one looks for in plastic deformation (Sec. 8-2-6); thus, desirable properties now include a low strain-hardening exponent (n), a low resistance to void formation and thus a low reduction in area (q), and a low fracture toughness.

2. To minimize cutting energy, the shear strength or—what is more practicably measured—the strength (TS) and hardness of the material should be low.

3. A strong metallurgical bond between tool and workpiece, usually expressed as adhesion (Sec. 4-9-2), is undesirable when it also promotes diffusion and weakening of the tool material by depletion of alloying elements. However, when diffusion does not take place, high adhesion helps to stabilize the secondary shear zone.

4. Very hard compounds (such as some oxides, all carbides, many intermetallic compounds, and elements such as silicon) embedded in the workpiece material act as cutting tools themselves and accelerate tool wear. They are particularly damaging when in the form of platelets with sharp edges.

5. Second-phase particles that are soft or softened at the high temperatures reached in the shear zone are beneficial because they promote localized shear and contribute to chip breaking, making the material free-machining. Because of their low shear strength, such inclusions also reduce the energy expended in the secondary shear zone, and some even act as internal lubricants by smearing on the rake face. Thus, cutting force and energy also decrease.

6. High thermal conductivity is helpful by keeping cutting temperatures low (Sec. 16-1-6).

7. A low melting point of the workpiece material means that cutting temperatures will also remain low, below the temperature at which the tool softens or reacts with the workpiece.

The above properties must be examined over a range of temperatures. A higher temperature lowers the shear strength of the material, thus making possible the machining of some very difficult materials. Indeed, in special cases metal removal rates can be greatly increased by localized heating of the workpiece just ahead of the cutting zone. To prevent dissipation of heat, the rate of heat input must be high, usually provided by induction heating or with the aid of a plasma torch or laser. High temperatures do, however, have the undesirable side effect of increased adhesion and accelerated diffusion, and tool life can drastically drop. If this is the case, every effort is made to keep the work zone cool with large quantities of cutting fluid.

Some of the requirements are seldom satisfied simultaneously. Some of the most ductile materials favored for plastic deformation are difficult to machine because of their ductility. Even more difficult are the ductile but also high-strength materials. Two-phase materials are often desirable because ductility is impaired by the presence of plate-like or, in general, sharp second-phase particles, especially if they are also brittle and of low strength (Sec. 6-3-2). In many instances it is economical to bring the material into a more machinable condition through metallurgical control (usually by a heat treatment) and then heat-treat it again after machining to impart the required service properties.

16-2-3   Ferrous Materials

The full range of machinability is encountered in ferrous materials.

Carbon Steels   The term *plain carbon steel* is applied to a great variety of materials, ranging from very low-carbon iron to hypereutectoid steel. These steels are commercially available in three different forms (Fig. 16–16):

1. In the fully annealed condition; strength increases while ductility decreases with increasing amounts of carbide present in the lamellar pearlitic form.

2. Heat-treated to bring the carbide into a spheroidal form; a spheroidized steel has lower strength and higher ductility (consider Fig. 6–15, 3*a* versus 2*c*).

3. Cold-worked (usually cold-drawn); strength is higher and ductility is depressed, while surface finish and tolerances are improved.

On this basis, one can readily choose the optimum treatment that ensures the best machinability for a given carbon content (Fig. 16–16). At low carbon levels (typically below 0.2% C), the annealed material is much too ductile, and the cold-worked material with its low ductility offers the best machinability. At intermediate carbon levels (typically up to 0.45% C) the strength of the cold-worked material would give rise to excessive cutting forces and the lamellar pearlite with its lower ductility and moderate strength is preferable. At yet higher carbon levels, the large quantities of carbide present in the lamellar pearlite act as minute cutting tools and cause premature abrasive wear of the cutting tool proper. Thus, the spheroidal condition with its relatively harmless globular carbides and lower strength is preferable even though the ductility is higher.

Free-Machining Steels   Vast quantities of carbon steels are machined, and efforts directed at improving their machinability have led to the development of free-machining grades. They contain an insoluble, soft element, primarily lead (leaded steels) or have an increased sulfur content (resulfurized steels) which forms MnS inclusions of controlled, globular shape. From the service point of view, an undesirable consequence is reduced ductility (see Fig. 6–16) and fatigue strength and slightly reduced tensile strength. The wear of cutting tools can be reduced without impairing the mechanical properties of the steel by the use of calcium

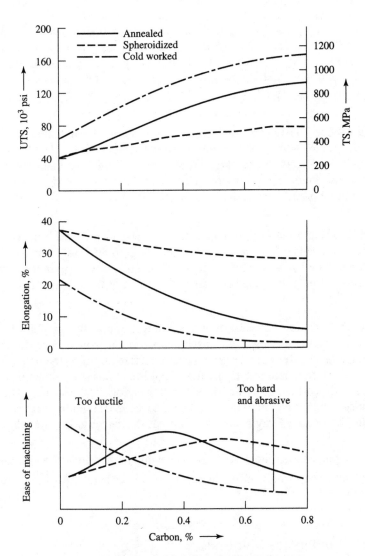

Figure 16–16   Carbon steels, like many other materials, are most machinable when brought into a condition that gives minimum strength combined with minimum ductility.

as a deoxidizing agent; when cutting such steels, a complex, low-shear-strength oxide forms on the rake face.

Alloy Steels   The greater hardness of alloy steels increases tool wear, especially if carbides are present in larger quantities. For dimensional control, these steels are often machined in the fully heat-treated (quenched-and-tempered) condition (*hard cutting*), and then cutting parameters are chosen to ensure surface integrity. For maximum productivity, large depth of cut is taken in roughing, and high speeds combined with small depth of cut and feed per tooth are used in finishing, usually with coated carbide or PCBN tools. Increasing quantities of powder-metallurgy steels are also produced. Machinability is reduced because porosity creates discontinuous cutting conditions, higher temperatures, and more wear. Parts with over 90% density are closer to wrought materials.

Stainless Steels   The higher strength and lower thermal conductivity of stainless steels (Table 4–1) results in higher cutting temperatures. The high strain-hardening rate of austenitic steels (AISI 300 series, Table 8–2) makes them more difficult to machine. Cutting fluids must contain EP compounds. If necessary, free-machining properties can be imparted by alloying.

Cast Irons   The presence of primary cementite makes white cast irons very difficult to machine, and white zones (chill zones) in graphitic cast irons are responsible for much tool wear and breakage. The machinability of graphitic cast irons is a function of graphite shape and distribution and of the microstructure of the matrix.

   1. Gray irons are basically free-machining because the graphite lamellae break up the chip. However, the machined surface is rough because graphite particles break out. Refining the graphite particle size improves the finish without impairing the free-machining properties. Tool life decreases with an increasing proportion of pearlite in the matrix and is lower for finer pearlite. The same factors also contribute to increased hardness, hence machinability decreases with increasing hardness. Gray irons are often cut dry because the fine chips clog filters.

   2. Compacted graphite cast iron is somewhat more difficult to machine.

   3. Nodular cast iron is more ductile and stronger but, surprisingly, can give a longer tool life.

16-2-4   Nonferrous Materials

In keeping with the convention adopted in Secs. 7-3 and 8-3, nonferrous materials will be discussed in order of increasing melting point.

Low-Melting Materials   Only zinc alloys are machined in significant quantities. Their low strength and limited ductility make them highly machinable.

Magnesium Alloys   The low ductility imparts free-machining properties, making magnesium a highly machinable material. Finely divided chips ignite spontaneously, therefore, finish cutting with chip thicknesses below 25 μm is always done with an oil-based cutting fluid. Mg–Al alloys form a BUE and must be cut with PCD or with oil.

Aluminum alloys   Pure aluminum and its ductile alloys are best machined in the cold-worked condition because their high ductility makes them "draggy" in the annealed condition: Cutting forces are higher than would be expected from their hardness, and high adhesion leads to a poor surface finish. Precipitation-hardened alloys can be readily machined in the fully heat-treated (solution-treated and aged) condition in which their ductility is low yet their strength is not unduly high. The high thermal conductivity and low melting point allow high machining speeds even with HSS tools, provided that a cutting fluid—which contains boundary-lubricating additives—is applied in a flood. Very high speeds (70 m/s) are possible with carbides and, especially, PCD; the limit is set by the capabilities of the machine tool. SiC cannot be used because of the solubility of Si in Al. Free-machining properties may be imparted by the addition of lead, bismuth, or tin. Castings that contain elementary silicon (hypereutectic Al–Si alloys) give rapid tool wear and must be cut with PCBN or PCD tools.

Beryllium   Beryllium is easily machinable, dry, but fine particles are toxic.

Copper-Based alloys   Pure copper, like pure aluminum, is best machined in the cold-worked condition. This applies also to most single-phase alloys which, nevertheless, can often be cut with less energy than pure copper. Chip disposition is difficult. In contrast, $\alpha + \beta$ brasses machine very well. Free-machining additions, usually lead, make all brasses more machinable, and the leaded $\alpha + \beta$ brass serves as a reference base in the machinability scale (Sec. 16-2-1). Free-machining coppers contain lead, sulfur, or tellurium; the chip may still be continuous but cutting force is greatly reduced and surface finish is improved. Lead is being replaced in applications where contact with food is possible.

Nickel-Based Alloys and Superalloys   For lower ductility, it would be desirable to cut these alloys in the cold-worked or fully heat-treated condition. However, their high adhesion and low thermal conductivity is often combined with high strength, and this dictates their cutting in the annealed or overaged condition. Sulfur must be avoided in cutting fluids because it forms a low-melting eutectic with nickel.

Titanium   The high reactivity and hence high adhesion of titanium, combined with its low thermal conductivity, make chip-formation discontinuous at most speeds and machining is difficult. For low speeds, HSS tools are used with a heavily compounded oil or emulsion. At higher speeds (30–60 m/min), cemented carbides or cermets are preferred. Heavier feeds are preferred because frictional heat is reduced and more heat is taken away in the chip.

16-3   CUTTING TOOLS

Specific features of cutting tools are varied to suit the process, but some basic characteristics are common to all.

16-3-1 Tool Materials

Machining can be, in general, regarded as competition for survival between workpiece and tool material. There have been improvements in the machinability of metals, but the main agents of recent advances in machining have been the developments in tool materials. They have not only allowed faster material removal but have also facilitated progress in machine tool design and control.

One can expect that the tool material should have properties just opposite to those of the workpiece:

1. The tool should be harder than the hardest component of the workpiece material, not only at room temperature (Table 16–3), but also at operating temperatures. High *hot hardness* prevents plastic deformation, ensures that the tool geometry is maintained under the extreme conditions presented by the chip formation process, and it also aids in resisting wear. Some feel for the wide range of hot hardnesses may be gained from Fig. 16–17.

Table 16–3   Hardness of typical tool materials or their constituents*

Material or Constituent	Hardness, HV
Martensitic steel	500–1000
Nitrided steel	950
Cementite (Fe_3C)	850–1100
Silicon	1150
Hard chromium coating	1200
Garnet	1360
Alumina	2100–2400
WC (Co-bonded)	1800–2200
WC	2600
W_2C	2200
$(Fe, Cr)_7C_3$	1200–1600
Mo_2C	1500
VC	2500
TiC	3200
Si_3N_4	1800
TiN	3000
SiC	2400
B_4C	3700
Cubic boron nitride	5000
Diamond	8000

*From various sources; for most materials, a wide range of hardness values is reported.

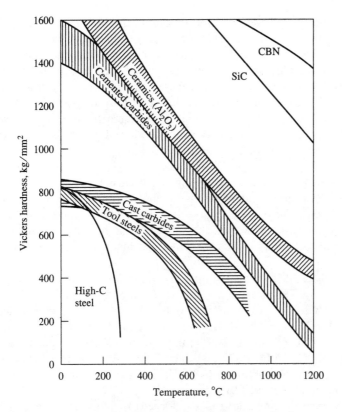

Figure 16–17   High temperatures developed in cutting hard materials at high speeds are better resisted by some tool materials. (*From J.A. Schey, Tribology in Metalworking, ASM International, 1983, p. 113. With permission.*)

2. Toughness is needed to survive mechanical shocks (impact loading) in interrupted cuts. Shocks occur even in continuous chip formation processes, when the tool encounters a localized hard spot.

3. Thermal shock resistance is needed when rapid heating and cooling take place in interrupted cuts. High thermal conductivity is preferable because it keeps temperatures low in the contact area.

4. Low adhesion to the workpiece material helps to avoid localized welding. Paradoxically, high adhesion is desirable when a secondary shear zone is to be stabilized; however, a *diffusion barrier* is then needed.

5. Diffusion of constituents of the tool into the workpiece material results in rapid wear; therefore, solubility of the tool in the workpiece material must be low.

Low hardness and high adhesion are undesirable because they allow distortion of the tool profile, rounding of the tool nose, gradual flank wear and, combined with diffusion, crater wear. Inadequate toughness and thermal shock resistance lead to edge chipping and even total fracture. Unfortunately, the hardness and heat resistance of materials can, in general, be increased only at the expense of toughness; therefore, there is no absolute best tool material available. Recent advances are often based on combining the desirable properties of a substrate and a coating. In the following, the most important tool materials will be discussed in order of rising temperature resistance.

Carbon Steels   Carbon steels derive their hardness from the martensitic transformation. Martensite softens (tempers) above 250°C; therefore, carbon steels are suitable only for machining soft materials such as wood, and then only at low production rates. However, they are hard and hold a keen edge, therefore, high-carbon steel hand reamers are sometimes made for metal cutting.

High-Speed Steel (HSS)   The vast majority of tool steels is in the *high-speed steel* (HSS) category. The two main groups are the molybdenum (M1, M2, etc., typically with 0.8% C, 4% Cr, 5–8% Mo, 0–6% W, and 1–2% V) and tungsten (such as T1, with 0.7C-4Cr-18W-1V) types. The carbides formed with the alloying elements constitute some 10–20% of the volume and allow repeated heating and cooling to 550°C without any loss in hardness. Even higher temperatures are permissible with the addition of 5–8% Co, sometimes coupled with an increased carbon content (M40 and T15 grades).

All these steels can be hot-rolled or forged to a dimension from which the cutting tool can be readily manufactured, in the annealed condition, by conventional machining techniques. Before final grinding, they are subjected to heat treatment which imparts great strength and high (HRC 63 and over) hardness coupled with reasonable toughness. They can be repeatedly reground. They remain important for the metal-cutting industry, especially for drills, reamers, broaches, and other kinds of form tooling (Fig. 16–18). Improvements in melting and casting techniques have improved their quality; some grades are made by consolidation of prealloyed atomized powder (Sec. 11-6), ensuring more uniform distribution of finer carbides and allowing higher alloying element concentration.

Surface coatings are playing an increasing role. Tempering in steam (*bluing*) creates a hard, porous Fe_3O_4 layer which increases tool life. More effective are nitriding and, especially, the gold-colored PVD TiN coating (Sec. 19-6) which reduces friction and minimizes built-up edge formation. The black-colored TiC coating is deposited at high temperature by CVD (Sec. 19-7) and the tool has to be re-heat-treated. Chromium carbide coatings are better for cutting Ti and are used also for Al. Coatings give two- to sixfold increases in tool life.

Cast Carbides   When the carbides reach very high proportions, the tool material is not hot-workable any more and must be cast to shape. The matrix of *cast carbides* (around 45%) is usually a cobalt alloy into which carbides of Cr and W,

Figure 16-18   Some high-speed steel (HSS) tools commonly encountered: (a) gear-tooth cutter, (b) shell-end mill, (c) slab mill, (d) side mill, (e) slotting mill, (f) combined drill and countersink, (g) countersink, (h) ball-end mill, (i) square-end mill, (j) single-angle cutter, (k) tap, (l) thread-cutting die, (m) reamer, and (n) angular cutter.

formed with 2–3% C, are embedded. Softening is gradual (Fig. 16–17) and higher cutting speeds are permissible, but ductility and toughness are much reduced.

Cemented Carbides   *Cemented carbides* produced by powder-metallurgy techniques (Sec. 11-6) have achieved a dominant position. The matrix is usually cobalt, 3–6% for greater hardness, 6–15% for greater toughness. Carbide grades are classified according to codes developed in various countries (e.g., the C system in the United States) and by the ISO (ISO 513-91). The carbide phase may be made entirely of WC for cutting nonferrous metals and gray cast iron (C1 and C2 grades; ISO group K), but diffusion would lead to rapid cratering in cutting steel (Fig. 16–19). Therefore, 10–40% TiC or TaC (or both), which form a carbide-rich diffusion-resistant interface, are added to grades destined for the machining of steel (C4 to C8; ISO group P). Malleable and spheroidal cast iron present the same diffusion danger and are cut with steel-cutting grades. General-purpose carbides (ISO group M) contain smaller quantities of mixed carbides. Cemented carbides soften only gradually (Fig. 16–17) and work best at higher temperatures (over 600°C).

Coated Carbides   Ideally, the tool should possess a very hard, nonreactive surface that also acts as a diffusion barrier, yet it should have a base of sufficient

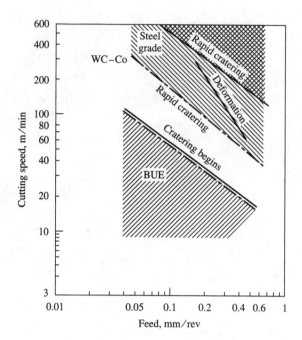

Figure 16–19   The dominant wear mechanism is a function of cutting speed and feed in cutting 0.4% C steel of HV 200. [*After E.M. Trent, Inst. Prod. Eng. J.,* **38**:*105–130,* *(1959). With permission of the Institution of Production Engineers, London.*]

fracture toughness to allow interrupted cuts. *Coated carbides* achieve this aim by combining the virtues of a cemented WC base with those of a thin (typically 5-μm) coating of a ceramic. TiC increases wear resistance, TiN reduces friction and adhesion, Al_2O_3 imparts oxidation and abrasion resistance, TiAlN has hot hardness and impact resistance. Deposition (Chap. 19) is by PVD at relatively low temperatures; CVD gives better bond to the substrate, but high temperatures cause the brittle η phase to form and this reduces transverse rupture strength. Several layers are usually deposited on top of each other to cater for various functions (e.g., a base layer of TiC, followed by Al_2O_3 and TiN). Multiple, alternating fine-grain layers of two or more coating types increase tool life at high speeds. Some feel for the benefits of coated carbides may be gained from Fig. 16–20. Coated carbides have captured some 80% of the market and are extensively used in production turning and milling of steels and cast irons.

Cermets   Cemented carbides are a subclass of *cermets*, ceramics bonded with a metallic phase (Sec. 11-6). For cutting steel and stainless steel, TiC bonded with nickel and molybdenum has gained acceptance. Better thermal conductivity

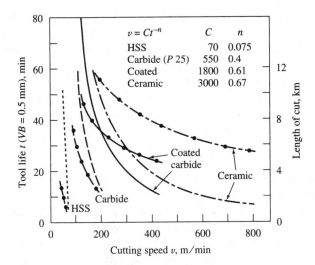

$v = Ct^{-n}$

	C	n
HSS	70	0.075
Carbide ($P\,25$)	550	0.4
Coated	1800	0.61
Ceramic	3000	0.67

Figure 16–20   Tool life is enhanced and the length of chip that can be cut increases greatly on using coated carbide or ceramic tooling for cutting steel. [R. Abel and V. Gomell, Ind. Anz., **102**:27–30 (1980). With permission of Verlag W. Giradet, Essen, Germany.]

and higher cutting speeds characterize the mixed TiC–TiN grades, often used for high-speed finishing and for machining near-net-shape parts.

Ceramic Tools   *Ceramics* such as Al_2O_3 may be used other than as coatings; they can be made, by sintering or hot pressing, into solid tool inserts. Since they are self-sintered (with the help of a sintering aid but without a metal binder), they are suitable for very high speeds, but only at light and continuous loads. However, great advances have been made in improving the reliability of these tools and their range of application is growing. Tools of Al_2O_3 reinforced with 25–40% SiC whiskers, and those made of silicon nitride (Si_3N_4) and Si–Al–O–N (sialon) ceramics, are tougher and more wear-resistant, and can be used for interrupted cuts. They are extensively used in cutting superalloys, gray cast iron, and sialon also for steel.

Polycrystalline Cubic Boron Nitride (PCBN)   Made by high-temperature, high-pressure techniques similar to those used for making synthetic diamonds, *cubic boron nitride* (CBN) has a hardness second only to that of diamond (Table 16-3). Its great advantage is that it does not suffer diffusive wear in cutting ferrous materials. It can be sintered into a 0.5-mm-thick layer onto a cemented carbide base, or made into inserts with or without a ceramic binder (for example, TiN for heat resistance). Inserts with high (> 70%) CBN content are hard and have high heat conductivity, making them suitable for cutting cast iron and superalloys. Higher binder contents increase toughness and reduce heat transfer, an advantage in hard turning of heat-treated steel where—if properly conducted—heat is taken away in the chip. It can replace grinding, producing a surface finish of $R_a = 0.4$ μm roughness or better. No coolant is normally used, especially in interrupted cuts where thermal shock would reduce tool life. It finds increasing application for near-net-shape parts; since the depth of cut is small, feeds need to be increased, because very thin chips do not take enough heat away and fail to break into short enough segments.

Polycrystalline Diamond (PCD)   The hardest material, *diamond* has long been used in the form of natural single crystals for high-speed finishing of aluminum and other nonferrous materials.   Natural diamond suffers from unpredictable early failure, and manufactured single crystals give more reliable performance.   Polycrystalline tool tips are available as self-sintered inserts or as 0.5-mm-thick layers sintered onto a carbide base.   Diamond is also applied by CVD as a < 50-μm-thick coating to drills.   Diamond outperforms all other materials on highly abrasive workpieces such as hypereutectic Al–Si alloys.   However, at high temperatures it changes into graphite which diffuses into iron; therefore, it is not suitable for cutting steel.

16-3-2   Tool Construction

High-speed steels have sufficient toughness to be made into *monolithic* (single-piece) tools (Fig. 16–18). Cemented carbides and cermets can be made into solid tools but the risk of total fracture is great and the cost can become high.   Therefore

Figure 16–21   Turning tools have coated carbide inserts clamped to toolholders (*Courtesy Kennametal, Latrobe, Pennsylvania.*)

their broadest application is in the form of tool *inserts*, which are either brazed or clamped (Fig. 16–21) to a tough steel body. Specially constructed cutters (*indexable cutters*, Fig. 16–22) permit moving the insert to compensate for wear, and can thus be used for extended periods of time. Ceramic tools are always made as inserts.

High-speed steel and many cemented carbide tools are reground several times in the machine shop. Some carbide and most ceramic tools are of the throwaway type, and are made so as to have several usable cutting edges (Figs. 16–8c and 16–21).

As discussed in Sec. 16-1-1, a large positive rake angle shortens the shear zone and reduces energy consumption. This also weakens the tool; therefore, large rake angles used to be permissible only for cutting lower-strength materials. The toughness of tool materials is, however, continually improving, and more turning and milling cutters are being made with positive rake angle. The more brittle tool materials are still made with a small positive, zero, or even negative rake angle. A three-cornered cutting insert can then have six usable cutting edges (as in Fig. 16–21). The edges of brittle tool materials may be chamfered to increase their strength, thus a negative angle develops even on a positive-rake tool (as in Fig. 16–14c). Positive-rake inserts in negative-rake tool holders combine advantages of both. In cutting with a negative rake angle, the force pushing the tool out of the workpiece is large and vibrations are easily generated; therefore, an extremely stiff machine tool is needed.

Figure 16–22   This indexable face mill holds peripheral carbide inserts for roughing and face inserts (replaced by dummy inserts during roughing) for finishing. Unequal insert spacing minimizes chatter. (*Courtesy Ingersoll Cutting Tool Company, Rockford, Illinois.*)

With the proliferation of tool materials and coatings, selection of the optimum tool is becoming more difficult, and tool manufacturers offer electronic catalogs which often incorporate some elements of an expert system.

16-3-3 Tool Holders and Fixtures

Tools are mounted in the machine tool in *tool holders*. For rotating tools, these typically have a tapered shank that fits into a matching tapered hole. A low-angle taper (e.g., the so-called Morse cone of 1/20 taper) is self-locking (friction holds the cone in place) and is difficult to remove; therefore, most tool holders designed for rapid, automatic tool change have a steeper taper (such as 7/24). A disadvantage is that in high-speed machining the centrifugal force opens up the seat and the grip on the toolholder loosens. Various designs have been developed for stiffness, concentricity, accurate axial location, and secure clamping of tool combined with ease of ejection. The cup-shaped, so-called HSK holders are finding increasing acceptance for higher speeds. *Integral tool holders* are in one piece and are stiffer but less flexible in application than *modular tool holders* in which only tool *cartridges* are changed.

Accurate machining requires that not only the tool but also the workpiece be accurately and rigidly located. In some machine tools (particularly, lathes) the construction of the machine provides this, but in other instances the part must be held—with mechanical or hydraulic clamps—against locating points in *fixtures*. Care must be taken not to distort thin parts. Dedicated fixtures are effective for mass production; general-purpose, reconfigurable fixtures are more suitable for medium- and small-lot production, and modular fixtures are between the two in terms of flexibility and ease of setup. To facilitate movement from station to station, parts may be held on *pallets*, precision metal slabs with holes or slots that engage on pins or tabs in the machine tool workspace.

16-4 CLASSIFICATION

The are innumerable ways of removing material by chip formation, and there are several ways of classifying processes. A useful grouping is by the number of cutting edges (Fig. 16–23); within this, two basically different approaches can be used: *forming* and *generating*.

16-4-1 Forming

A shape is said to be *formed* when the cutting tool possesses the finished contour of the workpiece. All that is necessary, in addition to the relative movement required to produce the chip (the *primary motion*), is to feed (*plunge*) the tool in depth.

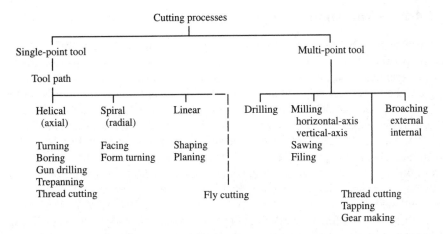

Figure 16–23   Classification of chipforming processes. *(Adapted from J.A. Schey, in ASM Handbook, vol. 20, Materials Selection and Design, ASM International, 1997, p. 695. With permission.)*

How the primary motion is generated is immaterial. The workpiece can be rotated against a stationary tool (*turning*, Fig. 16–24*a*), or the workpiece and tool can be moved relative to each other in a linear motion (*shaping* or *planing*, Fig. 16–24*b*), or the tool can be rotated against a stationary workpiece (*milling* and *drilling*, Fig. 16–24*c*) or against a rotating workpiece (as in cylindrical grinding). The accuracy of the surface profile depends mostly on the accuracy of the forming tool.

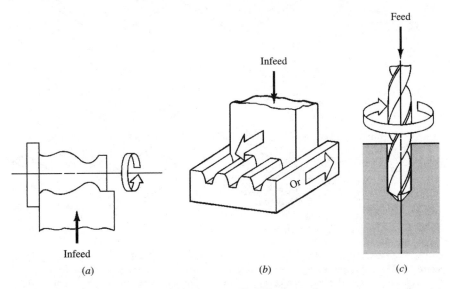

Figure 16–24   The cutting tool is made to the profile of the part in forming processes such as (*a*) form turning, (*b*) shaping or planing, and (*c*) drilling.

16-4-2   Generating

A surface may be *generated* by combining several motions that not only accomplish the chip-forming process (primary motion) but also move the point
of engagement along the surface (described as the *feed motion f*, Fig. 16–10).
Again, the workpiece may rotate around its axis, as in turning; the tool is set to
cut a certain depth and receives a continuous, longitudinal feed motion. When
the workpiece axis and feed direction are parallel, a cylinder is generated (Figs.
16–10*b* and 16–25*a*); when they are at an angle, a cone is generated (Fig. 16–25*b*).
If, in addition to the primary and feed motions, the distance of the cutting tool
from the workpiece axis is varied in some programmed fashion—e.g., by means
of cams, a copying device, or numerical control—a large variety of shapes can
be generated.

   When the tool (or the workpiece) is fed perpendicular to the primary linear
(shaping or planing) movement, a flat surface is generated (Figs. 16–10*a* and
16–25*c*). If the workpiece were given a feed motion by rotating it around its axis
parallel to the tool motion, a cylinder could be machined. The workpiece axis

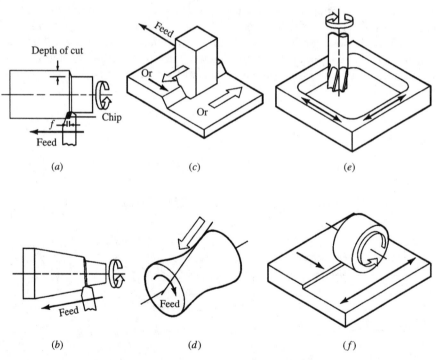

(a)   (c)   (e)

(b)   (d)   (f)

Figure 16–25   Programmed tool motion (feed) is necessary in generating a shape: (a) turning a
cylinder and (b) a cone; (c) shaping (planing) a flat and (d) a hyperboloid; (e)
milling a pocket; and (f) grinding a flat (principal motions are marked with
hollow arrows, feed motions with solid arrows).

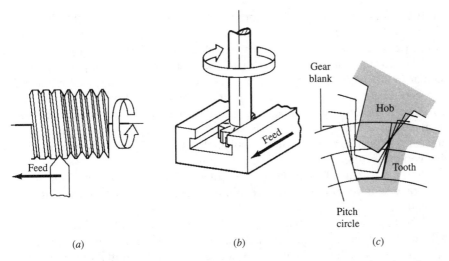

Figure 16–26   Forming and generation may be combined: (*a*) thread cutting, (*b*) T-slot milling, and (*c*) gear hobbing.

could be set at an angle and then a rotational hyperboloid would be generated (Fig. 16–25*d*). In principle, any surface that can be described by a straight generatrix may be produced by this technique. A tool of axial symmetry may rotate while the workpiece is being fed, leading to *milling* (Fig. 16–25*e*) or *grinding* (Fig. 16–25*f*).

Frequently, *combined forming and generating* offers advantages. Thus, a thread may be cut with a profiled tool fed axially at the appropriate rate (Fig. 16–26*a*). A slot or dovetail may be milled into a workpiece (Fig. 16–26*b*). A gear may be cut with a *hob* that gradually generates the profile of the gear teeth (Fig. 16–26*c*) while both hob and workpiece rotate.

16-5   SINGLE-POINT MACHINING

It is obvious from the previous discussion that one of the most versatile tools is a *single-point cutting tool* moved in a programmed fashion.

16-5-1   The Tool

The tool must accommodate not only the primary motion (as an orthogonal tool would, Fig. 16–1) but it must also allow for feeding and chip disposal. Therefore, the cutting edge is almost always inclined (oblique cutting, Fig. 16–9), and the chip is wound into a helix rather than a spiral. The tool is relieved both in the direction of feed and on the surface that touches the newly generated surface (Fig.

16–10*b*), and thus has major and minor flank surfaces (Fig. 16–27). Intersections of these with the rake face of the tool constitute the *major and minor cutting edges*, respectively. The nose is rounded with an adequate (typically, 1-mm) radius.

The all-important rake angle should really be measured in a plane perpendicular to the major cutting edge, but, for convenience, all angles are measured in a coordinate system that coincides with the major axes of the tool bit (Fig. 16–27). While this system appears simple, it creates various problems; these are resolved, however, by ISO 3002 (Basic quantities in cutting and grinding). In any case, it must be recognized that tool angles have meaning only in relation to the workpiece, after installation in the machine tool.

Some recommendations on cutting-tool angles are contained in Table 16–4. They represent a compromise to give minimum cutting force with maximum tool strength.

The single-point tool may be replaced with a *rotating tool*, which is a disk held at an appropriate angle. The disk may be rotated or it may rotate as a result of its contact with the workpiece; thus, all parts of the circumference are used.

16-5-2   Turning

The most widely used machine tool is the *engine lathe* (*center lathe*, Fig. 16–28), which provides a rotary primary motion while the appropriate feed motions are imparted to the tool.

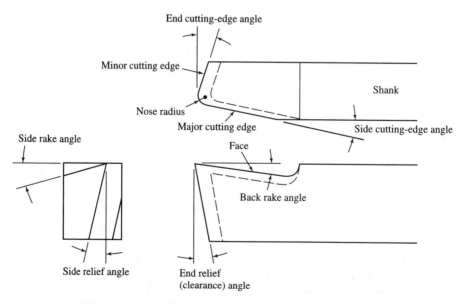

Figure 16–27   Nomenclature used for describing the geometry of single-point cutting tools (compare with Fig. 16–21).

Table 16–4   Typical single-point cutting tool angles*

Workpiece		Tool Material					
		High-Speed Steel		Brazed WC		Throwaway WC	
Material	BHN	Back Rake	Side Rake	Back Rake	Side Rake	Back Rake	Side Rake
Steels	< 225	10	12	0	6	−5	−5
	to 325	8	10	0	6	−5	−5
	to 425	0	10	0	6	−5	−5
	> 425	0	10	−5	−5	−5	−5
Stainless							
Ferritic		5	8	0	6	0	5
Austenitic		0	10	0	6	5	5
Martensitic		0	10	0	6	−5	−5
Cast iron	< 300	5	10	0	6	−5	−5
	> 300	5	15	−5	−5	−5	−5
Zn alloy	80–100	10	10	5	5	0	5
Al, Mg alloy		20	15	3	15	0	5
Cu alloy		5	10	0	8	0	5
Superalloy		10	5	6	0	5	5
Ti alloy		5	5	6	−5	−5	5
Thermoplastics		0	20–30	0	0	0	20–30
Thermosets		0	20–30	15	0	15	5

NOTE: End and side relief angles are typically 5° but higher values are usual for plastics. Edge angle is typically 15° but is less for nonferrous metals and plastics with HSS tooling.
*Extracted from *Machining Data Handbook*, 3d ed., Machinability Data Center, Metcut Research Associates, Cincinnati, Ohio, 1980.

The workpiece must be firmly held, most frequently in a *chuck* (Fig. 16–29a). Three-jaw chucks with simultaneous jaw adjustment are self-centering. Other chucks have two, three, or four independently adjustable jaws for holding other than round workpieces. Bars may also be held in *collets*, which consist of a split bushing pushed or pulled against a conical surface (Fig. 16–29b). Workpieces of awkward shape are often held by bolts on a *face plate* (Fig. 16–29c).

The *headstock* contains the drive mechanism, usually incorporating change gears and/or a variable-speed drive. Long workpieces are supported at their end with a *center* held in the *tailstock*. The tool itself is held in a *tool post* which allows setting the tool at an angle (horizontally and vertically). The tool post is mounted on a *cross slide* which provides radial tool movement. The cross slide is guided in a *carriage*, which in turn receives support from the *ways* machined in the *bed* that ensures rigidity and freedom from vibrations. An overhanging part, the *apron* of the carriage, may be engaged with the *feed rod* to give continuous feed motion, or with a *lead screw* for the cutting of threads. Very long workpieces

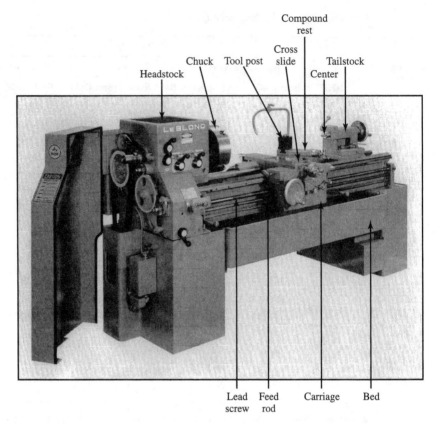

Figure 16–28 A typical engine lathe. Capacity: 380-mm- (15-in-) diameter swing; 1370-mm (54-in) length. (*Courtesy LeBlond Inc., Cincinnati, Ohio.*)

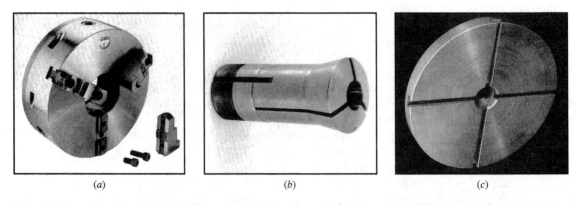

Figure 16–29 Workpieces may be held in a (*a*) chuck, (*b*) collet, or (*c*) face plate. [(*a*) and (*b*) *courtesy of DoALL Co.*]

are secured against excessive deflection by two fingers of a *center rest* or *steady rest* bolted to the lathe bed; a *follow rest* is clamped to the carriage.

Sometimes the tool post sits on a *compound tool rest* which incorporates a slide that can be set at any angle; thus, conical surfaces may be formed by hand feeding the tool. A *four-way tool post* can be rotated about a vertical shaft and allows quick changing of tools in preset positions, thus speeding up successive operations.

16-5-3   Boring

When the internal surface of a hollow part is turned, the operation is referred to as *boring* (Fig. 16–30a). For short lengths, the tool may be mounted on a cantilevered bar (*quill*) in the tool post. A long bar is prone to excessive vibration, and it is then preferable to have the workpiece secured to the lathe bed while the *boring bar*, clamped in jaws at one end and supported in the tailstock at its other end, is driven. A number of patented solutions exist that aim at reducing or damping out vibrations. The simplest is a *plug damper* (a heavy mass in the bore of the free end of the boring bar). Simultaneous cutting with two or three boring inserts equalizes the forces and reduces vibration. A special-purpose machine performing a similar operation, but with more firmly guided boring bars, is the *horizontal boring machine*.

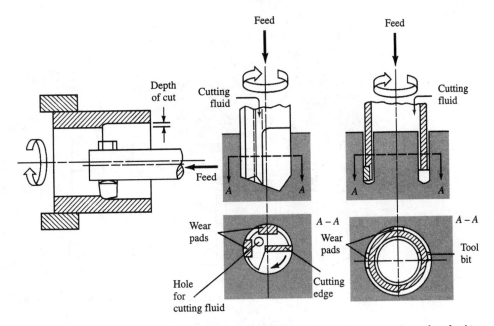

Figure 16–30   Machining of holes with single-point tools: (*a*) enlarging (improving the surface finish) by boring, (*b*) gun-drilling, and (*c*) trepanning.

Heavy and large-diameter workpieces that need to be machined on both inside and outside surfaces may be better supported on a lathe turned into a vertical position; called a *vertical turning and boring mill* or *vertical boring machine*, such a lathe can work on several surfaces of a workpiece fastened to the rotating, vertical-axis face plate of the machine (Fig. 16–31). (Alternatively, the tool rotates and the part is stationary.)   Additional live (driven) spindles allow secondary operations in one setup.

16-5-4   Gun Drilling and Trepanning

Holes may be produced in solid workpieces by single-point machining techniques resembling boring. In *gun drilling* the cutting forces are balanced by guide pads placed at angles of 90° and 180° to the cutting edge (Fig. 16–30*b*). To start a hole, a hardened steel guide (*boring bush*) is held against the face of the workpiece. Once the hole has started, the tool guides itself; the guide pads burnish the cut surface. Coolant fed through the tool keeps temperatures low and flushes out the chips. It can be used even for holes below 3-mm diameter. The tool is usually held stationary while the workpiece, clamped in a chuck and stabilized by steady

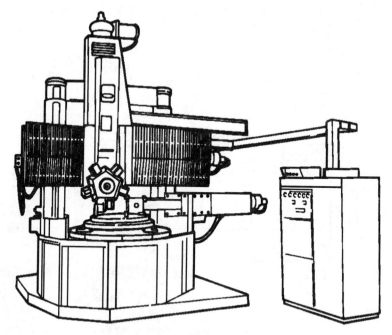

Figure 16–31   Large workpieces are often machined on vertical boring machines. The illustration is of a vertical CNC turret lathe. (*From The Tool and Manufacturing Engineers Handbook, 4th ed., vol. 1, p. 15.54. With permission of the Society of Manufacturing Engineers, Dearborn, Michigan.*)

rests, rotates. In a variant, the coolant is applied to the outside of the tool tube and flows back, together with the chips, in the hollow of the tube.

Larger holes (diameter of 20 mm and over) can be made by *trepanning*: The cutting tool bit is fastened on the end face of a tube, and the hole is machined by removing an annulus while leaving a center core (Fig. 16–30c). Again, greatly improved patented tool varieties exist.

Both techniques are suitable for making relatively deep holes, of a depth-to-diameter ratio of five and over. Force-fed cutting fluid lubricates and helps the removal of the chips, and is vital to success.

16-5-5 Facing

In *facing*, a plane perpendicular to the lathe axis is produced by moving the single-point tool in the carriage so that the feed motion is toward the center of the lathe (Fig. 16–32a). *Parting off* accomplishes the same task but two surfaces are now simultaneously generated (Fig. 16–32b). The cutting speed diminishes as the tool moves toward the center unless the rotational speed is increased in a programmed manner, using a variable-speed drive.

16-5-6 Forming

This method of producing complex rotational shapes (Fig. 16–24a) is fast and efficient, but cutting forces are high and the workpiece could suffer excessive deflection. On cantilevered workpieces the length of the *forming tool* is usually kept to 2.5 times the workpiece diameter. For longer lengths the workpiece is supported by a backrest or roller support, or, if possible, on a center.

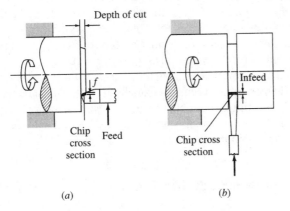

(a)   (b)

Figure 16–32   Flat end faces may be generated by (a) facing and (b) parting off.

16-5-7   Automatic Lathe

The hand operation of a lathe requires considerable skill. The talents of a highly skilled operator are poorly utilized in repetitive production; therefore, various efforts at automation have long been made. Unfortunately, the terminology has become somewhat confusing. In the context used here, an *automatic lathe* is similar to an engine lathe, but all movements of the carriage required to generate the workpiece surface are obtained by mechanical means.

Radial movement of the tool may be derived from a *cam bar* or a *tracer template*, or separate drives may be actuated by NC. Alternatively, the motions may be derived from a model of the workpiece using a *copying arrangement*.

All these machines may be supplied with material by hand, semiautomatically, or fully automatically.

16-5-8   Turret Lathe

When the surface can be generated or formed with relatively simple motions but requires a sequence of operations (such as turning, facing, boring, and drilling) for completion, the requisite number of tools can be accommodated by replacing the tailstock of a lathe with a *turret*. Equipped with a quick-clamp device, a turret brings several (usually six) tools into position very rapidly. All tools are fed in the axial direction, by moving the turret on a slide (*ram-type lathe*) or, for heavier work, on a saddle, which itself moves on the ways (*saddle-type lathe*). Axial feed movement is terminated when a preset stop is reached. Additional tools may be mounted in a turret on the cross slide and also on a rear tool post. The number of possible operations and the variety of combinations is very large, because several tools may be mounted at any one station for multiple cuts, or simultaneous cuts may be performed at several stations (*combined cut*). Once the machine is set up, it requires relatively little skill to operate.

The *CNC lathe* (Fig. 16–33) offers great flexibility of operation because all motions can be programmed by software. Equipped with a tool changer and linked to material-moving devices (automatic loaders), it becomes the center of a flexible manufacturing cell. To avoid the need for transfer to another machine, multifunction lathes (often called *turning centers*) can be fitted with a turret or turrets having driven tools that perform transverse milling or drilling on an arrested workpiece. When a part needs cutting from both ends, it must be released, turned, and rechucked, resulting in possible eccentricity; better accuracy is obtained with *dual-spindle machines* in which transfer to the second spindle (replacing the tailstock) is done within the machine.

16-5-9   Automatic Screw Machines

As the name suggests, these machines were originally developed for making screws at high production rates. Cold heading followed by thread rolling has

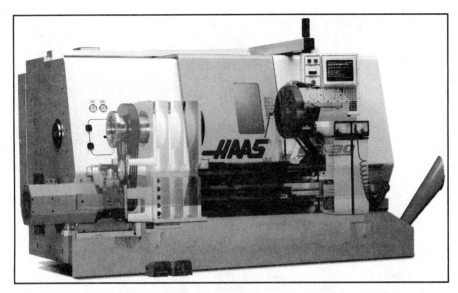

Figure 16–33   CNC lathe with 22-kW drive will turn 370-mm-diameter, 860-mm-long workpieces or 100-mm-diameter bars with tools held in its 12-position servo-controlled turret. Ghosted image shows the inner workings. Stiff cast-iron frame is filled with loose aggregate for vibration damping. Intuitive programming allows operator input. (*Courtesy Haas Automation, Oxnard, California.*)

almost eliminated this market, but machines have been developed to mass-produce more complex shapes.

Single-Spindle Automatics   *Single-spindle automatics* fall into two basically different groups:

Single-spindle automatic screw machines are based on the principle of the turret lathe, but operator action is replaced by appropriately shaped cams or CNC control that bring various tools into action at preset times. The stock (a bar cold drawn to close tolerances) is indexed forward, with cam-operated feed fingers, by the length of one workpiece, at the end of each machining cycle.

Swiss automatics are radically different in that all tools are operated in the same plane, extremely close to the guide bushing through which the rotating bar is continuously fed in a programmed mode. Individual tools are moved radially inward, mounted on slides that bring the appropriate tool into action. Since there is no workpiece overhang, parts of any length may be produced to unsurpassed accuracies and tolerances (down to 2.5 μm). More recent machines are numerically controlled (Fig. 16–34).

Even though several tools may be set to cut at the same time, the total machining time on single-spindle automatics is the sum of individual or simultaneous operations required to finish the part.

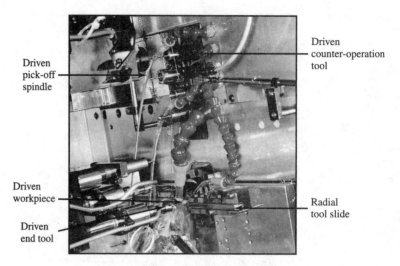

Driven
pick-off
spindle

Driven
counter-operation
tool

Driven
workpiece

Driven
end tool

Radial
tool slide

Figure 16–34   In a modern Swiss-type machine the bar is fed axially while a tool attached to the radial tool slide machines the surface. A driven end tool performs work at the free end. After cutoff, the part is grabbed by the pick-off spindle so that a counteroperation tool can finish the other end. (*Courtesy Tornos Technologies U.S. Corporation, Brookfield, Connecticut.*)

Multispindle Automatics   Productivity may be substantially increased if all operations are simultaneously performed. In *multispindle automatics* (Fig. 16–35) the head of the lathe is replaced by a *spindle carrier* in which four to eight driven spindles feed and rotate as many bars. The turret is replaced by a tool slide on which the appropriate number of tool holders (sometimes separately driven) are mounted. Additional tools are engaged radially, by means of cross slides; the number of these is often less than the number of spindles, because there may be insufficient room for them. The tool slide with the tool holders moves axially forward, and the cross slides move in radially under cam control, complete their assigned task, withdraw, and the spindle carrier indexes the bars to the next position. Thus, for each engagement of the tools, one part is finished.

Automatic screw machines produce mostly parts of axial symmetry (including threaded parts), but special attachments permit auxiliary operations such as milling or cross-drilling while the rotation of one spindle is arrested. Workpieces of irregular shape can be handled on so-called *chucking machines*. In CNC *multispindle turning centers* all movement is under computer control. *Rotary transfer machines* represent a radically different approach: an indexing rotary turret holds the parts which are worked upon by machining heads from the outside.

16-5-10   Shaping and Planing

As indicated in Fig. 16–25c, a surface can be generated with a linear primary motion. In the process of *shaping*, the primary motion is imparted to the tool and

Main drive motor Spindle drive shaft End toolslide Top slide Upper cross slide drums and cams Intermediate cross slide Intermediate cross slide cam and drums

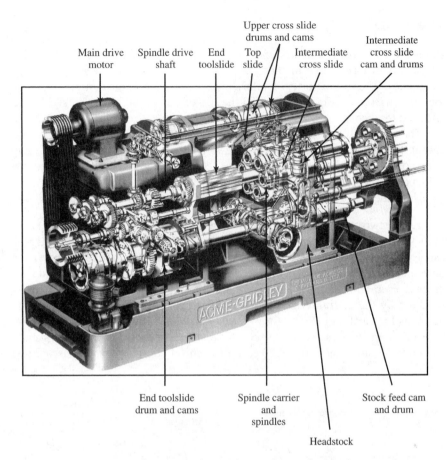

End toolslide drum and cams   Spindle carrier and spindles   Stock feed cam and drum

Headstock

Figure 16–35   Six-spindle automatic bar machine, without tooling. With every indexing of the head, a complete part is made. (*Courtesy National Acme, Cleveland, Ohio.*)

the feed motion to the workpiece (as in Fig. 16–10*a*). The tool is moved back and forth by an overhanging ram, the deflection of which limits the length of stroke. In the process of *planing*, a longer stroke (of practically unlimited length) is obtained by holding the workpiece on a long, horizontal, reciprocating table while attaching the tool to a sturdy column or arch or, rather, a cross rail with a lead screw that generates the feed movement. These processes have largely been replaced by milling, broaching, or belt grinding. For low production quantities, *slotting* (related to shaping) is still used for making internal keyways.

16-6 MULTIPOINT MACHINING

In multipoint machining at least two cutting edges of the same tool are simultaneously engaged at any one time.

16-6-1   Drilling

Holes are the most frequently encountered machined features. In Sec. 16-5-4 we already discussed two methods of making deep holes, based on single-edge techniques. The vast majority of holes, however, is made by the familiar two-edge tool, the *twist drill* (Fig. 16–36), and the life of this tool often has a significant effect on total cost.

The twist drill has several advantages: two cutting edges are more efficient; cutting forces are balanced; helical *flutes* allow access of cutting fluid and help to dispose of the chip; and small *margins* left on the cylindrical surface provide guidance.

Nevertheless, the twist drill also has its problems:

1. The two cutting edges must not come together into a point which, because of its small mass, would quickly overheat and lose its strength. A *chisel edge* is usually left and, because of the highly negative rake angle, no real cutting action takes place in the center of the hole. The material is plastically displaced by a process resembling piercing a semi-infinite body (Fig. 9–14*a*), to be subsequently removed by the cutting edges. The force required for this rotary indentation accounts for much of the total thrust (feed) force in drilling. If necessary, a pilot hole, of a diameter equal to the chisel-edge diameter of the larger drill, will greatly reduce the required feed force. Wear of the chisel edge further increases thrust

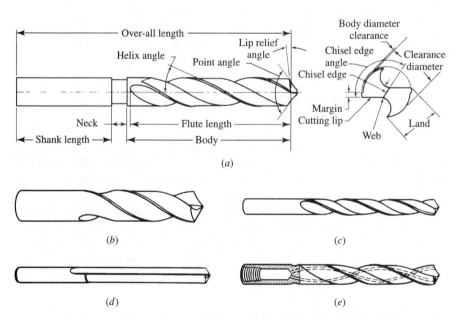

Figure 16–36   Twist drills are available in a great variety of shapes, including: (*a*) jobbers drill, (*b*) automatic screw-machine drill, (*c*) low-helix drill, (*d*) straight-flute drill, and (*e*) straight-shank oil-hole drill. (*From Metals Handbook, 8th ed., vol. 3, ASM International, 1967, p. 78. With permission.*)

force. Therefore, many variants of drill points (web-thinned points to reduce the chisel edge; helical points which cut on all edges; four- and six-facet points, etc.) have been developed for better concentricity, lower thrust force, and reduced wear. Some drills have chip-breaking grooves in the flutes.

2. When starting a hole, the chisel edge tends to wander. The drill must be kept in place by a *drill bushing*, or an indentation must be created with a *center punch*, or a *center-drill and countersink* is used.

3. The helix angle determines the rake angle at the periphery of the drill; the rake angle becomes smaller on moving along the edge to the center. *High-helix (fast-spiral) drills* with wide flutes aid in chip removal, and their increased bearing surface gives better guidance. When drilling thin sheet or materials such as free-machining brass, a reduced helix angle or even a *straight-flute drill* (zero helix-angle drill, Fig. 16–36*d*) is preferable.

4. A basic problem is cooling and removal of chips. Both are helped with internal coolant holes through which fluid is fed under pressure (Fig. 16–36*e*). Drilling in increments (*peckering* or *peck drilling*) also helps chip removal.

5. The surface finish of the hole is not as good as that of a bored hole, and the drill begins to drift at greater depths. Nevertheless, the quality is adequate for a great many purposes, in diameters ranging from 0.05–75 mm, at depth-to-diameter ratios of up to 5 (although deeper holes are often drilled).

6. Significant increases in drill life are obtained with coated (particularly, TiN-coated) HSS drills. Wear resistance is increased and longer tool life obtained with carbide inserts and solid (monolithic) carbide drills; for the latter, rigidity is critical, and three-flute drills are better in this respect. Holes of larger diameters may be cut with specially constructed drills equipped with several indexable inserts (Fig. 16–37).

Figure 16–37   Indexable insert drills for 16–170-mm diameter (*Courtesy Kennametal, Latrobe, Pennsylvania.*)

Spade drills of various configurations are suitable for drilling holes of all diameters and, when made of carbide, also for hard materials. Straight-shank, monolithic carbide spade drills have a larger cross section than twist drills of the same diameter and are preferred for small (from 0.013 mm up) holes. They cut slower and receive no guidance from flutes; hence, they are not suitable for deep holes.

Drilling Equipment   The simplest *drill press* has a single rotating spindle which is fed axially, at a set rate or under a constant feed force, into a workpiece held rigidly on a table (Fig. 16–38). A *radial arm drill* has a swinging arm which provides much greater freedom. When several holes have to be produced in a large number of workpieces, simultaneous drilling with a *multispindle drill head* or drill press assures better accuracy of relative hole location. Exceptional accuracy of hole location is achieved on the *jig borer*, which is really a drill press equipped with

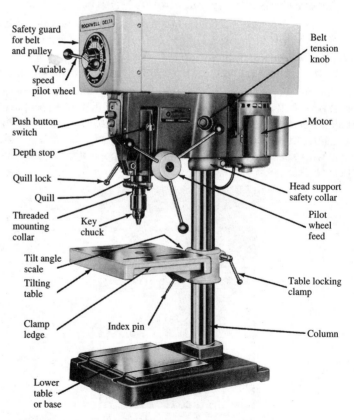

Figure 16–38   Components of a drill press of 380-mm (15-in) capacity (drill center to column distance). (*Courtesy Delta International Machinery Corp., Pittsburgh, Pennsylvania.*)

·a high-precision table movement in two directions. Numerical control is often employed.

Drills can be held in the tailstock of a lathe to machine holes of good concentricity, and drills are important tools for all automatics.

The quality of drilled holes is greatly improved by *reaming*, which could be classified as a milling operation (a reamer is shown in Fig. 16–18*m*). Seats for countersunk screws are prepared by *spot facing*, essentially an end-milling operation in the plunging mode (a countersink is shown in Fig. 16–18*g*).

16-6-2   Milling

Milling is one of the most versatile cutting processes, and it is indispensable for the manufacture of parts of nonrotational symmetry. There are innumerable varieties of milling cutter geometries, but, basically, they can all be classified according to the orientation of tool (or, rather, the orientation of the cutting edges and axis of rotation) relative to the workpiece, although everyday usage can be confusing.

Horizontal Mills   These machines have the axis of the cutter parallel to the workpiece surface, and in traditional mills the cutter axis is horizontal, with both ends of the cutter supported.

1. In *plain* or *slab milling* (Fig. 16–39*a*) the cutting edges define the surface of a cylinder and can be straight (parallel to the cylinder axis) or helical (see Fig. 16–18*c*). The milling cutter is wide enough to cover the entire width of the workpiece surface. The primary motion is the rotation of the cutter while feed is imparted to the workpiece. Both the primary and feed motions are continuous. The chips are thickest on the surface of the workpiece and diminish toward the base of the cut. In *down* or *climb milling* (Fig. 16–39*b*), feed motion is given in the direction of

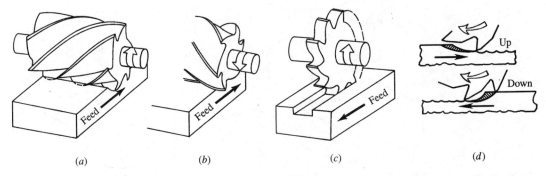

| (a) | (b) | (c) | (d) |

Figure 16–39   In horizontal-axis milling the cutter axis is parallel to the workpiece surface. Direction of feed and cutter rotation determine whether milling is (*a*) up or (*b*) down, resulting in (*d*) different undeformed chip thicknesses. A narrow cutter (*c*) performs slotting. [(*d*) *after M.C. Shaw, Metal Cutting Principles, Oxford University Press, 1984. With permission.*]

cutter rotation; thus, the cut begins at the surface with a well-defined undeformed chip thickness (Fig. 16–39*d*), and surface quality is good. However, the initial force is high and the machine must be of sturdy construction and equipped with backlash-free drives. In *conventional* or *up milling* (Fig. 16–39*a* and *d*) the tooth engages at a minimum depth, the surface may become smeared and more wavy, but starting forces are lower. Therefore, this was the preferred method before the arrival of more rigid machine tools. Note that chip thickness varies from maximum to zero, thus the power required is some 50% higher than calculated from Eq. (16-16). Mean chip thickness is usually around 0.1 mm.

2. When a cylindrical cutter is narrower than the workpiece, the cutting edges must be carried over the end faces of the cylinder (Fig. 16–39*c*; see also Fig. 16–18*d* and *e*). Because of their action, these mills are called *slotters* or *slitting cutters*. When only the side teeth are engaged, one speaks of *side-milling cutters*.

Vertical Mills   These machines have the axis of the cutter perpendicular to the workpiece surface (Fig. 16–40). The cutter axis used to be vertical and is still so in CNC *vertical machining centers*. Alternatively, CNC machines may have a horizontal-axis spindle, and then they are called *horizontal machining centers*; the workpiece surface is vertical and chip removal is easier. The cutter is always cantilevered (supported at one end only).

1. When the teeth are attached to the cutter face which is perpendicular to the axis (see Fig. 16–22), one speaks of *face milling* (Fig. 16–40*a*). In many ways,

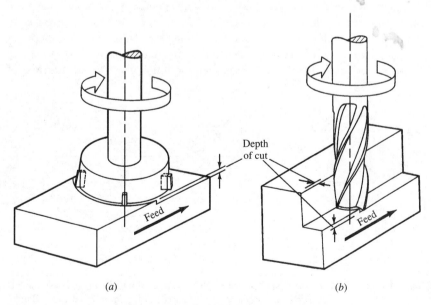

(*a*)   (*b*)

Figure 16–40   In vertical-axis milling the cutter axis is perpendicular to the workpiece surface: (*a*) face milling and (*b*) end milling.

this is similar to machining with many single-point tools moving in circles. Since the cut always starts with a definite chip thickness, face milling uses some 40% less power than slab milling. A variant frequently employed is *fly cutting*, that is, cutting with a single-point tool fixed to the end of an arm protruding from the perpendicular milling shaft.

2. Cutting edges carried over onto the cylindrical surface of the cutter create an *end mill* (Fig. 16–40b; see also Fig. 16–18b, h, i, and j). End mills are among the most versatile tools because they can be made to follow any path in the plane of and perpendicular (or at an angle) to the workpiece surface. Thus, pockets and contoured surfaces of almost any shape, depth, and size can be machined (Fig. 16–25e). Even huge surfaces are sometimes fully machined, as, for example, in making aircraft wing skins.

Milling machines The various feed motions of a milling machine may be controlled by hand, although the milling of complex shapes requires considerable skill. Milling machines can be mechanized or automated to various degrees:

1. *Copy millers* use a model of the finished part to transfer the movement from a copying head to the milling head.

2. *NC* and *CNC milling machines*, which move some of the skill of the operator into the programming stage (Sec. 2-5-3), have been rapidly developing. If properly utilized, they speed up production by eliminating much of the setup time and the trial-and-error procedures inevitable with manual control.

3. *Machining centers* (Fig. 16–41) are, as said earlier, CNC milling machines, often of extended capability, performing not only a variety of milling but also drilling, boring, tapping, and possibly also turning operations in one setup. More than one machining head may be used; *universal machining centers* have one horizontal and one vertical head. The *x-y* table may incorporate a rotary table. Some machining centers have a modular design: Tool heads can be changed for optimum production. After a reference surface has been prepared, sometimes on a separate machine, a machining center can work on the workpiece from five sides. The machines often form the core of flexible manufacturing cells (see Sec. 21-2-4).

One of the more challenging tasks in milling is the manufacture of dies and molds for metalworking and plastics processing. Often, the configuration is complex and surface finish requirements are high, and tools have grown to very large sizes (car body press tools may measure 5.5 m × 3 m and have a mass of 65 000 kg). Under manual control, toolmaking is extremely slow, with low metal removal rates, possible errors, and poor surface finish which requires extensive manual finishing. Four- or five-axis milling machines are now used extensively for cutting prehardened tool steel blocks at high rates and, increasingly, with finishing passes made in the same setup. Much of the work is done with ball-nose cutters; when the cutter axis is perpendicular to the cut surface, speed is zero at the center of the cutter. Therefore, mills now allow tilting the head at 15–30°

Figure 16–41   Machining centers are CNC milling machines. An automatic tool changer loads tools from a magazine into the vertical spindle. Note the chip-disposal chute on the left. (*Courtesy Bridgeport Machines Inc., Bridgeport, Connecticut.*)

so that all contact is with positive speed and feed. Feed rates in 3-D contouring are up to 0.2–0.5 mm/r at spindle speeds over 10 000 r/min (20 000 r/min is not exceptional any more and even 60 000 r/min has been reached), and acceleration of slides approaches and even exceeds 1.0 g (10 m/s^2). Cutting conditions must, however, be chosen to allow making the finishing cut with one tool, since tool change would leave a visible mark. On larger dies, this may require a tool life of several hours. An end mill or ball-nosed cutter will leave scallops which have to be removed. In advanced processes, scallop height is under 10 µm, and gradually the point is being reached where manual finishing is required only for dies that make exposed (visible) parts.

   An example of a large mold produced on a huge machining center is shown in Fig. 16–42.

16-6-3   Sawing and Filing

A very narrow slitting cutter becomes a *cold saw*. The precision-ground teeth need not go deep in the radial direction, and are usually made as HSS or carbide-tipped inserts attached to a larger saw blade. For less-demanding applications the very accurate formation of various tool angles may be relaxed, and the teeth are formed by bending into position. The basic cutting action of such *circular saws* is still closely related to milling.

   When the teeth are laid out into a straight line, one obtains a *hacksaw* or, if the saw blade is flexible and made into an infinite loop, a *band saw*. Machines now have controlled feed rates and some are run by programming software.

Figure 16–42   The mold for a pick-up truck bed liner is milled on a giant multiaxis machining center. (*Courtesy Ingersoll Milling Machine Company, Rockford, Illinois.*)

A fine-pitch slab mill laid out into a flat becomes a *file*. The individual cutting edges are broken up into a series of teeth in a *crosscut file*.

All these tools are form tools, and the cut progresses by a positive in-feed or by the pressure exerted on the tool.

16-6-4   Broaching and Thread Cutting

These processes differ from those discussed thus far in that the only motion is the primary motion of the tool. The feed is obtained by placing the teeth progressively deeper within the tool; thus, each tool edge takes off a successive layer of the material. Most of the material is removed by the roughing teeth which are followed by a number of finishing teeth designed to give the best possible surface finish. The shape of the tool determines the shape of the part (pure forming).

1. *Broaching* proceeds with a linear tool motion. A separate broach has to be made up for each shape and size; therefore, broaching is primarily a method of mass production. The workpiece must be rigidly held and the broach firmly guided. Rigidity of the machine tool is particularly important when a surface is broached with a *flat broach*, since the broach would be lifted out of the workpiece by the cutting forces. An *internal broach* is pulled through hollow parts (Fig. 16–43) and is capable of making holes with sharp corners. An *external (pot) broach* works on the outside surface of the part and the parts are pulled or pushed through. Both techniques are, to some extent, self-guiding. Large broaches may be of segmented construction which allows also some flexibility by using interchangeable broach elements. The machine tool resembles a hydraulic press of long stroke.

2. *Thread cutting* of a hole is an internal operation using a *tap* (Fig. 16–18*k*), which must be removed by reversing the spindle, unless a self-reversing tapping head is used. Threading of a shaft is an external operation using a *thread-cutting die* (Fig. 16–18*l*). For both, the primary motion is helical. Threads may also be cut on a lathe (Fig. 16–26*a*) or can be milled on a three-axis CNC machine. An alternative to thread cutting is thread forming (Sec. 9-7-4, Fig. 9–46). External thread rolling is, of course, of much higher productivity and is widespread (Fig. 9–45).

16-6-5   Gear Production

Gears are among the most important machine components whose importance has not diminished in recent years.

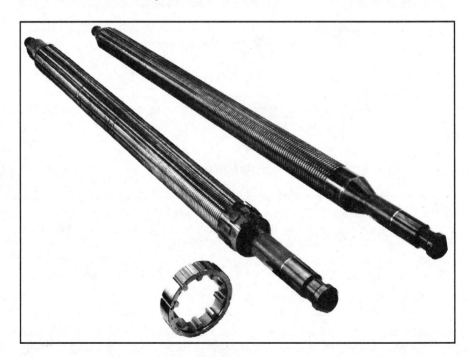

Figure 16–43   Two broaches finish in successive passes the internal profile of cam stator rings for automatic transmissions. (*Courtesy Apex Broaching Systems, Warren, Michigan.*)

Gear Making   All processes discussed hitherto are used:

1. Many gears used in more critical applications are still produced by metal-cutting processes.

a. *Form cutting* is conducted with a tool that has the profile of the space between two adjacent teeth. A form tool operated in a reciprocal (shaping) motion (Fig. 16–24b) or a form-milling cutter (Fig. 16–18a) installed on a horizontal milling machine may be used. Special gear cutting machines, working on either principle, are available. When the axis of the gear blank is set at an angle to the tool movement, helical gears are cut.

b. A multiplicity of cutting edges is engaged in *gear hobbing* (Fig. 16–44). The hob looks like a worm gear, the thread of which is interrupted to make several cutting teeth. In cutting spur gears, the hob axis is skewed relative to the gear axis by the helix angle of the hob. In cutting helical gears, the gear angle is added to the helix angle of the hob.

c. The cutting of bevel gears with straight or spiral teeth requires a combination of forming and generating and is performed on special-purpose machines that incorporate ingenious mechanisms to develop the required relative motions of tool and gear blank. Some machines use NC or CNC to guide the tool and blank.

d. Spur and helical gears may be cut by broaching, with the blanks pushed or pulled through pot broaches.

2. Lower costs and gear teeth of higher fatigue strength are often attainable by plastic deformation processes:

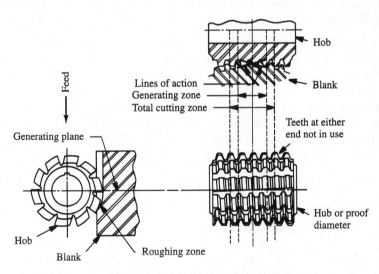

Figure 16–44   Hobbing is one of the many processes used for making gears. (*From The Tool and Manufacturing Engineers Handbook, 4th ed., vol. 1, p. 13.48. With permission of the Society of Manufacturing Engineers, Dearborn, Michigan.*)

a. Spur gears may be produced by cutting up a cold-drawn bar of the appropriate cross section.

b. Spur and helical gears may be cold extruded (Sec. 9-4-3).

c. Spur gears, bevel gears, and—with the use of more complex, rotating tooling—also spiral bevel gears may be hot forged (Sec. 9-3) to near-net shapes.

d. All forms of gears can be rolled. Indeed, transverse rolling (Sec. 9-7-4) is the standard method of production for worms of worm-gear drives and linear actuators, as well as for many spur and helical gears.

3. Spur gears are mass-produced by various forms of blanking, including fine blanking (Sec. 10-3-3).

4. Many gears are made by powder-metallurgy techniques, with or without restriking and resintering (Sec. 11-6).

Finishing   Gears produced by the above techniques are often suitable for immediate use in many applications. For smooth, noise-free running at high speeds and also for elimination of surface defects that would reduce fatigue life, many gears used in more critical applications are finished to close tolerances and a specified surface finish.

1. *Gear shaving* is, in some ways, related to broaching. The shaving tool is a meshing gear into which circumferential slots have been cut to make each tooth into a broach. The tool and gear are run in contact, slightly skewed, while imposing an axial oscillating motion, thus removing thin chips. A total of only 25–100 μm is removed. This is still the most frequently used technique.

2. Cold rolling (*burnishing*) between hardened meshing gears imparts a good surface finish and induces a residual compressive stress on the surface. This is a high-productivity process suitable for mass production.

3. Hardened gears are finished by form grinding, using a wheel dressed to the shape of the space between adjacent teeth, or by generating the tooth profile on special machines using straight-sided wheels. Ground gears are accurate but of high cost.

4. Hardened gears may be lapped against cast-iron lapping gears. In some gear assemblies, such as the hypoid gears of automotive drives, the assembly is lapped together by applying a lapping oil, containing a very fine abrasive, under controlled conditions.

16-7   CHOICE OF PROCESS VARIABLES

The decision to machine a component is usually part of a broader concurrent engineering process extending to the entire manufacturing sequence. Once the decision to machine the component is made, the sequence of machining steps is planned. This planning will take into account the machinability of the workpiece material, the shape, dimension, dimensional tolerances, and surface finish of the finished part, the characteristics of the machining process that is judged to be suitable for the purpose, the availability of machine tools at the plant or at outside vendors, and the economic aspects of production. In the simplest form, the result of these deliberations would be a decision to perform a sequence of operations, such as boring, milling, drilling, etc.

Metal cutting is somewhat unusual in that at this point of planning some important operational decisions have to be made: Even though the process is set, the speeds and feeds appropriate for the workpiece material and tool have to be chosen. When substantial volumes of material are to be removed, production is speeded up by taking one or more *roughing cuts* with large feeds and depths of cut. This applies also to rough castings and forgings which have scale, rough or hard spots, or inclusions in the surface. The required surface finish and dimensional tolerances are then obtained by taking a *finishing cut* with a small feed and depth of cut. A special situation arises with near-net-shape parts; only one cut may be needed but it has to cope with scale or hard spots in the surface, thus placing heavier demands on the tooling.

In a small-shop environment the choice of feeds and speeds may be based on the personal experience of the operator; almost always, such a choice will be conservative. In a competitive production environment the choice is more critical because low speeds and feeds result in low production rates, whereas excessive speeds and feeds reduce tool life to the point where the cost of tool change outweighs the value of increased production and, beyond a certain point, even production rates drop because of time lost in tool change. An initial choice of reasonable speeds and feeds is usually based on collective experience, gathered in many production plants and laboratories. Compilations have been prepared and are continually updated by various organizations, not only within large corporations, but also in specialized organizations such as the Machinability Data Center, Metcut Research Associates (a task now continued by the Institute of Advanced Manufacturing Science, Cincinnati, Ohio). Collected data are published in handbooks and are also available in computer data banks.

16-7-1   Cutting Speeds and Feeds

We saw in Sec. 16-1-8 that tool life decreases rapidly with increasing temperature which, in turn, depends on the energy expended per unit time and therefore on the cutting speed and shear stress (or, more generally, hardness) of the workpiece material. Hence, within any one material group, cutting speeds and feeds decrease with increasing hardness. However, large variations in speeds and feeds are noted between different material groups, and the general recommendations given in *Machining Data Handbook* (3d ed.) form the basis of Figs. 16–45 and 16–46. In using these figures, the following should be noted:

1. For some of the nonferrous materials speeds change little with hardness, and in Fig. 16–46 the speed is then simply indicated by the position of the alloy identification. For all materials, the feeds given apply to the hardness ranges indicated by the arrows.

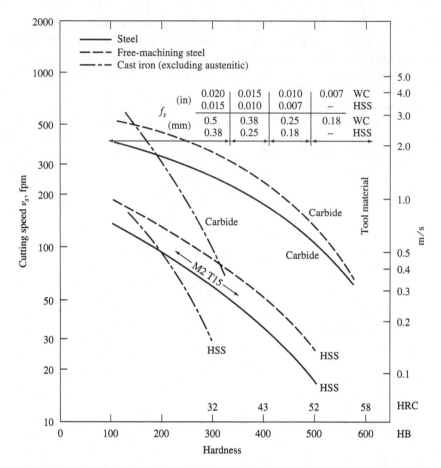

Figure 16–45   Typical speeds and feeds for roughing ferrous materials with a 3.8-mm (0.150-in) depth of cut. Increase speed by 20% for throwaway carbide inserts; reduce speed by 20–30% for austenitic stainless steels and for tool steels containing over 1% carbon.

2. The data provide a conservative starting point for rough turning with a maximum 4-mm depth of cut and a typical tool life of 1–2 h. The tool materials are identified generically. The HSS is typically M2 or, for heavier duties, T15; the symbol WC signifies cemented carbides, uncoated, of the grade appropriate for the workpiece material. Speeds may be increased by 20% for throwaway carbide inserts. If it is found that the actual tool life is much longer than 2 h, the operation may be speeded up. Remembering that heavier cuts are more efficient [Eq. (16-15)], the depth of cut and then feed (which determines the undeformed chip thickness) are increased. Higher cutting speeds generate more heat and should be used only when tool life is still excessive.

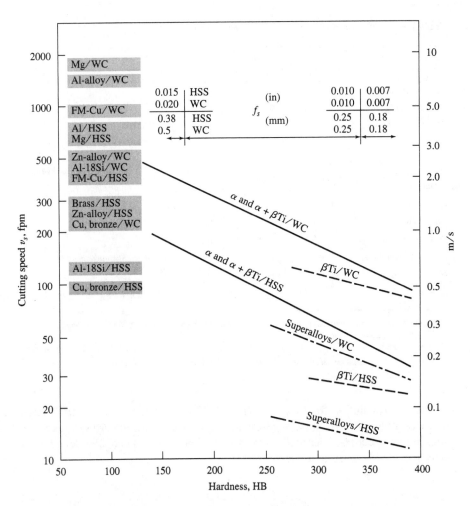

Figure 16–46   Typical speeds and feeds for roughing nonferrous materials with a 3.8-mm (0.150-in) depth of cut (FM stands for free-machining metal, WC for carbide tool).

3. While the data given are valid primarily for rough turning and boring, they can be used also as a guide for most other processes. The speed v and feed f for any particular process are found by multiplying v_s and f_s from Fig. 16–45 or 16–46 by the factors Z_v and Z_f, respectively (Table 16–5). It will be noted that the feed in shaping and planing is usually quite heavy; the lower feed values should be taken for harder materials.

4. Special conditions apply to drilling with HSS twist drills. The cutting speed is $v = 0.7v_s$ for ferrous and $v = 0.5v_s$ for nonferrous materials. Feed is typically a function of drill diameter D and is $0.02D$ per revolution for free-machining materials, $0.01D$ per revolution for tougher or harder materials, and

Table 16–5 Speeds and feeds in various metal-cutting operations*

Process	Z_v	Depth of Cut, mm	Z_f	Other
Rough turning	1	4	1	
Finish turning	1.2–1.3	0.65	0.5	
Form tools, cutoff	0.7			In-feed $0.1f$–$0.2f$
Shaping	0.7	4		Feed: HSS, 1.5–0.5 mm WC, 2–1 mm ($\times 2$ on Cu, Al, and Mg)
Planing	0.7	4		
Face milling	1	4	0.8–1[†]	
Slab milling	1	4	0.5[†]	
Side and slot milling	0.5–0.7	4	0.5[†]	
End mill, peripheral	1	1.2	0.5–0.25[†]	For 25-mm-diameter cutter
End mill, slotting	1	1.2	0.2[†]	
Threading, tapping	0.5–0.25			Slower for coarser thread

*Extracted from *Matchining Data Handbook*, 3d ed., Machinability Data Center, Metcut Research Associates, Cincinnati, Ohio, 1980.
[†]Feed per tooth.
NOTE: Speed $v = v_s Z_v$ and feed $f = f_s Z_f$; take v_s and f_s from Fig. 16–45 or 16–46.

$0.005D$ per revolution for very hard (HB > 420) materials. Coated drills have significantly longer life and may allow increased speed. Solid carbides take 2–3 times higher speed; heavier feed is often better than higher speed. For deep holes, feeds and speeds must be reduced:

Hole depth, D	Speed Reduction, %	Feed Reduction, %
3	10	10
4	20	10
5	30	20
6	35	20
8	40	20

5. For broaching with HSS tools, speeds range from 0.2 m/s (40 ft/min) on free-machining material to 0.025 m/s (5 ft/min) on hard material, while the undeformed chip thickness reduces from 0.12 to 0.05 mm per tooth.

6. Coated carbides allow higher speeds, and the recommendations of the manufacturer should be taken for first attempts. In general, speed can be increased by 25% with TiN, 25–50% with TiC, and 50–75% with Al_2O_3 coatings.

7. Ceramic tools can be tried at twice the recommended speed for cemented carbides in finishing cuts and then raised to higher speeds (up to 5 times higher).

8. With PCBN tools, speeds reach 2 m/s on high-Cr cast iron, 10–25 m/s on gray cast iron, and 2–5 m/s on superalloys and hardened steel of HRC 50–65 hardness.

9. Diamond tools are suitable only for finishing cuts of 0.05–0.2 mm depth at 0.02–0.05 mm feed and speeds of 4–15 m/s (800–2800 ft/min) for light nonferrous metals, although speeds of 70 m/s have been reached on 7075-T6 Al alloy.

The data given here are sufficiently accurate for initial planning purposes. Optimum conditions depend, however, on many factors, including the rigidity of the tooling, workpiece, workpiece holder, and machine tool. It is not uncommon for practical metal removal rates to reach twice the recommended values and, in specially constructed machines, very high-speed machining is possible with coated carbide or ceramic tools.

16-7-2   Cutting Time and Power

Once feeds and speeds are selected, details of the process can be planned and the appropriate equipment chosen. The following steps are usually followed:

1. The volume V to be removed is calculated. Handbooks contain numerous formulas but calculations from simple geometrical considerations are usually just as fast.

2. The chip removal rate V_t is calculated from cutting speed multiplied by chip cross-sectional area.

3. The net machining (cutting time) is simply

$$t_c = \frac{V}{V_t} \qquad \qquad \textbf{(16-23)}$$

(In some operations, such as turning, the total chip length l is easily calculated and then $t_c = l/v$.)

4. The adjusted specific cutting energy is taken from Eq. (16-15), the power of the machine tool from Eq. (16-16), and the cutting force from Eq. (16-17).

The constant C for Taylor's equation, Eq. (16-19), may be found by multiplying v_s by 1.75 for HSS and by 3.5 for WC. Since v_s in Figs. 16–45 and 16–46 is based on a tool life of 1–2 h or, on the average, $t = 90$ min, what is the implied value of n?

From Eq. (16-19), $v_s(90^n) = C$; for HSS, $v_s(90^n) = 1.75v_s$; $n = 0.125$. For WC, $v_s(90^n) = 3.5v_s$; $n = 0.28$.

Example 16-8

Example 16-9

The bore of a steel casting conforming to ASTM A27, 70-36 is to be machined out using disposable carbide insert tooling. The hole diameter is 130 mm in the as-cast condition; the finished diameter is 138 mm. Suggest cutting speed and feed, and calculate the power and cutting force.

From a handbook such as MHDE (2d ed., p. 249), TS = 485 MPa. This corresponds to HB = 3(485)/9.8 = 150 kg/mm^2.

Boring with a $w = (138 - 130)/2 = 4$ mm depth of cut is equivalent to rough turning. Hence, from Fig. 16–45, $v_s = 1.8$ m/s. For throwaway insert, increase by 20%: $v_s = (1.8)(1.2) = 2.16$ m/s. From Table 16–5, $Z_v = 1$ and $v = 2.16$ m/s.

From Fig. 16–45, $f_s = 0.5$ mm; from Table 16–5, $Z_f = 1$, and $f = 0.5$ mm/r.

To find the power required, the rate of material removal V_t must be calculated. Chip cross section $A = fw = (0.5)(4) = 2$ mm^2. $V_t = Av = 2(2160) = 4320$ mm^3/s.

From Table 16–1, $E_1 = 2.1$ W·s/mm^3. From Eq. (16-15), $E = E_1 h^{-a} = 2.1(0.5)^{-0.3} = 2.59$ W·s/mm^3. Thus, from Eq. (16-16a), the power, at an efficiency of 0.7 is 2.59(4320)/0.7 = 16 kW.

The cutting force, from Eq. (16-17a), $P_c = 15956/2.16 = 7.4$ kN.

Example 16-10

Holes of 10-mm diameter are to be drilled, with a twist drill, into a free-machining steel of HB 180. Determine the recommended cutting speed and feed.

From Fig. 16–45: for HSS, $v_s = 0.75$ m/s. From Table 16–5; $v = 0.7v_s = 0.5$ m/s.

Since the circumference of the drill is $10\pi = 31.4$ mm, the rotational speed is 500/31.4 = 16 r/s. The feed is 0.02D per revolution or 0.2 mm/r and the feed rate is 0.2(16) = 3.2 mm/s. Because there are two cutting edges, $f = 0.2/2 = 0.1$ mm/edge.

16-7-3   Choice of Machine Tool

The variety of commercially available equipment is immense, in terms of both size and type of operation. In addition to the more commonly encountered machine tools, special-purpose machines, some of enormous, others of minute size, are in operation. Thus, a lathe (made by Farrell Co.) for turning steam turbine rotors has a swing of 1.9 m and a bed 14 m long. There are vertical turning lathes for 4-m-diameter, 3-m-high parts of 400-Mg mass. A two-gantry NC milling machine (made by Cincinnati Milacron) for sculpturing airplane parts has a table 6 m (240 in) wide and 82 m (270 ft) long. It may turn out, of course, that no machine tool of sufficient power or stiffness is available and then the speed or feed, or both, must be reduced.

Example 16-11

Machine tools are generally built with solid beds and frames and all effort is made to minimize deflections. Machine tools of radically different construction have recently been proposed and built. Thus, in the *hexapod* the machining head is suspended on six telescoping struts. Motor-driven ball screws extend and retract the struts under computer control, giving great freedom of tool orientation. Since the struts are loaded only in tension or compression, stiffness is high. [Source: R.B. Aronson, *Manufacturing Engineering*, 1997(10):60–67]. Alternative designs have also been proposed, including ones for three-axis machining. [Source: M. Weck and N. Hennes, *Manufacturing Engineering*, 1998(9):54–62.]

16-7-4   Numerical Control and Automation

We have seen that the operation and control of machine tools requires considerable skill. This is particularly evident in milling three-dimensional contoured surfaces; it is not surprising that NC was first developed for machining such contoured aircraft components (Sec. 2-5-3). Parallel with the development of software, computer and control hardware developed at a rapid rate, and this was matched by advances in tool materials. Improved tools permitted material removal at higher rates; with the aid of CNC, the time during which actual cutting takes place could be increased and new demands could be set regarding tolerances. These developments also brought significant changes in machine tools and their operation.

1. Stiffness. We pointed out in Sec. 4-1-2 that all machinery suffers elastic deflections when forces act on it. In machining too, the machine tool, workpiece, tool, tool holder, and fixture form a dynamically interacting system which is subjected to dynamic loading and, in response to this, may develop vibrations. As discussed in Sec. 16-1-3, vibration results in variations in chip formation, poor surface finish, increased tool wear and, ultimately, tool breakage. Very large amplitude vibrations can be generated when the exciting frequency (as, for example, the number of tooth engagements in milling) is near the dominant natural frequency of the system. With higher speeds and increasing production rates achieved with CNC, stability of operation has become critical. Dynamic analysis of the machine tool is now an important step in equipment design.

2. Vibration control. Several measures can be taken. Stiffness is increased by appropriate design and by the use of materials of higher damping capacity (such as polymer-concrete, granite). Cavities are filled with vibration-damping material, and vibration isolation such as rubber pads, springs, and large masses (inertia blocks) are placed in strategic locations. Dynamic absorbers consist of a mass linked with springs to vibrating components; tuned to the natural frequency of the exciting force, they vibrate out of phase. Actively controlled dynamic absorbers use sensors and force actuators in a closed-loop system. The system may be detuned by oscillating the speed of spindle. Alternatively, spindle speeds are selected to be outside the vibration range while maintaining high output by taking the maximum depth of cut.

3. Drive spindle. This is a critical part of machine tools and has undergone substantial development. Concentricity and deflection under load (stiffness) are steadily improved. Drive is through a gearbox or with belts but, for higher speeds, integral drives are used in which spindles are extensions of the rotor. Centrifugal forces are very large at high speeds, and roller bearings made with low-mass ceramic balls, or hydrostatic, air, or magnetic bearings are used.

4. Thermal stability. With the quest for tighter tolerances, thermal expansion of the various elements of the machine tool becomes significant and various solutions are used to minimize expansion or, if this is not possible, to compensate for it.

5. Feed motion. To employ CNC, the handwheels of manually operated machines and the cams, tracer templates, and clutches of automatics had to be dispensed with. The table of the machine tool is fitted, for example, with *backlash-free lead screws*, actuated with *stepping motors* that move the table in small increments of, say, 2.5 μm; the number of increments depends on the number of pulses received from the NC control unit. For higher torque, *dc* or *ac motors* are employed, or the table is moved by *hydraulic actuators* or, most recently, for highest traverse speeds, by *linear motors*. Rapid acceleration requires ease of motion of tool slides, and hydrostatic bearings as well as linear ball bearings are extensively used for ways. Displacement transducers or encoders (Sec. 3-4) are installed for closed-loop control.

6. Chip management. Improved methods of *removing* large quantities of *chips* had to be developed. Partly for this reason, machining centers with horizontal spindles, but using cantilevered tooling, have found acceptance.

7. Tool management. *Automatic tool changers and magazines* holding 20 to 60 tools had to be installed (for FMS, Sec. 21-2-4, magazines holding as many as 200 tools are needed). Reduction of tool-change time is particularly critical in high-production machining centers. Tools are loaded in the sequence required in the operation, unless the machine is equipped with a double-arm swing system: this allows loading of one tool from a magazine into one arm while the active tool is cutting in the other arm. To eliminate setup time, tools are *qualified*; i.e., their dimension is premeasured and entered into the CNC program. Alternatively, tools are stored in the magazine in special, individual holders which ensure that they are always gripped in the same position relative to the machine tool.

8. Programming. This task has been greatly simplified. Two general trends have emerged:

First, programming languages—often based on APT—have been simplified for specific tasks such as two-axis contouring (ADAPT), and postprocessors have been added for specific processes such as drilling, milling, and lathe work (EXAPT).

Second, powerful software has been developed that performs the task of programming. Since the part description is already available in CAD, programs allow generation of the tool path directly. In conversational programming the operator has to enter only the dimensions of the part and the dimensions and material of the tool and workpiece, and the software executes programming. Advanced systems combine machine tool control with optimization for maximum productivity.

9. Unattended machining. Unattended and "lights-out" (supervised only by computers) operations have become possible with the introduction of automatic monitoring and adaptive control. Probes verify the presence of tools, or a camera is "taught" to recognize the tool setup and report on missing or broken tools or to initiate tool change. Tool force is sensed (e.g., from spindle deflection, using four position sensors) and torque and/or power is measured. Process optimization is possible; feed is often the controlled variable within the constraints set by maximum force, torque, and/or power. Techniques for in-process sensing of tool wear, vibration, and acoustic emission are utilized, and surface roughness is measured.

10. Automatic gaging. This is a vital element of automation. Some techniques are suitable for in-process measurement. At other times, a gaging station is set up within the machine tool or at an adjacent station, and the information gained is fed back to the control computer to make appropriate adjustments in cutting the next part.

11. Dynamic programing. With increasing speeds and feed rates, the programs controlling machine tool movements had to be changed too so as to keep up with the physical process of machining. For example, in milling molds and dies, the program has to look far ahead to decelerate feed on approaching a corner and accelerating again as the tool emerges from it.

The growth of CNC has been rapid; CNC machine tools account for the majority of current machine tool production, of which some 40% are machining centers and 40% are lathes (turning machines). In manually controlled machining, much time is wasted in waiting and material movement, and actual cutting takes place only about 20% of the time. Advances in integrating CNC machine tools with material-moving devices such as special-purpose robots and pallet changers raised machine-tool utilization to 40% and, in special cases, even to 70% of the time, greatly increasing productivity and reducing the number of machines required for a given output.

Advances in machine tools are illustrated by developments at the Dearborn Engine and Fuel Tank Plant of Ford Motor Company. In cooperation with the High Velocity Machine Division of Ingersoll Milling Machine Company and the linear motor maker Anorad Corporation, production cells were developed for high-speed machining of cast iron engine cylinder blocks and aluminum cylinder heads. They replace conventional transfer lines for flexible production of intermediate level (say, 200 000/year) quantities with the following results:

	Conventional Transfer Line	High-Velocity Machining Cell
Spindles in engine block line	550	18
Spindles in cylinder head line	371	6
Conversion to similar parts, weeks	40	8
Drilling rates for cast iron, mm/min	150	4 000
for Al, mm/min	330	6 300
Milling rate for cast iron, mm/min	800	9 500
for Al mm/min	4000	15 000

Example 16-12

[Data from J.V. Owen, *Manufacturing Engineering*, 1997(5):72–82.]

16-7-5   Optimization of the Cutting Process

The above-described procedure for selecting feeds and speeds is adequate for relatively small-scale production. In a competitive environment, *optimization of the process* becomes necessary. First, a management decision must be made regarding the criterion best applicable to the situation. If the backlog of orders is large and contractual obligations require fast delivery, the criterion will be *maximum production rate*. More usually, the criterion is *maximization of profit*; this requires a rather complex model of the entire production system. A less detailed analysis is sufficient to define production conditions for minimum cost per part, and only optimization according to the *minimum-cost criterion* will be discussed here. It will be assumed that constraints such as machine-tool capacity, surface finish, or surface integrity are not overriding, and that cost factors are known. Furthermore, it will be assumed that the constants in Taylor's equation are indeed constant; in reality, they have a certain statistical distribution and some analyses take into account the probabilistic nature of coefficients.

The cost of workpiece material and the loading and unloading time (and hence the associated cost C_l) are independent of cutting speed and do not enter into this optimization (Fig. 16–47). The cutting time per part is t_c (min) and is charged at a total rate R_t ($/min), which includes operator pay, overhead, and machine-tool charge. Cost must include an allowance for tool costs. If one tool costs C_t, takes time t_{ch} to change, and cuts N_t pieces before it has to be removed, the tool cost per piece C_{tp} is a $1/N_t$ fraction of the total cost

$$C_{tp} = \frac{1}{N_t}(t_{ch}R_t + C_t) \qquad \textbf{(16-24)}$$

The total speed-dependent production cost is

$$C_{pr} = t_c R_t + \frac{1}{N_t}(t_{ch}R_t + C_t) \qquad \textbf{(16-25)}$$

The cutting time t_c is simply the length of cut l divided by the cutting speed v

$$t_c = \frac{l}{v} \qquad \textbf{(16-26)}$$

The number of pieces cut with one tool (N_t) is equal to tool life t divided by the time t_c it takes to cut one part

$$N_t = \frac{t}{t_c} \qquad \textbf{(16-27)}$$

Substituting Eq. (16-26) and (16-27) into Eq. (16-25) gives

$$C_{pr} = \frac{l}{v}R_t + \frac{t_c}{t}(t_{ch}R_t + C_t) \qquad \textbf{(16-28)}$$

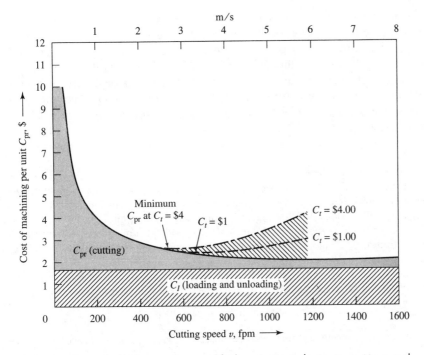

Figure 16–47   The optimum cutting speed for lowest-cost production is sensitive to tool cost and life.

From Taylor's equation, Eq. (16-19′), tool life is

$$t = t_{\text{ref}} \left(\frac{C}{v} \right)^{1/n} \tag{16-29}$$

Substituting into Eq. (16-28) gives

$$C_{\text{pr}} = \frac{l}{v} R_t + \frac{l}{t_{\text{ref}} C^{1/n}} (t_{\text{ch}} R_t + C_t) v^{(1-2n)/n} \tag{16-30}$$

The production cost C_{pr} is minimum where the first derivative with respect to v is zero

$$\frac{dC_{\text{pr}}}{dv} = 0 = -\frac{R_t l}{v^2} + \frac{1-n}{n} \frac{l}{t_{\text{ref}} C^{1/n}} (t_{\text{ch}} R_t + C_t) v^{(1-2n)/n} \tag{16-31}$$

Rearranging, we obtain the *cutting speed for minimum cost*

$$v_{c\ \min} = C \left[\frac{n}{1-n} \left(\frac{t_{\text{ref}} R_t}{t_{\text{ch}} R_t + C_t} \right) \right]^n \tag{16-32}$$

Even this simple model shows that the most economical speed increases with increasing C and n (as one might expect) and also with R_t, i.e., increasing labor and equipment cost. Conversely, there is not much point in increasing speeds if much time t_{ch} is wasted on tool changing or, as evident from Fig. 16–47, if the nonproductive time t_l of loading and unloading is a large fraction of the total time. All too often, this point has been overlooked and the effort spent on increasing metal removal rates could have been better spent on improving material movement and providing workpiece-locating jigs and fixtures. The importance of these factors is recognized in the application of the computer to machining processes, as was discussed in Sec. 16-7-4.

Example 16-13

Bars of free-machining steel of diameter $d_0 = 4$ in and length $l_0 = 40$ in are rough-turned with throwaway cemented carbide inserts. Calculate the variation of production cost with cutting speed if $R_t = \$12.60/h$, $t = \$4.00/edge$, $t_l = 8$ min, and $t_{ch} = 10$ min.

First, find cutting speed. From Fig. 16–45, $v_s = 500$ ft/min with 0.150-in depth of cut and feed of 0.020 in/r. From Sec. 16-1-8 (prediction of tool life), take $n = 0.25$ and $C = 3.5v_s = 1750$ ft/min.

Next calculate the cutting time. The diameter decreases by twice the depth of cut or $2(0.15) = 0.3$ in. The new diameter is $d_1 = 4.00 - 0.3 = 3.7$ in. The volume removed is

$$V = 40(4^2 - 3.7^2)\pi/4 = 72.6 \text{ in}^3$$

On dividing the volume by the cross-sectional area of the chip, the length of cut is obtained:

$$l = \frac{72.6}{0.15(0.02)} = 24\,190 \text{ in} = 2016 \text{ft}$$

[More simply, there are $40/0.02 = 2000$ turns, each of approximately $\pi(4 + 3.7)/2 = 12.095$ in $= 1$-ft length.] Thus, cutting time is

$$t_c = \frac{l}{v} = \frac{2000}{v} \quad \text{min}$$

Net cost of cutting

$$C_c = t_c R_t = 2000\frac{12.60}{60v} = \frac{420}{v} \quad \$$$

The cost of loading and unloading

$$C_l = t_l R_t = \$1.68$$

The total production cost per piece, C_{pr}, is next calculated from Eq. (16-28) and is plotted in a solid line in Fig. 16–47. Recalculation with $C_t = \$1.00/edge$ shows (broken line in Fig. 16–47) the magnitude of speed increase one could afford if the cost of tooling were lower. Obviously, the loading time t_l is the most likely target in a production-organization effort.

Computerized Optimization Systems   The machining system has an unusually high level of complexity and problems can be more effectively solved with the power of the computer. Several levels of optimization are possible:

1. Machining conditions are selected from stored databases according to the chosen optimization strategy and the tool loading sequence is determined.

2. Building on this first level, the process sequence is optimized for the specified workpiece shape, size, tolerances, and surface finish. If several machines are involved, the analysis aims at balancing workloads and minimizing investment.

3. Expert systems and process modeling are used to optimize the entire system.

16-8   ABRASIVE MACHINING

The term *abrasive machining* usually describes processes in which material is removed by a multitude of hard, angular abrasive particles or grains (also called *grits*) which may or may not be *bonded* to form a tool of some definite geometric form.

16-8-1   Classification

From the earliest times (Table 1–1), abrasive processes have been of greatest importance because they are capable of producing very well controlled—and if required, smooth—surfaces and, if the process is properly conducted, also very tight tolerances. Furthermore, the hardness of the abrasive grit makes it possible to machine hard materials, and, until the development of nontraditional machining processes, abrasive machining was often the only process for the manufacture of parts in certain materials. If properly conducted, abrasive machining can produce a surface of high quality, with a controlled surface roughness combined with a desirable residual stress distribution and freedom from surface or subsurface damage. Therefore, abrasive machining is often the finishing process for heat-treated steel and other hard materials. The number of processes has grown (Fig. 16–48), particularly in the last 50 years (Table 1–1). Some basic principles have general applicability and we will first explore these with reference to metals.

16-8-2   The Process of Abrasive Machining

In contrast to metal-cutting processes, in abrasive machining the individual cutting edges are *randomly distributed* and more or less *randomly oriented*, and the *depth of engagement* (the undeformed chip thickness) is small and not equal for all abrasive grains that are simultaneously in contact with the workpiece. These factors change the character of the process substantially.

Abrasive grit is manufactured to present sharp edges; however, the orientation of individual grains is mostly a matter of chance, and a grain may encounter the workpiece surface sometimes with a positive, zero or, in most instances, a negative rake angle. There is thus no guarantee that cutting will actually take place:

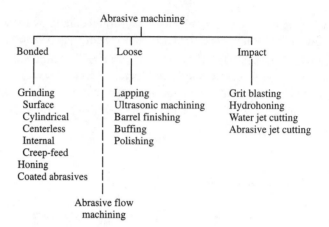

Figure 16–48   Classification of abrasive processes. (*Adapted from J.A. Schey, in ASM Handbook, vol. 20, Materials Selection and Design, ASM International, 1997, p. 695. With permission.*)

1. With a very slight engagement (Fig. 16–49*a*), only elastic deformation of the workpiece, grit, and—if present—binder takes place. No material is removed, but substantial heat is generated by elastic deformation and friction.

2. With a larger depth of engagement, events depend on process conditions. At highly negative rake angles, the grit may simply plow through the workpiece surface, pushing material to the side and ahead of the grit in the form of a prow (Fig. 16–49*b*). Occasionally the prow is removed, but the process is inefficient. Friction on the grit is significant.

3. Only at less negative rake angles, higher speeds, and with less-ductile workpiece materials will typical chip formation take place (Fig. 16–49*c*). The clearance angle is often zero and may even be negative, thus introducing large energy losses in overcoming friction and elastic deformation.

4. Much work is expended in a small space, therefore, temperatures rise substantially. When materials such as steel are machined in air, combustion takes place and sparks are seen to fly.

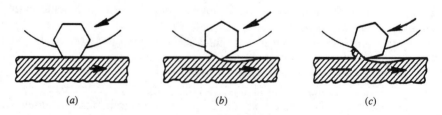

(a)   *(b)*   *(c)*

Figure 16–49   Depending on depth of engagement and attack angle, contact with an abrasive grit may result in (*a*) elastic deformation, (*b*) plowing, or (*c*) chip formation.

Because only a fraction of grit–workpiece encounters results in actual material removal and because there are many sources of friction, abrasive machining is inefficient; the energy required for removing a unit volume (Table 16–6) may be more than 10 times higher than in cutting.

Table 16–6   Approximate specific energy requirements for surface grinding*
Depth of grinding: 1mm (0.040 in)

Material	Hardness, HB	Specific Energy E_1	
		$W \cdot s/mm^3$	$hp \cdot min/in^3$
1020 steel	110	13	4.6
Cast iron	215	11	3.9
Titanium alloy	300	13	4.7
Superalloy	340	14	5.0
Tool steel (T15)	67 HRC	15	5.5
Aluminum	150 (500 kg)	6	2.2

*Extrapolated from data in *Machining Data Handbook*, 3d ed., Machinability Data Center, Metcut Research Associates, Cincinnati, Ohio, 1980.

Most of the energy is converted into heat, only some of which is removed with the chips; much heat remains in the workpiece with a number of undesirable consequences. Discoloration of the surface is due to oxidation and the color is indicative of temperatures reached. The surface of heat-treated steels may reach high enough temperatures to cause a transformation to austenite. After the grit has passed, sudden chilling causes the formation of martensite and associated cracks. A deep blue color is indicative of such "burning."

Thermal cycling leads to residual tensile stresses; these are superimposed on the residual compressive stresses generated by local deformation due to plowing and cutting with a highly negative rake angle. Hence, a properly controlled abrasive process can ensure a controlled surface roughness combined with high surface integrity, whereas an improperly controlled process may result in substantial surface damage (see Fig. 16–50*b*).

16-8-3   Abrasives

Abrasive grits must fulfill a number of often contradictory requirements.

1. High hardness at room and elevated temperatures helps to resist abrasion by hard particles in the workpiece.

2. Controlled toughness or, rather, ease of fracture (*friability*) allows fracture to occur under imposed mechanical and thermal stresses. Thus, new cutting edges are generated on a worn grit, but at the expense of a loss of abrasive material.

3. Low adhesion to the workpiece material minimizes BUE formation, re-deposition of grinding debris on the workpiece, and dislodging of grains from a bonded structure. At the same time, adhesion to the bond ensures the strength of a bonded abrasive.

4. Chemical stability increases wear resistance and resistance to corrosion by oxygen and cutting fluids.

5. The grain must have a shape that presents several sharp cutting edges. Because in-feed rates are low, grain size is several times larger than the depth of engagement. Grit size is specified as for other particulates (Sec. 11-2-2).

Only a few naturally occurring abrasives, primarily various forms of SiO_2 and Al_2O_3, find use and then only for softer workpiece materials or in finishing processes. The vast majority of industrial abrasives are manufactured (Sec. 12-3-2). *Alumina* (Al_2O_3) is made in grades of varying hardnesses; the hardest grades are friable, suitable for light-duty precision and finishing operations, whereas the less hard, tough grades (such as sintered alumina and alumina–zirconia abrasives) are suitable for heavy-duty stock removal. *Silicon carbide* (SiC) is harder than alumina (Table 16–3) but wears rapidly on low-carbon steel in which the carbon dissolves. The hardest grits are CBN and diamond, and are collectively called *superabrasives*. CBN is preferred for hardened steels, superalloys and cast iron. Diamond would dissolve in iron at high temperatures, therefore, it is used primarily for cemented carbide tools, ceramics, and glass. As with other abrasives, more friable grains fracture with sharp facets and cut cooler.

16-8-4   Grinding

Grinding is the most widespread of all abrasive machining processes.

Grinding Wheels   The abrasive is bonded, with an appropriate bonding agent, into a wheel of axial symmetry, carefully balanced for rotation at high speeds. The *grinding wheel* is a sophisticated tool made in a strictly controlled manufacturing environment.

1. The *strength of the bond*, which is governed by the quantity, kind, and distribution of bonding agent, is chosen so that the grit is held firmly, well supported, yet is still allowed to fracture. At some point, wear would result in unacceptably high forces on the grit, excessive heating, and poor surface quality; the grit must then be released to allow new grit to come into action. A natural consequence of this is that a wheel volume Z_s is lost in unit time. During this time, a volume Z_w is removed from the workpiece. It is usual to express the ratio of the two volumes as the *grinding ratio G*

$$G = \frac{Z_w}{Z_s} \tag{16-33}$$

Since grit wears more rapidly when grinding harder materials, a general rule states that a *softer* (*less strongly bonded*) *wheel should be used for grinding harder materials*. Grinding ratios are typically in the tens on tool steels, around 100–200 on hardened steel, can reach the hundreds or thousands with CBN or diamond, but can drop below unity on very hard materials.

2. Most frequently, the wheel is not solid but has a controlled porosity, a more or less open *structure*. Grains are bonded to each other by *bond posts* (Fig. 16–50*a*). A more open structure has slimmer posts, accommodates chips until they are washed away by the grinding fluid, and allows grinding fluid to move through the wheel.

The majority of grinding wheels is bonded with glass. Wheels with such *vitrified bonds* are the strongest and hardest, and the composition of the glass can be adjusted over a wide range of strengths. The so-called *silicate wheel*, bonded with water glass, is the softest.

Organic bonding agents are of lower strength but are available in a wide range of properties. *Resinoid wheels* are bonded with thermosetting resins and can be readily reinforced with steel rings, or fiberglass or other fibers, to increase their flexural strength. With the more flexible polymers such as shellac or rubber, very thin *cutoff wheels* can be made. Metals (usually, a sintered bronze) are used for superabrasives, although the cost of solid wheels is very high. Both CBN and diamond may be coated with a metal (typically, Ni or Cu) for better bonding to the matrix; part of the improvement comes from mechanical interlocking with the spiky coating. On diamond, a thin Ti coating forms a TiC interface which also improves crystal retention. For higher speeds, wheels have metal or fiber-reinforced composite centers and the abrasive (or superabrasive) is bonded into a 3–5-mm thick layer or is attached in segments. Alternatively, a steel wheel (often cut to shape) is coated with a single layer of superabrasive by brazing or electroplating. Such *coated wheels* have a more open grit structure and cut cooler.

The size and size distribution of the grit and the openness of the structure all contribute to determining grinding performance; therefore standard grinding wheel designations refer to all these factors (Fig. 16–51).

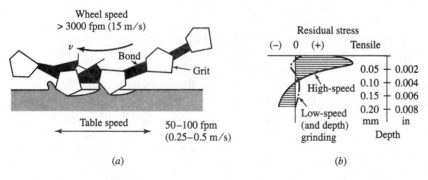

Figure 16–50   (*a*) A grinding wheel contains abrasive grit in a bond. (*b*) Grinding results in residual surface stresses that can become highly tensile in abusive grinding.

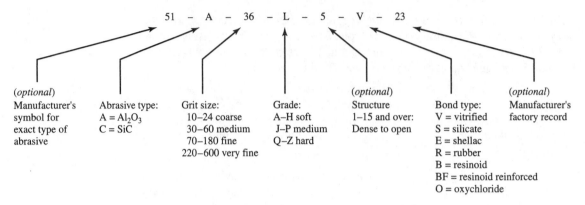

51 – A – 36 – L – 5 – V – 23

| (*optional*) Manufacturer's symbol for exact type of abrasive | Abrasive type: A = Al$_2$O$_3$ C = SiC | Grit size: 10–24 coarse 30–60 medium 70–180 fine 220–600 very fine | Grade: A–H soft J–P medium Q–Z hard | Structure 1–15 and over: Dense to open | Bond type: V = vitrified S = silicate E = shellac R = rubber B = resinoid BF = resinoid reinforced O = oxychloride | (*optional*) Manufacturer's factory record |

Figure 16–51   Grinding wheels are described by standardized nomenclature. (*From ANSI B74.13-1977, American Society of Mechanical Engineers, New York; also ISO 525-1975E.*)

Process Variables   *Operating (surface) speeds* are usually between 20 and 30 m/s (4000–6000 ft/min). Speeds up to 150 m/s (30 000 ft/min) are achieved in high-speed grinding with specially constructed wheels and superabrasives. There is usually an optimum speed at which the *G* ratio is highest. Material removal rate increases with the force imposed on the wheel but, beyond a certain limit, the *G* ratio drops steeply. Practical conditions represent a compromise, to give a high material removal rate with an economically tolerable grinding ratio. High rotational speeds impose large stresses on the wheel and all grinding wheels must be inspected for flaws prior to installation in balanced flanges, and equipment must be properly guarded to protect operators.

Truing and Dressing   Even the best-chosen grinding wheel undergoes gradual changes. It may develop a periodic wear pattern which leads to chatter and the emission of a howling sound, or it may lose its geometry even if it remains balanced. In grinding hard materials, wear causes *glazing* of the surface. Alternatively, the wheel surface may become clogged (*loaded*) with the workpiece material. For all these reasons, *truing* (restoration of desired geometry) and *dressing* (restoration of cutting ability) become necessary. This is usually done in the grinder itself so that alignment and wheel balance are not lost. Two basic methods are used. In one, the wheel is dressed by cutting with a *diamond point* (in the case of CNC grinders, under CNC control) or with a diamond-studded rotary disk. In the other, a wheel with a brittle bond is *crush-dressed* by pressing a high-strength steel roller against its surface; this is very fast and particularly economical for dressing forming wheels. Unnecessary dressing increases cost, especially with superabrasives, therefore, various sensing techniques (force, torque, or acoustic emission) are increasingly employed to monitor grinding conditions and determine the condition of the wheel. Surface-plated wheels require neither dressing nor truing, but can be replated.

Lubrication and Cooling   An indispensable part of the grinding system is the *grinding fluid* (Table 16–2). It fulfills a triple function: first, it keeps the ground surface cool and may prevent burning and cracking of the surface of hard materials; second, it affects the cutting process and reduces loading and wear of the grinding wheel (and thus increases the grinding ratio); third, it reduces friction and thus greatly reduces heat generation. Therefore, as in cutting, both cooling and lubrication are important; however, the lubricating function is much more important in grinding than in cutting. Greatest advantages are gained with oils and aqueous fluids that incorporate lubricating additives. It is not unusual to find that specific power requirements drop by a factor of 4 or more; metal removal rates and the volume that can be removed before the onset of chatter increase fivefold or more; grinding ratios may increase by a factor of 10 or more. For best results, lubricant is supplied under high pressure, if possible, through the wheel. Disposal of grinding sludge may be a serious problem, especially if the sludge is classified as a hazardous material. One reason for the growing use of CBN is that the slow wear of the wheel reduces the quantity of sludge formed.

Grinding Processes   The geometry of grinding can be as varied as that of other machining processes (Fig. 16–52).

1. *Surface grinding* is practiced with the cylindrical surface of a wheel (as in slab milling), but the wheel is normally narrower than the workpiece and must be cross-fed (usually by giving the workpiece a transverse feed motion, Fig. 16–52a).

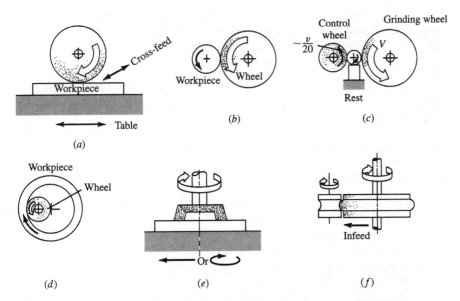

Figure 16–52   Various grinding processes: (*a*) surface grinding, horizontal spindle; (*b*) cylindrical grinding; (*c*) centerless grinding; (*d*) internal grinding; (*e*) surface grinding, vertical spindle; and (*f*) form (plunge) grinding.

2. *Cylindrical grinding* (Fig. 16–52b) is similar in its results to turning, except that the fast-rotating grinding wheel now works on the surface of the more slowly rotating part, and individual cuts are short. A comparison to surface grinding can be made with the equivalent diameter D_e

$$D_e = \frac{D_g}{1 + (D_g/D_w)} \qquad \textbf{(16-34a)}$$

where D_g is diameter of grinding wheel and D_w is diameter of the workpiece.

3. Cylindrical parts of great accuracy are obtained at high rates in *centerless grinding* (Fig. 16–52c): the workpiece is lightly supported on a workrest while the grinding pressure is taken up by the *regulating wheel* that rotates at about 1/20 the grinding wheel speed.

4. In *internal grinding* (Fig. 16–52d) a small wheel works on the cavity of the workpiece and individual cuts are longer than in cylindrical (external) grinding:

$$D_e = \frac{D_g}{1 - (D_g/D_i)} \qquad \textbf{(16-34b)}$$

where D_i is inner diameter.

5. The entire width of a flat workpiece may be ground with the annular end face of a *cup wheel* (Fig. 16–52e); this resembles face milling. Smaller pieces may also be ground on the end face of a cylindrical wheel; this is called *side grinding*. Parallel surfaces of high accuracy are formed by *double-disk grinding* between the end faces of two wheels.

6. Apart from generating basic geometric surfaces, grinding is used for finishing many parts of complex shape, including threads, gears, and most cutting tools. As in other machining, both forming (Fig. 16–52f) and generation of surfaces (Figs. 16–24 to 16–26) are practiced.

For a given process geometry, the undeformed chip thickness and length of cut increase with increasing depth of wheel engagement, increasing feed rate, and decreasing wheel speed.

Grinding Objectives Grinding was originally regarded as a finishing process; however, its scope has been expanded considerably. Processes can be categorized according to undeformed chip thickness:

1. *Precision grinding.* Originally most grinding was performed to improve tolerances and surface finish. Undeformed chip thickness is small and specific energy requirements are high (Table 16–6). The process is sometimes controlled by applying a constant force rather than constant feed. Until the advent of precision machining, grinding was the principal method of making precision parts, and it still occupies a dominant position.

2. *Coarse grinding.* More recently, grinding has become a material removal process. Wheels are made to release worn grit without dressing, yet without excessive wear. Speeds are increased because higher metal-removal rates are then obtained; chip formation dominates, and specific energy drops to 5–10 W·s/mm^3. Rough grinding (*snagging*) of castings and forgings—to remove ingates or flash—has long been practiced.

3. *Creep-feed grinding.* The full depth is removed in a single pass but at a low feed rate. Heat buildup ahead of the wheel accelerates metal removal but without harmful effects, because the heat-damaged material is ground away.

4. *High-efficiency deep grinding.* HEDG uses single-layer CBN form wheels to remove material at high rates, with speeds of 100–150 m/s and feeds of 700–2500 mm/min. Most of the material is removed in fine chips, as in milling, hence the process is also called *superabrasive machining*.

A fine surface finish is no guarantee of good surface quality; the surface integrity of hardened components is ensured by the use of soft wheels, low in-feeds per pass, high wheel speeds, frequent dressing, and a copious flow of grinding fluid. CBN-ground parts often show higher fatigue life, although the surface tends to be not as fine as with alumina.

Grinding lends itself to CNC control, and grinders have undergone development similar to machine tools (Sec. 16-7-4). Adaptive control is possible, and CBN wheels with their long life are particularly suitable for producing precision parts.

Example 16-14

A heat-treated H13 steel block of HRC 55 hardness and 50 × 100 mm surface area is ground on a vertical-spindle, reciprocating-table grinder, using a cup wheel of 150-mm diameter (as in Fig. 16–52e). *Machining Data Handbook* (3d ed., vol. 2, p. 8.45) recommends a wheel speed of 1500 m/min, a table speed of 30 m/min, and a downfeed of 0.05 mm, using a grade A80HB wheel. A total of 0.4 mm is to be removed. Calculate the wheel r/min and the total grinding time, assuming that reversal of the table takes 0.5 s.

It is usual to grind in the length direction. To obtain a uniform surface finish, the wheel must clear the workpiece at the end of the stroke. Thus the total stroke is $100 + 150 = 250$ mm. Surface speed is 1500 m/min $= D\pi N = 0.15\pi N$. Thus, $N = 3183$ r/min. The number of passes, at 0.05 mm/pass, is $0.4/0.05 = 8$ passes. Time/pass = stroke/table speed $= 0.25(60)/30 = 0.5$ s/pass. Total time for 8 passes $= 8(0.5 + 0.5) = 8$ s.

16-8-5   Other Bonded-Abrasive Processes

In addition to grinding, there are other processes that use bonded abrasive grit.

Coated Abrasives   Traditionally, abrasive grains attached to a flexible backing such as paper or cloth have been used for low-speed finishing of surfaces. However, with the development of stronger adhesives and backings, *coated belts* operating at high speeds (typically 15–30 but up to 70 m/s) have become important production tools, capable of high metal-removal rates. Up to 6 mm of material

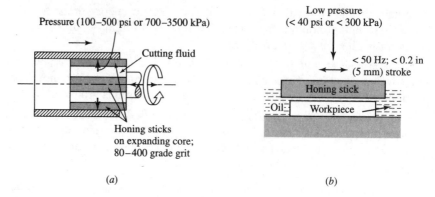

Figure 16-53   Schematic illustration of alternating-motion abrasive machining with bonded abrasives: (*a*) honing and (*b*) superfinishing.

can be removed in one pass, at rates of 200 cm³/min·cm width, replacing turning, planing, or milling in mass production (e.g., in machining the gasket surfaces of engine blocks and cylinder heads).

Grains are deposited on a layer of adhesive (*make coat*) applied to the backing and are held in place by a second layer (*size coat*). Bond strength is balanced to prevent stripping of new grains while allowing release of worn grains. Grains with sharp edges and of elongated shapes are electrostatically aligned to give cutting edges of low negative rake angles, spaced some 10 times farther apart than in grinding wheels. Some belts have multiple layers of grit for longer life.

Chip formation is the dominant metal-removal mode, and specific energy requirements and surface temperatures are relatively low. Grinding fluids serve to lubricate, cool, flush away the debris, and reduce clogging. A slotted contact wheel or platen is used behind the belt. Therefore, cutting is intermittent; heat can flow into the workpiece during the noncutting period, allowing high removal rates without burn. Even more aggressive cutting occurs with a backing disk.

Abrasive wire cutting utilizes CBN or diamond grit bonded to a wire surface.

Honing   In *honing* the abrasive is made into a slab (*stone, stick*). In the most frequent application (Fig. 16–53*a*), an internal cylindrical surface is finished with a number of honing sticks carried in an expanding, axially oscillating head (with axial speeds of 2–4 m/s), while the workpiece is rotated at 2–15 m/s. Thus, a crosshatch pattern is produced. A honing fluid, usually an oil, is applied to wash out abrasive particles. Water-based fluids are also used, especially for superabrasives. Filtration is essential in all cases.

Example 16-15   Oil carried up by the piston rings provides lubrication of the cylinder wall in internal combustion engines. A very smooth finish would not hold the oil and the engine would promptly seize, and the best results are obtained with a crosshatch finish. For this, a two-step process is used. In the first pass, 70–100-grit stones produce the crosshatch pattern. In the second pass, 600-grit stones knock off the peeks of the previous finish. Thus, deep grooves are left which trap lubricant and then feed it out to the flattened plateaus. (The finish is similar to that in Fig. 3–20, top profile, but with larger plateaus.)

Superfinishing is a variant of honing in which oscillating motion is imparted to a fairly large stone and surface pressure is kept very low. Thus, as the surface becomes flatter, it builds up its own hydrodynamic lubricant film which terminates the action of the abrasive (Fig. 16–53b).

Abrasive flow machining is a finishing process representing a transition to unbonded abrasives. A semisolid, abrasive-filled medium is forced through a hole or across a surface; it is particularly effective in removing burrs from inner surfaces.

16-8-6   Loose-Abrasive Processes

In a large number of processes, the abrasive is not bonded but is supplied as a powder.

Lapping   The process is based on a form of three-body abrasive wear (Fig. 4–23c). The abrasive is introduced as an oil-based slurry between the workpiece and a counterformal surface called the *lap*.

In the best-known form, the lap is a relatively soft, somewhat porous (e.g., cast iron) table, rotated in a horizontal plane. The workpieces are loaded onto the surface (sometimes in cages driven by a gear from the center of the rotating table, Fig. 16–54a). The workpiece describes a planetary movement and acquires a very uniformly machined, random finish of excellent flatness. When the lap is made into a three-dimensional shape, curved surfaces (e.g., glass lenses) may be lapped. Lapping is also a very fast process for breaking in mating gears or worms (thus eliminating the need for a running-in period, for example, for the hypoid gears of an automotive rear axle) and for removing surface protuberances from chromium-plated piston rings.

Less frequently, the lap is made of a bonded abrasive and the process is then similar to finish grinding but at a low speed.

Ultrasonic Machining   A piezoelectric transducer is used to generate ultrasonic (about 20 000-Hz) vibrations of small (about 0.04–0.08-mm) amplitude which drive the *form tool* (*sonotrode*) made of a reasonably ductile material such as soft steel. Abrasive grit is supplied in a slurry to the interface, and the workpiece is gradually eroded (Fig. 16–54b). The process is particularly suitable for machining less-ductile materials.

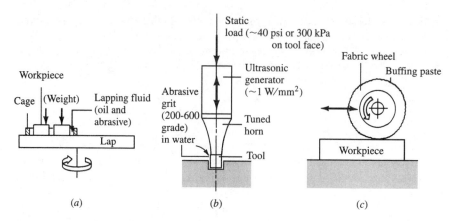

Figure 16–54   Machining with loose (unbonded) abrasives; (a) lapping, (b) ultrasonic machining, and (c) buffing.

Power Brushing   The tool is now a brush made up of flexible wires, usually rotated at high (15–30 m/s) speed and applied at a controlled force, with the distance selected to make a cut of 0.5–0.8-mm depth and remove material by impact. Most often, the aim is to create a specific surface finish, for esthetic or technical purposes. If not done properly, surfaces become smeared and loaded with reattached particles or bits lost from the wire. In any event, the wire should be of a material that will not cause galvanic corrosion if bits of wire are left on the surface. Plastic (nylon or PP) fiber brushes loaded with bonded abrasives (Al_2O_3, SiC, or sometimes CBN or diamond) can be very effective. The process is widely used for burr removal.

Barrel Finishing   Contact with a tool is replaced by contact between workpieces in *barrel finishing* or *tumbling*. In principle, the workpieces are placed into a barrel of many-sided cross section, so that the workpieces drop when the barrel is rotated. Mutual impact removes surface protuberances. Much improved finish is obtained when a *tumbling medium* is added, either in a liquid carrier (*wet medium*) or by itself (*dry medium*). The medium is chosen according to the intended purpose, and may range from soft materials such as nut shells to metallic or nonmetallic balls, chips, stones, and conventional abrasives. Similar results are obtained in *vibratory finishing*, with the barrel vibrated by some mechanical means. These completely random abrasive processes are of great value in removing burrs and fins from workpieces and, generally, in improving their surface appearance.

Buffing, Polishing, and Burnishing   The hard surface is abandoned and the abrasive is applied to a soft surface, for example, the cylindrical surface of a wheel composed of felt or other fabric (Fig. 16–54c). For *buffing*, the abrasive is in a semisoft binder. For *polishing*, the abrasive may be used dry or in oil or other carrier/lubricant. Both buffing and polishing are capable of producing surfaces which are highly reflective not because of greater smoothness but rather because of smearing (plastic deformation) of surface layers. Localized deformation of the surface induces compressive residual stresses and improves the fatigue strength of the components.

Impact Processes   In these processes all contact with a tool or workpiece is eliminated.

1. In *grit* or *shot blasting* particles of abrasives or of some other material (*shot*) are hurled at the workpiece surface at such high velocities that surface films—such as oxides—are removed, and a uniformly matte, indented appearance is imparted to the surface (Fig. 3–23*b*). Excessive impact velocities could cause damage, but properly controlled shot blasting promotes slight plastic surface deformation and residual compressive stresses which, as noted in Sec. 9-2-2, improve fatigue resistance. The required velocities may be produced by compressed air or by dropping the shot at the surface of a wheel rotating at very high speeds. The wheel may or may not have paddles; contact with the wheel accelerates the grit particles.

2. A special form of grit blasting is *hydrohoning*. The abrasive medium is suspended in a liquid which is then directed onto the surface in the form of a high-pressure jet.

3. In *abrasive jet machining* a tightly controlled air or CO_2 gas jet, loaded with dry abrasives, is applied at pressures of 0.5–0.8 MPa through sapphire or WC nozzles. Impacting at a speed of 150–300 m/s, it cuts slots or holes into very hard materials.

4. Softer materials (YS < 80 MPa) can be cut through the thickness with a high-pressure *water jet*. The jet stream is pressurized to 28–440 MPa (40–64 kpsi). The high-velocity (600–900 m/s) jet emerging from a sapphire or diamond nozzle of 0.08–0.4-mm diameter produces a clean, damage-free cut surface. The process is particularly advantageous when melting of the material is to be avoided.

5. Hard materials can be cut by introducing abrasives (such as garnet of 16–150 mesh, at 0.2–0.6 kg/min) into the water stream. Such *abrasive water-jet cutting* is highly favorable for heat-sensitive materials since there is no heat-affected zone. Installations with multiaxis control or integrated with a robot can make cuts on sheet, plate, and 3-D parts. Most are equipped with control software that selects (or helps in selecting) optimum processing parameters, such as travel speed and abrasive feed rate.

All these processes are extensively used also for deburring. Burrs are formed in many operations, including die casting, forging, blanking, injection molding, and machining, and their removal represents one of the high-cost stages of processing.

16-9   PROCESS CAPABILITIES AND DESIGN ASPECTS

General characteristic of processes are given in Table 16–7. Even though individual processes may be limited in the shapes they can produce, a succession of operations can develop highly complex shapes. Indeed, except for totally enclosed or very small-necked hollows (shapes T5 and F5 in Fig. 3–1), all shapes can be made from a solid block, even though the cost may be exorbitant. Complex shapes may be justified if they allow part integration, thus reducing problems with fits and cost of assembly. Size is limited only by the available equipment; as shown in Sec. 16-7-3, large machine tools are available, sometimes in plants that take work from outside. There is little limitation on minimum size, and very slender rotational bodies can be made on Swiss-type machines. There are, however, practical limits on minimum wall thickness, set by elastic deflections under the influence of tangential and normal forces developed in the process. Effective supports can eliminate the limit.

Frequently, machining is among the last operations on cast, forged, powder-metallurgy, and even ceramic and plastic parts (Figs. 5–2 and 5–3), and then its aim is to create shape features not obtainable by the preceding process, or to improve tolerances and surface finish. The scope of possible improvements is evident from Fig. 3–22: Machining processes can produce the tightest tolerances and smoothest surface finish (which is sometimes transferred to products of other processes as, for example, in the injection molding of plastics and cold rolling of metals). However, increased accuracy and smoother finish come at a price; frequently, they require extra processing steps and, even if they can be achieved in the same process, feed and depth of cut have to be reduced and productivity drops.

16-9-1   Design Aspects

In the design of parts for machining, principles of concurrent engineering must again be followed. In particular, material choice and part configuration must be taken into account in planning the machining process and process sequence. Some points to be considered:

1. Within the limits imposed by service requirements, choose the material of highest machinability. Thus, heat-treated steels are generally machined in the annealed (or normalized or spheroidized) condition, then heat-treated and finished, usually by grinding. Sufficient grinding allowance must be made, including an allowance for distortion in heat treatment. Alternatively, however, fully heat-treated material may be finished directly by hard cutting, competitively and perhaps with savings.

2. Machining can release residual stresses in castings and forgings and can introduce internal stresses on its own. Therefore, parts subject to distortion are heat-treated, either before machining or after completion of the bulk of machining.

Table 16-7 General characteristics of machining processes

	Machining Process					
Characteristics	Lathe Turning	Automatic Screw Machine	Drilling	Milling	Grinding	Honing, Lapping
Part						
Material	All	All	All	All	Hard	Hard
Preferred		Free-machining	Free-machining	Free-machining		
Shape*	R0-2,7; T0-2; Sp; U1-4	As turning	T0, 6	B; S; SS; F0-4, 6, 7; U7	As turning, milling	R0-2; B0-2, 7; T0-2, 4-7; F0-2; Sp
Min. section, mm	< 1 diam.	< 1 diam.	0.1 (hole diam.)	< 1	< 0.5	< 0.5
Surface configuration	Axially symmetrical	As turning	Cylindrical	Three-dimensional	As turning, milling	Flat, cylindrical, 3-D
Cost[†]						
Equipment	B–D	A–C	D–E	A–C	A–C	B–D
Tolling	D–E	A–D	D–E	A–D	E–D	A–E
Labor	A–C	D–E	B–D	A–B	A–E	B–D
Production						
Operator skill[†]	A–C	D–E	B–E	A–B	A–D	C–E
Setup time[†]	C–D	A–C	C–E	A–C	B–D	C–E
Rates (pieces/h)	1–50	10-500	10-500	1–50	1–1000	1–1000
Min. quantity	1	1000	1	1	1	1

[†]Comparative ratings with A indicating the highest value, E the lowest; ratings are greatly influenced by the degree of automation. For example, lathe turning involves high to low equipment cost, low to very low tool cost, high to medium labor cost, requires very high to medium operator skill, and has low setup time.
*From Fig. 3–1.

3. The workpiece must have a *reference surface* (an external or internal cylindrical surface, flat base, or other surface suitable for holding it in the machine tool or a fixture).

4. If at all possible, the part shape should allow finishing in a single setup; if the part needs gripping in a second, different position or on a different machine, one of the already-machined surfaces should become the reference surface.

5. The shape of slender parts should permit adequate additional support against deflection.

6. Radii (unless set by stress-concentration considerations) should accommodate the most natural cutting tool radius: the nose radius of the tool in turning and shaping, the radius of the cutter in milling a pocket (Fig. 16–55a), the sharp edge of the cutter in slot milling, or the rounded edge of a slightly worn grinding

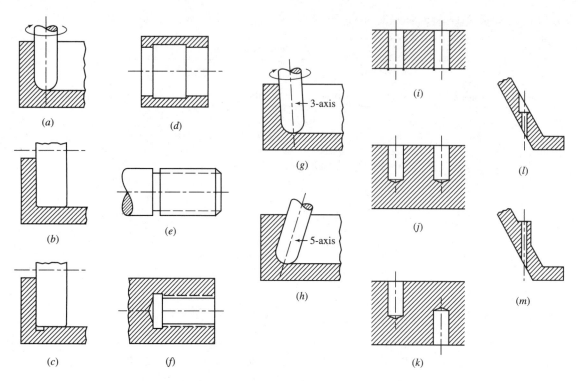

Figure 16–55   Some features of design for machining. See text for interpretation.

wheel (Fig. 16–55b). An undercut allows a sharp corner in grinding with the side of the wheel, but deflection of the wheel could make the side surface inclined (Fig. 16–55c).

7. Undercuts (Fig. 16–55d) can be machined if not too deep, but they increase costs. Undercuts are essential when runout is needed for threading tools (Fig. 16–55e and f).

8. Deflections of the tool in boring or milling of internal holes and recesses limit the depth-to-diameter (or depth-to-width) ratio. Deflection of the tool in machining deep recesses (Fig. 16–55g) necessitates special (and more expensive) techniques, or a sacrifice in tolerances.

9. Features at an angle to the main machine-movement direction call for a more complex machine, transfer to a different machine, interruption of the main machining action, and special tooling or special attachments, and should be avoided. CNC control of turning centers or multiaxis machining centers removes some of the limitations (Fig. 16–55h) but cost will be higher.

10. Holes of large depth-to-diameter ratio call for special processes (including nontraditional techniques, Chap. 17). When there is freedom of choice, through-holes (Fig. 16–55i) are preferable to blind ones (Fig. 16–55j). Blind holes are to be avoided on two sides of the part (Fig. 16–55k) since they make relocation of the part necessary.

11. Features placed at an angle to the workpiece surface deflect the tool and necessitate a separate operation such as spot-facing (Fig. 16–55l) or redesign of the part (Fig. 16–55m).

12. We saw in Sec. 10-3-3 that sheared parts often have burr. This is true also of machined parts. Everyone is familiar with the burr produced by a twist drill on the exit side; similar burrs form at the exit point of milling cutters and other tools. Burr removal is one of the most expensive operations. The thickness of the burr is the determining factor in the effort required for its removal, and burrs thicker than 0.4 mm have to be removed by machining, following the contour of the part. Both part and process design are (or should be) directed to minimize the formation of burrs or, if it is unavoidable, place burrs into an easily accessible location.

16-9-2 Precision Machining

The concept of *precision manufacturing* has several connotations. It can refer to the best dimensional accuracy achievable; in casting, plastic deformation, and particulate processing it is used rather loosely to denote processes that make parts to an order of magnitude tighter than usual tolerances, whatever they may be. More quantitative criteria are applied in machining and, as shown in Fig. 16–56, the development of machining processes has led to a gradual redefinition of what is "normal" machining accuracy. Each step from normal to precision, high-precision, and ultra-high-precision machining represents an order of magnitude tightening. While it may appear that the trends shown in Fig. 16–56 will continue, there is an absolute limit set by the size of atoms. With interatomic distances on the order of 0.3 nm, this limit is around 1nm. There are machine tools capable of producing to a tolerance of nanometers, and in Chap. 20 we shall see that there are solid-state electronic devices produced to similar tolerances. (The group of technologies capable of producing to such accuracy is sometimes collectively referred to as *nanotechnology*, although this may cause confusion with manufacturing on the nanoscale, Sec. 20-5-2.)

Example 16-16

Some everyday products meet the present definition of nanotechnology. For example, the hard disk of a PC contains a 130-mm-diameter, 1.9-mm-thick aluminum alloy disk machined to 0.007 μm (7 nm) R_y. The magnetic head is 3.2-mm long and rides with a clearance of 0.15 μm above the disk rotating at 1600 r/min. We can get a feel for these dimensions by scaling up the head to the 70-m length of the Boeing 747: It would have to fly 3.3 mm above a surface that has bumps of maximum 0.1-mm height over a circular area of 1820-m diameter. (Quoted by H. Nakazava in *Principles of Precision Engineering*, Oxford University Press, 1994, p. 9.)

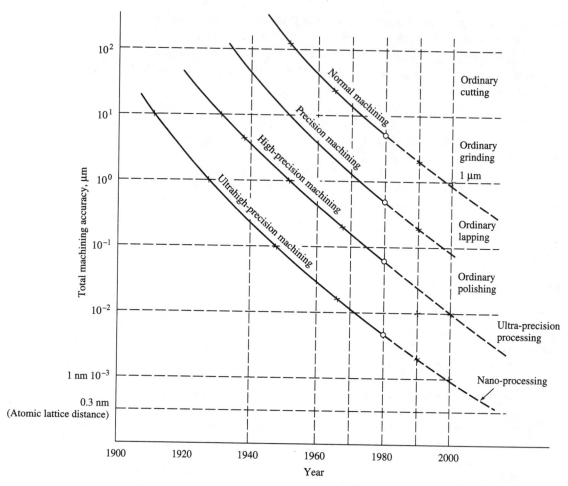

Figure 16–56 Progress in machining accuracy in the twentieth century has led to the continual redefinition of precision. [*From N. Taniguchi (ed.): Nanotechnology, Oxford University Press, 1996, p. 1, Fig. Int. 1. By permission of Oxford University Press.*]

The possibility of meeting a given accuracy depends, obviously, on the size of the part. The American Society for Precision Engineering takes this into account when it defines precision as ±50 parts per million (ppm); thus, the tolerance band is 100 ppm or 10^{-4}. In Fig. 3–22, this lies on the abscissa, and it would appear unreachable. To achieve such accuracy on a CNC lathe, grinder, or machining center, the machine tool itself is built to be extremely stiff and of high precision, with vibration-free spindles of high concentricity (low runout) and with ways having linear bearings or hydrostatic or air bearings (Sec. 16-7-4). Optical linear scales are used to sense position and provide feedback. Precision fixtures with controlled gripping forces avoid distortion of the part. Machining conditions are chosen to avoid chatter with appropriate tooling (most often, diamond; if the part is steel, the diamond is Ni-coated to prevent destruction of the tool). Temperature variation is the greatest enemy: According to an old saying, on a machine tool everything is a thermometer. Therefore, machine tools are housed in climate-controlled rooms with temperature control to $\pm 1°C$ or better; coolant temperature is controlled to $\pm 1°C$ or better; the part is allowed to equalize prior to machining; and CNC thermal-growth compensation programs are used to correct axis positioning. When the thermal expansion coefficient of the part is different from that of the machine tool (e.g., Al, Table 4–1), compensation is made for heat buildup in the part during machining. To bring the machine to equilibrium, it is "exercised" for some time before beginning actual cutting. Process monitoring detects worn or broken tooling, and precision measurement of finished parts is used to feed back information to the process.

Obviously, the achievable precision is set also by the ability of measuring instruments to resolve minute differences. Current laser measurement systems can measure up to 40-m lengths with a resolution of 10 nm at a precision of 0.1 ppm. Thus, for the 25-mm dimension that serves as the basis of Fig. 3–22, the precision is 2.5 nm and, since the precision of measuring instruments must be 5–10 times better than the accuracy to be determined, the part can be made to a value of 12–25 nm.

Example 16-17

The precision achievable by existing methods is illustrated by the Large Optics Diamond Turning Machine (LODTM) at Lawrence Livermore National Laboratory of the United States. Parts of up to 1620-mm diameter and 1360-kg mass can be machined. Multiple laser beams are used to position the tool with a precision of 28 nm, and parts are made to better than 5 ppm tolerance. Such precision—and its cost—can be justified only for exceptional products. There are, however, products made in the thousands that easily meet the 10^{-4} definition of precision. For example, femoral heads for hip replacements (Fig. 1–5) have typically 30-mm diameter and are made to a sphericity of 1.3 μm. [Source: J.R. Koelsch, *Manufacturing Engineering*, 1994(3):67–72.]

16-10   SUMMARY

Material removal is, in a sense, admission of defeat: One resorts to it when other processes fail to provide the requisite shape, tolerance, or surface finish, or when the number of parts is too small to justify a more economical process. How often we fail is shown by the fact that more operator hours are spent on machining than on all other unit processing methods. Machining is still indispensable for finishing the surfaces of some shafts; journal bearings; bearing balls, rollers, and races; accurately fitting machine parts; and bores of engine cylinders; and for creating shape features that a preceding process could not produce. Even though valuable material is converted into low-value chips, machining can still be economical, but some basic points must be observed for both economy and quality.

1. Machinability of materials is highly variable, and a material condition ensuring low TS, n, and q should be chosen whenever possible.

2. Machining involves a competition for survival between tool and workpiece; process conditions and tool materials must be chosen to prolong tool life while maximizing material removal rates. Tool life drops steeply with increasing temperatures. Temperatures in the cutting zone rise with cutting speed; therefore, material-removal rates are best increased by increasing first the depth of cut, then feed, and only finally speed.

3. Finishing must be directed toward producing a surface that is within tolerances, has the proper roughness and roughness texture, and is free of damage (tears, cracks, smeared zones, excessive work hardening, and changes in heat-treatment conditions) and of harmful residual stresses.

4. Machining, like all other manufacturing processes, must be treated as a system in which the workpiece and tool materials, the lubricant (coolant), process geometry, and machine tool characteristics interact. This approach is essential because some of the interactions may easily go undetected, to the detriment of quality, production rates, and economy.

5. Substantial increases in productivity result when time lost in loading, unloading, checking, and tool change is minimized. NC and CNC machine tools, in combination with part and tool handling systems, can achieve very high capacity utilization. The advantages of CNC machining can be fully realized only if the capabilities of advanced tool materials are exploited; for this, the machine tool must have sufficient rigidity, a high-precision spindle of sufficient speed and power, and ways of the highest quality and fast movement.

6. Large quantities of chips and grinding sludge are produced and the process must be conducted to make recycling feasible and economical.

7. Moving parts, flying or entangled chips, tool breakage, disintegration of grinding wheels, and other dangers must be recognized and protection devised. Some machine tools are totally enclosed; others are relatively accessible, and safe operation requires training and responsibility of operators. Eye protection is essential and protective clothing is necessary if exposure to coolants and lubricants is unavoidable.

PROBLEMS 16A

16A-1   (a) Make a sketch of ideal orthogonal cutting with a positive rake angle. Mark the (b) rake angle, (c) clearance angle, and (d) shear angle.

16A-2   Make sketches of orthogonal cutting with (a) positive and (b) negative rake angle. In each sketch, identify the shear angle and chip thickness; point out the differences.

16A-3 (*a*) Make a sketch of realistic orthogonal cutting with a positive rake angle, assuming a strain-hardening material and low friction on the rake face. (*b*) Repeat for high friction on the rake face; show the change in shear angle and chip thickness.

16A-4 Make a sketch of orthogonal cutting (*a*) without a chip breaker and (*b*) with a groove-type chip breaker. Show the difference in chip formation.

16A-5 State the most likely sources of (*a*) forced vibration and (*b*) regenerative chatter.

16A-6 (*a*) Make a perspective sketch of a turning tool, identifying the rake face and the flank face. Show the location of (*b*) crater wear, (*c*) flank wear, and (*d*) notch wear.

16A-7 (*a*) List the two principal functions of cutting fluids at low cutting speeds and (*b*) discuss at least two mechanisms by which they fulfill these functions.

16A-8 (*a*) List the two principal functions of cutting fluids at high cutting speeds and (*b*) discuss at least two mechanisms by which they fulfill these functions.

16A-9 (*a*) List three methods of applying cutting fluids. (*b*) Indicate where in a turning operation the fluid should preferably be applied.

16A-10 Make sketches to show the essential features of (*a*) turning, (*b*) boring, (*c*) facing, and (*d*) parting off. Always show the direction of feed and the cross section of the undeformed chip, distinguishing undeformed chip thickness h and depth of cut w.

16A-11 Make a distinction between (*a*) surface roughness and (*b*) surface integrity.

16A-12 (*a*) Draw a diagram showing tool life as a function of cutting speed. (*b*) Identify the axes and their scale. Draw lines showing the relative tool life for (*c*) HSS, (*d*) carbide, and (*e*) ceramic tooling.

16A-13 Draw three sketches to show the possible events that may take place between a grit in a grinding wheel and the workpiece.

16A-14 Define grinding ratio.

16A-15 Make a sketch of centerless grinding, identifying the main components of the system.

16A-16 Make sketches of surface grinding with a (*a*) horizontal spindle and (*b*) vertical spindle. (*c*) and (*d*) Show schematically the finishing (grinding) marks on the surface of parts produced by each method.

16A-17 Describe open-loop control as applied to a milling machine.

16A-18 Describe closed-loop control as applied to a milling machine.

16A-19 Distinguish between CNC and DNC as applied to a group of machining centers.

PROBLEMS 16B

16B-1 To explain the effects of tool geometry, make three sketches, showing orthogonal cutting with (1) highly positive, (2) zero, and (3) negative rake angle. In a tabular form, indicate the relative levels of (*a*) shear angle; (*b*) chip thickness; (*c*) chip speed; (*d*) thrust force; (*e*) cutting force; and (*f*) power.

16B-2 (*a*) Discuss the conditions that lead to BUE formation; make a sketch to illustrate. (*b*) List the advantages and (*c*) disadvantages of cutting with a BUE. (*d*) Suggest ways in which the BUE can be avoided.

16B-3 (*a*) Make a sketch of feed and chip thickness in turning with a single-point tool with zero lead angle. (*b*) In another sketch, use a positive lead angle and the same feed. State whether an increasing lead angle results in larger or smaller (*c*) undeformed chip thickness, (*d*) cutting force, and (*e*) thrust (radial) force.

16B-4 (*a*) Make a sketch of a ball-end milling cutter. (*b*) Indicate schematically the variation of undeformed chip thickness on moving toward the point of the cutter. (c) State what the undeformed chip thickness is at the center of the point (at the apex), and what effect this has on cutting efficiency. (d) Suggest a way of increasing cutting efficiency.

16B-5 (a) Draw a sketch showing the essential features of ideal orthogonal metal cutting with a positive rake angle. (b) Identify the rake angle α, the clearance angle θ, and shear angle ϕ. (c) Indicate the undeformed chip thickness h and deformed chip thickness h_c. In separate sketches, show the changes due to (d) strain hardening, (e) cutting with a secondary shear zone, and (f) the formation of a built-up edge. (g) State, generically, at what speeds, if any, (d), (e), and (f) apply to the cutting of low-carbon steel. (h) State which of the above cases give the highest and lowest energy requirements. Justify. (i) State which of the above cases give the best and worst surface finish.

16B-6 (a) Draw a sketch of orthogonal cutting with a BUE. State what effect the following variables have on the likelihood of BUE development: (b) positive rake angle; (c) nose rounding; (d) lubrication; (e) rougher rake face; (f) lower adhesion between tool and workpiece; (g) free-machining material. Briefly justify your choices.

16B-7 An unidentified carbide insert is used for cutting AISI 1045 steel at the recommended speed. The tool wears extremely rapidly. Suggest at least two possible causes of the problem and suggest remedies.

16B-8 Chips are found to clog up the machining space. (a) Identify the possible causes of the problem. Suggest remedies when cutting (b) ductile and (c) less-ductile materials.

16B-9 (a) Make a sketch of the ideal orthogonal cutting process with a positive rake angle, showing the rake angle α, clearance angle θ, and shear angle ϕ, undeformed and deformed chip thickness. (b) Make another sketch, with a negative rake angle. (c) State whether shear angle, chip-thickness ratio, chip speed, cutting force, and energy increase from (a) to (b). (d) Give at least two reasons why highly positive rake angles are seldom used. (e) State why one would use a tool with a negative rake angle.

16B-10 (a) Make a sketch of a facing operation. (b) Clearly show the feed, depth of cut, and undeformed chip thickness on a blow-up of the cutting zone. (c) Show, in a new sketch, the effect of an increased side-cutting edge angle, leaving the feed unchanged. State what r/min should be chosen for (d) maximum tool life and (e) maximum productivity.

16B-11 In principle, what kind of machine tool is likely to be most economical for producing rotationally symmetric parts of the following characteristics: (a) requires turning, drilling, boring, and parting off, at production rates of 10 000 parts per month; (b) as (a), but only 10 parts per month; (c) a very slender high-precision part requiring turning and parting off, at production rates of 1000 parts per month; (d) as (c), but only 10 parts per month; (e) as (a) but with a transverse hole; (f) as (b), but with a transverse hole.

16B-12 To explain the main features of plain (slab) milling, (a) make a sketch of up milling and (b) another for down milling. (c) In an enlarged view, show the feed, undeformed chip thickness, and depth of cut for both cases. (d) State which of the two will give better surface finish and why. (e) State which requires a stiffer milling machine and why.

16B-13 A machinist is turning annealed 1020 steel with a HSS tool of positive rake angle. (a)

What chip should one expect and why? (*b*) What is the likely cause if, on increasing the speed, the chip thins out, its underside shows occasional pickup particles and the machined surface is poorer? (*c*) Apart from slowing down, suggest at least two measures that could be taken to improve the quality of the machined surface, still using an HSS tool.

16B-14 To explain the effects of tool-face friction, (*a*) make sketches of orthogonal cutting with a positive rake angle tool. Show the shear angle and chip thickness for (*b*) low friction and (*c*) high friction.

16B-15 Energy consumption is a function of undeformed chip thickness. To demonstrate it, (*a*) draw a diagram of energy consumption versus undeformed chip thickness. (*b*) Draw a straight line to show the trend and (*c*) choose scales that will give such straight-line relationship. (*d*) Justify your choice of the slope of the curve.

16B-16 In a lathe turning operation, chatter sets in during machining a particular part. (*a*) What will be its manifestation (list at least two) and (*b*) indicate the steps you would take to identify the cause of chatter.

16B-17 A surface turned with a tool of 1-mm nose radius has unacceptably high roughness. To improve the finish, suggest a change to the (*a*) process and (*b*) tool, and indicate the unintentional consequences of such changes (assume that no tool pickup affects roughness).

16B-18 Suggest the best metallurgical condition for cutting 1010, 1045, and 1080 steel (in your answer, define the composition and structure of each material and justify your choice).

16B-19 A hardened steel part is ground with a vitrified alumina wheel. The surface of the part is discolored and minute cracks are visible. (*a*) Identify the events that lead to the problem and (*b*) suggest at least two remedies.

16B-20 The cast-iron flange of Example 7-9, Part *b* is to be finish machined. Design, in principle, the sequence of operations. For each operation, show how the part will be mounted and show the position of the tool working on the part. Consider alternatives, applying critical judgement to each. (Note: take dimensions from text, not from illustration.)

16B-21 (*a*) List the geometric factors that affect the mechanisms typical of abrasive grain/workpiece encounters. Indicate what relative levels of these factors result in (*b*) elastic deformation, (*c*) plowing, (*d*) chip formation. (*e*) Identify those that contribute to metal removal. (You may also answer in tabular form.)

16B-22 In grinding, the transition from elastic deformation to chip formation has great effect on power and material removal rate. (*a*) What are these effects? To promote the transition, what should be done in the choice of (*b*) workpiece material, (*c*) grinding speed, (*d*) grinding fluid?

16B-23 Make sketches to show the events that may take place when a grinding wheel grit encounters the workpiece surface. From the energy consumed and the material removed in these encounters, make a judgment whether the specific energy requirement should be greater in grinding or in cutting.

16B-24 From considerations of grinding geometry and speed relationships, indicate the variation of the following factors for otherwise identical conditions: (*a*) length of cut in cylindrical grinding versus internal grinding; (*b*) length of cut in up grinding and down grinding; (*c*) undeformed chip thickness at increased wheel speeds; (*d*) undeformed chip thickness at increased feed rates. (Use sketches, if desired.)

16B-25 The grinding ratio is controlled by a number of factors, among them the force acting on grains (which, when excessive, can rip out a grain) and temperature (which leads to more rapid wear of grains). (*a*) Define the grinding ratio. (*b*) Consider the effect of wheel speed on force and (*c*) temperature. (*d*) Deduce whether a high or low speed will give the largest grinding ratio.

PROBLEMS 16C

16C-1 (*a*) Make a sketch of ideal orthogonal cutting with a positive rake angle and a shear plane of unit length. Mark the shear and rake angles and the undeformed and deformed chip thickness. (*b*) Derive Eq. (16-3).

16C-2 A lathe is driven by a 5-hp motor. Estimate the possible metal removal rates (in units of in^3/min) if the workpiece material is steel (HB 270), an aluminum alloy, and a nickel-base superalloy. Express the answer also in SI units.

16C-3 The full form of Taylor's equation, Eq. (16-19′), is valid in any measurement system. In SI units, v and v_r will be in m/s, and t and t_{ref} in s. However, a reference tool life of $t_{ref} = 1$ s would be quite unreasonable; therefore, $t_{ref} = 60$ s is used. (*a*) Develop a general formula for converting C into v_r (in SI units). (*b*) If $C = 500$ ft/min, what is the value of v_r?

16C-4 Find the conversion factor for K_1 in Eq. (16-14) from SI to USCS units.

16C-5 If the life of an HSS cutting tool is 60 min at the recommended cutting speed, what will be the life at (*a*) 20% and (*b*) 40% higher speed? (Assume $n = 0.1$).

16C-6 The ring of Example 9-23 is made of O1 oil-hardening tool steel at HRC 58. It was hot forged, and the external and internal flash were removed in the forge shop where the part was allowed to cool slowly. (*a*) What metallurgical condition do you expect in the forging? (*b*) Consider the options available in designing the machining sequence, assuming that surface finish on parallel surfaces and ID must be better than 0.2 μm R_a.

16C-7 The threaded bushing shown in the illustration is to be made in large quantities (over 10 000 pieces/month). Design the optimum screw-machine operation sequence by going through the following steps: (*a*) Clarify missing dimensions, tolerances, and surface finish specifications (in the absence of consultation with the designer, make commonsense assumptions), (*b*) choose the starting material, stock dimensions, and metallurgical condition, (*c*) determine which way the part should be oriented to allow finishing from the bar, (*d*) select the basic operations required, establishing the appropriate speeds and feeds and machining times, and (*e*) determine the operation sequence, keeping in mind the possibility of simultaneous machining and of spreading lengthy operations over several positions. (You will need access to a data base such as MHDE.)

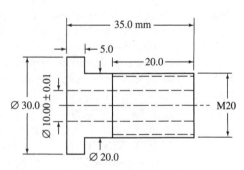

Figure Pr. 16C-7

16C-8 (*a*) Draw a sketch of facing on a lathe. Show the direction of rotation, the depth of cut, the feed, and the undeformed chip thickness. Assume that cast iron of BHN 200 is finish cut with carbide tools. (*b*) Find cutting speed and feed. (*c*) Calculate, from first principles, the time required to finish a disk of 200-mm diameter.

16C-9 An annealed 1045 steel tube is bored out, a layer of 3-mm thickness being removed in a single cut. The tube OD is 150 mm, the wall is 15-mm thick, and the tube length is 300 mm. (*a*) Make a sketch of the operation, then calculate (*b*) the recommended cutting speed and feed for a carbide tool, (*c*) the power requirement, (*d*) the time required to finish the cut, (*e*) the cutting force acting on the boring bar, if one tool is used, and (*f*) assuming that two tools are engaged simultaneously, the torque acting on the boring bar.

16C-10 (*a*) List the material properties that influence machinability. (*b*) Using the appropriate equilibrium diagrams, indicate what structure to expect, and from numerical data contained in this volume, (*c*) rank the following materials in order of their anticipated machinability: 1008 steel, 1045 steel, 302 stainless steel, 60Cu-40Zn brass, 1100 Al, and nickel, all in the annealed condition. (*d*) Show the reasoning you applied to arrive at your ranking.

16C-11 Discuss what material conditions should be specified for ease of machining (*a*) 1100 aluminum, (*b*) 6061 aluminum, (*c*) pure copper, (*d*) free-machining brass, (*e*) 1018 steel, (*f*) 1045 steel, (*g*) 1080 steel, (*h*) Hastelloy X.

16C-12 Holes of 6-mm diameter and 20-mm depth are to be drilled into 304 stainless steel, using carbide-tipped twist drills. Determine the recommended (*a*) cutting speed, (*b*) drill r/min, (*c*) feed (mm/r), (*d*) feed rate (mm/min), and (*e*) power requirement (W).

16C-13 Holes of 12-mm diameter and 36-mm depth are to be drilled into 6061-T6 aluminum alloy with HSS twist drills. Determine the recommended (*a*) cutting speed, (*b*) drill r/min, (*c*) feed (mm/r), (*d*) feed rate (m/min or m/s), and (*e*) power requirement (kW).

16C-14 If $n = 0.25$ and $C = 1000$ fpm, what should be the speed v for a life of 1 hour?

16C-15 A 100-mm-wide, 600-mm-long, 100-mm-thick slab of ASTM No. 20 cast iron is cleaned up, to a depth of 3 mm, on the 100 × 600 mm surface, by slab milling, using a HSS milling cutter of 120-mm diameter, with seven cutting edges. (*a*) Make a sketch of the process of down milling, showing the cutter rotation and the feed direction. (*b*) Find the hardness of the material. (*c*) Select the cutting speed, feed, and feed rate. (*d*) Calculate the time required. (*e*) Calculate the power requirement.

16C-16 The workpiece of Prob. 16C-15 is now of aluminum alloy. State whether the motor power increases or decreases. Justify.

16C-17 A rectangular slab of $4 \times 10 \times 1$ in dimensions, of 332.0 aluminum alloy, is to be face-milled on the two larger surfaces. The face-milling cutter has a diameter D_m and, for uniformity of surface finish, it is run clear off the ends of the workpiece. (*a*) Make a sketch to clarify the problem. (*b*) Make a reasonable assumption regarding milling cutter diameter and number of carbide inserts. Then determine, for a roughing cut, the (*c*) speed, (*d*) feed, and (*e*) net cutting time t_c.

16C-18 A steel is cut on a lathe with HSS tooling. It is now proposed to increase production rates by 30%. There is argument whether increased production should be obtained with increased (*a*) speed, (*b*) feed, or (*c*) depth of cut. Settle the argument by determining the change in tool life for the three cases, using the constants given with Eq. (16-21). (For clarity, take $K = 100$ arbitrary units, and initial values of $v = f = w = 1$.) (*d*) Make a recommendation on which variable should be increased first and which last.

16C-19 The cutting speed for minimum-cost machining v_c is given by Eq. (16-32). (*a*) Substitute v_c into Eq. (16-19′) and express the cutting time t for minimum-cost operation; denote it t_c. (*b*) Using $n = 0.1, 0.35,$ and 0.6 for HSS, carbide, and ceramic, respectively, express the relative magnitudes of t_c

(assuming identical tool and tool-changing costs).

16C-20 Establish the lowest-cost cutting speed for rough turning a cast iron of HB 200 with carbide tooling. Cost data are the same as in Example 16-13.

16C-21 A 1008 steel bar, cold drawn to 30% reduction, is machined. (*a*) Calculate the hardness. (*Hint*: for a cold-worked material, TS $= \sigma_f$, which can be found from K and n.) (*b*) Find the recommended cutting speed for HSS tools.

16C-22 A cast-iron cylinder block (TS $= 420$ MPa) is cleaned up by face milling. The surface subject to machining is 600 mm long and 150 mm wide. The face mill has seven coated carbide cutting inserts on 200-mm diameter. (*a*) Make a sketch of the process, showing the workpiece as a simple rectangular block. (*b*) Find the recommended cutting speed, feed, and depth of cut for a roughing cut. (*c*) Calculate the time required for the cut. (*d*) Estimate the power requirement, if efficiency is 70%.

16C-23 A hardened tool steel (HRC 55) surface of 30×200 mm dimensions is to be ground to a surface finish of 0.4–0.5 μm R_a. Past experience shows that a suitable wheel (A-46-H-V) will produce the requisite finish

with a cross feed of 0.5 mm/pass and a grinding depth of 0.050 mm. Table speed is 0.1m/s. The wheel is 20 mm wide, of 150-mm diameter, and clears the workpiece with a 50-mm overtravel. (*a*) Make a sketch to clarify the task. Calculate (*b*) the number of passes needed (count table movement back and forth as separate passes, and take into account that the wheel starts and finishes outside the workpiece width), (*c*) the total time taken, remembering that the wheel must run out at each end of the workpiece (make a sketch to calculate wheel position at end), (*d*) the material removed, total and per minute, and (*e*) the horsepower requirement.

16C-24 A 30-mm-diameter bar of low-alloy steel, pre-heat-treated to HRC 28, is finish turned over a 200-mm length, using a carbide tool insert. (*a*) Make a sketch of the process, showing clearly the undeformed chip thickness h, the feed f, and the depth of cut w. Then calculate, for the first (roughing) cut, (*b*) the cutting speed, (*c*) feed, and (*d*) depth of cut. Determine (*e*) the lathe r/min, (*f*) the rate of material removal, (*g*) the energy and power requirement (take $a = -0.4$), (*h*) the cutting force acting on the tool holder, (*i*) the time required for completion of the cut.

FURTHER READING

ASM Handbook, vol. 16, *Machining*, ASM International, 1989.

Drozda, T., and C. Wick (eds.): *Tool and Manufacturing Engineers Handbook*, 4th ed., vol. 1, *Machining*, Society of Manufacturing Engineers, 1983.

Machining Data Handbook, 3d ed., Machinability Data Center, Metcut Research Associates (now available from Institute of Advanced Manufacturing Science), Cincinnati, Ohio, 1983.

Armarego, E.J.A., and R.H. Brown: *The Machining of Metals*, Prentice-Hall, 1969.

Boothroyd, G., and W.W. Knight, *Fundamentals of Machining and Machine Tools*, Dekker, 1989.

Brown, J.: *Advanced Machining Technology Handbook*, McGraw-Hill, 1998.

Campbell, P.: *Basic Fixture Design*, Industrial Press, 1994.

Davis, J.R. (ed.): *Tool Materials*, ASM International, 1995.

DeVries, W.R.: *Analysis of Material Removal Processes*, Springer, 1992.

Gorczyca, F.E.: *Application of Metal Cutting Theory*, Industrial Press, 1987.

Hoffman, E.G.: *Jig and Fixture Design*, 4th ed., Delmar, 1996.

Hoffman, E.G.: *Setup Reduction through Effective Workholding*, Industrial Press, 1996.

Komanduri, R.: *Tool Materials*, in *Kirk-Othmer Encyclopedia of Chemical Technology*, 4th ed., vol. 24, Wiley, 1997, pp. 390–455.

Konig, W., I.K. Essel, and I.L. Witte: *Specific Cutting Force Data for Metal Cutting*, Verlag Stahleisen, Dusseldorf, 1982.

Madison, J.: *CNC Machining Handbook*, Industrial Press, 1996.

Malkin, S.: *Grinding Technology: Theory and Applications*, Wiley, 1989.

Metal Cutting Tool Handbook, 7th ed., Industrial Press, 1989.

Meyers, A.L., and T. Slattery: *Basic Machining Reference Handbook*, Industrial Press, 1988.

Nakazawa, H.: *Principles of Precision Engineering*, Oxford University Press, 1994.

Oxley, P.L.B.: *The Mechanics of Machining, an Analytical Approach to Assessing Machinability*, Wiley, 1989.

Roberts, G.A., G. Krauss, and R. Kennedy: *Tool Steels*, 5th ed., ASM International, 1998.

Salmon, S.C.: *Modern Grinding Process Technology*, McGraw-Hill, 1992.

Shaw, M.C.: *Metal Cutting Principles*, 3d ed., Oxford University Press, 1984.

Shaw, M.C.: *Principles of Abrasive Processing*, Oxford University Press, 1996.

Sluhan, C. (ed.): *Cutting and Grinding Fluids: Selection and Application*, Society of Manufacturing Engineers, 1992.

Smith, G.T.: *CNC Machining Technology*, 3 vols., Springer, 1993.

Schneider, A.F.: *Mechanical Deburring and Surface Finishing Technology*, Dekker, 1990.

Stephenson, D.A., and J.S. Agapiou: *Metal Cutting Theory and Practice*, Dekker, 1997.

Stout, K.J., E. J. Davis, and P.J. Sullivan: *Atlas of Machined Surfaces*, Chapman and Hall, 1990.

Szymanski, A., and J. Borkowski: *Technology of Abrasives and Abrasive Tools*, Ellis Horwood, 1992.

Taniguchi, N. (ed.): *Nanotechnology*, Oxford University Press, 1996.

Trent, E.M.: *Metal Cutting*, 3d ed., Butterworth-Heineman, 1991.

Walsh, R.A.: *McGraw-Hill Machining and Metalworking Handbook*, McGraw-Hill, 1994.

Webster, J.A., et al.: *Abrasive Processes: Theory, Technology, and Practice*, Dekker, 1996.

United States Cutting Tool Institute: *Metal Cutting Tool Handbook*, 7th ed., Industrial Press, 1989.

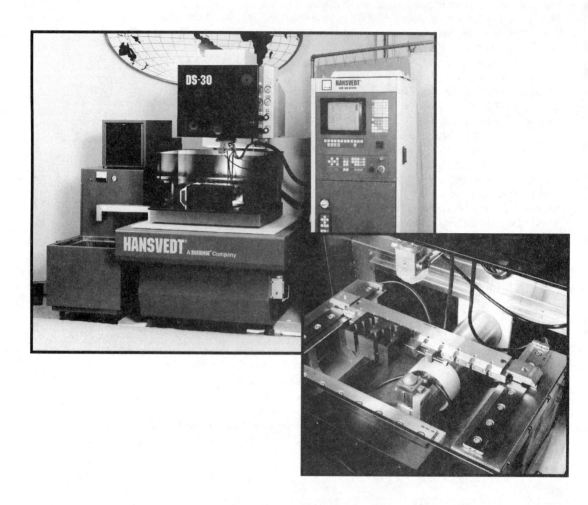

Wire-EDM is a powerful tool for cutting complex shapes into hardened steels on machines like the five-axis unit shown here. In the inset photo, fluid flow had been turned off to reveal the workspace with wire guides and a steel component with stepped cutouts. (*Courtesy Hardinge Inc., Elmira, New York.*)

17

Nontraditional Machining Processes

While most machining is done by removing material in the form of more or less well defined chips, there are other methods that offer unique capabilities. We will explore the basics of removing material through:

Chemical dissolution

Chemical dissolution aided by electric current

Controlled electrical discharge

Electron beams and laser beams of high energy intensity

Processes in direct competition with the above processes

All machining processes discussed up to this point were characterized by the mechanical removal of material in the form of chips, even though chip formation may have been imperfect. There are a number of processes that remove material by melting, evaporation, or chemical and/or electrical action; collectively they are often denoted as *nonconventional* or *nontraditional* processes. Some of these processes are not really new but have gained wider industrial application primarily because of the demands set by the aerospace and electronics industries. As a group they are characterized by insensitivity to the hardness of the workpiece material, hence they are suitable for shaping parts from fully heat-treated materials, avoiding the problems of distortion and dimensional change that often accompany heat treatment.

17-1   CLASSIFICATION

Most conveniently, processes can be classified according to the mode of action (Fig. 17–1). Some processes are exclusively for material removal, but high-energy beam processes can be also used for joining and will be discussed again in Sec. 18-7-2.

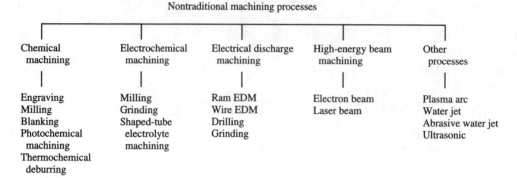

Figure 17–1   Classification of nontraditional metal removal processes.

17-2   CHEMICAL MACHINING (CM OR CHM)

It has been known for many years that most metals (and also some ceramics) are attacked by specific chemicals, typically, acids or alkalies. The metal is dissolved atom by atom and converted into a soluble compound over the entire exposed surface. In metal-removal applications, only part of the surface is etched away and the remaining parts must be protected by a substance such as wax, paint, or polymer film (the *maskant* or *resist*). Thick films are deposited by dipping or spraying over the entire surface; the pattern to be etched is cut manually with a knife along a template or by laser, and the resist is peeled off. (The process is repeated when stepwise etching is used to produce parts of varying thickness.) Greater accuracy is obtained by applying the resist through a silk or stainless-steel screen, using a stencil. Highest accuracy (better than 1 μm) can be obtained with photoresists used in semiconductor technology (see Sec. 20-3-5). There are several applications:

1. *Engraving* has been practiced for hundreds of years by the artist and printer, and is now used for nameplates and instrument panels.

2. *Chemical milling* serves to remove pockets of material, as in thinning down integrally stiffened wing skins and other aircraft components of often very large dimensions. The etchant dissolves material in all directions; therefore, it undercuts to approximately the same width as the depth of cut (Fig. 17–2a). The undercut can be exploited to increase the stiffness of panels with capped ribs (Fig. 17–2b). Surface roughness depends on the metallurgical structure; second-phase particles may etch at different rates.

3. *Chemical blanking* is used to cut through thin sheet. When the mask is made by photochemical techniques (Sec. 20-3-5), the process is called *photochemical machining*. Printed wiring boards and parts made of thin sheet are fabricated by this technique.

Metal removal rates are given in Table 17–1.

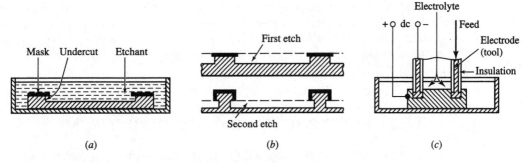

Figure 17–2   The undercut developed in chemical etching (a) can be exploited for making capped ribs (b). Material removal is accelerated on electrically conductive materials by electrochemical machining (c).

Table 17–1   Chemical and electrical machining processes*

	Chemical Machining	Electrochemical Machining	Ram-EDM	Wire-EDM
Metal removal rate	0.012–0.07 mm/min	$1.5 \text{ cm}^3/1000 \text{ A·min}$	$0.15–400 \text{ cm}^3/\text{h}$	$200 \text{ cm}^2/\text{h}$ steel $450 \text{ cm}^2/\text{h}$ Al
Surface finish, μm R_a	2 on Al 1.5 on steel 0.6 on Ti	0.1–1.4	0.2 at $0.15 \text{ cm}^3/\text{h}$ 5 at $8 \text{ cm}^3/\text{h}$ 10 at 50 cm/h	0.1 skim cut 1–2 roughing cut
Electric current				
Volts		4–24	< 300	45–60
Amperes		50–40 000	0.1–500	1–32
Frequency		DC	0.5 MHz–200 Hz	180–300 000

* Extracted from *Machining Data Handbook*, 3d ed., Machinability Data Center, Metcut Research Associates, Cincinnati, Ohio, 1980.

An aircraft component is chemically milled from a 500×800 mm 2024-T3 alloy plate of 20-mm thickness. A ribbed pattern is made by removing 80% of the area to a depth of 18 mm. (*a*) Assuming medium material removal rate from Table 17–1, calculate the time required. (*b*) To reduce assembly, the pattern is repeated and part integration results in a plate of 500×800 mm; calculate the time required.

| **Example 17-1**

(*a*) From Table 17–1, removal rate is $(0.012 + 0.07)/2 = 0.041$ mm/min. Depth to be removed is 18 mm; machining time $= 18/0.041 = 439$ min $= 7.3$ h. (*b*) Surface area does not enter into the calculation, and machining time remains the same. This shows the great advantage of the process for machining large parts (or several small parts in a larger tank).

---

4. *Thermochemical deburring* can be regarded as a high-temperature version of chemical machining. Parts are enclosed in a chamber into which an explosive mixture of natural gas, hydrogen, and oxygen is introduced. Upon ignition, very high flash temperature removes burrs by melting, vaporization, and oxidation. The parts get barely warm, provided the thinnest section is at least 10 to 15 times thicker than the burr. The process is suitable for metals and thermosetting plastics.

17-3 ELECTROCHEMICAL MACHINING (ECM)

The rate of material dissolution is greatly increased when a dc current is applied. The process is the reverse of electroforming (Sec. 11-8). The workpiece, which must be conductive, is submerged, together with the cathode, into the electrolyte (a dilute Na-chloride, Na-nitrate, sulfuric acid, or proprietary solution) and is made the anode. Metal removal rates W_c (g/s·m^2) can be calculated from Eq. (11-4). It is usual to express the volume removed in unit time

$$V_t = \frac{W_c}{\rho j} \qquad \left(\frac{m^3}{A \cdot s} \right) \tag{17-1}$$

where ρ is density (g/m^3) and j is current density (A/m^2). For convenience of calculation, V_t is given in Table 17–2 in units of mm^3/A·min, with valences shown in parentheses. The electrode feed rate is then

$$v_e = V_t j \tag{17-2}$$

On some metals an insulating oxide film may build up and can be broken down by intermittent spark discharges produced by an ac or pulsed dc circuit. Several versions of the process are used:

1. *Electrochemical milling* serves to remove material from large surfaces; the cathode is in the form of a distantly mounted flat plate.

2. *Electrochemical machining* (ECM) uses a metal or graphite cathode which is the negative of the shape to be produced. The cathode is fed into the workpiece at a controlled rate (Fig. 17–2c). The electrolyte is circulated—often through the cathode—to wash out the metal-hydroxide sludge and to ventilate the hydrogen formed in the course of electrolysis. The machine tool is rigidly constructed to prevent vibration and consequent inaccuracies. Rough guidelines on process variables are given in Table 17–1.

3. *Shaped-tube electrolyte machining* (STEM) uses an insulated titanium tube as the cathode through which an acid electrolyte is pumped to make holes of 0.25–6-mm diameter with aspect ratios (depth-to-diameter ratios) up to 300:1. Tolerances of ±0.08 mm have been achieved, with a finish of 0.4–3.2 μm R_a. The process is slow (0.75–3 mm/min) but many holes, of different diameters, can be simultaneously made. *Electrostream drilling* and *capillary drilling* use glass tubes to direct the electrolyte and make small holes in internally cooled turbine blades and similar parts.

Table 17–2   Specific material removal rates in electrochemical machining* *

Metal	Specific Removal Rate, $mm^3/A \cdot min$
Iron (2)	2.21
Iron (3)	1.47
Aluminum (3)	2.06
Copper (1)	4.39
Copper (2)	2.20
Nickel (2)	2.11
Nickel (3)	1.36
Titanium (3)	2.19
Titanium (4)	1.65

*Compiled from *Machining Data Handbook*, 3d ed., Machinability Data Center, Metcut Research Associates, Cincinnati, Ohio, 1980.

4. *Electrochemical grinding* uses a conductive grinding wheel (copper-bonded Al_2O_3 or metal-bonded diamond) as the cathode. Most of the material is removed by electrolysis, since the grit removes oxides and interfering films. In contrast to conventional grinding, the workpiece remains cool and pressures are low. The process is suitable for sharpening carbide cutting tools and for grinding turbine blades and vanes and delicate parts such as hypodermic needles and honeycomb structures.

Since metal removal occurs in the ionic state, the hardness of the material is of no consequence in either CHM or ECM; surface integrity is excellent; there is no heat damage; residual stresses are minimal or absent; surface finish is nondirectional. Superalloys, fully heat-treated steels, and aluminum alloys are often cut. Alloys susceptible to hydrogen embrittlement should be heated at 200°C for a few hours after CM.

Example 17-2

The component of Example 17-1 has equilateral triangular pockets of 100-mm sides. Find the machining time per pocket if the applied current is 5000 A.

The frontal area of the electrode is an equilateral triangle of 100-mm side. Area = $100(50\sqrt{3})/2 = 4330$ mm^2. Current density = 5000 A/4330 mm^2 = 1.155 A/mm^2. From Table 17–2, $V_t = 2.06$ $mm^3/A \cdot min$. Penetration rate, from Eq. (17-2) = 2.06/1.155 = 1.784 mm/min. Machining time = 18/1.784 = 10.1 min. The process has to be repeated to work on the entire surface of the part, therefore, even though machining time for one pocket is much less than the time required for chemical machining, total time would not be favorable especially because of the need for electrodes of different configurations. ECM is, however, highly competitive when smaller areas are to be removed.

17-4   ELECTRICAL DISCHARGE MACHINING (EDM)

In this process, chemical action is abandoned and metal is removed by the intense heat of electric sparks. There are several processes, all sharing the same mechanism.

The workpiece and the cathode (tool), made of metal or graphite, are submerged in a *dielectric fluid*. A direct current at a potential of up to 300 V is applied to the system; if a non-solid-state power supply is used, a capacitor is included in parallel with the spark gap (Fig. 17–3). At low voltages the fluid acts as an insulator; as the voltage builds up, the fluid suffers *dielectric breakdown* (large numbers of electrons appear in the conduction band) and a spark passes through the gap. Temperatures increase sufficiently to cause local vaporization of some of the workpiece material. After a discharge of controlled duration, the voltage is dropped to a low value for a short time (waiting time) to reestablish the insulating film by deionization of the dielectric. Pulse generators repeat the controlled cycle at a rate of 200–500 000 Hz. The dielectric must fulfill additional functions: It is supplied to the tool–workpiece interface to provide cooling and to flush out debris. It must be continuously filtered to remove debris that would cause short circuits. Discharge always takes place at the closest gap; therefore, the electrode is fed continuously to cut the desired shape. Optimum conditions and spark gaps are maintained by servocontrol. Overall process control is now frequently performed by CNC.

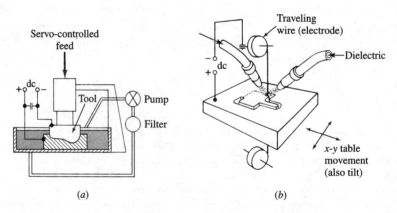

(a)   (b)

Figure 17–3   Electrodischarge machining is practiced with (a) form tools in ram-EDM and (b) traveling wire in wire-EDM.

Metal removal rate is a function of current density (Table 17–1). Empirically it is found to decrease for metals of higher melting point

$$V_{EDM} = 4T_C^{-1.23} \quad (cm^3/kA \cdot min) \qquad \textbf{(17-3a)}$$

where T_C is the melting point in degrees Celsius. In conventional volume units

$$V_{EDM} = 0.243T_C^{-1.23} \quad (in^3/kA \cdot min) \qquad \textbf{(17-3b)}$$

Surface finish is governed by a number of factors and becomes rougher with higher current densities (which give higher discharge energies), more viscous dielectric, and lower frequency. The same factors that contribute to a rougher finish also result in more *overcut* (typically, 0.005–0.5 mm/side) and in a deeper *heat-affected* (damaged) *zone* (typically of 2–400 μm depth). Some of the molten metal is redeposited (*recast layer*) and softening may also take place. Therefore, it is customary to end the cut at a low current density, or to finish the surface by other techniques. This is particularly important for workpieces subject to fatigue loading.

Material removal is not limited to the workpiece: The tool is eroded too. Under optimum conditions, the *wear ratio* (ratio of workpiece volume removed to tool volume lost) is 3:1 with metallic electrodes and from 3:1 to 100:1 with graphite electrodes. The process is insensitive to the hardness of the material; therefore it has found wide application for making forging, extrusion, sheet-metalworking, die-casting, and injection-molding dies into hardened steel die blocks, by various techniques:

1. *Ram-EDM.* Complex cavities are formed by the controlled penetration of a shaped electrode into the body (Fig. 17–3a). The electrolyte is low-viscosity oil. Electrode usage is improved by restricting the new electrode to finishing, and then using it to rough out the next workpiece. Roughing is sometimes done with a no-wear EDM process in which the polarity is reversed and the graphite anode suffers no weight loss. With the aid of CNC, the workpiece may be given controlled lateral motion (planetary motion of 10–100-μm amplitude) to improve accuracy and surface finish and increase metal removal rates. It becomes possible also to machine complex shapes with simple electrodes moved in a complex path, rather like multiaxis contour milling, provided that removal rates remain fast enough. Even though the process is fairly slow, CNC control with feedback is generally used and automatic electrode and pallet changers can be added; thus, the process lends itself to unattended operation. Electrodes are made by copymilling or CNC. Special abrasive processes have also been developed for making graphite electrodes.

2. *Electrical discharge wire cutting.* EDWC or, most often, *wire*-EDM has become an important production process. The electrode is now a brass, copper, tungsten, or molybdenum wire of 0.08–0.3-mm diameter (Fig. 17–3b). The wire acts like a band saw, but sparks instead of teeth do the cutting. The slot (*kerf*) formed is somewhat wider (by about 25 μm) than the wire. The wire is fed continuously from the take-off spool under controlled tension (about 60% of tensile strength) at a rate of 2.5–150 mm/s. To aid expulsion of the material from the slot, *stratified wire* is used: Brass wire is coated with Zn or a high-Zn brass, which boils off before the core wire loses strength. Thus, higher current densities can be used. The gas pressure causes the wire to bow, and travel speed must be reduced, especially in corners, if a straight cut through the thickness is required. As indicated by Eq. (17-3), cutting progresses faster in metals of lower melting point (Table 17–1). The electrolyte is oil or water with additives; bacterial growth is prevented by ozone treatment since bactericides destroy dielectric properties. One or more shallow (typically 0.04-mm) *skim cuts* may be taken with high-frequency ac current to remove the damaged surface. Multiaxis controls make it possible to cut 3-D shapes, for example, extrusion dies with tapered entry. CNC controls with feedback are used to monitor gap conditions, adjust cutting speeds, and rethread in case of wire breakage. Many parts have holes for which a pilot hole must be made by some process; the wire is then threaded and the slug removed manually or automatically. Very tight corners can be cut and the quality of cut is adequate for many sheet metalworking dies and other applications, including WC dies. Thicknesses up to 400 mm have been cut.

3. *Electrical discharge drilling.* Tungsten wire is used as the electrode in conjunction with an aqueous dielectric to make holes of small diameter (between 0.05 and 1 mm) to great depths, as for cooling holes in turbine blades made of superalloys. Pumping the electrolyte through a rotating electrode increases removal rates but at the expense of tolerances.

4. *Electrical discharge grinding.* Material removal takes place by discharges between a rotating graphite wheel and the workpiece.

Example 17-3 | Find the relative metal removal rates by EDM for 7075 Al, 1045 steel, Hastelloy X superalloy, and Ti-6Al-4V alloy. Take melting points from Table 8–3 and Fig. 6–11.

Alloy	Liquidus, °C	$T^{-1.23}$
7075 Al	640	0.354×10^{-3}
1045 steel	1500	0.124×10^{-3}
Hastelloy X	1290	0.149×10^{-3}
Ti-6Al-4V	1660	0.109×10^{-3}

Note that the metallurgical condition of the material does not enter into the relation. The 1045 steel could be quenched and tempered or fully annealed. It is cut at about the same rate as the superalloy or Ti alloy, which are normally regarded as much more difficult to machine.

Example 17-4 | Stainless steel rocket engine rotors of 50-mm diameter used to be made by 5-axis milling. Each rotor contains over 100 blades of complex shape with a tolerance of ±0.03 mm, and machining took 2000 hours per part. The rotors were then made by EDM; material was first removed from one side to obtain the leading edges, then the part was flipped over and machined again to obtain the trailing edges. Machining time was 400 hours, only a fifth of the time it took by milling. [Source: *Manufacturing Engineering*, 1998(3):160.]

17-5   HIGH-ENERGY-BEAM MACHINING

Materials can be machined—mostly cut or drilled—by melting and/or vaporizing the substance in a controlled manner. Processes are offshoots of joining processes to be discussed in Chap. 18. They are useful not only for metals but also for materials that are otherwise difficult to machine, for example, plastics, ceramics, and composites.

17-5-1   Electron Beam Machining (EBM)

The energy source for EBM is an electron gun (Fig. 17–4) similar to a vacuum tube. The W or Ta cathode, heated to 2500°C, emits masses of electrons that are accelerated and focused to a 0.25–1-mm-diameter beam of high energy density (in excess of 200 kW/mm^2). Upon impact, the kinetic energy of electrons is transformed into thermal energy, sufficient to melt and partially vaporize the workpiece material; the vapor ejects the melt. Pulses of 0.05–100 ms duration are used. The cavity is of small diameter and deep (hence the term *keyhole*), the heat-affected zone is very narrow, and energy-conversion efficiency is high, about 65%. Since an electron beam can be deflected with an electromagnetic coil, cuts of high quality can be made in complex patterns, on practically any material.

The electron gun is always in high vacuum. Deepest penetration is obtained when the workpiece too is enclosed in *high vacuum* (10^{-4} to 10^{-1} Pa), but

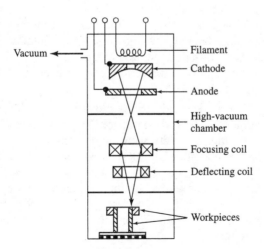

Figure 17–4   Very high energy densities are obtained in electron beam cutting.

pumping down the chamber takes several minutes. *Medium vacuum* (0.1–3 kPa) still permits operation on many metals with a pumping time of less than 1 minute. With specially constructed vacuum traps, the electron beam can emerge from the gun into a shielding gas, but power intensities are much lower in such out-of-chamber or *nonvacuum* operation.

With individual pulses, small holes (0.1–0.3-mm diameter in 0.1-mm-thick sheet, and 0.6–0.85-mm diameter in 8-mm-thick material) are made extremely rapidly; thousands of holes are made in only minutes. Depth-to-diameter ratios range up to 15. For through-holes, temporary backing is needed to maintain vapor pressure; the backing is often brass powder in silicone rubber or epoxy. Jet engine components as well as spinnerets for glass and polymer fibers are often made by the technique.

17-5-2   Laser Beam Machining (LBM)

The word laser stands for *light amplification by stimulated emission of radiation*. Some materials (*lasing media*) emit a highly collimated, coherent, monochromatic light beam when excited (*pumped*) by some appropriate energy source. Applications have mushroomed because very high energy densities are attainable, no vacuum is needed, and the beam is readily and rapidly directed by appropriate optics.

The first lasers utilized ruby (an Al_2O_3 crystal with Cr ions) as the lasing medium and such lasers are still useful for tasks such as alignment and measurement. For manufacturing purposes, three kinds of lasers have found application, each of which gives light of different wave length. The spot size is a function of wave length and optics but generally decreases with wave length. Thus, the width of cut (kerf) also decreases with decreasing wavelength.

1. *Gas lasers.* Most widely used are CO_2 *lasers*; these contain a mixture of gases in which CO_2 is the lasing medium, excited by an electric discharge between electrodes placed in the discharge tube. The emitted light is of 10.6-μm wavelength, in the far-infrared range, and can be directed with mirrors. Large units can develop over 40 kW in the continuous mode at 15% efficiency; in the pulsed mode, energy is under 1 J/pulse.

2. *Solid-state lasers.* The industrially most important *Nd:YAG lasers* contain small concentrations of neodymium ions in an yttrium aluminum garnet (YAG). Pumped with high-intensity white light from a xenon or krypton lamp (Fig. 17–5a), they emit radiation of 1.06-μm wavelength (near infrared) which, can be directed by fiber optics to great distances without excessive loss. The beam can be shared between several stations, justifying the higher investment. Operated in either pulsed or continuous mode, lasers develop up to 500 J/pulse of 0.1–20-ms duration, or several kW continuously. Overall energy conversion efficiency is low, around 2%.

3. *Excimer lasers.* These devices have much lower output and are used mostly for micromachining or semiconductor processing (and also for corrective eye surgery). In an electrical discharge, a noble gas atom (Ar, Kr, Xe) and halogen

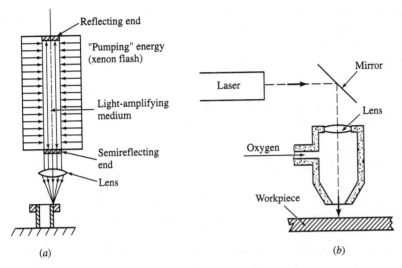

Figure 17-5   The high energy densities generated by (a) laser beams can be used for cutting, and cutting is greatly accelerated when (b) oxygen is supplied to the cutting zone.

gas (Cl_2 or F_2) form a dimer (hence the name *excited di*mer). As the electric discharge decays, the dimers split and emit UV light. The energy absorbed by the target material is large enough to break chemical bonds and eject small molecules without melting. A mask is used to define the pattern (*heatless laser etching*).

Laser beams can be focused with lenses made of materials transparent to the particular wavelength (zinc selenide or germanium for CO_2, and conventional glass optics for YAG lasers). Because beams are highly collimated, to a spot size of 0.2–0.3 mm, peak energy densities of 5–200 kW/mm^2 are reached. Depending on energy density, some material is melted (ablated), some is evaporated, creating a keyhole, as in EB cutting. Machining to a controlled depth is possible to some extent. Frequently, oxygen is supplied to the surface to increase energy absorption (Fig. 17–5b). The exothermic oxidation reaction also supplies heat and accelerates melting; furthermore, the oxide melts at a lower temperature and is blown away. Heat-treated steels may be cut; edge quality can be suitable even for punches and dies used in sheet metalworking.

Lasers provide one solution to the difficult problem of drilling holes in hard materials. In *percussion drilling* with intense bursts, the hole is the size of the beam, which is varied by focusing; tolerances are ±0.03 mm. In trepanning a circle is cut out with a tolerance of ±0.013 mm. The minimum hole size is typically 0.5 mm for CO_2 lasers, 0.08 mm for Nd:YAG lasers, and 1 μm for excimer lasers. Blind holes cannot be made to exact depth.

At Harley-Davidson, stretch-drawn sheet-metal parts such as motorcycle heat shields, fuel tanks, and fenders used to be shimmy-trimmed in complex press dies: The two dies close

Example 17-5

and then move in two lateral directions to trim the part in a horizontal plane. Setup times were 20–90 min per die and the trimmed edge had to be sanded. Now a 3-D robotic laser system requires setup times of only 1–10 min and trims as fast as the dies did, but with a directly acceptable edge finish and without die cost. Since the part is not touched by any tool, there are no blemishes to spoil the surface finish of these exposed parts. [After J.V. Owen, *Manufacturing Eng.*, 1992(6):35–41.]

Example 17-6	EB and laser drilling are often competitive. Spinnerets for glass and polymer fiber manufacture are prime examples. In one instance, 25 600 holes of 0.6-mm diameter were drilled by EB into a superalloy ring of 3.3–5.5-mm wall thickness in 16 min. A YAG laser would take some 12 times longer. [Source: R.R. Schreiber, *Manufacturing Eng.*, 1992(5):37–41.]

17-6   MACHINING OF NONMETALLIC MATERIALS

The emphasis in the preceding discussion and in Chap. 16 was on metals, but many of the basic principles discussed can be applied to other materials if allowance is made for their properties.

17-6-1   Machining of Ceramics

Most ceramics are hard and act as abrasives themselves. Therefore their machining is very often limited to abrasion by a yet harder ceramic. Thus diamond can be used to dress grinding wheels or to finish tool bits or ceramic (e.g., Al_2O_3) components. All abrasive processes, including grinding, lapping, polishing, ultrasonic machining, hydrohoning, grit blasting, and abrasive water-jet cutting are employed for both overall finishing and for localized shaping of ceramic (including glass) parts. Ceramics that are susceptible to chemical attack (as glass is to HF acid) can be chemically machined (etched). Mechanical as well as *chemical-mechanical abrasive processes* have been developed for finishing brittle materials to a very high level of surface quality.[1]

Creep-feed grinding can be economical for developing the shape from a simple preform. Compared to metals, speeds are low (45 m/s); depth of cut is 2.5–6 mm, and feed 250–600 mm/min. Heat is taken away in the particles; therefore, stresses, power, and temperatures are lower than in surface grinding.

Many nontraditional processes find application. EDM is feasible if resistivity is below 300 $\Omega \cdot$cm. Laser-beam vaporizing is suitable for drilling ceramics and silicon crystals and is extensively used for scribing electronic substrates: partial perforation creates a snap line along which parts can be safely separated.

Example 17-7	For long life, ball bearing balls made of Si_3N_4 must be smooth and free of surface defects. They are finished by polishing; chemomechanical polishing removes material rapidly by combining

---

[1] R. Komanduri, D.A. Lucca, and Y. Tani, *CIRP Annals*, **46**:545–596 (1997).

abrasion with chemical reactions in an aqueous environment. The surface is converted into brittle SiO_2, which has significantly lower hardness than Si_3N_4 and can be removed with a relatively soft abrasive such as CeO_2. Thus, a scratch-free surface of $R_a = 4$ nm ($R_t = 40$ nm) finish can be achieved. [Source: M. Jiang, N.O. Wood, and R. Komanduri, *J. Eng. Mat. Techn.*, **120**:304–312 (1998).]

17-6-2   Machining of Plastics

Even though plastics have a molecular rather than atomic structure, chip-forming processes can be applied to them if allowance is made for the differences in properties.

1. Compared to metals, plastics have a low elastic modulus and deflect easily under the cutting forces; therefore, they must be carefully supported.

2. Because of the viscoelastic behavior of thermoplastics, some of the local elastic deformation induced by the cutting edge is regained when the load is removed. Therefore, tools must be made with large relief angles and tools must be set closer than the finished size of the part.

3. In general, plastics have low thermal conductivity (Table 4–1); therefore, heat buildup in the cutting zone is not distributed over the body and the cut surface may overheat. In a thermoplastic resin the glass-transition temperature T_g may be reached and the surface smeared or damaged, while thermal breakdown and cracking may occur in thermosetting resins. Therefore, friction must be reduced by polishing and honing the active tool faces and by applying a blast of air or a liquid coolant (preferably water-based, unless the plastic is attacked by it). Best surface finish is produced at high speed and low feed. Saw teeth should be widely spaced to avoid overheating.

4. Since the shear zone is shortened and the cutting energy is reduced with a large rake angle (Sec. 16-1-1), cutting tools are made with a large positive rake angle (Table 16–4). This is permissible because the strength of plastics is low compared to metals. However, at excessive rake angles the cutting mechanism changes into *cleaving* in which coarse, disjointed fragments are lifted up and a very poor surface is produced.

5. Twist drills should have wide, polished flutes, a low ($<30°$) or even zero helix angle, and a 60–90° point angle, particularly for the softer plastics, although some plastics such as PP can be drilled with standard drills.

6. Plastics can be surprisingly difficult to machine when reinforced by fillers. Glass fibers are particularly hard on the tool and it is not uncommon that only carbide or diamond tools can stand up.

7. Trimming of plastic parts is often done with nonconventional processes. Water jet suffices for unfilled plastics but abrasive water jet is necessary for filled plastics; the processes cannot be used if the plastic absorbs water (as aramids do). In cutting with high-energy beams, polymer chains break down, thermoplastic polymers melt, and thermosets decompose (char). Clean edges are produced.

In general, molding and forming methods (Chap. 14) produce acceptable surface finish and tolerances (Fig. 3–22), and design usually aims at avoiding subsequent machining. Occasionally, however, machining is a viable alternative to molding (e.g., for PTFE, which is a sintered product and not moldable by the usual techniques).

17-6-3   Machining of Composites

A common problem in cutting composites is delamination, poor edge finish, and fiber or resin pull-out. Nontraditional processes are, therefore, widely used for making holes, trimming finished parts, and cutting blanks for further forming. A great advantage is that there is no fraying and the edges are sealed. Robots with multiaxis control are needed when the part has 3-D configuration.

Abrasive water jet can cope with all plastic-matrix composites but is best avoided for aramid-fiber composites. Plastics can be cut with laser, although some will decompose or char. Lasers are employed for glass/epoxy and Kevlar/epoxy composites, but not for carbon-fiber reinforced plastics because carbon does not evaporate at a low enough temperature. Glass/epoxy composites char and the edges become conductive; therefore, if isolation is important, the damaged edge must be removed.

As pointed out in Sec. 15-1-2 , wood is a natural polymer of highly directional structure. Its relatively low strength allows working with highly positive rake angles, however, cleavage (splitting) ahead of the tool is a problem when the grain direction encourages cleavage at an angle that produces an increasing undeformed chip thickness. High-quality cuts are made by lasers if the energy density is high enough to cause vaporization.

Metal-matrix composites, such as carbide tool bits, can be conventionally ground with diamond, or the electrical conductivity of the matrix can be exploited by electrodischarge or electromechanical machining and grinding.

17-7   PROCESS CAPABILITIES AND DESIGN ASPECTS

Process capabilities are determined by the nature of the process. Most processes are primarily for the generation of shapes in two dimensions, and there are few limitations on the complexity of these shapes. In the third dimension, simple steps can be made by chemical machining, and shapes without undercuts can be produced by advancing shaped tools in ECM and ram-EDM. Limited undercuts can be formed with ram-EDM if the electrode (or workpiece) is set at an angle and moved in a programmed manner. Wire-EDM is capable of making any shape that has a straight-line generatrix.

Ultrasonic machining is, to a small extent, competitive with ram-EDM and ECM. Wire-EDM, electron beam, and laser machining are in competition with abrasive water-jet (AWJ) cutting and with some welding processes, especially

Table 17–3   Characteristics of cutting processes

	Laser	Wire-EDM	Plasma	Abrasive Water Jet
Max. thickness, mm	12 steel, 4 Al	200	200	200
Kerf, mm	0.2–0.5	0.1–0.5	1–4	0.8–1.5
Tolerance, ± mm	0.02–0.4	0.01–0.03	1	0.2–0.4
Edge roughness, μm R_a	1–2	0.2–1	1–10	1–2
Edge inclination, degree	0.5–1	0	1–3	0
Heat–affected zone, mm	0.03–1	0.05–0.5	1–5	0
Metal	Yes	Yes	Yes	Yes
Ceramic	Yes	(No)*	No	Yes
Plastic	Yes	No	(Yes)†	Yes

*Yes if conductive.

†Yes only if edge burning is acceptable.

plasma arc cutting. Table 17–3 gives a comparative evaluation in terms of cost, productivity, and technical characteristics. Inferior tolerances and surface finish are acceptable for many applications and do not necessarily eliminate a process from consideration. There is usually some dross of resolidified metal at the exit side of plasma cuts and, to a lesser extent, of electron beam and laser cuts, but dross as well as heat-affected zone are totally absent in AWJ cutting. Water-jet cutting is a potential alternative for soft materials, primarily plastics, rubber, leather, and fabrics.

All cutting processes are eminently suitable for making contour cuts, using tracer mechanisms, NC, or CNC. Plasma or laser cutting is often incorporated into CNC punching centers (Sec. 10-3-4), greatly increasing the versatility of these machines. With high positioning rates and CNC control, productivity can be very high, making the laser competitive despite the high capital cost.

17-8 SUMMARY

A common characteristic of nontraditional machining processes is that material removal takes place not by chip formation but by chemical dissolution, melting, evaporation, or breaking of bonds. Because of this, the processes are insensitive to the hardness of the workpiece material and occupy a special position in machining fully heat-treated tool steels, superalloys, ceramics, and composites. In many cases, closed-loop CNC control allows processes to be operated unattended or in lights-out manner.

1.   Chemical machining is slow but large surfaces can be treated, hence productivity is acceptable, especially if the task of masking is speeded up.

2. ECM offers much increased material removal rates in electrically conductive materials. EC grinding is valuable for shaping cemented carbide tools.

3. EDM has reached a significant position in tool and die making. Ram-EDM produces cavities of complex shape and wire-EDM has become the process of choice for many two-dimensional die shapes.

4. Electron beam and, more frequently, laser beam cutting are suitable for making parts in small and medium production. Since laser beams do not require vacuum, they can be incorporated into cutting centers with mechanical punches and x-y tables. Lasers are used also for scribing and marking all materials.

5. Nontraditional processes often offer the only practical way to making small, deep holes in very hard materials.

6. Many processes present health hazards. Chemicals used in CM and ECM are corrosive and require fume exhaust and protective clothing for handling. Many EDM fluids are flammable and automatic fire extinguishers must be installed. High-energy-beam installations need to be enclosed and eye protection is essential.

PROBLEMS 17A

17A-1 Define (*a*) CM and (*b*) ECM. (*c*) Make a sketch showing the essential features of a setup for ECM.

17A-2 Define (*a*) EDM and (*b*) make a sketch showing the essential elements of ram-EDM.

17A-3 Define (*a*) electrochemical grinding and (*b*) electrical discharge grinding.

17A-4 What does the term *laser* stand for?

17A-5 Is it necessary that the workpiece be electrically conductive for (*a*) CM, (*b*) ECM, (*c*) EDM, (*d*) electron beam, and (*e*) laser beam machining? (Answer simply yes or no.)

17A-6 The circuitry of a printed wiring board is made by photochemical etching of a copper strip. State how much longer it will take to etch through a fully hard copper relative to a dead soft copper of the same thickness.

PROBLEMS 17B

17B-1 An internally-cooled turbine blade has 30 holes of 0.5-mm diameter. The material is a superalloy of high hardness. Suggest at least three processes that could be considered for making the holes.

17B-2 A fiber-reinforced plastic component of 1-m length and 0.5-m width is pressed into a shallow but complex shape by matched-die molding. The edges of the finished part must be trimmed. List all processes that can be used if the plastic is a thermoplastic polymer and the fiber is (*a*) glass or (*b*) carbon. List advantages and limitations.

17B-3 Blanks of a given shape are to be cut from a 0.5-mm-thick thermoplastic polymer sheet. For increased productivity, it is proposed to stack 10 sheets of the plastic and cut them together. List processes that could be used for cutting (*a*) individually and (*b*) in a stack, justifying your choice. (List also processes that might be considered but should be rejected.)

17B-4 A 3-D body is made by hand lay-up of fiberglass-reinforced epoxy prepreg. Several pieces are used, each of a different shape. List three methods (traditional or nontraditional) for cutting the shape, and make judgments on (a) applicability, (b) suitability for one-off production, (c) suitability for medium-lot production, (d) investment in equipment, and (e) accuracy of cut.

17B-5 Low-carbon steel pressings of 1-mm thickness and complex 3-D shape need to be trimmed. List at least five conventional and nontraditional methods that can be considered and rank them in order of suitability (include processes that have to be rejected). Justify your rankings by giving the main advantage and disadvantage of each process.

17B-6 A hole of 20-mm diameter is to be made in Al sheets by either CM or ECM. Sheet thickness is 1 mm or 3 mm. Choose for each sheet thickness the diameter of (a) cutout in the maskant for CM and (b) electrode for ECM. (Absolute values will be determined from experience; here our primary interest is in relative magnitudes.)

17B-7 A closed hole is to be made into a 30-mm-thick D2 tool steel of HRC 60 hardness by wire-EDM. Suggest at least three options for making the starting hole, with judgments of their suitability.

PROBLEMS 17C

17C-1 A hole of 80-mm diameter is trepanned by ECM (as in Fig. 17–2c) into an H13 hot-working die steel plate of 50 mm thickness. The inner diameter of the tube-shaped electrode is 60 mm. The available power supply delivers 10 000 A. Ignoring overcut, estimate the time required to finish the cut (assume that removal rate is halfway between bi- and trivalent iron in Table 17–2).

17C-2 A Ni-based superalloy part has a 3-D surface configuration with shallow cavities. It will be subjected to fatigue loading in service. Surface finish is 0.2 mm R_a. Processes under consideration include (a) creep-feed grinding, (b) CNC milling, (c) CNC milling followed by grinding, (d) CNC milling followed by electrochemical grinding, and (e) ram-EDM. Apply a critique, evaluating each process for the likelihood of producing the requisite surface finish without harmful residual stresses. (*Hint*: Start with Fig. 3–22 and refine your judgment by referring to process descriptions in this text.)

17C-3 A 304 stainless steel sheet of 1-mm thickness has 20 holes of 21.7-mm diameter in an irregular pattern. Diametral and locational tolerance is ±0.1 mm. Maximum 0.1-mm burr height is allowed. Review all methods you can think of for making (a) one sheet in total and (b) 150 sheets per week. (c) Apply a critique to the design of the part; suggest how the choices would change if the diameter of the hole were 20 mm. Justify your choices, including processes rejected.

17C-4 A 304 stainless steel sheet of 1-mm thickness is to be made with 1200 holes of 0.2-mm diameter. Diametral and locational tolerance is ±0.03 mm and no burr is allowed. Review all methods you can think of for making (a) one sheet in total and (b) 12 sheets per week.

17C-5 The forging dies needed for making the parts of Example 9-23b are made by ram-EDM. Take dimensions from Fig. Ex. 7-9b and, to complete the design of the die, add internal and external flash gutters as in Fig. Ex. 9-23b. (a) Make a dimension sketch of the lower die and (b) estimate the machining time if 99% of the volume is removed at the highest rate and, for improved surfaces finish and minimum recast layer, 1% is removed at the lowest rate.

FURTHER READING (see also Chap. 16)

Guitran, E.B.: *The EDM Handbook*, Hanser-Gardner, 1997.

Hablanian, M.H.: *High-Vacuum Technology*, 2d ed., Dekker, 1997.

Harris, W.T.: *Chemical Milling: The Technology of Cutting Materials by Etching*, Oxford University Press, 1976.

Hoare, J.P., and M.A. LaBoda: *Electrochemical Machining*, in *Comprehensive Treatment of Electrochemistry*, vol. 2, Plenum, 1981.

Ikeda, M.,I., Miyamoto, and T. Miyazaki: *Energy-Beam Processing of Materials*, Oxford Science Publications, 1989.

Powell, J.: *CO_2 Laser Cutting*, Springer, 1991.

Steen, W.M.: *Laser Material Processing*, Springer, 1991.

Taniguchi, N.: *Energy-Beam Processing of Materials*, Oxford University Press, 1989.

Weller, E.J. (ed.): *Nontraditional Machining Process*, 2d ed., Society of Manufacturing Engineers, 1984.

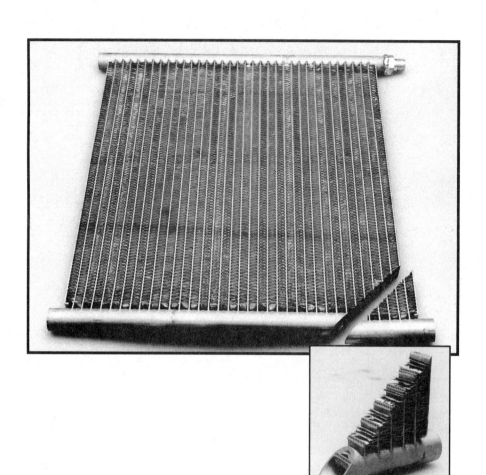

Hundreds of aluminum components of this automotive heat exchanger are simultaneously joined by brazing; louvered fins visible in the cut-off section are only 0.1-mm thick. (*Courtesy Modine Manufacturing Co., Racine, Wisconsin.*)

18

Joining Processes

Parts produced by any of the previously discussed processes can be made into larger, more complex bodies. We shall explore the basics of techniques such as:

Mechanical fastening

Creating a metallurgical bond by adhesion and diffusion

Joining by fusion with the use of various heat sources

Brazing or soldering with a lower-melting metal

Applications to plastics and ceramics

Production of laminates

Building of parts without the use of a mold

Joining differs from previously discussed processes in that it takes parts produced by other unit processes and unites them into a more complex part; therefore, it could also be regarded as a method of assembly. The product may replace a part that could have been made by other techniques (e.g., a cast machine-tool frame may be replaced by a welded frame) or it may be of a kind that can be produced only by joining processes (e.g., an automobile unibody, an automotive radiator, or a bicycle frame). While most joining methods are practiced on metals, ceramics and polymers may also be joined by similar techniques.

18-1   CLASSIFICATION

Some joints are purely mechanical, and within this category those devices that establish semipermanent joints (such as screws and bolts) are properly regarded as means of assembly. The emphasis of our discussions is on unit processes;

hence, we will concentrate on methods for establishing permanent joints (Fig. 18–1). We have much to build on. Mechanical joining methods are derived from metalworking processes, solid-state techniques are based on adhesion and deformation, fusion welds are related to casting processes, and liquid/solid processes draw on solidification, adhesion, and polymer technologies. There are, however, substantial differences in the way these processes are carried out, and it is these differences that will be at the focus of our attention. Where applicable, processes will be identified by the names and abbreviations given by the American Welding Society.

Joining processes often demand considerable skill. Harmful fumes, high electric voltages, and high temperatures require operator protection. For these reasons, and also for higher productivity, efforts aim at minimizing operator involvement, and we shall point out opportunities for mechanization (open-loop control of machinery) and true automation (closed-loop control, without or with artificial intelligence).

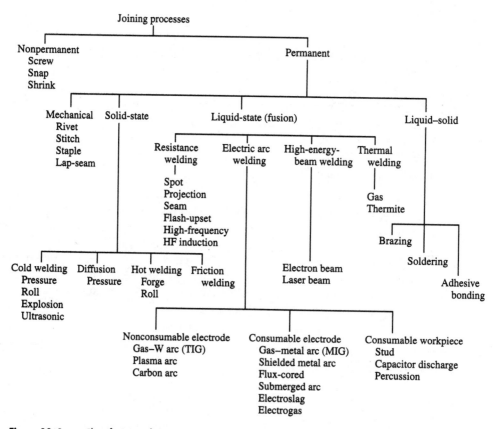

Figure 18–1   Classification of joining processes.

18-2   MECHANICAL JOINING

In addition to the semipermanent screw joint, there are several techniques for establishing a joint by *mechanical* means.

1. The most common mechanical fastener is the *rivet*. Whether it be solid or hollow (Fig. 18–2a and b), it makes a joint by clamping the two parts between heads. One head is usually formed in a prior operation; the resulting rivet is fed through predrilled or punched holes, and the second head is produced by upsetting, either cold or hot (for upsetting, see Sec. 9-2-1). On a hollow rivet, the head is formed by flaring, an operation related to flanging a tube (Fig. 10–34c). *Blind rivets* are tubular rivets that can be inserted from one side. They incorporate their own forming tool, usually a mandrel which is pulled by a special tool to expand the blind side (Fig. 18–2c). The notched stem of the mandrel then breaks off (pop rivets).

The hole represents a discontinuity in the structure and could cause fatigue failure. Therefore, the edges are deburred to remove stress raisers; for more critical applications, the hole is reamed or, to induce residual compressive stresses, the hole is slightly expanded by passing a larger pin through it. Riveting has lost its dominance in building construction and in the manufacture of automotive frames, but it is still of great importance. For example, tens of thousands of riveted joints are made on many airplanes. For greater consistency, riveting can be mechanized or entrusted to robots.

2. Thin sheets can be joined without preliminary drilling by *stitching or stapling* (Fig. 18–2d). Stapling is extensively used to fasten sheet to wooden backing.

3. *Seams* (Fig. 18–2e) are produced by a sequence of bends on tight (one-half sheet thickness) radii (Sec. 10-4). Lock seams can be made impermeable without or with fillers such as adhesives, polymeric seals, or solder. Some seams are along straight lines, as the seams of radiator tubes and the side seam of three-piece beverage cans; others are along the edges of circular parts, such as can lids

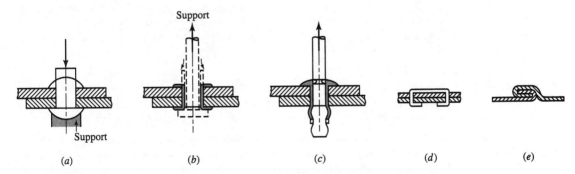

(a)	(b)	(c)	(d)	(e)

Figure 18–2   Permanent mechanical joints: (a) rivet, (b) tubular rivet, (c) blind rivet, (d) staple, and (e) seam.

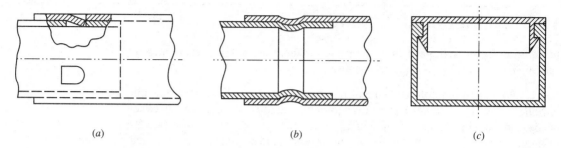

(a) (b) (c)

Figure 18-3   Permanent mechanical joints: (*a*) lanced tab, (*b*) crimped joint, and (*c*) snap joint.

(Fig. 10–35*d*), and then the process is related to flanging. A *hem* is formed when the edges of one part are turned over the other part (Fig. 10–35*a* and *b*).

4. Joints are produced also by creating mechanical interference with the aid of plastic deformation, as in twisting or bending of *lanced protrusions* (Fig. 18–3*a*) and *crimping* (Fig. 18–3*b*).

5. *Shrinking* a sleeve onto a core is applicable mostly to round parts. The compressive stress necessary to maintain a permanent joint is attained either by heating the sleeve (and/or cooling the core), by swaging (Fig. 9–24), or by pressing together two parts with a low-angle conical interference fit.

6. *Snap-fit joints* rely on the elastic springback of cantilevered elements (Fig. 18–3*c*).

7. *Clinching* forms the equivalent of a sheet metal rivet. Special tools are used to form the joint in a single press stroke.

Example 18-1 | **T**runk lids of automobiles are made of an outer sheet, formed by stretch-drawing to the desired curvature (Fig. 10–18*b*) and an inner stiffening member of deeper profile with various cutouts. The two are joined by hemming, usually with the inclusion of an adhesive bead which increases strength and contributes to sound deadening. A 90° flange is formed on the outer panel, the inner member is positioned, and the flange turned down for a tight seal. High lateral compressive strains are imposed on corners (essentially, a continuation of shrink flanging, Fig. 10–35*b*); to prevent wrinkling, the width of hem is reduced there. The same technique is used for joining door outer and inner panels.

Example 18-2 | **A** process related to clinching is used for fastening the tear tab to the lid of aluminum beverage cans. A dome is first stretched (*a*), gathered into a deeper, smaller-diameter dome (*b*), the tear tab placed over it, and the dome flattened (*c*) to form an integral, leak-proof riveted joint.

Example 18-3 | **S**hape-memory alloys allow joining of critical parts such as titanium-alloy hydraulic tubing in jet aircraft. The joining sleeves are made of a Ni–Ti alloy (of approximately 52% Ni) which is austenitic at room temperature but changes to a low-strength martensite below 130°C.

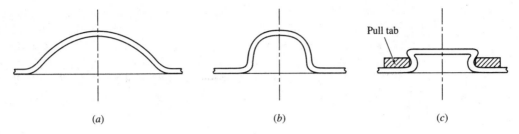

Figure Ex. 18-2

The coupling is machined in the austenitic condition with an inner diameter smaller than the outer diameter of the tubes. It is then cooled in liquid nitrogen into the martensitic range and mechanically expanded to a diameter larger than the tube OD. It is stored in this state in liquid nitrogen until installation. Once placed in position over the tube joint, it heats to room temperature, returns to its original machined diameter (shape memory) and establishes a leakproof shrink joint. [Source: *Adv. Mater. Proc.*, 1995(4):23–26.)

18-3   SOLID-STATE WELDING

In discussing adhesion we mentioned that interatomic bonds may be established by bringing the atoms of two surfaces into close proximity (Sec. 4-9-2). It is absolutely essential that the surfaces to be joined (also called *faying surfaces*) should be free of contaminants (oxides, adsorbed gas films, or lubricant residues) that might prevent the formation of interatomic bonds. This condition is difficult to satisfy in the atmosphere of the earth (see Fig. 4–21), and then measures must be taken to neutralize the effects of surface films:

1. *Relative movement* between surfaces helps to break up surface films. Roughening of the surface by wire brushing is helpful because, on joining, the ridges deform.

2. Plastic deformation of the contacting bodies causes a growth, *extension of the interfacial surface*, and, if surface films are unable to follow the extension, new, fresh surfaces are exposed which then form solid-state welds.

3. While theoretically no pressure would be required for bonding perfectly mating and clean surfaces, in practice a certain *normal pressure* is necessary to ensure conformance of the contacting surfaces and to break up surface films.

4. *Heat* is not an essential part of the basic bonding process, but softening of the materials promotes intimate contact and diffusion of atoms helps to achieve bonding. Diffusion is objectionable, however, when two dissimilar metals form intermetallic compounds that embrittle the joint.

In principle, any two materials can be bonded and, indeed, solid-state bonding is often applied when other techniques fail. Nevertheless, best bonds are obtained between metals when there is atomic registry (i.e., atoms of the two components are similarly spaced and crystallize in the same lattice structure). This means that metals bond best to themselves and to other metals with which they form solid solutions.

18-3-1   Cold Welding (CW)

The term *cold welding* (CW) is used loosely to describe processing at room temperature.

1. *Lap welding* relies on a 50–90% expansion of surfaces when indenters penetrate the sheets to be joined (Fig. 18–4*a*). Shoulders on the indenters limit distortion and promote welding.

2. *Butt welding* of wires establishes the joint by upsetting the wire ends to cause surface expansion (Fig. 18–4*b*). Welding is further aided when a twist is given.

3. *Roll bonding* or *roll welding* (ROW) (Fig. 18–4*c*) is highly effective because large reductions (50–80% in a single pass) result in great extension and break-up of surface films. Bonding can be locally prevented by depositing a parting agent (*stop-off*) such as graphite or a ceramic in a predetermined pattern. Inflation by pressurized air or fluid then yields parts such as refrigerator evaporator plates (Fig. 18–4*d*).

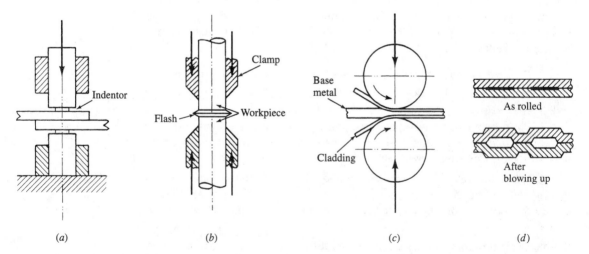

(a)   (b)   (c)   (d)

Figure 18–4   Parts may be joined in the solid state by: (*a*) cold lap welding, (*b*) butt welding, and (*c*) roll bonding. (*d*) Passages may be created by inflating roll-bonded assemblies.

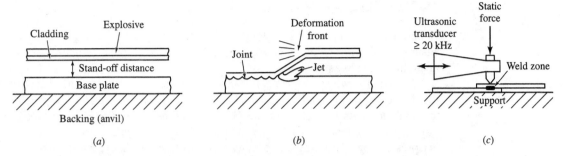

Figure 18–5   Solid-state joints are established in (a) explosion welding by (b) driving the cladding sheet toward the base material and in (c) ultrasonic welding by a rubbing motion at the interface.

4. In *explosion welding* (EXW) the joint is established by intense deformation of the interface. A mat of an explosive is placed on top of the cladding sheet or plate (the *flyplate*), held at some critical distance from the surface of the base material (Fig. 18–5a). When the explosive is detonated from one end, the pressure developed accelerates the flyplate at an angle to the surface. Under the intense pressure, a fluid jet forms which ejects surface contaminants; turbulent flow results in the formation of tight whirls or vortexes at the interface (Fig. 18–5b). In most instances there is no melting, but the combination of adhesion and mechanical interlocking assures a strong bond. The technique is used for cladding large plates for the chemical industries and also for the in-situ expansion of tubes into the header plates of boilers and tubular condensers.

5. Relative movement of the interface is induced in *ultrasonic welding* (USW) by tangential vibration (Fig. 18–5c). There is no heavy deformation, and the process is suitable for lap welding foils and delicate instrument and electronic components. When the welding tip is replaced by a roller, seam welds can be produced.

Example 18-4

In replacing silver-alloy quarters and dimes, nickel was unacceptable in the United States, where vending machines test for magnetic properties. Therefore, a copper core is clad with cupronickel (75Cu-25Ni), giving the desired whiteness and lack of ferromagnetism. A sandwich of 7.5-mm thickness is rolled to 1.36 mm; surface extension, together with the solubility of nickel in copper (Fig. 6–4), ensures a permanent bond. Calculate the proportion of new, atomically clean surface, assuming that preexisting surface films do not expand at all.

From constancy of volume (Eq. 4-2)

$$V = h_0 A_0 = h_1 A_1$$

$h_0/h_1 = A_1/A_0 = 7.5/1.36 = 5.515$, thus the new surface occupies $(5.515 - 1)/5.515 = 81.9\%$ of the interface.

18-3-2   Diffusion Welding

Generally, better bonding is obtained when the temperature is high enough to ensure diffusion—typically, above $0.5T_m$ (Sec. 8-1-6; typical temperatures are given in Tables 8–2 and 8–3).

Diffusion welding (DFW) is not new; for centuries the goldsmith has made filled gold by placing gold face sheets over a silver or copper core and compressing the sandwich with a weight (Fig. 18–6a). On holding in a furnace for a prolonged time, a permanent bond is obtained. The required pressure may be generated also in a press, or by restraining the assembly with a fixture made of a lower-expansion material (frequently, molybdenum). The term *hot pressure welding* (HPW) is also used.

In the 1970s, the technique was extended to airframe construction. Simultaneous deformation greatly helps in the development of a sound joint, therefore, *diffusion welding combined with superplastic forming* (Fig. 18–6b) has proven most successful. Using two, three, or four sheets, patterns are printed with boron nitride stop-off. The assembly is placed into an evacuated box (retort) held in a press and integral structures are formed while the cavities are blown up.

Because titanium dissolves its oxides (Sec. 11-4-1) and alloys such as Ti-6Al-4V have the small grain size required for superplastic forming (Sec. 8-1-6, Fig. 8–12) even without special processing, alloys of titanium are naturals for the technique. Large-scale part consolidation becomes possible. For example, in high-performance aircraft such as supersonic fighter planes, the fuselage is constructed of a thin skin strengthened by ribs of considerable complexity. Conventionally, sheet-metal parts, extrusions, and pocket-milled plates are joined by thousands of rivets. Diffusion bonding with superplastic forming reduces part number, labor, lead time, and cost. In applying the techniques to aluminum alloys, deformation is vital to break up the stable, brittle oxide.

Diffusion welding is applicable also to the manufacture of metal-matrix composites (e.g., titanium or aluminum reinforced with boron fiber). When bonding

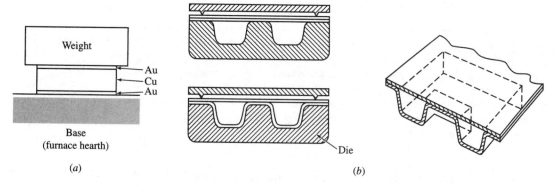

Figure 18–6   Elevated-temperature bonding is possible without substantial deformation in (a) diffusion welding which may be (b) combined with superplastic forming to make complex structural parts.

would not take place because of lack of solubility or would result in the formation of a brittle intermetallic compound, interlayers that are mutually soluble can be used in the form of foils or thin sheets. For example, superalloys are bonded after electroplating with a Ni–Co alloy.

18-3-3   Hot Welding

In a general sense, the term *hot welding* can be used to describe welding by deformation in the hot-working temperature range.

1. *Forge welding* (FOW) refers to the oldest industrial welding process. The bond is created by substantial local deformation of the joint. The hot, preshaped workpieces, usually of iron or steel, are forged together to squeeze out oxides, slag, and contaminants, and ensure interatomic bonding (Fig. 18–7a). The technique was used not only for joining (e.g., forging of chain links) but also for the welding of tubes, and for building up—layer by layer—medieval swords and even very large objects, such as anchors. Forge welding in which the ends of workpieces are pressed together axially (*butt welding*) is possible, but the joint quality tends to be poor.

2. In more recent variants of the process, heat is provided by *induction heating* to minimize oxidation (Fig. 18–7b). Much less deformation is sufficient, therefore butt welding of workpieces becomes practicable.

3. The heat may also be generated by passing a current through the compressed faces (Fig. 18–7c). *Electric butt welding* has now largely been replaced by flash butt welding (Sec. 18-5).

4. *Hot roll-bonding* (*roll welding*, ROW, the high-temperature version of Fig. 18–4c) has been used extensively to create composites of low cost or high performance. Thus Alclad combines the corrosion resistance of a pure aluminum cladding with the high strength of a precipitation-hardened aluminum core. Stain-

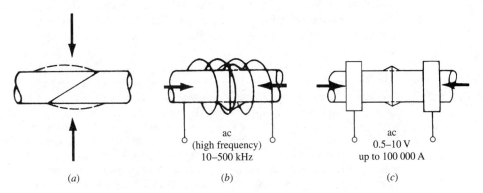

Figure 18-7   Elevated-temperature solid-state joining is obtained in (a) forge welding, (b) induction welding, and (c) electric butt welding.

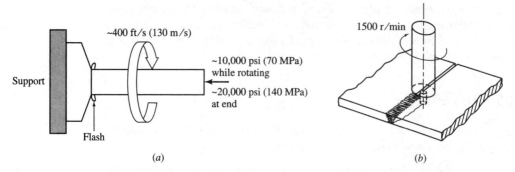

Figure 18–8   (a) Heat is generated in friction welding by rotating the workpieces against each other, and the bond is established by upsetting. (b) A rotating tool generates heat in friction stir welding.

less steel is clad to mild steel for corrosion protection, and alloys of different heat expansion are bonded to make *bimetallic strips* for thermostats. The sandwich consisting of base plate and cover sheets is enclosed and evacuated if oxidation must be avoided.

18-3-4   Friction Welding (FRW)

The frictional work generated when two bodies slide on each other is transformed into heat; when the rate of sliding is high and the heat is contained in a narrow zone, welding occurs.

In *continuous-drive* FRW (Fig. 18–8a) one part is firmly held while the other (usually of axial symmetry) is rotated under the simultaneous application of axial pressure. Temperature rises, partially formed welded spots are sheared, surface films are disrupted; rotation is suddenly arrested and a further upset force is applied when the entire surface is welded. Some of the softened metal is squeezed out into a flash, but it is not fully clear whether melting actually takes place. The heated zone is very thin; therefore, dissimilar metals are easily joined. In *inertia-drive* FRW, rotation is imparted by a flywheel, the energy of which is calculated so that the weld is completed when rotation stops.

Example 18-5   Continuous-drive FRW is performed in captive shops of automotive and appliance manufacturers and also in specialized job shops. For example, the aluminum cables of underground electrical distribution lines must be joined to copper switchgear. To prevent corrosion, impact-extruded aluminum barrels are crimped onto the cable and cold-headed copper terminal blocks are then friction-welded to these barrels. The process can also be used to increase shape complexity: piston rod ends are attached to piston rods of hydraulic cylinders; two forged gear blanks are united without losing properties; heavy-walled tubing is attached to transport trailer axle beams, and so on. (Source: J.R. Huber, ARD Industries Ltd., Cambridge, Ontario.)

Friction stir welding is suitable for joining sheets and plates. A rotating tool penetrates the joint area and, in its travel along the seam, generates enough heat to make the metal flow plastically to fill the joint (Fig. 18–8*b*).

18-4   FUSION WELDING

In the great majority of applications, the interatomic bond is established by melting. When the workpiece materials (*base* or *parent materials*) and the *filler* (if used at all) have similar but not necessarily identical compositions and melting points, the process is referred to as *fusion welding* or simply *welding*.

Welding is closely related to casting processes. Heat is provided to melt the base metal and filler. The melt is physically contained in the melt zone where, through its contact with the surrounding base metal of high thermal conductivity, it is rapidly chilled (typically, at tens or hundreds of degrees per second). Thus, cooling rates are between those prevailing in permanent-mold casting (Sec. 7-5-6) and atomization (Sec. 11-2-1), and proper control of processes calls for a thorough familiarity with nonequilibrium metallurgy (Sec. 6-1-6). Here we will have to compromise and be satisfied with a cursory examination of principles, based to a great extent on concepts discussed in Chap. 6, which could be usefully reviewed at this time. Some acquaintance with material effects is nevertheless indispensable if capabilities and limitations of various processes are to be appreciated.

18-4-1   The Fusion Joint

A fusion joint is far from homogeneous. The degree of inhomogeneity and complexity increases from pure metals to multiphase alloys, and is also a function of heat input per unit distance. Greater heat intensity gives deeper penetration, more concentrated heat zone, less change in metallurgical structure, and lower stresses in the weld. The following discussion is, therefore, generalized and does not necessarily apply for all rates of heat input.

Single-Phase Materials   A section through the joint in a pure metal (Fig. 18–9), such as aluminum or copper, welded with a rod of identical composition, shows that the applied heat has melted not just the welding rod but also some of the base material to form the *weld bead* (or, simply, the *weld*). The base material adjacent to the melt boundary was exposed to high temperatures, and the properties and structure are changed within this *heat-affected zone* (HAZ).

If the workpiece material was originally cold-worked and therefore of highly elongated grains, the HAZ will show recrystallization. For a given cold work, grain size increases with longer time at high temperature (Fig. 8–10); therefore, the very coarse grains found at the melt boundary gradually change to finer ones until, at the edge of the HAZ, only partial recrystallization is evident. If the workpiece was of annealed material, the further heat input during welding just

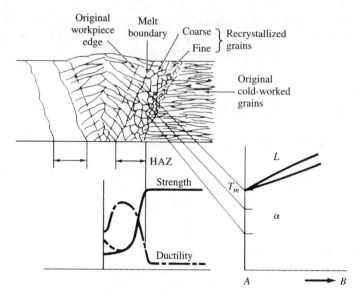

Figure 18–9   Fusion melting of a cold-worked pure metal or solid-solution alloy results in decreased strength in the heat-affected zone (HAZ).

coarsens the grains. In either case, a coarse-grained structure of lower strength (Sec. 6-3-6) exists at the melt boundary. Solidification begins at this boundary by *epitaxial growth*, i.e., by the deposition of atoms in the same crystal orientation as the surface crystals (from the Greek *epi* = upon, *teinen* = arranged, and axis). This leads to the development of coarse, columnar grains in the weld material itself.

Viewed from the top, there is a weld pool at the point of maximum heat input (Fig. 18–10). When making a line weld, the heat source is moved along at a set travel speed. In a pure metal or dilute alloy, at low travel speed, the solidifying grains turn to follow the heat source, and the center of the weld bead often has independently nucleated grains, especially in the presence of nucleating agents (Fig. 18–10a). At high travel speeds the weld pool becomes elongated

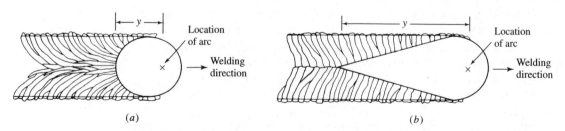

Figure 18–10   In making line welds, heat input per unit length determines the shape of the weld pool and the structure of the solidified weld bead: (*a*) at low travel speeds grains are interwoven, but (*b*) at high speeds a weak center plane often forms if low-melting constituents are present.

and solidification proceeds almost perpendicular to the melt boundary toward the center (Fig. 18–10b). If there are low-melting constituents, they segregate at the boundaries of cast grains and hot cracking may result.

Complicating factors enter in more highly alloyed materials. Dendrites form in the weld metal, just as they do in a casting (Sec. 6-1-2), but, as a result of faster cooling, the secondary dendrite arm spacing is smaller. Furthermore, nucleation of new grains often leads to a columnar to equiaxed transition in the center of the bead. Because melting occurs over the T_L–T_S temperature range (Fig. 6–4), there will be a partially melted, mushy zone at the melt boundary. Minor concentrations of low-melting alloying elements or contaminants can migrate to this boundary to cause hot-shortness and cracking during cooling.

Two-Phase Materials   Most technically important alloys have a two-phase or multiphase structure, and their suitability for welding is determined by the events taking place in both the liquid and solid states.

1. Materials of eutectic composition present little problem because of their favorable solidification mode (Sec. 6-1-3).

2. Precipitation-hardening alloys (Sec. 6-4-2) can be readily welded in the annealed state; if their freezing range is short, the effects of welding are mostly washed out on subsequent solution treatment and aging of the whole weldment. If the weldment is too large for this, fusion welding is possible in either the solution-treated or aged condition, but then the strength advantages of heat treatment are lost in the weld zone (Fig. 18–11). The weld itself may contain the intermetallic constituent in a coarse form, resulting in low strength and ductility. The immediately adjacent HAZ has been rapidly heated and quenched and would thus be in the solution-treated condition; however, heat conducted from the weld zone (back-heating) during cooling may cause overaging. Therefore, the HAZ is fully heat-treated but of a coarse grain, with high strength and moderate ductility. Farther away, the original structure becomes overaged and soft. Apart from the poor strength of the joint, the locally varying composition can also lead to corrosion problems, e.g., in aluminum and magnesium alloys. Sometimes a compromise solution is to use a different filler rod to avoid hot cracking or corrosion problems.

3. Solid-state phase transformations lead to complex changes. To take a medium-carbon steel in the annealed state as an example (Fig. 18–12), the parent metal has a microstructure consisting of pearlite colonies alternating with ferrite grains. The weld itself has the usual coarse, cast structure. Next to it the material has been heated high into the austenitic temperature range and then cooled, resulting in coarse grains. With decreasing temperatures, the austenite grains also become finer; thus, finer-grain ferrite and pearlite are usually found in the transformed structure, although high cooling rates may produce bainite or martensite. The edge of the HAZ has been heated just above the eutectoid temperature. Fast cooling rates will convert any austenite into martensite, and a large drop in ductility may result. With higher (above approximately 0.5%) carbon content, martensite will inevitably form, as it will in alloy steels.

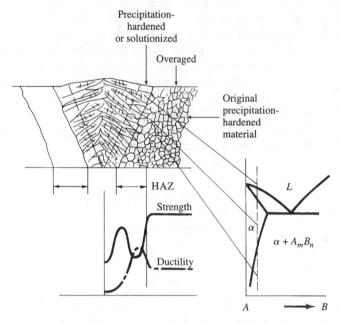

Figure 18–11   In welding a precipitation-hardened material, the weld bead is both weaker and less ductile.

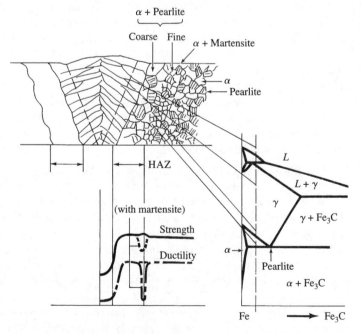

Figure 18–12   Heating into the austenitic temperature range can lead to martensite formation and resultant loss in ductility in welding carbon steels.

Dissimilar Materials   The situation becomes further complicated when two dissimilar materials are used, either in the workpiece or the filler. This is frequently the case because nonmatching fillers make it easier to obtain crack-free welds. The heating and cooling history is compounded by the effects of alloying, and events in the weld zone are determined by the equilibrium diagrams relating to both materials. Nonequilibrium solidification complicates the matter further.

As might be expected, alloys that form solid solutions present few problems. Eutectics tend to be more brittle, although they are favorable when both phases of the eutectic are ductile. Intermetallic compounds invariably embrittle the structure to make it useless; thus, for example, copper cannot be joined to iron. It is possible, however, to use a mutually compatible interface, in this case nickel, which forms solid solutions with both copper and iron.

In many instances the difference between melting points is very large, so that the parent metal does not melt, and then it is customary to classify the process as brazing or soldering (Sec. 18-8).

18-4-2   Weldability and Weld Quality

It is obvious from the foregoing discussion that the term *weldability* denotes an extremely complex collection of technological properties and is also a function of the process.

Welding Defects   If any of the requirements for producing a sound joint are not met, various defects may occur. They can exist in welds of any geometry and origin, but for illustrative purposes a fusion weld with filler metal may be used (Fig. 18–13*a*).

1. Fusion welding is a melting process and must be controlled accordingly (Sec. 7-4). Melting temperature determines—together with specific heat and latent heat of fusion—the required heat input. High thermal conductivity of the base metals allows the heat to dissipate and therefore requires a higher rate of heat input and leads to more rapid cooling. For any given parent material, the rate

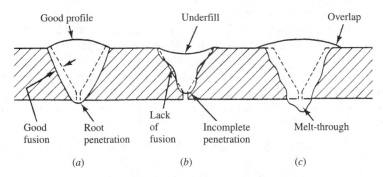

Figure 18–13   Fusion welds made with filler metal may show several
characteristics: (*a*) a well-formed bead; (*b*) lack of fusion, lack
of penetration, underfilling; or (*c*) melt-through, overlap.

of heat input (usually expressed as heat input per unit length) must be matched to the thickness of the workpieces, the rate of weld metal deposition, and the travel speed (the speed of moving along the weld bead). Insufficient heat input causes *lack of fusion* and, in thicker cross sections, *incomplete penetration* (Fig. 18–13*b*); the remaining gap then becomes an incipient crack. Insufficient rate of weld metal deposition causes *underfilling*. Excessive heat input can lead to *melt-through*, and high deposition rates result in *overlap* (Fig. 18–13*c*), the edges of which give a notch effect with reduced fatigue strength.

2. Surface contaminants, including oxides, oils, dirt, paint, metal platings, and coatings incompatible with the workpiece material, result in *lack of bonding* or lead to *gas porosity*, and must be kept out by mechanical and/or chemical preparation of the surfaces.

3. Undesirable reactions with surface contaminants and with the atmosphere are prevented by sealing off the melt zone with a vacuum, a protective (inert) atmosphere, or a slag. As in melting for casting, the slag is formed by dissolving oxides in a flux. Care must be taken to minimize *slag entrapment* which would reduce the ductility of the weld.

4. Gases released or formed during welding (e.g., CO) can lead to porosity which weakens the joint and acts as a stress raiser. High fluidity of the melt is helpful in allowing slag and gases to rise to the surface. Particularly dangerous is hydrogen which originates from atmospheric humidity or damp flux. When combined into the molecular form, it causes porosity in aluminum alloys. In the atomic form it diffuses to crack tips and causes *hydrogen embrittlement* of steels in which martensite forms upon cooling.

5. *Solidification cracks* appear under the influence of stresses in the weld when a low-melting liquid is rejected during dendritic solidification. *Liquation cracks* along grain boundaries are due to solid-state segregation of low-melting elements.

6. Solidification shrinkage coupled with solid shrinkage imposes internal tensile stresses on the structure (Fig. 18–14) and may lead to *distortion* and gross *cracking*. The problem is magnified when the structure is not free to shrink, in other words, when mechanical restraints are imposed. The properties of the base and filler materials are important. High thermal expansion results in greater distortion and residual stresses. The danger of cracking is greater in alloys of composition that gives the maximum freezing range in a system. The problem can be alleviated with a less-alloyed, more ductile filler that reduces hot-shortness (or a more-alloyed filler that avoids the maximum freezing range). Design is often based on the fracture-mechanics approach (Sec. 4-2) and the allowable flaw size is determined experimentally or theoretically. Residual stresses can also lead to stress-corrosion cracking.

7. Metallurgical transformations, discussed earlier, are of great importance especially when they lead to the formation of brittle phases such as martensite. It is then essential to preheat the base metal. The weld material is subject to further heat treatment if deposited in more than one pass (*multipass welding*).

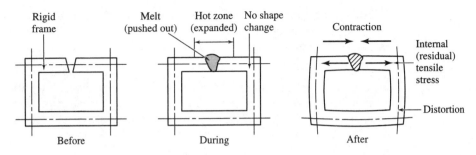

Figure 18-14   Expansion during welding followed by contraction on cooling can lead to distortion and residual stresses.

8. The absolute and relative thickness of parts to be joined and the design of the joint have a powerful influence on heating and cooling, and thus on weldability.

Metal Treatment   Some of the above-mentioned difficulties can be alleviated by various measures:

1. *Preheating* the weld zone or the entire structure reduces the energy input required for completion of the weld (important for materials of high diffusivity such as aluminum or copper); reduces cooling rates in the weld and HAZ (allowing the welding of hardenable steels and other materials in which fast cooling results in the formation brittle phases); aids in diffusing H away; and reduces differential shrinkage, distortion, and residual stresses.

2. *Peening* (hammering or rolling) of the weld bead improves the strength of welds. In multipass welding, peening between passes removes slag that might become entrapped and induces recrystallization of earlier layers (however, the last pass is not peened because ductility would be lost).

3. *Postwelding heat treatment* of the entire welded structure is often essential for a variety of reasons:

a. Stress-relief anneal (Sec. 6-4-1) reduces residual stresses to acceptable levels, making the structure dimensionally stable and not susceptible to stress-corrosion cracking. It tempers martensite and eliminates the danger of cold cracking in steels. Thus it also tends to increase ductility and fatigue strength with only minor loss in strength. It may, however, lead to overaging in precipitation-hardening materials.

b. Normalizing a steel (Sec. 6-4-3) wipes out most undesirable effects of welding. Recrystallization of the weld and parent material in the austenitic temperature range is followed by controlled cooling to give a structure consisting of ferrite and pearlite.

c. Full heat treatment (quenching and tempering of steels, solution treatment and aging of precipitation-hardenable alloys) gives highest properties in the entire structure, but may result in distortion.

d. An aging treatment of a precipitation-hardening material is sufficient if the weld bead ends up in the solutionized condition because of very fast cooling rates (as in electron beam welding).

Quality control is a vital part of all welding processes. Destructive techniques are useful for establishing process parameters, but production quality can often be controlled only by nondestructive inspection. Visual inspection is always mandatory and is supplemented by the whole arsenal of NDT techniques (Sec. 4-8). The most critical welds are often 100% inspected by radiography.

Process Factors   The heat required for melting may be provided by a number of heat sources. The shape of the weld zone and the width of the HAZ depend on the intensity of this heat source. For a given power, heat intensity increases with a decrease in the size of the heat source and ranges over several orders of magnitude. At low heat intensity the time required to reach melting temperature is long, heat is lost to the environment, and much of the heat reaching the workpiece is dissipated into the surrounding metal. Thus, in oxyfuel welding (with a heat intensity of $1-10$ W/mm^2) only some 10% of the heat reaches the metal and only a fifth of this does the actual melting. In consequence, efficiency is low, the weld zone is several times wider than it is deep, and the HAZ is broad. At heat intensities of $10-1000$ W/mm^2 typical of arc welding processes, some half of the total heat reaches the metal and about 40% of this is used to fuse metal; the weld pool and HAZ have the shape shown in Figs. 18–9 to 18–12. The high energy intensity (10^3-10^5 W/mm^2) of laser beam and EB welding gives high efficiency and the deep, cylindrical *keyhole* weld pool is several times deeper than wide. Because heating rates are so high, metallurgical changes occur primarily during cooling and the HAZ is about the width of the beam.

In many processes, heat is generated by electric power. Success depends on proper voltage, current, polarity, and the variation of these with time. Selection of the most suitable *power source* is a precondition of success. The power source may be as simple as a transformer, but highly sophisticated power sources—often based on solid-state electronics—are increasingly used. Closed-loop in-process control of parameters then becomes possible.

18-4-3   Weldable Materials

Generalizations are more dangerous for welding than for other processes, but some guidelines can be formulated. In actual production situations specialized reference volumes, computer databases, and industry standards should be consulted.

Ferrous Materials

Ferritic steels.   These are readily welded, but martensite formation is a danger in pearlitic steels. In general, increasing hardenability means that martensite is formed at lower critical cooling rates; thus, it indicates an increasing danger of martensite formation and a decreasing weldability. Martensite is not only hard and brittle, but its formation proceeds with a volume increase which imposes further stresses on the structure, reducing the strength of the weld. *Preheating* and, if possible, *postheating* are necessary when martensite or bainite formation is unavoidable.

Alternatively, the structure may be heated into the austenitic range, cooled to a temperature above M_S (Fig. 6–20), and welded before transformation begins. The completed structure is then cooled. Such *step welding* makes even tool steels amenable to welding. It should be noted that fully heat-treated alloy steels may be welded but the weld must be rapidly quenched to obtain a tempered martensite.

We already mentioned the danger of hydrogen embrittlement. Sulfur creates porosity and brittleness and, while the welding of resulfurized free-cutting steels is possible, they are often brazed in preference.

Coated sheets.   The widespread use of coated (primarily galvanized, but also prelubricated) sheet has posed new challenges. In resistance welding, zinc evaporates, creates a plasma, and leads to sputtering and porosity, and the life of copper electrodes is reduced by alloying with zinc. Coating thickness must not be excessive and heat densities must be adjusted. Some processes are inapplicable.

Stainless steels.   These steels always contain chromium, which forms an extremely dense Cr_2O_3 film. Welding conditions must be chosen to prevent its formation. Apart from this, austenitic steels (containing both Cr and Ni) are weldable but chromium carbides formed at high temperatures reduce the dissolved chromium content below the level required for corrosion protection, and subsequent corrosion (*weld decay failure*) is a danger. To avoid this, the carbon content should be very low, or the steel must be stabilized (Ti, Mo, or Nb added to form stable carbides), or the structure must be heated above 1000°C after welding and then quenched to retain the redissolved carbon and chromium in solution.

Stainless steels containing only chromium are either ferritic or martensitic in structure. Ferritic steels (over 16% Cr) can be welded but the coarse grain will weaken the joint. Martensitic steels form a martensite with a hardness depending on carbon content; careful preheating is followed by postheating to above 700°C to change the martensite into ductile ferrite with embedded chromium carbide precipitates.

Cast iron.   The weldability of cast irons (Sec. 7-3-1) varies greatly, but many cast irons are welded, especially by arc welding. A high-nickel filler is frequently used to stabilize the graphitic form. Preheating and slow cooling are also helpful. In welding gray iron the welding rod is enriched in silicon, and—to ensure spheroidal graphite formation—Mg is incorporated in the rod for welding nodular cast iron. Malleable iron reverts to brittle white iron, reducing the toughness of the weld. When toughness is important, the weldment is heat treated or the joint is made by brazing.

Nonferrous Materials

Low-melting materials.   Tin and lead are easily welded, provided that the heat input is kept low enough to prevent overheating. Zinc, on the other hand, is one of the most difficult materials to weld, because it oxidizes readily and also vaporizes at a low temperature (at 906°C). It can be resistance-welded and stud-welded, although soft soldering is more usual.

Aluminum and magnesium.   These materials share a number of characteristics. Most alloys are readily weldable, particularly with an inert gas envelope. Otherwise the oxide film must be removed with a powerful flux, which in turn may have to be washed off after welding to prevent corrosion. Moisture (H_2O) must be kept out, because it reacts to give an oxide and hydrogen (which embrittles the joint by causing porosity).

The high thermal conductivity and specific heat yet low melting point of these alloys require high rates of heat input and adequate precautions against overheating. The high thermal expansion coefficient necessitates preheating of hot-short materials. Because of the difficulties encountered with precipitation-hardening materials, such alloys are often heat treated after welding or, if this is not possible, a different filler is used (very often, Al–Si for aluminum alloys).

Copper-based alloys.   Deoxidized copper is readily welded, especially if the filler contains phosphorus to provide instant deoxidation. Tough-pitch copper cannot be welded because its oxygen content (typically 0.15%) reacts with hydrogen and CO to form water and CO_2, respectively, both of which embrittle the joint by generating porosity. Brasses can be welded but zinc losses are inevitable; therefore, either the filler is enriched in zinc, or Al or Si is added to form an oxide that reduces evaporation. Tin bronze has a very wide solidification range and is thus exceedingly hot-short. Phosphorus in the welding rod prevents oxidation, while postheating is necessary to dissolve the brittle nonequilibrium intermetallic phase. Aluminum bronzes present no problem but the oxide formed must be fluxed, just as with pure aluminum.

Nickel.   This metal and its solid-solution alloys are readily welded. The precipitation-hardening superalloys contain Cr, Al, and Ti, and the oxide must be fluxed or its formation prevented. All nickel alloys are very sensitive to even the smallest amount of sulfur, which forms a low-melting eutectic and results in hot cracking. Some precipitation-hardening alloys have a low-ductility temperature range and may crack too.

Titanium and zirconium.   Alloys are also weldable but an inert atmosphere is essential to prevent oxidation; therefore, they are often enclosed in inert-atmosphere welding chambers or are welded by electron beam.

Refractory metal alloys.   W, Mo, and Nb can be welded but the volatility of the oxides makes special techniques (e.g., electron beam welding) mandatory.

18-5   RESISTANCE WELDING

Electric resistance welding represents, in some ways, a transition from solid-state welding to fusion welding. The two parts to be joined are pressed together and an alternating current is passed through the contact zone. If the correct pressure is applied, this zone will present the highest resistance in the electric circuit and power losses will be concentrated there. The energy is converted into heat. The current is left on until melting occurs at the interface between the two parts and pressure is kept on until the weld solidifies. According to Joule's law, the heat generated (in joules) is

$$J = I^2 Rt \qquad \textbf{(18-1)}$$

where I is the current (A), R is the resistance (Ω) and t is the duration of current application in seconds. The voltage can be low, typically 0.5–10 V, but currents are very high (for examples, see Fig. 18–15).

Since heat must be concentrated in the weld zone, resistance away from this zone should be low, especially at the points where the current is supplied to the workpieces by the electrodes. Materials of high heat conductivity and specific heat (such as aluminum or copper) call for very high currents to prevent dissipation of heat.

Surface cleanliness is important but not quite as vital as in solid-state joining because some of the contaminants are expelled from the melt. Nevertheless, scale, thick oil films, and paint must be removed, but relatively simple surface preparation is adequate and zinc-coated steel can be welded too. Quality control is most important. Welds are destructively tested to establish optimum process

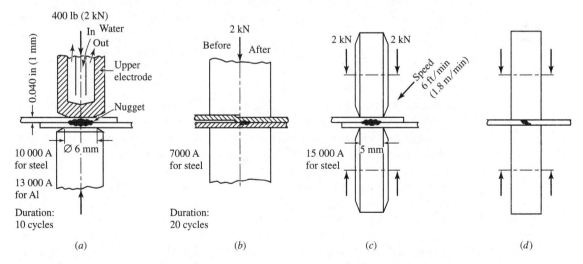

Figure 18–15   Only partial melting is aimed for in (a) resistance spot welding, (b) projection welding, (c) resistance seam welding, and (d) mash seam welding. (For illustrative purposes, process conditions are given for 1-mm-thick low-carbon steel sheet.)

parameters; thereafter, in-process inspection includes surface temperature measurement (from which weld zone temperatures can be extrapolated), ultrasonics, and acoustic emission techniques. Closed-loop and adaptive control are possible.

Resistance Spot Welding (RSW)   Because of the widespread application of sheet metal parts, *resistance spot welding* has acquired a prominent position, from attaching handles of cookware to assembling whole automobile bodies (there are several thousand spot welds per car). Two, usually water-cooled, electrodes press the two sheets together (Fig. 18–15a). Electrodes are made of materials of high conductivity and hot strength, such as copper with some Cd, Cr, or Be additions, or copper–tungsten or molybdenum alloys. The current is then applied for a predetermined number of cycles (in the automotive industry, 10–30 cycles for a current of 60 Hz), whereupon the interface heats up and within a fraction of a second a molten pool (*weld nugget*) is formed. The pressure is released only after the current has been turned off and the nugget has solidified. The sheet surface shows a slight depression and discoloration.

The electrodes may be incorporated into a fixed machine or a portable welding gun. Multiple electrodes (numbering sometimes into the hundreds) are used for the welding of large assemblies, with groups of electrodes brought into contact in a programmed sequence. A series of welds can be made accurately and in rapid succession by welding robots. Weld quality is ensured by in-process measurements based, for example, on the resistance change while the nugget is formed, or on the acoustic emission occurring during metal expulsion.

Example 18-6	Two low-carbon steel sheets of 0.5-mm thickness are joined by spot welding. Electrodes of 5-mm face diameter are pressed under a force of 2 kN. The weld draws a current of 8500 A for 7 cycles from a 60-Hz power transformer. Joint resistance is 100 $\mu\Omega$. How much heat is generated? From Eq. (18-1), heat = $(8500^2)(0.0001)(7/60) = 843$ J.

Projection Welding (RPW)   The extent of the weld zone is better controlled, and several welds can be made simultaneously with a single electrode, when small dimples or *projections* are embossed (Sec. 10-5-1) or coined (Sec. 9-3-2) on one of the sheets. When the current is applied, the projections soften and are pushed back in place by the electrode pressure as the weld nuggets form (Fig. 18–15b). Projections forged or machined onto solid bodies allow welding to a sheet or other solid body. A form of projection welding is practiced when grids formed by crossed wires are resistance welded.

Resistance Seam Welding (RSEW)   A series of spot welds may be made along a line much more rapidly if the electrodes are in the form of rollers (Fig. 18–15c). The current is switched on and off in a planned succession, giving uniform spacing of spot welds. When an alternating current is left on, a spot weld is made every time the current reaches its peak value, and the welds are spaced close enough to give a gas- and liquid-tight joint. Such *resistance seam welding* is one of the methods of producing the body of a can and is used for the manufacture of beams and box sections. In *mash-seam welding* (Fig. 18–15*d*) the overlap is only 1–2 times sheet thickness and the welding rollers apply sufficient pressure to reduce the seam to only about 10% over parent sheet thickness. This is one of the processes used for tailored blanks (Sec. 10-7).

High-Frequency Resistance Welding (HFRW)   An important application of resistance welding is to the manufacture of pipe, tubing, structural members, and wheel rims. Tubes are formed by roll forming (Sec. 10-4-4), and the longitudinal seam is then made by *high-frequency resistance welding* (HFRW) in which the electric current is applied through sliding or rolling electrodes (Fig. 18–16*a*). In *high-frequency induction welding* (HFIW) the tube is surrounded by an induction coil and the operating frequency is chosen to give optimum penetration; higher frequencies penetrate to a lesser depth. In both HFRW and HFIW there is a localized molten zone formed and then immediately squeezed out by compression. The excess metal is usually trimmed off.

Flash Upset Welding   Upset welding in the general sense means joining two bodies by pressing them together with sufficient force to cause deformation. In the narrower sense, the term is applied to resistance welding in which, just as in spot welding, the pressure is applied prior to switching on the current (Sec. 18-3-3). Much more widespread today is *flash butt welding* or *flash welding* (FW), in which the current is applied during the approach of the two parts; thus, extremely

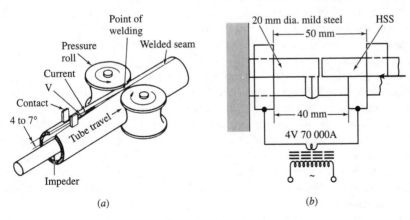

(a)   (b)

Figure 18–16   (*a*) High-frequency resistance (longitudinal butt seam) welding. (*b*) Flash-butt welding, illustrating critical process parameters. [*Part (a) from Metals Handbook, 9th ed., vol. 6, Welding. Brazing and Soldering, p. 760, ASM International, 1983. With permission.*]

rapid heating takes place when surface irregularities first make contact. Molten metal is violently expelled and burns in air, and some arcing occurs; hence the name *flash*. A substantial length may be burnt off to ensure a good weld, but all liquid metal is expelled and the weld is formed by upsetting the hot, solid metal surfaces (Fig. 18–16*b*) (*upset welding*, UW). Thus, the strength of the joint is not impaired by the presence of residual as-cast weld metal. Therefore, the process can be regarded as a transition between solid- and liquid-phase processes. A good weld will be produced only if the rate of approach, current and voltage, total travel, and upsetting pressure are closely controlled. Manual control is now often replaced with automatic, adaptive control.

Displacement of metal creates a flash which must be removed. The end faces to be joined are often chamfered so that melting moves from the center outward to squeeze out contaminants into the flash. Preheating is possible by switching on the current after the faces have been pressed together, but the faces are then again slightly separated to induce flashing. For uniform heating the two parts should have equal cross-sectional areas, but their composition can be different, for example, in joining a low-carbon steel shank to a HSS tool bit. Bars and sections bent into the shape of rings (Fig. 10–15*b*), tubes, and sheet structures are often welded end to end, and the process is extensively used for joining the ends of wire and sheet coils to allow the continuous operation of processing lines.

18-6   ELECTRIC ARC WELDING

Electric arc welding differs from electric resistance welding in that a sustained arc generates the heat for melting the workpiece (and, if used, the filler rod) material. Events are best followed on the example of gas tungsten-arc welding (GTAW).

When the tungsten electrode is connected to the negative terminal of a dc power source (*straight polarity* or *direct-current electrode negative*, DCEN mode), it becomes the cathode; the workpiece, connected to the positive terminal, becomes the anode (Fig. 18–17*a*). An inert gas shields both electrodes. The cathode is heated by the welding current until the *work function* (the energy required to strip off electrons) of tungsten is reached. The thermally induced (*thermionic*) emission creates a *space charge* (a cloud of electrons) in which electrons flow to the workpiece (anode), where most heat is generated (electron flow accounts for some 85% of heat transfer). In the space between electrode tip and workpiece, the high temperature ionizes some of the gas: Electrons are stripped off and an electrically conductive *plasma* (a neutral mixture of electrons and positive ions) forms. The energy of impinging electrons heats the workpiece. The weld zone is often deep and narrow (Fig. 18–17*b*).

When polarity is reversed, with the electrode connected to the positive terminal (*reverse polarity* or *direct-current electrode-positive*, DCEP mode), the workpiece becomes the cathode. The weld zone is wider and shallower (Fig. 18–17*c*); therefore, this mode of operation is more suitable for thin-gage material

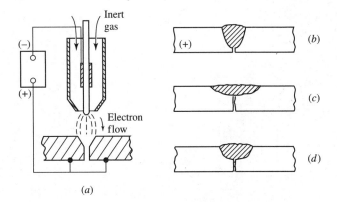

Figure 18–17   Gas-tungsten arc welding (a) and the shape of the
weld pool with dc current of straight (b) and
reverse (c) polarity and with ac current (d).

which would be burnt through in the DCEN mode. Reverse polarity has one
further effect: Oxide films on Al and Mg workpiece surfaces are stripped off;
thus, the surface can be cleaned, either by briefly reversing polarity in the DCEN
mode or by using an ac current (Fig. 18–17d).

Nominal heat input H is power divided by speed of travel v; for an electric
arc, power $= E \times I$, where E is voltage (V) and I is current (A). Thus,

$$H = \frac{EI}{v} \qquad \left(\frac{\text{J}}{\text{mm}} \right) \qquad \textbf{(18-2)}$$

As pointed out in Sec. 18-4-2, not all of this heat reaches the workpiece (*arc
efficiency* is less than 1) and more heat is lost to the zone adjacent to the weld.
Permissible heat input may be limited by metallurgical considerations and is
smaller when the workpiece is preheated. High temperatures are maintained for
some time; therefore, complete protection from the atmosphere is essential. In
some processes and with some materials, there is also need for a flux that dissolves
oxides and removes them from the melt zone. Very broadly, arc welding processes
include both nonconsumable- and consumable-electrode methods.

18-6-1   Nonconsumable-Electrode Welding

In these processes the electrode does not melt and the weld metal is supplied by
the flow of the parent metal (*autogenous weld*) or, for thicker (> 3-mm-thick)
sheet, from a separate filler rod.

Gas Tungsten-Arc Welding (GTAW)   As indicated, the arc is maintained between
the workpiece and a tungsten electrode protected by an inert gas (hence the older
name *tungsten inert gas* or TIG *welding*, Fig. 18–18a). The protective atmosphere

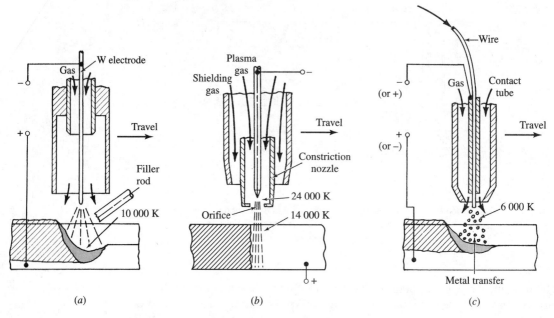

Figure 18–18   The arc is shielded by gas in the (a) gas tungsten-arc, (b) plasma-arc, and (c) gas metal-arc welding processes. Note that the depth of penetration increases with increasing arc temperature.

may be given by argon, which has a lower ionization potential and thus requires lower voltage (around 10 V) but gives a less hot arc and less deep penetration than helium. Polarity is DCEN except for Al and Mg where ac is helpful in stripping the oxide.

To strike an arc, electron emission and ionization of the gas are initiated by withdrawing the electrode from the work surface in a controlled manner, or with the aid of an initiating arc. High-frequency current, superimposed on the alternating or direct welding current, helps to start the arc and also stabilizes it.

Both hand and automatic operations are possible. The process demands considerable skill but produces very high-quality welds on almost any material, in any welding position, and also on thinner gages (below 6 mm). The weld zone is visible, and there is no weld spatter or slag formation, but electrode particles may enter the weld if the electrode overheats or touches the weld pool.

Plasma-Arc Welding (PAW)   If the arc and, within it, the plasma is constrained by an orifice, heat intensity increases (Fig. 18–18b). The arc is first drawn between electrode and nozzle by applying a high-frequency voltage. The torch is then brought close to the workpiece, the welding current is applied, and the arc is transferred to the workpiece (*transferred plasma arc* method of operation). At lower current densities, in the *melt-in mode*, the weld zone is similar in shape to arc welding; at high current densities the *keyhole mode* (Fig. 18–18b) prevails and the metal resolidifies behind the moving plasma beam. In the *nontransferred arc* technique the constriction nozzle is connected to the positive terminal; the arc is drawn between electrode and nozzle, and the arc heats the workpiece by radiation. This technique is used also for plasma spraying (coating of surfaces, Sec. 19-4-4). PAW is particularly useful for the welding of thin sheets. The small spot size requires careful control of path. Both manual and mechanical forms are practiced, and filler metal may again be used if an extra material supply is needed.

Carbon-Arc Welding (CAW)   was the forerunner of present-day nonconsumable electrode processes. The arc is drawn between the workpiece and a carbon electrode, or between two carbon electrodes.

18-6-2   Consumable-Electrode Welding

In this group of processes the *consumable electrode* is a metal which melts to become part of the weld seam. Its composition is often different from that of the base metals and recommendations can be found in the reference works cited at the end of this chapter. The weld zone is protected by a gas or a flux.

Gas Metal-Arc Welding (GMAW)   The consumable metal electrode, fed through the welding gun, is shielded by an inert gas, thus the older acronym MIG (*metal inert gas*) welding (Fig. 18–18c). It is suitable for most metals. As in GTAW, no slag is formed and several layers can be built up with little or no intermediate cleaning. Argon is a suitable gas for all materials; helium is sometimes preferred—because of its higher ionization potential and, therefore, higher rate of heat generation—for the welding of aluminum and copper; Ar with 2–30% CO_2 or pure CO_2 are generally used for carbon steels; specialty gases are also being introduced, tailored to specific tasks.

The electrode is usually connected to the positive terminal (DCEP or reverse polarity). Current density is the primary determinant of metal transfer mode. At low welding current, *short-circuiting* transfer takes place: The electrode touches the workpiece, current increases, the wire tip melts, and a droplet is transferred. The shielding gas is chosen to minimize spatter (CO_2 for steel, Ar–He for nonferrous metals). At higher current intensities (and with CO_2 almost always) *globular transfer* prevails: Particles larger than the electrode drop by gravity. The glob must be detached before it reaches the weld pool, and only horizontal welding is possible. Above a critical current density, metal is transferred in a fine spray (*spray transfer*) by arc forces, hence all welding positions can be used. More consistent welds are obtained with the use of pulsed welding current: Low background current maintains the arc, and metal transfer occurs when current pulses exceed the level required for spray transfer. Shielding gas is Ar or an Ar mixture. In a variant, wire is fed in a plasma torch (*plasma-MIG welding*).

The wire electrode can be supplied in long, coiled lengths which allow uninterrupted welds in any welding position. In semiautomatic welding the welder guides the gun and adjusts process parameters; in automatic welding all functions are taken over by the welding machine or robot. On-site welding can be difficult because drafts blow the shielding gas away from the weld zone.

Example 18-7

Butt joints are produced by short-circuiting DCEP GMAW between two stainless steel sheets of 2-mm thickness. Electrode wire of 1.1-mm diameter is fed at a rate of 4.4 m/min. Current is 140 A, arc voltage 21 V. Calculate the (*a*) power (kW) assuming an arc efficiency of 0.85, (*b*) deposition rate (mm^3/s), and (*c*) the speed of travel if the average width of the weld zone is 2.4 mm and half of this comes from the base metal.

(*a*) Power $= E \times I \times \eta = (21)(140)(0.85) = 2.5$ kW.

(*b*) Deposition rate $= (1.1^2 \pi /4)4400/60 = 70$ mm^3/s.

(*c*) Speed of travel $= 70/[(2 \times 2.4)/2] = 29$ mm/s.

Shielded Metal-Arc Welding (SMAW)   Again, the arc is struck between the filler wire or rod and the workpieces to be joined (Fig. 18–19*a*) but protection is now provided by a coating applied to the outside of the filler wire (*coated electrode*). The coating fulfills several functions: Combustion and decomposition under the heat of arc creates a protective atmosphere; melting of the coating provides a molten slag cover on the weld; the Na or K content of the coating readily ionizes to stabilize the arc. Also, alloying elements may be introduced from the coating. Choice of the electrode (which determines also polarity) is critical for success. During welding, the electrode melts at a rate of approximately 250 mm/min while the coating melts into a slag which must be removed if more than one pass is required to build up the full weld thickness.

Since the coating is brittle, straight sticks of typically 450-mm length are generally used, making this process suitable only for hand operation, at relatively slow rates, but still at a low cost. The process is versatile and suitable for field application, but requires considerable skill. Welding in all positions, including overhead welding is possible if the metal and slag solidify fast enough.

Flux-Cored Arc Welding (FCAW)   Basically the same result, but deeper penetration is obtained with the flux inside a tube. The welding wire can now be coiled, and automatic, continuous welding becomes possible. Sometimes additional shielding is provided with a gas, and then the process resembles gas metal-arc welding.

Submerged Arc Welding (SAW)   The consumable electrode is now the bare filler wire and the weld zone is protected by a granular, fusible flux supplied quite independently from a hopper (Fig. 18–19*b*) in a thick layer that covers the arc. The flux shields the arc, allows high currents and great penetration depth, assures

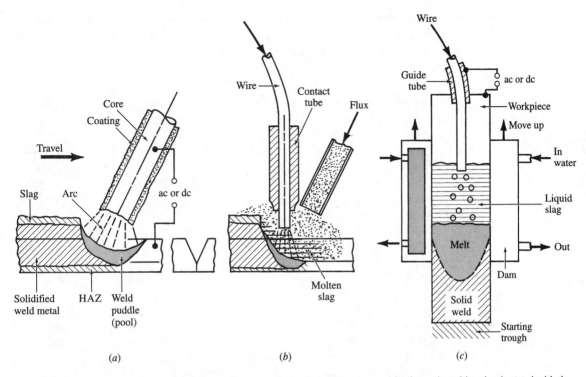

Figure 18–19   A sustained arc, shielded by molten slag, is maintained in consumable-electrode welding by the (a) shielded metal-arc, (b) submerged arc, and (c) electroslag methods.

high efficiency, acts as a deoxidizer and scavenger, and may contain powder-metal alloying elements. SAW is primarily an automatic welding process with high deposition rates. Tandem electrodes can be used to deposit large amounts of filler material.

The weld position must be horizontal; thus, it is suitable for steel line pipes, cylinders, and also for circular welds if the workpiece is rotated. Double submerged-arc welding (with one weld from the inside, the other from the outside) is used in making spiral-welded pipelines. It can also be used with welding robots, with the workpiece manipulated into appropriate positions.

Electroslag Welding (ESW)   The process is used extensively for welding vertical joints in thick (25-mm or over) plates and structures, such as oil rigs, bridges, ships, and press and rolling mill frames. Electrode wire is fed into a molten slag pool (Fig. 18–19c); an arc is drawn initially but is then snuffed out by the slag, and the heat of fusion is provided by resistance heating in the slag. Water-cooled copper *shoes* (*dams*) close off the space between the parts to be welded to prevent the melt and slag from running off. The welding head must be raised as the weld deposit builds up. In a variant of the process, a consumable, steel guide tube melts into the weld pool, thus the welding head need not be moved. If the part is rotated, circumferential welds can also be made.

Electrogas Welding (EGW)   This is an outgrowth of electroslag welding but is related to gas metal-arc welding. The electrode wire is solid or flux-covered, and protection is provided by a gas (typically 80% Ar, 20% CO_2). The molten pool is again retained with copper dams.

Automation of Welding Processes   We already remarked on the possibilities of mechanization and automation for various arc welding processes. Developments have taken two directions.

First, the process itself is increasingly controlled automatically. The shape of current pulses is controlled by electronic devices. The optimum current, travel rate, electrode feed rate, etc., are set and controlled, and CNC and adaptive control—utilizing noncontacting temperature sensors, electrode-to-workpiece distance sensors, etc.—are introduced.

Second, manipulation of the workpiece and/or the welding gun relative to each other is increasingly automated. The task is relatively simple when welding along a straight-line, spiral, or other readily defined path, and mechanization is often possible. A prerequisite is proper edge preparation, fit-up of parts, and fixturing, so that a sound weld bead is made in the correct location. Welding along complex spatial lines became possible only with the appearance of multiaxis, teachable robots. Welding robots achieve an arc time of 80%, whereas a manual welder seldom maintains 30%. Repeatability is much higher too, and it is not surprising that welding is the single largest field of robot application.

Ideally, seam preparation and accurate location by fixturing should allow the welding torch to follow a predetermined path. If this is not possible, some form of *seam tracking* becomes necessary. In the simplest form, a probe or wheel is in front of the torch and guides it in the groove. Through-the-arc tracking relies on sensing the change in current when the groove-to-torch distance changes. For better definition of the groove, the torch is moved in a weaving or orbital pattern. The most sophisticated is laser tracking in which the location of the groove is found by triangulation. Such vision systems can be further developed to help plan and program the welding sequence, locate the joint, position the electrode, begin the cycle, control the electrode during welding, stop the cycle, and inspect the resulting weld.

18-6-3   Consumable-Workpiece Welding

In some special cases of autogenous welding the workpiece becomes the electrode.

Stud Arc Welding or Stud Welding (SW)   The arc is maintained between a *projection* (reduced cross section) on a *stud* (typically, a threaded or smooth rod) and the surface of the workpiece (typically, a plate, Fig. 18–20a). When the projection and the surface of the plate melt, pressure is applied to join the stud to the plate. Polarity is usually DCEN for steel and DCEP for aluminum. An expendable ceramic *shielding ferrule* concentrates the heat of the arc, protects against oxidation, and confines the melt. Flux, embedded in the stud, may be

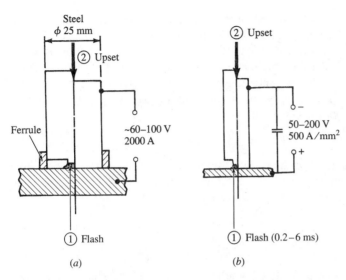

Figure 18-20   An arc is drawn in (a) stud and (b) percussion welding.

used. Stud diameter is chosen so that the joint should fail in the stud and not the sheet. Millions of studs, often threaded, are used in building construction, shipbuilding, auto industry, electric panel construction, and for the attachment of handles and feet to appliances.

Capacitor-Discharge Stud Welding   In attaching smaller (2–6-mm-diameter) studs, the energy stored in a condenser is used for heating (Fig. 18–20b). Discharge takes place just before or during approach to the surface. The intense, localized heat allows the joining of widely differing cross sections and also of dissimilar materials. Timing and motion control are critical. Studs can be welded to thin sheets, even to those coated with paint or PTFE on the other side, allowing the fastening of instrument panels, nameplates, and auto trim.

The term *percussion welding* (PEW) is used to describe capacitor-discharge welding applied to joining wires to terminals and other flat surfaces. Since the two terminals must be separate prior to impact, rings cannot be welded.

18-7   OTHER WELDING PROCESSES AND CUTTING

There are a number of welding processes that do not fit into the previous classes.

18-7-1   Chemical Heat Sources

The heat required for fusion may be provided by a chemical heat source.

Gas Welding *Oxyfuel gas welding* (OFW) is the most widespread form of gas welding. Heat is produced by the combustion of acetylene (C_2H_2) with oxygen (*oxyacetylene gas welding*, OAW). Both are stored at high pressure in gas tanks and are united in the welding torch. After ignition, a temperature of approximately 3700 K (3400°C) is generated in the flame. Three zones may be distinguished (Fig. 18–21a). Primary combustion takes place in the *inner zone* and generates two-thirds of the heat by the reaction

$$2C_2H_2 + 2O_2 \rightarrow 4CO + 2H_2 \tag{18-3}$$

These reaction products predominate in the *second zone*, and thus provide a reducing atmosphere favorable for the welding of steel. Complete combustion takes place in the *outer envelope* by the reactions

$$4CO + 2O_2 \rightarrow 4CO_2 \tag{18-4}$$

$$2H_2 + O_2 \rightarrow 2H_2O \tag{18-5}$$

The flame protects low-carbon steels, lead, and zinc sufficiently, but a flux is needed for most other materials. The relatively low flame temperature, the ability to change the flame from oxidizing to neutral and even reducing, and the flexibility of manual control make the process suitable for all but the refractory metals and reactive metals such as titanium and zirconium. The process has the advantage of portability and is suitable for all welding positions.

The welder (assumed to be right-handed) holds the torch at an angle to the surface, and may pull the filler rod (with its flux coating) away from the *weld puddle* in *leftward* or *forehand welding* (Fig. 18–21b), thus preheating the joint area and obtaining a relatively wide joint. When the filler is moved above the weld bead (*rightward* or *backward welding*, Fig. 18–21c), the weld puddle is kept hot for a longer time and a narrower weld, often of better quality, results.

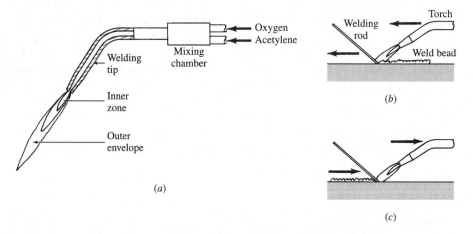

Figure 18–21   A thermal heat source is employed in (a) oxyfuel gas welding. A wider bead is produced in (b) forehand than in (c) backhand welding.

Other gases, including propane, natural gas, and hydrogen are also used as the heat source, particularly for aluminum and lower-melting metals.

Thermite Welding (TW)   When a metal oxide of low free energy of formation is brought in intimate contact with a metal of higher free energy of oxide formation, the metal oxide is reduced in an exothermic reaction, named, generically, a thermite reaction. *Thermit powder* (a registered trademark of Th. Goldschmidt AG, Essen, Germany) is a mixture of a metal and an oxide, e.g., aluminum and iron oxide. The reaction, initiated with a special ignition powder, liberates iron

$$3Fe_3O_4 + 8Al \rightarrow 4Al_2O_3 \text{ (slag)} + 9Fe \qquad (18\text{-}6)$$

Iron alloy pellets are added to reduce the reaction temperature to about 2500 °C.

The process finds application for joining heavy (minimum 60-cm^2-area) sections such as rails or reinforcing rods in the field. A sand or semipermanent mold—complete with sprue, gate, and riser—is built around the joint area, the thermite powder is ignited in a crucible placed on top of the mold, and the resulting iron is tapped, through a bottom hole of the crucible, directly into the mold. After solidification, the mold is destroyed and the still-hot excess steel is chiseled off. Large electrical bus bars are similarly welded, using aluminum and Cu_2O.

18-7-2   High-Energy-Beam Welding

The heat of fusion may be provided by converting the energy of impinging electron or light beams into heat; we already saw the application of these processes to cutting (Sec. 17-5). The same processes can be used for joining, except that energy densities are seldom over 10 kW/mm^2 because vaporization and melt-through would ruin the weld.

Electron Beam Welding (EBW)   The electron gun (Fig. 17–4) now melts the parent metal (Sec. 17-5-1); molten metal ahead of the vapor hole flows around to fill the gap, thus narrow gaps can be welded without a filler (although filler rods may be used). The heat-affected zone is very narrow. Most welding is done in high vacuum (EBW-HV) or medium vacuum (EBW-MV) but nonvacuum (EBW-NV) operation also yields high-quality welds in many materials. The processes are extremely adaptable and excel in welding both thin gages and thick sections, parts of dissimilar thickness, hardened or high-temperature materials, and dissimilar materials. EBW lends itself to automatic control.

Example 18-8

A 12.7-mm-thick 2219 aluminum plate was welded by electron beam and gas tungsten-arc welding. The weld bead was about 1.3-mm wide in the EB weld and on the average about 7-mm wide in the GTAW joint; the latter was made in two passes. The following conditions applied:

	EB Weld	GTAW 1st Pass	GTAW 2d Pass
Voltage, V	30,000	11.7	13.0
Current, A	0.2	270	270
Welding speed, mm/s	40	2.75	3.18

Calculate the power, energy/mm, and total heat input for the two processes.

	EB Weld	GTAW 1st Pass	GTAW 2d Pass
Power, kW	6	3.16	3.51
Energy, kJ/mm	0.15	1.15	1.10
Heat input, J/mm	150	1149	1104
Total heat input, J/mm	150	2253	

Thus, 15 times more heat was needed to weld by GTAW than EBW. In GTAW, 5 times more metal was melted, thus, efficiency was about one-third that of EB welding. This demonstrates the importance of rapid heat transfer which minimizes heat losses into the environment and surrounding metal. (Data from W.J. Farrell, *ASTME Paper SP 63-208*.)

Laser Beam Welding (LBW)   The energy of a laser (Sec. 17-5-2) may be used to heat from the surface of the material (*conduction-limited mode*) or to penetrate the full depth of the joint (*deep-penetration mode* or *keyhole welding*). Because heating is a function of surface emissivity, the shorter-wave Nd:YAG lasers are more suitable for highly reflective materials but cannot be used on glass or polymers. The laser has the advantage that vacuum is not necessary. The workpiece usually needs protection by a gas, except for spot welding in which exposure time is very short. Oxygen blown on the surface reduces light reflection and increases removal rates for steel; inert gas (N_2 for aluminum, Ar or Ar–He for titanium) increases heat transfer for nonferrous metals.

The laser is finding growing application, particularly for thin-gage metals. Welding speeds of about 7 m/min are achieved on steel sheet 1.5 mm thick. The process is suitable for automation, and either the workpiece, the laser, or the beam may be moved along prescribed paths (Fig. 18–22). Relative to resistance welding, one-sided access is an advantage, for example, in welding hydroformed tubes. It is one of the principal methods of making tailored blanks (Sec. 10-7).

Example 18-9   In Example 5-1 we indicated that the ULSAB auto body has an all-welded structure. The number of spot welds is greatly reduced by the use of laser-welded tailored blanks. Six highly stressed nodes, where vital elements meet, are also laser welded.

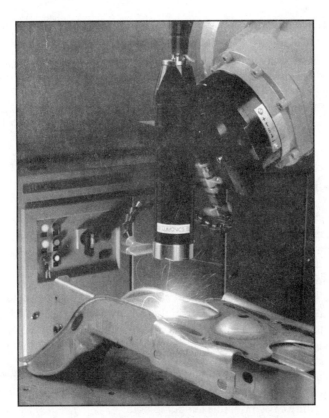

Figure 18–22 In welding a zinc-coated automotive support member, the beam of a 2-kW continuous-wave Nd:YAG laser is delivered through a robot-manipulated optical fiber. (*Courtesy Lumonics, Livonia, Michigan.*)

18-7-3 Cutting

A very important application of fusion welding processes accomplishes an exactly opposite function: headers, risers, and flash are removed from castings, forgings, or moldings, or workpieces of varying shape are cut out from sheet, plate, and even heavy sections. There are applications beyond those shown in Sec. 17-5.

1. Steels, preheated to 850 °C, burn if oxygen is blown on them. *Oxygen cutting* is widespread in steel mills, for cleaning up surfaces (*scarfing*) and cutting up billets while the steel is still hot.

2. For general cutting applications, the steel is first preheated with an oxyfuel gas flame, and a stream of high-pressure oxygen is then directed onto the heated spot. Oxidation generates more heat, melting the steel. The melt is flushed away by the flame and oxygen (*flame cutting* or *oxyfuel gas cutting*, OFC). Plates of 5–1500-mm thickness are cut. When iron powder is added to the gas stream *(powder oxyfuel cutting* or *powder metal cutting*, POC), oxidation of the powder provides the heat to melt oxidation-resistant materials.

3. In general, cuts of better surface quality are produced by variants of welding processes. *Plasma-arc cutting* (PAC) with a transferred arc has acquired great importance in cutting all metals in thicknesses up to and even over 25 mm. The quality of the cut edge is better than in flame cutting but not as good as with EB or laser beam cutting (Table 17–3).

18-8 LIQUID–SOLID-STATE BONDING

When the joint is established without melting the base metal, the main source of strength is adhesion between filler and base metal, developed in the absence of contaminant surface films. The strength of the joint is higher than the bulk strength of the filler. Dissimilar materials can be joined, as can be parts with greatly differing wall thickness. A great advantage is that there is no need for access to all parts of the joint; hence, complicated assemblies, including those consisting of many parts, can be simultaneously joined. The techniques are applied to the manufacture of millions of automotive radiators, plate-and-tube heat exchangers, impellers, fans, and appliance parts, and to the joining of pipes and wires. Processes are somewhat arbitrarily divided into *brazing* and *soldering*: When the *filler metal* melts below 425°C (800°F), one speaks of soldering. In both processes, the filler metal is drawn into the joint by capillary action (Fig. 18–23) and some basic principles apply to both.

18-8-1 The Bond

The liquid–solid bond has, in many ways, unique properties which can be developed only if a number of conditions are satisfied. Frequently, the filler metal is

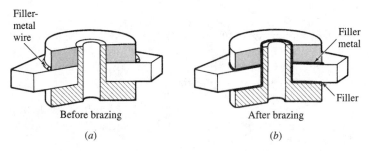

Figure 18–23 Several components are simultaneously joined in brazing; the filler metal is distributed through capillary action. (*From Metals Handbook, 8th ed., vol. 6, AMS International, 1971, p. 607. With permission.*)

substantially weaker than the base metal, and the joint derives its strength from the support it receives from the base metal. For this, two criteria must be met:

1. *Joint thickness.* In a thick layer, the bulk properties of the filler would dominate. For greater strength, the layer must be so thin that the restraint given by the parent metal prevents necking in tension. The joint is weaker in shear but still stronger than the bulk filler because the thin layer is subjected to very high shear strain.

2. *Bonding.* Full strength is developed only if there is perfect interatomic bonding between parent and filler metal. As discussed in Sec. 4-9-2, surfaces are normally covered with oxides and adsorbed layers from the atmosphere, and lubricant residues may also be present from prior processing steps. To secure bonding, surfaces must be wetted by the filler (Sec. 6-3-2); otherwise, the joint becomes a crack which would fail at very low loads.

These two conditions dictate the practice of brazing and soldering.

Clearance   The gap between mating surfaces is critical. If too small, the filler cannot penetrate, and imperfect joins form; if too large, the support received from the parent metal is gradually lost. Only within a narrow range of clearances are maximum properties developed (Fig. 18–24). This clearance must exist at the joining temperature; therefore, differential expansion must be taken into account when dissimilar metals are joined. With the optimum clearance, the filler penetrates the gap by *capillary action* (Fig. 18–23). Surface roughness must be controlled too; slight roughness opens passages for the melt but on a very rough surface only the asperities would be joined.

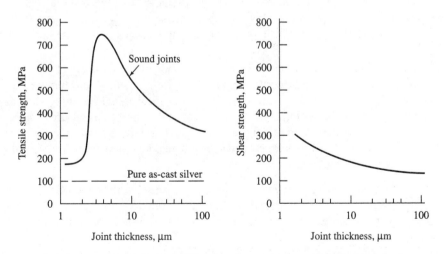

Figure 18–24   The (a) tensile and (b) shear strength of brazed joints depends critically on joint thickness. (Silver-brazed butt joint on 12.7-mm-diameter steel rods; data from *Brazing Handbook*, 4th ed., American Welding Society, 1991.)

Filler Metal   Capillary action is possible only if the filler wets the base metal [typically, the contact angle in Eq. (6-6) should be between 15° and 45°]. The filler metal must have high fluidity to penetrate crevices (but not so high that it would run out of the joint) and it should preferably have a narrow melting range (pure metal or eutectic). In alloys of long freezing range, the low-melting phase may melt prematurely and such *liquation* may leave the higher-melting constituent behind. In general, wider-melting alloys require larger clearance. The filler is applied in the form of wire, strip, preforms, powder, or paste to the joint area, which is then heated locally or by heating the entire assembly. It may be placed outside the future joint (*local*, Fig. 18–23a) or inside the joint (*preplaced*). Alternatively, the filler metal is preapplied to the surface of one of the contacting parts as a coating (cladding), often by rolling (Fig. 18–4c), electrolytic deposition, or hot dipping. In choosing the filler metal, service conditions must be taken into account. In particular, the filler must not cause galvanic corrosion.

Fluxing   Surface films that would prevent wetting must be removed by degreasing; heavy oxides must be removed by mechanical or chemical means. To prevent oxidation during joining, a protective atmosphere or vacuum may be applied, or heating at temperature is kept very short. In many cases all this is still insufficient to ensure wetting, and fluxes are then used. A good flux melts at a low enough temperature to prevent oxidation of the base and filler materials; it has a low viscosity so that it is replaced by the molten filler metal; it may react with surfaces to facilitate wetting; it shields the joint while the filler is still liquid; and it is relatively easy to remove after solidification of the filler. Fluxes are in the form of powder, paste, or slurry. In general, strongly reducing atmospheres or more active fluxes are needed when stable oxides form, such as on aluminum, magnesium, heat-resistant alloys, and stainless steels. Ideally, residues can be left on the finished assembly. If they would cause corrosion, they have to be removed, a task that can be difficult if there are tight spaces in the assembly.

Heating   The parts to be joined are most often made of sheet metal but may also be forged, extruded, or cast components. They are assembled and temporarily held together in the correct position by fixtures or mechanical fastening such as expanding, staking, swaging, etc. The design of some assemblies ensures correct positioning and makes them self-fixturing. A great variety of heating techniques can be employed, each with some special advantages. Most techniques are readily mechanized or automated.

    1. *Furnace heating* is a mass-production process, with assemblies placed into a box furnace or, for highest production rates, conveyed through a continuous furnace on belts, roller gangs, or suspension hooks. A protective atmosphere can be applied.

    2. *Torch heating* allows selective heating of the joint area by an acetylene, propane, or natural gas flame. It is suitable for manual operation but is easily mechanized.

3. *Resistance heating* utilizes the same equipment as resistance welding (Sec. 18-5) but a filler is placed in the joint. Rapid heating minimizes oxidation.

4. *Induction heating* with a typically 10–460-kHz power supply has the same advantages as resistance heating.

5. *Infrared heating* with high-intensity quartz lamps is suitable for conveyor-type operation but, if required, heat can be concentrated to the area to be heated. This applies also to *microwave heating.*

6. *Laser beam* and *electron beam heating* are justified mostly for precision assemblies of high-value, relatively high-temperature materials.

18-8-2   Brazing

Typical filler metals are listed in Table 18–1 for various classes of base metals. In general, filler metals of higher melting point give higher strength, but the high brazing temperature may affect the strength of the base metal. In brazing steel with copper, the clearance is zero or even negative; with other materials, brazing proceeds with the usual small, positive clearance. With aluminum or magnesium alloys, the melting point of the filler metal is close to that of the base metal, hence induction heating cannot be used. Small Mg additions to aluminum alloys are beneficial because Mg acts as a getter (captures the oxygen) and modifies the oxide film. The filler is often applied as a cladding of 5–10% thickness to one or both sides of a sheet (an Al–Si alloy to aluminum, Cu to steel and stainless steel). Surface carbon would prevent wetting of cast irons and must be removed by electrolytic or molten salt-bath treatment of the parts.

Table 18–1   Filler metals used for brazing*

AWS Designation	Brazing Filler Metal Composition, wt %	Brazing Temperature, °C	Base Metal
BCu-1	99.9Cu-0.075P	1100–1150	Steel (< 0.3% C); low-alloy steel
BCuP-2	92.75Cu-7.25P	820–870	Copper alloys
BAg-1	45Ag-15Cu-16Zn-24Cd	620–760 ⎱	Carbon steel; low-alloy steel; cast iron;
BAg-5	45Ag-30Cu-25Zn	735–845 ⎰	stainless steel; copper alloys
RBCuZn-D	48Cu-42Zn-10Ni	950–970	Low-carbon steel; low-alloy steel; WC
BNi-2	74Ni-3B-7Cr-4.5Si-3Fe-0.06C	1010–1175	Stainless steel, low-alloy steel; tool steel
BAu-1	37Au-63Cu	1015–1090	Stainless steel, low-alloy steel; tool steel
BAlSi-3	Al-10Si-4Cu-0.15Cr-0.2Zn-0.15Mg	570–605	Aluminum alloys (flux brazing)
BAlSi-7	Al-10Si-0.25Cu-0.2Zn-1.5Mg	590–605	Aluminum alloys (vacuum brazing)

*Compiled from *ASM Handbook*, vol. 6, *Welding, Brazing, and Soldering*, ASM International, 1993.

Higher temperatures accelerate oxidation; therefore, fluxes have to be more aggressive. Fluxes are composed of borates, fluorides, chlorides, and similar materials in various proportions. Tailor-made for specific applications, they assure perfectly wetted joints (Fig. 18–25). In *vacuum brazing*, pressure in the furnace is reduced to prevent oxidation, and, in some instances, no flux is needed. Specially constructed furnaces allow degreasing by vaporization prior to initiating the vacuum brazing process.

Dip brazing derives the heat from immersion of the assembly into molten salt. Heating rates are high and the salt may perform the fluxing function. *Braze welding* differs from brazing in that a much wider gap is filled up with the brazing metal (mostly brass) with the aid of a torch, and thus capillary action plays no part. Besides assembly, it is used also for repairs of steel and iron castings. *Diffusion brazing* differs from diffusion welding in that a liquid layer forms while diffusion also takes place.

Controlled cooling of brazed assemblies is necessary to ensure rapid solidification yet without causing distortion of the assembly or cracking of the joint.

18-8-3   Soldering

Soldered joints are of lesser strength than brazed ones, but parts can be joined without exposing them to excessive heat. The lower temperatures make good wetting more critical than in brazing. Therefore, surface preparation by mechanical and chemical means and the use of fluxes are essential.

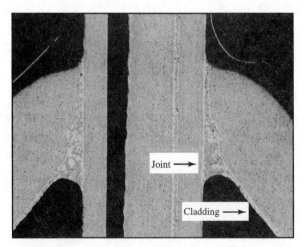

Joint →

Cladding →

Figure 18–25   With the aid of a special flux, braze metal flows from the cladding on the base metal to fill the joint in an all-aluminum heat exchanger. (*Courtesy Modine Manufacturing Co., Racine, Wisconsin.*)

The filler metal (*solder*) may be chosen by reference to the relevant phase diagrams. Solid solubility is a sign of good wettability, although the presence of surface films greatly modifies the behavior. A surface that appears uniformly coated by the solder may still de-wet on reheating. Thus, *solderability* is a technological property that can be determined only experimentally. In general, precious metals and copper are readily soldered; iron and nickel require more aggressive fluxes; the tenacious oxides of aluminum and chromium make soldering of aluminum alloys and high-chromium steels more difficult. Cast iron, titanium, magnesium, and ceramics (including graphite) require preplating.

The most widely used solders used to be tin-lead alloys (Table 18–2). A low (< 5%) tin content gives higher strength and is still extensively used for automotive radiators and tubes made of tinplate. The wide freezing range of the 35% Sn alloy made it ideal as a wiping solder for the joining of copper tubes. The eutectic composition (63% Sn) has, by definition, the lowest melting point and solidifies at a constant temperature, making it most suitable for electric connections. The toxicity of lead has prompted its complete elimination from joints of domestic water pipes, food processing equipment, and containers for food products; Sn–Ag and Sn–Sb solders are used for food applications and stainless steel. Because of the danger to operators, lead solders have been or are being replaced with lead-free solders in many other applications. Coated metals, especially tinplate, facilitate soldering. A large variety of other solders are used in specific applications. In particular, Sn–Zn and Zn–Al alloys have been developed for the soldering of aluminum in conjunction with special fluxes.

Table 18–2   Selected solders*

	Melting Range	
Composition	Liquidus, °C	Solidus, °C
Sn-5Pb	240	233
Sn-30Pb	193	183
Sn-37Pb	183	183
Sn-50Pb	216	183
Sn-58Pb-2Sb	231	185
Sn-95Pb	312	308
Pb-5.5Ag-0.25Sn	380	304
Sn-4Ag	221	221
Sn-5Sb	240	233
Sn-9Zn	199	199
Sn-50In	125	117
Zn-5Al	382	382

*Compiled from *ASM Handbook*, vol. 6, *Welding, Brazing, and Soldering*, ASM International, 1993.

For general work with tin-based solders, the *flux* is a water solution of zinc (or Zn + Na + ammonium) chloride and must be washed off to prevent corrosion. Noncorrosive organic rosins are essential for electrical connections where corrosion would create high local resistance and even loss of conduction.

Soldering may be carried out by all the techniques previously described in Sec. 18-8-1. Because of the low temperatures involved, a heated copper or iron-plated copper bit (*soldering iron*) is used extensively. Low-temperature solders can be *pumped* to flood the joint area. Agitation of the molten solder bath by ultrasonic waves improves wetting in *dip soldering*, a technique particularly useful in soldering aluminum, as are techniques in which the surface to be bonded is rubbed (e.g., by an ultrasonic transducer) while the solder is flowed onto it. Special methods are used in the electronics industry (Sec. 20-4-3).

18-9   ADHESIVE BONDING

All processes discussed hitherto used a metal to establish a joint between two metallic or ceramic parts. *Adhesive bonding* differs in that the material of the joint is a polymer or, less frequently, a ceramic. Alloying is not possible, and bond strength relies entirely on the adhesion of the adhesive to the metal or polymer substrate and on the strength of the adhesive itself. Adhesion between metal and polymer surfaces is not fully understood but it is clear that only secondary bonding forces can come into play. Bond strength is, in general, lower than when primary (metallic, covalent, or ionic) bonds are established, and the design of joints must take this into account.

Adhesive bonding has many advantages: Only low temperatures are involved, thus no undesirable changes occur in most substrate materials; the exterior surface remains smooth; dissimilar materials and thin gages can be joined; the adhesive contributes to energy adsorption in shock loading and in the presence of vibrations; and, not least, complex assemblies can be made at low cost.

Adhesive bonding technology has advanced to the point where load-bearing structures can be built reliably. Applications include such critical structures as control surfaces in aircraft, entire aircraft bodies, and countless applications in the automotive, appliance, and consumer goods fields. In addition, adhesives are used also for sealing, vibration damping, insulating, and other nonstructural applications. Our concern here is with load-bearing applications, for which a class of adhesives termed *structural adhesives* are used.

18-9-1   Characteristics of Structural Adhesives

The terminology is somewhat imprecise but a structural adhesive is one used to establish a *permanent joint* between two higher-strength parts (*adherends*). The joint is expected to retain its load-bearing ability over a long period of time, under a wide variety of—and often hostile—environmental conditions. Thus, a structural adhesive must possess a number of desirable attributes; many of these are linked to properties of polymers, and a review of Chap. 13 may be useful at this point.

1. The adhesive must have sufficient *cohesive strength* at service temperatures. Shear strength is governed by composition (Fig. 18–26) but, for long-term applications, creep strength—which is generally lower than short-term strength—is of great concern. Many adhesives are thermosets in which the degree of cross-linking can be increased to increase creep resistance, although usually at the expense of bond flexibility. For example, in epoxies an average cross-link separation of < 2 nm gives a strong but brittle bond; some toughness is obtained by increasing the separation to approximately 3 nm, and at > 3 nm the bond becomes soft and flexible. Creep resistance can be increased also by the addition of fillers. These are particularly important in thermoplastics but are also used to improve the impact properties of thermosets.

2. The adhesive must be able to *distribute stresses* imposed in service. Flexibility helps to reduce stress concentrations arising from mechanical loading, makes the adhesive joint more resistant to fatigue, and prevents sudden catastrophic separation in the event of joint failure. Even in the absence of mechanical loading, stresses are generated during thermal cycling because polymers have, in general, higher thermal expansion coefficients than metals (Table 4–1). Flex-

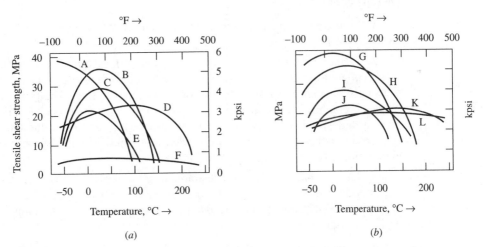

Figure 18–26 The shear strength of adhesives declines rapidly at some temperature which depends on composition, limiting their range of application. (*a*) Paste and liquid adhesives, (*b*) tape, film, and solvent-based adhesives. A: two-part urethane + amine; B: one-part rubber-modified epoxy; C: one-part epoxy, general-purpose type; D: one-part epoxy, heat-resistant type; E: two-component room-curing epoxy–polyamide; F: silicone sealant; G: nylon–epoxy; H: nitrile–epoxy; I: nitrile–phenolic; J: vinyl–phenolic; K: epoxy–phenolic; L: polyimide. (*Reprinted from J.C. Bolger, in Adhesives in Manufacturing, Dekker, 1983, p. 142, by courtesy of Marcel Dekker Inc.*)

ibility is again useful, and fillers can be used to reduce thermal expansion by as much as 75%.

3. The adhesive must not suffer *degradation* such as splitting of chains by water (hydrolysis) or ultraviolet radiation; oxidation, burning, cracking, loss of coherence at operating temperature (thermal degradation), or stress-corrosion cracking or dissolution in certain media.

4. Many adhesives are applied as liquids and must then have a low enough *viscosity* to flow into joint areas but not so low as to be lost from the joint. With thermoplastics, this requires heating above T_g; with thermosets, a prepolymer is employed which cross-links and polymerizes after application. A temporary decrease of viscosity on shearing (*thixotropy*) is useful.

5. Development of the adhesive bonds requires *wetting* of the adherend surfaces. Wetting is a surface phenomenon, depending greatly on the nature of the adherend surface, the presence of adsorbed surface films, and the migration of certain components of the adhesive to the surface. In some instances, sudden, catastrophic debonding may take place after some time.

18-9-2   Adhesive Types and Their Application

High-quality structural joints can be established only if all phases of the process are fully controlled.

1. *Joint design* must recognize the limitations of adhesive bonding (Sec. 18-11).

2. *Surface preparation* is crucial. Organic surface films are harmful with most (but not all) adhesives. Oxide films are harmful if loosely bonded, whereas a strongly bonded, porous oxide film or conversion coating is helpful. Special surface preparation techniques have been developed to promote adhesion and to prevent undesirable reactions that can lead to bond separation in the long term. Primers and adhesion promoters are needed when there is delay between surface preparation and the application of the adhesive.

3. *Surface roughness* can be desirable because it gives a larger contact area and provides some mechanical interlocking. However, surface films are difficult to remove from valleys; air may become entrapped too, and bonding is then limited to asperities.

Adhesives are often classified by their mode of action or chemical grouping; for our purpose, grouping by application method is more relevant.

Application Methods   A good adhesive bond is, in general, stronger than the strength (cohesive strength) of the weaker body, which may be one of the adherends or the adhesive itself. Therefore, maximum strength is usually obtained with very thin—but not starved—adhesive films. This is ensured by the proper application method.

1. *Hot-melt adhesives* are thermoplastic polymers. To assure wetting and high fluidity, the polymer is heated well above T_g and applied in beads or webs, by appropriate equipment, to the surfaces to be joined. Rapid cooling quickly establishes a bond. The flexibility of the bond is controlled by choosing an appropriate polymer. Water resistance is high but heat and creep resistance tend to be low; polyamides and polyesters are suitable for high-temperature applications. There are many applications in the construction, packaging, furniture, and footwear industries.

2. *Tapes and films* of thermosets are applied at room temperature and cured at elevated temperature. They are composed of a high-molecular-weight backbone polymer such as a flexible vinyl, neoprene, or nitrile rubber and of a low-molecular-weight cross-linking resin such as an epoxy or phenolic cured by a catalyst. To prevent starvation, a support mesh of reinforcing material such as fiberglass may be used. Some pressure is applied and reliable joints of good toughness are produced. Several thousand kilograms of such adhesives are used in the construction of a large aircraft. The technique is used also for the bonding of brake linings, lamination of windshield glass, and a variety of construction applications.

3. *Pastes* consist of the polymer and often also contain fillers. The most frequently encountered classes are room-temperature-curing, two-component epoxies, heat-curing one-component epoxies, acrylics, or flexible polyurethanes, with the catalyst added just before application. They are often used to bond elastomers, fibers, fabrics, and fiber-reinforced plastics to steel in the automotive and appliance industry, as well as for metal-to-metal bonds. They are applied by systems designed to handle high-viscosity fluids.

4. *Anaerobic adhesives* are monomeric liquids that have the unusual property of curing on the exclusion of air. They are stored in air-permeable containers and cure when applied to a narrow gap that excludes oxygen. Hence, they are extensively used as sealants and for holding threads and other fasteners (e.g., Loctite, a registered trademark of Loctite Corporation).

5. *Moisture-curing adhesives* include cyanoacrylates, low-viscosity fluids that cure on contact with moist or oxidized surfaces (hence they instantly bond human skin). They have low peel strength and temperature resistance but are useful in the electronics industry and for joining rubber or plastic to metal. Some polyurethanes and silicones are also cured by moisture.

6. *Liquids* usually contain a solvent. Thus, PVC plastisols, sometimes fortified with an epoxy, are used for assembling car hoods, trunk lids, roofs, furniture, and appliance cabinets. They can be applied by a brush, spray, printing, etc., but the solvent can create an environmental problem.

7. *Water emulsions* of polymers, such as polyvinyl acetate (white glue) need a porous substrate to allow removal of the water.

8. *Conductive adhesives* represent a separate class. For electrical conduction, the plastic (usually epoxy) is filled with up to 85% silver flakes; for thermal conduction, up to 75% ceramic (alumina) powder is added.

It is evident from the foregoing discussion that the adhesive joint is a system composed of adherends, metal oxide, primer, adhesive, and the environment. Therefore, in all instances, total control of application is crucial. In critical applications, such as aircraft construction, the presence and continuity of bonds is verified by NDT techniques.

18-10 JOINING OF PLASTICS AND CERAMICS

In many instances, special considerations apply when nonmetals are to be joined.

18-10-1 Joining of Plastics

Parts made of plastics may be joined to each other or to metals by variants of the techniques discussed hitherto.

1. *Mechanical fastening* employs metal or plastic screws, driven into bosses or holes molded (sometimes cut) in the part, or into metallic inserts molded into the plastic part. The joint must be designed to minimize the effects of creep; thus, for example, countersunk screws are to be avoided. Snap fits are widely used.

2. *Thermal sealing and bonding* of thermoplastic polymers relies on heat and pressure. As in the pressure welding of metals, surfaces must be clean, and localized deformation that breaks up adsorbed surface films is helpful. However, in contrast to metals, melting of the polymer usually occurs. Several techniques are used:

a. *Hot-tool welding* with heated tools or rollers is suitable mostly for thinner parts since polymers are poor heat conductors. For example, LDPE is heat-sealed with PTFE-coated, electrically heated tooling.

b. *Friction joining* relies on heat generation at the interface, either by rotation (*spin welding*, as in Fig. 18–8a) or by *vibration*. Vibration may be *low-frequency oscillation* (100–500 Hz) imposed by mechanical means, or *ultrasonic* (20–40 kHz, as in Fig. 18–5c).

c. *Hot wire welding* provides localized heating by incorporating a resistance wire into the joint area. The wire ends must stick out so that current may be passed through. A closed wire loop may be heated inductively (as in joining knife blades to handles). Once the polymer melts, the parts are pressed together.

d. *Hot-gas welding* is the equivalent of oxyfuel gas welding (Fig. 18–21), except that hot air (or inert gas) is the heat source. The joint is prepared, beveled as on metals, and a filler rod, of the same thermoplastic as the parts, is used.

e. *Focused infrared welding* relies on the heat of an infrared lamp focused by reflectors into a narrow (1.5–3-mm) beam.

f. *Dielectric heating* results when polymers, which are insulators, are placed into an electromagnetic field. Frequencies of 13–100 MHz (most often

27.12 MHz, the frequency allocated by the Federal Communications Commission) are used; therefore, one speaks also of *radio-frequency* (rf) or *high-frequency* (hf) heating. Polymers of high dielectric loss factor, such as PVC, ABS, nylon, polyurethane, and rubber can be through-heated and are most suitable.

g. *Electromagnetic bonding* and *magnetic heat sealing* are made possible by embedding very small (of about 1-μm-diameter) magnetic particles into the polymer. When a high-frequency field is applied, the polymer is heated and melted by induction heating of the particles. Particles are often limited to a bead of thermoplastic, placed at the future joint.

3. *Adhesive bonding* is the most versatile of all joining methods, suitable for thermoplastics, thermosets, dissimilar materials, polymer-metal and ceramic-metal combinations, and even for joining manufactured and natural structures (e.g., metal or ceramic crowns on teeth, or implants into bone). The cement may be a monomer (as for PMMA), an elastomer, or a thermoset. The optimum adhesive is chosen with regard to service requirements, and the method of application (brush, roller, screen printing, spraying, tape, etc.) is chosen according to manufacturing considerations. Surface preparation is critical.

4. *Solvent welding* gives a strong bond when a thermoplastic polymer (especially, an amorphous one) has a specific solvent as, for example, ABS, PVC, PMMA, polycarbonate, and polystyrene do.

5. *Cocuring* establishes the joint at the same time as when the plastic (or polymer-based composite) parts are cured.

The above methods, with some modifications, are also applicable to the joining of polymer-based composites.

18-10-2 Joining of Ceramics

An important and rapidly developing field is the joining of ceramics to themselves and to metals. Special problems arise from the structure, inertness, and very high melting point of many ceramics.

Carbides or, more accurately, sintered carbides are routinely brazed with Ag, Cu, or Ag–Ni filler metal, since the presence of the Co or Ni binder allows an essentially metal-to-metal joint to form.

Glasses are readily bonded to each other simply by heating to the softening temperature and then applying pressure, as in the solid-state welding of metals. Glass–metal seals are feasible because metal ions enter the glass network; the main concern is a matching of the thermal expansion coefficient (hence the use of Invar and Kovar, Table 4–1).

Ceramics can be bonded to metals or to each other after *metallizing*. For alumina, the *moly–manganese* (Mo–Mn) *process* is most widely used. A slurry of Mo, MoO_3, Mn, MnO_2, and glass-forming compounds is applied to the surface. Firing near 1500°C results in the formation of a glassy layer bonded to the glassy grain boundary phase in the alumina. After Ni plating, brazing with filler metals is possible. In *active brazing* the filler contains chemically reactive elements, chiefly Ti, and is applied directly to the ceramic. *Diffusion bonding* is also applicable to both metal–ceramic and ceramic–ceramic joints, sometimes with a metal interlayer. Many processes are highly specialized and beyond the scope of our discussions.

18-11   PROCESS CAPABILITIES AND DESIGN ASPECTS

As in all design, the potentials and limitations of joining processes must be taken into account together with service requirements. Limitations imposed by the chosen process will be sensed from an understanding of the processes themselves, and general guidance is given in Table 18–3. Mechanical joints were touched upon in Sec. 18-2 and remarks here are limited to fusion welds and brazed, soldered, and adhesive bonds which, as a group, are subject to similar conditions.

Loading Modes   In service, assemblies are usually exposed to complex loading patterns but, from the design and testing point of view, some basic loading modes can be recognized:

1. *Normal loading.* The joint is subjected to pure tensile or compressive loading (Fig. 18–27a).

2. *Shear loading.* The joint is subjected to pure shear. This is possible only in a double lap joint (Fig. 18–27b). Lap strength is the force sustained by a joint of unit width (N/mm). A single lap joint is subjected to a moment that, if large enough, causes deformation (Fig. 18–27c) which in turn leads to a more complex stress state, in particular, an increase in stress at the ends of the joint. Lap strength does not increase beyond a certain overlap length L and the strength of the assembly can be increased only by increasing the width of the lap joint.

3. *Cleavage.* The joint is subjected to balanced tensile forces that open up the joint perpendicular to its plane (Fig. 18–27d). Cleavage strength is the force required to separate a specimen of unit width (N/mm).

4. *Peeling.* The joint is separated by a tensile force that bends and progressively separates one of the adherends (the flexible adherend, Fig. 18–27e). Peel strength is the average force required to peel back a joint of unit width at the rate of 2.5 mm/s, with a peel angle of 180°. Brazed, soldered, and, in particular, adhesive joints are most vulnerable. For example, a rigid epoxy may have a lap shear strength of 21 MPa, but a peel strength of only 3 N per millimeter lap width.

Joint Design   The are few limitations on the size of assemblies, but there may be limits on material and thickness. For example, cross sections to be joined must be equal when heat generation is a function of cross section, as in butt welding and flash butt welding, whereas they may be quite dissimilar in projection

Table 18-3 General characteristics of welding processes

	Welding process							
	Arc welding							
Characteristics	Gas W	Gas Metal	Shielded Metal	Flux-Cored	Submerged	Oxyacetylene	Electron Beam	Laser Beam
Part (Assembly)								
Material	All but Zn	All but Zn	All but Zn	All steels	All steels	All but refractory metals	All but Zn	All but Zn
Preferred	All but Zn	Steels; Cu; non-HT Al	Steels	Low-C steels	Low-C steels	Cast iron; steels	All but Zn	All but Zn
Thickness, min. mm	0.2	0.5	(1.5) 3	1.5	5	0.6	0.05	0.05
Single pass, max.	5	5	8–10	3–6	40	10	75	25 (steel) 12 (Al)
Multiple pass max.	> 6	> 25	> 25	> 15	> 200	> 20		
Distortion*	B–C	B–C	A–B	A–C	A–B	B–D	C–E	C–E
Jigging needed	Variable	Variable	Minimum	Minimum	Full	Minimum	Full	Full
Deslagging for multipass	No	No	Yes	Yes	Yes	No		
Current								
Type	DCEN, ac for Al	DCEP	ac or dc	DCEP	ac or dc			
Volts	60–150	20–40 or 70	40 or 70	40–70	25–55		30–175 kV	
Amperes	100–500	70–700	30–800	30–800	300–2500		0.05–1	
Cost*								
Equipment	B–C	B–C	C–D	B–D	B–C	D–E	A	A–B
Labor	A–C	A–C	A	A–D	B–D	A	A–D	A–D
Finishing	B–E	B–D	A–B	A–C	A–C	A	C–E	C–E
Production								
Operator skill*	A–D	A–D	A	A–D	C–D	A	A–D	A–D
Welding rate, m/min	0.2–1.5	0.2–15	(1–6 kg/h)	0.02–1.5	0.1–5	(0.3–0.6 kg/h)	0.2–2.5	0.2–10
Operation	All	All	Manual	All	Automatic	Manual	Automatic	Automatic

*Comparative ratings, with A indicating the highest value of the variable, E the lowest. For example, shielded metal-arc welding results in fairly high distortion, has fairly low equipment cost, high labor and finishing costs, and requires high operator skill.

welding, electric arc welding, brazing, soldering, and adhesive bonding. The joint configuration is governed by several considerations:

1. *Butt joints* produce the smallest joint area between two parts and are thus the weakest joints in tension. Nevertheless, they are completely adequate when the joint itself is strong, as in solid-state and fusion welding. In fusion welding the

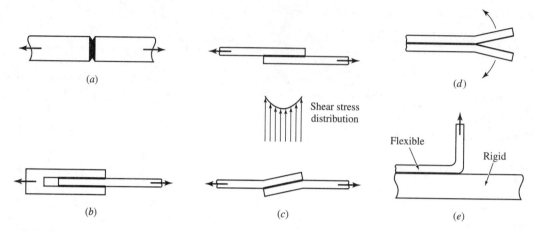

Figure 18–27   Basic loading modes on joints: (a) pure tension, (b) pure shear, (c) unbalanced shear and resultant deformation, (d) cleavage, and (e) peeling.

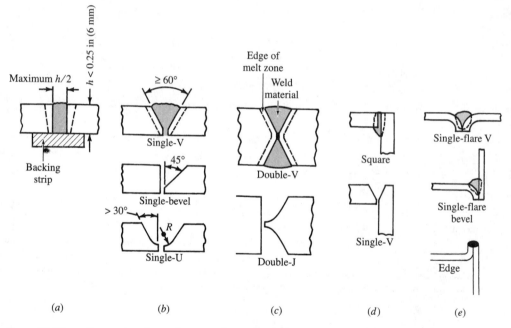

Figure 18–28   Preparation of joints for arc and gas welding by providing: (a) square, (b) single-V, (c) double-V, (d) corner, or (e) edge grooves.

soundness of the weld is ensured by appropriate preparation (Fig. 18–28) which vitally affects the fatigue performance of the joint.

    a. Square grooves are adequate for thinner stock (Fig. 18–28a) but grooves shaped for greater weld penetration and controlled bead formation are essential for gas and arc welding of thicker gages (Fig. 18–28*b* and *c*). No groove is needed (square preparation) for electron beam welding.

b. Corner joints with square grooves (essentially, a butt joint) are suitable only for thinner material; for thicker gages, groove preparation is essential (Fig. 18–28*d*). Cleavage fracture may occur in brazed, soldered, and adhesive joints and reinforcement is then necessary.

2. Butt welding may not give sufficient strength in thin sheet. Several solutions are available:

a. *Flared joints* give larger surface area (Fig. 18–28*e*). In fusion welding the weld bead provides the strength; in other processes, including brazing and adhesive joining, the joint area is increased by bending the sheet. However, such joints are susceptible to failure by peeling separation.

b. A better joint is usually obtained in nonfusion processes by creating a large contact area loaded in shear (*lap joint*, Fig. 18–29*a*) Better stress distribution in the joint area is achieved by offsetting the lap (Fig. 18–29*b*) or tapering the sheets (Fig. 18–29*c*). Lap joints are strengthened with *straps* (Fig. 18–29*d*). If a smooth surface is required, the edges must be prepared (Fig. 18–29*e*) or a *scarf joint* made (Fig. 18–29*f*). A tongue-and-groove configuration converts butt joints into lap joints of greater strength (Fig. 18–29*g*). In corner or *T* joints, the contact surface may be increased by fillers and covers (Fig. 18–29*h*).

c. For highest strength, a mechanical (riveted or lock-seam) joint is brazed, soldered, or adhesively bonded. In the process of *weld bonding*, sheets coated with an adhesive are spot-welded through the adhesive (Fig. 18–29*i*), and then the adhesive is cured. Alternatively, a spot-welded joint is infiltrated with the

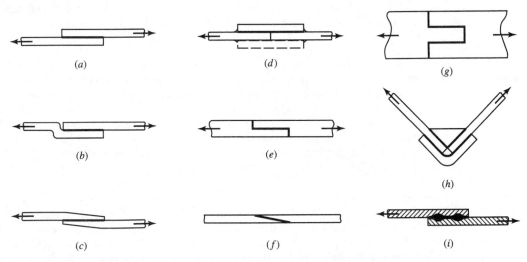

(a)   *(d)*   *(g)*

(b)   *(e)*

(h)

(c)   *(f)*   *(i)*

Figure 18–29   Brazed, soldered, and adhesive joints are designed to give maximum adhesive joint area: (*a*) simple lap joint, (*b*) offset lap joint, (*c*) tapered lap joint, (*d*) reinforced lap joint, (*e*) smooth-surface lap joint, (*f*) scarf joint, (*g*) tongue-and-groove butt joint, (*h*) corner joint with inserts, and (*i*) weld-bond joint.

liquid adhesive which is then cured. The spot weld increases shear strength and the adhesive increases fatigue strength.

A critical aspect of joining processes is quality assurance. Process conditions must be strictly monitored and documented. Periodic destructive testing guarantees that the process is under control. The strength of joints is tested in tension and that of lap joints also in peel tests. Additionally, NDT techniques are used to check the soundness of joints. Ultrasonic, eddy current, and x-ray techniques find general application. In processes where cracks may form, surfaces are inspected by dye-penetrant and magnetic flaw detection. Many products see critical service (e.g., nuclear vessels, ships, boilers, etc.) and then special codes and inspection techniques apply.

18-12   LAMINATES

Laminates represent a special class of products. As the name implies, two or more *laminae* (thin sheets or plates) are joined over their entire interface. The product is a composite, but not in the sense defined in Chap. 15, because there is no continuous matrix in which the other phase is enclosed. Laminates differ also from coatings in that they are made by joining solid laminae to each other. One lamina may be much thinner, and then it is usual to speak of clad products.

We have already encountered methods for making laminates, such as calendering for plastic-coated paper and fabric, coextrusion of plastic film (Sec. 14-3-2), fiber reinforcement (Sec. 15-1-2), and diffusion, explosion, and roll welding for bimetallic strip and metal cladding (Sec. 18-3). Brazing, soldering, and adhesive bonding are also employed in a vast variety of products.

Example 18-10 | **W**indshields of automobiles must protect the occupants in a collision from a disintegrating windshield yet must not be so strong as to injure—possibly fatally—an occupant hurled against them. Therefore, they are made as laminates: two pieces of pre-curved annealed glass are joined with a central layer of polyvinyl butyral thermoplastic. Upon impact, the glass breaks into small fragments, but most of the shards are retained by the plastic which is capable of taking large deformation. A further plastic layer may be added to the inner surface to prevent laceration. (Bulletproof glass is a laminate of tempered glass layers.)

Example 18-11 | **A** laminate consisting of a plastic sheet sandwiched between two adhesively bonded, thin metal (steel or aluminum) sheets has several advantages. The metal cover sheets make the sandwich more creep resistant and much stronger in bending than a plastic alone; the plastic makes it much lighter than a solid metal sheet; formability is comparable to a ductile metal; and the sandwich has excellent sound-deadening properties. We saw in Example 10-16 that the ULSAB design uses high-strength steels for much of the structure. The deep spare tire well in the trunk cannot be formed in such a steel. The designers chose an insert, deep drawn from a steel/PP laminate. Since this cannot be welded, it was adhesively bonded into a cutout in the trunk.

Honeycombs form a special class of laminates. They are exceptionally light for their strength: The cover plates (*face plates*) provide tensile and compressive strength and the *core* gives resistance to shear and through-thickness compression (Fig. 18–30a). As the name suggests, the basic core configuration is hexagonal (Fig. 18–30b), but other patterns are also used; in particular, a *flex core* (Fig. 18–30c) allows shaping into a three-dimensional form. The core is made by applying an adhesive to plastic or metal foil in a pattern to build up a block (Fig. 18–31a). After curing, the block is expanded (Fig. 18–31b). A metal core is ready at this point but a polymer core would not hold its shape and is dipped into thermosets which, upon curing, fix the shape. The face sheets are then joined with adhesives.

For high-temperature application of metal honeycombs, the adhesive joints are replaced by brazed ones. Metal cores can also be made from strips corrugated between rolls resembling gears; the core is built up by brazing or welding.

Honeycombs have many structural applications, for helicopter rotor blades, control surfaces of aircraft, and all applications where high bending strength to mass ratio is desired. In addition, cores are placed in ducts to straighten air flow, are used for radio-frequency shielding, or are filled with fiberglass batting to provide exceptional sound insulation. Honeycomb structures offer efficient energy absorption.

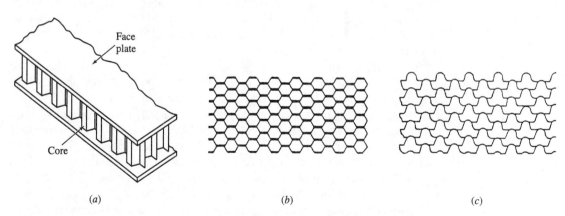

(a)   (b)   (c)

Figure 18–30   Honeycombs are lightweight yet strong structures (a); among core configurations, the (b) hexagonal and (c) flex-core are frequently used. (*Adapted from Engineered Materials Handbook, Desk Edition, ASM International, 1995, pp. 546 and 547. With permission.*)

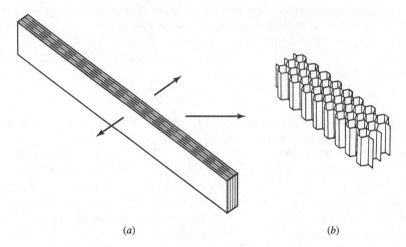

(a)   (b)

Figure 18–31   Fabrication of honeycomb core: (a) assembly of block and (b) expansion of core. (As Fig. 18–30.)

18-13   SOLID FREEFORM FABRICATION (SFF)

In all our discussions to this point, shapes were developed by solidification, deformation, or consolidation into a die, or by removing material from a previously produced shape. The alternative is to build up a shape incrementally, free of the confines of a die. Since the mid-1980s, there have been many processes developed that employ the principle and there will no doubt be many others yet to come. Terminology is not fully established, but the term *solid freeform fabrication* seems to be gaining general acceptance. It is useful because it expresses the main physical characteristic of the processes without reference to the intent or ultimate use.

The enabling technology is solid modeling: The solid model created in the CAD file is "sliced" in the computer into thin 2-D layers and the physical part is then built—layer by layer—from various materials by a variety of techniques. The thickness of layers is typically 0.1–0.5 mm, but can be as much as 2.5 mm. In a sense, the process is an automated version of the activity of the pattern maker who often built the pattern from thin layers of wood, glued together; smoothed by sanding, the 3-D body was suitable for inspection and could eventually be molded into sand. Similarly, a part built from 2-D layers will show steps. Thinner layers need less smoothing but take longer to make. As we shall see, an alternative is to build a body by laying down coils of material in a helical, spiral, or other pattern. This too has its precedent in the coil forming process used for ceramic products in antiquity and occasionally even today. The whole field is still in a flux but several trends are already discernible.

18-13-1   Purposes of Freeform Fabrication

The wider adoption of concurrent engineering gave the impetus for freeform fabrication: if time to market is to be reduced and costly mistakes avoided, the soundness of design must be proved at an early stage, with the full use of opportunities offered by CAD/CAM.

Rapid Prototyping   The first application of freeform fabrication was to building physical prototypes for the classical reasons: They give a better appreciation of shape and esthetic appeal. If dimensions and tolerances can be brought within the range of actual products, fits and assembly can also be checked. For these purposes, the prototype can be built of any material of dimensional stability such as a plastic or even paper. The process is a powerful aid also in reverse engineering; the necessary computer file is already available from measurements made by CMMs. Somewhat ironically, developments in solid modeling have cut into the predicted growth rate for these applications.

Functional Models   The next step is to build working models. If the purpose is only to check kinematic functions, the material can again be a plastic. Since layer-by-layer fabrication allows parts within parts (a sphere in a cage, or gears in a frame), working models can be built. If, in addition, performance is to be checked too, the model must be made of the real material, very often a metal or ceramic, and this presents additional challenges.

Prototype Tooling   Tools and dies for casting, metal deformation, and ceramic and plastic molding are expensive and may take several months to make. We saw in Sec. 10-7 that castable zinc alloys are used for die tryout in the sheet-metalworking industries, but even these take months to make, try out, and correct. Freeform fabrication can reduce lead times by various routes. A wax model can be directly used as the pattern for investment casting (Sec. 7-5-5). If more patterns are needed, the model is made of a plastic; from this, a room-temperature-curing silicone rubber mold is made into which several wax patterns can be molded. The model can be made conductive and then a nickel shell mold, suitable for RTM or SRIM, is created by electrodeposition. If the model has higher temperature capability, a low-melting Al–Zn alloy can be sprayed on it. The computer is capable of generating the negative of the part and the SFF can directly produce a mold.

Automated Fabrication   The term is sometimes used as an inclusive term for the above freeform fabrication purposes but, in the most general sense, it is applied to all processes that require no human intervention for the manufacture of parts once the computer program is created. This includes, of course, both lights-out machining (*subtractive processes*) and SFF (*additive processes*). The two can be in competition even for rapid prototyping: With the continued development of CNC machining, prototypes can often be made—especially in free-machining Al alloys—competitively with SFF processes. The term *desktop manufacturing* is sometimes applied to SFF processes when, in reality, the only machines that fit on a desktop are CNC mills specially built for prototype production.

18-13-2 SFF Processes

As indicated, most processes build a part layer by layer. The differences are in the often proprietary approaches to this goal.

1. Photopolymer fabrication. Specially formulated liquid polymers have been developed that can be cross-linked to a controlled depth under the influence of UV (or, sometimes, visible) light. The polymer is in a vat that has a vertically movable platform (Fig. 18–32). To begin with, the platform is below the surface of the liquid by one slice thickness and the first slice is solidified by controlled exposure to light. Processes differ in the way a light is applied.

a. The light is generated by a laser and the beam is scanned across the surface. The platform is then dropped and the process is repeated until all slices are made. Overhanging parts need support (Fig. 18–32).

b. The polymer layer is exposed to UV light through a glass mask on which the pattern is made by copier-like techniques. The nonexposed liquid is vacuumed off and replaced with wax, the layer is milled flat, and the platform is lowered for the repetition of the sequence. The wax provides support and, upon completion of the part, is removed by melting or dissolution.

2. Selective laser sintering. In principle, any powder that can be sintered may be used. A thin layer of the powder is spread and sintered by the passage of a laser beam, the platform lowered, a new layer spread and sintered, etc. The nonsintered powder provides support for the structure and falls away when the part is removed.

a. When the powder is investment-casting wax, the part thus formed becomes a pattern for investment casting.

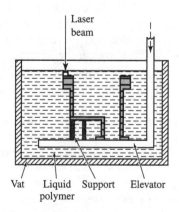

Figure 18–32   Rapid prototyping by photopolymerization. Note the supports needed for overhanging parts.

b. Thermoplastic polymers or B-stage thermosets can be made into proto-types or molds for wax patterns. If made with hollow walls, they can serve as patterns for investment casting.

c. Metal powders, coated with a polymer, can be sintered to establish polymer bonds. The binder is removed and the part sintered as in metal injection molding (Sec. 11-3-2).

d. Polymer-coated ceramic powders may be similarly treated to produce ceramic parts or ceramic molds for investment casting.

e. With higher energy intensities, metal powders can be directly sintered but the part may need final sintering or HIP to develop acceptable mechanical properties.

3. Three-dimensional printing. Metal or ceramic powder is deposited in layers and an ink jet sprays a liquid binder in the pattern defined by the computer. Further treatment is as in powder injection molding.

4. Liquid-state deposition. The material is heated into the melt regime and deposited in a pattern by various techniques.

a. A filament of thermoplastic polymer or wax is heated and extruded through a nozzle, the movement of which is controlled in x-y coordinates. The table drops when a layer is completed. Rapid solidification gives sufficient strength to obviate the need for outside support.

b. In Sec. 11-5 we referred to direct spray deposition of atomized metal. By spraying through masks cut out by laser, a shaped part can be built up in layers. In a variant, a laser beam is directed through the center of a powder-delivery nozzle; thus, the part is built directly from fused layers.

c. When the platform (or the nozzle) is given a programmed z-axis movement, 3-D shapes can be built up directly by the coil-building method. The part may also be built as a continuous weld bead. Surface quality will have to improve before these processes can be used for functional parts.

5. Laminated object manufacturing. The term is used for a process in which polymer-coated paper (butcher paper) is fused in layers. Into each layer, the contour of a slice of the solid model is cut by laser, and surrounding areas and internal cavities are cut into squares. On completion of the part, the small cubes can be peeled away. The part can also serve as an investment-casting core. This is, in the strict sense, not SFF but a special case of subtractive manufacturing.

6. Laminated fabrication. The principle of building a part from several layers can be taken further. The solid model is divided in the computer into relatively thick slices that can be rapidly made by relatively simple 3-axis CNC machining. The parts are then united by some joining process.

The prototype of a five-valve cylinder head was made by vacuum brazing six slabs machined by three-axis milling. This process, named *precision stratiform machining*, allowed completion of a working prototype in 3 months, about one-third the time it would have taken by conventional methods. [Source: *Adv. Mater. Proc.*, 1995(1):16.]

Example 18-12

18-14 SUMMARY

Joining expands the scope of all manufacturing processes; castings, forgings, extrusions, plates, sheet metal, and machined parts can all be joined to make more complex shapes or larger structures. Welded constructional girders, machine frames, automobile bodies, tubing and piping of all sizes, containers, and cans are all around us. Welding is often the most economical and practical repair for broken machinery. Brazed bicycle frames, heat exchangers, and soldered radiators and plumbing joints abound. Adhesive joining is increasingly used in the aircraft, automotive, appliance, and other industries.

Joints, by their very definition, provide a transition between two not necessarily similar materials. Quality control is even more important than in other processes because oxidation, surface films, slag inclusions, porosity, gaps, undercuts, hot cracks, cold cracks (embrittlement), and residual stresses could cause dangerous delayed failures. Nevertheless, good-quality joints, sometimes equal to the parent (workpiece) material in strength, can be obtained through a variety of means:

1. Solid-phase welding relies entirely on adhesion; while it is extremely sensitive to surface contaminants, it does allow joining of a very wide range of similar and dissimilar materials.

2. Highly localized melting in resistance welding represents a transition from solid- to liquid-phase processes; it is still sensitive to contaminants but the heat-affected zone is small.

3. Deep or through-the-thickness melting in fusion welding processes broadens the heat-affected zone in the parent metal in all but the high-energy-beam (EB and laser) processes. Surface preparation, weld geometry, protective atmospheres, and/or slag and fluxes, and the rate of heating and cooling must be simultaneously controlled. Localized heating and cooling makes welded structures susceptible to distortion and cracking under the influence of internal stresses.

4. In brazing and soldering, solid workpieces are joined—without melting the parent metal—with a lower-melting metal. The joint again relies on adhesion, making surface cleanliness and fit the critical factors.

5. Polymers can be formulated to suit the particular task in adhesive joining. Control of surface preparation and adhesive application are crucial to success.

6. Joining processes are additive: they build up a body from smaller units. This principle can be applied to making parts by freeform fabrication. These fabrication processes have made rapid prototyping possible; further developments have reduced the lead time necessary for the construction of molds and dies. In the ultimate development, the techniques may lead to small-batch production of end-use items.

7. Joining processes offer ample scope for mechanization, automation, and the application of robotics. Mathematical modeling has made great strides and artificial intelligence is applied.

8. Many joining techniques present considerable hazards. Light emitted by high-temperature arcs and laser beams call for eye protection or total enclosure; heat, molten metal, and sparks flying from a weld zone require face shields, goggles, gloves, and other protective clothing; irritating or toxic fumes emanating from solvents and monomers and from welding, brazing, soldering, and adhesive bonding operations necessitate exhausts and other safety and health measures.

PROBLEMS 18A

18A-1   (*a*) Make a sketch of roll bonding. (*b*) List (or mark on sketch) three essential conditions for achieving a good bond.

18A-2   (*a*) Make a sketch showing friction welding of two bars on their end faces. (*b*) List (or mark on sketch) the essential conditions for obtaining a good joint.

18A-3   (*a*) Draw a sketch showing the structure of a butt fusion weld between two copper plates cold rolled to 50% reduction. (*b*) In a sketch underneath, indicate the relative strength and ductility of different zones.

18A-4   (*a*) Make a sketch of the cross section of a fusion-welded joint between a low-carbon steel and a normalized 0.8% C steel plate. (*b*) Assuming no preheat and rapid cooling, show the characteristic features of the microstructure, identifying the various phases to be expected.

18A-5   Define (*a*) stud welding and (*b*) percussion welding (use sketches if desired).

18A-6   State what is the main energy transfer medium in (*a*) plasma arc and (*b*) electroslag welding.

18A-7   Make simple sketches to show the main features of (*a*) resistance seam and (*b*) mash seam welding (including the energy source).

18A-8   Make sketches showing the essential elements of (*a*) gas tungsten-arc, (*b*) gas metal-arc, and (*c*) shielded metal-arc welding.

18A-9   Two low-carbon steel plates of 10-mm thickness are butt-joined by fusion welding. Make sketches to show the expected weld bead geometry and the width of the HAZ with (*a*) oxyacetylene, (*b*) electron beam, (*c*) shielded metal arc, and (*d*) plasma arc welding.

18A-10 State the common characteristics of (*a*) brazing, (*b*) soldering, and (*c*) braze welding, and define the major difference between the three.

18A-11 Draw a sketch to show the basic design for preparing the edges of two 20-mm-thick plates to be butt-joined by shielded metal arc welding with (*a*) one-sided and (*b*) two-sided access. (*c*) Which of the two will be free of distortion?

18A-12 Draw sketches to show at least two joint designs that increase the strength of an adhesive lap joint.

PROBLEMS 18B

18B-1   Your company produces copper-clad steel sheet by hot roll bonding. A customer complains that the cladding separated during deep drawing. You are asked to investigate your company's procedure to make the sheet and identify possible problems. List the likely process steps and problem areas.

18B-2   It is claimed that elevated temperature is essential for creating a metallic bond in rolling a sandwich of base metal plate and two cover plates. Do you agree? Justify.

18B-3   A design requires that two austenitic (304) stainless steel sheets of 0.6-mm thickness be butt-joined at their edges in a straight line. (*a*) List the processes that could be considered. (*b*) List additional processes that would become feasible if the sheets were allowed to overlap.

18B-4   (*a*) Make a sketch showing the shape of a good weld bead produced by gas metal-arc welding two 20-mm-thick steel plates. (*b*) Suggest inspection methods for ascertaining the quality of weld.

18B-5   A HSS twist drill is to be joined to a low-alloy steel bar. Suggest processes that are applicable if the HSS drill steel is already fully heat-treated and must not soften at the cutting edges.

18B-6   Explain why an off-eutectic solder is often preferred to a eutectic one.

18B-7   When the bodies of three-piece steel cans are seam-welded, the welding current is switched off just before the end of the seam is reached. Suggest a reason, keeping in mind the subsequent operations.

18B-8   Many bicycle frames are constructed of steel tubing joined to hollow fittings. (*a*) Suggest a way of joining, (*b*) define the process, (*c*) make a sketch of a joint indicating critical features.

18B-9   Three-piece cans are made of two lids and a body (the cylindrical part, rolled up of a sheet and joined on the cylindrical surface). (*a*) Make sketches of processes you can think of for making the body, then make judgments on their applicability to (*b*) tin plate, (*c*) black plate (thin steel

sheet, not coated), and (*d*) aluminum alloy.

18B-10   Steel rings of 1.2-m OD, of the configuration shown in Fig. Ex. 7-9*a*, are made by bending an angle section on pyramidal rolls (Fig. 10–15*b*). Make judgments whether the following processes are feasible for establishing a joint equal in strength to the parent metal: (*a*) forge welding, (*b*) friction welding, (*c*) flash butt welding, (*d*) brazing, (*e*) shielded metal-arc welding, (*f*) soldering, and (*g*) gas metal-arc welding. Briefly justify each judgment.

18B-11   Modern railroad tracks are made by welding rolled rails into long lengths. (*a*) Suggest a method for making joints in the field. (*b*) In hot weather the rails expand. From basic considerations, suggest a way of minimizing the consequences to prevent buckling.

18B-12   Inspect as many pressure vessels of two-piece construction as you can identify (propane gas cylinder for barbecue, oxygen cylinder for scuba-diving equipment, disposable cylinder for paint-stripping torch, etc.) for signs of methods of manufacture, including the methods for making the half-vessels and the method of joining.

PROBLEMS 18C

18C-1   Parts of the fuselage of an aeroplane are joined by 2024 aluminum alloy rivets. The operator of a riveting machine reports that many rivets crack; the riveting force also seems to be inadequate as judged by the many incompletely formed rivet heads. The rivets are in the solution-treated condition, to attain their full strength by natural (room-temperature) aging after riveting. (*a*) Determine what could have gone wrong and (*b*) suggest, step-by-step, what remedial action should be taken.

18C-2   A small, complex aircraft part is to be manufactured of 6061 aluminum alloy.

The highest strength obtainable with this alloy must be maintained in the entire part. (*a*) Check (in Table 8–2 or, preferably, in reference works such as *Metals Handbook*) what metallurgical condition will give the highest strength. (*b*) Determine whether two parts made of material in this condition could be welded without much loss of properties. (*c*) If a joining process can be found, specify the best welding process (in justifying your choice, work by the process of elimination). (*d*) Suggest a suitable filler, if used, and (*e*) describe what postwelding treatment, if any, is needed.

18C-3   It is proposed to join aluminum to low-carbon steel. After reviewing the equilibrium diagram, consider the feasibility of establishing a reasonably ductile joint by: (*a*) cold-welding, (*b*) diffusion bonding, (*c*) explosion welding, (*d*) spot welding, (*e*) friction welding, and (f) flash butt welding. Justify your judgments.

18C-4   A single-phase resistance-welding machine is used to join two steel sheet parts with 20 projection welds made simultaneously. One assembly is joined every 10 s. If the steel sheets are 1-mm thick and individual projection welds are similar to the one shown in Fig. 18–15*b*, calculate (*a*) the welding current required; (*b*) the duty cycle, expressed as the fraction (or percentage) of time the current is on; (*c*) the required kVA for each welding cycle, if the desired current is attained at a voltage of 4 V; (*d*) the kVA rating of the transformer, taking into account the duty cycle; (*e*) the kW rating, allowing for a power factor of 0.5 (for phase shift due to inductive loading in the transformer); (*f*) the electrical energy consumption for each weld in kWh; and (*g*) the force requirement.

18C-5   5052 Al alloy plates of 3.2-mm thickness are butt-joined by DCEP GMAW using 1-mm-diameter electrode wire. Current is 120 A, arc voltage 18 V, and arc travel speed 14 mm/s. Average width of joint is 2 mm. Calculate the (*a*) power (kW), (*b*) deposition rate (mm^3/s), and (*c*) electrode feed rate (mm/s).

18C-6   The design of the floor pan of an automobile calls for joining two 5052 Al-alloy sheets of 1-mm thickness by spot welding. (*a*) Review the appropriate phase diagram and deduce if the operation is feasible. (*b*) If the answer to (*a*) is yes, calculate the transformer rating and power consumption as in Prob. 10C-4, but with one spot weld (8 cycles at 30 000 A per electrode) made every 2 s. (*c*) Suggest an appropriate welding process if the design is changed to butt welding at the edges.

18C-7   Two carbon steel plates of 0.25-in (6.35-mm) thickness are to be joined by oxyacetylene welding in a single-V groove (Fig. 18–28*b*). Using data from Table 18–3, determine (*a*) whether this is feasible, and, if the answer is yes, (*b*) what welding rate (mm/min or in/min) one might expect at the top metal deposition rate of 0.6 kg/h.

18C-8   A steel tube of 25-mm OD is to be adhesively joined to another steel tube of 25-mm ID with a general-purpose epoxy. The assembly will operate at room temperature (15–30°C) and an axial tensile load of 12 kN will have to be supported with a factor of safety of 2. Calculate the minimum length L of the lap joint (i.e., the depth of interpenetration of the tubes).

18C-9   A cold-rolled pure copper bar and an annealed pure nickel bar of 10-mm thickness and 20-mm width are to be butt-joined together without a filler metal. The joint must be ductile and free from inclusions. (*a*) Review the relevant phase diagram to determine whether this is possible. (*b*) If the answer is yes, use Fig. 18–1 to suggest at least two suitable processes, using the process of elimination. (c) Make a sketch of the structure of the resulting joint, indicating grain shape, grain size, and composition. (d)) Indicate, in relative terms, the width of the HAZ for the two processes.

18C-10 Welding transformers used for spot welding are usually rated in kVA for a 50% duty cycle. This means the transformer is designed to withstand the heat generated when it is switched on for 50% of the base time cycle (usually 1 min). Since heat generated is proportional to time, Eq. (18-1), higher kVA can be drawn if the duty cycle is shorter. Calculate the permissible loading for duty cycles of 1, 5, 10, 20, 30, 40, 70, and 100%.

18C-11 In Sec. 18-4-3 it is stated that gray iron is welded with a silicon-rich filler. Explain the reasons for this, drawing on information from Chapter 6.

FURTHER READING

ASM Handbook, vol. 6, *Welding, Brazing, and Soldering*, ASM International, 1993.

Engineered Materials Handbook, vol. 3, *Adhesives and Sealants*, ASM International, 1990.

Brazing Handbook, 4th ed., American Welding Society, 1991.

Welding Handbook, 8th ed. (4 vols.), American Welding Society, 1987–1998.

Handbook of Plastics Joining: A Practical Guide, Plastics Design Library, 1996.

Bickford, J.H., and S. Nassar, *Handbook of Bolts and Bolted Joints*, Dekker, 1998.

Blodgett, O.W.: *Design of Welded Structures*, James F. Lincoln Arc Welding Foundation, Cleveland, OH, 1982.

Blodgett, O.W.: *Design of Weldments*, James F. Lincoln Arc Welding Foundation, Cleveland, OH, 1982.

Cary, H.B.: *Modern Welding Technology*, 4th ed., Prentice Hall, 1998.

Davies, A.C.: *The Science and Practice of Welding*, 10th ed., Cambridge University Press, 1993.

Dawes, C.T.: *Laser Welding*, Woodhead, Cambridge, 1992.

Evans, G.M., and N. Bailey: *Metallurgy of Basic Weld Metal*, Woodhead, 1997.

Grong, O.: *Metallurgical Modeling of Welding*, The Institute of Materials, London, 1994.

Halmshaw, R. (ed.): *Introduction to the Nondestructive Testing of Welded Joints*, Woodhead, 1997.

Hicks, J.G.: *Welded Joint Design*, 2d ed., Abington, 1997.

Humpston, J., and D.M. Jacobson: *Principles of Soldering and Brazing*, ASM, 1993.

Karlsson, L. (ed.): *Modeling in Welding, Hot Powder Forming and Casting*, ASM International, 1997.

Lancaster, J.: *Handbook of Structural Welding*, Woodhead, 1992.

Lancaster, J.F.: *The Metallurgy of Welding*, Allen and Unwin, 1986.

Linnert, J.E.: *Welding Metallurgy*, 4th ed., vol. 1, American Welding Society, 1994.

Lucas, W. (ed.): *Process Pipe and Tube Welding*, Woodhead, 1991.

Manko, H.H.: *Solders and Soldering Materials, Design, Production and Analyses for Reliable Bonding*, McGraw-Hill, 1992.

Marshall, P.W.: *Design of Welded Tubular Connections*, Elsevier, 1992.

Messler, R.W., Jr.: *Joining of Advanced Materials*, Butterworth-Heinemann, 1993.

Olson, D.L., R. Dixon, and A.L. Liby, (eds.): *Welding Theory and Practice*, North Holland, 1990.

Parmley, R.O. (ed.): *Standard Handbook of Fastening and Joining*, 3d ed., McGraw-Hill, 1997.

Pecht, M.G.: *Soldering Processes and Equipment*, Wiley, 1993.

Radaj, D.: *Design and Analysis of Fatigue Resistant Welded Structures*, Halstead Press, 1990.

Rahn, A.: *The Basics of Soldering*, Wiley, 1993.

Schulz, H.: *Electron Beam Welding*, Woodhead, 1994.

Schwartz, M.M.: *Ceramic Joining*, ASM International, 1990.

Schwartz, M.M.: *Joining of Composite Matrix Materials*, ASM International, 1994.

Swenson, L.-E.: *Control of Microstructures and Properties in Steel Arc Welds*, CRC Press, 1994.

Tres, P.A.: *Designing Plastic Parts for Assembly*, 3rd ed., Hanser-Gardner, 1998.

Walker, J.M. (ed.): *Mechanical Joining*, 1996.

Wick, C., and R. Veilleux (eds.), *Tool and Manufacturing Engineers Handbook*, vol. 4, *Quality Control and Assembly*, Society of Manufacturing Engineers, 1987.

Solid Freeform Fabrication

Burns, M.: *Automated Fabrication*, PTR Prentice Hall, 1993.

Jacobs, P.F.: *Rapid Prototyping and Manufacturing*, SME, 1992.

Jacobs, P.F.: *Stereolithography and Other RP&M Technologies*, ASME, 1995.

Wood, L.: *Rapid Prototyping: An Introduction*, Industrial Press, 1993.

Biocompatible coatings (GPX^{TM} Bio Coatings) are applied to enhance fixation of prosthetic implants to bone. (*Courtesy Engelhard Corporation, East Windsor, Connecticut.*)

19

Surface Treatments

Many manufactured products are coated with layers designed to impart
corrosion or heat resistance or to satisfy esthetic requirements. We shall
explore the principal manufacturing methods, including:

Removal of unwanted surface films

Conversion of surface layers by chemical reaction

Deposition of metallic layers from the melt or from solution

Creation of metal, ceramic, or organic coatings by vapor deposition

Deposition of coatings by high-temperature chemical reaction

Changing surface composition by ion implantation

Applying organic coatings

Surface treatments aim at creating a composite structure by changing only the properties of the surface. Thus the products differ from composites in which bulk properties are affected (Chap. 15) and from laminates (Sec. 18-12) which are made by uniting two separate entities. In previous chapters we already saw numerous examples of surface treatments; here we examine them systematically.

19-1 CLASSIFICATION

There is a vast and ever growing variety of surface treatments, and we shall have to limit our discussion to the most essential features of them. As shown in Fig. 19–1, some processes are subtractive, others are additive, and yet others convert the surface by reactions involving the base material. Some processes (such as painting) are used to treat a finished product, others (such as phosphating) are an essential

Surface treatments

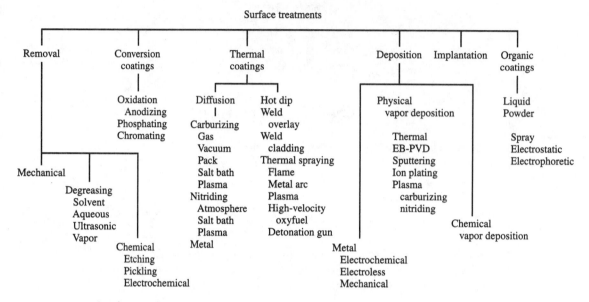

Figure 19–1   Classification of surface treatments.

part of a production sequence, yet others (such as CVD) are production processes in their own right and will play a significant role in the manufacture of semiconductor devices (Chap. 20). Coatings may serve a large variety of purposes. For processing, they provide a controlled surface topography or composition for lubricant retention, for service they improve wear or corrosion resistance, or they may be applied for purely esthetic reasons. Treatments designed to change surface residual stresses are sometimes included among surface treatments; we already dealt with shot peening in Sec. 9-2-2, roller burnishing in Sec. 9-7-6, tempering of glass in Sec. 12-5-2, and surface integrity of machined parts in Secs. 16-1-9 and 16-8-4. Therefore, they will not be treated here further; neither will be the ceramic and glassy coatings (enamels) discussed in Sec. 12-5-3.

19-2   REMOVAL PROCESSES

The purpose is usually that of removing unwanted surface layers or changing the topography of the surface.

Mechanical   Coarse or strongly adherent contaminants are removed by mechanical means, including bonded-abrasive (Sec. 16-8-5) and loose-abrasive processes (Sec. 16-8-6), particularly, grit blasting, power brushing, and barrel finishing.

Degreasing   Organic fluids (oils) are extensively used in many phases of manufacturing, including deformation processing and machining, but may also be applied to metals to provide corrosion protection in storage and shipment. The surface may also pick up oil, dirt, and other organic and inorganic contaminants (*soil*). Before final joining, assembly, or coating operations, all residues must be removed by solvent, aqueous, or vapor processes referred to in Sec. 8-2-4. Applying the degreasing fluid in a *high-pressure jet* helps in removing heavy soil. A higher level of cleanliness is assured when the fluid is agitated with an *ultrasonic* transducer: The vibrations set up *cavitation* (the formation of tiny bubbles). When the bubbles hit the surface of the part, they collapse into a high-speed (up to 400-km/h) jet and exert a powerful mechanical cleaning action. The *water-break test* offers a quick check on the efficiency of degreasing: When water flowed on the surface remains in a thin film, the surface is clean; when it gathers into pools, oily residues are present.

Chemical   We saw in Sec. 17-2 that aqueous solutions of chemicals can be used for material removal. The chemical may be chosen to remove only a surface layer, usually an oxide; it can be an acid (as in the *pickling* of steel) or an alkaline solution (as in stripping the oxide of aluminum). An inhibitor may be necessary to prevent attack on the substrate metal. Thorough degreasing is often necessary to ensure uniform attack (a grease patch can act as a maskant). After etching, the part must be rinsed to prevent further attack. Just as chemical machining is accelerated by the application of an electric current (Sec. 17-3), so cleaning can also be accelerated by *electrolytic cleaning*, with the part serving as the cathode, anode, or at alternating polarity, depending on material and soil to be removed.

19-3   CONVERSION COATINGS

In conversion coating, the surface layer of metals is transformed through a chemical reaction.

Oxidation   Many metals oxidize in service, and uncontrolled oxidation can lead to undesirable changes in surface appearance or in localized or overall corrosion. Chemical or elevated-temperature treatments can, however, be found that result in desirable, corrosion-resistant forms of oxidation. Thus, the blackened surface produced by chemicals on steel resists rusting, and copper and brass can be treated to various shades of black, brown, blue, and green. A hard oxide of controlled thickness and color is built up on aluminum (and also on Mg and Ti) by *anodizing*. After stripping the natural oxide, the part is immersed into an aqueous electrolyte and connected to the positive terminal of a dc power supply (and thus becomes the anode). An oxide film, structurally bonded to the substrate, is built up. A porous oxide can be infiltrated with dyes for decorative color or color may be produced during deposition. Immersion in boiling water converts the alumina to monohydroxide and seals the surface. Thicker (25–100 μm), more dense coatings made in *hard anodizing* are wear-resistant, and PTFE can be included for reduced friction and increased wear resistance.

Phosphate Coating   In Sec. 8-2-4 we saw that steel surfaces are phosphate-coated to retain the lubricant for heavy deformation. Immersion in a bath of Zn phosphate and dilute phosphoric acid causes the growth of a crystalline zinc phosphate layer in which interconnected channels help to entrap the lubricant; if applied in a thinner layer, the coating serves as a paint base. Similar treatments are available for other metals, including aluminum and galvanized steel.

Chromate Coating   A very thin (under 2.5 μm), clear or colored coating is formed by immersion in a bath of dilute chromic acid and other chemicals. Some food cans have, for example, such a coating as a base layer on the steel surface.

19-4   THERMAL TREATMENTS

Some thermal treatments change the properties of a surface layer; others apply a coating by solidification.

19-4-1   Surface Heat Treatment

The prime application is to steel.

Surface Hardening   As shown in Sec. 6-4-4, a surface layer can be hardened if the steel has sufficient carbon. For this, the surface must be heated into the austenitic temperature range and rapidly quenched to form martensite. Several heating methods are applicable:

1. *Flame hardening* relies on the heat of an oxyfuel torch; the surface is then water quenched. It can be manual or mechanized, for local or overall hardening, in both small and medium-to-high production situations.

2. In *induction hardening*, a water-cooled copper coil surrounds the part. High-frequency current supplied to the coil heats the surface to a controlled depth, followed by water quench.

3. High-energy-beam hardening is based on the high energy intensity of electron beams (defocused to prevent melting) and laser beams (Sec. 18-7-2). *Electron beam* and *laser beam hardening* allows rapid heating of the steel surface in a controlled pattern. If the part is large enough, the nonheated bulk cools the surface fast enough to provide the requisite quench.

19-4-2   Diffusion Coating

When the steel has insufficient carbon, the surface can be enriched by diffusing the required atomic species (C or N) into the surface.

Carburizing   Carbon is diffused into steel held in the austenitic temperature range (typically, 850–950°C) by various techniques. *Gas carburizing*, the most widely used technique, is carried out in a furnace with carbon-rich gases (natural gas, propane, or methane) and a carrier gas (usually a mixture of N_2, CO, and H_2). For *vacuum carburizing*, the part is heated in a moderate (rough) vacuum and the hydrocarbon gas is then admitted. Carburizing in a solid compound (*pack carburizing*) or in a *salt bath* is now less used. Most effective is plasma carburizing, to be discussed in Sec. 19-6-4. The part is quenched and tempered.

Nitriding   Nitrogen can be diffused into alloy steels at temperatures of 500–560°C to form a thin, hard surface layer by forming nitrides with Cr, V, W, and Mo. Nitrides formed with Al give a very hard but brittle case. The process can be performed in a gas or liquid; plasma nitriding (Sec. 19-6-4) is gaining popularity. There is no need for quenching and, if the steel is heat-treatable, it must be quenched and tempered before nitriding. *Carbonitriding* is a variant of gas carburizing, with ammonia mixed in the furnace atmosphere. The resultant case is thin and hard.

Metal Diffusion   In *chromizing* chromium is diffused into the surface of steels and superalloys to increase corrosion resistance, but the carbide formed with carbon steels decarburizes the surface. *Codiffusion* of silicon with chromium eliminates the problem. *Aluminizing* is applied to increase high-temperature corrosion resistance of steel and superalloys.

Petrochemical companies have to cope with highly corrosive aqueous fluids in their refineries, especially if they process acidic or heavy crude oils. Carbon steel suffers severe corrosion, ferritic or austenitic stainless steels (Table 8–2) have poor heat conductivity (Table 4–1) and are not free of corrosion problems either. One solution is to make heat-exchanger tubes of a carbon steel (AISI 1018) coated with a codiffused Cr–Si layer on the inside. A powder, made up of Cr and Si (source alloys), halide salts (activators), and an inert filler (such as alumina), is placed in contact with the surface. Upon heating to 1060–1200°C, the activator reacts with the source alloys and the halides thus formed react with the tube surface. Chromium and silicon are deposited and, over a cycle time of 12–45 h, diffuse to a depth of 0.25–0.5 mm. [Source: G.T. Bayer, *Adv. Mater. Proc.*, 1998(2):25–28.)	**Example 19-1**

19-4-3   Hot-Dip Coatings

Finished steel parts as well as steel sheet for further working (Sec. 10-1-1) are coated by these techniques.

Hot-Dip Galvanizing   Parts are dipped into a molten zinc bath which may be protected with a flux. Steel strip—welded to make a coil of infinite length—is guided through a molten bath with the aid of rollers. Residence time and bath composition are controlled to minimize the formation of a brittle intermetallic at the Fe–Zn interface and to develop or suppress *spangle* (visible crystal formation). Coating layer thickness is controlled by blowing off excess zinc with the aid of air or nitrogen. Many sheet products (especially in auto bodies) are spot welded, and alloying of zinc with the copper electrodes reduces electrode life.

Galvannealing   When the strip emerging from the zinc bath is reheated to around 500°C, the coating is transformed into Fe–Zn intermetallic phases of about 9–12% Fe. This is more favorable for spot welding because alloying with the copper electrode is slowed down. The brittle coating is, however, prone to cracking and powdering during forming.

Zn–Al Coatings   There is virtually no brittle interface in coatings of the near-eutectic Zn-5Al alloy, and spangle is minimized. A Zn-55Al coating provides better corrosion resistance, but the surface shows spangles.

Aluminum Coatings   There is no significant intermetallic layer with thin (20–25-μm) coatings of Al–Si alloy (with 5–11% Si), hence the sheets are readily formable and are suitable for high-temperature applications such as automobile exhaust. A pure Al coating forms a brittle intermetallic and is used primarily for lightly formed outdoor structures.

19-4-4   Weld Overlay Coatings

Several welding processes can be used to deposit relatively thick (up to 25-mm) coatings.

Hard Facing   To improve the wear resistance of a lower-strength but tough body, the part is coated with a very hard and sometimes brittle surface layer which is bonded by melting the substrate. *Hard facing* is used not only for repair but also for the initial manufacture of cutting tools, rock drills, cutting blades, parts for earth-moving equipment, valves and valve seats for diesel engines, saw guides, forging dies, and screws for the extrusion of plastic and food products. Iron- or cobalt-based alloys with up to 4% C and a high proportion of carbide-forming elements (Cr, Mo, W) are used; nickel-based alloys contain both carbides and borides (up to 5% C + B). They often have such a high alloying-element concentration that they cannot be manufactured into welding rods. The ingredients are then incorporated in the flux coating or packed inside tubular rods, and the alloy is formed in the welding process itself. Hard particulates such as WC can also be deposited in a metal matrix. Deposition is by fusion welding techniques (Secs. 18-6 and 18-7). Heat input is controlled to prevent excessive melting of the bulk, which would dilute the deposit, but there is a heat-affected zone. When the processes are used to build up worn gears, shafts, dies, and other components, one speaks of *weld overlays*. The rate of metal deposition is increased two- to fivefold by supplying metal powder to the heat zone or by substituting strip for the welding wire (*strip* *overlay welding*). Coatings, mostly of stainless steel, are applied for corrosion resistance in *weld cladding*. Deposition of WC by an electric arc is called *spark hardening*, useful for cutting tools.

Thermal Spraying (Molten Particle Deposition)   This differs from hard facing in that the substrate does not melt and there is little if any HAZ; indeed, the substrate can be cooled by gas or liquid. Any material whose melting point is at least 300°C below its vaporization or decomposition temperature can be sprayed. A metallic or nonmetallic material in the form of wire or powder is fed into a heat source such as an oxyfuel flame (*wire* or *powder flame spraying*, Fig. 19–2a), an arc drawn between two consumable electrodes (*arc spraying*), or a nontransferred plasma arc (*plasma spraying*, Fig. 19–2b) which may also be operated in vacuum (*vacuum plasma spraying*). In a unique process, combustion of an oxyfuel in a confined space produces supersonic, high-pressure gas flow either continuously [*high-velocity oxyfuel* (HVOF) *powder spray*, Fig. 19–2c] or intermittently (*detonation gun*), and the powder is fed into this flow. The molten or softened material strikes the surface, flattens into platelets (*splats*) to build up a lamellar coating. Velocities range from 30 to 1200 m/s (Table 19–1) and can exceed 2000 m/s. Coating thickness is up to 3 mm for ferrous and 5 mm for nonferrous alloys and is typically 0.4 mm for ceramics and carbides.

Since the bond is largely mechanical, the surface must be roughened by grit blasting or, if the body is cylindrical, by threading. Metallic (Zn, Al, Zn–Al alloy), metal-matrix (Al-10Al$_2$O$_3$), and ceramic coatings (as, for example, on rocket motor nozzles and on dies for the hot extrusion of steel, Sec. 9-4-2) are used. Metal powders are often made by rapid solidification techniques (Sec. 11-2-1). The possibilities are endless; for example, alternating metal–ceramic layers are deposited, as are ceramic-coated fluoropolymer powders. Coatings may be applied for corrosion resistance at atmospheric and high temperatures (land-based turbines and jet engines); wear resistance in the textile, paper, and metal industries; and as thermal barrier coatings for jet and diesel engines.

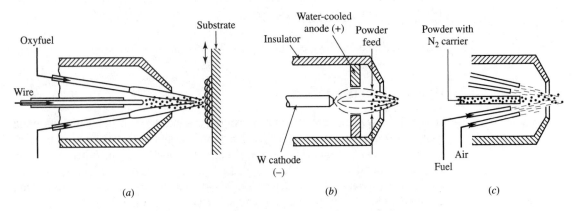

Figure 19–2   (a) Flame spraying, (b) plasma spraying, and (c) high-velocity oxyfuel spraying.

Table 19–1   Characteristics of thermal spray processes*

Process	Temperature, °C	Impact Velocity, m/s	Spray Rate, kg/h	Energy, kWh/kg	Oxide Content, %	Adhesive Strength[†]	Cohesive Strength[†]	Cost
Oxyfuel, powder	2200	30	7	11–22	6	E	E	3
Oxyfuel, wire	2800	180	9	11–22	4	D	C	3
Wire arc	5500	240	16	0.2–0.4	0.5–3	C	B	1
Plasma	5500	240	5	13–22	0.5–1	C	B	5
Vacuum plasma	8300	240–600	10	11–22	ppm	A	A	5
Detonation gun	3900	900	15	220	0.2	B	A	4
HVOF	3100	600–1200	25	20–200	0.1	B	A	10

*Chiefly from data by M. L. Thorpe, *Adv. Mater. Proc.*, 1993(5):50–61.
[†]Relative ratings, with A indicating the highest value.

Example 19-2

Stationary gas turbines used in electric power generation are not only exposed to high temperatures but also to corrosion by contaminated fuels. Therefore, coatings are essential on all hot-stage components. The aluminum diffusion-type coatings applied to jet engines do not provide adequate protection because operating times are longer and fuel and air are more contaminated. For an oxidation barrier, Al_2O_3 forms early in the cycle if there is sufficient Al present in the alloy. Otherwise, a platinum-aluminide coating is deposited by vacuum plasma spray. On top of this, a thermal barrier coating is applied which consists of a thin bond coat followed by an insulating oxide layer. Plasma-sprayed Co-based coatings also may be applied for wear resistance. [After P.W. Schilke, A.D. Foster, J.J. Pepe, and A.M. Beltram, *Adv. Mater. Proc.*, 1992(4):22–30.]

19-5   METAL COATINGS

Metallic surface coatings can be applied by several techniques.

19-5-1   Electroplating

In Sec. 11-8 we saw that, in electroforming, a metal layer is deposited and then stripped from a tool (mandrel) to make a part. The same electrochemical process can be used for *electroplating* (*coating*) a part produced by some other technique, but now the deposit stays in place for esthetic or technical reasons, primarily wear and corrosion resistance. Therefore, our efforts will be directed toward increasing the bond strength between coating and substrate.

Extreme care is exercised in cleaning the surface by all techniques of Sec. 19-2. Metal deposition is slow (typically up to 75 μm/h) but coating thickness

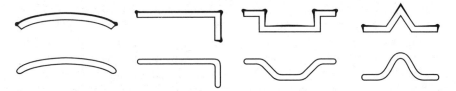

Figure 19–3   Differential buildup of coatings (top row) and design changes for its mitigation (bottom row).

(usually 10–500 μm) can be closely controlled. Parts with irregular surface configurations can be reasonably uniformly covered (*throwing power* is adequate) especially if the anode is well placed and, if necessary, shaped. A thicker layer builds up on edges and if this is objectionable, the excess is removed by abrasive machining, or the part shape is changed to compensate for differential buildup. Thus, corners are given generous radii and the walls of recesses are tapered (Fig. 19-3). Deposition can be locally prevented by masking the surface.

Tin and zinc are deposited continuously on steel sheet for further working (Sec. 10-1-1). Zinc and cadmium are deposited on parts for corrosion resistance, but cadmium cannot be used for food applications because of toxicity. Copper is deposited for corrosion protection and for making electrical circuits (Secs. 20-3-4 and 20-4). Nickel is extensively used for corrosion resistance. Chromium is applied in a thin (about 1.3-μm) layer on top of nickel coatings for appearance. Thicker (2.5–500-μm) layers deposited directly on the base metal (*hard chromium coating*, HV 1000) may need diamond polishing, but then impart wear resistance to dies and reduce their adhesion to many workpiece materials, including aluminum and zinc. Precious metals are used for decorative purposes and, primarily in electronic devices, for corrosion protection.

In the automotive industry, dies used for stretch-drawing (Sec. 10-7) are expected to produce hundreds of thousands of sheet metal parts. Local lubricant breakdown leads to cold welding and metal transfer (tool pickup) and the surface of parts becomes scored. In drawing of galvanized sheet, zinc transferred to the dies would frequently force a shutdown of the press and the pickup would have to be manually removed. For increased productivity, dies are hard chromium plated. Tests have shown that friction is not necessarily reduced but tool pickup is virtually eliminated and the process becomes much less sensitive to lubrication. [Source: J.A. Schey, *Lubric. Eng.*, **52**(1996):677–681.]

Example 19-3

19-5-2   Electroless Coatings

The part—submerged into an aqueous bath containing metal salts, a reducing agent, and various additives—acts as a catalyst for the reduction of metal ions to form the coating. Deposition progresses at a uniform rate over the whole surface, even on parts of complex configuration and in cavities. The coating thins out on sharp edges, and edges must be given radii of 0.4 mm.

The widest application is to nickel with hypophosphite as the reducing agent. The coating incorporates 2–10% P; hardness drops but corrosion resistance increases with increasing P content. The hardness of low-P coatings can be further increased (to HV 1000) by heat treatment. PTFE, SiC, and diamond particles can be incorporated. Electroless deposition of copper and gold is used in printed circuit boards.

Mechanical Plating   When parts are tumbled together with glass beads, metal powder, and accelerator chemicals in water, a metal coating is deposited under the impact of glass beads. Bonding is essentially by cold-welding mechanisms. A great advantage is that there is no hydrogen embrittlement, so often encountered with chemical and electrolytic methods.

19-5-3   Metallizing of Plastics and Ceramics

Electrodeposition on nonmetallic surfaces has long been practiced, but poor adhesion made complete enclosure necessary (as for flowers and baby shoes). Technical applications became possible by improving adhesion through mechanical interlocking and possibly also chemical bonds. Coatings are applied for decorative purposes (plumbing fixtures, automotive parts), reflectivity (headlamps), electrical conduction (printed circuit boards), and electromagnetic shielding (electronic devices).

Preparation for electroless plating involves a number of steps, designed to remove surface films and improve adhesion. Thus, in preparing ABS, a mixture of chromic and sulfuric acid is used to remove the butadiane phase from the surface. All residues of the acid are then removed, the surface is catalyzed with a very thin layer of palladium, and nickel or copper are deposited. On top of the plating thus produced, various metals can be electroplated. The design considerations shown in Fig. 19–3 again apply.

Thermal spraying can be applied to all plastics provided that temperatures are not too high. This limits it to low-melting metals, primarily, zinc. The surface is often roughened, or coated with a paint, to improve adhesion.

We already discussed some processes for metallizing ceramics in Sec. 18-10-2. Metallization by vacuum deposition techniques, to be discussed next, is widespread, especially for semiconductor devices (Sec. 20-3-4).

19-6   PHYSICAL VAPOR DEPOSITION (PVD)

A substance such as a metal or oxide may be evaporated by the application of sufficient heat. The atoms or molecules liberated move away from the source in all directions; when they come into the range of atomic or molecular attraction

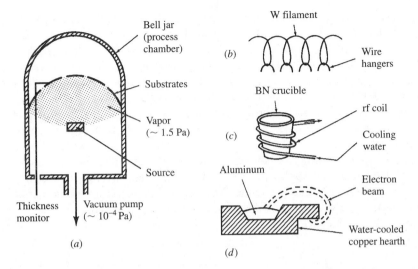

Figure 19–4   Physical vapor deposition (PVD): (a) basic arrangement, in which the source may be a (b) wire evaporated on a tungsten filament, (c) melt heated by radio-frequency induction, or (d) metal impinged by an electron beam. [After D. B. Fraser, in VLSI Technology, S.M. Sze (ed.), McGraw-Hill, 1983, p. 357. With permission of Bell Telephone Laboratories.]

of the workpiece (substrate), they condense on it. To facilitate bonding, the workpiece is often heated, below its melting point. Several methods of operation are practiced.

19-6-1   Thermal PVD

In the basic form of PVD (also called *vacuum deposition*, Fig. 19–4a), the substance such as a metal is evaporated by one of several techniques: placing small wire hangers onto a refractory-metal (typically, W) filament (Fig. 19–4b); heating in a crucible by induction (Fig. 19–4c); or, for higher evaporation rates, by the impingement of electron beams on the surface of the metal (EB-PVD, Fig. 19–4d). In EB-PVD, heating is confined to a small zone of the metal and no contaminants are introduced from crucibles. A new variant is *pulsed-laser deposition*, in which short (10–30-ns) laser pulses evaporate the substance (which may be a metal or a ceramic such as a high-temperature superconductor).

An unusual combination is found in the GE90 turbofan (Fig. 1–2); the large fan blades are made of a graphite-epoxy composite and the leading edges are titanium for protection against hard-body impacts. (Source: GE Aircraft Engines, Cincinnati, Ohio.) | **Example 19-4**

Collision with atoms of air or other gas would reduce the efficiency of the system; therefore, PVD is conducted in a bell jar, evacuated to typically 10^{-4} to 10^{-2} Pa. Thickness is usually monitored by placing a quartz crystal into the bell; as the deposit builds up, the natural frequency of the crystal and thus the frequency of a quartz-controlled oscillator changes. Deposition rates are a few micrometers per hour but increase to over 100 μm/h in EB-PVD. Gases (oxygen, nitrogen, or hydrocarbon gases) may be metered into the vacuum chamber to produce oxides, nitrides, or carbides by *reactive evaporation* (a transition to chemical vapor deposition).

For better bonding, the substrate is often heated. The rate of deposition is accelerated by the application of a dc electric field, making the source the cathode and the substrate the anode. Material is removed from the source in the form of negatively charged ions which are accelerated toward the substrate by the substrate's positive charge.

Example 19-5

Jet engine components are exposed to the most exacting conditions during operation. At cruising speed, parts rotate at 12 000 r/min, with clearances of 0.025 mm, at temperatures in excess of 1100°C, while 960-km/h air, laden with contaminants, rushes past to react with hydrocarbon combustion products at 1900 km/h. Coatings are extensively used and EB-PVD is one of the techniques for coating turbine blades. To increase productivity, several blades are racked and rotated during coating. After evacuation to 2 Pa, the blades are resistance heated to 960–980°C (depending on alloy), the EB guns are operated, and a 0.1–0.2-μm-thick film of MCrAlY is built up (M is base metal and can be Fe, Co, or a Co–Ni alloy). Coating alloy is continuously supplied to the crucible to keep the melt level constant. The parts are then allowed to cool to 300°C before the chamber is opened. Further improvements are obtained with a 0.25-mm-thick TBC (thermal barrier coating) of $ZrO_2 + Y_2O_3$. [Data from G. Simmons and R.C. Seanor, *Adv. Mater. Proc.*, 1994(12):35–37, and from H. Lammermann and G. Kienel, *Adv. Mater. Proc.*, 1991(12):18–23.]

19-6-2   Sputter Deposition

The method of operation changes when the vacuum chamber is partially backfilled (to 3–8 Pa) with a heavy inert gas such as argon (Fig. 19–5).

Gases are not conductors at ambient temperature, but electrons may be removed from atoms (the gas can be ionized and made conductive) by supplying energy in the form of electron impact, x-rays, UV light, or heat. Indeed, in Sec. 18-6 we already saw that a *plasma*—a neutral, gaseous mixture of electrons and positive ions—is formed in the intense heat of a welding arc. Such high-temperature plasma would be useless for other purposes, but a plasma can be obtained also by imposing on a gas an electric field of some critical magnitude. Many of the electrons released by the energy input are captured by positive ions, but some survive and are elastically scattered to collide with the surrounding atoms to excite them into higher quantum states. When the atoms drop back to

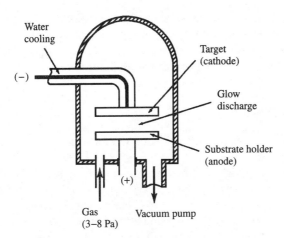

Figure 19-5   Arrangement for sputtering. (*After D. Roddy, Introduction to Microelectronics, 2d ed., Pergamon Press, Oxford, 1978, p. 137. Reprinted with permission.*)

the stable state, the energy released is emitted as photons, leading to the characteristic *glow discharge*. If the energy of electrons is high enough, they ionize other atoms, therefore, above a critical voltage (the *breakdown potential*) the gas becomes ionized and conductive. Typically, of the 10^{16} atoms/cm^3 that are present at atmospheric pressure, only 10^9–10^{12} atoms/cm^3 are ionized. Hence, even though the mean electron temperature is between 10^4 and 10^5 K, the gas itself remains cool, a great advantage for many processes.

For PVD, the plasma is formed by imposing a high (2–6-kV) dc potential on the two electrodes placed in the backfilled envelope. The heavy positive argon ions are accelerated in the electric field and are hurled against the cathode (*target*, the substance to be deposited) at such velocity that the impact dislodges (*sputters*) atoms from the surface; the atoms thus released are deposited on the anode (*substrate*, the part to be coated). If the gas is oxygen instead of argon, atoms dislodged from the cathode are immediately oxidized, and an oxide is deposited on the substrate (*reactive sputtering*). Sputtering is useful in other ways too: If the substrates are given a negative charge, their surface will be cleaned of all contaminants and adsorbed films (*ion-beam etching*).

Insulators cannot be treated with dc PVD because a positive charge builds up on the cathode, repelling the Ar ions. Therefore, PVD is then conducted with ac in the radio-frequency (rf) range (the frequency of 13.56 MHz is internationally assigned for industrial and scientific use). A great advantage of such *rf sputtering* is that, because the cathode is discharged during each reverse-polarity half-cycle, the electrode can be covered even with an insulator, allowing the processing of SiO_2 and other ceramics. In *magnetron sputter deposition* the target (cathode) is surrounded by a magnetic field which captures electrons, increasing their ionizing efficiency, and thus increasing the rate of sputtering.

Atoms travel in a straight line (*line-of-sight deposition*), and reentrant shapes cannot be coated. Deposition rate is constant for points lying on the inside surface of a sphere, therefore, substrates are attached to rotating, spherical sections. Typical deposition rates are 1 μm/min for aluminum.

19-6-3   Ion Plating

Ion plating is a combination of thermal PVD and sputtering: The evaporation rate is kept higher than the rate of sputtering from the substrate. In consequence, a dense, adherent coating is formed while the surface is continually cleaned by sputtering ions. In *reactive ion plating* a reactive gas is admitted into the argon gas; the process thus represents a transition to chemical vapor deposition. The familiar gold-colored TiN coating is formed on cutting tools by evaporating titanium in an Ar–N_2 atmosphere. Coatings are smooth, dense, and have high compressive residual stresses. Other compositions are being added; in particular, TiAlN has high oxidation resistance.

Ion-beam-assisted deposition combines PVD with the bombardment by ions produced independently in an ion-assisted gun. Coatings are dense and have high adhesion, and, if the evaporant or ion beam is reactive, a compound layer (e.g., Si_3N_4) can be deposited.

19-6-4   Plasma Carburizing and Nitriding

Also called *ion carburizing* and *ion nitriding*, these processes take advantage of the ionization of gas in a plasma. Thus, the ions bombard the surface and develop deep diffusion layers. Carbide layers are formed in methane gas and nitride layers with nitrogen.

19-7   CHEMICAL VAPOR DEPOSITION (CVD)

The term *chemical vapor deposition* (CVD) refers to the deposition of an element (or compound) produced by a vapor-phase reaction between a compound of the element and a reactive vapor or gas, with the formation of by-products that must be removed from the reactor. Provisions are often made for the introduction of various reacting compounds in succession.

1. In the basic form of CVD (*thermal* CVD), the reactant gases are introduced into a chamber (reactor) and by-products are removed through an exhaust system (Fig. 19–6). It can be conducted at atmospheric pressure (AP-CVD) or at a reduced, lower (10^2 Pa) pressure (LP-CVD). Temperature in the reactor is kept high enough (typically 900°C or higher) to initiate the reaction. Because of the high temperature, the coating will crack upon cooling if there is a large difference

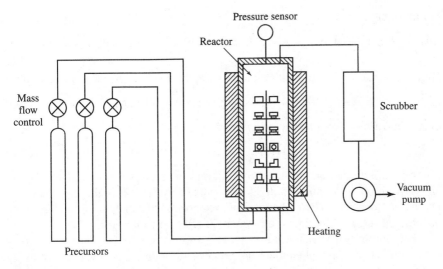

Figure 19–6 Typical arrangement for chemical vapor deposition.

in coefficient of thermal expansion between base and coating. Furthermore, tool steels have to be re-heat treated after coating.

2. Some of the above disadvantages are minimized in *plasma* CVD (*plasma-assisted* CVD or PA-CVD): The reaction is activated by a plasma, thus temperatures are much lower (300–700°C). The process is gaining popularity for tool coatings. The energy of laser beams can also be utilized (*laser* CVD).

3. In *closed-reactor* CVD the parts are enclosed with the reactants in a box and heated. Thus, pack carburizing and nitriding, chromizing, aluminizing, and siliconizing belong in this category.

CVD is used extensively not only in semiconductor manufacture (Chap. 20) but also for making very fine ceramic powders and for coatings on tools. Some typical reactions are:
For boron nitride:

$$BF_3 + NH_3 \rightarrow BN + 3HF \quad \text{(at 1100–1200°C and 103 kPa)} \quad \textbf{(19-1)}$$

$$B_2H_6 + 2NH_3 \rightarrow 2BN + 6H_2 \text{ (in plasma at 300–400°C and 0.13 kPa)} \quad \textbf{(19-2)}$$

For silicon nitride:

$$3SiCl_4 + 4NH_3 \rightarrow Si_3N_4 + 12HCl \quad \text{(at 850°C and 103 kPa)} \quad \textbf{(19-3)}$$

For titanium nitride:

$$TiCl_4 + NH_3 + \tfrac{1}{2}H_2 \rightarrow TiN + 4HCl \quad \textbf{(19-4)}$$

For alumina:

$$2AlCl_3 + 3H_2 + 3CO_2 \rightarrow Al_2O_3 + 3CO + 6HCl \quad \text{(at 1050°C and 0.13 kPa)}$$

$$\text{(19-5)}$$

High-temperature decomposition yields elements such as nickel from carbonyl:

$$Ni(CO)_4 \rightarrow Ni + 4CO \quad \text{(at 180–200°C and 103 kPa)} \quad \text{(19-6)}$$

or silicon carbide from methyl trichlorosilane:

$$CH_3SiCl_3 \rightarrow SiC + 3HCl \quad \text{(19-7)}$$

CVD diamond coatings are made by thermal decomposition of a hydrocarbon and deposition on carbide tools at 700–900°C. Adhesion is the critical issue. *Diamond-like carbon* (DLC) has a structure between that of diamond and graphite but still has a useful hardness (around 5000 HV) for tool coatings.

The number of possible reactions and applications is very large. In common with many chemical engineering processes, modeling is advanced for many reactions and computer control of processes can be based on valid models.

19-8   ION IMPLANTATION

The process is based on the observation that, if ions are sufficiently accelerated, they will become lodged in the target and the foreign atom species *dope* the surface. For the process to be industrially viable, an economical ion source must be found.

1. In the original process (Fig. 19–7), the *ion source* consists of a gas feed and hot oven. The vapor or volatile compound generated in the oven enter a noble-gas discharge plasma where they become ionized. Thus, ions of the doping element and gas are generated. After *acceleration*, ions are subjected to a strong magnet (*analyzing* or *mass-separating magnet*) which deflects ions so that only the desired species passes through a slit (*resolving aperture*). The ions are accelerated to a high (30–200-keV) energy level so that on impact they penetrate some 10–1000 nm below the surface of the substrate, even if a thin oxide film is present (indeed, such a film is often intentionally grown to protect the surface from unwanted reactions). Voltages applied to *deflection plates* cause the beam to sweep over the target area (or the beam is fixed and the target is mechanically moved).

2. In a more recent process, a plasma arc is the source of ions, greatly reducing the cost of implantation.

The ions lose their energy in collisions with target nuclei and electrons, and come to rest some distance below the surface. Thus, surface composition is modified without dimensional change and with minimum heating (usually below 200°C). Wear-resistant coatings are made on tools and implants such as hip-joint replacements, but the major application is to solid-state devices (Sec. 20-3-3).

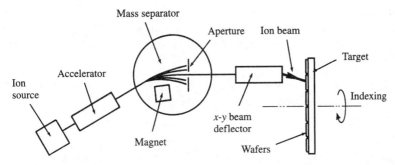

Figure 19–7   Ion implantation is the dominant technique in VLSI and ULSI technology.

If the ion enters in perfect alignment with a major crystallographic direction, it penetrates deeply. To avoid such *channeling* effect, the ion beam is directed slightly obliquely to the crystal axis, and a fairly uniform depth distribution from the surface is then obtained.

19-9   ORGANIC COATINGS

The topmost coating of many products is a clear or pigmented organic layer (*paint*). It is a multicomponent system, consisting of a film former (*binder*, which may be a natural oil or natural or synthetic polymer), solid, finely disbursed pigments to give color and opacity (less frequently a soluble die), various additives (curing agents, plasticizers, UV stabilizers, etc.) and, in most instances, a solvent or dispersant. Depending on the type of solvent, the paints are classified as:

1. *Enamels* (not to be confused with vitreous enamels, Sec. 12-5-3) form films by solvent evaporation, oxidation, or polymerization. High-solid paints contain more than 70% solids by volume to reduce VOCs.

2. *Lacquers* form films by solvent evaporation.

3. *Waterborne paints* can be true solutions, colloidal dispersions, or emulsions.

4. *Powder paint* is a fully formulated paint ground into a fine powder.

Drying of a paint may be as simple as standing in air but may involve a controlled heating, UV, or microwave *curing cycle*. Paint formulation is a well-developed art and science, and paints suitable for most purposes and application methods can be found. A major concern is impact on the environment and health, and paint formulations and application methods have undergone rapid change in the last two decades. For both environmental and cost considerations, every effort is made to increase *transfer efficiency* (percentage of paint that lands on the product), to recover overspray, and to minimize air pollution.

Application   The first step is always thorough cleaning (Sec. 19-2). For better paint adhesion, a conversion coating (Sec. 19-3) may be applied. Paint application may be manual, dip, mechanized, or by specialized paint robots with five or more axes of control.

1. *Brushing* is purely manual and is seldom applicable in production to high standards except for repair in nonexposed areas.

2. *Roller application* is suitable for mechanization in the coating of flat surfaces.

3. *Spray coating* is widely used under both manual and automatic control. In manual spray coating, transfer efficiency is poor; paint is applied in line of sight and the gun must be manipulated around the part. In electrostatic spraying the part and the paint are connected to a high-voltage (up to 130-kV) dc power source. Transfer efficiency increases dramatically. Paint particles wrap around the part, reaching into recesses but, unless properly controlled, they build up a thicker coating on edges (Faraday-cage effect). Atomization may be by air at about 2 MPa (for best surface finish), by pressurizing the paint to 30 MPa and dispensing it through a small nozzle (*airless electrostatic spray*), by developing a lower (2–7 MPa) pressure and assist atomization by auxiliary air (*air-assisted airless spray*), or by centrifugal forces developed in a fast-rotating (30 000-r/min) *bell* or *disk*.

4. The paint may be applied by submerging the part in a tank. *Autophoretic paints* are deposited by the catalytic action of the metal (ferrous alloys). *Electrophoretic paints* are deposited on the charged metal body and may be formulated to form a film on either a cathodic or anodic surface.

5. In *powder deposition*, the powder is dispensed by compressed air through a gun, most often electrostatically. Since the paint is fully formulated, it has to be heated prior to application (thermoplastics) or cured after application (thermosets).

Example 19-6   Automobile bodies are endowed with the most advanced corrosion protection system that also satisfies esthetic requirements. Most bodies are still made of sheet metal, and the corrosion protection system has several elements. Steel sheets are coated (e.g., by one of the galvanizing processes, Secs. 19-4-3 and 19-5-1). The welded body then goes through a number of spray stations and tanks where it is cleaned, repeatedly rinsed, phosphated (for corrosion protection and paint adhesion), rinsed, neutralized, rinsed again (it may also receive a rinse of chromic acid), rinsed, and dried. The paint itself is usually applied in three or four coats. An electrophoretic dip deposits up to 35 μm of *resin base coat* which is baked around 180°C. A sound-deadener plastic (PVC) coating is applied to selected areas (wheel well and underbody) and an elastomeric stone-guard coating to the lower part of the body. *Primer* is then applied, often as powder, in a 40-μm-thick layer, baked at 170°C. Some products then receive a finish coat, but in most instances a *color base coat* is applied followed by the topmost *clear coat*, baked at around 140°C. Between each layer, defects must be corrected by hand; great efforts are made to minimize this work by controlling the paint and its application and also by design features that make uniform paint application easier.

19-10 PROCESS CAPABILITIES AND DESIGN ASPECTS

The number of processes is very large and several processes may satisfy a given need. Therefore, processes must be chosen within the concurrent engineering exercise.

Dimensional constraints are few, although size will often determine the applicability of alternative approaches. Thus, for example, entire car bodies can be dipped into a coating tank, but this will obviously be impracticable for a jet liner. Shape limitations are few, although access to recesses and reentrant shapes varies greatly by process.

1. Several processes are characterized by line-of-sight operation. This is generally true of processes in which deposition is by electron beam, laser beam, electric arc, or high-pressure spray. It is then necessary to manipulate the part and/or the deposition device to reach surfaces, although a limit is always set by the size of openings relative to the size of the deposition device.

2. Better coverage is obtained if an electrical potential can be imparted to the substrate, as in electrodeposition and electrostatic spray deposition. An unwanted side effect is thickening of coatings on edges and, apart from process control, part design can help in this respect (Fig. 19–3).

3. Total coverage of surfaces (all but the closed shape T4 in Fig. 3–1) is possible in dipping, provided that excess fluid can be drained and residues flushed.

4. Processes vary greatly in their ability to fill in surface details (leveling), and surface preparation must be directed accordingly. If the coating relies on mechanical interlocking for adhesion, it may be necessary to smooth the surface of the coating by some secondary (usually abrasive) finishing process.

19-11 SUMMARY

Parts and assemblies often need to be endowed with surface coatings that increase corrosion or heat resistance. Because the surface of products is often visible, esthetics also must be satisfied. The coating may actually be a sophisticated multilayer system, as exemplified by the finish of automobiles and appliances.

1. The first step in most cases is that of removing unwanted (contaminant) surface films by mechanical or chemical means.

2. The surface of metals may be converted by chemical reactions for protection and appearance (anodizing of aluminum) or to facilitate further processing (phosphating of steel).

3. Surfaces of great hardness or corrosion resistance can be produced by thermal diffusion of carbon, nitrogen, or metallic species into metallic (ferrous) substrates.

4. Metal, ceramic, and plastic coatings are deposited by the melt route—with or without melting the substrate—for corrosion and wear resistance (hot dip, weld overlay, and thermal spray coatings) and higher temperature. capability (thermal barrier coatings).

5. Metal coatings are applied for corrosion protection (nickel coating), hardness (hard chromium coatings), electrical conduction (copper circuitry), or appearance (chromium on nickel), mostly on metal but also on ceramic and plastic substrates by electroplating and other techniques.

6. Principally metal, but also ceramic and polymeric coatings can be deposited by physical evaporation and condensation at moderate temperatures (PVD), sometimes aided by a plasma. In conjunction with a reaction, ceramic compounds can be deposited (as are TiN coatings on tools). Sputtering maintains atomically clean surfaces. Plasma is also used to facilitate diffusion, as for carbon and nitrogen into steel.

7. Chemical vapor deposition (CVD) at high temperatures is the principal method for forming thin ceramic layers on tools and semiconductor devices.

8. Ion implantation provides a means of changing surface composition without dimensional change.

9. Organic coatings are the last and often visible layers of a surface protection system.

10. Many coating processes generate significant amounts of gaseous, liquid, or solid by-products during operation, and there may also be residues that must be disposed of. Protection of workers and the environment is of paramount concern and may add substantial cost.

11. Processes are often amenable to computer control and automation, and process modeling is an important element in quality and cost control.

PROBLEMS

19-1 (*a*) Define anodizing of aluminum and (*b*) indicate two reasons for anodizing.

19-2 (*a*) Define phosphating of steel and (*b*) indicate two applications and the reason for doing it.

19-3 (*a*) Define carburizing and (*b*) nitriding of steel. (*c*) State two methods for each process and (*d*) indicate the process temperature and the necessary heat treatment.

19-4 (*a*) Define galvanizing and (*b*) state two methods of doing it.

19-5 Explain the difference between (*a*) hard facing and (*b*) thermal spraying.

19-6 Draw a diagram showing the essential parts of an electroplating setup. Indicate polarity.

19-7 (*a*) Is it possible to deposit a metal coating without electricity? (*b*) If the answer is yes, suggest two methods.

19-8 (*a*) Define PVD and (*b*) make a sketch showing thermal PVD.

19-9 Define the terms (*a*) plasma and (*b*) sputtering.

19-10 (*a*) Define CVD and (*b*) make a simple sketch showing the essential elements of a CVD system.

19-11 Define ion implantation.

19-12 We saw in Fig. 18–25 that the braze alloy was supplied as a cladding on the 3003 Al base metal. Suggest a way of making aluminum strip of 0.3-mm thickness with 10% cladding of braze alloy on both sides. Considering the base and cladding composition, list the likely processing steps.

19-13 (*a*) Define clad metal and (*b*) name at least four processes by which clad sheets (of unspecified base and cladding metal) can be produced.

19-14 Define galvannealed sheet.

19-15 A brass cup of 50-mm OD, 46-mm ID is to be chromium plated. The designer specified zero outside radius at the base. (*a*) Draw a sketch to show the likely coating thickness distribution. (*b*) Suggest a design modification to obtain uniform thickness.

19-16 A ceramic thermal barrier coating is to be deposited on a superalloy turbine component. Suggest at least two methods for doing it.

19-17 It is desired that an alumina part be coated with a chromium layer. From basic principles, suggest a processing sequence that results in good adhesion.

19-18 A low-carbon steel cooking vessel is enameled on the inside. Rapid heating causes uneven temperature distribution and can lead to cracking of the enamel. Calculate the maximum allowable temperature difference for the cases when (*a*) only the steel heats up and (*b*) both steel and enamel are gradually brought to temperature. (Assume that the enamel fractures at the flexural modulus; you will find it, the elastic modulus, and coefficient of thermal expansion in this text.)

19-19 Ice skates are required to hold a keen edge. They must also be corrosion-resistant and have esthetic appeal. Critically review at least three potential solutions available to the designer and choose the one most likely to satisfy the requirements at reasonable cost. (List advantages and disadvantages.)

19-20 We have repeatedly come across the terms coating and cladding. Review the use of the terms and formulate definitions for them.

FURTHER READING

ASM Handbook, vol. 5, *Surface Engineering*, ASM International, 1994.

Wick, C., and R. Veilleux (eds.): *Tool and Manufacturing Engineers Handbook*, vol. 3, *Materials, Finishing, & Coating*, Society of Manufacturing Engineers, 1985.

Edwards, J.: *Coating and Surface Treatment Systems for Metals: A Comprehensive Guide to Selection*, ASM International, 1997.

Kodas, T.T., and M.J. Hampden-Smith: *The Chemistry of Metal CVD*, Wiley, 1994.

Lambourne, R., and T.A. Strivens: *Paint and Surface Coatings: Theory and Practice*, 2nd ed., Woodhead, 1999.

Lindsay, J.H. (ed.): *Coatings and Coating Processes for Metals*, ASM International, 1998.

Rausch, W.: *The Phosphating of Metals*, 2d ed., ASM International, 1990.

Stern, K.H. (ed.): *Metallurgical and Ceramic Protective Coatings*, Chapman and Hall, 1996.

Sudarshan, T.S. (ed.): *Surface Modification Technologies*, Dekker, 1989.

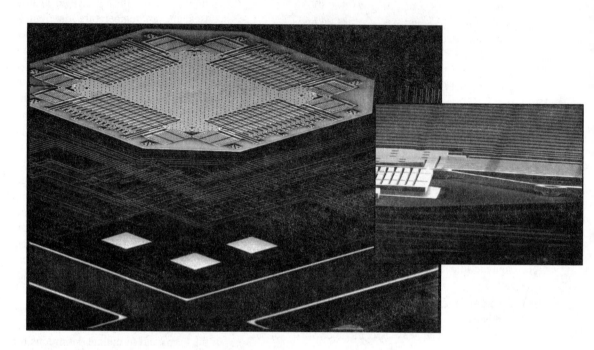

The ADXL202 accelerometer has a 400 × 400 µm micromachined polysilicon spring mass system integrated with all signal conditioning on a single 2 × 2 mm IC chip. It gives digital outputs for ±2g acceleration in two axes and is widely used for tilt sensing in vehicle security systems. (*Courtesy Analog Devices, Wilmington, Massachusetts.*)

20

Manufacture of Semiconductor Devices

We are now ready to delve into an entirely new field, based on unique electronic properties of materials. After reviewing elements of semiconductor devices, we shall explore new processes and the application of previously discussed processes to:

Growing silicon crystals and making wafers

Treating wafers by PVD, CVD, and other processes to make integrated circuits

Using lithography to develop patterns

Establishing interconnections and protecting devices by encapsulation

Integrating devices into circuit boards

Creating microelectromechanical systems

Fabricate parts on the micro- and nanometer scale

In Sec. 1-1-3 we referred to the immense changes brought about by the development of microelectronic devices. Since the invention of the *transistor* by Bell Telephone Laboratories scientists John Bardeen, Walter Brattain, and William Shockley in 1947–48, the field has grown by leaps and bounds and is still in a phase of rapid development. This rapid growth is attributable to concentrated efforts in developing both the physical, chemical, and materials science background *and* the manufacturing technologies needed for making solid-state products.

Transistors were first made as individual devices, essentially as straightforward replacements for electron tubes. They were then wired, with other components, into complete circuits. A revolutionary change came after 1958–59, when the monolithic, *integrated circuits* (ICs) were invented by J. St. C. (Jack) Kilby of Texas Instruments and R. N. (Bob) Noyce of Fairchild Camera. All components of a circuit are now placed on a single chip cut from a thin (perhaps 200-μm-thick) wafer of doped silicon or other semiconductor material. With advances in manufacturing technology, more and more components—making up circuits of increasing complexity—have been accommodated on a single chip. This allowed the development of that most powerful of devices, the microprocessor, in the early 1970s. The minimum feature width gradually decreased and the size of chips increased (although most still measure only a few millimeters on their edges), and the industry moved from small-scale integration (SSI) to medium-scale (MSI), large-scale (LSI), very large-scale (VLSI), and now ultra-large-scale (ULSI, with over 1 million devices on a chip) integration. As expressed by Moore's law, the number of components per chip has doubled every 18 months (Fig. 20–1) and is still increasing; gigascale integration is not too far into the future.

Almost unparalleled in the history of industrial development, these advances came at a lower price; the cost of one bit of memory has been halved every 2 years. Thousands of millions of integrated circuits are now manufactured every year, to be used not just in calculators and computers, radios, and television sets, but also in automobiles and all kinds of machinery and toys, articles of production and consumption. All this could not have been possible without the development of mass-manufacturing techniques. Mass production, as discussed up to now, was mostly based on making many parts in sequence; in contrast, in IC technology it is based mostly on *batch manufacture*, with hundreds or thousands of units processed simultaneously.

A study of manufacturing processes would be incomplete without at least a general outline of the most important techniques. Some processes are modifications or further developments of processes previously used in other fields; many others have been developed specifically for the purpose and have, over the years,

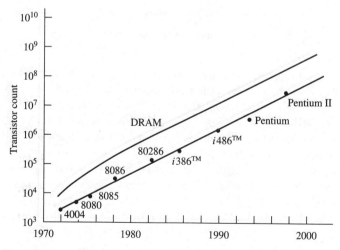

Figure 20–1   The number of transistors on a chip has doubled every 18 months. (*Data for Intel microprocessors from Intel Internet site.*)

found applications in other fields and are also important in microfabrication, a field of tremendous promise.

Compared to most other manufacturing processes, the production of microelectronic devices places extreme demands on the absence of unintentional elements (impurities) in the starting materials, and on cleanliness and strictest control of process parameters during all phases of manufacturing. To understand the necessity for these measures, it is important to have at least a nodding acquaintance with the physical principles underlying the technology.

20-1 ELEMENTS OF SEMICONDUCTOR DEVICES

Within the scope of this book, it is impossible to cover all semiconductor devices serving various purposes. The emphasis will be on devices used in logic (digital) circuits, with only passing reference to analog devices.

20-1-1 The Semiconductor

It will be recalled from physics that electrons in an isolated atom can occupy only discreet energy levels and that only two electrons can share the same energy level. Therefore, in a body consisting of many atoms, the outermost (*valence*) electrons must occupy very slightly different energy levels, and the very closely spaced individual levels will broaden into *energy bands*. Some of these bands overlap, while a *forbidden energy gap* separates others.

Current flow may be visualized as the physical movement of electrons under the influence of an imposed electric field. The electron is a negative charge carrier, and the electrical conductivity of a material increases (resistivity decreases) if an increasing number of charge carriers is available. Three classes of materials are commonly distinguished:

1. *Conductors* are typified by metals. As mentioned in Sec. 4-9-3, they have very low resistivities (e.g., aluminum has a resistivity of $\rho = 3 \times 10^{-8}\Omega\cdot$m; in the electronic industry, the old cgs unit of $\Omega\cdot$cm is frequently used, and resistivity is quoted as $3 \times 10^{-6}\Omega\cdot$cm). We saw that metals may be visualized as positively charged ions glued together by valence electrons. These electrons are free to move and can be set in motion on imposing the slightest voltage, because the conduction bands are only partially filled (Fig. 20–2a). All disturbances of the lattice—including dislocations, solute atoms, vacancies, and thermal excitation of atoms—present obstacles to electron motion; therefore, the resistivity of metals *increases* with temperature.

2. *Insulators* are typified by ceramics such as SiO_2, which has a resistivity of 10^{14} $\Omega\cdot$cm. We already saw (Fig. 12–3) that all valence electrons of Si are used up in completing the outer shell of oxygen atoms, and are thus firmly held in the covalent bond. The valence band is completely filled (Fig. 20–2c) and an

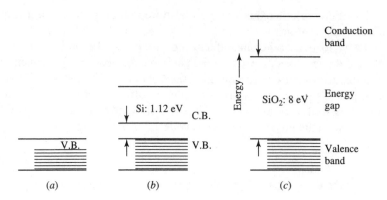

Figure 20–2   Energy bands: (a) unfilled valance bands in metals, (b) filled valance bands separated by a small energy gap from the conduction band in semiconductors, and (c) large energy gap in insulators.

electron can be brought into the conduction band only if an energy of 8 eV is supplied so that an electron can jump the band gap. For an electron to acquire this 8 eV, a very large energy must be applied to the solid.

3. *Semiconductors*, as the name implies, stand between conductors and insulators. The most frequently used semiconductor is still Si. Silicon is a tetravalent element and all electrons are again taken up in completing the outer electron shell of neighboring atoms, creating covalent bonds. The valence band is filled as in insulators, but the bond energy is lower, and the energy gap is only 1.12 eV (Fig. 20–2b). When external energy is supplied, a number of electrons can acquire enough energy to break away (jump into the conduction band), therefore, the resistivity of semiconductors *decreases* with increasing temperature. Very pure Si is a poor conductor because at room temperature only one electron in 10^{13} has the probability of acquiring enough energy to jump the gap; hence, the conductivity of silicon (an *intrinsic semiconductor*) is low (resistivity is approximately $3.2 \times 10^5 \, \Omega \cdot \text{cm}$).

However, the properties of Si can be dramatically altered by introducing very small quantities, on the order of 1 part per million (1 ppm) foreign elements, impurities, also called *dopants*. Two possibilities exist:

a. Pentavalent elements (from group V of the Periodic Table) such as N, P, As, or Sb can be introduced to form a substitutional solid solution in Si. One electron per atom will now be surplus, available to conduct electricity; hence, these are called *donor* elements. Since the electron is a negative charge carrier, one speaks of an *n-type extrinsic semiconductor*.

b. Trivalent elements (from group III of the Periodic Table), such as B, Al, Ga, or In also form solid solutions but there is now one electron missing in the valence band. Such a *hole* can be regarded as a positive charge carrier; hence, one speaks of *p-type extrinsic semiconductors*. The hole can accept an electron; therefore, the trivalent elements are also called *acceptors*.

When an electron (*n* carrier) finds a hole (meets a *p* carrier), the carriers recombine and disappear. However, it takes time for carriers to recombine, and one speaks of *carrier lifetime*.

Even a very small amount of donor or acceptor element provides many charge carriers. For example, 1.0 ppm P amounts to 4.5×10^{16} atoms/cm^3 of Si; by injecting the same number of electrons, it reduces resistivity to about $1.5\,\Omega$·cm (Fig. 20–3). Hence, the silicon must be extremely pure if unintended conducting paths (and leakage currents) are to be avoided.

Crystallographic defects (vacancies and dislocations) also interfere with the intended operation of devices because they reduce charge mobility; furthermore, dopants migrate to them and provide low-resistivity leakage paths. Hence, the

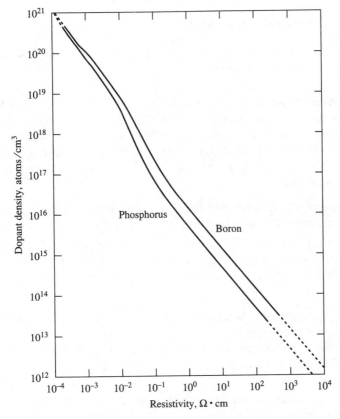

Figure 20–3   The resistivity of pure silicon drops rapidly even with small concentrations of dopant (impurity) atoms. (*From W.R. Turber, R.C. Mattis, and Y.M. Liu, in Semiconductor Characterization Techniques, Electrochemical Society, Pennington, N.J., 1978, p. 89. This figure was originally presented at the Spring 1978 Meeting of The Electrochemical Society, Inc. held in Seattle, Washington. With permission.*)

silicon must be not only very pure but must also be free of defects (unless defects are intentionally introduced to capture, *getter*, contaminants). Therefore, a great majority of semiconductor devices are made on silicon single crystals, although polycrystalline silicon (often called *polysilicon*) is used for specific purposes.

Germanium, the substrate for early transistors, has an energy gap of only 0.7 eV, and is now of limited industrial significance. Semiconducting devices are, to a still limited extent, made on group III–V substrates, particularly GaAs. Compared to silicon devices, GaAs devices are more expensive; however, they operate at higher speeds, have higher resistance to radiation, lower power consumption, and function at temperatures from -200 to $+200°C$ (as compared to the -55 to $+125°C$ of silicon devices). Our discussion will concentrate on silicon semiconductors.

20-1-2   Semiconductor Devices

The variety of devices based on semiconduction is very large and is still growing; here we will mention only a few.

Diodes   When adjacent regions of a silicon single crystal are doped with *p*- and *n*-type dopants, a *p-n* junction is formed (Fig. 20–4a). There will be excess electrons in the *n* region and excess holes in the *p* region; in the *n* region, electrons are *majority carriers* and holes are *minority carriers*. The reverse is true of the *p* region. When a *diode* is connected to an electric current source, one of two situations may exist:

1. *Reverse bias* (Fig. 20–4b) causes the charge carriers to move away from the interface, an insulating zone forms, and practically no current can flow.

2. *Forward bias* (Fig. 20–4c) results in the movement of majority charge carriers into the opposing halves. Current flows as in a conductor, although some electrons and holes are annihilated in the recombination zone (around the interface).

In the described form the diode can serve as a rectifier. In other diodes, made of III–V-type semiconductors, recombination may result in the emission of light (*luminescence*). Among the *light-emitting diodes* (LEDs) GaAs emits infrared light and GaP emits green light.

Transistors   A *transistor* is a device formed by a coupled pair of *p-n* junctions. They are used in two basic forms:

1. *Bipolar transistors* are two junctions in series. The *p-n-p* transistor was originally more widespread because of easier fabrication, but the *n-p-n* transistor (Fig. 20–5) is now preferred because the mobility of electrons is some 3 times higher than the mobility of holes, making the *n-p-n* transistor faster and more suitable for operation at high frequencies and fast switching rates.

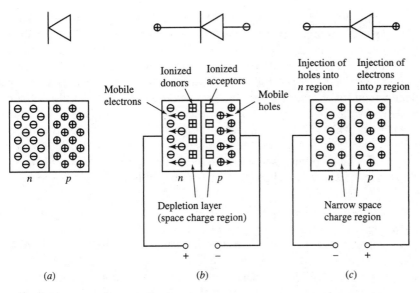

Figure 20-4 caption image 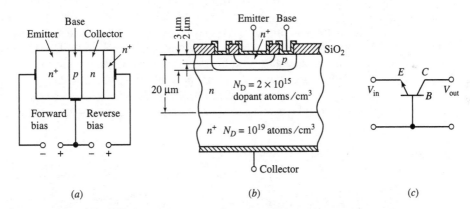

Figure 20–4   The simplest semiconducting device is the diode which (a) consists of n and p zones. When a current source is connected, (b) virtually no current flows when reverse bias is applied but (c) the diode conducts when forward bias is applied.

Figure 20–5   A bipolar junction transistor consists of (a) n-p-n (or p-n-p) zones which, in integrated circuits, are made (b) in a planar form. (c) Symbolic presentation (*From E.S. Yang, Microelectronic Devices, McGraw-Hill, 1988, p. 118. With permission.*)

The three regions are called *emitter*, *base*, and *collector* (Fig. 20–5a). In the n-p-n transistor a very thin, lightly doped p region (base) is sandwiched between a moderately doped n region (collector) and a more heavily doped, hence marked n^+, region (emitter). The emitter junction is forward-biased to drive electrons into the narrow *p* region, where some electrons recombine with holes but the majority passes through and is injected into the collector. A small change in base current results in a large change in collector current, hence this transistor acts as an *amplifier* or it can be used as a *switch*. Since both *n* and *p* charge carriers are involved, one refers to *bipolar junction transistors*. Metal connections to *n*-doped silicon would make other transistors. To avoid this, the crystal is heavily doped (indicated by the plus sign) where connections (*ohmic contacts*) will be made.

In the early phases of development the transistor looked physically somewhat similar to Fig. 20–5*a*; however, the fabrication method soon changed to *planar devices*, in which connections are brought to the wafer surface (Fig. 20–5*b*). Irrespective of its physical appearance, the transistor is symbolically shown as in Fig. 20–5*c*.

2. The *insulated-gate field-effect transistor* (IGFET) or *metal-oxide-semiconductor field-effect transistor* (MOSFET) is a *unipolar* device because only one charge carrier plays a role. In the *n*-channel MOSFET shown in Fig. 20–6*a*, a channel of *L* length is formed, in a lightly doped *p* substrate, between the heavily doped (and hence marked n^+) *source* and *drain*. The magnitude of current is controlled by the *gate*, which is insulated by a thin (< 100-nm-thick) SiO_2 layer. No current flows until a positive voltage is applied to the gate; the electric field repels holes from the substrate and, when the applied voltage exceeds a threshold value, a mobile electron layer is set up: An *n*-type conductive channel exists now in the zone between source and drain. If the drain is made positive, an electron current will flow from source to drain, the magnitude of which is controlled by the gate voltage. Since the structure consists of *m*etal, *o*xide, and *s*emiconductor, and the current is carried by electrons (*n*) and is controlled by the field set up under the gate, one speaks of an *n-channel* MOSFET. The metal gate is most often replaced by a highly doped, *n*-type polysilicon stripe.

MOSFETs are slower than bipolar transistors, but they dissipate much less power; they are self-insulating, thus occupy less space; they are easier to manufacture; they can be used as memory elements; and the channels act as resistors,

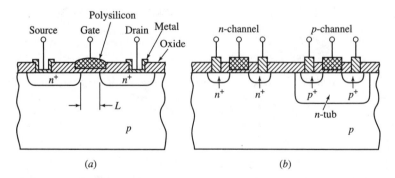

Figure 20–6   Only one kind of charge carrier is involved in the (*a*) metal-oxide-semiconductor field-effect transistor (MOSFET) and (*b*) complementary MOS (CMOS), transistor; in the latter, the *p* channel is formed in an *n* tub (or *well*).

eliminating the need for separate resistors. Because of their many advantages, MOSFET circuits were used in the first hand-held calculators and in microprocessors, and are used in most VLSI and ULSI devices. *Complementary* MOS (CMOS), in which an *n* channel and a *p* channel are paired in adjacent positions (Fig. 20–6*b*), draw less current and became the dominant devices in the 1990s.

Less frequently used are *junction field-effect transistors* (JFET, also simply called FET). Bipolar transistors (which are faster and can drive higher current) are integrated with CMOS (BiCMOS) for high-performance applications such as very high-speed *static random-access memories* (SRAMS). *Application-specific ICs* (ASICs, of varying degrees of sophistication) are dedicated to a single end user and are often mass-produced.

20-1-3   Integrated Circuits

We mentioned that transistors were first made as discrete devices which were then wired into complete circuits. With the introduction of the *integrated circuit* (IC), all circuit elements—including transistors, resistors, and capacitances—are fitted on a single silicon chip, at an ever increasing density. In general, ICs are designed to avoid the need for inductors although inductors can be made as integrated components on Si chips, chiefly for wireless communication devices.

1. We saw in Figs. 20–5*b* and 20–6 some typical cross sections of transistors. In plan view, they appear as small, electrically isolated areas on the wafer surface, interconnected by metallic patterns, referred to as *metallization.*

a. Bipolar transistors need isolation by dielectric insulating regions or by reverse-biased *p-n* junctions. An example is shown in Fig. 20–7*a*; the *n-p* junction is reverse-biased and provides the isolation. Because of the planar arrangement, the active part of the transistor is far from the collector, and a highly doped n^+ *buried layer* is introduced to provide a low-resistance path. Other technologies are available that allow higher density of devices.

b. We mentioned that MOSFETs are self-isolated; in plan view they appear as in Fig. 20–7*b*.

2. Resistors can be formed as thin sheets (stripes) of doping elements diffused into a silicon substrate. The resistance of one square portion (Fig. 20–8*a*) is

$$R = \rho\frac{a}{ha} = \frac{\rho}{h} \qquad \textbf{(20-1)}$$

and is independent of a, the side of the square. Hence it is quoted in units of $\Omega/\square$ (ohm per square). Resistivity ρ depends on doping element concentration (Fig. 20–3) and, for a given doping depth h, it can be increased by diffusing over a greater length l, i.e., by creating the equivalent of several squares in series. For a given resistance, the length of the resistor can be reduced by reducing a. It should be noted that a *p-n* junction is also formed. As mentioned, MOS transistors can also be used as high-value resistors (the channels have a sheet resistance on the order of $10\ \mathrm{k}\Omega/\square$).

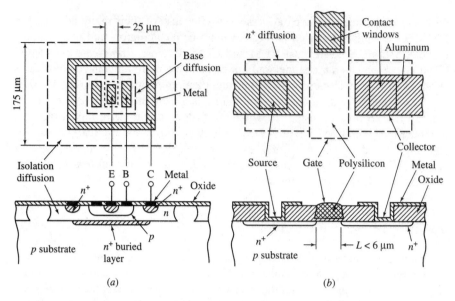

Figure 20–7   The physical appearance of an (a) insulated bipolar junction transistor and (b) MOSFET. (*Adopted from E.S. Yang, Microelectronic Devices, McGraw-Hill, 1988, p. 162. With permission.*)

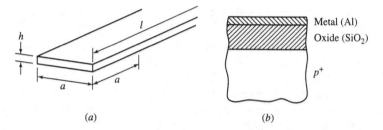

Figure 20–8   Passive devices: (a) typical dimensions of diffused resistors and (b) capacitor.

3. Capacitors are formed by depositing a dielectric (insulating) layer such as SiO_2 on top of a heavily doped p^+-type Si substrate. The other plate of the capacitor is formed by a deposited metal (aluminum) film (Fig. 20–8*b*). Obviously, an MOS structure can also serve as a capacitor.

Miniaturization   The first integrated circuits contained only a few devices and were quite large. There are powerful incentives to reduce the size of devices so that more of them can be fitted on a single chip. Signals propagate at the rate of about 5 ns/m; hence, speed of operation increases with decreasing device size. Power consumption drops too and thus the problem of cooling is also alleviated.

With larger-scale integration, the number of external interconnects and the failures associated with them decrease and the cost of complete circuits drops.

Miniaturization hinges on reducing the feature size. The minimum feature size is defined as the average of the minimum line width and spacing. It is anticipated that certain limitations imposed by physical phenomena will be reached when feature sizes are reduced below 0.1 μm. Before these limits are reached, various difficulties have to be overcome. First of all, the design of very complex circuits becomes a formidable task; fortunately, the concurrent development of CAD has made this task possible. Computer programs are used not only for the design of circuit layout but also for the modeling of circuit performance. Manufacturing challenges are also of great magnitude, and in the remaining part of this chapter we will explore some solutions. Problems multiply with increasing integration because the likelihood of a defect falling within a given circuit becomes higher. Nevertheless, it is believed that gigascale integration—with several hundred million components on a single chip—will be reached with three-dimensional (multilevel) chips.

There are innumerable and ever newer devices designed, and there are corresponding developments in manufacturing technology. Because the technology is still fluid, our discussion will be limited to some of the basic technologies employed for the manufacture of integrated circuits, in particular, *monolithic integrated circuits* (ICs). As shown by the Greek derivation (monolithic = single-stone), these contain all circuit components on a single *chip* (also called *die*, plural: *dice*). In most instances the chip is cut from a wafer of a silicon crystal.

20-2   MANUFACTURE OF SILICON WAFERS

We saw that a semiconductor must be free of accidental doping elements and that the tolerable concentration of impurities is very low. In general engineering applications, a 99.99% pure metal (e.g., superpurity aluminum or electrolytic zinc) is regarded as very pure, yet it contains 0.01% or 100 ppm impurities. In *electronic-grade silicon* (EGS) the concentration of doping elements is on the order of *parts per billion* (ppb), so low that their presence can only be inferred from resistivity measurements.

20-2-1   Production of EGS

Production of EGS involves the following steps:

1. Quartzite (pure, natural SiO_2) is reduced with carbon in a submerged-arc electric furnace. The resulting 98% purity silicon is suitable for metallurgical alloying but must be further refined for electronic purposes.

2. Silicon is hard and brittle. It can be broken up, pulverized, and reacted at 300°C with anhydrous HCl

$$\text{Si (solid)} + 3\text{HCl (gas)} = \text{SiHCl}_3\text{(gas)} + \text{H}_2\text{(gas)} \qquad \textbf{(20-2)}$$

Trichlorsilane (SiHCl_3) and chlorides of impurities condense to a liquid at room temperature. The boiling point of SiHCl_3 is 32°C; it can be separated from the chlorides of impurities by fractional distillation.

3. EGS is obtained by CVD (Sec. 19-7), in this case by reaction with H_2

$$2\text{SiHCl}_3\text{(gas)} + 2\text{H}_2\text{(gas)} = 2\text{Si (solid)} + 6\text{HCl (gas)} \qquad \textbf{(20-3)}$$

Deposition takes place on thin (4-mm-diameter) resistance-heated silicon rods (*slim rods*, Fig. 20–9a), over several hours, until polycrystalline silicon rods of up to 200-mm diameter grow.

The processes are power-intensive but the entire world demand for EGS amounted to only 13 000 Mg in 1995, hence the microelectronics industry consumes very little power per unit value produced.

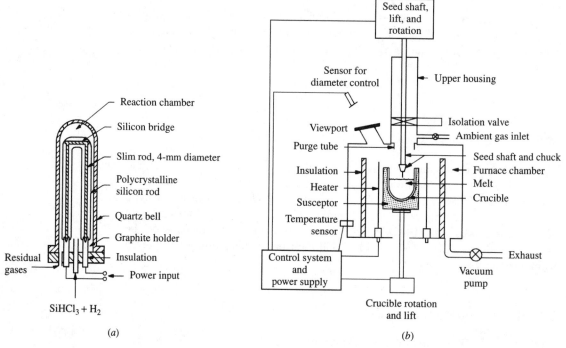

(a)   (b)

Figure 20–9   The growth of (a) polycrystalline silicon rods by CVD and of (b) single-crystal ingots by the Czochralski technique. [*Part (a) after L.D. Crossmann and J.A. Baker, in Semiconductor Silicon 1977, Electrochemical Society, 1977, Pennington, N.J., p. 18. This figure was originally presented at the Spring 1977 Meeting of The Electrochemical Society, Inc., held in Philadelphia. With permission. Part (b) after C.W. Pearce, in VLSI Technology, S.M. Sze (ed.). McGraw-Hill, 1983, p. 25. With permission of Bell Telephone Laboratories.*]

20-2-2   Crystal Growing

Silicon crystallizes in the diamond lattice (Fig. 12–1*a*). Microelectronic devices generally require that lattice defects be minimum; therefore, the aim is to grow near-perfect single crystals, usually in the ⟨111⟩ or ⟨100⟩ orientation.

Crystals are grown in a scaled-up version of the laboratory technique known as the *Czochralski method* (Fig. 20–9*b*). Broken pieces of EGS are loaded into a crucible (usually made of pure SiO_2) supported by a graphite susceptor. Heat is provided by induction or resistance heating. Silicon melts at 1414°C. To grow a crystal of fixed orientation, a single crystal of Si (the *seed crystal*) is partly immersed in the melt, and then pulled slowly upward with simultaneous rotation, while the crucible also rotates. Solid silicon deposits on the seed in the same orientation to form a cylindrical single crystal. A melt of 60 kg yields a 3-m-long, 100-mm-diameter *crystal* (also called *boule* or *ingot*). Most crystals are of 150- or 200-mm diameter (and 300-mm diameters are anticipated). Oxygen would burn up the graphite susceptor and would enter the single crystal; hence, growth takes place in vacuum or in an inert gas (He or Ar). Operating conditions are closely monitored by instrumentation, and computers are used for closed-loop control.

A smaller quantity of single crystals is produced by placing a polycrystalline rod onto a single-crystal seed and then passing a melt zone upward, by moving a high-frequency induction coil at a controlled rate along the length of the rod.

20-2-3   Wafer Preparation

After cooling, the ingot is subjected to a number of steps:

1. It is inspected for crystal structure, resistivity, and defects. Defective parts and the ends of the ingot are removed and the scrap (up to 50% of the ingot) is returned to melting. Since silicon is very hard, it is cut with rotary diamond saws.

2. The surface of the remaining parts is ground perfectly cylindrical with diamond wheels, and then flats are ground along the length to identify crystal orientation and doping type.

3. The crystal is cut into thin (0.6–0.7-mm-thick) wafers. Reasonably straight cut is obtained in ID cutting (Fig. 20–10), in which thin (0.325-mm-thick) stainless steel annular blades are used, coated with diamond powder on their ID. Here some one-third of the silicon is again lost. Blades are rotated at 2000 r/min, and are fed into the crystal at 0.5 mm/s.

4. Flatness and parallelism are improved by simultaneous, two-sided lapping of the surfaces, using Al_2O_3 in glycerine as the lapping medium. Some 20 μm per side is removed.

5. The edges are rounded with diamond wheels to prevent chipping.

6. The wafers are chemically etched to remove the mechanically damaged surface layers.

7. The wafers are polished in a slurry of fine (10-nm) SiO_2 particles in an aqueous solution of NaOH. Some 25 μm is removed per side by a combination of

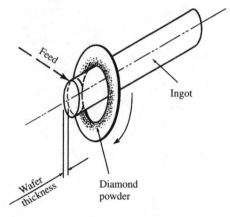

Figure 20–10   Single-crystal ingots are cut into wafers by the ID cutting process.

mechanical action and the oxidation of silicon by NaOH. At this point, the wafer has a surface roughness of about 2 nm R_a and this must be maintained throughout the cleaning steps.

8. In ULSI devices gate oxide thickness is below 10 nm, hence the surface must be very clean, smooth, and free of native oxide. The wafers are cleaned in a special solution (*slice clean*) to remove all residues. This step is critical for ULSI circuits and is aided by high-power *megasonic* (850–900-kHz) cleaning. Dry cleaning techniques have also been developed. After a rinse with deionized water, wafers are rigorously inspected.

20-3 DEVICE FABRICATION

The silicon wafers serve as the substrate on which planar devices are made by a sequence of process steps. Since an IC contains hundreds of thousands or millions of devices, it is necessary to build test circuits or conduct computer simulations. Once the design is fixed, the manufacturing sequence is determined.

20-3-1 Outline of Process Sequence

As shown in Fig. 20–7, circuits consist of intricate patterns of semiconducting, insulating, and conducting features which must be created in several superimposed layers. This can be achieved either by *deposition in the requisite pattern*, or by deposition over the entire surface and *removal of the unwanted portions*.

The major steps of processing are shown in Fig. 20–11. The substrate of all devices is either an *n*-type or a *p*-type wafer (Fig. 20–7). The doping element concentration is set either during crystal growing or, after wafer preparation, by overall diffusion. The substrate is now ready for developing circuit features. For each layer of the device, a pattern is made which is optically reduced in size to make an optical mask. To transfer the pattern onto the surface of the wafer, the surface of the wafer is oxidized and a photosensitive resist is deposited and exposed through the optical mask. Unwanted portions of the photoresist are dissolved, and the oxide film thus exposed is etched away. Thus, an oxide mask is generated which then controls the diffusion of doping elements into the substrate. This sequence is repeated several times until the device is constructed. Metallic conducting paths are deposited on the surface and the circuit thus created is provided with connections to the outside world. The IC is then packaged to protect it against damage and the effects of the environment.

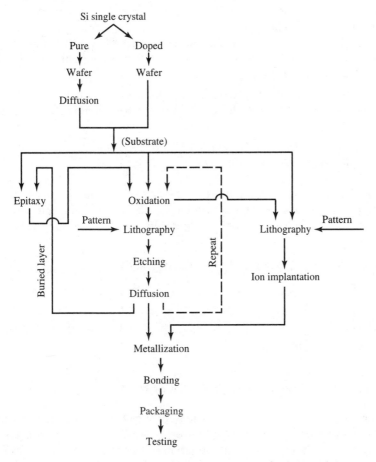

Figure 20–11   General outline of fabrication sequence for integrated circuits.

In the following, each processing step will be discussed from the manufacturing point of view, and then the integration of processes will be indicated.

20-3-2   Basic Fabrication Techniques

It will be useful to review some of the equipment and techniques that find a variety of applications in semiconductor manufacture. It should be noted that basically all processing is done in batches. However, by the simultaneous processing of tens and even hundreds of wafers (each containing up to several hundred ICs), and by a high degree of automation, the cost of individual devices can be kept very low.

Heating   Much of wafer processing takes place at elevated temperatures. Two approaches are usual:

1. In *cold-wall processing* the wafers rest on a susceptor which is heated by resistance, induction, or radiant heat. Gases (or vacuum) necessary for processing are contained in a chamber with cooled walls. In *rapid thermal processing* (RTP) individual wafers are on a holder that contacts in only a few places; thus, they can be heated rapidly by lamps through a quartz window. RTP is gaining importance as the line width diminishes.

2. In *hot-wall processing* the wafers are placed into a furnace which is heated to the processing temperature. In RTP they are moved rapidly in and out of the heat zone.

When thermal or chemical reactions take place in the cold-wall enclosure or furnace, it is usual to speak of a *reactor*.

Pyrolysis   The atomic species required is obtained by the thermal decomposition of a compound (often a halide) of the element. The compound is introduced into the furnace where decomposition and deposition on the substrate (wafer) takes place.

PVD   We already encountered it in Sec. 19-6. All its forms are used in IC fabrication.

CVD   The process (Sec. 19-7) was developed in response to the needs of the semiconductor industry. It is used in the production of EGS (Sec. 20-2-1) and it has many other applications.

20-3-3   Changing the Composition of the Surface

The composition of the wafer may have to be changed either over its entire surface or only in locations determined by the resist pattern. Basically, two kinds of changes may be made: Impurity atoms are introduced into the surface (diffusion or ion implantation), or reaction with oxygen is induced (thermal oxidation).

Diffusion   We already discussed in Sec. 6-1-2 the role of diffusion in metals. Diffusion of doping elements into silicon has been one of the most important methods of changing the composition of the wafer in a controlled manner, and it is unavoidable whenever the wafer is heated for any length of time. Diffusion can serve one of two purposes (Fig. 20–11):

1. Overall diffusion to change the characteristics of the wafer and thus of the entire substrate. Silicon crystals can and are being grown with controlled amounts of doping elements, particularly As, P, Sb, or B, to give resistivities from 0.0005–50 $\Omega \cdot$cm. However, pure EGS crystals are also grown, and dopants are then introduced into the wafers at the beginning of processing.

2. Localized diffusion may serve several purposes: Develop p and n regions in active devices; create stripes for resistors; and develop highly doped regions for interconnects and for pads to which metallic contacts will be applied (ohmic contacts).

The most common technique is thermal diffusion, conducted at 800–1200°C, using a gaseous, liquid, or solid source of the dopant. For example, B can be diffused from boron tetrabromide. The first step is oxidation

$$4BBr_4(\text{liquid}) + 3O_2 = 2B_2O_3 + 16Br \qquad \textbf{(20-4)}$$

Boron is incorporated into the silicon surface with the formation of a thin SiO_2 film

$$2B_2O_3(\text{gas}) + 3Si(\text{solid}) = 4B + 3SiO_2 \qquad \textbf{(20-5)}$$

Alternatively, BN slices can be placed between the silicon wafers; at elevated temperatures, the concentration gradient drives B into the silicon. Good control is obtained by depositing a doped oxide and then driving the dopant into the silicon at elevated temperatures.

We saw in Sec. 6-1-2 [Eq. (6-2)] that diffusion is greatly accelerated with increasing temperature, hence diffusion is conducted at high temperatures. In general, smaller atoms diffuse more rapidly, and differences in diffusion rates can be used to control the depth of impurity penetration in various steps of processing. All atoms diffuse more readily and preferentially along grain boundaries and at other disturbances of the crystal structure, and this is one of the reasons for using single-crystal wafers, free of defects.

Diffusion is omnidirectional; hence, diffusion away from the edges of masks cannot be avoided, and this puts a lower limit to the length of features (width of lines) that can be produced. To prevent unwanted diffusion, thin layers of *diffusion barriers* are deposited. These may, however, reduce adhesion, and several different layers may have to be deposited to balance requirements.

The concentration of dopant is highest at the surface and may reach the limit of solid solubility (Sec. 6-1-2). Concentration falls off rapidly and, if a more uniform distribution is desired, deposition (now called *predeposition*) is followed by subsequent heating (*drive-in diffusion*). To prevent the escape of dopants through the surface during drive-in diffusion, a thin oxide film is formed after predeposition. Much of the diffusion takes place interstitially, and holding at temperature allows dopants to enter into substitutional positions and thus become electronically active.

As mentioned, diffusion—wanted or unwanted—takes place every time the temperature is raised in the course of processing, and this must be taken into account in designing the process sequence.

Ion Implantation   This process (Sec. 19-8) too was developed for the semiconductor industry to introduce dopants in a highly controlled manner. It has many advantages: The dose can be readily controlled, usually in the range of 10^{14}–10^{18} atoms/cm^3 (but even to concentrations exceeding the solubility limit); a very steep concentration gradient is obtained at the edge of masks, thus line width can be reduced; and the temperature of the wafer remains low.

The high-energy impacts greatly disturb the wafer surface. Silicon atoms are knocked out of their lattice locations, vacancies form, and substantial structural damage occurs. In effect, the implanted layer becomes amorphous. This is undesirable since it reduces carrier lifetime by accelerating recombination; therefore, ion implantation is followed by annealing at 200–800°C to reestablish the crystalline structure. Coincidentally, impurity atoms are driven to greater depths. This has the undesirable effect of omnidirectional diffusion (both lateral and in-depth); therefore, very rapid annealing by laser beam is often favored.

Thermal Oxidation   Silicon dioxide (SiO_2) has several properties desirable for IC fabrication. It is an insulator, hence it can provide dielectric isolation of devices; it is an essential component in MOS devices (Fig. 20–6a and b); and it isolates conductors in multilevel structures. It adheres well to silicon but has a lower (0.8×10^{-6}) thermal expansion than silicon (2.6×10^{-6}), and large thermal stresses can be set up. Diffusion of P, Sb, As, and B is very slow in it; hence, even a thin film (on the order of 0.1–1 μm, more usually, around 0.5 μm) can serve as a diffusion barrier. It is used as a mask for doping by diffusion, and also as a barrier against the loss of previously deposited dopants, as in the drive-in stage of two-stage diffusion. However, Ga, Al, Zn, Na, and O diffuse fast in the oxide and, for these elements, a silicon nitride layer must be used as a barrier, often with a thin oxide interface on the silicon surface.

The thermal oxidation of Si may be conducted in dry oxygen

$$Si + O_2 = SiO_2 \qquad \text{(20-6)}$$

but steam oxidation is more suitable for thicker (over 0.5-μm-thick) films

$$Si + 2H_2O = SiO_2 + 2H_2 \qquad \text{(20-7)}$$

Growth rates are greatly accelerated by moderately high pressures; increasing the pressure from atmospheric to 20 atm (2 MPa) causes a tenfold increase in oxidation rate.

Plasma oxidation in a pure oxygen discharge has the advantage of keeping temperatures below 600 °C.

20-3-4   Deposition of Surface Films

These techniques differ from the previous ones in that layers, films, are formed entirely by material deposited on the surface, with no intentional reaction with the surface.

Epitaxy   We saw in Sec. 18-4-1 that epitaxial growth refers to the growth of a crystal in the same orientation as the substrate: essentially, the new layer becomes a continuation of the substrate. This was the case also in growing a silicon crystal from the liquid on a seed crystal (Sec. 20-2-2), and also in annealing after ion implantation. In *homoepitaxy* the deposited layer is identical to the substrate; in *heteroepitaxy* it is of a different material (fairly substantial mismatch in structure can be accommodated in very thin films). Most frequently, doped silicon is grown on the substrate, at temperatures below T_m. The layer may be grown over the entire surface prior to forming local features (as the n layer in Fig. 20–5b), or after a local feature had been formed (Fig. 20–11); this local feature then becomes a *buried layer* (as the n^+ layer in Fig. 20–7a).

Cleanliness of the substrate surface is critical; therefore, the previously cleaned wafers are first etched, in situ, with anhydrous HCl at 1200°C. Deposition then follows in the same equipment (reactor). The *growth reactor* is a high-purity quartz bell jar in which wafers are placed on a graphite or SiN- or SiC-coated quartz susceptor, heated inductively (rf) or by radiant heating. The most frequently used technique is CVD by hydrogen reduction of the halides of Si. Thus, silicon is deposited by passing, for example, silicon tetrachloride ($SiCl_4$) over the surface of wafers at 1150–1250°C

$$SiCl_4 \text{ (gas)} + 2H_2 \text{ (gas)} = Si \text{ (solid)} + 4HCl \text{ (gas)} \qquad \textbf{(20-8)}$$

The layer grows at a rate of 0.2–0.3 μm/min. Doping elements are codeposited from halides, such as arsine (AsH_3) which decomposes on the hot surface and becomes entrapped in the epitaxial silicon layer. Some of the compounds (such as arsine) are highly toxic; therefore, reactors are constructed and operated with the utmost care. Dopant atoms will also diffuse from the substrate into the epitaxial layer (*autodoping*) and this sets a lower limit on the thickness of epitaxial layers that can be grown with a controlled impurity concentration.

For this reason, *low-temperature epitaxy* (LTE) processes have been developed. Silicon can be deposited below 700°C if a bare (ultraclean) Si substrate can be maintained. One of the techniques is *ultrahigh vacuum CVD* (UHV/CVD): the reactor chamber is pumped down to 10^{-7} Pa and the reactants are then admitted. The other is *molecular beam epitaxy* (MBE). This too is conducted in ultrahigh vacuum (10^{-6} to 10^{-8} Pa): Si and dopants are evaporated with electron guns at 500–900°C. The evaporated species are transported at a very high velocity to the surface of the substrate. In *selective epitaxial growth*, Si is deposited only on the bare Si substrate without the growth of an amorphous Si layer on the SiO_2 or SiN masks.

Deposition of Dielectrics and Polysilicon   *Dielectrics* (insulators) and polycrystalline silicon (or, simply, *polysilicon*) do not take an active part in the semiconductor action but are essential for the functioning of devices.

Deposition normally takes place by CVD with reactant gas flowing over the surface of wafers in cold-wall reactors. Deposition is more uniform, at reduced pressures (30–250 Pa), in hot-wall reactors (low-pressure CVD or LPCVD). Lowest temperatures (100–400°C) are typical of plasma-assisted deposition (or, simply, plasma deposition), because the energy for the chemical reaction is supplied primarily by the glow discharge (Sec. 19-6-2).

1. Dielectrics are deposited for various purposes: To provide electrical insulation between successive layers; to present barriers to diffusion; to serve as masks for diffusion and ion implantation; to passivate the surface (tie down the free surface bonds); and to protect devices from environmental and mechanical damage. The most frequently used dielectrics are as follows:

a. Silicon dioxide (SiO_2) is usually deposited in the pure form, although phosphor-doped silica (P glass) is also frequently grown. With 4–7% P, it has the advantage of flowing in a viscous manner at 950–1100°C, giving good step coverage with a smooth surface.

b. Silicon nitride (Si_3N_4) is deposited in stoichiometric form by CVD, from a reaction of silane (SiH_4) with ammonia (NH_3), at 700–900°C. Plasma-assisted deposition produces a nonstoichiometric nitride (SiN). Both kinds of nitride are excellent barriers to water and sodium diffusion. In plasma deposition the temperature is low enough to allow deposition over the completed device (i.e., encapsulation for protection against the environment and mechanical damage).

2. Polysilicon is deposited by LP-CVD, at 570–650°C and 25–130 Pa, by the pyrolytic decomposition of silane

$$SiH_4 = Si + 2H_2 \qquad \textbf{(20-9)}$$

Deposition proceeds at the rate of some 10 nm/min. The layer may be doped by adding dopant gases during deposition, or after deposition by ion implantation or diffusion. Doped to high concentrations, polysilicon can serve as: gate elements in MOS devices; conductive contacts (ohmic contacts) to crystalline silicon; high-value resistors; and as an intermediate product, namely as a diffusion source for the formation of shallow junctions. In MOSFET devices it is used to self-align source and drain diffused regions (Sec. 20-3-7) and enables a high level of integration. Annealing at 800°C in wet or dry oxygen produces an insulating *polyoxide* film.

Film deposition processes require stringent controls. Most reactants, including silane, are toxic and flammable, and the properties of films—including composition, resistivity, and internal stresses—are greatly affected by process conditions.

Metallization   Metallic elements replace the wiring used in conventional electronics, and as such have made integrated circuits practical. Their purpose is to provide highly conductive current paths between devices. *Metallization* plays a much more important role than might first appear. Even though currents are small, on the order milliamperes, they result in very high current densities in the fine wires and can accelerate wear-out by *electromigration* (diffusion under the influence of operational currents); for wires to conduct without significant loss, resistance must be low. Also, conductors create capacitive and inductive loads, which are among the main factors limiting the switching speed of circuits. In VLSI and ULSI devices it would be difficult to route wires around devices; therefore, *multilevel* (2–6-level) *metallization* (Fig. 20–12a) is employed. This has the further advantage of reducing the size of the chip, but metal conductors (plugs) are now needed to connect successive levels.

The most frequently used metal is aluminum; it adheres well to SiO_2 and silica glasses, and it makes low-resistance contacts with the highly doped p^+ and n^+ regions of transistors and with polysilicon. Most metallization used to be carried out by PVD (Fig. 19–4). For VLSI and ULSI circuits, sputter deposition at over $0.5T_m$ is preferred because it produces dense films (Al is deposited at 200–300°C). Magnetron deposition allows a lower argon concentration, thus little or no Ar is included in the films. Deposition is followed by annealing to form an alloy interface with Si. To limit electromigration, aluminum is alloyed by low levels of copper and is then referred to as Al(Cu). The gaps between the metal lines must be filled with a dielectric. A smooth surface is needed for lithography of the next layer: *Planarization* of the interlevel SiO_2 is by etching or chemical/mechanical polishing with fine ceramic (alumina or silica) powder and an etching/oxidizing chemical.

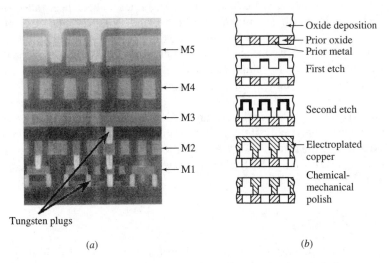

Tungsten plugs

(a)   (b)

Figure 20–12   Cross section of semiconductor device with 5 layers of Al(Cu) metallization with W studs (a) and process sequence for dual Damascene process of copper metallization (b). [Part (a) from Intel Web site; part (b) courtesy IBM Corporation.]

Plugs are usually made of tungsten which is not only a reasonably good conductor but also resists high temperatures. PVD, being a line-of-sight process, gives poor coverage on the walls of connecting holes (*vias*); therefore, CVD finds wide use. Tungsten plugs may be formed by the hydrogen reduction of WF_6

$$WF_6 + 3H_2 = W + 6HF \tag{20-10}$$

or by pyrolytic decomposition

$$WF_6 + \text{energy (thermal, plasma, or optical)} = W + 3F_2 \tag{20-11}$$

A TiN barrier around the W plug is necessary to prevent attack of the lower-lying Al by the WF_6. A thin Ti cladding is added under and/or over the Al(Cu) wires; upon subsequent sintering, a $TiAl_3$ alloy is formed which serves as a redundant current strap that maintains circuit continuity in the event of electromigration of the Al(Cu). Some aluminum plugs are made by *reflow* (melting); for this, the melting point must be reduced (e.g., by alloying with Ge).

A recent breakthrough at IBM and Motorola makes it possible to use copper metallization. At IBM, trenches are etched in the SiO_2 layer and copper is used for both wires and studs (Fig. 20–12*b*). Using proprietary additions and process control for electrodeposition, the trenches are coated on three sides from the bottom up and void-free deposits are made. The technique is named *dual Damascene process* (after the ancient decorative technique in which grooves engraved in a metal surface are filled with fine Cu, Ag, or Au wire). To prevent Cu diffusion, the trenches are first coated with a barrier (in the case of Motorola, Ti nitride). A thin Si_3N_4 cap serves as diffusion barrier and adhesion promoter on the top surfaces of the Cu wires. There are several advantages. Copper interconnects have 40% lower resistivity than the clad Al(Cu) ones. The aspect ratios of Cu wires can be reduced (thus reducing capacitance and cross talk) and finer pitches and line widths become possible. At the same time, upper Cu wiring levels can be made thicker than Al(Cu) ones but still at a tighter pitch, thus reducing the delay caused by long lines on the chip. These scaling trends significantly improve the performance of large, high-speed chips with multilevel Cu wiring.[1]

A disadvantage of W (and Mo) is rapid oxidation. Therefore, silicides such as WSi_2, $MoSi_2$, $TiSi_2$, $CoSi_2$, and $TaSi_2$ are used as MOSFET gate electrodes, alone or with heavily doped polysilicon. Deposition is by simultaneous evaporation (or sputtering) of the refractory metal and silicon, or by CVD. Relative to doped polysilicon, these polysilicides (*polycides*) have the advantage of lower resistivity, important for VLSI devices.

20-3-5   Lithography

Having reviewed the basic techniques of film formation and deposition, we are ready to see how the spatial distribution of features is controlled. It is evident from Fig. 20–7 that an IC containing thousands or millions of devices on a small

[1]D.C. Edelstein: "Copper Chip" Technology, *SPIE*, **3508**:8–18 (1998); L. Geppert: Solid State, *IEEE Spectrum*, 1998(Jan):23–28.

chip must have a very intricate pattern indeed. To control deposition and removal in such fine detail, a modern version of an old technique is used.

At the end of the 18th century, *lithography* (from the Greek lithos = stone) was developed for printing. The design is put on a flat stone (or metal) surface with grease. First water and then an oil-base ink is applied; the ink is repelled by the wet parts but is adsorbed by the greasy surfaces. Thus an image of the pattern can be transferred to paper. For microcircuits, the pattern is transferred photographically, hence the term *photolithography*. Because it controls the geometry of the IC, it is the most critical step in fabrication.

Generation of Masks   The first step is the preparation of a *photomask* through which a photosensitive film can be exposed. For this, a pattern is made for each layer of the IC.

For circuits of moderate complexity, the patterns can be large-scale drawings (also called *artwork*), magnified some 100–2000 times. The drawing is photographically reduced to 4× or 5× magnification onto a glass plate, which in turn is reduced again to 1× size, reproduced repeatedly (*step-and-repeat*) in exact locations, until a plate corresponding to the wafer surface area is completely covered with tens or hundreds of identical patterns.

For VLSI and ULSI circuits, the complexity of design demands that the pattern be generated and verified by CAD, and the digital data can be used to drive directly a 1× or 5× pattern generator. Best definition is obtained with an electron beam (such as is found in a TV tube, but highly collimated to give a spot size below 1 μm).

A mask prepared by photographic techniques consists of a glass plate covered by a photographic emulsion, which is soft and susceptible to damage. For this reason, masks are most often made with a thin (100–200-nm) film of a hard material such as chromium metal or iron oxide, again patterned by the photoresist technique.

Pattern Transfer   The pattern is then transferred to the surface of the wafer that has been coated with a resist. A *resist*, in the most general sense, is a substance in which wave energy produces chemical or physical changes, making it resistant to acids. A portion of the resist is removed with a solvent, either in the exposed areas, creating a positive of the pattern (*positive resist*) or in the unexposed areas, creating a negative image (*negative resist*).

A *photoresist* is a light-sensitive substance, typically a polymer. It is applied in a thin layer: A few drops of the resist are placed on the wafer which is then spun (at up to 12 000 r/min) to spread out the resist. Residual solvent is driven off by baking; the resulting film is typically 0.3–1.0 μm thick. Negative resists are polymers that cross-link upon irradiation, hence the exposed area becomes insoluble. Positive resists contain a photoactive dissolution inhibitor; where it is destroyed by light, the exposed area becomes soluble. A high-energy electron beam causes chain scission. Negative photoresists are faster and wafers can be produced in seconds, but many polymers swell during development and this limits resolution. Positive resists are slower, but do not swell and give a finer line width, and are the choice for ULSI circuits.

If there already are step-like features on the wafer from previous processing stages, the steps may have to be filled up with a thicker polymer film until a level surface is produced on which a thin photoresist can then be deposited.

Several techniques of exposure are used:

1. In *optical lithography*, line width is limited to the wavelength of light. In UV-light lithography, exposure is made with a mercury lamp, emitting light of 310–450 nm wavelength. The mask is made of chromium deposited on glass. Phase-shift masks have a second layer which creates an interference pattern to increase resolution and depth of focus. Exposure with deep UV light of 200–300 nm wavelength gives resolutions of 0.25 μm and a resolution of 0.18 μm is sought with 193-nm wavelength; this represents the lower limit attainable by light lithography. To minimize light absorption at short wavelengths, a quartz plate is used instead of glass.

a. In some ways the simplest is the *contact method* in which the mask is laid upon the wafer. However, a single trapped silicon dust particle can damage the mask and all subsequently exposed wafers; therefore, extremely high standards of cleanliness must be maintained. Resolution is around 0.35 μm and is limited by diffraction of light.

b. The chance of damage is less when the mask is held at a distance of 20–50 μm from the surface (*proximity method*), but resolution drops to 2–4 μm.

c. No damage occurs at all when the mask is held away from the wafer and the image is projected on the mask (*projection method*). The mask is either full-size or enlarged (typically, 10× or 4×). By exposing only a small part of the wafer at a time, submicron resolutions can be achieved.

2. Proximity printing with x-rays of 0.4–5-nm wavelength reduces diffraction effects and allows higher resolutions, below 0.2 μm. A *synchrotron* is a powerful but expensive x-ray source. The mask must be made on material transparent to x-rays (such as polyimide, Si, SiC, Si_3N_4, or Al_2O_3), with the pattern made of Au, W, or Ta. Only a few square millimeters can be exposed at a time and the step-and-repeat process must be used. It takes typically 1 min to process a 125-mm wafer.

3. Patterns can be directly written with an electron beam, focused to 0.01–0.5-μm diameter. The resists are the same as in x-ray lithography. Resolutions of 0.1 μm (100 nm) are possible with resists such as PMMA. Further resolution is limited by backscatter. The main drawback is the relatively low output rate; typically, it takes minutes to hours to write a 125-mm wafer.

Techniques have been continually developing, allowing a reduction of feature size with all techniques, and some more exotic techniques, not discussed here, are also being developed.

Since all ICs are multilayer devices, several lithography steps are involved in succession. Registry relative to previously developed layers is most important. Masks are aligned mechanically or with the aid of alignment features.

The photoresist is destroyed by high heat; therefore, for high-temperature processing steps such as diffusion, the pattern must be developed in a heat-resistant film such as SiO_2 or Si_3N_4.

Irrespective of the nature of the surface film (high-temperature mask or metallization), the pattern can be obtained by one of two methods (Fig. 20–13). In the *lift-off method* the lithographic mask (the photoresist pattern) is made first; the surface layer (most often metallization) is deposited on the wafer, and then the unwanted portions of the film are lifted off by dissolving the mask (Fig. 20–13a). The deposited features have rounded tops and the resist limits processing temperature to below 300 °C. Therefore, most patterns are transferred by the *subtractive* or *etching method* in which the film is first deposited over the entire surface and unmasked portions of the film are then removed by etching (Fig. 20–13b).

20-3-6   Etching

Etching is the most frequently performed process step. It may serve one of several purposes:

1. Remove (strip) a film such as an oxide from the entire wafer surface.

2. Remove material over selected areas, as defined by a photoresist. Thus, a silicon oxide or silicon nitride film formed on the wafer is locally etched away in

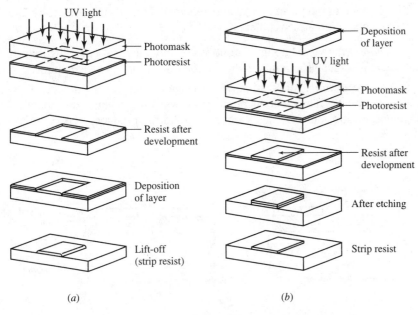

(a) (b)

Figure 20–13   Typical photolithography by (a) lift-off and (b) deposition techniques.

preparation for diffusion, ion implantation, or surface layer deposition. Alternatively, a film such as aluminum deposited over the entire surface is locally etched away to leave only the metal needed for interconnections. Two basic approaches are used.

Wet Etching   The wafers are submerged in aqueous solutions of chemicals that selectively dissolve one or the other material. Thus, HF dissolves SiO_2 but not aluminum, whereas phosphoric acid dissolves Al without attacking SiO_2 or Si. Photoresists are then removed (stripped) with a solvent such as acetone.

Wet etching has lost favor for several reasons. Reaction products build up in the etchant, and this creates the problem of periodic disposal. A further problem is that wet etchants are *isotropic*; i.e., they remove material at the same rate in all directions, causing an undercut under the resist (Fig. 20–14*a*). In designing a photomask this can be taken into account, but features still must have relatively large dimensions. Single-crystal silicon is anisotropically etched, but dry etching has still become the dominant technique for VLSI and ULSI devices.

Dry Etching   Material removal relies on the energy created by a plasma. A great advantage is that no large quantities of effluents are produced. Furthermore, if the bombardment is directional, a high degree of *anisotropy* of etching can be achieved: The rate of etch becomes much greater in depth than in lateral directions, hence undercuts are minimized and fine features, with walls perpendicular to the wafer surface (Fig. 20–14*b*) can be produced. Such directional bombardment is obtained in the plasma of *planar reactors* (Fig. 20–15). Basically, three methods of operation are possible:

1. *Sputter etching.* Material is removed by the impact of energetic ions of a nonreactive gas such as argon, and the bombardment may induce damage.

2. *Plasma etching.* Ionization of some molecular gases produces highly reactive fragments. Thus, chemical etching is obtained when gases containing halogen atoms are ionized; e.g., carbon tetrafluoride produces very aggressive

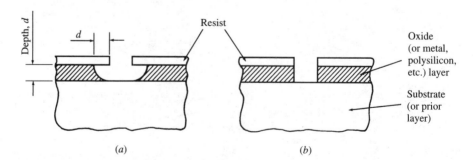

Figure 20–14   Patterns of etching: (*a*) fully isotropic material removal in wet (chemical) etching creates much less-well-defined features than (*b*) anisotropic plasma etching.

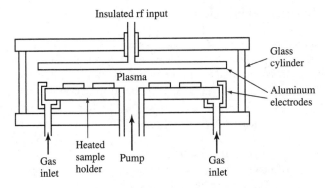

Insulated rf input

Glass
cylinder

Plasma

Aluminum
electrodes

Gas
inlet

Heated
sample
holder

Pump

Gas
inlet

Figure 20–15   Highly directional, anisotropic etching is obtained
in planar plasma reactors in which wafers lay flat
on the lower electrode, with the plasma directly
above them. [*From A.C. Adams, in S.M. Sze (ed.):
VLSI Technology, 2d ed., McGraw-Hill, 1988,
p. 237. With permission of Bell Telephone
Laboratories.*]

free F

$$CF_4 + e^- = CF_3^+ + F + 2e^- \qquad \textbf{(20-12)}$$

CF_4 does not react with Si but F forms the volatile SiF_4. Etching is selective, isotropic, and free of surface damage. To produce submicron features and to protect the ozone layer, fluorocarbons are being replaced with Cl-, F-, or Br-base gases.

3. *Reactive ion etching.*   This method is a hybrid of sputter etching and plasma etching. The system is arranged so that the benefits of both are retained. In sputter etching gas pressure is low and rf energy high; in plasma etching gas pressure is higher and rf energy lower; in reactive ion etching both parameters are intermediate. In contrast to wet etching, reactive ion etching (RIE) is highly anisotropic because ion bombardment is perpendicular to the surface; bombardment helps to remove etch-inhibiting species from the bottom but leaves them in place on the side walls; thus, deep, straight-sided grooves can be made. Gases can be chosen to ensure selective etching. Pure Si etches rather slowly and etching rates can be greatly accelerated by adding some 12% O_2 to CF_4. Etch rates of SiO_2 increase greatly with the addition of H_2 to the plasma. Chlorine etches polysilicon several times faster than it does SiO_2. High etch-rate ratios relative to photoresists are also required. Etch rates range between 10–1000 nm/min. It is this technique that made VLSI and ULSI possible, since grooves of under 1-μm width and of several micrometers depth can be etched. Because gas temperatures are low, conventional photoresists can be used as masks. After plasma etching is completed, the reactive gas is replaced with oxygen and the plasma is used to strip the resist.

20-3-7   Process Integration

As indicated in Sec. 20-3-1, techniques described in Secs. 20-3-3 to 20-3-6 are employed in succession to generate several superimposed layers. Not surprisingly, microprocessor and computer control of individual steps is widely practiced, and great advances have been made in the complete closed-loop automation of processing sequences. This has the benefit of increased productivity and, since contamination of wafers through human contact is avoided, the yield is also greatly improved. In the following, a typical process sequence will be given for a CMOS device (as in Fig. 20–6b).[2]

A number of masks are needed for defining various regions. The p-type substrate (Fig. 20–16a) is cleaned, and mask 1 defines the area of n-type dopant in the n-tub region (Fig. 20–16b). Heavy oxide regions are formed with mask 2 (Fig. 20–16c); to obtain the exact dopant concentration in the p region, additional implanting may be needed with mask 3 (Fig. 20–16d). The thin gate oxide is then grown (mask 4, Fig. 20–16e) and the polysilicon gates are developed (mask 5, Fig. 20–16f). The gate oxide prevents diffusion of dopant into the underlying channel; hence, channel and gate are automatically aligned without the need for masking (*self-aligned silicon gate process*). The source and drain regions are made (mask 6, Fig. 20–16g and, with the mask reversed, Fig. 20–16h). After growing an insulating oxide and cutting windows in the oxide (mask 7, Fig. 20–16i), aluminum metallization is deposited over the entire wafer. Mask 8 allows selective removal of the aluminum (Fig. 20–16j). (The sequence would be different for metallization with copper, Fig. 20–12b.)

The IC is now complete but vulnerable. To protect it, it is *passivated* (protected from outside mechanical and environmental damage) by the deposition of a low-melting P glass layer or silicon nitride (Fig. 20–16k). A photoresist is again needed to open contact windows where connections to the Al metallization will be made. A plasma-deposited silicon-nitride film may be used to provide further protection.

In detail, many more operating steps are involved in VLSI and ULSI chip manufacture, but the basic steps are the same. They are just repeated several times to form the many layers of these structures. An advanced example is shown in Fig. 20–17a for a 32-bit RISC CPU; the six-level copper metallization is revealed by etching away the silicon dioxide (the connecting pads are on top in Fig. 20–17b).

Process Control   The ever-increasing circuit density has been achieved by tightening up on process control and eliminating sources of contamination. Dust particles of only one-tenth of the feature geometry cause fatal defects; therefore, processing takes place in *clean rooms* (or enclosures). Detailed specifications have been developed for air cleanliness. For example, U.S. Federal Standard 209E allows maximum 10^1 particles/m^3 of 0.5-μm size in class M1, 10^2 particles/m^3 in class M2, etc. The number of smaller particles is also specified. Temperature is controlled to $22 \pm 0.1°$C, humidity to $43 \pm 2\%$. Ultrapure water is used for rinsing

---

[2]After P.R. Shepherd, *Integrated Circuit Design, Fabrication and Test*, McGraw-Hill, 1996.

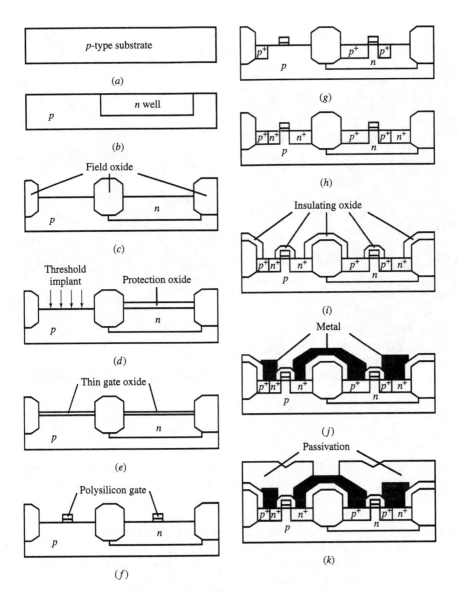

Figure 20–16   Simplified sequence of CMOS device fabrication. See text for process steps. (From P.R. Shepherd, *Integrated Circuit Design, Fabrication and Test,* McGraw-Hill, 1996, Fig. A1.12, p. 207. With permission.)

after slice clean. Human contact is avoided as far as possible. Process conditions, film thicknesses, and doping levels are continuously monitored.

Test chips designed to measure electrical properties are located at several sites (usually in the kerf areas between product chips) on each wafer. Circuits are tested on completed wafers, using often very complex test patterns designed to check all aspects of the operation of devices. Access is gained through windows

Figure 20–17   (a) 32-bit RISC CPU measures only 5.5 × 7.5 mm yet contains 6.4 million transistors. (b) The 6-level copper metallization is revealed by etching away the silicon dioxide. (*Courtesy Tom Way, IBM Corporation.*)

etched in the glass layer. Defective chips are identified and marked. On some chips, especially the more complex ones, redundant circuits are provided and, by blowing fuses in the metallization, can be activated to replace defective circuits.

The back of the wafer is ground off (typically, some 250 μm is removed) and, unless connections are made by beam-lead bonding, to be described later, individual chips are now separated. Because silicon is hard and brittle, it can be scribed with diamond-tipped scribers, *diced* with diamond-impregnated disks, or scribed with a pulsed laser beam, and then snapped apart. Complete separation with diamond-coated wheels is the method of choice for ULSI devices. Sorting may take place at this point. Defective chips are discarded: Good chips may be attached to a backing to facilitate automatic processing. The chips are now ready for packaging.

20-3-8   Packaging

The delicate chip must be electrically connected to the outside world and must be protected from mechanical damage and environmental influences. The techniques used are adaptations of processes described in earlier chapters. This does not mean that packaging is free of problems or that it is inexpensive; indeed, often the cost of packaging exceeds that of chip fabrication.

Physically, the package may appear in various forms. One of the most popular ones used to be the *dual in-line package* (DIP, Fig. 20–18*a*). The IC, measuring only a few millimeters on its sides, sits on or in a recess of the substrate to which it is attached (*bonded*) by a metallic or polymer layer. Contact with the outside world is made with *pins* or *terminals*, typically, on 2.54-mm (0.1-in) centers, which in the example of Fig. 20–18*a* would be fed through holes in a printed wiring board (Sec. 20-4-3). For higher pin counts (especially on logic and microprocessor devices), leads are on all four sides (*quad package*) or over the entire back face (*pin grid array*) on much closer (0.8–0.3-mm) centers. Pins are of J (Fig. 20–18*b*) or gull-wing (Fig. 20–18*c*) configuration for surface mounting. Significantly smaller footprints are secured by the *lead-on-chip* (LOC) structure (Fig. 20–18*b*).

Leads (terminals) are connected to the pads provided on the metallization of the chip (die), and the whole package is hermetically sealed against the outside environment. Thus, there are three areas of major interest to us: attachment of the chip to the substrate (*die bonding*), attachment of leads (*interconnecting*), and protection of the assembly (*encapsulation*).

Die Bonding   The chip must be attached to a sturdier substrate which also helps to remove heat. The substrate may be alumina, which is an electrical insulator but has low heat conductivity (Table 4–1). Beryllia (BeO) has high heat conductivity but is more expensive and its dust is highly toxic. Alternatively, the substrate is a metal, such as Kovar (Table 4–1) or, for higher heat conductivity, a copper alloy. Metal substrates are often made up in the form of 0.1–0.15-mm-thick, etched or stamped *lead frames* that provide conducting paths to pins. With increasing chip

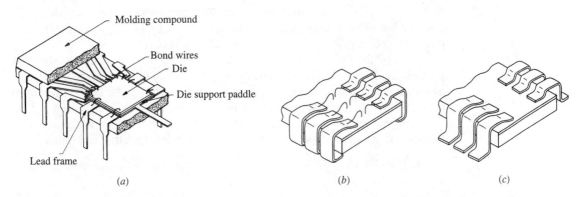

Figure 20–18   (*a*) Sectional view of a DIP package shows the chip (die) bonded to a lead frame and encapsulated in plastic. Surface-mount packages have (*b*) J leads or (*c*) gull-wing terminals. [*Part (a) from J. R. Howell, in Proc. Int. Rel. Phys. Symp. p. 105 ©1981, IEEE. With permission.*]

size, mismatch in thermal expansion becomes a serious concern and influences process choice. Basically, three techniques are available:

1. When the thermal expansion of the substrate is much different from that of the die, a ductile, high-lead solder (of a liquidus between 300 and 315°C) is used.

2. Bonding to Kovar lead frames or ceramics is usually done with pure Au or Au-2Si foil, of less than 50-μm thickness, cut to size (*preform*). To ensure good wetting, the die and the substrate are plated with Au or, for metal lead frames, also with Ag. In the Au–Si system there is a eutectic at 3.6% Si. On heating above the eutectic temperature of 370°C, the eutectic forms by alloying between die and preform (*eutectic bonding*). For ceramic packages, the conductor can be deposited by screen printing. As in all joining processes, surface films must be broken through and mechanical scrubbing is helpful.

3. For most less-demanding applications, attachment is by polymers, primarily epoxy or polyimide loaded with Ag flakes (*epoxy bonding*). The expansion coefficient is high (Table 4–1), but thermal stresses are low because the elastic modulus of the polymer is low. Polymers are easily dispensed pneumatically or by printing, processes which lend themselves to automation. The low (125–175°C) curing temperature prevents damage to delicate active devices. The cured epoxy can stand temperatures of 320°C for short times, as are needed for some wire bonding techniques.

Interconnection   The chip emerges from the fabrication process with aluminum (or copper) connecting pads. It is now necessary to make connections to the terminals (or connecting pins) of the package. In principle, the highest density of interconnection (50–1000 mm^{-2}) can be achieved, at low cost, on the chip itself. Within the package (*first-level connection*) the density is lower (typically, 1 mm^{-2}) but this is still higher than can be made outside the package (*second-level connection*). Hence the trend is to put more complex circuits within a single package, even if this necessitates more pins to interconnect with the outside world. The number of outside connections can be reduced by placing two or more ICs on a common circuit base (*multichip module*, MCM).

To keep the die size small, pads of the metallization are made as small as possible, and are typically spaced only 0.1–0.2 mm apart. Therefore, techniques had to be developed to make delicate connections to fingers of lead frames or to pins. Aluminum is always covered with a tenacious oxide; as discussed in Sec. 18-3, reliable bonds can be established only if surface films are broken up, and methods of achieving this goal are central to all interconnecting techniques.

1. *Wire bonding.* Individual, thin (25-μm-diameter) Al or Au wire is bonded to the pads by one of two techniques:

a. *Thermocompression bonding* is an example of solid-state welding (Fig. 18–4) by hot (300–350°C) flattening a wire. Obviously, no lubricant can be used and, as in all unlubricated upsetting, sticking friction prevails over the contact surface (Fig. 9–4*d*). The oxide remains undisturbed in the dead-metal zone, and heavy deformation is needed to increase the *d/h* or *L/h* ratio; this ensures sliding and thus breaks up the oxide, at least away from the central dead-metal zone. To reduce the required pressure and facilitate diffusion, bonding is performed with a heated tool (*thermode*) and the pressure is kept on for 0.5–2 s.

b. *Thermosonic bonding* is an example of ultrasonic joining (Fig. 18–5*b*). The substrate and die are heated to 150–250°C, and the wire is compressed while the bonding tool is ultrasonically vibrated. The ultrasonic energy input heats and softens the wire, rubbing is effective in breaking up oxides, and a good quality joint results at a relatively low temperature.

The application of these techniques is shown in Fig. 20–19.

2. *Tape automated bonding.* TAB is a technique of making all joints simultaneously. The pattern of connectors is etched, by photolithography, into 33- or 66-μm-thick Cu foil, which is usually Au-plated to improve bonding. It may be backed with a polymer such as polyimide. Alternatively, the pattern is plated on polymer tape. Either way, it is supplied in a long tape, allowing automation. The bond is established against *bumps*; these are tiny Au- or Cu-plated columns of approximately 25-μm height. They are deposited either on the Al pads of the die or at the corresponding locations on the tape. Bond is then established by one of two techniques:

a. Thermocompression bonding, with a thermode in the shape of the connector pattern, heated to 450–550°C, applying a pressure of 275–480 MPa (the joint does not reach this temperature).

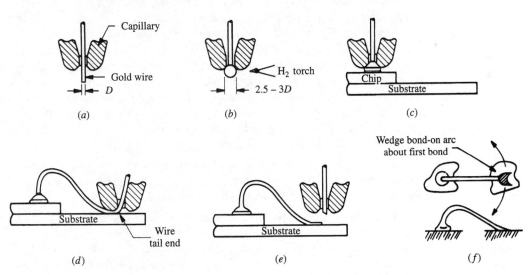

Figure 20–19   Thermosonic ball-wedge bonding involves: (*a*) feeding gold wire, (*b*) forming ball with hydrogen torch, (*c*) bonding the ball to the pad on the chip, and (*d*) bonding the wire to the lead frame. (*After J. S. Stafford, Semiconductor International, May, 1982, p. 82. With permission Cahners Publishing Co., Chicago.*)

b. Eutectic bonding to a tin-plated copper tape. The Au–Sn eutectic forms at 280°C. Under the applied pressure, rapid alloying takes place.

3. *Flip chip technique.* This also involves the formation of Au-plated bumps on the die, but a layer of Pb–Sn is now evaporated (or, more recently, electroplated) on top of the bumps. Heating makes the solder flow into a ball. The chip thus prepared is then flipped over the tape, and the bond is established by heating. In the *controlled collapse chip component* (C^4) technique the distance between chip and substrate is maintained.

4. *Beam-lead bonding.* This differs from all above techniques in that the gold leads to the outside world are deposited on the wafer, and the chips are then separated by etching away the silicon. The leads overhang the chip, appearing like tiny beams (hence their name); connections to the Al pads can be made by wire bonding.

Packaging Once the connections are established, the IC is complete and needs only a hermetic package to protect from the external environment. The degree of protection depends on the application; for example, packages must be mechanically stronger and more temperature-resistant in an industrial or military application than in a home computer. Apart from the required electrical characteristics, the package must also aid in removing heat. The following are the major packaging techniques used:

1. *Plastic molding.* The completed assembly is placed into the cavity of a transfer-molding die (Fig. 14–12*b*) and is enveloped with a thermosetting (e.g., epoxy, epoxy–silicone, or silicone) or thermoplastic (e.g., polyphenylene sulfide) polymer. For reduced shrinkage and thermal expansion and for increased strength, the polymer is filled with SiO_2 or Al_2O_3 (Table 4–1). As indicated in Sec. 14-3-4, shear heating reduces the viscosity of the polymer so that it does not damage the delicate connecting wires. Molding pressures can be kept low (around 6 MPa) and exposure to temperature (typically, 175°C) is limited to 1–5 min. Such *postmolding* is relatively inexpensive, fast, and is the dominant technique for mass-produced ICs. However, the difference in thermal expansion between polymer, silicon, and gold sets up stresses during molding and in service.

2. *Premolding.* The chip is spared the exposure to molding stresses and temperatures when the mold is premade with an appropriate cavity (*premolds*). Pin connections are usually made with a lead frame.

3. *Ceramic packaging.* We saw in Secs. 12-4-2 and 12-4-6 that ceramics, mostly Al_2O_3, are tape-cast into thin strips from which substrates can be blanked out. Holes through which electric connections will be made (*via holes*) are punched out, wiring paths are printed on the surface in the form of a refractory metal (usually W) powder slurry, and the via holes are filled with metal. Several ceramic blanks are assembled into a sandwich which is then fired. The W is nickel plated to facilitate brazing of Kovar lead wires or lead frames, using a eutectic Cu–Ag brazing alloy. The process requires strictest control of shrinkage. It is expensive but essential for highest performance requirements. Beryllia substrates offer higher heat conductivity but at a higher price.

4. *Glass-sealed refractory packages.* These too use ceramic substrates but the lead frame is sealed into a glass (Table 12–2) above 400°C. To prevent alloying of Au and Al (which would lead to the formation of a brittle intermetallic compound), an all-aluminum construction is used.

5. *Glass-ceramic flip-chip multichip modules.* These are the mainframe computer packages. The glass ceramic has a thermal expansion similar to Si, a much lower dielectric constant than alumina, and allows higher speed in interchip communication.

20-4   PRINTED WIRING BOARDS

The completed IC package or other semiconductor device may be installed directly in some industrial product, and then it only needs a socket or other means of connection. Frequently, however, it is part of a yet larger circuit, and then it is joined to a *printed wiring board* (PWB). Connections are made on the bottom of the board either by soldering the pins (Fig. 20–18a) protruding through holes (*through-hole mounting*) or by joining J (Fig. 20–18b) or gull-wing (Fig. 20–18c) terminals by *surface-mount* technology (Sec. 20-4-3) to make up a *hybrid circuit*.

The board may have *thick-film* or *thin-film* circuitry. The distinction between the two is rather arbitrary and is based more on the method of manufacture than on actual thickness, although *thin films* are typically $0.1–1$ μm thick whereas *thick films* are $10–25$ μm thick. These films are used to provide resistors and capacitors which would use too much chip surface, as well as for the interconnection of several discrete components (including inductors, diodes, transistors, etc.) and ICs, usually in mass production where discrete components assembled on a printed circuit board would be too expensive.

20-4-1   Thin-Film Fabrication Methods

The technology is essentially the same as for ICs, but a variety of film materials are used. The circuit is designed as for ICs and, if features are fine, photolithography is used to develop an SiO_2 mask (Fig. 20–13). Metals are then deposited by PVD or sputtering. Alternatively, metals are deposited over the entire surface and etched away with the aid of a photomask. Various devices can be formed:

1. For low-value *resistors*, a Nichrome (Ni–Cr) alloy, tin oxide, or tantalum nitride is deposited for resistivities of $10–1000$ $\Omega/\square$. For high-value resistors, cermets (usually Ta–Cr silicides) of $100–20\,000$ $\Omega/\square$ are used. Stability is achieved by heating (baking).

Exact values are obtained by *trimming*, i.e., subjecting the resistor to a treatment while its resistance (or the intended operation of the circuit) is monitored.

The method of trimming depends on the resistor material. Oxides can be further oxidized until the desired resistance value is reached. Nichrome is mechanically trimmed; the line width is locally reduced by a diamond scribe, spark erosion, or high-energy beam (electron beam or laser).

2. *Capacitors* are constructed by the successive deposition of metal (usually Al or Au), insulator (SiO_2, Al_2O_3, or, for higher dielectric constant and capacitance, HfO_2), and again metal films. For an insulating film of 100-nm thickness, the capacitance is 800 pF/mm^2 with Al_2O_3 and 3500–7000 pF/mm^2 with HfO_2. Adjustment is possible by providing tabs (spurs) which can be trimmed off.

3. Wiring is produced by techniques used for metallization (Sec. 20-3-4). Electrodeposition allows relatively thick but narrow (submicron-width) connections for high-density multichip modules. Indeed, electrodeposition is a vital element in all microfabrication.[3]

20-4-2   Thick-Film Circuits

The basic difference between thick-film and thin-film techniques is that, when the fine definition given by thin-film fabrication is not needed, thick-film circuits can be produced at a lower cost. This comes from the replacement of photolithography with *silk-screen printing*, a technique adopted from the textile and graphic arts industries. In the original form, a dye or ink is squeezed through the holes of a silk screen to form a predetermined pattern on fabric or paper. In thick-film technology the silk is replaced with a metal screen. The following steps are followed:

1. Circuits are designed as in IC fabrication, except that circuits are often less complex and, because they contain much coarser features (typically, 50–500-μm line width), the artwork is only 5–20 × magnified. It is drawn on paper or cut out of a plastic film by hand or computer. A photomask is then prepared.

2. The screen is usually stainless steel of typically 8 to 325 mesh (woven of wires of 0.94–0.028-mm diameter, giving openings of 0.118–0.051 mm). In one of the processes, the so-called *direct emulsion steel screen process*, a photosensitive emulsion is deposited on the screen, exposed through the photomask, and the unexposed areas are dissolved. The circuit pattern is now defined by the nonblocked holes of the screen.

3. The screen is held, with controlled tension, some 0.5–1.0 mm off the ceramic substrate, and an ink is applied with a squeegee (Fig. 20–20). The *ink* is formulated to have appropriate rheological properties: It must not flow under gravity (hence it must have a well-defined yield point, as a Bingham fluid does, Fig. 7–5, line *D*), yet it must flow at a low shear stress when it is spread (hence it must be pseudoplastic, Fig. 7–5, line *C*). The ink contains several powders dispersed in a polymer–solvent system. The ingredients are: for a conductor, metal powder (usually, Pt–Au, Pd–Ag, or similar alloy) with some oxides; for

---

[3] *IBM J. Research and Development,* **42**(5):561–722 (1998).

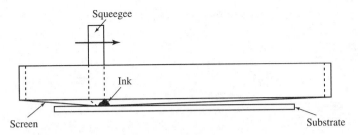

Figure 20–20   Thick-film circuits are deposited by silk-screen printing.

resistors, also a metal powder–oxide mix but with a higher proportion of oxide; for dielectrics, only ceramic (such as $BaTiO_3$) powder. Inks of glass dispersions are used for encapsulation. To establish bonding between the metal and oxide powders and the substrate, a low-melting glass powder (*glass frit*) is always incorporated in the ink.

4. The printed substrates are dried at 125°C to drive off the solvent, then fired, usually in a continuous furnace, where the polymer is first burnt off. The temperature is then raised to about 850°C to sinter the metal and oxide particles and to fuse them, with the aid of the glass, to the substrate. The circuits are then allowed to cool in a programmed manner to prevent cracking and oxidation.

5. Circuit performance can be optimized by trimming resistors as in thin-film technology. Air-abrasive trimming is also possible: the width of the resistor is adjusted by abrasive blasting the edge of the resistor stripe, using fine (approximately 50-μm) powder (typically Al_2O_3) in high-pressure air.

20-4-3   Soldering

ICs and other components are finally joined to the PWB by soldering. Potentially all techniques described in Sec. 18-8-3 could be used, but some techniques have been specifically developed for the purpose.

Tin-based solders are generally used and are chosen for corrosion resistance, fluidity, and lack of (or controlled) reaction with leads. Melting range is particularly important when the board is subjected to several soldering operations in sequence (*step soldering*). Lead-rich solders are used for higher temperature, the eutectic Sn–Pb composition flows readily, and off-eutectic compositions (such as Sn-40Pb) fill larger holes (Table 18–2). For lowest temperature, Sn–In solders are used (Sn-50In). Sn–Ag solders minimize attack on Ag coatings. Leads of devices often need coating to assure wetting by the solder, and ceramics are metallized. Fluxes are typically of rosin base or contain an organic acid in water or alcohol. There are basically two approaches:

1. Through-the-hole mounting requires the solder to be applied to the leads protruding through holes in the PWB. Of great value is *wave soldering*: Molten solder is pumped through a narrow slot (nozzle) so that fresh solder moves in a gently flowing wave (Fig. 20–21).

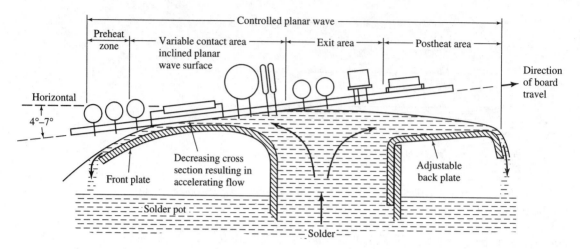

Figure 20–21   Planar wave soldering showing adjustable backplate to control wave configuration and flow pattern. (*Courtesy Electrovert Consulting Services, Elmsford, New York.*)

2. Surface mount technology has become the preferred method for VLSI and ULSI circuits because higher densities are obtainable and the problems associated with making holes are avoided. The solder is applied as a paste (*ink*), consisting of Sn–Pb (or, to prevent loss of Ag from conductors by diffusion, Sn–Pb–Ag) solder powder and wetting and fluxing compounds. Application is by screening, stenciling, or in dots by *x-y* coordinate dispensers; electrodeposition has recently been introduced for high-density flip-chip interconnections. After placing the parts, the boards are baked to remove the binder (if present) and the solder is reflowed. In mass production this is done by passing the board through a furnace or by infrared heat. A special form is *vapor-phase soldering*: the board is passed through a chamber in which the saturated vapor of a liquid of selected boiling point (100–265 °C) assures rapid heat transfer, so that the solder is melted in a few seconds.

20-5   MICROFABRICATION

As the name implies, *microfabrication* involves, in the wider sense, all processes that make products on the submillimeter scale. Many traditional and nontraditional processes are capable of doing so: Wires are drawn to a few micrometers (microns) in diameter; metal foils are rolled to a few microns in thickness; miniature cold-formed products are forged or extruded; submillimeter shafts are machined on Swiss-type lathes; and many plastics processing and all abrasive and nontraditional machining processes are capable of operating on this scale. In a sense, a mechanical wristwatch is a microfabricated product. However, from the perspective of people involved in microfabrication, these are all traditional or conventional processes and products. Microfabrication, in the narrower sense, is taken to apply to IC technologies for the manufacture of mechanical or electromechanical devices. Indeed, one of the alternative (and more restrictive) terms is *microelectromechanical systems* (MEMS). The products may be purely mechanical, such as gear trains; electromechanical, such as actuators and motors; or they may combine different forms of operation in a vast variety of sensors.

Sensors and actuators, which convert energy from one form into another, currently constitute the largest application of microfabrication. Such devices are not new. A bimetallic strip made from two metals of different thermal expansion (Sec. 18-12) deflects with rising temperature, actuates a switch when the desired temperature is reached, the toast is ejected, and the toaster is turned off. Piezoelectricity of crystals (Sec. 4-9-3) is exploited to convert the force exerted on a load cell into electric current and, vice versa, to drive an ultrasonic tool by imposing an alternating current (Sec. 16-8-6). Such devices are regarded as microfabricated products only when based on IC technology. This technology allows integration with IC signal conditioning and processing capability (IMEMS), and it is then usual to speak of *smart sensors*.

20-5-1   Techniques of Microfabrication

For electromechanical devices, it is necessary to extend IC technology since, in constructing ICs, layers are often but a fraction of a micrometer thick. In contrast, even a membrane for a pressure sensor will have to be up to several micrometers thick, and much thicker layers are needed for a gear. Pattern definition will have to be of high quality and walls must be perpendicular and smooth. Basically, two approaches can be taken to silicon-based microfabrication:

1. *Bulk micromachining* is the term applied to *top-down* (removal) processes in which a crystalline or noncrystalline material is etched in a pattern defined by lithography. High-sensitivity photoresists, in conjunction with oxide masks, assure the required definition. Wet, or more frequently, dry etching techniques (especially RIE) are used to develop features. With suitable etching techniques, deep (> 200 μm) and narrow trenches (with aspect ratios over 10:1) can be developed. Undercuts and closed spaces are created by placing several thin micromachined slices on top of each other and uniting them by diffusion bonding.

2. *Surface micromachining* is a term applied to *bottom-up* (additive) approaches. Successive layers of polysilicon are deposited and unwanted areas are removed by lithography and etching. If a base (sacrificial) layer is deposited on the substrate and is dissolved upon completion, freestanding parts can be developed. Since the structure is built up layer by layer, it is possible to create multipart assemblies such as gear trains with integral shafts without the need for assembly (*in-situ assembly*).

Detailed reviews of micromachining techniques are given by Kovacs and Madou.[4]

Silicon, even though brittle, has attractive properties for mechanical components: density (2.3 g/cm^3) is below that of aluminum, elastic modulus is the same as for steel (210 GPa), yield strength is high (1–2 GPa) and, because there is no hysteresis, it makes excellent elastically loaded structures. With a melting point of 1414°C, it can be diffusion bonded at 1000°C.

In sensors, many physical and chemical principles can be exploited. For example, pressure sensors based on 2–20-μm-thick membranes with implanted strain gages or piezoelectric sensors are used in automobiles (for active suspension hydraulics, lumbar seat support, and barometric pressure sensors). Accelerometers are usually based on displacement of a mass, attached to a piezoelectric crystal or to one side of a capacitor.

Example 20-1 | Accelerometers are indispensable parts of automobiles and many are installed for the deployment of airbags. In one example, accelerations up to $50g$ are measured with ±5% accuracy through the capacitance change caused by shifting the interpenetrating combs of a polysilicon capacitor. The chip contains all electronics needed for excitation, signal conditioning, and self-test; the micromachined sensing capacitor occupies only 5% of the surface area.

Some unusual problems are encountered because of scaling effects. Detailed consideration is beyond the scope of our inquiry, but it will be readily appreciated that, as linear dimensions diminish, effects linked to mass shrink by the third power but those linked to surface shrink only by the square. Thus, friction and viscous drag increase steeply. This creates difficulties in reducing friction and increases the danger of wear.[5]

Example 20-2 | High friction and rapid wear are among the problems micromotors have to overcome. An alternative is offered by the wobble motor. The stator is constructed of isolated conductive segments which are sequentially energized and exert a pull (image) force on an insulated but conductive (or dielectric) rotor. The torque thus created causes the rotor to roll toward the next stator segment. Friction is now beneficial, the surface may even be made wavy or gear-like, and rotary motion ensues. [After S.C. Jacobsen et al, *Proc. IEEE MEMS*, 1989(2):17–24.]

In an entirely different approach, lithography is used to generate the negative of the desired pattern in a thick (from microns- to centimeters-thick) photoresist (usually PMMA) deposited on a *conductive* subplate. After development, the

---

[4]G.T.A. Kovacs: *Micromachined Transducers Sourcebook*, McGraw-Hill, 1998; M. Madou: *Fundamentals of Microfabrication*, CRC press, 1997.

[5]A.D. Romig, Sandia National Laboratories, Albuquerque, New Mexico.

resist is filled with metal by electrodeposition (this is the same as the Damascene process for metallization, Sec. 20-3-4, except that trenches in the photoresist are now filled). When the resist is removed, the metal structure can be used as a permanent mold for injection or compression molding of a plastic; this plastic part is an end-use item or it can be used in a lost-mold process for ceramic molding or for a second electroforming. With synchrotron x-ray lithography, features as small as 0.1 μm can be made with smooth ($R_{max} = 0.02$ μm), vertical walls. The technique is then referred to as LIGA, from the German acronym for Lithographie (x-ray lithography), Galvanoformung (electrodeposition), and Abformung (molding).

Sensors and actuators may convert magnetic to electrical energy. A prime example is the read/write head in magnetic storage devices (hard disks, etc.), originally made as a horseshoe magnet wound with insulated copper wire. Miniaturization became possible with the development of a thin-film head.[6] On a ceramic substrate (originally SiO_2, now $Al_2O_3.TiC$) a magnetically soft material (Ni-20Fe) is deposited to form a yoke with a small read/write gap. The magnet is energized by a copper coil in which relatively thick (2–3 μm) wires are electrodeposited through a polymeric mask. This technology is the basis of microfabrication by electrodeposition and LIGA. In the most recent devices, a separate magnetoresistive head is used for readback and the thin-film head only for recording. This has allowed a tremendous increase in areal densities, so that even commodity hard disks have capacities measured in gigabytes.[7] Hundreds of millions of heads are made every year.

Example 20-3

The ubiquitous CD (or CD-ROM) disk is an example of microfabrication. It combines IC and conventional techniques. The disk itself is a mass-produced plastic part. The pattern is created by IC techniques on a glass substrate through a sequence of steps:

The glass disk is (a) polished to 10-nm R_a roughness and 10-μm flatness, (b) cleaned and dried. (c) A positive photoresist is deposited and (d) patterned with a laser beam. [The pattern is in a spiral with 1.6-μm pitch. Pits of 0.5-μm width and of nine different lengths between 0.83 and 3.56 μm carry the information. Depth is equal to one-quarter the wavelength of laser used for reading the information, so that there is total reflection or total extinction (see Fig. 3–17); each level change (transition) corresponds to 1 in the binary code; the distance between level changes determines the number of zeroes.] The resist is (e) developed and the (positive) glass master is (f) inspected and (g) silver-coated to check the quality. The glass master is then (h) coated with a sputtered Ni layer which permits (i) electroforming to a thickness of 250–300 μm. This (negative) nickel master (called the stamper) is (j) polished on its back to 0.3 μm R_a, (k) separated from the glass master and, (l) after removing the resist, is ready for (m) molding the (positive) polycarbonate plastic disks, which are then (n) coated with PVD aluminum for reflectivity and (o) protected by a transparent plastic coating.

Example 20-4

---

[6]L.T. Romankiw, I. Croll, and M. Hatzakis, *IEEE Trans. Magn.*, **6**:729 (1970).
[7]P.C. Andricacos and N. Robertson, *IBM J. Res. Develop.* **42**(5):671–680 (1998).

The medical field represents vast opportunities for microfabrication. Very small instruments for minimally invasive surgery have been produced for years, at high cost, by conventional techniques; microfabrication has the promise of mass producing them cheaply. One example is the endoscope, a tube fitted with an imaging fiber, light source, and a working channel through which microtools are operated. Such tools can be made by microfabrication. The tip of the endoscope is bent with sheathed cables; these can be replaced by shape-memory alloy wires. Microfabrication also allows developments for new applications.

20-5-2   Nanotechnology

The term *nanotechnology* is somewhat ambiguous. Sometimes it is applied to processes capable of making parts to *tolerances* expressed in nanometers (Sec. 16-9-2). More frequently, it refers to processes that make parts with *dimensions* in the submicron range. A further subdivision is sometimes made by separating out the range between 1 and 0.1 μm as the *mesoscopic domain*. Nanorange then refers to 10–100 nm. This corresponds to only 30–300 interatomic distances and is thus on the molecular scale.

1. *Nanofabrication* produces parts *top-down*, largely by reduced-scale versions of IC processes. Some mass-produced products are already on the market.

Example 20-5 | We saw in Example 20-4 that CD disks are mechanical devices on which information is embedded in the form of tiny pits. This information is read by lasers, and some of these are devices falling within the definition of nanostructures. Molecular beam epitaxy and metalorganic chemical deposition are used to create layers of only 100 nm thickness; in these, electrons are confined and emit light only at defined wavelengths. Such *quantum well lasers* emit light at much lower current densities than conventional lasers.

---

2. *Molecular engineering* is a *bottom-up* approach that aims to build structures the way nature does. The notion seems farfetched but some beginnings have already been made. Some medicines work by *self-assembly*: Molecules of the drug fit into receptors in the body like key in a lock. *Self-organization* of molecules results, for example, in molecular crystals, such as the liquid-crystal polymers (Sec. 13-2-3). DNA is tailor-made by processes of *self-replication*. All these techniques are potential candidates for manufacturing not only replacements for natural products but also for devices, such as computers, sensors, and actuators, now made by conventional techniques and microfabrication.

Example 20-6 | We saw in Sec. 12-1-1 that carbon can exist in a stable form as diamond. We came across another stable, hexagonal, form in Sec. 12-4-6 as graphite, a solid lubricant. In 1985 a third stable form was discovered: In the most well-known version, 60 atoms produce perfect hollow spheres, in a crystal form similar to the patches of a soccer ball. Spatial curvature in these *buckminsterfullerenes* (or, colloquially, buckyballs, after Buckminster Fuller of the geodesic dome fame) and larger, not perfectly round molecules (collectively called *fullerenes*), is obtained by combining 12 pentagons with $m = (n - 20)/2$ hexagons, where n is the total number of carbon atoms in the C_n molecule. Fullerenes can be made in quantity by vaporizing carbon in a helium atmosphere, in electric arcs, flames, and plasmas. Techniques have been developed to open the crystal and trap (cage) other atoms or molecules in it. The composition can be manipulated by other means too, and insulators, conductors, semiconductors, and superconductors can be made. Nanotubes are related structures: A hexagonal sheet is folded into a tube and closed by conical ends (many nanotubes consist of several concentric tubes). These tubes are potentially very strong and, by trapping other atoms inside, nanoscale wires can be made. Intense research will no doubt result in practical applications.

20-6 SUMMARY

Microelectronic devices are the agents of the Second Industrial Revolution. They control machines from robots to microwave ovens, from automotive engines to the landing of aircraft; they aid in computation, from hand-held calculators to supercomputers; they allow communication, from telephones to satellites to fiber-optic devices; they help to entertain and teach, from radio to television to computer-aided instruction; and they are at the heart of CAD/CAM and CIM.

Microelectronic devices are based primarily on electrical phenomena taking place in semiconductor materials such as doped Si and GaAs. Analog and digital circuits are formed, by batch production methods, in a planar arrangement on the surface of wafers. By shrinking the size of individual features, integrated circuits containing millions of components can be fitted on a single chip, the sides of which measure only a few millimeters. Advances in design and manufacturing have allowed increased performance at steeply reduced costs, at a rate unparalleled in other manufacturing fields.

The enormous complexity of the circuits characteristic of VLSI and ULSI can be mastered with the aid of CAD. The density of components demands that features be fabricated with minimum sizes below 1 μm, and a host of manufacturing techniques had to be developed—some from laboratory techniques, others from more generally used manufacturing processes.

The operation of devices depends on the incorporation of impurity (doping) atoms in a highly controlled manner. Unintended impurities, crystal defects, localized damage to the circuit or crystal all defeat operation. Therefore, manufacturing is conducted in an exceptionally clean environment, with a high degree of automation, using techniques that minimize chance events. Because many chemicals used are toxic, workers need protection; conversely, the high standards of cleanliness demand protection of the product from human contact. If workers must be present, they need to be dressed in complete protective suits.

A complex sequence of manufacturing steps is needed to develop more complicated devices. Miniaturization hinges on the development and strictest control of lithographic and film-deposition and implantation techniques.

Some of the techniques can be applied to microfabrication, the batch manufacture of submicron-size devices based on mechanical, electromechanical, and other principles. Hundreds of millions of sensors are already made every year and actuators are not far behind. The frontiers of research and production are being extended to yet smaller-scale products by various techniques of nanofabrication and molecular engineering.

PROBLEMS

20-1   State what is the difference between (*a*) epitaxial silicon and (*b*) polysilicon.

20-2   Give the name of a process for the deposition of polysilicon.

20-3   (*a*) Define the term doping in semiconductor manufacture and (*b*) give at least two processes for performing it.

20-4   (*a*) Define the term metallization and (*b*) give two processes for performing it.

20-5   Describe the basic steps involved in creating the metallization pattern on a Si wafer.

20-6   Describe the basic steps in making a glass mask for photolithography.

20-7   Draw sketches to show the results of (*a*) wet etching and (*b*) dry etching. (*c*) Define reactive ion etching and explain why it is suitable for making deep trenches.

20-8   Make sketches to show the difference between (*a*) through-the-hole and (*b*) surface mount techniques for mounting chips on a board.

20-9   Define MEMS.

20-10   Define (*a*) top-down and (*b*) bottom-up processes for fabricating a micromechanical system.

20-11   Suggest the processing sequence for making a miniature Si gear.

20-12   Suggest a processing sequence for making a miniature metal mold for plastic molding a part defined by a straight generatrix (similar to a part made by blanking a sheet, but of complex shape and micrometer dimensions).

FURTHER READING

Semiconductor Devices

Electronic Materials Handbook, vol. 1, *Packaging*, ASM International, 1989.

Chang, C.Y., and S.M. Sze (eds.): *ULSI Technology*, McGraw-Hill, 1996.

Doane, D.A., and P.D. Franzon (eds.): *Multichip Module Technologies and Alternatives*, Van Nostrand Reinhold, 1993.

Lee, T.W. (ed.): *Microelectronic Failure Analysis Desk Reference*, 3d ed., ASM International, 1993.

Mahajan, S.: *Principles of Growth and Processing of Semiconductors*, McGraw-Hill, 1999.

Murarka, S.P.: *Metallization: Theory and Practice for VLSI and ULSI*, Butterworth-Heinemann, 1993.

Rao, G.K.: *Multilevel Interconnect Technology*, McGraw-Hill, 1993.

Shepherd, P.R.: *Integrated Circuit Design, Fabrication and Test*, McGraw-Hill, 1996.

Sze, S.M. (ed.): *VLSI Technology*, 2d ed., McGraw-Hill, 1988.

Tummala, R.R., and E.J. Rymaszewski (eds.): *Microelectronics Packaging Handbook*, Van Nostrand Reinhold, 1989.

Yang, E.S.: *Microelectronic Devices*, McGraw-Hill, 1988.

Zorich, R.: *Handbook of Quality Integrated Circuit Manufacturing*, Academic Press, 1991.

Micro- and Nanotechnology

Andonovic, I., and D. Uttamchandani (eds.): *Principles of Modern Optical Systems*, Artech House, 1989.

Bouwhuis, G., et al: *Principles of Optical Disk Systems*, Adam Hilger, 1986.

Crandall, B.C., and J. Lewis (eds.): *Nanotechnology*, MIT Press, 1996.

Dresselhaus, M.S., G. Dresselhaus, and P.C. Eklund: *Science of Fullerenes and Carbon Nanotubes*, Academic Press, New York, 1996.

Drexel, K. E.: *Nanosystems*, Wiley, 1992.

Fatikow, S., and U. Rembold: *Microsystem Technology and Microrobotics*, Springer, 1997.

Fujimasa, I.: *Micromachines*, Oxford University Press, 1996.

Kovacs, G.T.A.: *Micromachined Transducers Sourcebook*, McGraw-Hill, 1998.

Madou, M.: *Fundamentals of Microfabrication*, CRC Press, 1997.

Nicolini, C. (ed.): *Molecular Manufacturing*, Plenum, 1996.

Regis, E.: *Nano: The Emerging Science of Nanotechnology*, Little, Brown, 1995.

Taniguchi, N. (ed.): *Nanotechnology*, Oxford University Press, 1996.

Trimmer, W.S. (ed.): *Micromechanics and MEMS: Classic and Seminal Papers to 1990*, IEEE Press, 1997.

Vaidya, R., G. Lopez, and J.A. Lopez: *Nanotechnology*, in *Kirk-Othmer Encyclopedia of Chemical Technology*, 4th ed., Suppl., Wiley, 1998, pp. 397–437.

Two horizontal CNC machining centers, each with its own automatic workchanger (on the right), are linked into an automated pallet cell in which a rail-guided vehicle (in the foreground) shuttles pallets to and from machines and to off-line load/unload stations. A cell controller tracks all activities. (*Courtesy Cincinnati Milacron, Cincinnati, Ohio.*)

21

Manufacturing Systems

Individual production processes require organization into a
functioning manufacturing system. We shall review:

Material movement, including robots

Organization for batch and mass production

The application of group technology to flexible manufacturing cells

Mechanization and automation of assembly

The all-important topic of quality management

Statistical process control

The role of manufacturing engineering in a company organization

In Sec. 2-1 we already outlined the major activities involved in the totality of manufacturing, and in Chaps. 6–20 we investigated the principles related to individual processes. Armed with this knowledge, we may now proceed to explore elements of some vital technological and organizational features of manufacturing systems, including the movement of material within a plant, organization of production facilities for mass and batch production, quality assurance, and manufacturing management.

21-1 MATERIAL MOVEMENT

The movement of materials, parts, and tools is an essential element of all manufacturing operations, including the production of parts and the assembly of parts into subassemblies or finished products. Studies of batch-type shop operations have shown that, for some 95% of the total production time, parts are being transported from one place to another or are just waiting for something to happen. Even of the 5% of time that they spend on a machine tool, they are actually worked upon only some 30% of the time, while the rest of the time is absorbed in loading and unloading, positioning, gaging, or idling for some extraneous cause (Fig. 21–1). If productivity is to be increased, first the methods of material movement, loading, positioning, clamping, and unloading must be improved, and only then will it make sense to worry about speeding up the process itself. Conversely, if in-process time is already high, there is little incentive to improve upon material movement. There are several ways of moving material.

21-1-1   Attended Material Movement

Operators can move objects with the least capital expenditure, but this is usually also the least efficient and most costly technique. Efficiency can be increased by loading smaller parts into baskets, but this has the disadvantage that parts have to be picked out for the next operation. Parts can be arranged on *pallets* (or platforms or trays) and, if desired, oriented so that they become more accessible for the next operation, and this practice is spreading because it also protects the parts from accidental damage.

Forklift trucks facilitate material movement while retaining flexibility, but need unobstructed passageways. *Cranes* also ensure flexibility and need no floor space but may interfere with each other.

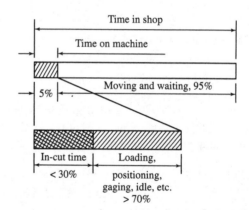

Figure 21–1   In batch production, the average workpiece is actually worked upon only a small fraction of the total time it spends in the shop. (*After C. F. Carter, in Proc. 2d Int. Conf. on Product Development and Manufacturing Technology, pp. 125–141, Macdonald, London, 1972. With permission.*)

21-1-2   Mechanized Material Movement

The term *material movement* usually refers to transportation between production units.

1. Roller gangs, endless belts, carousels, overhead conveyors, towline carts (moved by below-floor chains), and similar devices can be highly efficient for moving parts as well as pallets but can be reorganized only by changing the physical layout of the system.

2. Flexible transportation is ensured with *automated guided vehicles* (AGVs). The vehicles move, like forklift trucks, on the factory floor, but follow any one of several paths, for example, under inductive guidance provided by wire (cable) guides embedded in the floor. The paths of individual vehicles can be readily programmed and reprogrammed for changing production requirements, avoiding collisions while optimizing the path of each vehicle. Sensors stop the vehicles when they encounter an obstruction. In a further development, vehicles are equipped with sensors and interactive programs that allow them to find their own way.

Material movement to and from machine tools often requires that the part be turned, oriented, gripped, and placed into a predetermined position. Manual operation is the most flexible but it is also most prone to error from operator fatigue, especially if the task is repetitious or involves the movement of very small or very large (and heavy) parts. Heat, smoke, fumes, gases, or particulates may make the environment unpleasant or unhealthy. Therefore, there are powerful incentives to mechanize and automate the loading and unloading of parts. Simple and relatively low-cost automation has long been employed.

1. Purely *mechanical devices* of varying degrees of complexity are, in general, highly efficient but inflexible. They may combine material movement with machine loading and unloading. We already saw examples of mechanized transportation between successive stages of progressive machine tools such as cold headers, transfer presses, and automatic screw machines. Instrumentation may be added to provide some simple feedback, such as the presence or absence of the part.

2. Parts can be individually *palletized*. Alignment in the machine tool is automatically obtained if: The pallet is made with the precision of a fixture, the part is held (clamped) in exact position (usually with the aid of locating holes), and the pallet is accurately located (with the aid of locating pins) on the machine-tool bed. The cost of pallets is reduced by the use of *modular fixtures* constructed on precision base plates.

3. Small parts are often handled effectively with simple mechanical devices. Vibratory belts and bowls, reciprocating forks, rotary disks, or magnetic devices are combined with simple but ingenious work-orientation devices from which the parts progress through feed tracks to the machine tool, where a metering device (such as a mechanically or electronically actuated escapement) releases the part at the proper time.

4. Loading and unloading can often be done with mechanical arms, generally referred to as *manipulators*. They can be divided into several subgroups. Classification is possible by the method of control:

a. *Manipulators*, in the more restricted sense of the term, are mechanical arms under manual control, with the aid of push buttons, joysticks, or devices that take the motions of the operator's hand and transform it into equivalent motions of the mechanical arm. Their lifting capacity ranges from a few grams to hundreds of tons. Examples are *remote manipulators* used in dangerous environments (e.g., in the atomic industry) and *forge manipulators* used in the open-die forging of large ingots. In a sense, programming is totally flexible in response to the operator's commands.

b. *Fixed-sequence manipulators* advance to preset positions in a preset sequence. Position and sequence are set by limit switches, proximity sensors, light interrupts, and relays. Limit switches are used also to sense whether an action—by the manipulator or the machine it serves—has indeed been taken, thus providing a primitive version of feedback. Typical examples are manipulators that move sheet-metal parts between presses, and unloading shuttles used with die-casting and injection-molding machines.

c. *Robots* are defined by the Robot Institute of America as "programmable multifunctional manipulators designed to move material, parts, tools, or specialized devices through variable programmed motions for the performance of a variety of tasks." Thus they differ from fixed-sequence manipulators only in their variable programming. The so-called *pick-and-place robots* are reprogrammable, for example, with a programmable logic controller (PLC) or personal computer (PC), but often lack a feedback system, and thus could be classified as programmable fixed-sequence manipulators.

21-1-3   Robots

Robotic devices consist of two elements:

1. A *mechanical structure* which includes:

a. A *base* with movable parts, articulated in such a way that one or more (usually up to six) degrees of freedom are attained. Many devices comprise simple shuttles moving along guide rods; each shuttle has one degree of freedom. In its most familiar form, the robot has an arm which may be articulated in various ways (Fig. 21–2). A rigid arm moving up and down and swiveling around a column has two degrees of freedom. An arm that moves (or tilts) up and down, rotates (swivels), moves radially in and out, and has a wrist with twist (swivel), bend (pitch), and yaw movements, possesses six degrees of freedom.

b. The *gripper* (*hand, jaw*, or, more generally, the *work-holding device* or *end effector*) which holds and moves the part or tool.

c. *Drive elements* that provide the motive power for the various motions. Drives are usually pneumatic, hydraulic, or electric, and sometimes combinations

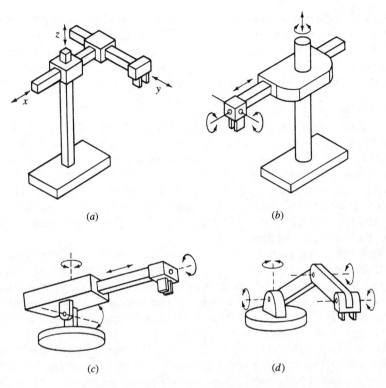

Figure 21–2   The mobility of robots depends on prismatic (P) and revolute (R) joints. Their combination gives (a) Cartesian (PPP), (b) cylindrical (RPP), (c) polar (RRP), and (d) revolute (RRR), also called articulated, configurations. (After L.L. Toepperwein et al., Technical Report AFWAL-TR-80-4042, Air Force Wright Aeronautical Laboratories, Dayton, Ohio.)

of these. Mechanical devices such as cams, levers, and linkages are less frequently encountered because of their relative inflexibility of programming. Air or, more frequently, hydraulic cylinders or servo-driven ball screws, attached to precision ways, generate linear motion. Air, hydraulic, or stepping motors, or ac or dc servomotors provide rotation.

d. Most robots are *fixed* to the floor, but there are some that move on ground or overhead rails or pneumatic tires (*mobile robots*).

It should be noted that a robot may either move a part relative to the tooling or move the tooling relative to the part (both versions are used, for example, in painting).

2. A *control system.* True robots are driven by servomechanisms which incorporate closed-loop control (as in Fig. 2–3c). Sensors measure displacements and feed a signal back to the controller, so that the gripper is positioned accurately, usually within 1.0 mm or better. In their simpler forms, robots move from point to point without following a defined path. Continuous-path robots follow a defined path, thus can be used for such operations as spray painting and arc welding. Robots can be used in the greatest variety of applications, including the loading of machine tools and presses, inspection, and assembly.

Robots are programmed in various ways:

Playback robots can be programmed or "taught" using the "walk-through" method in which a robot arm (or a substitute "training arm") is manually moved through the required path, with control commands inserted whenever some particular action (switching a tool on or off, or waiting for a machine tool to perform a given action) is needed. In the "lead-through" method a control panel ("teach pendant") is used to position the arm. Either way, the controller stores the instructions thus received and plays them back, but without the interruptions, delays, and hesitations typical of manual control.

Other robots are programmed the same way as CNC machine tools (Sec. 2-5-3). In the most advanced form, the database established in CAD/CAM is used to preprogram all robot motions and actions. Several programs may be stored and called upon when the appropriate part is presented. Part identification may be given by a previous workstation or by reading bar codes (similar to those used in supermarkets) applied to the part or pallet.

An *intelligent robot* or *sensory robot* is a CNC robot equipped with some form of artificial intelligence which allows it to cope with nonfixed situations (randomly oriented parts, parts not presented in exact positions) and to perform adaptive control of operations. Some forms of sensing are already fairly widely available:

1. *Visual sensing* requires cameras that usually contain CCD light-sensitive elements (picture elements or pixels). These are rapidly scanned to acquire information on the distribution of light intensities. This information, when converted into the required digital form, can be processed in a computer for image recognition (*image processing*). More complex tasks call for the processing of images obtained from several cameras simultaneously.

2. *Tactile sensing*, in the simplest form, requires force-sensing elements built into the end effector. There are many possibilities for further feedback. For example, infrared light may be ducted through fiber-optic bundles into the jaws of the end effector; when the jaws come close enough to the part for light to be reflected, jaw movement is slowed and the part is gripped with a preset force. Systems more closely approaching human senses are being continually developed.

3. *Adaptive control* links the actions of the robot to the information obtained by sensors. Among others, force or torque sensing elements are involved in adaptive control. For example, a deburring robot may move along the edge of a part at a high rate while searching for a burr. Increased deflection of the tool (increased force on the toolholder) indicates the presence of burr, whereupon feed rate is reduced until the burr is removed. Similarly, in a polishing operation the correct polishing pressure can be maintained, irrespective of part shape, by feedback from a force sensor.

Although robots can often be introduced into an existing plant, some changes are most likely needed. The robot is less tolerant of variations in shape and dimension of parts than a human operator, and it often pays to redesign parts to suit the limitations of the robot. However, robots can perform tasks tirelessly and, if adequately protected and maintained, reliably, even in hostile, dangerous, or unpleasant environments.

21-2   PRODUCTION ORGANIZATION

We already saw in Sec. 2-1 that problems of manufacturing are best treated as a system. Within this system, parts production is organized for maximum efficiency and minimum cost, consistent with the required quality standards. There is no single form of organization that satisfies all requirements, and the choice depends on the characteristics of production.

21-2-1   Production Characteristics

Two important factors in the choice of processes and their organization are the *total number* of parts to be produced and the *rate of production* (i.e., the number of units produced in a time period such as an hour, day, month, or year). Total production quantity and production rate together define the justifiable expenditure on special machinery and tooling.

The total production quantity is often insufficient to keep a production unit continuously occupied, and production proceeds in lots representing a fraction of the total number of parts. The *batch size* or *lot size* is the number of units produced in an uninterrupted run. There are no strict definitions, but it is customary to speak of *small-batch* (1–100 units), *batch* (over 100), and *mass* (over 100 000 or even a million units) *production*. In general, a larger batch size justifies the choice of processes with inherently higher production rates and thus more favorable economies.

Lot size is not determined by purely technical considerations. The cost of setting up (changing over) must be weighed against the cost of stocking (warehousing) parts between production runs. The move to *just-in-time* delivery has had the effect of reducing lot sizes, resulting not just in lower cost of stocking but also lower floor space requirements.

In evaluating the number of parts produced and the rates of production, it is best to consider all parts that show any similarities in features and operating sequences (Sec. 3-1-2). Close similarities may allow grouping of parts for processing by more productive techniques; absence of similarities will require that great flexibility of operation be retained.

21-2-2   Optimum Manufacturing Method

The *optimum manufacturing process* is selected from a knowledge of process capabilities and limitations, tempered by restraints imposed by the requisite production rates and batch sizes. The choice of machine tool depends on cost factors, and break-even charts (similar to Fig. 21–3) can be constructed to show where one machine tool becomes more profitable than an other.

1. Stand-alone machines with manual control require the smallest capital outlay, but their operation is labor-intensive. Labor costs do not drop significantly with increasing batch size (Fig. 21–3); thus, such machines are best suited to one-off and small-batch production. The operator may be a highly skilled artisan or, in repetitive production, may be semiskilled, with a setup person providing the necessary skills.

2. Properly chosen stand-alone NC or CNC machines are most suitable for small-batch production, although with the trend toward increasingly user-friendly programming devices and with the application of group technology, they become competitive with manually operated machinery. Once the workpiece is clamped in place on the CNC machine tool table and a reference point is established, machining, bending, welding, cutting, etc., proceed with great accuracy and repeatability. Nonproductive setup time is practically nil; therefore, CNC can become economical even for very small lots (Fig. 21–3) widely separated in time. The operator may again be highly skilled, this time with some programming knowledge, or the programs may be provided to the machine by a programmer who may be working from the database of a CAD/CAM system; in this case the operator performs machine supervision and service functions such as pallet loading.

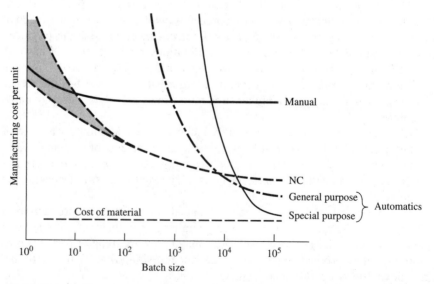

Figure 21–3   The most economical approach to production depends on batch size.

3. In large-batch production programmable automatics are most economical, while special-purpose (and often hard-programmed) automatics are limited to the mass production of standard parts.

Automobiles, appliances, and consumer goods generally fall into the large-batch category and have been relatively efficiently produced with traditional methods. There has been, however, a marked shift towards the customization of many products and this has forced greater flexibility even in mass production. Machine tools, off-the-road and railroad equipment, heavy machinery, and aircraft usually fall into the batch-production categories. Products of the latter industries are characterized by large expenditures for the main components of complex shape, and a relatively small expenditure on the much more numerous, mass-produced, often purchased components (Fig. 21–4). Obviously, greatest economy will be ensured by organizing effective production of the complex main parts. Thus, the organization of manufacturing has traditionally followed different philosophies for mass production and batch production, but the differences are gradually diminishing.

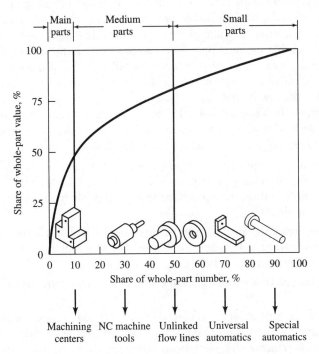

Figure 21–4   A large proportion of the total value added originates in the batch manufacture of complex main parts, whereas standard parts, usually of smaller size, are mass-produced. [*H. Opitz and H. P. Windahl, Int. J. Prod. Res.* **9**:*181–203* (*1971*). *With permission.*]

21-2-3   Organization for Mass Production

The large lot sizes typical of mass production make the installation of special-purpose machines economical. When parts are of identical, simple shapes, they can be easily held in the correct position relative to the tool. Parts of irregular shape can be worked upon after establishing a reference surface, hole, or boss (*qualifying the part*), for example, by high-speed machining. The part is clamped to a base, pallet, or fixture, or moved on its own from machine tool to machine tool.

1. In *transfer lines* parts are moved by fixed means (such as conveyors, carousels, mechanical arms) between machine tools organized in the sequence of operations. Each machine performs only one (or a related group) of operations, and is controlled by fixed (hard) automation (cams, levers, or relays). Setup is manual, time-consuming, and calls for a highly skilled setup person. Product changes cannot be accommodated without substantially rebuilding the production line.

Lines must be carefully balanced to equalize the output of various stages, otherwise the most time-consuming station would slow down the entire line. Tooling and process conditions are selected so that all tools can be changed at the same time, avoiding costly, random shutdowns. Sensors, gaging heads, and probes are built into the line at appropriate places to confirm that an operation has indeed taken place and that the next operation can proceed. In the event of difficulty, one of several actions may be provided: Signal lights or alarms alert the line attendants; defective parts are automatically marked with paint or ink; the line is slowed down; the line is shut down entirely.

Fixed production lines are of very high productivity but of virtually no flexibility. Because of this, input material and in-process inventories must also be large so as to provide *buffers* against unexpected disturbances.

2. *Flexible transfer lines* have developed in response to increasing global competition, rapidly changing customer demands, and the high cost of money. Mass-production facilities have been made more flexible by several approaches, alone or in combination:

a. The production line is grouped into *sections* of 5–12 stations, with a smaller buffer storage in between, so that breakdown, tool change, or setting in one group does not stop the whole line.

b. Operations that would upset the line balance are performed on *branch lines*.

c. Fixed machine tools are replaced by *powerhead production units* which consist of a base with feed mechanism, a drive unit (power spindle), and several interchangeable attachments, so that various operations (drilling, tapping, turning, milling, etc.) can be carried out according to need. These units are examples of *modular* (sometimes also called *metamorphic*) hardware. Sometimes even CNC machining centers are incorporated.

d. *Quick-change tool holders* allow rapid tool change or the change of the tool holder complete with a preset tool.

e. Parts belonging to the same family and differing only in the presence or absence of some feature (such as a hole) can be processed on the same line if the parts are identified on entering the line (e.g., by a boss provided for sensing, or a bar code on the part or pallet); the appropriate station is then activated or deactivated.

Such flexible transfer lines are operated under logic control, increasing their flexibility by the ease of reprogramming.

To make a family of transmission cases for Chrysler's LH cars, aluminum castings are machined at the rate of 200/h on a flexible transfer line. Locating pads are milled and holes drilled in the raw castings so that they can be transferred on precision palletized fixtures. To prevent distortion of the thin-walled castings, they are held by servo-driven wrenches rather than hydraulic clamps. In addition to several milling operations, numerous holes are drilled and bored in patterns corresponding to different engines. Splines are machined to tight tolerances and shaft holes are bored at right angles. Depth sensors have feedback controls and temperature is closely controlled to hold tolerances. (Source: *Manufacturing Engineering*, 1994(4):73–78).

| **Example 21-1**

21-2-4   Organization for Batch Production

Batch production differs from mass production not only in batch size but also in the speed of response to changing demands. The epitome of batch production used to be the job shop which makes its living by providing service to a large number of customers; recent developments have endowed production facilities with the same flexibility.

Functional Layout   Batch production has traditionally taken place in shops organized around individual machine tools. Parts are moved by some flexible means (manually, by overhead conveyors, cranes, forklift trucks) from machine to machine. This results in complex and often disorganized, time-consuming material movement.

Separating machine tools of one kind into groups (*functional layout*) hardly improves the situation, since different parts are made in different production sequences (Fig. 21–5a) and material movement remains chaotic.

Management of such a plant is also highly demanding. Each machine is attended by one operator; production plans must be drawn up that ensure full utilization of machine and operator time while at the same time assuring that parts are produced in the correct number for scheduled delivery. This usually turns out to be impossible. Frequent setups would take up most of the production time, therefore, batch sizes are increased whenever possible, even at the expense of increasing the in-process inventory. This, however, increases processing time and reduces the plant's ability to respond to changing customer needs.

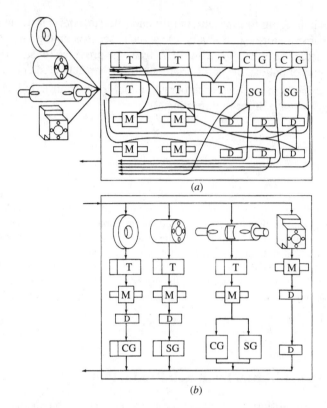

Figure 21-5   Comparison of material flow in plants with (a) functional and (b) group layout. T, turning; M, milling; D, drilling; SG, surface grinding; CG, cylindrical grinding. (*C. C. Gallagher and W. A. Knight, Group Technology, Butterworths, London, 1973. p. 2. With permission.*)

Group Technology   Many problems can be resolved if parts to be produced can be classified according to the principles of *group technology* (GT, Sec. 3-1-2). The potential of GT can be fully exploited only if production flow is analyzed and the plant is reorganized. All the equipment necessary to produce a family of parts is grouped into a *cell*. In a more modern plant a cell may comprise, for example, a complex (and expensive) machine such as a CNC machining center, supported by several special-purpose and, therefore, lower-cost machines.

The parts are transferred with minimum movement and wasted time from one unit to the other. For larger batch sizes, machines are laid out along a line (or U or L shape) in the sequence of operations (Fig. 21–5b), creating a transition between cell and modular-construction transfer line.

Typically, such cells are still attended, but one operator may take care of several machines. Thus productivity increases while the task of the operator is made much more varied and interesting.

There are multiple benefits, many of them flowing directly from the application of GT principles:

1. The variety and quantity of starting material as well as in-process inventory are reduced.

2. Production planning is simplified, and better information can be collected for production control and planning.

3. Tooling costs can be reduced by standardization, and setup times are minimized.

4. Total processing time is reduced, lead times are shorter, response to customer needs is faster, and competitiveness increases.

Flexible Manufacturing Cells (FMC)   A further increase in flexibility is obtained if several operations are combined into one (or more) highly flexible CNC machine which is served by some flexible means of material movement, such as a robot or pallet changer (Fig. 21–6). In such a *flexible manufacturing cell*, FMC, the task of the operator is reduced to: loading the racks from which the robot will pick up the parts, clamping parts on pallets, removing finished parts, and changing the tools and other supplies in magazines. Most FMCs are for machining and comprise CNC lathes, milling machines, machining centers, grinders, etc. Sheet-metalworking cells comprise sheet stacking, CNC punching, laser or electron beam cutting, bending, and parts stacking units. Many FMCs incorporate automatic inspection. (The principle of FMC is not limited to the manufacture of hardware; one of the early applications was for the making of cakes).

Compared to attended cells, FMCs are more demanding: machine tools must be more rigid, foundations more substantial for greater stability and better alignment, and preventive maintenance must be strictly enforced. Still, FMCs offer several advantages:

1. Machine tools of greater flexibility are more expensive but may replace several conventional tools. Therefore, investment may be 70–130% of manned cells of similar capacity.

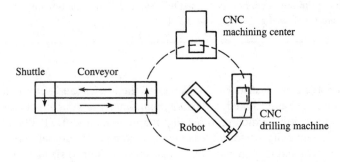

Figure 21–6   Typical layout of a flexible manufacturing cell.

2. In-process time increases from the less than 5% typical of stand-alone attended machine tools to 75 or 80%; thus productivity, expressed as output per machine, is much higher, and deliveries are much faster.

3. Productivity is higher also in terms of output per operator hour. Still, the requisite investment can often be justified only if the FMC is operated 24 hours a day, seven days a week, with operators in attendance only in one shift (or, sometimes, two). Five- to tenfold productivity increases have been realized. For unattended operation, the operator loads or palletizes numerous parts; the robot then identifies the parts and calls up the appropriate program from the memory of the control computer. Provisions must be made to detect malfunctions or impending malfunctions, and appropriate strategies must be formulated for dealing with them. For example, sensors built into a milling chuck indicate whether the cutter has been chucked correctly and, if not, give a command for rechucking.

4. The greater flexibility allows reduction of in-process part inventory, often to one-quarter of the usual amount. Production can, if required, proceed in random order; in a sense, the advantages of flow-line production are attained in batch production and small runs become profitable. In principle, manufacturing of at least the smaller parts can be geographically distributed to small centers, creating jobs in many locations.

5. Quality improves because human error is eliminated as a source of problems. Unattended operation requires though that quality be routinely (and often 100%) checked by automatic inspection. Cleanliness becomes of paramount importance: fluids, chips, dust create problems in fixturing and also interfere with the proper functioning of sensors.

Example 21-2 | **A**ircraft wings are complex structures in which wing posts act as stiffeners. In each wing of the Boeing 737 and 777, there are 100–200 posts, similar in shape but each one of different dimensions, with heights of 1.5 m near the fuselage tapering to 150 mm at the wing tip. By installing a flexible manufacturing cell with five 4-axis horizontal machining centers served by 44 pallets, 600 part types can be machined in any order. Compared to stand-alone machines, total machining time for a rib post decreased by 33% and total cycle time by 60%. [Source: *Manufacturing Engineering*, 1998(3):114–124.]

Flexible Manufacturing System (FMS)   When all the FMCs (and automatic inspection) of a plant are interlinked, a *flexible manufacturing system* is created. This is a huge undertaking which requires that many elements of CIM be already in place. The complexity of computer control becomes substantial and, to ensure real-time control and response to situations, several (four to six) levels of hierarchical control are usually needed. An essential feature of an FMS is the *automatic storage and retrieval* (ASR) warehouse.

An FMS is often implemented by installing several FMCs first. New plants can be designed as FMSs but experience has shown that complete cooperation between user and supplier of the FMS is essential and that the hardware and software must be developed as a joint effort. Once an FMS is up and running, it outperforms attended operations by a substantial margin.

A feel for the ranges of application for different manufacturing systems may be gained from Fig. 21–7. It is important to remember that flexibility is relative and that it costs money. Therefore, many FMSs are designed to deal with only 10 or so products of the same family; they are flexible only relative to the fixed automation they replace. One of the difficulties is communication between machines and controllers of different manufacture, although great progress has been made by the worldwide introduction of a standard set of communication specifications (*manufacturing automation protocol*, MAP), using General Motor's MAP document as a starting point.

In principle, it is possible—but not necessarily desirable—to build a fully automated factory in which all unit processes, tool changing, material movement, and inspection are accomplished without operator assistance. This is true also of assembly, although some assembly operations remain difficult to automate.

21-2-5   Organization of Assembly

In the final phase of manufacturing, individual components are assembled into the end product. This presents a wide range of problems, depending on production quantities.

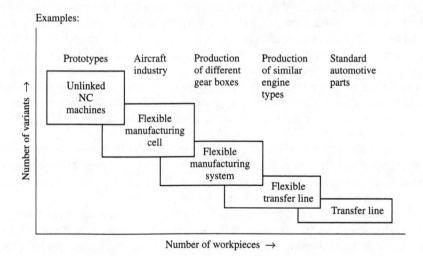

Figure 21-7   The suitability of various manufacturing systems for specific production tasks depends on the variety of products and the number of workpieces. [*After W. Eversheim and P. Herrmann, J. Manufacturing Systems* **1***(2):139–148 (1982.) With permission of the Society of Manufacturing Engineers, Dearborn, Michigan*].

Manual assembly is still the preferred method for small-batch production. In larger production quantities, the repetitive nature of work, the danger of making errors when hundreds of parts are involved, and the low overall efficiency have led to early attempts at organizing and mechanizing assembly operations.

Assembly Lines   In assembling a complex machine, great progress can be made by breaking down the operation into smaller units; this also facilitates material handling by ensuring that all parts can be supplied in their proper place and sequence.

This concept led to the *synchronous assembly line*, pioneered by Henry Ford in 1913. The units to be assembled move on a conveyor at a preset rate while operators stationed along the conveyor perform their assigned tasks. This assembly method, more than any individual advance, has made possible the mass production of consumer goods that previously were regarded as a luxury. However, the monotony of work has led to some dissatisfaction with the system, resulting in attempts at replacing it with alternative yet similarly productive methods.

Nonsynchronous assembly lines permit operator judgment; units are passed on when finished. Another potentially attractive solution entrusts an entire assembly (e.g., an automobile engine) to a group of operators. In both cases, operators are given considerable freedom in organizing themselves and also perform the quality-control function. The alternative is, of course, mechanization and automation.

Mechanization of Assembly   Some types of assembly operations lend themselves to fairly simple mechanical methods of assembly. Thus, screws or bolts can be driven and parts placed, crimped, or riveted with mechanical devices (particularly advantageous are snap joints, Fig. 18–3c). Cost is reduced while productivity and consistency of product increase, but only if the reliability of mechanization is very high. The cost of off-line repairs can quickly cancel all savings. A crucial factor of success is *in-line inspection* to pinpoint and remove imperfect assemblies, either during or at the end of assembly operation. Many of the elements used in automatic assembly are the same as in mechanized production and workpiece handling, and may be purely mechanical, electromechanical, or numerically or computer controlled.

During assembly the unit may move continuously; *indexing workheads* move with it and retract after completing their task, backtrack, and repeat the operation on the next unit to come along. Alternatively, the line itself indexes and *stationary workheads* perform the operation while the line is at rest.

In all instances, assembly may be performed in line, that is, along a conveyor on which parts move (if necessary, on pallets that ensure accurate positioning). The line may be laid out in a straight line or U or L shape. Pallets are returned below the line. The assembly line may also be oval or circular, so that pallets return to their starting position. When the total number of stations is not too large, a rigid carousel may be used to carry the unit from station to station.

Operation of a line may be *synchronous*, in which case each unit is moved at the same time and any local holdup affects the entire line. Breaking down the line into smaller modules, with storage buffers between modules, makes operation of the line less critical.

Greater freedom is secured by *nonsynchronous* movement of units. Each unit is moved on command at the completion of operation; the total number of units is larger than the number of assembly stations so that there is always a buffer between each station.

Branch lines feeding into the main line help in maintaining output when a subassembly operation is more time consuming.

Flexible Assembly System (FAS)   The same pressures that have forced the evolution toward the FMC and FMS have contributed also to the development of the *flexible assembly systems*. Such systems bring the economy of mechanized assembly to batch production. Many of the techniques used for manned and mechanized assembly are incorporated. The difference is that many—and sometimes all—operators are replaced by flexible assembly machines, usually robots. These robots range from pick-and-place devices to complex, fully articulated robots. By definition, they are under computer control, and sometimes they have artificial intelligence, particularly pattern recognition, to pick randomly oriented parts from a bin or conveyor belt and to locate parts in the assembly in the correct position. The robot itself may perform the assembly operation, or it may present the part to an assembly machine such as a press or nut driver.

Automatic assembly is successful only if existing product designs are modified to take into account the limitations and capabilities of automatic assembly.

21-2-6   Scheduling of Assembly

In the traditional mode of operating an assembly line, large stocks of parts are *warehoused* for several reasons. First, the supplier of parts achieves economy of production by shipping larger batches. Second, these stocks provide a buffer in the event of a disruption of supply for whatever reason (technical problems, labor disputes, transportation breakdown). Third, if there is no assurance that all parts received actually meet requirements, parts can be rejected as assembly proceeds.

Many industries find the cost and quality implications of this approach unacceptable and have turned to the *just-in-time* (JIT) *delivery* system developed in Japan. Deliveries are made to the assembly line frequently (daily or several times a day) so that only the parts immediately needed are stocked directly at the line. The system *pulls* supplies as required by consumption on the assembly line, instead of *pushing* on the basis of forecasts. This presents several advantages as well as challenges:

1. Investment in factory space and inventory drops drastically, increasing the economy of production.

2. Assembly becomes more flexible because production schedules can be changed in the absence of large inventories. However, assembly also becomes more sensitive to any malfunction on the assembly line.

3. Reliability of supply becomes of particular concern because, in the event of an interruption in supplies, the parts required for a temporary change in production profile may simply not be available.

4. Suppliers of parts must adopt flexible manufacturing techniques so that they can respond to demand. Quality problems must be corrected immediately: equipment maintenance and good labor relations become critical; failure to supply on time may mean a permanent loss of business.

5. There is often neither time nor inclination to inspect incoming material, and there is no buffer that would help to tide over a bad batch. Responsibility for quality assurance is shifted, to a large extent, to the supplier.

6. Frequent deliveries necessitate that suppliers be within reasonably close geographic locations (with distance depending on the transportation mode). Parts are shipped in carefully designed, often reusable packaging to ensure that parts arrive in perfect condition and, frequently, also in the correct orientation. In the ultimate development, parts are produced on branch lines feeding the assembly line on demand; this represents a true zero-inventory system.

JIT delivery can function only if there is the closest cooperation between producer and user of parts. This requires joint efforts in quality control, technology transfer, and even in the design of parts. Competitive bidding based purely on price is not used as the basis of awarding contracts.

21-3   QUALITY MANAGEMENT

The aim of manufacturing is the creation of reliable products, i.e., products that will perform their intended function, under stated conditions, for a specified period of time. *Reliability* is the probability that a product will do so, and is usually expressed as a percentage. Thus, when we say that a product has 98% reliability for a 1000-h service, we mean that 98 out of 100 units will perform without breakdown.

Assuming that the product had been correctly designed and that the proper manufacturing operations were chosen, steps must be taken so that the planned reliability will indeed be achieved. For this, the quality of the product must be controlled. *Quality* is usually defined as conformance to written specifications, but it should be recognized that there are aspects of quality that are difficult to define exactly yet can be readily judged subjectively.

The approach to quality control has changed over time. In preindustrial days quality was assured by the pride of the artisan and the control exerted by guilds. In earlier days of the industrial revolution, responsibility for quality was divided between operator and management. Following the work of Fredrick W. Taylor in the early 1900s, tasks were organized with clearly identifiable responsibilities, and the task of quality control shifted to quality-control departments. Beginning with the 1920s, statistics was applied: W.A. Shewhart of Bell Telephone Laboratories introduced statistical process control and control charts and H.F. Dodge and H.G. Romig developed acceptance sampling. The huge increase of production during the Second World War made a systematic, unified approach necessary, and a number of military specifications were drafted, some of which are still used.

The need for a broader view had been recognized by leaders such as W. Edwards Deming and Joseph M. Juran in the United States but their concepts were first adopted in Japan. Thus, Juran's *quality road map* leads from the customer to a proven process and quality end product; Deming's *14 points* stress cooperation between departments and all personnel in the service of quality and the customer. Genichi Taguchi contributed the concept of a *quality loss function*: Loss of quality represents a loss to the producer through the cost of rejects, repairs, and customer dissatisfaction, and to society through the higher cost of operation, repair, and maintenance. This loss increases according to a quadratic function with deviation from the normal (target) value, even if variables are within the tolerance range (Fig. 21–8). This view is in contrast to the traditional concept of quality which accepts of equal value everything within the tolerance band and becomes concerned only when the upper or lower tolerance limits are exceeded. Thus, the Taguchi approach aims at producing to target value, instead of just staying within specifications, resulting in a measurable increase in quality. The philosophy is supported by fractional factorial design of experiments for evaluating the major factors that affect the process. This is one of several *design of experiment* (DOE) techniques and it has proven successful for initial evaluation. In a mature process, where the last percents of improvement are sought, other techniques, including multiple regression analysis, may serve better.

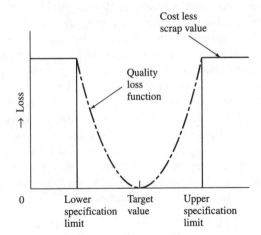

Figure 21–8   Traditionally, products are regarded to have equal value as long as they are within specification (solid line); according to Taguchi, there is a loss when the product deviates from the target value (dash-dot line).

21-3-1   Quality Assurance

Quality assurance has developed its own language and acronyms. *Quality control* (QC) is concerned primarily with inspection and the analysis of defects; it aims simply at maintaining quality standards. *Quality assurance* (QA) is defined as the totality of planned and systematic actions necessary to provide adequate confidence that a product or service will deliver the required quality. *Total quality control* (TQC) is a broader activity, beginning with interaction with design, and spreading over most aspects of manufacturing, including the formulation and auditing of quality-control programs; it requires a company-wide effort and total devotion to the concept and its realization. *Total quality management* (TQM) aims at continuous improvement with the aim of satisfying the customer's needs and desires. One tool is *quality function deployment* (QFD), a methodology for determining the customer's needs and translating them into specific steps to make products that meet these requirements. It will be recognized that all these techniques readily fit into concurrent engineering.

Continuous improvement is central to higher quality. One path is through the ISO 9000 series of standards. They are generic; they describe what elements should be part of a quality system, but do not impose any specific method. They create *models for quality assurance*: broadest is ISO 9001 (for all activities of the enterprise, including design/development, production, installation, and servicing); more limited are ISO 9002 (for production and installation) and ISO 9003 (for final inspection and testing). Guidance for setting up quality systems is given in ISO 9004. A company desiring registration has to document its quality assurance system, including document control, communication, operational procedures, training programs, measurement and calibration methods, methods of dealing with nonconforming products, corrective actions, and, importantly, management involvement and responsibility. In preparation for registration, the company conducts an internal review of its operations and procedures, involving all employees; this usually leads to uncovering areas of weakness and prompts improvements that have long-term productivity and cost benefits. Registration is by plant or may be company-wide. Third-party, accredited registrars are chosen to perform an audit. If the registrars find that the quality system operates as documented, they issue a *certification* which has to be periodically renewed (note that it is not a product but the company that is certified). Many companies now require that their suppliers should also be certified, and the ISO standards are often made part of industry-wide quality standards, such as the QS-9000 standard adopted by North American auto and truck manufacturers.

21-3-2   Statistical Aspects of Manufacturing

Quality assurance starts from the recognition that all properties of manufactured products are subject to random variations. Allowance for this is made when the designer specifies dimensional tolerances (Sec. 3-2-2) or requests some minimum strength, knowing well that the strength of individual parts will show some spread above that minimum. In dealing with random variations, quality assurance draws heavily on concepts of statistics; hence, one often speaks of *statistical quality control* (SQC).

In Sec. 3-4-1 we already reviewed statistical aspects of dimensional measurements. They also apply to other measurable *variables* (such as surface finish or mechanical or electrical properties) and *attributes* (qualitative data such as surface blemishes, dents, defects in welded seams, etc.).

Variations are changes in the value of measured characteristics. They are of two kinds: *Assignable (attributable, special) causes* can be followed up to remedy the situation. Even then, there will remain small, random variations. Such *common-cause variations* are inherent in any process and as long as only these random variations exist, the process is stable and is said to be *in statistical control*.

Analysis of variations is best illustrated by returning to Fig. 3–7a, in which the measurements taken on a cylindrical part (a shaft) are plotted. Data points can be plotted as a function of time (*run chart*); this will immediately reveal if something is drastically wrong and corrective action can be taken. Assuming a smooth production run, a sample of 100 consecutive parts is taken: Eccentricity in the lathe spindle, variations of cutting force and hence tool deflection, etc., reflect in random variations in the diameter of the shaft. The data points can be organized by grouping them into narrow dimensional ranges and plotting a bar graph (Fig. 3–7a). For a process in statistical control, the distribution approximates normal distribution (Fig. 3–7b) and is characterized by the statistical average $\bar{x}$ and the range R or, more commonly, the standard deviation σ (see Sec. 3-4-1). It may, of course, happen that the samples taken show a nonstandard distribution: The process is not in statistical control because a nonrandom variable is affecting the results. Such special-cause variations may usually be attributed to faulty machines, operator error, wrong material, worn tool, etc. A substantial proportion of parts produced may then fall outside the specification limits.

There are many attributes that cannot be measured but are no less important. For example, machining may reveal slag inclusions in some of the shafts, or a paint finish may show blemishes. We can then count the number of defects in one part (unit), or the number of defective parts (units) in a production lot.

21-3-3   Acceptance Control

Ideally, specifications would assure that no defective parts are delivered for assembly and that no defective assemblies are delivered to the final customer. *Acceptance control* is based on tests applied to parts and completed products by the producer and/or customer. The tests may involve:

1. Hundred-percent inspection. When inspection is conducted by humans, there is no guarantee that 100% of the defective parts will be found; because of fatigue and boredom, human inspection is only 80–85% effective. With the wider availability of automatic inspection techniques (Secs. 3-4-4 to 3-4-6 and Sec. 4-8), 100% inspection can be fully effective and, in many instances, also economical. Integrated with the process, it allows instant feedback for process control. Analysis of the data provides important clues regarding the effects of process variables, allowing the tightening of control limits.

2. Acceptance sampling. Hundred-percent inspection may be uneconomical and, if inspection involves destructive testing, impossible. A limited number of samples will then be tested. The number of samples and the methods of obtaining them are prescribed on the basis of probability theory. The sampling plan is drawn up so that the risk of rejecting good lots (the producer's risk) is minimized without an undue increase in the risk of accepting bad lots (the consumer's risk).

This approach is essentially *reactive*. Inspection is made after the fact; if an unacceptably high proportion of nonconforming parts or assemblies is detected, the sampling rate is increased and, if necessary, out-of-specification parts are separated by sorting. Correction involves rework and reinspection, or scrapping. The origin of rejects may be difficult to trace and, by the time the information gets back to the producer, further lots—of good or bad quality—may already have been produced. In a sense, *quality is inspected into the product*. Even though quality control may be performed by highly trained people employing the best equipment and inspection plan, nonconforming parts will be allowed to go through; the purpose of inspection is only to ensure that the proportion of nonconforming parts does not exceed a predetermined value. Assemblies consisting of many parts may turn out to be defective in a significant proportion of cases, destroying the competitiveness of the product. Tighter quality standards can usually be satisfied only at higher cost. Quality now has a cost.

21-3-4   Statistical Process Control

A much greater probability of success is attained if quality control is applied in the course of production itself. As mentioned, the beginnings of *statistical process control* (SPC) go back to the 1920s, but its execution was usually assigned to quality-control departments, with inspectors drawing samples during production runs. This often led to an adversarial relationship with the operators and reduced the effectiveness of control.

Beginning with the 1950s, American experts, notably W.E. Deming, introduced the technique in Japan in a modified form, with the execution of quality control entrusted to the operator. This approach recognizes that without appropriate guidance, the operator is not able to control quality; he or she may take actions when they are not needed and, conversely, take no action when action is needed. Therefore, management (usually through the quality assurance department) must provide the appropriate tools: not just measuring instruments, but also the plan of control based on principles of statistics. Thus, the operator can perform the measurements, evaluate the significance of results, and take immediate corrective action. The key role played by the operator is often recognized by the name *operator process control* (OPC). The phenomenal success of this approach has prompted its acceptance also in North America and elsewhere. The principle of OPC is straightforward:

1. Quality is not regarded as a separate issue, to be controlled in isolation from the process. Instead, the entire production system is reviewed, and factors affecting quality at various workstations are evaluated.

2. From this review, critical variables or attributes are identified and control limits are set. Measurements and inspections are prescribed at time intervals (or number of production units) that ensure statistical significance.

3. The operator performs the prescribed measurements and immediately plots the data on appropriate charts (or enters them into a computer program). If trends suggest that deviations may reach values approaching the control limits, remedial action is taken before any nonconforming parts would be produced.

Thus, even though a chance event may lead to the production of a defective part, this will be extremely rare; the vast majority of parts will be acceptable without further inspection: *quality is built into the product.* The producer benefits because all parts that are manufactured can also be shipped. The purchaser benefits too; the need for reinspection and for the rework or repair of assemblies disappears. Competitiveness increases through higher quality, and costs drop because of higher productivity. The technique even allows improvements to the process. By identifying variations that are attributable to a definite source (assignable cause), means of removing these sources can be found, control limits tightened, and parts of higher quality produced. Higher quality now results in lower cost.

Example 21-3

The basic elements of the approach can be illustrated by returning to the example of producing a shaft on a lathe (Fig. 3–7). The operator controls the diameter of the shaft and aims to keep it well within the specified limits of 9.90 mm and 10.00 mm. A micrometer is provided for measurement. Before evaluating the capability of the machine tool or process to produce the part, we need to establish whether the measuring instrument has the required accuracy.

1. Evaluate the measuring device. This task is usually performed by the quality assurance department, for example, by measuring two calibration-accuracy plug gages, of 9.900- and 10.000-mm diameter. The measuring device must have graduations corresponding to one tenth of the tolerance or better (in this case, graduations of 0.01 mm).

2. Evaluate reproducibility. Once the micrometer is found to be accurate, the combined reproducibility and repeatability of measurement is evaluated in the plant (separation of the two requires more extensive testing). Two operators (A and B) measure, only once, five shafts selected at random. Estimates of readings are rounded to the nearest half graduation. Gage error is then calculated as shown in Table 21–1 (the multiplier 4.33 comes from statistical theory and is a function of the number of operators and samples). If the error is greater than 10% of the tolerance range, the micrometer is adjusted, changed, or—if the gage error cannot be reduced—other means of measurement must be found. In less critical applications, an error of 20% is tolerable. It should be noted, however, that a less accurate gage will reject parts that, in reality, are within tolerance.

Table 21–1   Checking gage error

Part No.	Operator		Range, mm
	A	B	
1	9.945	9.960	0.015
2	9.925	9.930	0.005
3	9.940	9.940	0.000
4	9.935	9.930	0.005
5	9.955	9.945	0.010

Sum of ranges $\Sigma R = 0.035$

Average range $\bar{R} = \Sigma R/n = 0.035/5 = 0.007$ mm

Tolerance range $= 10.00 - 9.90 = 0.10$ mm

Gage error (gage repeatability and reproducibility, GRR) $= 4.33\bar{R} = 0.0303$

GRR as percent of tolerance $= (0.0303/0.100)100 = 30.3\%$.

3. Determine process capability (i.e., whether the process is capable of meeting specifications at all). To obtain a snapshot in time, *short-term capability* is established by sampling. To ensure that the results will be statistically significant, the sample size should be large; for economy, it must be kept small. A good compromise is $n = 5$ pieces (or, more frequently, 4 pieces) per *sample (subgroup)*. At least 10, but preferably 20 samples (100 consecutive pieces) are taken; no adjustments must be made during this period. The measured diameters (Table 21–2) are plotted in a histogram (Fig. 3–7a). If the distribution is close to normal, proceed with the analysis. If normal distribution is centered off 9.95 mm, reset the tooling and repeat sampling. If distribution is bimodal, find the source of the effect (e.g., if the shaft is turned down in two stages, analyze each stage). If the distribution is skewed, find the source (e.g., a drift due to tool wear). If skew cannot be eliminated, proceed with the analysis but note the cause of drift.

Table 21–2   Evaluation of averages and ranges*

		Sample number									
		1	2	3	4	5	6	7	8	9	10
Data	1	9.92	9.95	9.98	9.99	9.93	9.96	9.97	9.94	9.95	9.96
from	2	9.94	9.96	9.94	9.92	9.95	9.97	9.94	9.96	9.93	9.92
sample	3	9.93	9.96	9.95	9.93	9.99	9.94	9.92	9.95	9.97	9.92
x,	4	9.95	9.97	9.97	9.94	9.93	9.98	9.92	9.96	9.96	9.94
mm	5	9.93	9.94	9.93	9.92	9.94	9.99	9.93	9.92	9.98	9.93
Total	Σx	49.67	49.78	49.77	49.70	49.74	49.84	49.68	49.73	49.79	49.67
Average	$\bar{x}$	9.934	9.956	9.954	9.940	9.948	9.968	9.936	9.946	9.958	9.934
Range	R	0.03	0.03	0.05	0.07	0.06	0.05	0.07	0.04	0.05	0.04

*Only 10 samples out of 20 shown. $\bar{\bar{x}} = 9.947 = 9.95$ mm; $\bar{R} = 0.049$ mm.

From the data, calculate the average diameter $\bar{x}$ [Eq. (3-2)] and range R for each 5-piece sample (Table 21–2; note that the measurements are shown only to the nearest hundredths and that only 10 of the 20 samples are included in the table, hence the averages and standard deviation are not quite the same as indicated in Fig. 3–7). Calculate the average of averages (*grand average*)

$$\bar{\bar{x}} = \frac{\sum \bar{x}}{(n)}$$ **(21-1)**

and the average $\overline{R}$ of ranges R for the $(n) = 20$ samples (again, only 10 samples are shown in Table 21–2). An estimate of the standard deviation may now be obtained from

$$\sigma = \frac{\overline{R}}{d_2}$$ **(21-2)**

where d_2 is taken from Table 21–3. Since a dispersion of $\pm 3\sigma$ is expected in manufacturing, it is usual to compare 6σ to the tolerance to express *machine capability ratio*

$$C_r = \frac{6\sigma}{\text{tolerance}}$$ **(21-3)**

It is desirable that 6σ should occupy only a fraction of the tolerance band; in practice, $C_r = 0.75$ is acceptable. At $C_r = 1.00$ the process is barely acceptable, and at $C_r > 1.00$ it is not acceptable at all (or the parts produced would have to be sorted). *Process capability* is often expressed by the reciprocal

$$C_p = \frac{\text{tolerance}}{6\sigma}$$ **(21-4)**

An acceptable process then has the capability of $C_p = 1/0.75 = 1.33$ or better. This expresses only the potential of the process; if, for example, the tool was misset, the bell curve would shift and rejects produced (Fig. 3–7d). It is, therefore, necessary to compute also

$$C_{\text{pk}} = \frac{\text{USL} - \bar{x}}{3\sigma}$$ **(21-5a)**

Table 21–3   Factors for control limits

Number of observations in sample, n	d_2	A_2	D_3	D_4
2	1.128	1.880	0	3.267
3	1.693	1.023	0	2.575
4	2.059	0.729	0	2.282
5	2.326	0.577	0	2.115
6	2.534	0.483	0	2.004
8	2.847	0.373	0.136	1.864
10	3.078	0.308	0.223	1.777
15	3.472	0.223	0.348	1.652
20	3.735	0.180	0.414	1.586
25	3.931	0.153	0.459	1.541

$$C_{\text{pk}} = \frac{\bar{x} - \text{LSL}}{3\sigma} \tag{21-5b}$$

where USL and LSL are the upper and lower specification limits, respectively. If either C_{pk} is below 1.33, some parts will be outside tolerance, but resetting the tool will correct the matter.

4. Calculate control limits. If the process is capable, *control limits* can be calculated for the averages:

$$\text{UCL}_{\bar{x}} = \bar{x} + A_2\overline{R} \tag{21-6}$$

$$\text{LCL}_{\bar{x}} = \bar{x} - A_2\overline{R} \tag{21-7}$$

and for the ranges:

$$\text{UCL}_R = D_4\overline{R} \tag{21-8}$$

$$\text{LCL}_R = D_3\overline{R} \tag{21-9}$$

where, for 3 standard deviations from the mean, A_2, D_3, and D_4 are taken from Table 21–3. In our example

$$\text{UCL}_{\bar{x}} = 9.95 + (0.577)(0.049) = 9.976 \text{ mm}$$
$$\text{LCL}_{\bar{x}} = 9.95 - (0.577)(0.049) = 9.919 \text{ mm}$$
$$\text{UCL}_R = (2.115)(0.049) = 0.1036 \text{ mm}$$
$$\text{LCL}_R = (0)(0.049) = 0 \text{ mm}$$

These limits are then plotted on so-called $\bar{x}$–R (x bar–R) charts, Fig. 21–9; it should be noted that these charts *do not* show the specification limits. For a process in control, the points are randomly distributed within the limits.

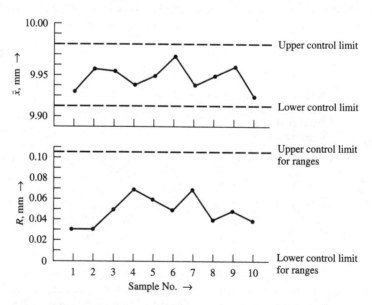

Figure 21–9   In turning a simple shaft of 9.95 ± 0.05-mm diameter, the process is in statistical control because none of the sample averages or ranges fall outside the established control limits.

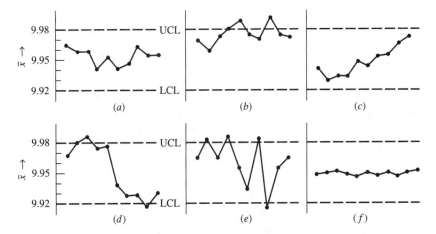

Figure 21-10   Statistical process control gives valuable clues: (a) process is in statistical control, (b) improperly set tooling, (c) rapid tool wear, (d) mixed material lot of two hardnesses, (e) process out of control, and (f) something is going very well and is worth investigating.

5. Operator quality control. At this point, the operator takes over and plots, on the control charts provided, $\bar{x}$ and R for samples taken at predetermined intervals (Fig. 21–9). As long as variations are random and stable, the process remains in a state of statistical control; the sampling interval can be lengthened with the agreement of the quality control department. The $\bar{x}$–R charts alert the operator to conditions that lead to greater than normal variations, so that remedial action can be taken *before* out-of-tolerance parts are produced; some examples are shown in Fig. 21–10.

When statistical process control is practiced, the purchaser of parts is assured that no out-of-specification part will be delivered. Increasingly, purchasers will buy parts only from producers who maintain statistical process control and are willing to cooperate with suppliers in installing a meaningful process-control procedure.

The example shown here applies to many but by no means all processes and products. There is a vast literature on the subject, and many software products are available to help introducing quality control and analyze the data. When processes are run under statistical control, a very small spread of R indicates that either something is wrong with the measurement or that the process is better controlled than normally; by following up clues, improvements can again be made.

In some instances tolerances are so tight that no existing process is capable of satisfying them. Then parts must be separated by 100% inspection into subgroups for *selective assembly*, and parts in the assembly are no longer interchangeable.

In many applications direct measurement of a variable is not possible, and process control must be based on the number or ratio of defectives in a sample or

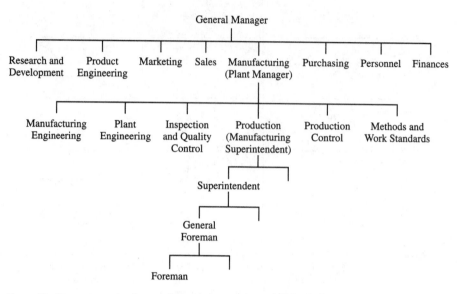

Figure 21–11   Example of a possible organizational structure.

on the number of defects in a unit. Quality control charts can be set up for such attributes too.

21-4 MANUFACTURING MANAGEMENT

A plant equipped with efficient machine tools and served by the best means of material movement and assembly can still lose money. For profitable production, the technology and physical means of production must be effectively managed. We will come back to this theme in Chap. 22; here we look only at some typical organizational schemes and the role of manufacturing engineering within them.

21-4-1 Company Organization

Neither company organizations nor the terminology used to describe organizational elements is standardized. Nevertheless, the *organization chart* in Fig. 21–11 is fairly typical of North American practice.

All manufacturing concerns have, at the company level, finance, personnel, purchasing, and sales departments. *Marketing, market development*, and *research and development* departments are essential for growth. Depending on the nature of operations there may also be *product engineering* which takes care of the development, design, testing, and evaluation of new products, ideally with the full involvement of the manufacturing group of the company and outside vendors. These groups together provide the talent for concurrent engineering.

At the plant level, all facets of production are under the direction of the plant manager. Actual production is headed by the manufacturing superintendent, with the assistance of superintendents, general foremen, and foremen. In many plants several levels of supervision are eliminated and highly trained technicians or associates are given broad responsibilities and authority. To keep production going at peak efficiency, support is provided by several other departments.

1. *Production planning and control* determines economical lot sizes and establishes schedules for manufacturing and assembly, using various techniques. *Computer-aided process planning* (CAPP) is aided by *material requirements planning* (MRP), based on the master production schedule, the bill of materials, and inventory records relating to raw materials, purchased components, parts to be delivered, in-process materials and parts, finished products, and tools and maintenance supplies. With this information as the input, MRP manages the inventories with due regard to the timing of material requirements. In detail, it provides notices for releasing orders, scheduling and rescheduling, cancellation, inventory status, performance reports, deviations from schedules, etc. A good MRP system results in reduced lead times, minimum inventory, faster response to customer requests, and increased productivity. A further development is *manufacturing resource planning* (also called MRP-II or closed-loop MRP), which integrates a complete manufacturing control system. Production planning, master scheduling, capacity requirements planning, and functions necessary for executing the production plan (including vendor schedules and dispatch lists) are incorporated, and provisions are made for continuous updating. All these techniques are often regarded as elements of a management information system (MIS). More recently, MIS is taken to stand for manufacturing information system.

2. The primary task of the *methods and work standards* group used to be provision of time and motion studies and norms. Now the emphasis is often on the analysis of the whole process, sometimes in cooperation with the *process engineering department.*

3. *Plant engineering* is responsible for preventive maintenance of equipment, replacement of machinery, and provision of services (power, heating, lighting, etc.). Because downtime results in large economic losses, maintenance is particularly critical with computerized equipment and robots, and *total productive maintenance* (TPM) aims at keeping the equipment running at all scheduled times.

4. *Inspection and quality control* are a vital activity, reporting to a high level of management while also increasingly integrated into production.

5. *Manufacturing engineering* is central to our considerations and will be discussed separately.

It should be noted that the organization shown in Fig. 21–11 is by no means universally adopted and is not necessarily the best. It tends to encourage separation of departmental efforts and, for this reason, there is a growing trend to a *horizontal organization* scheme in which all departments report to the plant manager, thus ensuring greater interdepartmental cooperation.

21-4-2 Manufacturing Engineering

The *manufacturing engineering* group, also called *process engineering*, is usually headed by the *chief engineer*. It is in this group that technological awareness resides, and the competence of its people determines whether the company will be competitive and profitable. In the present environment, people in this group must understand processes and their control, and must be versed in materials, mechanics, electronics, computers, and systems analysis.

Typical tasks include evaluation of manufacturing feasibility and cost; selection of optimum processes and process sequences; production equipment; tooling, jigs, and fixtures (their design and manufacture, and control of the tool room); material movement methods and equipment; and plant layout. In addition to unit processes discussed in this book, the group also issues specifications for auxiliary and finishing processes and, in general, treating of individual parts and finished assemblies. Cost analysis (Sec. 22-2) is often performed here, and the group also has a central function in value analysis. The group cooperates closely with research and development specialists and delegates its members to concurrent engineering teams.

With increasing mechanization and automation, and particularly with the special demands set by numerical and computer control, the activities of the manufacturing engineering department account for a substantial share of the total production cost, and there are convincing reasons why at least some of these activities should be regarded as direct rather than indirect labor (Sec. 22-2). The true cost-effectiveness of new manufacturing technologies can then be judged and the need for a thorough reorganization of production facilities revealed. This is particularly true of computer control, which often results in no savings unless the entire manufacturing concept is changed.

21-5 SUMMARY

The unit processes discussed in Chaps. 6–20 can lead to competitive production only if they are integrated into a manufacturing system. Of special importance are the following points not previously discussed:

1. Material movement can be most time-consuming and may limit the productivity of machine tools. Therefore, parts and tools must be transported—manually or with mechanical devices—so that they are immediately available as the need arises for them. Programmable devices such as robots (especially those fitted with elements of artificial intelligence) and automated guided vehicles combine flexibility and productivity. Fixturing or palletization of parts allows quick and accurate setup.

2. Production quantities, production rates, and similarities in features and operational sequences determine the most appropriate method of production organization. Small-batch production of highly varying units

requires stand-alone machine tools; as the similarity of features and operational sequences increases, FMC or even FMS becomes feasible and economical for both small- and large-batch production. Mass production is still often performed in dedicated automatic machines and production lines, but flexibility is being introduced even in these, and the distinction between flexible transfer line and FMC is becoming blurred. Group technology is a most important aid in the organization of production.

3. Assembly too may be manual or automatic; automation may be flexible or fixed, depending on the number and variety of products. Some flexibility is often incorporated even into mass production, aided by the introduction of just-in-time delivery of smaller batches of components.

4. Lowest-cost production is attained when each and every part produced satisfies quality requirements. Therefore, statistical process control—which ensures that quality is built, and not inspected, into the product—can be most profitable and also helps to secure the market in a competitive environment.

5. The best process will fail unless supported by strong management; conversely, the best management will do nothing for a fundamentally noncompetitive process.

PROBLEMS

21-1 Define the difference between a (a) manipulator and (b) robot.

21-2 What is the minimum equipment in a FMC?

21-3 Define the difference between (a) transfer line and (b) flexible transfer line.

21-4 Define (a) synchronous and (b) nonsynchronous assembly lines.

21-5 Define the term process capability.

21-6 Draw a sketch of an $\bar{x}$–R chart for a cutting process (a) in statistical control, (b) subject to gradual tool wear, and (c) with an improperly set tool.

21-7 Give examples of (a) assignable and (b) random causes for variations in the diameter of a shaft turned on a lathe.

21-8 We indicated that $C_p = 1.33$ is desirable. (a) Express this in terms of standard deviation. (b) Find the number of defectives that would be produced in a lot of 10 000 parts.

21-9 If a process has $C_p = 1.00$, how many parts will be outside the tolerance (a) per million and (b) in a lot of 10 000?

FURTHER READING

Tool and Manufacturing Engineers Handbook, vol. 4, *Quality Control and Assembly*, 1987; vol. 5, *Manufacturing Management*, 1988; vol. 7, *Continuous Improvement*, 1993; vol. 9, *Material and Part Handling*, 1998; Society of Manufacturing Engineers.

Badiru, A.: *Industry's Guide to ISO 9000*, Wiley, 1995.

Clausing, D.P.: *Total Quality Development*, ASME, 1994.

Clements, R.B.: *Handbook of Statistical Methods in Manufacturing*, Prentice Hall, 1991.

Craid, R.J.: *The No-Nonsense Guide to achieving ISO 9000 Registration*, ASME, 1994.

Deming, W.E.: *Out of the Crisis: Quality, Productivity and Competitive Position*, MIT Press, 1982.

Doty, L.A.: *Statistical Process Control*, 2d ed., Industrial Press, 1996.

Griffith, G.: *Statistical Process Control Methods*, 2d ed., ASQC Quality Press, 1996.

Hodson, W.K.: *Maynard's Industrial Engineering Handbook*, McGraw-Hill, 1992.

Hoyle, D.: *ISO 9000 Quality Systems Handbook*, Butterworth-Heinemann, 1994.

Juran, J.M.: *Juran on Leadership for Quality*, Free Press, New York, 1989.

Juran, J.M. and A.B. Godfrey: *Juran's Quality Handbook*, 5th ed., McGraw-Hill, 1999.

Koenig, D.T.: *Manufacturing Engineering: Principles for Optimization*, 2d ed., Taylor and Francis, 1994.

Kolarik, W.J.: *Creating Quality: Process Design for Results*, McGraw-Hill, 1999.

Nof, S.Y., W.E. Wilhelm and H.-J. Warnecke: *Industrial Assembly*, Chapman and Hall, 1998.

Novack, J.L.: *The ISO 9000 Documentation Toolkit*, Prentice Hall, 1995.

Oakland, J.S.: *Statistical Process Control*, 3d ed., Butterworth-Heinemann, 1996.

Rabbitt, J.T. and P.A. Bergh: *The QS-9000 Book*, Quality Resources, New York, 1998.

Schoonmaker, S.J.: *The Complete ISO 9001 Reference for Engineers and Designers*, McGraw-Hill, 1996.

Salvendy, G.: *Handbook of Industrial Engineering*, 2d ed., Wiley, 1991.

Taguchi, G.: *Introduction to Quality Engineering*, Asian Productivity Organization, Tokyo, 1986.

Tempelmeier, H., and H. Kuhn: *Flexible Manufacturing Systems*, Wiley, 1993.

Singh, N., and D. Rajamani: *Cellular Manufacturing Systems*, Chapman and Hall, 1995.

Askin, R.G., and C.R. Standridge: *Modeling and Analysis of Manufacturing Systems*, Wiley, 1993.

Nee, A.Y.C., K. Whybrew, and A.S. Kumar: *Advanced Fixture Design for FMS*, Springer, 1994.

Williams, D.J.: *Manufacturing Systems*, 2d ed., Chapman and Hall, 1994.

Wu, J.-K.: *Neural Networks and Simulation Methods*, Dekker, 1994.

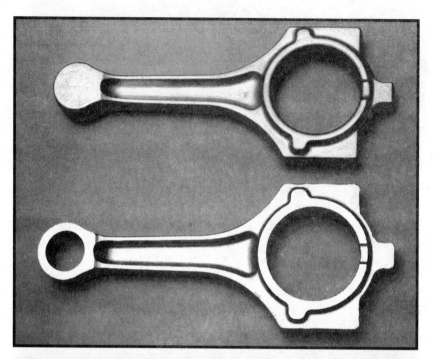

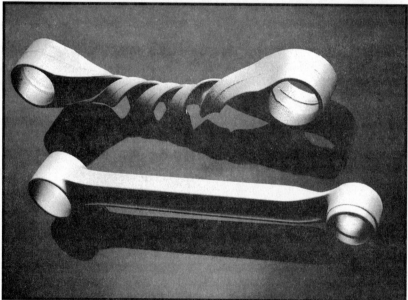

An example of competition between processes: Internal combustion engine connecting rods are usually hot-forged (top) or powder-forged (center; note the closer definition of features and the hole at the small end), but vacuum casting of steel can also produce thin-walled, highly ductile rods, as shown by the twisted part (*Top: courtesy Ford Motor Co., Dearborn, Michigan, bottom: courtesy American Foundrymen's Society, Des Plaines, Illinois.*)

22

Competitive Aspects of Manufacturing Processes

Manufacturing is a dynamic, constantly developing activity, and, for survival, any enterprise must be highly adaptive. Elements of success are:

Doing more with less: lean manufacturing

Responding to rapid change: agile manufacturing

Control of costs, including indirect costs

Choosing the optimum process for a given task

Recognizing the competition between materials

We argued in Sec. 1-2 that the economic well-being of a nation is critically dependent on a competitive manufacturing industry, and that productivity is a key issue in securing a competitive position in the world.

In this book the emphasis has been on unit processes that are needed to make discrete parts. We showed in Chaps. 2 and 21 that these processes must be organized into a well-coordinated manufacturing system, operating on the principles of concurrent engineering. In the narrower sense of the term, the functions of design and manufacturing must be interwoven; whenever appropriate, these points were also emphasized in Chaps. 3–20. There is, however, a danger that the inevitable mass of details obscures the larger picture; it is now necessary to take a bird's eye view and examine the competitive aspects of manufacturing.

22-1   COMPETITION IN THE GLOBAL ECONOMY

Competition in the everyday sense often refers to competition between corporations and nations. Indeed, even without any technical knowledge, it would be difficult to ignore global competition: The goods we buy originate from all around the world, and newspapers are full of stories of successes (and failures) of companies in meeting competition. The structure of global competition is extremely complex. There certainly is competition between nations, and a highly educated workforce is a basic requirement in an age when wealth is created by knowledge. There are also other factors determining competitiveness and multinational (or, as they are often called, transnational) corporations will move production to countries where conditions are more favorable. Ultimately, however, competition is between products.

Much is heard about *world-class products*. It must be realized that these are moving targets; what appears on the market today is already yesterday's technology and design. Therefore, a company striving for world-class status will employ all techniques described in Secs. 2-1, 2-3, 2-4, and Chap. 21 and make them, in their broadest interpretation, part of the company culture. This is the subject of *management science*, a multifaceted discipline outside our scope of inquiry. It is also subject to fast changes and the birth of new techniques, described by various acronyms. Indeed, the acronym MBF has been coined for *management by fad*, a reference to the many highly touted techniques that may prove of little value in the long run.[1] There are, nevertheless, some techniques and trends that have stood the test of time. Thus, concurrent engineering will always be practiced to bring new products to market in minimum time. Benchmarking will be a continuous process of comparing processes and practices against the best in the industry, in a cooperative effort between two partners. JIT will not be simply scheduling for just-in-time delivery, but a continual effort to eliminate non-value-added activity. Only *what* is needed will be produced *when* it is needed, leading to the shortest possible lead time with the fewest possible mistakes. The requirement for fast movement through the production process with minimum inventory leads to changes with social consequences such as a clustering of suppliers in geographic locations close to the final assembler. The result is *lean manufacturing*, a set of practices that removes all waste from the manufacturing system and produces maximum value with minimum resources (a concept originally advocated by Juran).[2]

Manufacturing must also respond to new demands. A clearly discernible trend is customization of mass production: In products such as automobiles, airplanes, and even computers, customers now expect products tailored to their wishes, yet at mass-production prices. One response is *flexible manufacturing* which, in its broader sense, relates to the ability to make a variety of products in a cost-effective and timely manner. *Agile manufacturing* is more: a top-down

---

[1] *Enhancing Organizational Performance*, National Academy Press, Washington, D.C., 1997.
[2] J.M. Juran: *Juran on Leadership for Quality*, Free Press, New York, 1989.

enterprise-wide effort to adjust operations and processes in an environment of continuous and unpredictable change, capable of fast response to consumer demands, creating and producing products economically in small lots. Achieving agility is helped by *virtual manufacturing*, the integrated application of simulation, modeling, and analysis to enhance manufacturing design and production decision and control.

In the narrower sense, there is also competition between technical approaches. There is no absolute and eternal solution to any manufacturing problem. Only the function of the product is defined, and even this function may be continually redefined. For products of well-defined function, there is not only a never-ending competition between various processes and process sequences, but there is also a competition between various materials and the processes implied in the material choice. Competition is based on serving the function at minimum cost, hence cost considerations will always enter into the choice of manufacturing processes.

22-2 MANUFACTURING COSTS

In choosing a manufacturing process, no decision—not even a preliminary one—can be made without considering at least the major cost elements. A complete discussion of *manufacturing cost analysis* is outside the scope of this book and only an outline of the approach can be given here.

22-2-1 Cost and Productivity

In manufacturing, *cost* is usually expressed in a monetary unit (such as the dollar) per unit output. *Productivity* is a more difficult term to define. In the simplest form, it is taken as the value of production per employee. This *labor productivity* is useful as a very general measure, but it does not take into account the capital employed in generating the output. This aspect has become particularly significant with the increase of automation. The *productivity of capital* (holding the quality and quantity of other inputs fixed) increases only if the efficiency of new capital goods grows faster than the price of these goods. Capital productivity is much more difficult to measure and, therefore, more controversial than labor productivity. What is more easily seen is that there must be an *economic limit* on automation where further investment yields *diminishing returns* in terms of cost reduction.

To illustrate these points, it will be instructive to look at an example taken from the automotive industry, with symbolic dollar values that have no relation to present reality. More than half the total mass of a production-type automobile is still steel, much of it in the form of pressworked sheet-metal parts. To take one part as an example, the outer door panel is a fairly sophisticated product with a shallow curvature, complex outer profile, and several cutouts.

Example 22-1

Several car models of a manufacturer share the same door, and the production quantity over a typical 4-year model period will account to several million units for mass-produced cars. This justifies a special-purpose production line (a transfer line) in which several presses perform the required sequence of operations, with mechanical means of passing the part from one press to another. With the increasing need for fast response to customer demands, the line will be equipped with quick-change die sets. The line will be tended by a few (say, three) operators in each of three shifts, and their productivity will be high (Table 22–1). A more generally applicable measure of labor productivity is value produced, and if we assume (without further justification) that the door panel is worth $3.00 for the manufacturer, productivity can be expressed in value produced per operator.

Table 22–1   Comparison of productivities and costs in a sheet-metalworking operation

	Mass Production	Batch Production
Production, units	6×10^6	1.2×10^4
Value generated, $	1.8×10^7	3.1×10^5
Number of operators	9	5
Number of employees	27	7
Productivity, units/operator	6.7×10^5	2.4×10^3
$/operator	2×10^6	6.25×10^4
Productivity, units/employee	2.2×10^5	1.7×10^3
$/employee	6.4×10^5	4.4×10^4
Capital outlay, total $	10×10^6	4×10^5
Amortization, $/annum	4×10^6	8×10^4
Tooling cost, total $	5×10^5	1×10^5
Amortization, $/annum	3×10^5	4×10^4
Cost of tool changes, $/annum	1×10^4	1×10^4
Labor cost, $/employee	2×10^4	2×10^4
Total $/annum	5.4×10^5	1.4×10^5
Production cost, $/annum	4×10^6	8×10^4
	$+ \quad 3 \times 10^5$	$+ \quad 4 \times 10^4$
	$+ \quad 1 \times 10^4$	$+ \quad 1 \times 10^4$
	$+ 5.4 \times 10^5$	$+ 1.4 \times 10^5$
	4.85×10^6	2.7×10^5
Production cost, $/unit	0.86	22.50
Material cost, $/unit	1.50	2.00
Total cost, $/unit	2.36	24.50

NOTE: All values are fictitious and bear no relation to any existing manufacturer's operation.

At the other end of the scale, some luxury cars are made by specialized companies in small numbers. Equipment and tooling costs have to be reduced, for example, by stretch-forming and low-cost blanking (such as rubber forming) techniques. The plant operates perhaps on only one shift and we can assume that five operators produce the requisite 4000 door panels per year. This will, of course, be split up into several lots spread over the year, and we assume that equivalent parts are produced for the rest of the time for a total of 1.2×10^4 units per year. The door panel, in the luxury market, is now worth $26.00 to the manufacturer, and productivities can again be calculated (Table 22–1).

When productivities are compared on either a unit or dollar basis, there would seem to be no room for the batch-produced automobile even in the luxury market. However, our calculation ignored several important factors. First of all, the nine operators of the transfer line are backed up by a team of, say, 18 people in maintenance, supervision, quality control, programming, and various levels of management. The five operators of the batch production plant perform some of these functions themselves and are supported by only two people. Productivities now change (Table 22–1).

These numbers reveal nothing about *costs* or *profits*. For these, the means of production must be also considered. Capital is invested in the purchase of presses and auxiliary equipment, their installation, and the requisite buildings and services. This capital must be regained through some time period that formed the basis of investment decision. The actual annual repayment (amortization) depends on interest rates, tax treatment, and accounting procedures. The same argument applies to the cost of tooling, except that it has to be repaid over a shorter time period.

The *production cost* per unit is still vastly different (Table 22–1) between mass and batch production and the difference remains when the cost of material (somewhat higher in batch production because of the greater losses in rubber blanking and stretching) is added. After deducting other costs not accounted for in Table 22–1 (e.g., carrying an inventory of parts), the profit for the two operations is obtained.

This example, while fairly crude, illustrates that several cost elements must always be considered. It also shows that before any meaningful cost calculation can be made, the starting material and the major production sequence must be identified. It is also obvious that value is not simply related to utility; the market has different segments, each with its own values, and product design must address these values.

22-2-2   Operating Costs (Direct Costs)

Direct costs can be clearly allocated to the product, and they are proportional to the number of units produced.

Material Costs   The *net material cost* is the cost of purchasing the *raw material* (whether it be a casting, forging, rolled section or plate, metal or ceramic powder, polymer, or any other starting material), less the value of the *scrap* produced. Thus the weight of the purchased starting material, W_0 (including the weight loss in cutting off, etc.), and of the finished part, W_f, is determined. If the unit price of the starting material is C_0 and that of the scrap is C_S (separated into heavy and light scrap if their resale value is widely different), the material cost per piece C_P

is

$$C_P = W_0 C_0 - (W_0 - W_f) C_S \qquad \textbf{(22-1)}$$

In general, a process that generates less scrap, or generates it in a more valuable (separated by composition and usually heavier) form, is more economical.

Labor Costs   With a possible process sequence and the appropriate equipment settled, the *number and skill of operating personnel* can also be predetermined. From experience, time studies, or a breakdown of operator functions into identifiable physical and mental action elements, the time required for completion of one piece can be calculated.

Strictly speaking, the *net production time* is the time period during which the material is actually shaped or processed; thus in machining (Sec. 16-7-5) the net cutting time t_c is the time during which material is actually removed. However, in other processes where the truly productive time is only a fraction of a longer but essential cycle, the total cycle time is usually taken as the productive time. For example, in pressworking the time during which the press actually moves (and not just forms) is taken as the productive time.

Whether the time required for moving the material from the floor to the machine and off again is regarded as productive or nonproductive is a matter of philosophy; a truer picture of productivity is obtained if it is classified as nonproductive. In the cost analysis, it usually appears in the total or floor-to-floor time, and is charged at direct labor rates, especially if the movement is performed by the machine-tool operator personally.

When *energy costs* are a significant fraction of the total cost, they too are allocated directly to production units, as are tools and dies if used only for that particular purpose.

22-2-3   Indirect Costs

As seen in the example of loading and unloading, the distinction between direct and indirect cost may be a fuzzy one.

Indirect costs arise from the functions and services that contribute to the efficient performance of the actual production process. Traditionally, they include: indirect labor (including material movement, cleaning services, etc.); repair and maintenance; supervision (from foreman to plant superintendent); engineering (manufacturing and industrial engineering, quality control, laboratory, etc.), research and development; sales; the entire management hierarchy of the company; lighting and heating (and sometimes all energy supplies and materials not directly used in production); office and sales expenses, etc. Some of these are somewhat flexible and can vary with the volume of production, but the relationship is never as direct as with operating costs.

If properly controlled, indirect activities represent an indispensable part of the total production effort. In the ultimate development of a fully automated manufacturing process, there would be no direct labor cost in the classical sense. Yet, many of the indirect costs—including programming and other activities required for the operation of a computer-integrated manufacturing system—obviously have to be regarded as directly related to production. Similarly, development—and even long-range research if properly managed—is a vital part of the production process, and it could be argued that even these should be regarded as direct costs.

Indirect costs are also called *overhead* or, sometimes, *burden*. Unless properly controlled, indirect cost can indeed become a burden that finally nullifies the productivity gains in direct production. Therefore, breakdown of indirect costs into their elements is essential, and this is one of the purposes of CIM, discussed in Sec. 2-4. All too often, the actual production process is carefully analyzed, improved, and made more productive, while the indirect cost sector is allowed to grow out of proportion. This finally impairs the competitive position of the company; for that matter, unchecked growth of indirect costs can destroy a national economy.

In manufacturing cost estimates the indirect costs may appear as a multiplying factor applied to direct labor costs, as a fixed cost per unit product, or a fixed cost per hour worked on a machine. This can, however, be highly misleading because it ignores the cost of activities not directly related to production. As a result, simple, mass-produced products are typically overburdened while the cost of complex products is underestimated. This is corrected in *activity-based costing* which collects the costs of all activities related to a product.

22-2-4   Fixed Costs

Fixed costs include the costs of equipment, buildings, and of total facilities in general, taking into account depreciation, interest, taxes, and insurance. For a given capital outlay, the fixed costs depend on the interest rates prevailing at the time of purchase and during the life of the equipment, the tax treatment accorded to investments, and the useful life of the production facility. In general, the life of equipment can be extended, but only at the expense of rapidly rising maintenance costs and at the danger of obsolescence, and a replacement decision must sooner or later be made.

In manufacturing cost estimates, fixed costs are allocated on the basis of anticipated equipment utilization. Thus, if a press was purchased on the premise that it will be used in a two-shift operation, the fixed cost per unit production doubles if only a one-shift operation can be sustained. Conversely, the capital outlay for flexible manufacturing systems can usually be justified by the three-shift, seven-day operation of such systems. In the simplest estimating procedure, the fixed cost appears as a machine-hour rate or burden.

22-3   COMPETITION BETWEEN MANUFACTURING PROCESSES

We have repeatedly seen that there is also competition between manufacturing processes. In discussing various processes, we already saw that each process has its particular capabilities and limitations (see Tables 7–3, 9–1, 10–2, 11–2, 12–1,

14–1, 16–7, and 18–3). The shape of the part (Sec. 3-1-1) is, obviously, a prime factor, and the grouping in Fig. 3–1 gives some feel for the degrees of complexity.

We also saw that the maximum size producible by any one technique is most often limited simply by the availability of large-size equipment. In some processes, however, we found limitations due to process conditions themselves. Thus, a casting mold may not stand up to the excessive solidification times imposed by very heavy walls, a welding process may be limited to a maximum metal thickness if only single-pass welding is permissible, or the thickness of plastic parts may be limited by the low heat conductivity of the plastic.

More frequently, however, the limitation was on the minimum producible size or wall thickness. Since the difficulty of filling a mold or deforming a part increases with the distance over which material must be moved, minimum attainable wall thickness was a function of the extent (width) of the thinnest section. Thus, the wall thickness of a casting was limited by the fluidity of the metal (Fig. 7–28), and that of a forging by the die pressures developed with increasing w/h ratios (Fig. 9–25). Thinner, smaller, and larger parts may well be made, but usually under special circumstances and at extra expense.

We already stated in Sec. 3-5-2 that unnecessarily tight tolerances and surface finish specifications are a major cause of excessive manufacturing costs. In Fig. 3–22 we saw that various manufacturing processes are capable of producing parts to a certain surface finish and tolerance range without extra expenditure. If tighter than usual tolerances or smoother surfaces are required, the cost generally rises. To take a hot-forged part as an example, the process can be tightened up by more careful preforming in several die cavities, using the finishing die only for a limited amount of work, or by subsequent additional operations such as machining or grinding. The cost-conscious designer will specify the loosest possible tolerances and coarsest surfaces that still fulfill the intended function. It is often necessary to specify both maximum and minimum surface roughness for the proper functioning and/or ease of manufacturing of the part.

Manufacturing industries have responded to competitive pressures by tightening tolerances, reducing minimum wall thicknesses, and generally improving quality without necessarily raising the cost of their products.

22-4 COMPETITION BETWEEN MATERIALS

The designer always has the opportunity to consider alternative materials, as previewed in Chap. 5. Substantial economies are sometimes possible. For example, machining costs can vary over several orders of magnitude as shown by the data in Table 22–2 (the actual ratios have changed with the arrival of coated tools, but the point to be made remains valid). If there are no other offsetting costs or constraints imposed by service requirements, great savings can result from switching to another metallic material.

Table 22-2 Relative machining costs for a given part*
(Lathe turning, 60-min tool life)

Material	Hardness	Machining Cost
7075-T6 Al		10
1020 steel	111 BHN	25
410 stainless steel	163 BHN	40
310 stainless steel	168 BHN	55
Ti-6Al-4V		75
4340 steel	52 R_C	100
Inconel X		170
Inconel 700 (aged)	400 BHN	340

*From *Profile Milling Requirements for Hard Metals 1965–1970*.
Report of the Ad Hoc Machine Tool Advisory Committee to the
Department of the U.S. Air Force, May 1965.

A new dimension entered into the competition between materials with the appearance of engineering (high-performance) plastics. Plastics first presented challenges to zinc alloys, although the zinc die-casting industry succeeded in holding onto much of the market by developing thin-wall casting methods. More recently, plastics have become attractive alternatives for sheet-metal parts, zinc- and aluminum-alloy castings, and cast iron. The competition is particularly fierce in the automotive field where mass reduction is one of the means of reducing fuel consumption. Within only a decade (from 1975 to 1985), the average mass of American cars was reduced from almost 1720 kg to 1230 kg; the mass of aluminum rose from 10 kg to 60 kg per car and that of plastic parts from 40 to 100 kg per car. The first applications were in obvious directions: fascia panels, interior fittings, and other non-load-bearing components. Now structural parts such as bumpers, brake-fluid tanks, and some body parts are also made of plastics. Even within the share of steel, shifts occurred in the direction of more extensive HSLA and galvanized sheet usage.

Example 22-2

In Chap. 1 we pointed out that manufacturing is one of the main sources of the wealth of nations. Its development results in vast social changes, and one product that has had enormous influence is the automobile. It is, therefore, fitting that U.S. industry and government formed a Partnership for a New Generation of Vehicles (PNGV, Example 5-6) with the aim of developing a midsize car that, compared to a typical 1993 model, uses much less fuel, costs the same, and is 80% recyclable. Higher fuel efficiency, low emissions, ease of recycling, and safety are technical goals but of obvious social implications. Even though the program concentrates on a typical midsize vehicle, the findings will have much broader applicability and will ultimately benefit all humankind.

A key factor in reaching the goals is a 40% reduction in total vehicle mass. Throughout this text we saw examples of how this goal can be achieved with the application of both established and new technologies. There is limited opportunity for reducing the mass of power plant and transmission; thus, a significant part of mass reduction must come from the "body-in-white" (the basic body without doors, trunk lid, and hood). We have already seen several approaches. The ULSAB initiative has achieved the required mass reduction (Ex. 5-1) by using high-strength steel (Ex. 10-16), material substitution (Ex. 18-11), and techniques such as pressing of tailored blanks (Ex. 10-12 and Ex. 18-9) and hydroforming (Ex. 10-16). The goal was also reached by Ford with an aluminum-intensive unibody and by General Motors with a glass-composite unibody. Another approach is to manufacture a drivable space frame (a complete load-bearing, crashworthy structure) on which metal or plastic panels are fastened (this technique has been used in the production of General Motors' Saturn vehicles). Armco Steel reduced the mass with a stainless-steel space frame, and an all-aluminum construction is used in General Motors' EV1 electric vehicle and Porsche's Boxter. Further mass reduction can come from other components. Thus, an increasing number of engine blocks and almost all cylinder heads are made of aluminum. Chrysler is using aluminum castings for engine mounts in their minivans; dimensional accuracy is assured by automatic straightening, freedom from porosity by 100% x-ray inspection. Aluminum alloy wheel rims are widely used, and magnesium alloy is gaining application for wheel rims and some parts that had been made of plastics.

22-5   IDENTIFYING THE OPTIMUM APPROACH

In the ideal manufacturing situation, concurrent engineering assures that the designer cooperates with the manufacturing engineer. Together they ensure that the part should have features that make it eminently producible by a selected, optimum process (*design for manufacturability*). Process selection then follows a formalized procedure.

In one approach, group members get together in a *brainstorming session*. The basic rule is that no idea is regarded as too silly; no criticism is allowed until all ideas emerge, and only then does the process of whittling down alternatives begin. Those processes that stand concentrated criticism are then evaluated in detail and the optimum is finally chosen, often after detailed scrutiny by estimators and designers working with CAD.

The same function is performed, but in a much more organized form, by *value engineering groups* within the concurrent engineering team. The task of such groups is to review the functional requirements and ease of manufacturing of the part, and attach a value to functions and processes. Feedback from the group is used to initiate design changes that make manufacturing (including assembly) easier. In addition, economic aspects are also fully weighed. It is not unusual that, as a result of value analysis, a part is entirely redesigned, or several parts forming an assembly are made as a single unit. With the growth of CAD/CAM, some of these functions are incorporated within the activities of the design group.

In many smaller companies the situation is quite different. The manufacturing engineer or technologist may have to rely on his or her own resources, without the benefit of interactions with specialists in other fields. There is then real danger of settling on the first obvious solution, conditioned by experience with a given process. As a minimum, the manufacturing engineer must consider alternatives in a "brainstorming session" with himself or herself. The availability of low-cost CAD packages and increased access to computer databases greatly improves the situation.

In many instances, a company performs only assembly operations, or possesses only machining facilities of its own. The raw material is purchased in the form of forgings, castings, sheet-metal stampings, plastic moldings, and semifabricated products such as bars, tubes, sections, plate, and sheet. A vital function is then fulfilled by the purchasing department that conveys the information to suppliers and often also acts as an intermediary between designer (or manufacturing specialist) and supplier. Identification of alternative production methods and location of suppliers of appropriate capacity is essential. Without proper coordination between purchasing and manufacturing, the part is often "hacked out of the solid." While there are instances in which this approach is indeed the most economical, alternatives must always be explored.

First, one would assume that the designer has taken full advantage of the innumerable standardized and semistandardized components obtainable from specialized producers who employ mass-production techniques and can guarantee highest quality at lowest cost. Screws, rivets, clips, springs, packings, seals, bearings, gears, etc., are available in almost any justifiable size and material, and should seldom if ever have to be custom-made.

Second, geometrically similar parts can often be made identical without sacrificing functional performance, or they may be made into a family suitable for group-technology treatment (Sec. 3-1-2).

Third, in specifying materials, function as well as ease and cost of manufacturing must be considered. A sound decision is possible only if the true function of the part is known and no unnecessary constraints are imposed.

Example 22-3

The foregoing discussion has shown that, more often than not, several processes are potential candidates for making any one part. Indeed, many of these processes coexist for the manufacture of parts that meet the product specification (design) requirements. This is particularly true of parts of relatively little shape complexity, as demonstrated on the example of a spool-shaped part shown in Fig. 22–1 (shape T2, Fig. 3–1). Options will be a strong function of material, size, tolerances, and production volume.

For illustrative purposes, assume that the part is made from a ferrous material, with a minimum tensile strength of 280 MPa (40 kpsi); ductility is of no major concern. Take first the following dimensions: bore diameter 50.00 mm, wall thickness 2.5 mm, flange diameter 75.0 mm, flange width 10 mm, and total length 100.00 mm. Tolerances of ±0.025 mm apply only to the bore diameter and total length. For a tolerance range of 0.05 mm over a 25-mm dimension, Fig. 3–22 indicates that only a machining operation can provide the required quality. This applies even more to the larger dimensions of the part. Since finish machining is needed irrespective of the method of manufacture, there is great freedom in choosing the initial process for creating the shape.

1. The shape is suitable for turning on a lathe (Table 16–7). Machining from a solid bar (Fig. 22–1a) results in a cutting-off loss of 6 mm, and material utilization is only 18%. Starting

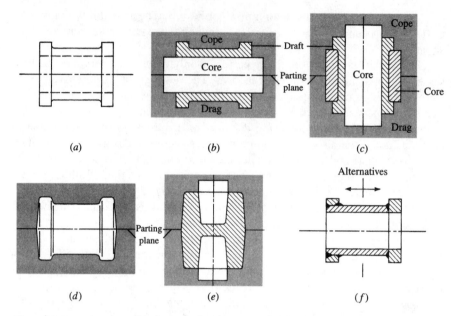

Figure 22–1   Some possible methods of making a spool-shaped part: (*a*) machining from bar or tube, (*b*) casting in a horizontal position, (*c*) casting in a vertical position, (*d*) forging in a horizontal position, (*e*) forging in a vertical position and (*f*) welding.

with (a much more expensive) tube of 46 mm ID, material utilization increases only slightly to 25%.

2. Casting is feasible by all expendable-mold processes (Table 7–3). Casting in a horizontal position requires an inexpensive cylindrical core (Fig. 22–1*b*). With a machining allowance of 2 mm on all surfaces, wall thickness becomes 6.5 mm; this allows for possible eccentricity of the core. This thickness has be to be fed over a distance of 76 mm; according to Fig. 7–28, this should be possible even for a steel casting but, in view of the modest strength requirement, a nodular cast iron will do. Material utilization increases to 37%.

3. Casting in a vertical position (Fig. 22–1*c*) complicates the mold but allows bottom feeding. In view of the small size of the part, the benefit is hardly worth the expense.

4. If the strength requirement were raised to the point where cast iron is insufficient and, especially if the part is exposed to fatigue loading, a higher grade of nodular cast iron or cast steel could be considered. Forging also becomes an alternative. Forging in a horizontal position (Fig. 22–1*d*) saves little material and the flash forms at a position where the flow lines create a potential weakness under an axially rotating load.

5. Forging in a vertical position (Fig. 22–1*e*) gives poor material utilization and fiber orientation is still not optimum. Forging on a horizontal forging machine creates, however, a part with favorable material flow, very close to the finished shape, and machining allowances could be reduced. Fairly expensive tooling with a sliding insert would, however, be required.

6. Powder metallurgy by die pressing could be used to produce a tube, with an OD equal to flange diameter and an ID equal to bore less machining allowance. Since there are flanges at both ends, the reduced OD cannot be directly pressed, and the economy of the process is doubtful. Isostatic pressing in a deformable mold would come much closer to final shape. Powder injection molding could produce a net-shape part but the economy is doubtful for such a large and simple part.

7. In principle, the part could be made by welding a ring (Fig. 22–1f, left) or flange (Fig. 22–1f, right) but finish machining would still be required and it is highly unlikely that this method would be competitive.

The final choice would most likely come down to machining from the solid for small-lot production and casting or horizontal upsetting for larger production volumes. Economies would change if the part were much longer, but with the same flange: upsetting of tube (pipe) ends or welding flanges onto a tube would become much more attractive.

Increasing the flange diameter to 100 mm would change the evaluation. Machining from a bar again gives only 18% material utilization and a thick-walled tube is not worth considering. Casting becomes a clear favorite with 56% material utilization. Horizontal upsetting of the flange from 56-mm to 100-mm diameter is possible but stretches the capabilities of the process (material has to be gathered in several die cavities). If the length of a part with the larger flange were increased, the machining option would become even less favorable. Casting is still attractive. Horizontal upsetter work is feasible only to a certain length (typically, a length-to-hole-diameter ratio under 4), although a solid spool could be forged. Assembly by welding may be the most economical, depending on the cost of tubing. Cold-drawn tube would eliminate the need for machining the bore, provided that welding were performed without causing distortion or loss of properties.

In general, the range of process options is narrowed with increasing shape complexity and, as shown by the above example, with increasing differences in dimensions between various portions of the part. It may then be necessary to return to the design stage and consider alternative geometries and material options. The number of possible processes would increase greatly if the restriction on the material (ferrous alloy) were removed and the strength criteria were relaxed, since both nonferrous materials and reinforced plastics would become viable candidates.

Example 22-4

A nonpressurized cylindrical oil storage tank of $D = 750$-mm diameter and $L = 2000$-mm length is to be made of 5-mm-thick mild steel plate. The design (Fig. 22–2a) prescribes welding flat disks onto the ends of a cylindrical shell. (We will ignore the fill and drain fittings.)

The cylindrical shell can be made by three-roll bending with a longitudinal welded seam. The weld needed for closing the ends is unfavorable since through-thickness penetration is difficult and the remaining gap would act as a stress raiser (one could, however, question the need for the prescribed wall thickness, depending on the support structure). Chamfering (Fig. 22–2b) helps but requires very large lathes and involves extra expense. Changing the design to disks fitted into the ends of the shell makes welding easier but high stress concentrations still result (Fig. 22–2c). Welding of parts of equal thickness is always preferable; if production quantities warrant it, dished ends (produced by spinning or drawing without a blankholder) could be pressed into the ends to secure equal wall thickness (Fig. 22–2d; note, however, that a gap still exists). Domed ends allow circumferential welds and, if penetration is complete, there are no stress raisers (Fig. 22–2e).

The task becomes much more open-ended if the design statement calls simply for a storage container; let us assume that this time it is for water. Depending on water quality, corrosion of mild steel could become a serious issue, aggravated by potential crevice corrosion if the welds leave small gaps on the inside. Protection with an enamel layer is possible but could be too costly, unless justifiable by other requirements. For cold water storage, a plastic liner could

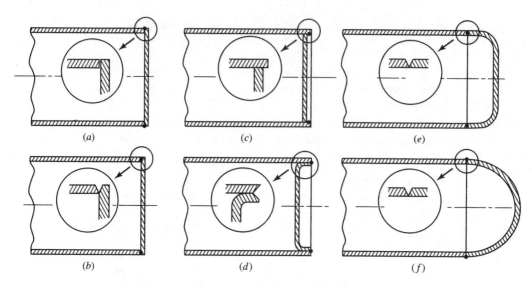

Figure 22–2   The design of a container has implications for ease of manufacture.

be applied by rotary molding. Austenitic stainless steel would be justifiable if the tank were part of a food-processing system. Alternatively, an all-plastic tank deserves consideration; a nonreinforced tank would have to have thick walls, and a fiberglass-reinforced composite is a more likely choice.

The situation changes entirely if the contents are pressurized. For an internal pressure of 3 MPa, wall thickness with mild steel has to increase to 8 mm and the ends now have to be made domed (e.g., hemispherical, Fig. 22–2 f). At this thickness, the domes will most likely be spun and the edges trimmed and chamfered. The cylindrical shell must have the same thickness, and circumferential welding is feasible after edge preparation.

The very open-ended nature of the problem indicates that this is properly the task of a concurrent engineering team, especially if the tank is to be made in larger quantities.

Example 22-5 | In Ex. 7-9 we introduced a generic part (a flange) as a casting. We then examined it for production by forging (Ex. 9-23), sheet metalworking (Ex. 10-14 and 10-15), and powder metallurgy (Ex. 11-8 and Prob. 11C-6), and looked at machining it (Prob. 16B-20 and Prob. 16C-6). Collectively, these examples showed in some detail the steps to be taken. If production alternatives are to be considered by one person, without the benefit of consultation with others, there is danger of discarding profitable alternatives. The danger is best avoided by following the *process of elimination*; that is, one first considers all alternatives, throws out those that are obviously impossible, and then concentrates on the feasible ones. The task is actually more difficult for a part of such simple shape because the final choice will hinge on a rather detailed economic analysis.

Example 22-6 | The economies attainable by a radical change in processes are shown in Fig. 22–3. Sparkplug shells are made of low-carbon steel and are needed by the tens of millions. Traditionally they were machined of hexagonal bar stock, on automatic screw machines, with one shell finished every 4 to 6 seconds. Much of the material was turned into low-value chip (typically, worth

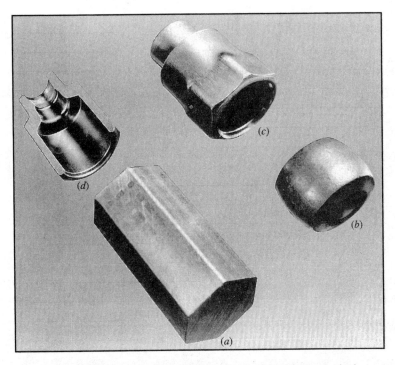

Figure 22-3   The savings achieved by changing the manufacturing method are illustrated by the sparkplug body. (a) Originally it was produced by machining from hexagonal bar stock; now (b) a round slug is upset and (c) cold-extruded into a shell of (d) thin wall, for a material saving of 70%. (*Courtesy Braun Engineering Co., Detroit, Michigan.*)

only one-twentieth the price of the bar). The process became totally noncompetitive when the introduction of the phosphate–soap lubrication system allowed the cold extrusion of steel. The starting material is round wire (bar), sheared and cold-upset (preformed) in a cold header at the rate of 250–400 per minute to form a slug which, after application of the phosphate–soap system, is cold extruded at the rate of 160/min in a single stroke. (Many shells are also progressively cold formed at similar rates). The shell is finished, at the rate of 60/min, with minimum machining, saving 70% of the material once used.

Example 22-7

The manufacture of molds and dies is one of the major cost elements in forming and molding processes. An example of time and cost savings achievable by process shortcuts is provided by the nickel vapor deposition (NVD) process. As stated in Secs. 11-2-1 and 19-7, pure nickel is obtained from its carbonyl according to Eq. (19-6). Instead of producing a powder, the nickel can be directly deposited on the positive (mandrel) of the shape to be molded. The mandrel can be made from any metal, preferably a highly machinable aluminum alloy. It is placed into a pressurized chamber at 177 °C; nickel is deposited atom by atom from carbonyl gas at the rate

Figure 22–4   Polishing the mold for injection molding the body of the Chrysler Composite Concept Vehicle, shown on the chapter-opening photo for Chap. 15. (*Courtesy: Nickel Tooling Technology, division of Weber Manufacturing, Midland, Ontario.*)

of 0.25 mm/h. The shell thus formed can be hardened to HRC 38–42 by the addition of diborane (B_2H_6) during deposition. It is readily separated from the mandrel, and multiple shells can be made on the same mandrel. The surface can be highly polished and, using special techniques, intricate and authentic surface details of leather and wood can be faithfully reproduced.

The stripped shell can be backed with a network of copper, steel or stainless steel tubing to provide heating or cooling. Metal-filled epoxies, concrete, steel weldments, aluminum castings have been used separately and in various combinations to provide structural support for the nickel shell (Fig. 22–4). A comprehensive design guide (*Design Manual for NVD Nickel Shell Molds*) is available from the Industrial Research and Development Institute, Midland, Ontario. (Source: Nickel Tooling Technology, division of Weber Manufacturing Limited, Midland, Ontario.)

22-6   SUMMARY

For a company to survive, it must adopt practices that allow it to compete in the global economy. It must learn to create maximum value with minimum input, respond rapidly to changing customer demands, and bring to the market new products in the shortest possible time with highest quality. The foundation to all of this is concurrent engineering. In the broadest sense, a concurrent engineering team will weigh all implications of engineering decisions. In the narrower sense, product design will proceed with the full appreciation of manufacturing processes.

In this book we concentrated on the principles that underlie unit processes of manufacturing. We also indicated, although in a much more general sense, how these processes can be brought together into a viable system. We stressed the importance of keeping an open mind, since a part of a given function can often be made by several techniques and from different materials. We showed how design can be adapted to the capabilities and limitations of processes, and saw that minor changes in design can greatly improve the feasibility and economy of manufacture. At its minimum, knowledge acquired here should allow active participation in a concurrent engineering team. In the less-than-ideal situation where a product designer is required to work without team support, it is still necessary that he or she should be aware of manufacturing implications even while striving to satisfy functional requirements. This demands at least some familiarity with processes and a willingness to discuss the design with manufacturing specialists. Increasingly, many of the alternatives can be weighed with the aid of specialized computer software, but some familiarity with processes is still essential for the proper use of these computer programs. Some basic rules are always worth considering:

1. Specify the broadest tolerance and coarsest surface finish that still satisfy functional requirements.

2. Subject the design to value analysis, formally or informally. Start with the real functions, strip off all unnecessary constraints, and seek alternatives with the early involvement of people with knowledge of various manufacturing techniques (as opposed to narrow specialists).

3. Remember that a seemingly minor change in shape, wall thickness, or radius can make the part suitable for manufacturing by a different, more economical technique.

4. Look at the relation of parts to each other; it may well happen that several parts are much easier to make as a single unit.

Manufacturing is a peculiar blend of science, art, and economics, with very broad social implications. It would be unreasonable to expect designers or manufacturing engineers to consider formally all these elements whenever a detail decision is made. The broad view must, nevertheless, guide the general approach to the profession; without sound scientific foundations, manufacturing remains but a collection of myriads of isolated rules; without practical and economic sense, the best theory and research will fail to make impact; and without conscious attention to quality and cost, the best-designed product and process will lose out against competition.

PROBLEMS

22-1 A cylindrical component, made of heat-treated steel, is found to fail in service by fatigue. Laboratory examination reveals the presence of residual surface tensile stresses. Describe the methods that could be considered for changing the residual stresses to compressive, assuming that the original cause of surface tensile stresses cannot be eliminated.

22-2 Trace the coordinate system of Fig. 3–22 on transparent paper, then construct a line corresponding to a tolerance of 20 R_a. From Fig. 3–22 judge whether a 0.4-μm R_a surface finish specified for a journal of 25.00 ± 0.01-mm diameter can be produced (a) by grinding and (b) by cold drawing. (c) If the answer to (a) and/or (b) is yes, is the chosen finish reasonable? (Refer to the line constructed above.)

22-3 Collect at least three types of metal cans, selected from among containers for fizzy drinks, fruit juices, baby food, canned meat, and sardines. Make sure you have samples of two- and three-piece containers. Section them carefully (protecting your hand) and investigate their structure. Write an essay describing their method of manufacture as deduced from the evidence available; to support your analysis, quote the relevant figure numbers from this text. (This is an extension of Prob. 10B-18, and now includes the methods of joining the can).

22-4 Carefully inspect the trunk lids of at least two cars. Write a brief report with your judgments on (a) how the lid and inner reinforcing member were made, (b) how they were joined, and (c) whether sound deadening was applied. (d) Press on the outer lid; make a reasoned judgment how the lid was formed for higher stiffness.

22-5 What processes and materials could you envisage for making (a) a high-pressure gas cylinder; (b) a CO_2 cartridge? (c) Suggest 3 other processes and materials one could consider if the part were of shape F5 (Fig. 3–1) without the need to sustain internal pressures.

22-6 Circles of 300-mm diameter are to be cut from 10-mm-thick low-carbon steel plate. (a) Make a numbered list of all potential processes you can identify from this text. (b) Rank the processes based on the quality of the cut surface (if two processes are similar, assign the same ranking). (c) Choose the process that appears the most economical for production rates of 10 000 pieces/month, with no special quality requirements regarding the cut surface. (d) Choose processes that appear to be economical for production rates of 500 pieces/month, and a requirement of a perpendicular edge surface; rank them in order of anticipated quality.

22-7 It has been said that the customer does not buy a drill but the expectation of a hole. (a) Make a numbered list of all hole-making methods you can think of (the quality and production quantity are not specified). (b) List (by number) the methods suitable for making through-holes of 50-mm diameter in 20-mm-thick 7075-T6 Al alloy plate and judge their suitability by assigning an order of preference (if two or more methods are equal, assign the same preference to them). Justify with brief statements (poor surface quality, expensive, rapid tool wear, etc.).

22-8 Repeat Prob. 20-7*b* for 2-mm-thick sheet but a hole of irregular shape, in small quantities.

22-9 Repeat Prob. 20-7*b* for 1-mm-diameter hole in 20-mm-thick superalloy (Hastelloy X).

22-10 Repeat Prob. 20-7*b* for 50-mm-diameter hole in (*a*) 2-mm-thick thermoplastic and (*b*) glass-fiber-reinforced thermoplastic.

22-11 Repeat Prob. 20-7*b* for 5-mm-diameter hole in 2-mm thick alumina.

22-12 List all materials and methods you can think of for making automotive wheel covers (the finished part must have a metallic appearance). If the method requires several manufacturing steps (including surface treatment), state the sequence.

22-13 Obtain two samples each of socket-head cap screws, and recessed- and slotted-head screws. With the aid of a magnifying glass or, preferably, a stereomicroscope, investigate the heads for evidences of the method of manufacture. Report your findings, including any defects discovered, in a professional manner.

22-14 Carry out the tasks described in Prob. 22-13, but collect six different self-tapping screws and include the threaded portion in your investigation. If facilities are available, mount screws in plastic and make longitudinal sections. Etch to reveal the flow lines. Note any defects that may be present and draw conclusions regarding their possible effects on the performance of the screws.

22-15 Carry out the tasks of Prob. 22-13, but on six wood screws.

22-16 Survey the products used for merchandizing liquids in approximately 1-liter units. Describe at least five products, identifying the materials of construction and the probable methods of manufacture.

22-17 Lightly loaded spur gears of 50-mm diameter and 3-mm thickness are to be mass produced with 50 teeth. Suggest all methods you can think of if the material is (*a*) steel, (*b*) aluminum alloy, (*c*) zinc alloy, (*d*) brass, (*e*) nylon, or (*f*) phenolic resin. For each material, indicate the most likely candidate process.

22-18 Recreational products such as golf clubs often embody materials and processes at the leading edge of technology. Suggest as many materials as seem applicable for making the shaft; for each material, list the likely sequence of manufacturing steps.

22-19 Small gears are made of an aluminum alloy by conventional blanking. List all processes you can think of for deburring.

22-20 From information available in this volume, discuss the method by which bone growth into hip joint implants can be encouraged.

22-21 Suggest at least three approaches to making the hull of a 3-m-long rowboat.

22-22 From information available in this text, suggest ways to increase the permissible operating temperature for turbine blades (consider materials, manufacturing processes, and surface coatings).

22-23 Ribbed sheets have desirable stiffness in the longitudinal direction. Make a table to list all applicable processes if the ribs are (in column 1) tapering, (in column 2) rectangular, and (in column 3) T-shaped. In separate rows, show the relevant processes for (*a*) 6061 Al alloy, (*b*) 4140 steel, (*c*) fiberglass-reinforced epoxy. (The ribs must be integral with the base.)

FURTHER READING

Brimson, J.A.: *Activity Accounting*, Wiley, 1997.

Brown, J.: *Value Engineering: A Blueprint*, Industrial Press, 1992.

Cosumano, M.A.: *Thinking Beyond Lean*, The Free Press, 1998.

Creese, R.C., M. Adithan, and B.S. Pabla: *Estimating and Costing for the Metal Manufacturing Industries*, Dekker, 1992.

Ernst and Young Guide to Total Cost Management, Wiley, 1992.

Mahoney, R.M.: *High-Mix Low-Volume Manufacturing*, Prentice Hall, 1997.

Meyers, F.E.: *Motion and Time Study for Lean Manufacturing*, Prentice Hall, 1998.

Miller, J.: *Activity-Based Management*, Wiley, 1995.

Miller, J.G.: *Benchmarking Global Manufacturing*, Irwin, 1992.

Montgomery, J.C., and L.O. Levine (eds.): *The Transition to Agile Manufacturing*, ASQC Quality Press, 1996.

Murphy, S.: *The Fat Firm*, McGraw-Hill, 1997.

Niebel, B., and A. Freivalds: *Methods, Standards, & Work Design*, 10th ed., McGraw-Hill, 1999.

Oleson, J.D.: *Pathways to Agility: Mass Customization in Action*, Wiley, 1998.

Shunta, J.P.: *Achieving World Class Manufacturing Through Process Control*, Prentice Hall, 1995.

Vollman, T.E.: *Manufacturing Planning and Control Systems*, Irwin, 1997.

Womack, J.P., D.T. Jones, and D. Roos: *The Machine That Changed the World*, Simon & Schuster, 1990.

Womack, J.P., and D.T. Jones: *Lean Thinking*, Simon & Schuster, 1996.

Conversion factors from U.S. conventional units to metric (SI) units

| Quantity | To Convert from USCS Units | | Multiply by | To Obtain SI Unit | |
	Symbol	Name		Symbol	Name
Length	in	inch	*25.4	mm	millimeter
	ft	foot	*0.3048	m	meter
Area	in^2	square inch	*6.4516×10^{-4}	m^2	
Volume	in^3	cubic inch	1.639×10^{-5}	m^3	
	gallon	US gallon	3.785×10^{-3}	m^3	
Time	min	minute	*60	s	second
Velocity	ft/min		*5.08×10^{-3}	m/s	
Mass	lb	pound	0.4536	kg	kilogram
	ton	short ton	0.9072	tonne	tonne (Mg)
Acceleration (gravitational)	ft/s^2 $(32 \ ft/s^2)$		*0.3048 *(9.80665)	m/s^2 m/s^2	
Force	lbf (or lb)	pound force	4.448	N	newton
	tonf	ton force (2000 lb)	8.9	kN	
	kgf	kilogram force	*9.80665	N	
Stress (pressure)	lbf/in^2	psi	6.895×10^3	Pa	Pascal (= N/m^2)
	ksi (or kpsi)	1000 psi	6.895	MPa	(or N/mm^2)
	torr	mm Hg	133	Pa	
Torque (work)	lbf · ft	foot-pound	1.356	N · m	newton-meter
Energy (work)	Btu	British thermal unit	1055	J	joule (= N · m)
	cal	gram calorie	*4.1868	J	
Power	hp	550 ft · lb/sec	*746	W	watt (= J/s)
Viscosity	P	poise ($dyne \cdot s/cm^2$)	0.1	Pa · s	(or N · s/m^2)
Temperature interval	°F	Fahrenheit degree	0.5555	C or K	Celsius degree or kelvin
Temperature	t_F	degree Fahrenheit	$(t_F - 32)\,5/9$	t_C	degree Celsius
	t_C		$t_C + 273.15$	t_k or T	absolute degrees

NOTES: Exact conversion factors are recorded with an asterisk. The Celsius degree is often written °C to avoid confusion with C (coulomb). Most frequently used multipliers:

	Prefix	Symbol
10^9	giga	G
10^6	mega	M
10^3	kilo	k
10^{-3}	milli	m
10^{-6}	micro	μ
10^{-9}	nano	n

The international Committee of Weights and Measures (CIPM) modernized the metric system in 1960. The resulting SI units are now used worldwide in the literature; all industrialized nations have already committed themselves to conversion to the international System (SI).

For a detailed discussion see, for example, *Standard for Use of the International System of Units (SI): The Modern Metric System* IEEE/ASTM SI 10-97, or *Canadian Metric Practice Guide*, CAM/CSA-Z234.1-89 (R1995).

B

Approximate Conversion of Hardness Values

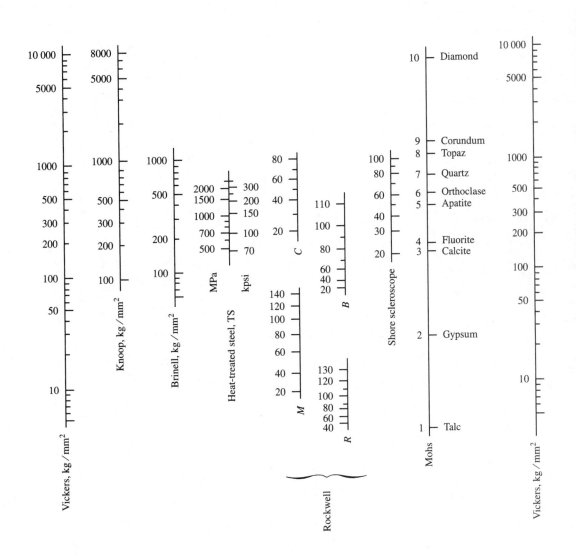

C

Solutions to Selected Numerical Problems

3C-2 Shaft: ring gage, diameter GO = 24.980 mm, NOT GO = 24.959 mm. Bearing: plug gage, diameter GO = 25.000 mm, NOT GO = 25.033 mm.

3C-4 Change = 0.0025 mm (12% of total tolerance).

3C-6 (a) Error = 0.0372 mm; (b) no; (c) R_t = max. 8 μm; R_a = 0.8 μm.

3C-8 Error = 0.0224 mm.

4C-2 Minimum hardness = HRC 39.

4C-4 $0.5T_m$ = 73°C for Zn, 405°C for Cu, 591°C for Ni: Zn will creep.

4C-7 (b) e_t = 82%; (c) e_c = 0.45; (d) e_c tension = 0.45, e_c compr = 0.87; (e) hydrostatic pressure.

4C-11 TS = 206 MPa.

4C-13 YS = 239.7±9.6 MPa; TS = 332.4±9.1 MPa; el. = 42.7±1.9%.

6C-2 Freezing range $(T_L - T_s)/T_S$ = 0.022, 0.025, 0.07 for the three alloys; coring in Al-5Mg.

6C-5 For d = 10^{-2}, 10^{-1}, 10^0, 10^1, 10^2; the value of σ = 11.4, 4.16, 2.0, 1.316, 1.1.

6C-8 (a) Al at room temperature is << $0.5T_m$: use fine grain. (b) Creep: use coarse grain.

6C-11 87.5%.

7C-5 (a) 93 cm³/100 g steel; (b) 87.86%; (c) 6.27 cm/min.

7C-7 (a) Strain e = 0.0046. (b) ε = 7.67(10^{-6}) s^{-1}. (c) σ_f = 36.9 MPa. (d) Plastic deformation.

7C-9 (b) 943 kN (106 tonf).

7C-11 837.5 kg 70/30 scrap; 1163.75 kg Cu.

7C-13 d = 77.1 mm, h = 92.5 mm.

8C-1 m = 0.174.

8C-3 $\varepsilon = \ln(1 + e_t) = \ln(1 - e_c)$

8C-5 r = 0.56.

9C-2 (a) p_a = 1282 MPa; P_a = 227 kN. (b) p_a = 3012 MPa; P_a = 532 kN.

9C-5 With m^* for sticking: p_p = 5967 MPa; lubricated: p_p = 3397 MPa.

9C-8 (a) At 500°C: p_a = 61 MPa, P_a = 420 kN. (b) At 400°C, p_a = 153 MPa, P_a = 1050 kN.

9C-11 h/L = 4.12: deformation is inhomogeneous. Use lower die angle: for h/L = 2, α = 20°.

9C-13 (b) p_e = 242 MPa; P_e = 4280 kN; p_l = 519.2 MPa.

9C-15 (a) 417MPa; (b) 5240 N, 5.24 kW; (c) 4866 N.

9C-17 (a) for the four draws, σ_{exit} = 100, 175, 223, 53 MPa; (b) P_{dr} = 1602, 2245, 2278, 528 N; (c) safe P = 4623, 5087, 4904, 4866 N; (d) yes.

9C-19 (a) No. (b) For the two passes, P_r = 31.0, 59.0 kN; (c) power = 1.57, 2.99 kW.

9C-22 σ_f = 572 MPa.

9C-24 P_r = 4190 kN; power = 296 kW.

9C-27 For rolls of 800, 300, 30 mm diameter, P_r = 6559, 3443, 907 kN; power = 359, 308, 257 kW.

9C-29 v = 0.033 mm/s.

10C-1 (a) P_s = 1442 kN.

10C-3 (a) $P_s = 18$ kN; (b) 44; (c) staggered, 283-mm wide, 67.95% material utilization.

10C-5 Single row: 61.7%, double row: 67.75%, triple row: 69.97%; with punchouts utilized: 68.0%, 74.45%, 76.89%.

10C-7 $R_f = 2.016, 10.405, 62.06, 163.56, 877.2$ mm for $R_b = 2, 10, 50, 100, 250$ mm.

10C-9 With Al ($\sigma_{0.2}/E = 0.00207$).

10C-11 For 410 SS, (b) necking at $n = 0.1$, (c) fracture $q = 65\%$; (d) for 1008, $n = 0.25$ (much better), $q = 70\%$ (slightly better).

10C-13 (c) Check n and el.; 302 SS will be much better.

10C-15 (a) $d_0 = 295.5$ mm; (b) 845 kN; (c) DR = 1.508; (d) $D_p = 196$ mm; $P_d = 647$ kN.

10C-17 (b) Yes; (d) $P_{dr} = 189.5$ kN.

11C-1 $l/d = 6.1$.

11C-4 (a) 6.932 g/cm^3; (b) 88.08%; (c) 11.92%.

11C-6 (b) Force = 1209 kN; (c) ejector force = 246 kN (min.).

11C-8 (a) Diameter 1.5684; (b) yes, on barreled surface; (c) enclose in can; (d) no.

12C-1 (a) 27.1%; (b) 13.33%; (c) $8.667 \times 17.334 \times 130$ mm; vol. 19.53 cm^3.

12C-4 (a) 40 kN; (b) 74 MPa; (c) 94.2 MPa.

13C-1 (a) When $m = 1$, $n = 1$; (b) when $m = 0$, $n \to \infty$.

13C-3 26.23 MPa.

13C-5 (a) $T_g = -18°C$; (b) yes; (c) use crystalline form.

13C-7 Polyimide ($T_g > 310°C$; heat-deflection. temp. 340°C at 1850 kP) Glass-filled nylon-6,6 (heat-deflection temp. 255°C at 1850 kP).

14C-1 (a) At the waist, 0.417 mm, at the bulge 0.265 mm. (b) Assuming that 0.265 mm is sufficient, make tube ID = 28.74 mm at waist.

14C-3 There is a saving of 13.4%.

14C-4 Press = 589 kN; clamp = 1080 kN.

14C-6 (a) 239 kg/h; (b) 40.8 cm/s; (c) 32.6 kW.

15C-1 (a) CTE = $17.92(10^{-6})/°C$; (b) transverse $E_c = 89.6$ GPa; (c) Ex. 15-4: CTE = $16.4(10^{-6})/°C$ (slightly less); $E_c = 96$ (slightly better).

15C-3 From Eq. (15-5), $E_c = 197.8$ GPa; from Eq. (15-6), $E_c = 95$ GPa. Measured value 110 GPa (better than transverse).

15C-5 $l_{cr} = 583$ μm.

15C-7 Length = 498.82 mm; width = 299.86 mm.

16C-2 Removal rates 5.55, 25, 3.13 in^3/min; 1554, 7460, 820 mm^3/min.

16C-4 Multiply mm^3/W·s by 2.73 to obtain in^3/hp·min.

16C-5 At 20% higher speed, 9.69 min; at 40% higher speed, 2.07 min.

16C-8 (b) $v = 1.8$ m/s, $f = 0.25$ mm; (c) 140 s.

16C-9 (b) $v = 1.8$ m/s, $f = 0.5$ mm; (c) 7978 W; (d) 126 s; (e) 4432 N; (f) 532 N·m.

16C-12 (a) $v = 0.35$ m/s; (b) 1114 r/min; (c) 0.06 mm/r; (d) 1.1142 mm/s; (e) 0.421 kW.

16C-14 $v = 360$ ft/min.

16C-17 (b) $D_m = 5$ in, 7 teeth; (c) $v = 1500$ ft/min; (d) $f = 0.112$ in/r; (e) $t_c = 0.117$ min.

16C-20 $v_{c\ min} = 345$ ft/min.

16C-22 (b) $v = 3.9$ m/s, $f = 2.8$ mm/r, depth of cut = 4mm; (c) $t_c = 0.768$ min; (d) power = 25.4 kW.

16C-24 (b) $v = 2.016$ m/s; (c) 0.25 mm/r; (d) depth of cut = 0.65 mm; (e) 1283 r/min; (f) 327.6 mm^3/s; (g) 1956 W; (h) $P_c = 679$ N; (i) 37.4 s.

17C-1 Time = 5.976 min.

17C-5 Machining time 0.9 h at 200 cm^3/min, 12 h at 0.15 cm^3/min.

18C-4 (a) 140 kA; (b) 3.3%; (c) 560 kVA; (d) 18.5 kVA; (e) 37kW; (f) 0.103 kWh/20 joints; (g) 40 kN.

18C-6 (a) Yes. (b) 6.7%; (c) 120 kVA; (d) 8.04 kVA; (e) 16.08 kW; (f) 8.93 W·h/weld.

18C-8 $L = 10.9$ mm.

19-18 (a) $\Delta T = 62.7°C$; (b) $\Delta T = 128.65°C$.

INDEX

No separate glossary is provided; however, definitions of terms will generally be found on the first referenced page. Page numbers in *Italics* indicate tables.

List of Recurrent Symbols

A	contact area; cross-sectional area (instantaneous)	R	resistance; universal gas constant
A_0	original (starting) cross-sectional area	R_a	average surface roughness
A_1	cross-sectional area after deformation	R_e	extrusion ratio; reduction ratio
C	strength coefficient in hot working	R_q	root-mean-square surface roughness
C	Taylor constant (cutting speed for 1 min tool life)	R_t	maximum surface roughness
C_E	composition of a eutectic	S	solubility
C_L	composition of liquid	T	temperature
C_p	heat capacity	T_E	eutectic temperature
C_S	composition of solid crystals	T_L	liquidus temperature
C_0	alloy composition	T_S	solidus temperature
D_p	punch diameter	T_g	glass-transition temperature
E	adjusted cutting energy	T_m	melting point (K)
E	Young's modulus	V	volume
E	voltage	V_t	rate of material removal
E_c	specific cutting energy	W_e	weight per unit area removed by an electric current
E_1	specific cutting energy for 1-mm undeformed chip thickness	Z_f	multiplying factor for feed
		Z_v	multiplying factor for cutting speed
F	shear force; frictional force	a	side dimension of hcp prism
F_s	shear force in shear plane	c	height of hcp prism
G	shear modulus; grinding ratio	c	crack depth; specific heat
I	current	d	average grain size
K	strength coefficient in cold working	d	instantaneous diameter of workpiece
K_{Ic}	plane-strain fracture toughness	d_0	original diameter of workpiece
L	length of contact zone between tool and workpiece	e_c	compressive engineering strain
L	barrel length in screw extrusion	e_f	tensile engineering strain at fracture
M	mass; atomic weight	e_t	tensile engineering strain
M_s	temperature at which martensite transformation begins	e_u	engineering tensile strain at necking
		f	feed (in cutting)
N	rotational frequency (r/min)	f	fiber fraction in composite
P	applied force	g	gravitational acceleration
P_a	force in axial upsetting of cylinder	h	instantaneous height
P_b	bending force	h_c	undeformed chip thickness
P_c	cutting force	h_m	mean or average height
P_d	deep-drawing force	h_0	original height or thickness
P_{dr}	drawing force (wire)	h_1	height after deformation
P_e	extrusion force	j	current density
P_i	indentation force	k	constant; heat conductivity
P_n	normal force on cutting-tool face	l	instantaneous length
P_r	rolling force	l_c	critical length of fiber
P_s	shearing force (maximum)	l_f	final length at fracture
P_t	thrust force on cutting tool	l_0	original length
R	universal gas constant	m	strain-rate sensitivity exponent
R_e	extrusion ratio	m^*	frictional shear factor
Q	activation energy	n	number of observations
	pressure-multiplying factor	n	strain-hardening exponent
	Q for axial upsetting of cylinder	n	Taylor exponent
	Q for extrusion	p	interface pressure
	Q for inhomogeneous deformation	p_a	average interface pressure in axial upsetting
	for plane-strain upsetting	p_c	specific cutting stress
	...s of tool	p_e	average extrusion (ram) pressure